Microbial Biodegradation and Bioremediation

Microbial Biodegradation and Bioremediation

Edited by

Surajit Das

Laboratory of Environmental Microbiology and Ecology (LEnME)
Department of Life Science
National Institute of Technology
Rourkela Odisha, India

ELSEVIER

AMSTERDAM • BOSTON • HEIDELBERG • LONDON • NEW YORK • OXFORD
PARIS • SAN DIEGO • SAN FRANCISCO • SINGAPORE • SYDNEY • TOKYO

Elsevier
32 Jamestown Road, London NW1 7BY
225 Wyman Street, Waltham, MA 02451, USA

First edition 2014

British Library Cataloguing-in-Publication Data
A catalogue record for this book is available from the British Library

Library of Congress Cataloging-in-Publication Data
A catalog record for this book is available from the Library of Congress

ISBN: 978-0-12-800021-2

This book has been manufactured using Print On Demand technology. Each copy is produced to order and is limited to black ink. The online version of this book will show color figures where appropriate.

Contents

Preface

The evolution of environment into its components of atmosphere, hydrosphere, and troposphere took place 600 million years ago. Due to rapid urbanization and emerging anthropogenic activities, the natural biodiversity of our environment is becoming disturbed. The introduction of contaminants into an environment causing disorder, unsteadiness, and distress to the physical and chemical systems, including living organisms, is called pollution. The intrusion of humans and their activities has placed a major pressure on our environment, possibly threatening the dynamics of nature, by producing certain xenobiotics. These compounds have a detrimental effect on the existing flora and fauna, alarming the natural ecosystem throughout our biosphere. Therefore, there is an immediate global demand for diminution of environmental pollution produced by local, national, and global processes. We should have a wide awareness and knowledge of the detrimental effects of these xenobiotic compounds and approaches to their remediation. Advanced techniques for the disposal and treatment of these xenobiotic compounds are the major concern but the recently developed treatment strategies are very costly and lead to production of toxic intermediates which can adversely affect the living organisms.

Over time, microbial remediation processes have been accelerated to produce better, more eco-friendly, safer, and more biodegradable measures for complete dissemination of these toxic xenobiotic compounds. Bioremediation is the process of the usage of living organisms such as plants (phytoremediation) and microbes such as bacteria, algae, and fungi (microbial remediation) and their enzymes to detoxify toxic xenobiotic compounds. Some toxic xenobiotics include synthetic organochlorides, such as plastics and pesticides, and naturally occurring organic chemicals, such as polyaromatic hydrocarbons (PAHs), and some fractions of crude oil and coal. The evolution of new metabolic pathways from natural metabolic cycles has enabled the microorganisms to degrade almost all the different complex and resistant xenobiotics found on Earth. This is an imperative, efficient, green, and economical new alternative to conventional treatment technologies.

The present book, *Microbial Biodegradation and Bioremediation*, comprises chapters dealing with various bioremediation strategies with the help of different groups of microorganisms, along with detailed diagrammatic representations. This book will be useful both for novices and experts in the field of microbial bioremediation. We hope to instill the present status, practicality, and implications of microbial bioremediation to academicians, students, teachers, researchers, environmentalists, agriculturalists, industrialists, and professional engineers, as well as to other enthusiastic people who are wholeheartedly devoted to conserving nature. I thank all the contributors who have expertise in this field of research for their advanced, timely chapters and their help in making this a successful endeavor.

Dr. Surajit Das
Laboratory of Environmental Microbiology and Ecology (LEnME),
Department of Life Science, National Institute of Technology,
Rourkela, Odisha, India

Biography

Dr. Surajit Das is an Assistant Professor at the Department of Life Science, National Institute of Technology, Rourkela, Odisha, India, where he has been on the faculty since 2009. Earlier he served at Amity Institute of Biotechnology, Amity University Uttar Pradesh, Noida, India as Lecturer. He received his PhD in Marine Biology (Microbiology) from the Centre of Advanced Study in Marine Biology, Annamalai University, Tamil Nadu, India. He is the recipient of the Australian Government's Endeavour Research Award for postdoctoral research on marine microbial technology at the University of Tasmania. This research paved the way for using marine actinobacteria as the biocontrol agent in aquaculture.

He has multiple research interests with the core research program on marine microbiology and is currently conducting research as the group leader of the Laboratory of Environmental Microbiology and Ecology (LEnME) on biofilm-based bioremediation of heavy metals and PAHs by marine bacteria; nanoparticle-based drug delivery and bioremediation; and the metagenomic approach for exploring the diversity of catabolic gene and immunoglobulins in the Indian Major Carps, with the help of research grants from the Ministry of Science and Technology and the Indian Council of Agricultural Research, Government of India.

In recognition of his work, the National Environmental Science Academy, New Delhi presented him with the 2007 Junior Scientist of the Year Award for his research on marine microbial diversity. He is also the recipient of the 2002–2003 Ramasamy Padayatchiar Endowment Merit Award, given by the Government of Tamil Nadu from Annamalai University, Tamil Nadu, India. He is a member of the IUCN Commission on Ecosystem Management (CEM), South Asia and a life member of the Association of Microbiologists of India, the Indian Science Congress Association, the National Academy of Biological Sciences, and the National Environmental Science Academy, New Delhi. He is also a member of the International Association for Ecology.

In addition to his research, Dr. Das is the reviewer of many scientific journals published by reputed publishers. He has written and edited three books, and authored more than 50 research publications in leading national and international journals as well as 11 book chapters and several popular articles on different aspects of microbiology.

List of Contributors

Devin Alvarez Environmental Biotechnology Laboratory, School of the Environment, Florida A&M University, Tallahassee, FL, USA

María Julia Amoroso Planta Piloto de Procesos Industriales y Microbiológicos (PROIMI), CONICET. Av. Belgrano y Pasaje Caseros. 4000. Tucumán, Argentina; Universidad del Norte Santo Tomás de Aquino (UNSTA). 4000. Tucumán, Argentina; Universidad Nacional de Tucumán (UNT). 4000. Tucumán, Argentina

Juan Daniel Aparicio Planta Piloto de Procesos Industriales y Microbiológicos (PROIMI), CONICET. Av. Belgrano y Pasaje Caseros. 4000. Tucumán, Argentina

Bastos A.C. Department of Biology and Centre for Environmental and Marine Studies (CESAM), University of Aveiro, Aveiro, Portugal

Claudia S. Benimeli Centro Científico Tecnológico (CCT). Tucumán, Argentina; Universidad Nacional de Catamarca (UNCA). 4700. Catamarca, Argentina; Universidad del Norte Santo Tomás de Aquino (UNSTA). 4000. Tucumán, Argentina

Beyer A. Friedrich Schiller University, Institute of Microbiology, Jena, Germany

Punyasloke Bhadury Integrative Taxonomy and Microbial Ecology Research Group, Department of Biological Sciences, Indian Institute of Science Education and Research Kolkata, Nadia, West Bengal, India

Chiranjib Bhattacharjee Chemical Engineering Department, Jadavpur University, Kolkata, West Bengal, India

Sutapa Bose Department of Earth Sciences, Indian Institute of Science Education and Research Kolkata, Mohanpur, Nadia, West Bengal, India

Brangsch H. Friedrich Schiller University, Institute of Microbiology, Jena, Germany

Anirban Chakraborty Department of Life Science and Biotechnology, Jadavpur University, Kolkata, India

Jaya Chakraborty Laboratory of Environmental Microbiology and Ecology (LEnME), Department of Life Science, National Institute of Technology, Rourkela, Odisha, India

Sudip Chakraborty Chemical Engineering Department, Jadavpur University, Kolkata, West Bengal, India; Department of Chemical Engineering and Materials, University of Calabria, Cubo-42a,87036 Rende(CS), Rende (CS), Italy

Ashvini Chauhan Environmental Biotechnology Laboratory, School of the Environment, Florida A&M University, Tallahassee, FL, USA

Archana Chuahan The University of Tennessee, The Center for Environmental Biotechnology, The Joint Institute for Biological Sciences, Knoxville, TN

Lowell Collins Environmental Biotechnology Laboratory, School of the Environment, Florida A&M University, Tallahassee, FL, USA

Chanchal K. DasGupta Department of Life Science and Biotechnology, Jadavpur University, Kolkata, India

Surajit Das Laboratory of Environmental Microbiology and Ecology (LEnME), Department of Life Science, National Institute of Technology, Rourkela, Odisha, India

Hirak R. Dash Laboratory of Environmental Microbiology and Ecology (LEnME), Department of Life Science, National Institute of Technology, Rourkela, Odisha, India

Jaysankar De Department of Biotechnology, UNESCO Chair—Life Sciences International Postgraduate Educational Center, Yerevan, Armenia

Kashyap Kumar Dubey Microbial Biotechnology Laboratory, Department of Biotechnology, University Institute of Engineering and Technology, Maharshi Dayanand University, Rohtak, Haryana, India

Nelson Duran Center of Natural and Human Sciences, Universidade Federal do ABC, SP, Brazil; Institute of Chemistry, Biological Chemistry Laboratory, Universidade Estadual de Campinas, Campinas, SP, Brazil

Paul H. Fallgren Advanced Environmental Technologies, LLC, Fort Collins, CO 80525, USA

Fragoeiro S. MSD Animal Health, Walton Manor, Milton Keynes, UK

Supratim Giri Department of Chemistry, National Institute of Technology, Rourkela, Odisha, India

Gouma S. School of Agricultural Technology, Technical Educational Institute of Crete, Stavromenos, Heraklion, Crete

Paramasamy Gunasekaran Thiruvalluvar University, Vellore, India

Neelam Gurung Department of Earth Sciences, Indian Institute of Science Education and Research Kolkata, Mohanpur, Nadia, West Bengal, India

Jahangir Imam Central Rainfed Upland Rice Research Station (CRRI), Hazaribag, Jharkhand, India; Enzyme Technology and Protein Bioinformatics Laboratory, Department of Microbiology, Maharshi Dayanand University, Rohtak, Haryana, India

Avinash P. Ingle Department of Biotechnology, Sant Gadge Baba Amravati University, Amravati, Maharashtra, India

Sathyanarayanan Jayashree Department of Genetics, Centre for Excellence in Genomic Sciences, School of Biological Sciences, Madurai Kamaraj University, Madurai, India

Song Jin Department of Civil and Architectural Engineering, University of Wyoming, Laramie, WY 82071, USA; Advanced Environmental Technologies, LLC, Fort Collins, CO 80525, USA

Rashmi Kataria Department of Forests Biomaterials and Technology, Swedish University of Agricultural Science (SLU), Umea, Sweden

Kothe E. Friedrich Schiller University, Institute of Microbiology, Jena, Germany

Punit Kumar Microbial Biotechnology Laboratory, Department of Biotechnology, University Institute of Engineering and Technology, Maharshi Dayanand University, Rohtak, Haryana, India

Jose Antonio Lopez-Elias Departamento de Investigaciones Científicas y Tecnológicas de la Universidad de Sonora, Hermosillo, Sonora, México

Madhu Priya Department of Environmental Science, Central University of Rajasthan, Bandar Sindri, Ajmeer, India; Department of Earth Sciences, Indian Institute of Science Education and Research Kolkata, Mohanpur, Nadia, West Bengal, India

Magan N. Applied Mycology Group, Cranfield Soil and AgriFood Institute, Cranfield University, Bedford, UK

Neelam Mangwani Laboratory of Environmental Microbiology and Ecology (LEnME), Department of Life Science, National Institute of Technology, Rourkela, Odisha, India

Luis Rafael Martinez-Cordova Departamento de Investigaciones Científicas y Tecnológicas de la Universidad de Sonora, Hermosillo, Sonora, México

Marcel Martinez-Porchas Centro de Investigacion en Alimentacion y Desarrollo, Hermosillo, Sonora, Mexico

Bibhuti Bhusan Mishra Department of Microbiology, Orissa University of Agriculture and Technology, Bhubaneswar, Odisha, India

Anindita Mitra Faculty of Arts and Science (FAS), Harvard University, Cambridge, MA 02138, USA

Swati Mohapatra Department of Microbiology, Orissa University of Agriculture and Technology, Bhubaneswar, Odisha, India

Koninika Mukherjee Department of Earth Sciences, Indian Institute of Science Education and Research Kolkata, Mohanpur, Nadia, West Bengal, India; KIIT School of Biotechnology, Patia, Bhubaneswar, Odisha, India

Arijit Nath Chemical Engineering Department, Jadavpur University, Kolkata, West Bengal, India

Tonya L. Peeples Department of Chemical and Biochemical Engineering, Seamans Center, The University of Iowa, Iowa City, IA

Nicole Perry Department of Biology, Wittenberg University, Springfield, OH

Marta A. Polti Planta Piloto de Procesos Industriales y Microbiológicos (PROIMI), CONICET. Av. Belgrano y Pasaje Caseros. 4000. Tucumán, Argentina; Universidad Nacional de Tucumán (UNT). 4000. Tucumán, Argentina

Marco Antonio Porchas-Cornejo Centro de Investigaciones Biológicas del Noroeste, Guaymas, Sonora, México

Muthuirulan Pushpanathan Department of Genetics, Centre for Excellence in Genomic Sciences, School of Biological Sciences, Madurai Kamaraj University, Madurai, India

Mahendra Rai Department of Biotechnology, Sant Gadge Baba Amravati University, Amravati, Maharashtra, India; Institute of Chemistry, Biological Chemistry Laboratory, Universidade Estadual de Campinas, Campinas, SP, Brazil

Ritu Raj Laboratory of Environmental Microbiology and Ecology (LEnME), Department of Life Science, National Institute of Technology, Rourkela, Odisha, India

Jeyaprakash Rajendhran Department of Genetics, Centre for Excellence in Genomic Sciences, School of Biological Sciences, Madurai Kamaraj University, Madurai, India

T. Subba Rao Biofouling and Biofilm Processes Section, Water and Steam Chemistry Division, BARC Facilities, Kalpakkam, India

Sujata Ray Department of Earth Sciences, Indian Institute of Science Education and Research, Kolkata, Mohanpur, West Bengal, India

Steven Ripp The University of Tennessee, The Center for Environmental Biotechnology, The Joint Institute for Biological Sciences, Knoxville, TN

Rohit Ruhal Department of Chemistry, Umea University, Sweden

Deviprasad Samantaray Department of Microbiology, Orissa University of Agriculture and Technology, Bhubaneswar, Odisha, India

Gary Sayler The University of Tennessee, The Center for Environmental Biotechnology, The Joint Institute for Biological Sciences, Knoxville, TN

Amedea B. Seabra Universidade Federal de São Paulo—Unifesp, Diadema, SP, Brazil

Pratyoosh Shukla Enzyme Technology and Protein Bioinformatics Laboratory, Department of Microbiology, Maharshi Dayanand University, Rohtak, Haryana, India

Sudhir K. Shukla Biofouling and Biofilm Processes Section, Water and Steam Chemistry Division, BARC Facilities, Kalpakkam, India

Puneet Kumar Singh Enzyme Technology and Protein Bioinformatics Laboratory, Department of Microbiology, Maharshi Dayanand University, Rohtak, Haryana, India

Stoiber-Lipp J. Friedrich Schiller University, Institute of Microbiology, Jena, Germany

Suneetha S. Department of Botany, Andhra University, Visakhapatnam, Andhra Pradesh, India

Swapna G. Department of Botany, Andhra University, Visakhapatnam, Andhra Pradesh, India

K. Uma Devi Department of Botany, Andhra University, Visakhapatnam, Andhra Pradesh, India

Weist A. Friedrich Schiller University, Institute of Microbiology, Jena, Germany

Tingting Xu The University of Tennessee, The Center for Environmental Biotechnology, The Joint Institute for Biological Sciences, Knoxville, TN

1 Microbial Bioremediation: A Potential Tool for Restoration of Contaminated Areas

Surajit Das and Hirak R. Dash

Laboratory of Environmental Microbiology and Ecology (LEnME), Department of Life Science, National Institute of Technology, Rourkela, Odisha, India

1.1 Introduction

More than 1 billion people around the world cannot get healthy air to breathe; 3 million die annually due to air pollution (WHO, 2006). This century has witnessed the warmest quarter in recorded history (from September 2003 to November 2003). Globally, over 1 million seabirds and thousands of sea mammals are killed by pollution every year. Each year 1.2 trillion gallons of untreated sewage and industrial waste are dumped into waterways by the United States alone. More than 3 million children under the age of 5 die annually from various environmental pollutions (EPA, 2000). These statistics not only give us a clear understanding of the precarious conditions at present, but warn us to think today about a better tomorrow. An unprecedented increase in population, anthropogenic activities, and urbanization in the name of modernization have increased pollutants to critical levels. In day-to-day life, we use more than 60,000 chemicals in the form of fuels, consumer products, industrial solvents, drugs, pesticides, fertilizers, and food additives. Though industrialization is essential for the faster growth of the developing nations, its concomitant pollution level should not come without sustainable management. Hence, proper management policy, suitable remedial strategies, and sustainable utilization of resources without altering the natural ecosystem should be the prime aim of all researchers and decision-making bodies.

In this context, microorganisms play an important role in the maintenance and sustainability of any ecosystem as they are more capable of rapid adjustment towards environmental changes and deterioration. Microorganisms are considered to be the first life forms to have evolved; they are versatile and adaptive to various challenging environmental conditions. Microorganisms are omnipresent, and they impact the entire biosphere. They play a major role in regulating biogeochemical cycles, from extreme environmental conditions like frozen environments, acidic

Microbial Biodegradation and Bioremediation. DOI: http://dx.doi.org/10.1016/B978-0-12-800021-2.00001-7

lakes, hydrothermal vents, bottoms of deep oceans, to the small intestines of animals (Seigle-Murandi et al., 1996). Microorganisms are responsible for carbon fixation, nitrogen fixation, methane metabolism, and sulfur metabolism, thus controlling the global biogeochemical cycling (Das et al., 2006). They produce diverse metabolic enzymes that can be employed for the safe removal of contaminants, which can be achieved either by direct destruction of the chemical or through transformation of the contaminants to a safer or lesser toxic intermediate (Dash and Das, 2012). Due to their versatility, microorganisms have provided a useful platform to be used for an enhanced model of bioremediation of heavy metals, hydrocarbons, polythenes, food wastes, greenhouse gases, etc. as discussed in the subsequent chapters.

1.2 Pollution: A Major Global Problem

The pollution crisis is a major problem all around the globe. It adversely affects millions of people every year, causing many health disorders and deaths. Although urban areas are usually more polluted than the countryside, pollution can also spread to remote places; for example, pesticides and other chemicals have been found in the Antarctic ice sheet. In the middle of the northern Pacific Ocean, a large collection of microscopic plastic particles are found in what is known as the Great Pacific Garbage Patch. Pollutants can be moved from one place to another through land, water, and the atmosphere, worsening the situation day by day (Figure 1.1). Air and water currents carry pollution, ocean currents and migrating fish carry marine pollutants far and wide, and the wind can pick up and scatter

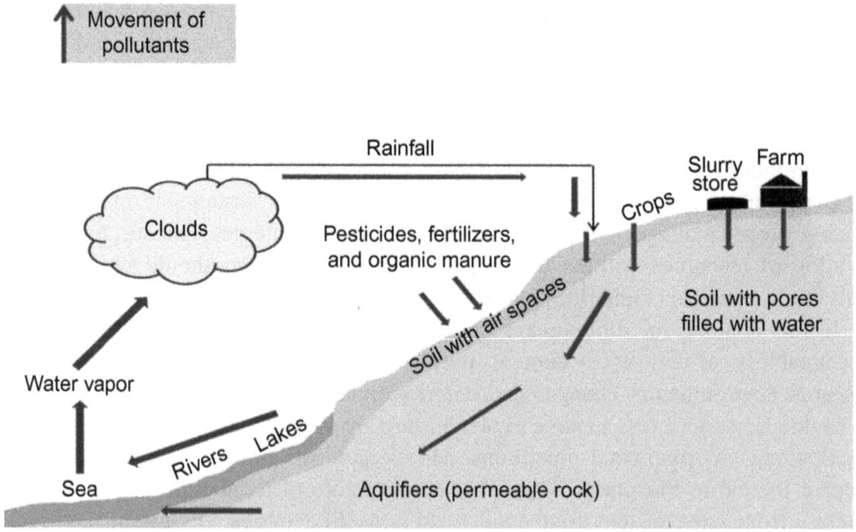

Figure 1.1 Movement of pollutants through land, water, and atmosphere.

radioactive material accidentally released from a nuclear reactor or smoke from a factory from one country into another (Doney et al., 2012). Thus, pollution does not believe in the limitations of geographical boundaries.

There are many major pollution-causing agents found around the globe, i.e., oil spills, fertilizers, garbage, sewage disposals, toxic chemicals. All of them contribute to global pollution in the form of soil, air, water, and marine environmental pollution. Fifty-three chemicals have been identified by United States Environmental Protection Agency as persistent, bioaccumulative, and toxic among 87,000 commercial chemicals (USEPA, 2000). Contamination may be defined as the presence of elevated concentrations of substances in the environment which may or may not be harmful for society. However, pollution is the deliberate introduction of the noxious elements by human beings into the environment, resulting in a toxic effect. These contaminants may be produced due to various natural as well as anthropogenic activities like the large-scale manufacturing of chemicals as well as their processing and handling. The pollution level has increased at a rapid pace throughout the globe and is a major concern for developed as well as developing nations. A comparative detail of air pollution in terms of CO_2 emission and generation of municipal waste has been given in Table 1.1.

Though the statistical data varies greatly among countries within years, most of the pollutants possess high residence time and may travel long distances. Subsequently, they transform to become highly toxic chemicals, thus worsening the situation. As they don't obey geographical boundaries, pollutants can travel easily across long distances, as in the case of the occurrence of xenobiotics in Arctic regions where no immediate sources of pollution-causing agents are found. Thus, pollution has become a global problem and it should be taken care of irrespective of political borders and geographical boundaries.

Table 1.1 A Comparative Account of CO_2 Emission and Generation of Municipal Waste Throughout the Globe in Subsequent Years

Country	CO_2 Emission (Million Tonnes)			Municipal Waste Generation (kg per capita)		
	1971	2000	2010	1980	2000	2010
Australia	144	339	383	700	690	–
Canada	340	533	537	–	–	–
Germany	979	825	762	–	640	580
Italy	293	426	398	250	510	540
Mexico	97	349	417	–	250	370
Norway	24	34	39	550	550	470
United States	4291	5698	5369	610	760	720
Russian Federation	–	1506	1581	160	190	480
China	824	3317	7428	–	210	250
India	200	972	1626	–	–	–

Source: OECD Factbook, 2013; http://dx.doi.org/10.1787/factbook-2013-en.

1.3 Current Remediation Practices

Removal of pollutants from contaminated sites is a great challenge for environmental management. Various processes based upon ion exchange resins or biosorbents have been employed for the removal of toxic chemicals, but these are sensitive to environmental conditions (Wang and Chen, 2009). The most promising conventional techniques for remediation involve chemical precipitation, conventional coagulation, adsorption by activated carbons, adsorption by natural materials, ion exchange, and reverse osmosis (USEPA, 2007). In the conventional coagulation process, coagulants such as aluminum sulfate, iron salt, and lime are employed for the removal of toxic elements. This process can be applied when pollutants are present at high concentrations in the waste materials. The main shortcoming of this procedure is that it does not reach desired levels of precipitation (Henneberry et al., 2011). The process based on adsorption by activated carbon is highly efficient and the water can be reused after regeneration. However, among the disadvantages of this process are the facts that the sorption reaction is slow, it is suitable for low pollutant concentrations only, and the regeneration costs are also much higher (Mohan et al., 2001). Adsorption processes using natural waste materials are more eco-friendly and economical since natural waste materials are used as adsorbents and these processes produce fewer secondary effluents, but the adsorption reaction is slow (Ramadevi and Srinivasan, 2005). The ion exchange processes are based on various ion exchange resins. The major advantages of these processes are that they are insensitive to variability, they can achieve zero level of contaminants, and the resins can be reused after regeneration (Chiarle et al., 2000).

Cellulose acetate or aromatic polyamide membranes are used for the reverse osmosis process at very high pressure, which results in the solvent being forced out through the membrane to the dilute solution (Zhu and Elimelech, 1997). Alternatively, low-cost biological processes such as phytoremediation have also been employed to remediate metal-contaminated sites (Narang et al., 2011). Use of bacteria for metal removal from contaminated sites is also a promising technology. However, the cons for the bacterial-based or plant-based processes may include production of large volumes of pollutant-loaded biomass, the disposal of which is problematic. Thus, in the current scenario, biological methods, i.e., bacterial mediated bioremediation, have the upper hand in terms of sustainability, cost effectiveness, and easy applicability *in situ*.

1.4 Characteristics of Microorganisms Suitable for Remediation

There are many characteristic features of bacteria which make them suitable for application in bioremediation practices. In this chapter, the unique characteristics of bacteria have been described, with marine bacteria considered as a model microorganism. The marine environment is the largest habitat on Earth, accounting for more than 90% of total biosphere volume; the microorganisms present in the marine environment are responsible for more than 50% of the global primary production and nutrient cycling (Lauro et al., 2009). These marine bacteria can be isolated from the marine water, sediments, and mangroves associated with marine habitats, the normal flora of the marine organisms, and deep-sea hydrothermal vents. They usually require sodium and potassium ions for their growth and to maintain osmotic balance of their cytoplasm (MacLeod and Onofrey, 1957). This requirement for the Na^+ ion is an exclusive feature of marine bacteria which is attributed to the production of indole from tryptophan (Pratt and Happold, 1960), oxidation of L-arabinose, mannitol, and lactose (Rhodes and Payne, 1962), as well as transport of substrates into the cell (Hase et al., 2001). Other physical characters imputed to marine bacteria include facultative psychrophilicity (Bedford, 1933), higher tolerance to pressure than their terrestrial counterparts (Zobell and Morita, 1957), the capacity to survive in seawater, mostly Gram-negative rods (Buck, 1982), and motile spore formers (Buerger et al., 2012) which distinguishes them from terrestrial bacteria. β-aminoglutaric acid or β-glutamate, which is rare in nature, is present in higher amounts in marine sediments and is utilized by the marine bacteria as osmolytes. Some of the thermophilic marine bacteria isolated from the deep-sea hydrothermal vents are also capable of producing nitrogen. The most unique feature of a photosynthetic marine bacterial genome is the presence of rhodopsin, which contains 2197 genes. In addition to that, marine cyanobacteria also harbor a similar pattern of gene contents which are correlated with their isolation sources. The sole cause behind the diverse genetic level in marine microbes is due to the acquisition of an alternative mechanism for obtaining carbon and energy. Copiotrophic marine habitats have higher genetic potential to sense, undergo transduction, and integrate extracellular stimuli. These characteristics are likely to be crucial for their ability to fine-tune and rapidly respond to changing environmental conditions like sudden nutrient influx or depletion.

1.5 Adaptation in Extreme Environmental Conditions

The vast diversity of microorganisms is significant to the functional role they play in their normal habitat. They respond very quickly to changing environmental patterns which makes them ideal for potential bioremediation and bioindicator purposes. There are various changes that occur periodically in the environment which include variation in temperature and pH of the surrounding environment, changing patterns of light, sea level rise, tropical storms, and terrestrial inputs. In marine environments, microorganisms get continuous exposure to changes in oceanic temperature; however, the level of exposure varies in different microbial niches. Some groups of microorganisms overcome this problem by shifting their physical locations beneath sediments or by symbiosis with other organisms, a trait mostly found in pathogenic microorganisms. Other reported mechanisms of adaptation towards elevated temperature in seawater are chemotaxis and adhesion to a β-galactoside receptor in the coral mucus (Banin et al., 2001), penetration into epidermal cells, differentiation into a viable-but-not-culturable state, intracellular multiplication, production of toxins that inhibit photosynthesis, and production of superoxide dismutase to protect the pathogen from oxidative stress. Though ocean acidification is mainly caused by accumulation of CO_2 gas in marine environments, pH has not gone below 6.0. Bacteria are more adapted to this variation of pH conditions. However, Takeuchi et al. (1997) showed that the oceanic pH will very soon go below 6.0 and may reach 5.5, which will create a serious problem. Furthermore, rainfall and river flooding add pollutants and xenobiotics into the seawater, which has the potential to alter the microbial community's structure and function. However, bacteria adapt to such situations by changing their pattern of growth rates, gene expression, and physiological or enzymatic activities, and undergoing changes in intimate or symbiotic associations with other organisms. Some groups of bacteria have also been reported to develop many unique mechanisms like the synthesis of bioactive compounds, biofilm formation in the environment, and production of biosurfactants when they are exposed to extremes in pressure, temperature, and salinity or depletion of micronutrients (Mangwani et al., 2012; Mangwani et al., 2014).

Any microorganism to be used for bioremediation practices has to possess the resistant genotype for the particular pollutant. Apart from that, microorganisms possess certain unique characteristics which make them more suitable for bioremediation practices. The bacteria used in bioremediation practices should possess metal-processing features as described in Figure 1.2. These processes include the uptake and reflux of contaminating metals; their bioabsorption, intracellular assimilation, immobilization, complexion, and precipitation; and their release (Stelting et al., 2010).

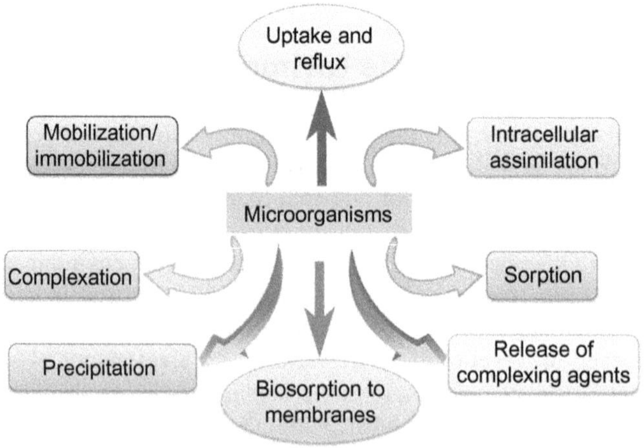

Figure 1.2 Metal-processing features of bacteria required to utilize in bioremediation.

1.6 Applications of Bacteria for Bioremediation

The use of bacteria for biodegradation of various natural and synthetic substances, and thereby reducing the level of hazards, is increasingly drawing attention. Bacteria possess a wide variety of bioremediation potentials which are beneficial from both an environmental and an economic standpoint. Bioremediation and biotransformation methods have been employed to tap the naturally occurring metabolic ability of microorganisms to degrade, transform, or accumulate toxic compounds including hydrocarbons, heterocyclic compounds, pharmaceutical substances, radionuclides, and toxic metals (Karigar and Rao, 2011). The goal in bioremediation is to stimulate microorganisms with nutrients and other chemicals that enable them to destroy the contaminants. The bioremediation systems in operation today rely on microorganisms native to the contaminated sites, encouraging them to work by supplying them with optimum levels of nutrients and other chemicals essential for their metabolism. However, researchers are still investigating ways to augment contaminated sites with non-native microbes as well as genetically engineered microbes suited for degrading the contaminants of concern at particular sites.

Microorganisms gain energy by catalyzing energy producing chemical reactions that involve breaking chemical reactions, thus transferring electrons away from the contaminant. The energy gained from these electron transfers is then invested, along with some electrons and carbon from the contaminant, to produce more cells. There are at least five critical factors that should be considered while evaluating the microbial bioremediation for site cleanup. These factors include:

1. Magnitude, toxicity, and mobility of contaminants: The site should be properly investigated and characterized to determine the (i) horizontal and vertical extent of contamination; (ii) the kinds and concentrations of contaminants at the site; (iii) the likely mobility of contaminants in future, which depends on the geological characteristics of the site.
2. Proximity of human and environmental receptors: Whether bioremediation is an appropriate cleanup remedy for any site is dependent upon the rate and extent of contaminant degradation.
3. Degradability of contaminants: The degradability of a compound is dependent on the occurrence of the compound in nature. In some instances, compounds with a high molecular weight, particularly those having complex ring structures and halogen substituents, degrade more slowly than simpler straight chain hydrocarbons or low molecular weight compounds. Thus, the rate and extent to which the compound is metabolized in the environment is determined by the availability of electron acceptors and other nutrients.
4. Planned site use: The critical factor deciding the appropriateness of bioremediation is the rate and extent of contaminant degradation.
5. Ability to properly monitor: There are inherent uncertainties in the use of bioremediation for contaminated soils and aquifers due to physical, chemical, and biological heterogeneities of the contaminated matrix.

1.6.1 Removal of Heavy Metals

Heavy metal pollution caused by various natural and anthropogenic activities is one of the most important environmental concerns. Though various physical and chemical methods have been proposed to remove such hazardous metals from the environment,

Table 1.2 Catabolic Genes Reported in Bacteria Responsible for Bioremediation

Potential Bacteria	Target Substance	Catabolic Genes	References
Pseudomonas aeruginosa	Organic and Inorganic mercury	*merA*, *merB*	De et al. (2008), Dash and Das (2012)
Cycloclasticus sp.	PAH	*phnA1*, *phnA2*, *phnA3*, and *phnA4*	Kasai et al. (2002)
Pseudomonas sp.	Phenol	*dmpN*	Selvaratnam et al. (1997)
Staphylococcus aureus	Chromate	*chrB*	Aguilar-Barajas et al. (2008)
Bacillus subtilis, *Bacillus cereus*	Cobalt–Zinc–Cadmium	*czcD*	Abdelatey et al. (2011)
Pseudomonas sp., *Bordetella* sp.	Nickel–Cobalt–Cadmium	*nccA*	Abou-Shanab et al. (2003)

they are least successful in terms of cost effectiveness, limitations, and generation of harmful substances (Wuana and Okieimen, 2011). Microorganisms solve these problems as they do not produce any by-products, and they are highly efficient even at low metal concentrations. *Vibrio harveyi*, a normal inhabitant of the saline environment, is reported to possess the potential for bioaccumulation of cadmium up to 23.3 mg Cd^{2+}/g of dry cells. In line with that, Canstein et al. (2002) reported a consortium of marine bacteria to efficiently remove mercury in a bioreactor in a disturbance-independent mechanism. A new combination of genetic systems in bacteria for the potential degradation of phenol and heavy metals was also described. Bacteria also possess the properties of chelation of heavy metals, thus removing them from the contaminated environment by the secretion of exopolysaccharides which have been evident from the reports of a marine bacterium *Enterobacter cloaceae* (Iyer et al., 2005). This bacterium has been reported to chelate up to 65% of cadmium, 20% copper, and 8% cobalt at 100 mg/L of metal concentration. In line with that, certain purple nonsulfur bacterial isolates, e.g., *Rhodobium marinum* and *Rhodobacter sphaeroides*, have also been found to possess the potential of removing heavy metals like copper, zinc, cadmium, and lead from the contaminated environments either by biosorption or biotransformation. Thus, the bacteria have been designated for assessing pollution through their tolerance and biosorption of heavy metals. However, the genetic mechanisms of bioremediation towards toxic metals have been reduced for a smaller number of bacteria (Table 1.2).

1.6.2 Degradation of Polyaromatic Hydrocarbons and Other Recalcitrants

Polyaromatic hydrocarbons (PAHs) are ubiquitous in nature and are of great environmental concern due to their persistence, toxicity, mutagenicity, and carcinogenicity in nature. However, many marine bacteria have been reported to have the potential for bioremediation of the same in the process of metabolism to produce CO_2 and metabolic intermediates, thus gaining energy and carbon for cell growth. The bioremediation potential of these bacteria can be increased, as was shown in an experiment by Latha and Lalithakumari (2001) when they transferred a catabolic plasmid of *Pseudomonas putida* containing hydrocarbon degradation genotype in a marine bacterium, which increased its efficiency. Some novel marine bacterial species like *Cycloclasticus spirillensus*, *Lutibacterium anuloederans*, and *Neptunomonas naphthovorans* have also been utilized in enhanced biodegradation of PAHs in a marine environment (Chung and King, 2001). Similarly, *Achromobacter denitrificans*, *Bacillus cereus*, *Corynebacterium renale*, *Cyclotrophicus* sp., *Moraxella* sp., *Mycobacterium* sp., *Burkholderia cepacia*, *Pseudomonas fluorescens*, *Pseudomonas paucimobilis*, *P. putida*, *Brevundimonas vesicularis*, *Comamonas testosteroni*, *Rhodococcus* sp., *Streptomyces* sp., and *Vibrio* sp. have been isolated from marine resources and are capable of degrading naphthalene by the process of mineralization. However, bacteria belonging to the genus *Cycloclasticus* play the major role in biodegradation of hydrocarbons. Bacterial isolates like *Sphingomonas paucimobilis* EPA505 have been found to utilize fluoranthene as their sole carbon source.

1.6.3 Petroleum and Diesel Biodegradation

Crude oil is the most important organic pollutant in the environment, as $1.7-8.8 \times 10^6$ tonnes of petroleum hydrocarbons are being released to the marine and estuarine environments annually (McKew et al., 2007). These organic pollutants can be degraded by the oil-eating microbes present in the environment which are used for their carbon and energy source. Some of the important genera of bacteria that are capable of degrading oil include *Acinetobacter*, *Marinococcus*, *Methylobacterium*, *Micrococcus*, *Nocardia*, *Planococcus*, and *Rhodococcus*. In terms of commercial applications, a consortium has been developed by Deppe et al. (2005) by using Arctic bacteria like *Agreia*, *Marinobacter*, *Pseudoalteromonas*, *Pseudomonas*, *Psychrobacter*, and *Shewanella* for significant degradation of crude oil and its components. In addition to that, a more potent bacterium has been isolated from the Arabian Sea sediments capable of degrading oil by 39% in 8 days in laboratory conditions. Recently, bioaugmented and biostimulated products of marine bacteria have been reported for oil remediation in marine environments.

1.6.4 Degradation of Plastic

Several broad classes of plastic include polyethylene, polypropylene, polystyrene, polyethylene terephthalate, and polyvinyl chloride are used in environments for fishing, packing, etc. which ultimately pollute the environment. However, microorganisms can develop the mechanism to degrade the plastic to nontoxic forms. Recent findings showed that *Rhodococcus ruber* degrades 8% of dry weight of plastic in 30 days in concentrated liquid culture *in vitro*. Similarly, bacterial isolates belonging to genera *Shewanella*, *Moritella*, *Psychrobacter*, and *Pseudomonas* isolated from the deep seas of Japan possess the potential of degrading ε-caprolactone in an efficient manner. Some mangrove-associated bacterial species like *Micrococcus*, *Moraxella*, *Pseudomonas*, *Streptococcus*, and *Staphylococcus* were also found to degrade 20% of plastic (Kathiresan, 2003).

Besides their bioremediation function, bacteria have also been used for biosurfactant production from *Acinetobacter anitratus*, *Bacillus pumilus*, *Bacillus subtilis*, *Myroides* sp., *Micrococcus luteus*, and *V. parahaemolyticus* which may be utilized in the process of enhanced bioremediation.

1.7 Factors of Bioremediation

The optimization and control of bioremediation process is a complex system of many factors. These include the existence of a microbial population capable of degrading the pollutants, the availability of contaminants to the microbial population, and the environmental factors (i.e., type of soil, temperature, pH, and the presence of oxygen or other electron acceptors and nutrients) (Table 1.3). Microorganisms can be isolated from almost any environmental conditions. Microbes can adapt and grow at subzero temperatures as well as in extreme heat, desert conditions, water, and anaerobic conditions; with an excess of oxygen; in the presence of hazardous compounds; or on any waste stream. The main requirements are an energy source and a carbon source for microbes and other biological systems that can be used to degrade or remediate.

Among the environmental factors, carbon is the most fundamental element of living forms and is needed in greater quantities than that of other elements. In addition to hydrogen, oxygen and nitrogen constitute about 95% of its weight. The type of bioremediation depends on the concentration of soil contaminants, presence and/or absence of other nutrients such as phosphorus and sulfur. For optimum microbial bioremediation, nutritional requirement of carbon to nitrogen ratio is 10:1 and carbon to phosphorus is 30:1.

For high concentrations of contaminants, the soil is agitated in a water solution containing an interface active agent to separate it from the soil. Then bioremediation

Table 1.3 Linked Factors of Efficient Bioremediation

Factors	Condition Required
Microorganisms	Aerobic or anaerobic
Natural biological processes of microorganism	Catabolism and anabolism
Environmental factors	Temperature, pH, oxygen content, electron acceptor/donor
Nutrients	Carbon, nitrogen, oxygen
Soil moisture	25–28% of water holding capacity
Type of soil	Low clay or silt content

Table 1.4 Optimum Environmental Conditions Required for Microbial Activity

Environmental Factors	Optimum Conditions	Condition Required for Microbial Activity
Available soil moisture	25–85% water holding capacity	25–28% of water holding capacity
Oxygen	>0.2 mg/L DO, >10% air-filled pore space for aerobic degradation	Aerobic, minimum air-filled pore space of 10%
Redox potential	Eh > 50 mV	
Nutrients	C:N:P = 120:10:1 molar ratio	N and P for microbial growth
pH	6.5–8.0	5.5–8.5
Temperature	20–30°C	15–45°C
Contaminants	Hydrocarbon 5–10% of dry weight of soil	Not too toxic
Heavy metals	700 ppm	Total content 2000 ppm

is started to efficiently clean the soil. In the experimental stage, bioremediation alone has been able to turn contaminated soil into soil suited for landscaping. In less contaminated soils, contaminants can be treated using bioremediation alone. It may take 6 months to a year to purify soil containing 2% heavy oils, but at a concentration of 0.8%, the job can be done in about 1–2 months. Thus, this environmentally friendly method makes it possible to recycle and reuse soil without much effort. The optimum conditions of the environmental factors are provided in Table 1.4.

1.8 Microbial Bioremediation Strategies

1.8.1 In situ *Bioremediation*

In situ bioremediation is the application of a biological treatment to clean up hazardous compounds present in the environment. The optimization and control of microbial transformations of organic contaminants requires the integration of many scientific and engineering disciplines. Some of the *in situ* bioremediation practices have been discussed below.

Biosparging: This involves the injection of air under pressure below the water table to increase ground water oxygen concentrations and to enhance the rate of biological degradation of contaminants by naturally occurring bacteria. Biosparging increases the mixing in the saturated zone and thereby increases the contact between soil and ground water. The ease and low cost of installing small-diameter air injection points allow considerable flexibility in the design and construction of the system.

Bioventing: Bioventing is a promising new technology that stimulates the natural *in situ* biodegradation of any aerobically degradable compounds by providing oxygen to the existing soil microorganisms. It uses low air flow rates to provide only enough oxygen to sustain microbial activity. Oxygen is most commonly supplied through direct air injection into residual contamination in soil by means of wells. Henceforth, the adsorbed fuel residuals are biodegraded and volatile compounds are also biodegraded as vapors move slowly through biologically active soil.

Bioaugmentation: Bioaugmentation is the introduction of a group of natural microbial strains or a genetically engineered strain to treat contaminated soil or water. Most commonly, it is used in municipal waste water treatment to restart activated sludge bioreactors. At sites where soil and ground water are contaminated with chlorinated ethanes, such as tetrachloroethylene and trichloroethylene, bioaugmentation is used to ensure that the *in situ* microorganisms can completely degrade these contaminants to ethylene and chloride, which are nontoxic in nature.

Biopiling: This is a full-scale technology in which excavated soils are mixed with soil amendments, placed on a treatment area, and further bioremediated using forced aeration. The contaminants are reduced to carbon dioxide and water. A basic biopile system includes a treatment bed, an aeration system, an irrigation/nutrient system, and a leachate collection system. Moisture, heat, nutrients, oxygen, and pH are also controlled to enhance biodegradation. The irrigation/nutrient system is buried under the soil to pass air and nutrients either by vacuum or positive pressure.

1.8.2 Ex situ *Bioremediation*

The most common *ex situ* bioremediation practice involves composting, which is a process by which organic wastes are degraded by microorganisms, typically at elevated temperatures. A typical compost temperature is in the range of 55–65°C. The increased temperatures result from heat produced by microorganisms during the degradation of the organic material in the waste.

1.8.3 *Bioreactors*

Slurry or aquatic reactors are used for *ex situ* treatment of contaminated soil and water pumped up from a contaminated plume. Bioreactors involve the processing of contaminated solid material or water through an engineered containment system. The slurry reactor is a containment vessel and apparatus used creates a three-phase (solid, liquid, and gas) mixing condition. It increases the bioremediation rate of soil bound water soluble pollutants by increasing the bioavailability of the target contaminants to the biomass. The various advanced methods applied in bioremediation practices are described in Table 1.5.

1.8.4 *Alternative Bioremediation Technologies*

Bioremediation technologies responding to the various contaminants may be divided into three discrete categories: (i) nutrient enrichment, (ii) seeding with naturally occurring microorganisms, and (iii) seeding with genetically engineered microorganisms (GEMs) (Table 1.6).

Table 1.5 Various Developmental Methods Applied in Bioremediation

Technique	Examples	Benefits	Applications	References
In situ	Biosparging	Most efficient and noninvasive	Abilities of indigenous microorganisms for biodegradation in presence of metals and inorganic compounds	Sei et al. (2001), Niu et al. (2009)
	Bioventing	Relatively passive	Environmental parameters	
	Bioaugmentation	Naturally attenuated process, treat soil and water	Biodegradability of pollutants	
			Distribution of pollutants	
Ex situ	Land farming	Cost effective, simple, self-heating	Surface application, aerobic process, application of organic materials to natural soils followed by irrigation	Antizar-Ladislao et al. (2008)
	Composting	Low cost, rapid reaction rate, self-heating		
	Biopiles	Can be done onsite		
Bioreactors	Slurry reactors	Rapid degradation kinetics and optimized environmental parameters	Faster treatment of polluted soils by biostimulation and bioaugmentation	Behkish et al. (2007)
	Aqueous reactors	Enhances mass transfer and effective use of inoculants and surfactant	Toxic concentrations of contaminants	

(*Continued*)

Table 1.5 (Continued)

Technique	Examples	Benefits	Applications	References
Precipitation/ flocculation	Nondirected physicochemical complexation reaction between dissolved contaminants and charged cellular components	Cost effective	Removal of heavy metals	Natrajan (2008)
Microfiltration	Microfiltration membranes are used at a constant pressure	Remove dissolved solids rapidly	Waste water treatment; recovery and reuse of more than 90% of original waste water	–
Electrodialysis	Uses cation and anion exchange membrane pairs	Withstand high temperature and can be reused	Removal of dissolved solids efficiently	–

Table 1.6 Principal Features of Alternative Bioremediation Approach

Nutrient Enrichment
• Intended to overcome the chief limitation on the rate of the natural biodegradation • Most suited of the three approaches as well as most promising approach for most types of spills • No indication that fertilizer causing algal blooms or other significant adverse impacts
Seeding
• Intended to take advantage of the properties of the most efficient species for degradation • May not be necessary at most sites as in few locations degrading microbes do not exist • Requirements for successful seeding are more demanding than those for natural enrichment • In some instances, seeding may help biodegradation getting started much faster
Use of GEMs
• Probably not needed in most of the cases due to the availability of naturally occurring microbes • In some cases, potential use for components of petroleum not degradable by naturally occurring microorganisms • Development and use could face major regulatory hurdles

1.8.5 Use of Microbial Consortia for Bioremediation

The use of microbial consortia instead of pure cultures has increased dramatically due to their known synergistic metabolism, which improves the efficiency of hydrocarbons and other chemical degradations. The biodegradation process becomes much more efficient and much faster with the use of consortia due to many factors: (i) the metabolic intermediate of one bacteria can be utilized by another for efficient degradation; (ii) the system becomes much faster from a bioremediation point of view; (iii) suitable trapping methods may be employed for efficient degradation, etc. The uses of some microbial consortia are shown in Table 1.7.

1.8.6 Improvement of the Strains by Genetic Manipulation for Enhanced Bioremediation

Microbial metabolic potential provides an effective mechanism for eliminating environmental pollutants. Anthropogenic pollution introduces some xenobiotic substances to which bacteria have not been exposed before. Upon exposure, resistant bacteria slowly change their metabolic pathway to survive with the stress. However, in order to increase the bioremediation potential and/or metabolic activity of any bacteria, insertion of certain functional genes into their genome is necessary. This phenomenon can be achieved by insertion of new genes into the genomic complexion, insertion of new plasmid, alteration of metabolic pathways like transport

Table 1.7 Examples of Some Potential Microbial Consortia for Bioremediation

Pollutants	Bacterial Consortium	Reference
Oil spills	*Brachybacterium* sp., *Cytophaga* sp., *Sphingomonas* sp., *Pseudomonas* sp.	Angelim et al. (2013)
n-alkane	*Micrococcus* sp., *Bacillus* sp., *Corynebacterium* sp., *Flavobacterium* sp., *Pseudomonas* sp.	Rahman et al. (2003)
Diesel hydrocarbon	*Moraxella saccharolytica, Alteromonas putrefaciens, Klebsiella pneumonia, Pseudomonas fragi*	Sharma and Rehman (2009)
Crude petroleum oil	*Acinetobacter faecalis, Staphylococcus.* sp., *Neisseria elongate*	Mukred et al. (2008)

and chemotaxis, and most importantly, adaption of features towards the environmental conditions. Due to significant developments in the field of molecular microbiology and genetics, there are success stories of the development of genetically engineered microbes for bioremediation of toxic substances. However, limited reports are available to date for the genetic manipulation of marine bacteria to achieve the goal of enhanced bioremediation. Insertion of the *bmtA* gene coding for metallothionein into a suitable vector and its transformation into marine bacteria has been conducted and successfully employed in highly metal-contaminated environments. Similarly, *Pseudoalteromonas haloplanktis*, possessing a shuttle plasmid-encoding suppressor for amber mutation, has been used for genetic manipulation for bioremediation (Kivela et al., 2008). Bacteria possessing plasmid with a *merA* gene, responsible for converting the toxic form of mercury to the nontoxic form, may be transformed into marine bacteria for better applications in field conditions of bioremediation of mercury (Dash et al., 2013; Dash et al., 2014). *Deinococcus radiodurans*, the most radio-resistant organism, has been modified genetically to consume and digest toluene and the ionic form of mercury from nuclear wastes (Brim et al., 2000). A list of bacteria possessing genetic alterations in their genome to increase their bioremediation potential is given in Table 1.8.

1.9 Pros and Cons of Using Bacteria in Bioremediation

Bacteria are found in a wide range of environmental conditions, from the sea floor to fish stomachs, and develop unique mechanisms of resistance in adverse and diverse conditions. Thus, there is ample opportunity to employ them as potential bioremediating agents. When a bacterium utilizes the contaminant as its food source, its number increases rapidly in the contaminated environments; on subsequent decontamination, the number decreases to produce harmless biomass. The process is cost effective in comparison to chemical processes and can be carried out onsite. Utilization of bacteria in bioremediation is highly specific; hence, the

Table 1.8 List of Some Engineered Bacteria Developed for Enhanced Bioremediation

Microorganism	Modification	Application	Reference
Vibrio harveyi	Conjugation with *E. coli*	Detection of mutagens	Czyz et al. (2000)
Staphylococcus aureus	Fusion of *arsB* gene with lux genes	Antimonite and arsenite sensing	Ramanathan et al. (1997)
Synechococcus sp.	Insertion of *smtA* gene	Heavy metal tolerance	Sode et al. (1998)
Sulfate reducing bacteria (SRB)	Consortium with other SRB	Chromate reduction	Cheung and Gu (2003)
Thalassospira lucentensis	Change in culture medium	Hydrocarbon degradation	Sutiknowati (2007)
Nocardia sp.	NA	Remediation of oil contaminated soil	Balba et al. (1998)
E. coli	Change in substrate specificity	PCB, benzene, toluene	Kumamaru et al. (1998)
E. coli FM5/pKY287	Regulation	TCE, toluene	Winter et al. (1989)

chance of forming harmful by-products is lower, which is the major advantage of utilizing these isolates. However, there are some disadvantages in the process of using bacteria. In the case of mixed contaminants, finding a suitable consortium becomes difficult. In the same case, the process is time consuming, sometimes taking years to finish. Though there are fewer chances of forming by-products, in some cases, lethal by-products may form. After the process is over, the bacterial biomass is degraded, and the serious problem of biofouling may occur. Another problem associated with the use of recombinant strains is the instability of the cloned genes in the contaminated environment due to change of habitat. These problems persist not only with the marine microorganisms but also with bacterial entities isolated from any environments. However, when the potential of the microorganisms is concerned in bioremediation, marine bacteria have been proven to be the most valuable and efficient candidates.

1.10 Conclusion and Future Prospects

One of the major problems that the twenty-first century is facing is environmental pollution, and research communities are paying close attention. The global requirement for a solution to this problem includes various remediation aspects, but bioremediation is one step ahead of all these due to its many advantages over other remediation protocols. Among various microorganisms employed in bioremediation, marine bacteria can adapt quickly to rapidly changing, noxious environments and may potentially be utilized to solve the problem by remediating toxic materials. Though many studies have been conducted and a large number of marine microbial entities have been discovered so far, the microbial diversity from different marine habitats is yet to be explored. Hence, by combining their molecular aspects with metabolic approaches, the microbial diversity of the oceanic environment can be explored. The treatment of environmental pollution by employing microorganisms is a promising technology; however, various genetic approaches to optimizing enzyme production, metabolic pathways, and growth conditions will be highly useful to meet the demand.

Acknowledgments

The research at the Laboratory of Environmental Microbiology and Ecology (LEnME), Department of Life Science, National Institute of Technology, Rourkela, Odisha, India is supported by research grants from the Department of Biotechnology, the Ministry of Science and Technology, the Government of India, and the Indian Council of Agricultural Research.

References

Angelim, A.L., Costa, S.P., Farias, B.C.S., Aquino, L.F., Melo, V.M.M., 2013. An innovative bioremediation strategy using a bacterial consortium entrapped in chitosan beads. J. Environ. Manage. 127, 10–17.

Antizar-Ladislao, B., Spanova, K., Beck, A.J., Russell, N.J., 2008. Microbial community structure changes during bioremediation of PAHs in an aged coal-tar contaminated soil by in-vessel composting. Int. Biodeterior. Biodegradation. 61, 357–364.

Balba, M.T., Al-Awadhi, N., Al-Daher, R., 1998. Bioremediation of oil contaminated soil: microbiological methods for feasibility assessment and field evaluation. J. Microbiol. Met. 32, 155–164.

Banin, E., Khare, S.K., Naider, F., Rosenberg, E., 2001. Proline-rich peptide from the coral pathogen *Vibrio shiloi* that inhibits photosynthesis of zooxanthellae. Appl. Environ. Microbiol. 67, 1536–1541.

Bedford, R.H., 1933. Marine bacteria of the northern Pacific Ocean. The temperature range of growth. Contrib. Cancer Biol. Fish. 8, 433–438.

Behkish, A., Lemoine, R., Sehabiague, L., Oukaci, R., Morsi, B.I., 2007. Gas holdup and bubble size behavior in a large-scale slurry bubble column reactor operating with an organic liquid under elevated pressures and temperatures. Chem. Eng. J. 128, 69–84.

Brim, H., McFarlan, S.C., Fredrickson, J.K., Minton, K.W., Zhai, M., Wackett, L.P., et al., 2000. Engineering *Deinococcus radiodurans* for metal remediation in radioactive mixed waste environments. Nat. Biotechnol. 18, 85–90.

Buck, J.D., 1982. Nonstaining (KOH) method for determination of Gram reactions of marine bacteria. Appl. Environ. Microbiol. 44, 992–993.

Buerger, S., Spoering, A., Gavrish, E., Leslin, C., Ling, L., Epstein, S.S., 2012. Microbial scout hypothesis and microbial discovery. Appl. Environ. Microbiol. 78, 3229–3233.

Canstein, H., Kelly, S., Li, Y., Wagner-Dobler, I., 2002. Species diversity improves the efficiency of mercury-reducing biofilms under changing environmental conditions. Appl. Environ. Microbiol. 68, 2829–2837.

Chiarle, S., Ratto, M., Rovatti, M., 2000. Mercury removal from water by ion exchange resins adsorption. Water Res. 34, 2971–2978.

Chung, W.K., King, G.M., 2001. Isolation, characterization, and polyaromatic hydrocarbon degradation potential of aerobic bacteria from marine macro-faunal burrow sediments and description of *Lutibacterium anuloederans* gen. nov., sp. nov., and *Cycloclasticus spirillensus* sp. nov. Appl. Environ. Microbiol. 67, 5585–5592.

Czyz, A., Jasiecki, J., Bogdan, A., Szpilewska, H., Grzyn, G.W., 2000. Genetically modified *Vibrio harveyi* strains as potential bioindicators of mutagenic pollution of marine environments. Appl. Environ. Microbiol. 66, 599–605.

Das, S., Lyla, P.S., Khan, S.A., 2006. Marine microbial diversity and ecology: present status and future perspectives. Curr. Sci. 90 (10), 1325–1335.

Dash, H.R., Das, S., 2012. Bioremediation of mercury and importance of bacterial *mer* genes. Int. Biodeterior. Biodegradation. 75, 207–213.

Dash, H.R., Mangwani, N., Chakraborty, J., Kumari, S., Das, S., 2013. Marine bacteria: potential candidates for enhanced bioremediation. Appl. Microbiol. Biotechnol. 97, 561–571.

Dash, H.R., Mangwani, N., Das, S., 2014. Characterization and potential application in mercury bioremediation of highly mercury resistant marine bacterium *Bacillus thuringiensis* PW-05. Environ. Sci. Pollut. Res. 21, 2642–2653.

De, J., Ramaiah, N., Vardanyan, L., 2008. Detoxification of toxic heavy metals by marine bacteria highly resistant to mercury. Mar. Biotechnol. 10, 471–477.

Deppe, U., Richnow, H.H., Michaelis, W., Antranikian, G., 2005. Degradation of crude oil by an arctic microbial consortium. Extremophiles. 9, 461–470.

Doney, S.C., Ruckelshaus, M., Duffy, J.E., Barry, J.P., Chan, F., English, C.A., et al., 2012. Climate change impacts on marine ecosystems. Ann. Rev. Mar. Sci. 4, 11–37.

Hase, C.C., Fedorova, N.D., Galperin, M.Y., Dibrov, P.A., 2001. Sodium ion cycle in bacterial pathogens: evidence from cross-genome comparisons. Microbiol. Mol. Biol. Rev. 65, 353–370.

Henneberry, Y.K., Kraus, T.E.C., Fleck, J.A., Krabbenhoft, D.P., Bachand, P.M., Horwath, W.R., 2011. Removal of inorganic mercury and methylmercury from surface waters following coagulation of dissolved organic matter with metalbased salts. Sci. Total Environ. 409, 631–637.

Iyer, A., Mody, K., Jha, B., 2005. Biosorption of heavy metals by a marine bacterium. Mar. Pollut. Bull. 50, 340–343.

Karigar, C.S., Rao, S.S., 2011. Role of microbial enzymes in the bioremediation of pollutants: a review. Enzyme Res. Available from: http://dx.doi.org/doi:10.4061/2011/805187.

Kasai, Y., Kishira, H., Harayama, S., 2002. Bacteria belonging to the genus Cycloclasticus play a primary role in the degradation of aromatic hydrocarbons released in a marine environment. Appl. Environ. Microbiol. 68, 5625–5633.

Kathiresan, K., 2003. Polythene and plastics-degrading microbes from the mangrove soil. Rev. Biol. Trop. 51, 629–634.

Kivela, H.M., Madonna, S., Krupovic, M., Tutino, M.L., Bamford, J.K.H., 2008. Genetics for *Pseudoalteromonas* provides tools to manipulate marine bacterial virus PM2. J. Bacteriol. 190, 1298–1307.

Kumamaru, T., Suenaga, H., Mitsuoka, M., Furukawa, K., 1998. Enhanced degradation of polychlorinated biphenyls by directed evolution of biphenyl dioxygenase. Nat. Biotechnol. 16, 663–666.

Latha, K., Lalithakumari, D., 2001. Transfer and expression of a hydrocarbon degrading plasmid pHCL from *Pseudomonas putida* to marine bacteria. World J. Microbiol. Biotechnol. 17, 523–528.

Lauro, F.M., McDougald, D., Thomas, T., Williams, T.J., Egan, S., Rice, S., et al., 2009. The genomic basis of trophic strategy in marine bacteria. PNAS. 106, 15527–15533.

MacLeod, R.A., Onofrey, E., 1957. Nutrition and metabolism of marine bacteria. III. The relation of sodium and potassium to growth. J. Cell. Comp. Physiol. 50, 389–401.

Mangwani, N., Dash, H.R., Chauhan, A., Das, S., 2012. Bacterial quorum sensing: functional features and potential applications in biotechnology. J. Mol. Microbiol. Biotechnol. 22, 215–227.

Mangwani, N., Shukla, S.K., Rao, T.S., Das, S., 2014. Calcium mediated modulation of *Pseudomonas mendocina* NR802 biofilm influences the phenanthrene degradation. Colloids Surf. B. 114, 301–309.

McKew, B.A., Coulon, F., Osborn, A.M., Timmis, K.N., McGenity, T.J., 2007. Determining the identity and roles of oil-metabolizing marine bacteria from the Thames estuary. Environ. Microbiol. 9, 165–176.

Mohan, D., Gupta, V.K., Srivastava, S.K., Chander, S., 2001. Kinetics of mercury adsorption from wastewater using activated carbon derived from fertilizer waste. Colloids Surf. A Physicochem. Eng. Asp. 177, 169–181.

Mukred, A.M., Hamid, A.A., Hamzah, A., Yusoff, W.M.W., 2008. Development of three bacteria consortium for the bioremediation of crude petroleum-oil in contaminated water. J. Biol. Sci. 8, 73–79.

Narang, U., Bhardwaj, R., Garg, S.K., Thukral, A.K., 2011. Phytoremediation of mercury using *Eichhornia crassipes* (Mart.) Solms. Int. J. Environ. Waste Manage. 8, 92–105.

Natrajan, K.A., 2008. Microbial aspects of acid mine drainage and its bioremediation. Trans. Nonferr. Met. Soc. China. 18, 1352–1360.

Niu, G., Zhang, J., Zhao, S., Liu, H., Boon, N., Zhou, N., 2009. Bioaugmentation of a 4-chloronitrobenzene contaminated soil with *Pseudomonas putida* ZWL73. Environ. Pollut. 57, 763–771.

OECD, 2013. OECD Factbook 2013: Economic, Environmental and Social Statistics. OECD Publishing. Available from: http://dx.doi.org/doi:10.1787/factbook-2013-en.

Pratt, D., Happold, F.C., 1960. Requirements for indole production by cells and extracts of a marine bacterium. J. Bacteriol. 80, 232–236.

Rahman, K.S.M., Rahman, T.J., Kourkoutas, Y., Petsas, I., Marchant, R., Banat, I.M., 2003. Enhanced bioremediation of n-alkane in petroleum sludge using bacterial consortium amended with rhamnolipid and micronutrients. Bioresour. Technol. 90, 159–168.

Ramadevi, A., Srinivasan, K., 2005. Agricultural solid waste for the removal of inorganics: adsorption of mercury (II) from aqueous solution by Tamarind nut carbon. Indian J. Chem. Technol. 12, 407–412.

Rhodes, M.E., Payne, W.J., 1962. Further observations on effects of cations on enzyme induction in marine bacteria. Antonie Van Leeuwenhoek. 28, 302–314.

Sei, K., Nakao, M., Mori, K.M.I., Kohno, T., Fujita, M., 2001. Design of PCR primers and a gene probe for extensive detection of poly (3-hydroxybutyrate) (PHB)-degrading bacteria possessing fibronectin type III linker type-PHB depolymerases. Appl. Microbiol. Biotechnol. 55, 801–806.

Seigle-Murandi, F., Guiraud, P., Croize, J., Falsen, E., Eriksson, K.L., 1996. Bacteria are omnipresent on *Phanerochaete chrysosporium* Burdsall. Appl. Environ. Microbiol. 62, 2477–2481.

Sharma, A., Rehman, M.B., 2009. Laboratory scale bioremediation of diesel hydrocarbon in soil by indigenous bacterial consortium. Indian J. Exp. Biol. 47, 766–769.

Stelting, S., Burns, R.G., Sunna, A., Visnovsky, G., Bunt, C., 2010. Immobilization of *Pseudomonas* sp. strain ADP: a stable inoculant for the bioremediation of atrazine. In: 19th World Congress of Soil Science, Soil Solutions for a Changing World, Brisbane, Australia, August 1–6, 2010.

Sutiknowati, L.I., 2007. Hydrocarbon degrading bacteria: isolation and identification. Makara Sains. 11, 98–103.

Takeuchi, K., Fujioka, Y., Kawasaki, Y., Shirayama, Y., 1997. Impacts of high concentrations of CO_2 on marine organisms: a modification of CO_2 ocean sequestration. Energy. Convers. Mgmt. 38, S337–S341.

US Environmental Protection Agency, 2000. Marine Litter—Trash That Kills. Available from: <http://www.epa.gov/owow/oceans/debris/toolkit/files/trash_that_kills508.pdf>.

U.S. EPA, 2007. Treatment Technologies for Mercury in Soil, Waste, and Water. Office of Superfund Remediation and Technology Innovation, Washington, DC, 20460.

Wang, J., Chen, C., 2009. Biosorbents for heavy metals removal and their future. Biotechnol. Adv. 27, 195–226.

WHO, 2006. Air Quality Guidelines for Particulate Matter, Ozone, Nitrogen Dioxide and Sulfur Dioxide-Global Update 2005. World Health Organization, Geneva.

Winter, R.B., Yen, K., Ensley, B.D., 1989. Efficient degradation of trichloroethylene by a recombinant *Escherichia coli*. Biotechnol. 7, 282–285.

Wuana, R.A., Okieimen, F.E., 2011. Heavy metals in contaminated soils: a review of sources, chemistry, risks and best available strategies for remediation. ISRN Ecol. Available from: http://dx.doi.org/doi:10.5402/2011/402647.

Zhu, X., Elimelech, M., 1997. Colloidal fouling of reverse osmosis membranes: measurements and fouling mechanisms. Environ. Sci. Technol. 31, 3654–3662.

Zobell, C.E., Morita, R.Y., 1957. Barophilic bacteria in some deep sea sediments. J. Bacteriol. 73, 563–568.

2 Heavy Metals and Hydrocarbons: Adverse Effects and Mechanism of Toxicity

Surajit Das, Ritu Raj, Neelam Mangwani, Hirak R. Dash and Jaya Chakraborty

Laboratory of Environmental Microbiology and Ecology (LEnME), Department of Life Science, National Institute of Technology, Rourkela, Odisha, India

2.1 Introduction

Rapid industrialization, population growth, and complete disregard for environmental health have led to global environmental pollution. The release of pollutants into the environment may occur accidentally or due to anthropogenic activities which ultimately results in soil, water, and air pollution, leading to many health hazards. From the galaxy of chemicals, heavy metals and polycyclic aromatic hydrocarbons (PAHs) are two of the major environmental pollutants (Dash et al., 2013). From the environment, these chemicals enter into the food chain. The flow of chemicals through the lower constituents of the food chain to different tropic levels imparts risk to the ecosystem, as chemicals tend to bioaccumulate and can be transferred from one food chain to another. Pollutants have been detected in various food chains where the results are usually detrimental to microorganisms, plants, animals, and humans alike (Vinodhini and Narayanan, 2008).

Contamination of the soil, water, and air with organic compounds and heavy metals has drawn much public concern as many of them are toxic and mutagenic as well as carcinogenic. Out of these, PAHs are considered to be the most acute toxic component; they are associated with many chronic diseases and disrupt the natural equilibrium between living species and their natural environment (Mangwani et al., 2014). In earlier days, heavy metals were present naturally in soils as natural components; however, currently their presence in the environment has been increased due to human activities. This is a widespread problem around the globe in areas where excessive concentration of heavy metals such as As, Cd, Hg, Pb, Cr, and Se can be found in soil. Contamination of soils is one of the most critical environmental problems as the contaminants infiltrate deep into the layer of underground water and pollute the ground water as well as the surface water (Perfus-Barbeoch et al., 2002).

Microbial Biodegradation and Bioremediation. DOI: http://dx.doi.org/10.1016/B978-0-12-800021-2.00002-9

Contamination of the environment with hydrocarbons and toxic metals constitutes a nuisance to the environment due to their persistent nature and tendency to spread into ground as well as surface water. There are around 30 chemical elements which play a pivotal role in various biochemical and physiological mechanisms in living organisms and are recognized as the essential elements of life. As the concentrations of toxic metals and metalloids have largely increased due to anthropogenic activities, they can disturb important biochemical processes, thus constituting an important threat to the health of plants and animals (Da Silva and Williams, 2001). Living organisms absorb these elements from soils, sediments, and water by contact with their external surfaces through ingestion and through inhalation of airborne particles as well as vaporized metals. These excess heavy metals aid neurological depositions, enhancing oxidative damage—the key component for chronic inflammatory disease as well as a suggested initiator of cancer. Further, potential pathological roles of metal ions are emerging in the form of premature aging and many other characteristic features of a wide range of diseases (Chowdhury and Chandra, 1986; Duruibe et al., 2007). In this chapter, the adverse effects of heavy metals and PAHs on the ecosystem and human health have been discussed in detail.

2.2 Source of Contaminants in the Environment

The major sources of environmental pollution originate from the environment itself. The anthropogenic uses of many substances lead to the formation and generation of these hazardous materials to pollute the environment. Though many natural sources are present, the majority of pollutant comes from human interference that triggers the amount of exposure. However, anthropogenic sources contribute higher amounts of toxic substances to the environment than do the natural sources (Figure 2.1). Detection of the origin points of pollutants is necessary to undertake any suitable remedial measures to decrease the pace of spread and limit further consequences of these toxic elements (Nriagu, 1979). The origins of contaminants may be entirely natural, such as from radon emissions, volcanic eruptions, earthquakes, floods, and storms. Other contaminants are produced by industrial activities.

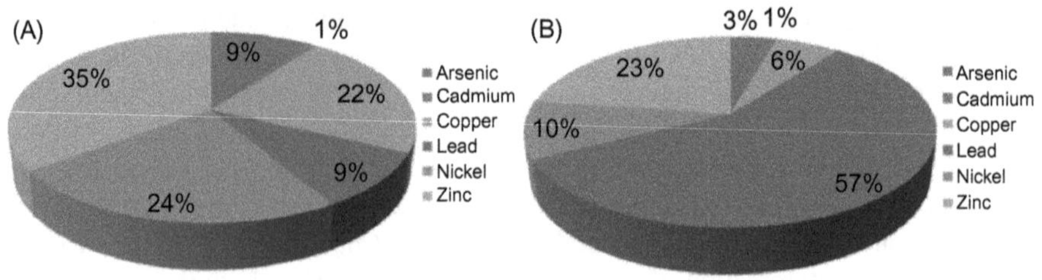

Figure 2.1 Comparative account of sources of pollutants in nature (A) natural sources; (B) anthropogenic sources.
Source: Pacyna and Pacyna (2001).

2.2.1 Natural Sources

The major group of substances like iron, arsenic, manganese, chloride, fluorides, sulfates, and radionuclides are present naturally in certain rocks and soils; these substances contaminate when dissolved in water, whereas, organic matter and natural substances move directly to ground water (Mukhopadhyay et al., 2005). Apart from this, the *in situ* conditions play a major role in contamination. Certain chemicals lead to potential health problems when consumed in large quantities, whereas others are responsible for the physical deterioration of materials by changing their taste, odor, or color. Water resources containing these harmful substances should be removed completely before going for their domestic use.

The major natural sources of pollutants include volcanic eruptions, forest fires, and cosmetic dusts that release H_2O, CO_2, SO_2, CO, H_2S, CS, CS_2, HCl, H_2, CH_4, and HF as well as many toxic heavy metals like mercury, lead, and gold (Timmreck, 2012). Burning of forests emits more pollutants back to the environment, mostly in the form of particulate matters, which is a major source of concern as the generated particulate matters are too tiny to be filtered by any means. In another natural process a photochemical pollutant the ozone is generated which is a potential toxic substance and the greater part of it is generated in the stratosphere due to the action of ultraviolet (UV) radiation on oxygen (Piver, 1989). Another major group of pollutants of plant origin is volatile organic compounds (VOCs), which cause more asthma and other allergic reactions than anthropogenic irritants. Other sources of irritants include pollen grains from plants, grass, and trees that trigger many allergic reactions; the levels of these are highest during the growing seasons of crops.

2.2.2 Anthropogenic Sources

Anthropogenic activities are a major source of environmental pollutants. Industrialization and the increase in population have boosted the release of pollutants into the environment at an alarming rate. Power plants, industries, and domestic and agricultural waste discharge introduce diverse classes of chemicals to the environment (Figure 2.2). Heavy metals are released into the environment from various anthropogenic activities like mining, energy and fuel production, electroplating, waste water sludge treatment, nuclear fuels and from agricultural wastes (Reimann and de Caritat (2005)). Smoke stacks of power plants, traditional biomass burning, and waste incineration are stationary sources of greenhouse gases and PAHs in the air, whereas motor vehicles, aircraft, and marine vessels are mobile sources. Many pollutants are present naturally in the environment; however, anthropogenic activities have increased their emission (Uherek et al., 2010).

2.3 Major Groups of Pollutants

A substance or energy introduced into the environment having undesired adverse effects and possibly causing long- or short-term damage by changing the growth rate of plants or animal species, or by interfering with human amenities and health,

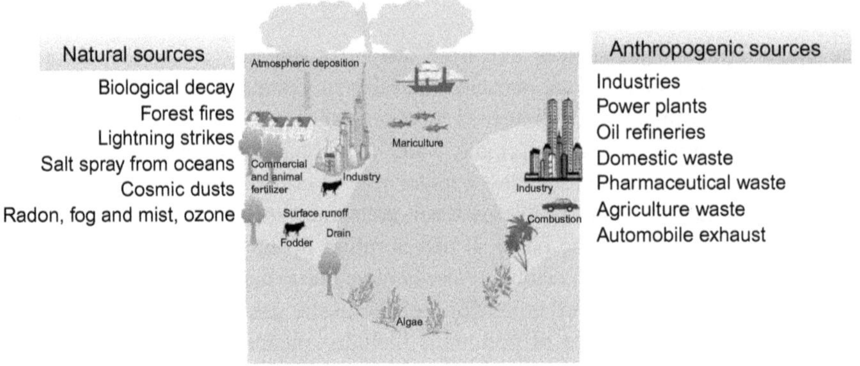

Figure 2.2 Sources of pollutants in nature.

is called a pollutant. There are many organic or inorganic compounds known as pollution based on their known or suspected carcinogenicity, mutagenicity, teratogenicity, or high acute toxicity. A few of them have been described in the following sections.

2.3.1 Heavy Metals

1. *Arsenic (As)*: Arsenic is a toxic element to multicellular life forms, whereas some groups of unicellular microorganisms can use these compounds as a respiratory metabolite. The major pollution problem associated with arsenic is the contamination of ground water. The toxicity level of arsenic depends upon its physical and chemical forms, valence state, the route of administration into the body, dose, and duration of exposure (Hughes et al., 2011). There are various commercial uses for this metal such as an alloying additive in lead solder, lead shot, battery grids, cable sheaths, and boiler piping. Paints and pharmaceutical industries also discharge huge amounts of arsenic as a waste material which ultimately meets oceanic inventories; it has been reported that the concentration of arsenic in seawater is around 0.002 ppm (Maher and Butler, 1988). The major concerns for arsenic toxicity include carcinogenicity and mutagenicity (Ratnaike, 2003).
2. *Cadmium (Cd)*: Though cadmium is mostly found in the Earth's crust, it is associated with zinc and copper, which are produced as a by-product of zinc smelting. Cadmium is absolutely not an essential trace element and is highly toxic to living systems (Bernhoft, 2013). Cadmium and its compounds come to the environment from geological or anthropogenic activities such as metal mining, smelting, and fossil fuel combustion. The permissible limit of cadmium in drinking water is 0.005 ppm (Hutton, 1987).
3. *Mercury (Hg)*: This is a rare element present in the universe but highly toxic in comparison to other heavy metals. Mercury in the environment comes from both natural and anthropogenic sources. Natural sources of mercury pollution include weathering, dissolution, and biological processes. There are various forms of mercury, but methyl mercury is the most toxic among all other forms of mercury and is a potential neurotoxin (Dash and Das, 2012). Most of the contamination to human beings comes from ingesting mercury-contaminated fishes; hence, most government agencies have regulated

the consumption of seafood containing mercury. The permissible limit of mercury in the aquatic environment is 1 ppb. The major sources of mercury pollution are industrial effluents connected with chlorine and caustic soda production, pharmaceuticals, mirror coatings, mercury lamps, and certain fungicides (Barkay and Wagner-Dobler, 2005).

4. *Lead (Pb)*: The most abundant heavy metal in nature, first explored for human use, was lead. It is highly toxic to living systems and has never been a nutritionally essential element; hence, its regular monitoring in the environment is of utmost importance from a human health point of view. The toxicity of lead increases when it is accumulated in the human body to worsen the Central Nervous system. The major source of lead pollution comes from batteries, cable sheathing, lead sheets, and pipes (Hernberg, 2000). Although in some cases it can be recovered and recycled, in most cases it is in compound form, as in paints and petrol additives, are lost to the environment, landing in the aquatic environment. The permissible limit of lead in drinking water is 0.01 ppm.
5. *Barium (Ba)*: Barium is present in the Earth's crust in a concentration of 0.50 g/kg. The most common source of barium in the environment is mineral barite and barium sulfates; there are also traces of barium present in soils (Jacobs et al., 2002). Foods such as Brazil nuts also contain traces of barium. Many industrial processes like the manufacturing of vacuum tubes, spark-plug alloys, Getter alloys, Fray's metal, and lubricants for anode rotors in X-ray machines add barium to the environment. However, the permissible limit for barium in drinking water should not be more than 0.050 ppm (Purdey, 2004).
6. *Selenium (Se)*: Levels of selenium in soil and vegetation vary greatly as a result of differences in natural geochemical cycles. The solubility and the chemical form of selenium is the decisive factor regarding its presence in drinking water. Though human beings and other animals require trace levels of selenium, more than 5 mg/kg of selenium in diet (food consumed per kg) causes chronic intoxication (Spallholz, 1997). Exceeding the permissible range of selenium intake causes health-associated problems in humans, followed by toxicity. The permissible limit of selenium in drinking water is 0.01 ppm. Manufacturers of electronics, TV cameras, solar batteries, computer cores, rectifiers, xerographic plates, and ceramics use selenium to color glass (Tinggi, 2003). In animal feeds selenium is also used as a trace essential metal.
7. *Silver (Ag)*: In nature, silver can be found in its elemental form as well as in the form of various ores. Many ores of other metals like lead, copper, and zinc also contain certain amount of silver. In certain countries silver oxide is used to disinfect water; the silver level in tap water is elevated in such instances. The permissible limit of silver in drinking water is 1 ppb (Gaiser et al., 2009). Silver is widely used in many processes like the manufacture of silver nitrate, silver bromide, and other photographic chemicals; water distillation equipment; mirrors; silver plating equipment; special batteries; table cutlery; jewelry; and dental, medical, and scientific equipment, including amalgams (Flegal et al., 2007).
8. *Chromium (Cr)*: Chromium is found in trace amount in most rocks and soils. However, in nature it is found in a highly insoluble form. All the common soluble forms are due to the contamination of industrial effluents (Megharaj et al., 2003). There are many uses of chromium that leads to contamination, including chrome alloys, chrome plating, oxidizing agents, corrosion inhibitors, pigments for the textile glass, ceramic manufacturing, and photography. Chromium in its hexavalent form is carcinogenic and the permissible limit is 0.05 ppm (Mohanty and Patra, 2011).
9. *Nickel (Ni)*: Though nickel is ubiquitous in the environment, it is an essential metal for nutrition in animals, and in some instances essential to humans. Although it is relatively nontoxic, certain nickel compounds have been reported to be carcinogenic to animals at higher concentrations.

10. *Tin (Sn)*: The most controversial element in the environment is tin and its chemical forms. In contrast to the other metals, all forms of tin are not equally biologically active. The inorganic form of tin derived from canned foods is less toxic than its organic form, known as organotin (Blunden and Wallace, 2003). Tin derivatives like tributyltin and triphenyltin, the constituents of antifouling paints, are highly toxic.

2.3.2 Organic Compounds

1. *PAHs*: PAHs are a large group of compounds formed during incomplete combustion of organic matter in electricity-generating power plants, automobiles, coal combustion etc. (Samanta et al., 2002). PAHs are widely distributed and relocated in the environment. They are highly hydrophobic in nature and tend to adsorb into the surface of soil (or sediments in a marine environment). A total of 16 PAHs (e.g., naphthalene, pyrene, phenanthrene, anthracene, and benzo[a]pyrene) have been listed as toxic pollutants by the US Environmental Protection Agency (EPA) (Perelo, 2010). They are potent carcinogens and mutagenic. Air and marine sediments are the chief sources of PAHs. PAHs bind to dust particles in air and sediments, which make it inert for degradation. However, it accumulates in the bodies of marine animals in lipid-rich organs when such individuals are routinely exposed to a mixture of PAHs (Bojes and Pope, 2007).
2. *Polychlorinated biphenyls (PCBs)*: PCBs are chlorinated aromatic compounds produced by chemical synthesis. There is no natural process that can generate PCBs. PCBs are good insulators, and thus they are commercially very significant (Hornbuckle and Robertson, 2010). PCBs have been used as coolants and lubricants in transformers, capacitors, and other electrical equipment. PCBs are present in the dust control formulations of dedusting agents and in used oil for dust suppression. Large volumes of PCBs have been introduced to the environment through the burning of PCB-containing products (ATSDR, 2001).
3. *Nitroaromatic compounds (NACs)*: NACs are relatively rare in nature. They are introduced to the environment from anthropogenic sources and synthesis. They are an essential class of chemical and used for the synthesis of diverse classes of chemicals. NACs and their derivatives are widely used as dyes, pesticides, explosives, and pharmaceuticals. NACs are recalcitrant due to the presence of a highly active nitro group, which offers functional and chemical diversity to NACs. Vast application of NAC has led to the environmental contamination of soil, ground water, and freshwater (Kulkarni and Chaudhari, 2007).
4. *Phthalates*: Phthalates are used as plasticizers and are used to make plastic flexible and resilient. Structurally they contain a phenyl ring with two attached and extended acetate groups. They are used in products like automobile parts, toys, cosmetics, automobile parts, and food packaging. Phthalates cause developmental and reproductive toxicity. Chronic exposure to them causes cancer (Shea, 2003).
5. *Polybrominated biphenyls (PBBs)*: PBBs are artificial chemicals. PBBs are used as plastic additives to make products like televisions, plastic foams, computer monitors etc. They are persistent environmental pollutants. PBBs are mixtures of brominated biphenyl compounds known as congeners (Alaee et al., 2003).

In addition to the above list, there exist many dissolved inorganic compounds associated with the esthetic and organoleptic characteristics of water. They may include aluminum, chlorides, dyes, copper, hydrogen sulfide, iron, manganese, pH, sodium, sulfates, zinc, petroleum hydrocarbons, and many more.

2.4 The Environmental Fate and Biogeochemical Cycle of Pollutants

All matter gets recycled; it can neither be created nor destroyed. Planet Earth is a closed system with respect to matter: all matter is continuously used, recycled, and reused although its forms and states change. Heavy metals and PAHs in the environment exist in different forms and states; due to their adverse effect on the environment it is essential to understand their movement in different natural systems, particularly terrestrial ecosystems and aquatic ecosystems. The environmental fate of any elements/chemicals is determined by the processes by which these elements/chemicals move and are transformed in the environment. These movements are known as biogeochemical cycles: the movement (or cycling) of matter through living and nonliving parts of a system.

2.4.1 *Biogeochemical Cycle of Heavy Metals*

The biogeochemical cycle of elements is a process where different biological, chemical, and geological interactions take place simultaneously; for example, the transport and mobilization of an element between different systems (living and nonliving), or the bioconcentration and biodegradation of elements by living organisms. The transport of elements by volatilization is much faster; here it enters the atmosphere through evaporation from soil or water (Figure 2.3). However, its tendency to remain in a particular environment in an unchanged form depends upon other interactions. Atmospheric residence time is defined as the ratio of the total mass of a chemical in a particular system to either the total emission rate or the total removal rate, under steady-state conditions. Some heavy metals (mercury and lead) have greater residence time in the atmosphere than others. Microbial diversity plays a very important role in biogeochemical cycles; microbes change the residence time and the half-life time of chemicals.

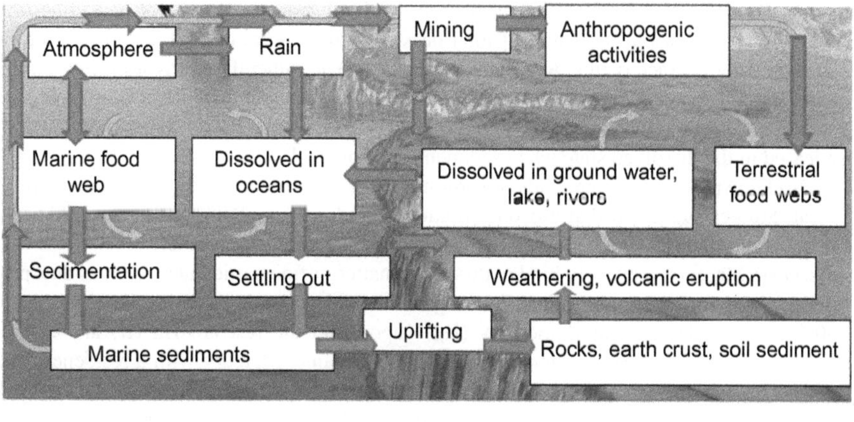

Figure 2.3 A generalized view of the biogeochemical cycles of metals in the environment.

Acclimatization is a process where continuous exposure of a microbial population to a chemical results in a more rapid transformation (biodegradation) of the chemical than initially observed. Bioconcentration and biodegradation are two antagonistic processes which involve living organisms, especially microorganisms. Bioconcentration is process leading to a higher concentration of a substance in an organism than in environmental media to which it is exposed whereas biodegradation is the transformation of a substance into new compounds through actions of microorganisms.

1. *Arsenic*: Arsenic is a complex element compared to other heavy metals because of its ionic character; it can form cationic and anionic compounds and also transform ions to neutral atoms. Arsenic is stable in several oxidation states (−III, 0, +III, +V), but the + III and + V states are the most common in natural systems among all arsine (−III), shows extremely high toxicity, but its occurrence in nature is rare. As arsenic is predominantly used in various industries and agriculture, its toxicity to the environment cannot be ignored. It is always in high concern list to understand arsenic circulation in natural system, its migration and transformation in the nature for protecting the environment against arsenic contamination (Figure 2.4). High concentrations of arsenic in ground water are often due to natural and anthropogenic activities which contribute to naturally occurring arsenic deposits. While the average abundance of arsenic in the Earth's crust is between 2 and 5 mg/kg, enrichment in igneous and sedimentary rocks, such as shale and coal deposits, is not uncommon. High arsenic concentrations in ground water are not necessarily directly linked to geologic materials with high arsenic content. Arsenic mobility is related not only to the amount of arsenic in the geological source material but also to the environmental conditions that control chemical and biological transformation of the material.
2. *Cadmium*: Natural events such as volcanic eruptions, ocean sprays, forest fires, and the release of metal-enriched particles from terrestrial vegetation are sources of atmospheric cadmium. Volcanoes are considered to be a major contributor. In the natural cycle of cadmium, weathering of crustal materials plays a major role but not sufficient to increase cadmium concentration in high range. The adsorption of cadmium into soils is strongly pH-dependent, since alkalinity favors adsorption, while low pH initiates desorption. Cadmium is expected to be more mobile in the environment than copper, zinc, and lead due to its lower affinity for the absorbents. Cadmium concentrations in bed sediments are generally at least an order of magnitude higher than in overlying water, where sorption and desorption are dependent on the nature of total heavy metal loading, type of sediment, and the surface water characteristics. Association with carbonate minerals and hydrous iron oxides make cadmium less available for mobilization by resuspension of sediments or biological activity. Cadmium absorbed by organic materials is more easily bioaccumulated or released in the dissolved state when sediments are solvated.

   No known volatile compounds of cadmium have been reported in the aquatic environment. Studies suggested that bioconcentration of cadmium (sulfate, nitrate, and chloride) in aquatic organisms can be quite high in some species, but low in others.
3. *Mercury*: Mercury appears to bind to dissolved matter or fine particulates, such as dust particles in the atmosphere or bed sediment particles in rivers and lakes. In the atmosphere, 50% of the volatile form is mercury (Hg) vapor, and the rest are Hg (II) and methyl mercury; 25−50% of Hg in the aquatic system is organic. Mercury with a residence time of few days in the atmosphere is deposited and revolatilized many times in the environment. Two characteristics of mercury, volatility and biotransformation, make it a unique

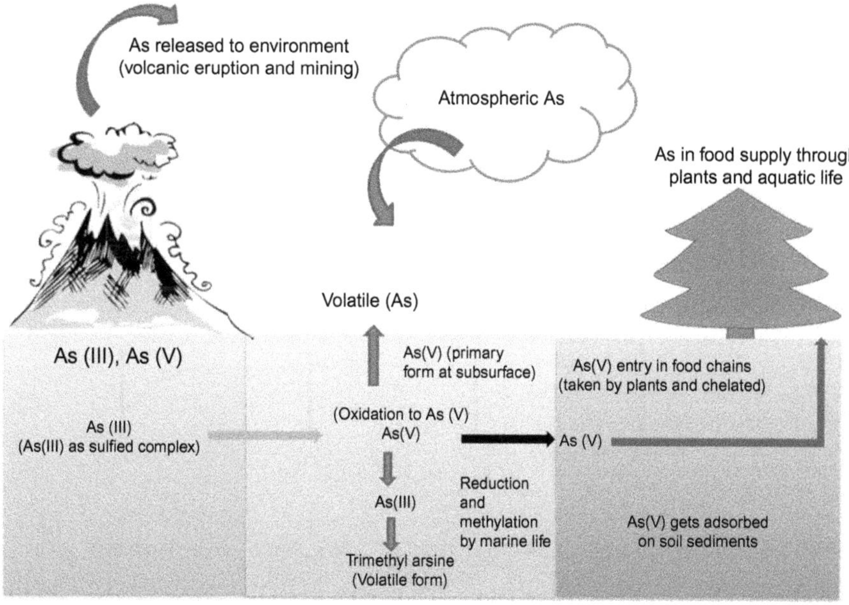

Figure 2.4 Biogeochemical cycle of arsenic in the environment.

environmental toxicant. Volatility accounts for atmospheric concentrations up to four times the level present in contaminated soils in an area. Inorganic forms of mercury could be converted to organic forms by microbial action in the biosphere (Figure 2.5).

The bioconcentration factor of mercury has been found to be 63,000 for freshwater fish, 10,000 for salt water fish, 100,000 for marine invertebrates, and 1000 for freshwater and marine plants. Mercury's accumulation rate depends on tissue, concentration of mercury at a steady-state, uptake rate, membrane transport rate, and/or depuration rate of the saturation of storage sites. Acidification of water might also increase mercury residues in aquatic animals if no new input of mercury occurs, most likely because lower pH increases ventilation rate and membrane permeability, accelerates the rates of methylation and uptake, affects partitioning between sediment and water, or reduces the growth and reproduction of marine animals.

4. *Lead*: Lead enters into the environment mostly due to various anthropogenic activities (mining, ore processing, smelting, refining, recycling, and waste disposal). Natural forces also add lead into the atmosphere (e.g., weathering of soil sediments and volcanoes). Both inorganic Pb(II), Pb(IV), and organic forms of lead exist in nature. One source of organometallic lead in the environment is the volatilization of gasoline additives. Usually, the topsoil of soil sediment retains lead when the organic matter content of soil is about 5% with a pH 5 or above. Leaching does not have a significant effect under normal conditions; most of the lead particles deposited on soil are retained and eventually become mixed into the topsoil or surface layer, whereas accumulated lead particles are taken up directly animals and by soil microorganisms. The solubility of lead decreases with time; eventually, it forms insoluble sulfate, sulfide, oxide, and phosphate salts. Lead enters into water from atmospheric runoff (largely anthropogenic), erosion (mostly

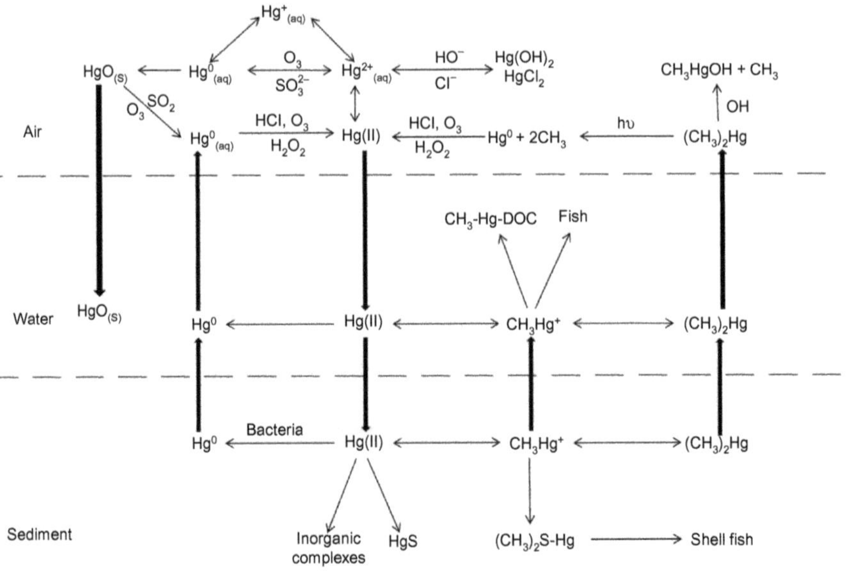

Figure 2.5 Biogeochemical cycle of mercury in the environment.

natural), and direct deposition from air (anthropogenic); little is transferred from natural ores. Interaction of metallic lead with pure water occurs in the presence of oxygen, but carbonates and silicates form protective films and prevent further interaction. Marine microorganisms help in methylation of certain inorganic lead compounds. Lead stability in natural water depends upon the pH and redox potential of the system. The least soluble common forms of leads are probably carbonate, hydroxide, and hydroxyl carbonate (in an oxidizing system); however, lead sulfide (PbS) is the stable solid (reducing system). The solubility of lead in water can be very low (10 ppb at pH ≥ 8) or very high (100 ppb at pH 6.5 depending upon pH). Except in some shellfish such as mussels, lead bioaccumulation is not prominent. Lead accumulation in fishes does not go deeper than the muscles; it remains in the epidermis and scales.

2.4.2 Biogeochemical Cycles of PAHs

PAHs are semivolatile compounds under environmental conditions. They move between the atmosphere and the Earth's surface in repeated, temperature-driven cycles of deposition and volatilization. In the presence of sunlight, PAHs undergo photooxidation in the ambient air, where they are present as vapors, or are absorbed into airborne particulate matter. Chemical breakdown of PAHs due to photooxidation can take several days to weeks; their presence in the atmosphere is subject to complex physicochemical reactions, photochemical transformations, and reactions with other pollutants. The widespread occurrence of PAHs is largely due to their formation and release in all processes of incomplete combustion of organic

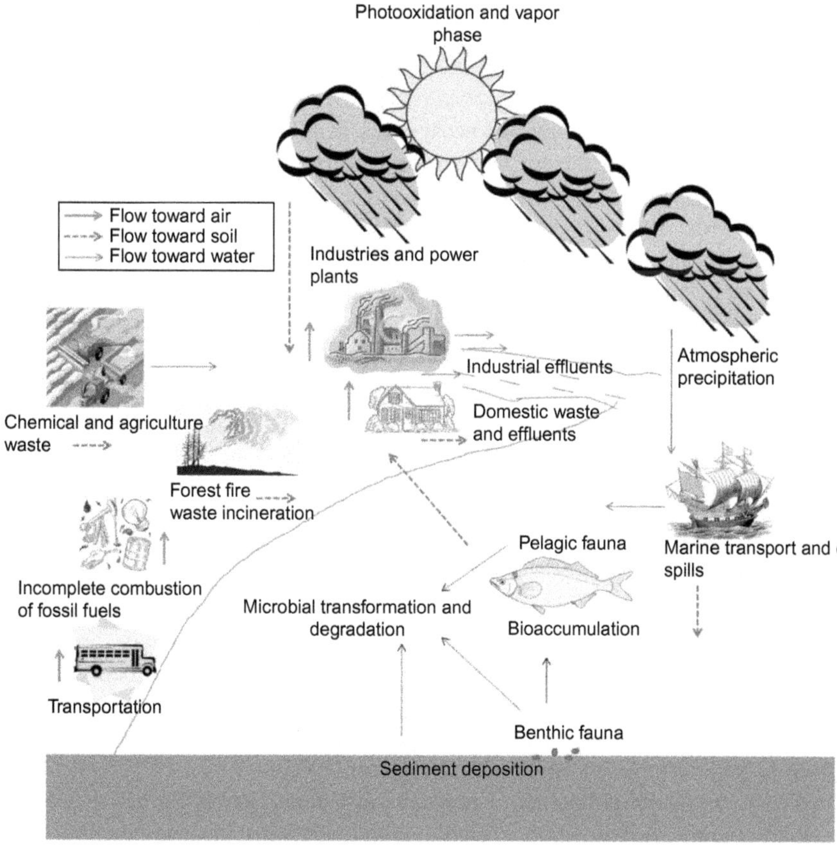

Figure 2.6 A generalized view of the biogeochemical cycle of PAHs in the environment.

materials. In the bottoms of aquatic systems, PAHs are adsorbed into particulates and precipitated or solubilized in any oily matter due to low solubility (Figure 2.6). These persistent adsorbed PAHs later contaminate soil and water systems. Microbial populations in sediment/water systems degrade some PAHs and reduce their toxicity over a period of time. Aquatic organisms are mostly affected by PAH metabolism and photooxidation; the presence of UV light usually increases toxicity. PAHs enter into terrestrial animals by various routes, for example, inhalation, dermal contact, and ingestion, while absorption is the route of entry into plants from soil, that is, through their roots. Bioaccumulation of PAHs occurs in plants, although certain plant species can synthesize PAHs that act as growth hormones. Since they are a moderate persistent contaminant in the environment, the concentration of PAHs in marine animals (fish and shellfish) is expected to be much higher than in the environment due to biomagnification.

2.5 Effect of Pollutants on the Ecosystem

2.5.1 Aquatic Ecosystems

Contamination of aquatic and marine ecosystems with heavy metals has become an issue internationally. The concern with metal toxicity in aquatic ecosystems started with the tragedy at Minimata, which was caused by consumption of mercury-contaminated fish. Although certain metals are essential for cellular machinery and development of the organism, others are toxic. In aquatic ecosystems, physical parameters such as temperature, pH, salinity and the presence of other metals affect a given metal's toxicity to aquatic organisms. Metal toxicity is well documented in fish. Mercury and cadmium toxicity is more prevalent in freshwater as compared to hardwater or seawater fish. The toxicity of mercury is masked by calcium, whereas chlorine combines with cadmium to form inert salt, which is not readily accessible to organisms (Wright and Welbourn, 2002). At a lower pH, mercury is converted to methyl mercury, which is more toxic than metallic or inorganic mercury. Mercury is more toxic to freshwater ecosystems. Cadmium causes skeletal deformities and impaired functioning of kidneys in fish. The effects of cadmium are directly or indirectly lethal to aquatic organisms. Cadmium also impairs aquatic plant growth, which in turn affects the entire ecosystem. Cellular damage has been observed in the hepatopancreas of marine crustaceans that experience prolonged exposure to cadmium. Pb does not show biomagnifications up the food chain. It concentrates in the skin, bones, kidneys, and liver of fish rather than in muscle. Pb is particularly toxic to aquatic plants and algae and inhibits the activity of enzymes involved in photosynthesis (Wright and Welbourn, 2002).

PAHs are persistent pollutants of aquatic ecosystems. PAHs have five fates in the environment: volatilization, leaching, degradation, bioaccumulation, and sequestration. They are ubiquitous in the marine environment. The occurrence is higher in urban areas (Meador et al., 1995). Although PAHs are not easily metabolized by a variety of microorganisms, they penetrate easily into the bodies of higher vertebrates by crossing the cell membrane, after which they are deposited into organs rich in lipids. Thus, lipids and organic carbon are important factors governing the accumulation of PAHs in sediment, water, and tissues. The aquatic ecosystem receives PAHs largely because of atmospheric precipitation and surface runoff. The marine ecosystem in particular is contaminated by a variety of PAHs. PAHs generally get deposited in lipid-rich tissues. Thus accumulation of PAHs occurs in almost all marine organisms. PAHs have been listed as priority pollutants by both the US EPA and the European Union (Maiti et al., 2012).

2.5.2 Terrestrial Ecosystems

Terrestrial ecosystems are characterized by low water availability and greater temperature fluctuations. There is greater availability of gases and light in terrestrial ecosystems as compared to aquatic ecosystem. Terrestrial biota are exposed to heavy metal by different means. The effect of metal and PAHs on terrestrial animals is more or less similar to that in humans. However, in plants the response to heavy metal and PAHs is diverse. Contact with Pb is largely through atmospheric pollutants, Hg through soil, and As through water. In terrestrial ecosystems, plants can accumulate many metals. The uptake process for toxic metal is the same as the uptake processes for essential micronutrient metal ions (Patra et al., 2004). At extreme concentrations, metals result in phytotoxicity through changes in membrane permeability, binding with sulphydryl (–SH) groups, reacting with phosphate groups and active groups of Adenosine di-phosphate (ADP) or Adenosine tri-phosphate (ATP), and replacement of essential. Metals like Pb can significantly affect the water content of plant tissues. In the presence of Pb, transpiration intensity, osmotic pressure of cell sap, water potential of xylem, and relative water content were significantly reduced in plants. Studies on alfalfa (*Medicago sativa*) show that PAHs can significantly reduce uptake of Cu and Zn (Carlo-Rojas and Lee, 2009). Elevated concentrations of heavy metals have been reported as harmful for function and growth of roots (Rautio et al., 2005).

2.6 Exposure, Metabolism, and the Fate of Environmental Pollutants in Humans

The most likely means of exposure to pollutants is by inhalation, oral, or dermal. The route of exposure is commonly through, food, drinking water, and air, or predominantly occupational in some cases (Figure 2.7). The metabolism of metals and PAHs in living systems is governed by their chemical state and their affinity toward biomolecules, and the response of metabolic machinery toward them.

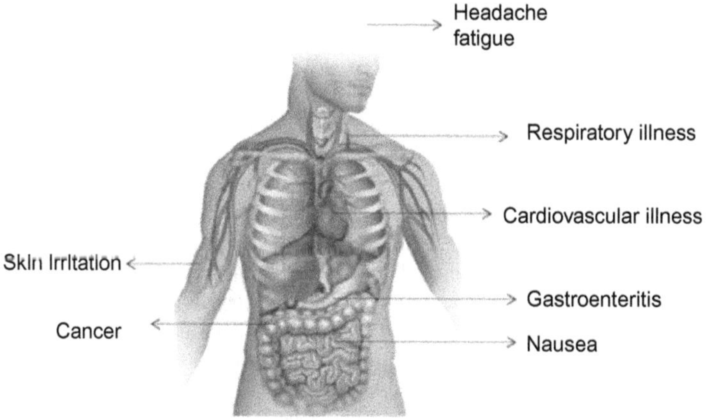

Figure 2.7 A generalized view of the effects of toxic metals and PAHs on the human body.

2.6.1 Routes of Exposure and Metabolism of Heavy Metals

Metals are present everywhere and exposure to them is unavoidable. The frequent route of exposure is through air, water, and food. However, the major route is specific to each metal. The source, major route of exposure, and associated health effects have been provided in Table 2.1.

1. *Arsenic*: The major route for As exposure is by consumption of As-contaminated food or water. Inhalation or dermal contacts are usually occupational. As is released naturally into rock, which accounts for arsenic contamination of food and ground water. This is a major global concern. The US EPS has enlisted As as the most toxic element. Metabolism of inorganic arsenic has been extensively studied in human and animal models. It can undergo oxidation/reduction and alkylation. Inside the body it can interconvert from As(III) to As(V). Upon methylation it forms monomethylarsonate (MMA) or dimethyarsinate (DMA). Arsenic exists in various forms in the human body. Out of total As in human body, 20% is present in methylated form (14% MMA, 6% DMA), whereas 78% remains as As(III) and 2% as As(V). Long-term exposure to As(III) and As(V) elevates the level of DMA in the body (Aposhian et al., 2004). A community study on children and women claims the poor metabolism of As in children. Children exposed to As in drinking water were able to eliminate it as DMA (47%) and inorganic As (49%) as compared to 32% inorganic As and 66% DMA in case of women (Concha et al., 1998). As(V) is reduced to As(III) by metallothioneins such as glutathione. *In vitro* studies assert the complex of glutathione with both arsenate and arsenite. Glutathione is oxidized (and arsenate reduced) in the glutathione-arsenate reaction Scott et al. (1993). The As(III) form of As can undergo methylation. In the body the principal site of methylation is the liver. In the liver, the enzymes utilize *S*-adenosylmethionine as cosubstrate mediates the methylation process (Buchet and Lauwerys, 1988; ATSDR, 1989).
2. *Cadmium*: The primary source of exposure to Cd is through food and tobacco smoking. Tobacco leaves can accumulate high levels of Cd. Individuals may also be exposed to Cd by water, if the water source is in the vicinity of industries that use cadmium. Cd exists in a +2 ion state and does not undergo metabolic conversions like oxidation, reduction, or alkylation. Cd^{+2} has a strong affinity toward anionic groups. Cd^{+2} binds to sulfhydryl groups in protein and other molecules. Cadmium is active toward proteins such as albumin and metallothionein in plasma (Pan et al., 2010). Once consumed or inhaled, cadmium is excreted in feces. It is not absorbed by the gut lining. Cadmium is readily absorbed by the liver, the kidneys, and the tissues of muscles, skin, and bones. The half-life of Cd in the human body is around 26 years, 6–38 years for kidney and 4–19 years for kidney (Shaikh and Smith, 1980; Kjellstrom and Nordberg, 1985).
3. *Mercury*: Mercury and its compounds are highly lipophilic and easily cross the blood–brain and placental barriers (Clarkson, 1993). It accumulates primarily in the kidney. Exposure to metallic mercury is by inhalation and it oxidizes to mercuric mercury inside the body. Contact with organic mercury is usually through oral exposure. The kidney contains metallothionein, a metal-binding protein that is also found in fetal and maternal livers and other organs. In the kidneys, the production of metallothionein is stimulated by exposure to mercury. The increased levels of metallothionein increase the amount of mercuric ion binding in the kidneys (Cherian and Clarkson, 1976). Once absorbed, metallic and inorganic mercury enter an oxidation–reduction cycle.

Table 2.1 Major Pollutants in Nature, their Source, Route of Exposure, Level of Contamination, and Associated Health Effects

Compound	Source of Exposure	Major Route of Exposure	Environmental Level	Health Effects
Arsenic	Food, air, and drinking water	Oral	Air: 1–3 ng/m^3(rural areas), 20–100 ng (urban areas); drinking water: 2 μg/L; soil: 0.1–97 mg/kg	Lung cancer, cardiovascular effects, and encephalopathy
Cadmium	Food, cigarette smoking, drinking water, and air	Oral, inhalation, and dermal	Air: 0.1–5 ng/m^3 (rural areas), 2–15 ng/m^3 (urban areas); water: <5 μg/L (drinking); soil: 0.06–1.1 mg/kg, 0.27 mg/kg (agricultural)	Glomerular damage, bone mineralization, and emphysema
Lead	Contaminated food, drinking water, lead-based paint	Inhalation	Air: <0.05 μg/m^3; water: 5–10 μg/L; soil: <10–30 g/kg	Elevated blood pressure, colic in children, neuropathy, reduced fertility
Mercury	Water, air, dental amalgam fillings, waste incinerators	Inhalation and oral	Air: 0.9–1.5 ng/m^3; water: 0.5–100 ng/L; soil: 6–17 mg/kg	Diarrhea and/or abdominal pain, kidney damage, and acrodynia
PAHs	Air and food	Inhalation and oral	Air: 0–1 ng/m^3; water: 0.001–0.01 μg/L; sediment: 0.01 μg/g	Cancer, mutation, and skin irritation

Source: WHO (2000), EPA (2005), and ATSDR (2013).

4. *Lead*: Another toxic metal of key concern is lead (Pb). Humans are exposed to Pb through food and water. However, exposure to Pb also happens through contact with dust or soil and products containing lead. Water is the most common source of exposure, as Pb can leach to water from Pb-coated water pipes. The primary route of exposure to Pb for the general population is by oral consumption of Pb-contaminated food or water, whereas, occupational exposure to Pb is through inhalation. Pb gets attached to dust and deposited in ciliated airways. Inside the body, inorganic Pb can intricate with protein and sulphur-rich ligands. Pb can form complex with proteins in cell nucleus and cytosol. Alkyl lead compounds are actively metabolized in the liver by oxidative dealkylation catalyzed by cytochrome P-450. Independent of the route of exposure, absorbed lead is excreted primarily in urine and feces; sweat, saliva, hair and nails, and breast milk are minor routes of excretion (ATSDR, 2007; Hsu, 1981; Stauber et al., 1994).

2.6.2 Route of Exposure, Metabolism, and Excretion of PAHs

PAHs exist throughout the environment. Humans are exposed to a variety of PAHs daily. Exposure to individual PAH is occupational. The primary source of exposure is air. Usually humans come into contact with PAHs by exposure to either dust or air, where they are present as air particulates. The primary sources of PAHs are vehicle exhaust, cigarette smoke, agricultural burning, wood burning, waste incineration, coal and coal tar. PAHs are highly hydrophobic and liophililic compounds. The lipophilicity permits trouble-free penetration through cellular membranes. They remain in the body indefinitely. Metabolism of PAHs occurs inside the body in all tissues with the help of varying enzymes and cofactors. The metabolites of PAHs are soluble in water to some extent. The lipophilicity of PAHs makes them persistent inside the body and difficult to excrete. The metabolism of PAHs has been studied extensively *in vitro* and *in vivo*; these studies have contributed significantly to the explanation of PAH metabolism. Benzo[a]pyrene metabolism has been extensively studied as a model for PAHs metabolism. It is also a major air pollutant, higly toxic to humans. Metabolism of benzo[a]pyrene is initiated by microsomal cytochrome P-450 systems. The initial breakdown of benzo[a]pyrene produces several arene oxides. These arene oxides rearrange spontaneously to phenols, undergo hydration to the corresponding trans-dihydrodiols in a reaction catalyzed by microsomal epoxide hydrolase, or react covalently with glutathione, either spontaneously or in a reaction catalyzed by cytosolic glutathione-*S*-transferases (IARC, 1983). *In vitro* studies on benzo[a]pyrene in human bronchial epithelial and lung tissue revels the metabolism to the 9,10-dihydrodiol, 7,8-dihydrodiol, 4,5-dihydrodiol and 3-hydroxybenzo[a]pyrene, (Kiefer et al., 1988). The rate of formation of the dihydrodiols was greater in the bronchial epithelium than in the lung (Autrup et al., 1980; Cohen et al., 1979). This may render some areas of the respiratory tract more sensitive to the effects of carcinogens. The sulfate conjugate of 3-hydroxybenzo[a]pyrene, benzo[a]pyrene-3-yl-hydrogen sulfate. This sulfate is very lipid soluble and, thus, would not be readily excreted in the urine (Cohen et al., 1979). Activation of benzo[a]pyrene has also been detected in human fetal esophageal cell culture (Chakradeo et al., 1993; Schwarz et al., 2001).

2.7 Effects of Heavy Metals and PAHs on Human Health

A healthy lifestyle requires certain heavy metals but in small quantities, which are nutritionally essential too. Among them few metals are also referred as the trace elements (e.g., iron, copper, manganese, and zinc). These heavy metals or their compounds are commonly present in foodstuffs, fruits and vegetables, and in commercially available multivitamin products.

Toxicity due to heavy metal occurs, when they are not being metabolized by the organism and accumulate in the different parts of body, such as soft tissues in humans. Medical conditions due to heavy metal toxicity are uncommon, but clinically significant, if unrecognized, uncharacterized, false diagnosis or inappropriately treated, heavy metal toxicity could be fatal. Since early days of metallurgy, human are exposed to heavy metals and after industrial revolution occupational exposure to heavy metal become prominent reason of various health effects. In early medical books, abdominal colic in a man who extracted heavy metals and the pernicious effects of arsenic and mercury among smelters were known. One of the most studied heavy metal toxicity syndrome is metal fume fever (MFF), or "brass founders' ague," "zinc shakes," "Monday morning fever" as it is variously known, which is characterized by fever, headache, fatigue, dyspnea, cough, and a metallic taste occurring within 3–10 h after exposure. This is self-limiting inhalation syndrome seen in workers exposed to metal oxide fumes (zinc oxide, magnesium, cobalt, and copper oxide fumes). A neutrophil alveolitis ensues, with hypoxia, reduced vital capacity, and diffuse bilateral infiltrates seen on radiographs. Other health effects, which are seen or clinically recognized during chronic exposure to heavy metal dust (cobalt); pulmonary fibrosis, pneumoconiosis, fibrotic and emphysematous lung damage, chronic renal failure and osteoporosis. In case of PAH, they have been associated with increased incidences of different types of cancers (lung, skin, and bladder cancers) from occupational exposures. However, PAHs do not have a high degree of acute toxicity in humans. Chronic bronchitis, chronic cough irritation, bronchogenic cancer, dermatitis, cutaneous photosensitization, pilosebaceous reactions and the most significant endpoint of PAH toxicity is cancer. Most exposures to PAHs are mixtures of PAHs, thus it is difficult to ascribe health effects for a specific PAHs through epidemiological studies. Certain PAHs have well-known carcinogenicity, which has been well established in laboratory animals, there are reports of increased incidences of skin, lung, bladder, liver, and stomach cancers, as well as injection-site sarcomas, in animals. Studies over animals have also revealed that PAHs account for hematopoietic, immune, reproductive, and neurologic systems and also cause developmental effects. PAHs metabolites or their derivatives can be potent mutagens even though they are having only slightly mutagenic or even non mutagenic property *in vitro*. Naphthalene, the most abundant constituent of coal tar, is a skin irritant, and its vapors may cause headaches, nausea, vomiting, and diaphoresis.

2.7.1 Diseases Caused by Heavy Metals Contamination

1. *Respiratory effects*: During occupational exposure to arsenic dust in air, the most common reaction is irritation to the mucous membranes of the nasopharyngeal ducts and

throat. This leads to laryngitis, bronchitis, or rhinitis; prolonged exposure and/or very high levels of exposure can cause perforation of the nasal septum (Morton and Caron, 1989; Sandstrom et al., 1989). Noncancerous respiratory disease (e.g., emphysema) due to arsenic exposure has shown increased mortality in arsenic-exposed workers, but no conclusive evidence has been produced (Welch et al., 1982; Lee-Feldstein, 1983; Saraiya et al., 1999). The relation between inhaled inorganic arsenic and respiratory disease is inconclusive due to lack of studies. Irritation and hyperplasia were observed in rats and hamsters when arsenic trioxide and/or gallium arsenide were instilled in their intratracheal systems (Webb et al., 1986; Webb et al., 1987; Goering et al., 1988; Ohyama et al., 1988). Short exposure studies on laboratory animals (rats and mice) have shown that dimethyl arsenic acid caused respiratory distress, and necropsy of animals that died revealed bright red lungs with dark spots (Stevens et al., 1979).

Respiratory tissues are highly sensitive to cadmium oxide fumes or dust; symptoms of toxicity are limited to coughing, and slight irritation of the throat and mucosa can be delayed (Beton et al., 1966). Prolonged acute exposure can cause a pulmonary response (cadmium pneumonitis) including chest pain/precordial constriction, severe dyspnea/wheezing, persistent cough, weakness/malaise, anorexia, nausea, diarrhea, nocturia, abdominal pain, hemoptysis, and prostration (Barnhart and Rosenstock, 1984; Townshend, 1982). Acute, high-level exposures can be fatal.

Respiratory symptoms including cough, dyspnea, and tightness or burning pains in the chest are a prominent effect of acute-duration high-level exposure to metallic mercury vapors in humans. Clinical analysis of the human lungs has primarily shown diffuse infiltrates or pneumonitis. Other prominent effects are impaired pulmonary function, airway obstruction, restriction, hyperinflation, and decreased vital capacity. Severe cases show respiratory distress, pulmonary edema (alveolar and interstitial), lobar pneumonia, fibrosis, desquamation of the bronchiolar epithelium, and bronchiolar obstruction by mucus and fluid resulting in alveolardilation, emphysema, pneumothorax, and possibly death. The effects of lead exposure in the respiratory system lack sufficient studies.

2. *Cardiovascular effects*: There are some epidemiological studies regarding inhalation of inorganic arsenic and its effects on the cardiovascular system. Studies on smelter workers have shown significantly increased incidences of Raynaud's phenomenon, increased vasospasticity, and increased systolic blood pressure in comparison to diastolic pressure (Jensen and Hansen, 1998). Several other researchers have reported increased risk of mortality from cardiovascular disease, specifically ischemic heart disease and cerebrovascular disease in the arsenic-exposed workers of cohorts (Qiao et al., 1997; Xuan et al., 1993). However, it is still not conclusive that the observed increase in risk was due to arsenic exposure.

   Cadmium exposure does not appear to have a significant association with the cardiovascular system. Whereas few reports had shown adverse cardiovascular effects of cadmium exposure in humans, Wisniewska-Knypl et al. (1971) reported rhythmic disturbances and ventricular fibrillation, and Kagamimori et al. (1986) reported disorders of the cardiac conduction system, lower blood pressure, and decreased frequency of cardiac ischemia. Peters et al. (2010) reported that a 50% increase in blood or urinary cadmium levels resulted in a significant increase in the risk of stroke and congestive heart failure.

   Increases in heart rate and blood pressure have been reported following inhalation exposure to metallic mercury in humans. High concentrations of metallic mercury vapor exposure (acute) have shown increased blood pressure, increased heart rate/palpitations, acrodynia, and hypertension. Chronic exposure may lead to death due to ischemic heart and cerebrovascular disease. Lead has been found to affect the cardiovascular system in a diverse manner; chronic exposure to lead has shown increased blood pressure, disruptions of the vasodilatory actions of NO, elevations in systemic blood pressure, and decrements in the glomerular filtration rate.

3. *Gastrointestinal effects*: Nausea, vomiting, and diarrhea in workers with occupational inhalation exposure (acute poisoning) were reported in several case studies (e.g., Muzi et al., 2001; Chakraborti et al., 2003; Uede and Furukawa, 2003; Vantroyen et al., 2004). Animal studies on the gastrointestinal effects of arsenic are very limited; reports on pregnant rats have shown severe hyperemia and plasma discharge into the intestinal lumen at autopsy (Holson et al., 1999).

   Studies of cadmium exposure in humans and animals have indicated that oral exposure to in high concentrations causes severe irritation to the gastrointestinal epithelium. Common symptoms in humans include nausea, vomiting, salivation, abdominal pain, cramps, and diarrhea (Andersen et al., 1988; Frant and Kleeman, 1941; Buckler et al., 1986; Frant and Kleeman, 1941; Nordberg et al., 1973; Shipman, 1986; Wisniewska-Knypl et al., 1971).

   One characteristic of mercury intoxication is stomatitis (inflammation of the oral mucosa) due to acute-duration exposure to high concentrations of metallic mercury vapors. Other symptoms include abdominal pains, nausea and/or vomiting, and diarrhea (Bluhm et al., 1992; Taueg et al., 1992). Studies in animals show cellular degeneration and necrosis of the colon (Ashe et al., 1953).

   The early symptoms of lead poisoning include colic, characterized by abdominal pain, constipation, cramps, nausea, vomiting, anorexia, and weight loss (Rosenman et al., 2003).

4. *Hematological effects*: Acute and chronic oral exposure of arsenic has shown common hematological effects (anemia and leukopenia) due to arsenic poisoning in humans. Arsenic may have a cytotoxic or hemolytic effect on the blood cells, which suppresses erythropoiesis and causes anemia and leukopenia as well (Bolla-Wilson and Bleecker, 1987; Ide and Bullough, 1988; Morton and Caron, 1989).

   Cadmium exposure (oral) reduces the gastrointestinal uptake of iron, which can result in anemia if dietary intake of iron is low (Kagamimori et al., 1986). Hypoproteinemia and hypoalbuminemia were reported in a male who ingested a high dosage of cadmium in his diet (Wisniewska-Knypl et al., 1971).

   Exposure to metallic mercury is characterized by fatigue, fever, chills, elevated leukocyte count, and leukocytosis. Other hematological effects related to mercury exposure include a decrease in hemoglobin and hematocrit (dental amalgam) and anemia a decrease in red cell count and rupture of erythrocytes were observed after intraperitoneal injection of mice (Solecki et al., 1991). The reports regarding prolonged exposure of humans to high levels of mercury, which may result in hematological changes, are very limited. Lead poisoning induces anemia which is microcytic and hypochromic, mostly due to inhibition of heme synthesis and shortening of the erythrocyte lifespan.

5. *Musculoskeletal effects*: No significant studies have been conducted related to musculoskeletal effects of inorganic and organic arsenic exposure in humans and animals.

   Bone is a sensitive target of cadmium toxicity. Elevated levels of cadmium exposure can result in osteomalacia, osteoporosis, bone fractures, and decreased bone mineral density. Itai-Itai, or "ouch-ouch" disease, is characterized by multiple fractures of the long bones, osteomalacia, and osteoporosis.

   Elemental mercury vapor exposure may cause tremors, muscle fasciculations, myoclonus, or muscle pains (Aronow et al., 1990; Karpathios et al., 1991; Taueg et al., 1992). High levels of exposure to lead have been reported to cause muscle weakness, cramps, and joint pain. Studies in animals have indicated that lead exposure may impair normal bone growth, decrease bone density and bone calcium content, decrease trabecular bone volume, increase bone resorption activity, and alter growth plate morphology.

6. *Hepatic effects*: Several studies of hepatic effects due to arsenic exposure (inorganic arsenic) in humans have symptoms of hepatic injury. Clinical analysis has shown swollen and tender liver and elevated levels of hepatic enzymes in blood. (Chakraborty and Saha, 1987; Franzblau and Lilis, 1989; Liu et al., 2002). Hepatic effects have also been reported in acute bolus poisoning; in chronically exposed persons there has been a consistent finding of portal tract fibrosis, leading in some cases to portal hypertension and bleeding from the esophagus (Vantroyen et al., 2004; Guha Mazumder, 2005). Liver cirrhosis has also been reported at an increased frequency in arsenic-exposed individuals (Tsai et al., 1999).

   Very high levels cadmium exposure can cause pronounced liver damage and can be fatal (Buckler et al., 1986). Another hepatic effect is hypertriglyceridemia (increase in liver triglyceride levels) (Larregle et al., 2008).

   Metallic mercury exposure may result in elevated serum glutamic pyruvic transaminase (SGPT), ornithine carbamyl transferase, and serum bilirubin and decreased synthesis of hepatic coagulation factors (Jaffe et al., 1983). Various case studies of mercury have reported that acute-duration, high-level exposure in humans may lead to hepatomegaly and hepatocellular vacuolation, jaundice, an enlarged liver, and elevated aspartate aminotransferase, alkaline phosphatase, lactate dehydrogenase, and bilirubin (Kanluen and Gottlieb, 1991). Studies related to the hepatic effects of lead exposure are few; case studies have shown that exposure to lead results in inhibition of the formation of the heme-containing protein cytochrome P-450 and decreased activity of hepatic mixed-function oxygenases.

7. *Renal effects*: Studies on the renal system have shown no significant clinical signs of injury due to acute and chronic arsenic toxicity, although in some case elevated serum levels of creatinine or bilirubin have been found (Moore et al., 1994b; Cullen et al., 1995). Studies of animals exposed to arsenic also indicate that the kidney is not a major target organ for arsenic (Woods and Southern, 1989).

   Kidney is the primary target organ in cadmium toxicity following extended oral exposure and/or inhalation exposure; several studies have shown associations between increased mortality and renal dysfunction in residents living in cadmium-polluted areas (Friberg, 1950). Several biomarkers such as β2 microglobulin, pHC, and retinol binding protein are of wide use for the detection of urinary cadmium excretion followed by early renal changes associated with renal damage due to cadmium exposure.

   Exposure to metallic, inorganic, and organic forms of mercury has adverse renal effects in both humans and experimental animals. An increase in excretion of urinary protein, hematuria, oliguria, urinary casts, edema, inability to concentrate the urine, and hypercholesterolemia may be observed during exposure to mercury (Anneroth et al., 1992). In the kidneys, degenerative change in the epithelial cells of the renal proximal tubules is the primary toxic effect of both inorganic and organic mercury (Bernard et al., 1992).

   Lead exposure-related renal symptoms include enzymuria, proteinuria, impaired transport of organic anions and glucose, depressed glomerular filtration rate, proximal tubular nephropathy, glomerular sclerosis, and interstitial fibrosis. Lead has been shown to induce nephrotoxicity, renal injury, intranuclear inclusion bodies, and cellular necrosis in the proximal tubule.

8. *Endocrine effects*: Endocrine glands targeted studies on arsenic exposure are very limited, arterial thickening in the pancreas in chronic exposure, increased incidence of diabetes mellitus are few clinical observance (Tsai et al., 1999; Tseng et al., 2000; Wang et al., 2003). In rats, hypertrophy of the thyroid epithelium was observed after chronic exposure to arsenic (Arnold et al., 2003). However, no biologically significant effects of arsenic were observed in other endocrine tissues.

   Animal studies on the toxicity of cadmium to the endocrine system showed pancreatic atrophy and pancreatitis (Wilson et al., 1941). Other major glands such as the pituitary, adrenals, thyroid, thymus, and parathyroid glands were found to be less affected or unaffected. However, cadmium has been found to disrupt the daily pattern of aspartate, glutamate, and glutamine content in the pituitary gland.

   There is no clear evidence of association of mercury exposure with endocrine glands; however, case studies have shown thyroid enlargement with elevated triiodothyronine and thyroxine, as well as low thyroid-stimulating hormone levels (Karpathios et al., 1991). Serum-free thyroxine (T4) and the ratio of free thyroxine to free 3,5,3′-triiodothyronine (T3) were found to be slightly, but significantly, higher in mercury exposed people (Barregard et al., 1994). Studies on lead exposure-related endocrine effects show an association between high exposures to lead and changes in thyroid, pituitary, and testicular hormones.

9. *Dermal effects*: Dermal effects are the most common visible effects of arsenic exposure; characteristics of dermal effects include a pattern of skin changes. Dermatitis, hyperkeratosis, general hyperpigmentation, hypopigmentation (small areas of on the face, neck, and back) are common features of dermal effects due to arsenic exposure. The most prominent dermal effect associated with chronic ingestion of inorganic arsenic is skin cancer (Chen et al., 2003).

   No significant studies on the dermal effects of cadmium exposure have been conducted. Exposure to elemental mercury vapors (inhalation) for acute and intermediate durations has resulted in various dermal effects, including erythematous and pruritic skin rashes (Bluhm et al., 1992; Karpathios et al., 1991; Schwartz et al., 1992), heavy perspiration (Fagala and Wigg, 1992), and reddened and/or peeling skin on the palms of the hands and soles of the feet (Aronow et al., 1990; Karpathios et al., 1991; Fagala and Wigg, 1992).

10. *Immunological and lymphoreticular effects*: Very few studies were located regarding immunological and lymphoreticular effects due to exposure to inorganic and organic arsenicals. No direct evidence of immunosuppression was detected in laboratory experiments on mice exposed to arsenate in drinking water.

   There is limited study of immunological effects due to cadmium exposure in humans. Cadmium exposure results in a slight decrease in the generation of reactive oxygen species by leukocytes. Studies in animals have revealed that cadmium intermediate-duration exposure increases resistance to viral infection and mortality from virally induced leukemia and suppresses the humoral immune response.

   Immune response due to mercury exposure varies with individual genetic susceptibility; it may result in elevated or suppressed immune activities. Several case studies and experimental evidence have suggested that mercury could alter the host's immune system and lead to increased susceptibility to infections, autoimmune diseases, and allergic manifestations (Anneroth et al., 1992; Warfvinge, 1995). Several studies on the immunological effects of lead have been carried out, but significant data are lacking. Lead may have an effect on the cellular component of the immune system.

11. *Neurological effects*: Epidemiological studies have suggested that arsenic (inorganic and organic) exposure can cause injury to the nervous system. Other reports have revealed that acute high-dose exposure of arsenic results in encephalopathy with general symptoms such as headache, lethargy, mental confusion, hallucination, seizures, and coma, whereas lower dosage levels of arsenic are typically characterized by a symmetrical peripheral neuropathy, diminished sensitivity to stimulation, and abnormal patellar reflexes (Cullen et al., 1995; Uede and Furukawa, 2003; Vantroyen et al., 2004). Intellectual deficit in children due to chronic exposure to arsenic in drinking water also became evident.

    Cadmium-induced neurotoxicity has not been clearly demonstrated in human studies. It has been observed in animal studies that intermediate-duration exposure of adult rats to cadmium resulted in significantly decreased motor activity, weakness, and muscle atrophy and induced aggressive behavior (Baranski and Sitarek, 1987).

    Metallic mercury exposure results in adverse neurological effects, especially damage to the central nervous system. Any exposure (acute-, intermediate-, and chronic-duration) elicits similar neurological effects, which may intensify and become irreversible as exposure duration and/or concentration increases. Neurological effects include cognitive, personality, sensory, and motor disturbances. Symptoms of metallic mercury exposure include tremors, emotional lability, insomnia, memory loss, neuromuscular changes, headaches, and polyneuropathy (Bluhm et al., 1992; Karpathios et al., 1991). Studies in animals show that intermediate-duration exposure of mercury vapor results in mild, unspecified, pathological changes to specified cellular degeneration and necrosis in the brain (Ashe et al., 1953). Lead shows most severe neurological disorder known as lead encephalopathy. Early symptoms include dullness, irritability, poor attention span, headache, muscular tremor, loss of memory, and hallucinations. Prolonged exposure may lead to delirium, convulsions, paralysis, coma, and death.

12. *Reproductive effects*: Some studies on exposure to arsenic in drinking water have found a relation between adverse reproductive outcomes and arsenic concentrations. A significant increase in spontaneous abortions and an increased risk of stillbirth during pregnancy were also reported (Ahmad et al., 2001; Milton et al., 2005). A possible relation between cadmium exposure and male reproductive toxicity has been examined by several researchers with respect to sex steroid hormone levels. It has been found that increasing blood cadmium levels and increasing levels of serum luteinizing hormone, follicle stimulating hormone, prolactin, and testosterone among infertile men have important significant associations. However, animal studies have shown adverse reproductive effects to male and female reproductive capacity from cadmium exposure such as testicular atrophy, necrosis, and concomitant decreased fertility (Andersen et al., 1988).

    There is insufficient data on effects on the reproductive system due to metallic mercury; however, reported studies found no effect on fertility in humans due to intermediate or chronic exposure to metallic mercury (Alcser et al., 1989; Cordier et al., 1991). High occupational exposure to lead may result in abortion and preterm delivery in women, and decreased fertility in men.

13. *Developmental effects*: Chronic exposure of females to arsenic in drinking water has been associated with low birth weights in infants, late fetal mortality, neonatal mortality, and postneonatal mortality (Yang et al., 2003). There are very limited data on the developmental effects of cadmium in humans, whereas studies in animals (rats and mice) indicate that cadmium can be fetotoxic from oral exposures prior to and during gestation.

    Defects or abnormalities due to cadmium exposure in animals during the gestation period include malformations, (primarily of the skeleton), sirenomelia (fused lower limbs), amelia (absence of one or more limbs), delayed ossification of the sternum and ribs, dysplasia of facial bones and rear limbs, edema, exenteration, cryptorchism, palatoschisis, and sharp angulation of the distal third of the tail.

No direct association of metallic mercury with development has been found (Alcser et al., 1989). Low levels of lead have not been reported to cause any major congenital anomalies, whereas minor anomalies include hemangiomas and lymphangiomas, hydrocele, skin anomalies, and undescended testicles.

14. *Cancer*: Evidence of cancer due to arsenic exposure in the human population varies from a large number of epidemiological studies and case reports. Reports have suggested that ingestion of arsenic increases the risk of developing skin cancer, multiple squamous cell carcinomas, and multiple basal cell carcinomas. Prolonged exposure is seen as a major contributor to skin cancer and can also result in the development of bladder cancer (transitional cell cancers being the most prevalent) and respiratory tumors and an increased incidence of lung cancer (Guo et al., 2001a; Beane Freeman et al., 2004; Chen et al., 2003). Cadmium toxicity is mostly related to lung cancer, prostate cancer (in males), bladder cancer, and breast cancer (in females) (Kellen et al., 2007).

    There is no clear evidence that metallic mercury produces cancer in humans (Cragle et al., 1984). The carcinogenicity of lead was established through studies in animals, which revealed that the most common tumors that develop are renal tumors.

2.7.2 Diseases Caused by PAH Contamination

1. *Respiratory effects*: PAH exposure may result in adverse noncancer respiratory effects such as bloody vomit, breathing problems, chest pains, and chest and throat irritation (Gupta et al., 1993). Studies in animals show benzo[e]pyrene induces hyperplasia of the tracheal epithelium, also observed with pyrene, anthracene, and benz[a]anthracene. Tracheas implanted with pyrene also exhibited a more severe mucocilliary hyperplasia. Almost all PAHs have induced acute inflammation (edema and/or granulocyte infiltration), subacute inflammation (mononuclear infiltration and an increase in fibroblasts), and fibrosis and hyalinization in prolonged exposure (Topping et al., 1978).
2. *Cardiovascular effects*: PAHs are among the risk factors in the development of atherosclerosis, which includes arterial smooth muscle cell proliferation, collagen synthesis, lipid accumulation, and cellular necrosis. *In vitro* studies have reported that benzopyrene affects the aforementioned processes as well as collagen secretion, and increases cellular toxicity (Stavenow and Pessah-Rasmussen, 1988).
3. *Gastrointestinal effects*: Gastrointestinal toxicity in humans due to PAHs has been reported. The consumption of laxatives containing anthracene increases the incidence of melanosis of the colon and rectum in humans (Badiali et al., 1985). PAHs have found to interfere with enzyme activity in the intestinal mucosa of animals which may produce reactive intermediates and lead to tissue injury.
4. *Hematological effects*: Studies in animals show that PAHs have adverse hematological effects. A single dose of benzo[a]pyrene given to mice resulted in a small spleen, marked cellular depletion, prominent hemosiderosis, and follicles with large lymphocytes, finally leading to the death of the animal (Shubik and Porta, 1957). PAHs also affect rapidly proliferating tissues such as bone marrow and blood-forming elements (EPA, 1988).
5. *Hepatic effects*: PAH exposure has no known adverse hepatic effects in humans. However, adverse effects have been observed in animals after acute oral, intraperitoneal,

or subcutaneous administration of various PAHs. Observed hepatic effects include the induction of preneoplastic hepatocytes, induction of carboxylesterase and aldehyde dehydrogenase activity, increased liver weight, and stimulation of hepatic regeneration (Danz et al., 1991). Hepatic injury is deemed the most serious adverse effect observed in animals after PAH exposure.

6. *Renal effects*: Reports on PAHs having adverse renal effects in humans are not available. In animal studies, renal effects were observed after administration of fluoranthene. Renal tubular regeneration and interstitial lymphocytic infiltrates and/or fibrosis are just a few of the observed effects in female mice (EPA, 1988).
7. *Endocrine effect*: PAHs may have adverse effects on the endocrine system, based on limited studies in animals. A decrease in thymic glucocorticoid receptors after treatment with benzo[a]pyrene was observed in rats (Csaba et al., 1991).
8. *Dermal effects*: Skin is sensitive to PAH exposure. A mixture of carcinogenic PAHs can cause skin disorders in humans and animals. However, the specific effect of individual PAHs, except for benzo[a]pyrene, is difficult to determine, The effects of others have not been reported. Despite their adverse effects, mixtures of PAHs are used in the treatment of some skin disorders in humans. Regressive verrucae (i.e., warts) and local bullous eruption are characteristics symptoms of benzo[a]pyrene exposure to skin (Cottini and Mazzone, 1939).
9. *Immunological and lymphoreticular effects*: Reports have illustrated that PAHs have adverse immunological effects. Humoral immunity was suppressed in workers exposed to benzo[a]pyrene (Szczeklik et al., 1994). It has been reported that benzo[a]pyrene inhibits T-cell–dependent antibody production by lymphocytes (Blanton et al., 1986).
10. *Neurological effects*: No report is available on neurological effects due to exposure to PAHs in humans and animals.
11. *Reproductive effects*: Reports regarding the reproductive effects of PAHs in humans are limited or absent, and studies in animals have limited information; the effect of only one of the PAHs (benzo[a]pyrene) was reported with respect to reproduction.
12. *Developmental effects*: Data regarding the developmental effects of PAHs are limited to pregnant animals exposed to PAHs. Exposure to benzo[a]pyrene produced resorptions and malformations in fetuses, as well as sterility (Mackenzie and Angevine, 1981). Other PAHs, such as dibenz[a,h]anthracene, produced fetolethal effects; chrysene produced liver tumors (Buening et al., 1979a).
13. *Cancer*: Occupational studies of workers exposed to a mixture of PAHs give primary evidence for the carcinogenicity of PAHs (Hammond et al., 1976). Studies in animals confirm the carcinogenicity of certain PAHs (benz[a]anthracene, benzo[a]pyrene, and dibenz [a,h] anthracene) (McCormick et al., 1981). From animal studies and reports on humans it has become obvious that the route of exposure and site of absorption of PAHs influence tumor induction, for example, stomach tumors – ingestion, lung tumors – inhalation, skin tumors – dermal exposure, and mammary tumors – intravenous injection.

2.8 Conclusion

Heavy metals and hydrocarbons are a huge threat to the environment when they are present beyond their permissible limits. However, most of the anthropogenic activities are responsible for their increased level in the environment. As there is no body part left unaffected by these noxious elements, responsible for many life-threatening diseases like cancer, a few success stories are there for the removal of these contaminants from the living systems. Hence, most likely the death is the ultimate. In this regard, care should be taken while handling these toxic substances. Sustainable development should promote minimal emission of these elements to the environment to limit contamination. Strict legislation, successful implementation, and public awareness may decrease the pollution load in the environment, which is essential to sustain any life form in the ecosystem.

References

Agency for Toxic Substances and Disease Registry, 1989. Toxicological Profile for Cadmium. Agency for Toxic Substances and Disease Registry, Life Systems, Atlanta, Georgia.

Ahmad, S.A., Sayed, M.H., Barua, S., 2001. Arsenic in drinking water and pregnancy outcomes. Environ. Health Perspect. 109 (6), 629–631.

Alaee, M., Arias, P., Sjodin, A., Bergman, A., 2003. An overview of commercially used brominated flame retardants, their applications, their use patterns in different countries/regions and possible modes of release. Environ. Int. 29 (6), 683–689.

Alcser, K.H., Birx, K.A., Fine, L.J., 1989. Occupational mercury exposure and male reproductive health. Am. J. Ind. Med. 15 (5), 517–529.

Andersen, O., Nielsen, J.B., Svendsen, P., 1988. Oral cadmium chloride intoxication in mice: effects of dose on tissue damage, intestinal absorption and relative organ distribution. Toxicology. 48, 225–236.

Anneroth, G., Ericson, T., Johannson, I., 1992. Comprehensive medical examination of a group of patients with alleged adverse effects from dental amalgams. Acta. Odontol. Scand. 50 (2), 101–111.

Arnold, L.L., Eldan, M., van Gemert, M., 2003. Chronic studies evaluating the carcinogenicity of monomethylarsonic acid in rats and mice. Toxicology 190, 197–219.

Aronow, R., Cubbage, C., Wisner, R., 1990. Mercury exposure from interior latex paint. Morb. Mortal. Wkly. Rep. 39 (8), 125–126.

Ashe, W., Largent, E., Dutra, F., 1953. Behavior of mercury in the animal organism following inhalation. Arch. Ind. Hyg. Occup. Med. 17, 19–43.

ATSDR. 2001. Polychlorinated Biphenyls. February 2001. Agency for Toxic Substances and Disease Registry Web page: <http://www.atsdr.cdc.gov/tfacts17.html>.

ATSDR, 2007. Toxicological Profile for Lead. US Department of Health and Human Services, 1-582, Atlanta, Georgia.

Autrup, H., Wefald, F.C., Jeffrey, A.M., Tate, H., Schwartz, R.D., Trump, B.F., et al., 1980. Metabolism of benzo[a]pyrene by cultured tracheobronchial tissues from mice, rats, hamsters, bovines and humans. Int. J. Cancer 25 (2), 293–300.

Badiali, D., Marcheggiano, A., Pallone, F., 1985. Melanosis of the rectum in patients with chronic constipation. Dis. Colon Rectum. 28, 241–245.

Baranski, B., Sitarek, K., 1987. Effect of oral and inhalation exposure to cadmium on the oestrous cycle in rats. Toxicol. Lett. 36, 267–273.

Barkay, T., Wagner-Dobler, I., 2005. Microbial transformations of mercury: potentials, challenges, and achievements in controlling mercury toxicity in the environment. Adv. Appl. Microbiol. 57, 1–54.

Barnhart, S., Rosenstock, L., 1984. Cadmium chemical pneumonitis. Chest. 86, 789–791.

Barregård, L., Lindstedt, G., Schütz, A., Sällsten, G., 1994. Endocrine function in mercury exposed chloralkali workers. Occupat. Environ. Med. 51 (8), 536–540.

Beane Freeman, L.E., Dennis, L.K., Lynch, C.F., 2004. Toenail arsenic content and cutaneous melanoma in Iowa. Am. J. Epidemiol. 160 (7), 679–687.

Bernard, A.M., Collette, C., Lauwerys, R., 1992. Renal effects of in utero exposure to mercuric chloride in rats. Arch. Toxicol. 66 (7), 508–513.

Bernhoft, R.A., 2013. Cadmium toxicity and treatment. Sci. World J. <http://dx.doi.org/10.1155/2013/394652>.

Beton, D.C., Andrews, G.S., Davies, H.J., 1966. Acute cadmium fume poisoning; five cases with one death from renal necrosis. Br. J. Ind. Med. 23, 292–301.

Blanton, R.J., Lyte, M., Myers, M.J., 1986. Immunomodulation by polyaromatic hydrocarbons in mice and murine cells. Can. Res. 46, 2735–2739.

Bluhm, R.E., Bobbitt, R.G., Welch, L.W., 1992a. Elemental mercury vapour toxicity, treatment, and prognosis after acute, intensive exposure in chloralkali plant workers: Part I. History, neuropsychological findings and chelator effects. Hum. Exp. Toxicol. 11 (3), 201–210.

Blunden, S., Wallace, T., 2003. Tin in canned food: a review and understanding of occurrence and effect. Food Chem. Toxicol. 41, 1651–1662.

Bojes, H.K., Pope, P.G., 2007. Characterization of EPA's 16 priority pollutant polycyclic aromatic hydrocarbons (PAHs) in tank bottom solids and associated contaminated soils at oil exploration and production sites in Texas. Reg. Toxicol. Pharmacol. 47 (3), 288–295.

Bolla-Wilson, K., Bleecker, M.L., 1987. Neuropsychological impairment following inorganic arsenic exposure. J. Occup. Med. 29 (6), 500–503.

Buchet, J.P., Lauwerys, R., 1988. Role of thiols in the in-vitro methylation of inorganic arsenic by rat liver cytosol. Biochem. Pharmacol. 37 (16), 3149–3153.

Buckler, H.M., Smith, W.D., Rees, W.D., 1986. Self-poisoning with oral cadmium chloride. Br. Med. J. 292, 1559–1560.

Buening, M.K., Levin, W., Karle, J.M., 1979a. Tumorigenicity of bay region epoxides and other derivatives of chrysene and phenanthrene in new born mice. Can. Res. 39, 5063–5068.

Carlo-Rojas, Z., Lee, W.Y., 2009. Cu and Zn uptake inhibition by PAHs as primary toxicity in plants. Proceedings of the 2007 National Conference on Environmental Science and Technology. Springer, New York, NY (pp. 41–46).

Chakraborti, D., Mukherjee, S.C., Saha, K.C., 2003. Arsenic toxicity from homeopathic treatment. J. Toxicol. Clin. Toxicol. 41 (7), 963–967.

Chakraborty, A.K., Saha, K.C., 1987. Arsenical dermatosis from tube well water in West Bengal. Indian J. Med. Res. 85, 326–334.

Chakradeo, P., Kayal, J., Bhide, S., 1993. Effect of benzo(a)pyrene and methyl (acetoxymethyl) nitrosamine on thymidine uptake and induction of aryl hydrocarbon hydroxylase activity in human fetal oesophageal cells in culture. Cell Biol. Int. 17, 671–676.

Chen, Y.C., Guo, Y.L., Su, H.J., 2003. Arsenic methylation and skin cancer risk in southwestern Taiwan. J. Occup. Environ. Med. 45 (3), 241–248.

Cherian, M.G., Clarkson, T.W., 1976. Biochemical changes in rat kidney on exposure to elemental mercury vapor: effect on biosynthesis of metallothionein. Chem. Biol. Int. 12 (2), 109–120.

Chowdhury, B.A., Chandra, R.K., 1986. Biological and health implications of toxic heavy metal and essential trace element interactions. Pro. Food Nut. Sci. 11 (1), 55–113.

Clarkson, T.W., 1993. Mercury: major issues in environmental health. Environ. Health Perspect. 100, 31–38.

Cohen, G.M., Mehta, R., Meredith-Brown, M., 1979. Large inter individual variations in metabolism of benzo(α)pyrene by peripheral lung tissue from lung cancer patients. Int. J. Can. 24 (2), 129–133.

Concha, G., Vogler, G., Lezcano, D., Nermell, B., Vahter, M., 1998. Exposure to inorganic arsenic metabolites during early human development. Toxicol. Sci. 44 (2), 185–190.

Cordier, S., Deplan, F., Mandereau, L., 1991. Paternal exposure to mercury and spontaneous abortions. Br. J. Ind. Med. 48 (6), 375–381.

Cottini, G.B., Mazzone, G.B., 1939. The effects of 3,4-benzpyrene on human skin. Am. J. Can. 37, 186–195.

Cragle, D., Hollis, D., Qualters, J., 1984. A mortality study of men exposed to elemental mercury. J. Occup. Med. 26, 817–821.

Csaba, G., Inczefi-Gonda, A., Szeberenyi, S., 1991. Lasting impact of a single benzpyrene treatment in pre-natal and growing age on the thymic glucocorticoid receptors of rats. Gen. Pharmacol. 22 (5), 815–818.

Cullen, N.M., Wolf, L.R., St Clair, D., 1995. Pediatric arsenic ingestion. Am. J. Emerg. Med. 13 (4), 432–435.

Da Silva, J.F., Williams, R.J.P., 2001. *The Biological Chemistry of the Elements: The Inorganic Chemistry of Life*. Oxford University Press, New York.

Danz, M., Hartmann, A., Otto, M., 1991. Hitherto unknown additive growth effects of fluorene and 2-acetylaminofluorene on bile duct epithelium and hepatocytes in rats. Arch. Toxicol. Suppl. 14, 71–74.

Dash, H.R., Das, S., 2012. Bioremediation of mercury and importance of bacterial *mer* genes. Int. Biodeterior. Biodegradation. 75, 207–213.

Dash, H.R., Mangwani, N., Chakraborty, J., Kumari, S., Das, S., 2013. Marine bacteria: potential candidates for enhanced bioremediation. Appl. Microbiol. Biotechnol. 97 (2), 561–571.

Duruibe, J.O., Ogwuegbu, M.O.C., Egwurugwu, J.N., 2007. Heavy metal pollution and human biotoxic effects. Int. J. Phy. Sci. 2 (5), 112–118.

EPA. (1988) US Environmental Protection Agency. Code of Federal Regulations. 40 CFR 372.65.

Fagala, G.E., Wigg, C.L., 1992. Psychiatric manifestations of mercury poisoning. J. Am. Acad. Child Adolesc. Psychiatry 31 (2), 306–311.

Flegal, A.R., Brown, C.L., Squire, S., Ross, J.R.M., Scelfo, G.M., Hibdon, S., 2007. Spatial and temporal variations in silver contamination and toxicity in San Francisco Bay. Environ. Res. 105, 34–52.

Frant, S., Kleeman, I., 1941. Cadmium food-poisoning. JAMA. 117, 86–89.

Franzblau, A., Lilis, R., 1989. Acute arsenic intoxication from environmental arsenic exposure. Arch. Environ. Health 44 (6), 385–390.

Friberg, L., 1950. Health hazards in the manufacture of alkaline accumulators with special reference to chronic cadmium poisoning. Acta Med. Scand. 138 (S240), S1–S124.

Gaiser, B.K., Fernandes, T.F., Jepson, M., Lead, J.R., Tyler, C.R., Stone, V., 2009. Assessing exposure, uptake and toxicity of silver and cerium dioxide nanoparticles from contaminated environments. Environ. Health. 8 (S2). Available from: http://dx.doi.org/doi:10.1186/1476-069X-8-S1-S2.

Goering, P.L., Maronpot, R.R., Fowler, B.A., 1988. Effect of intratracheal gallium arsenide administration on delta-aminolevulinic acid dehydratase in rats: relationship to urinary excretion of aminolevulinic acid. Toxicol. Appl. Pharmacol. 92 (2), 179–193.

Guo, X., Fujino, Y., Kaneko, S., 2001a. Arsenic contamination of groundwater and prevalence of arsenical dermatosis in the Hetao plain area, Inner Mongolia, China. Mol. Cell. Biochem. 222 (1-2), 137140.

Gupta, P., Banerjee, D.K., Bhargava, S.K., 1993. Prevalence of impared lung function in rubber manufacturing factory workers exposed to benzo(a)pyrene and respirable particulate matter. Indoor Environ. 2, 26–31.

Hammond, E.D., Selikoff, I.J., Lawther, P.O., 1976. Inhalation of B[a]P and cancer in man. Ann. N.Y. Acad. Sci. 271, 116–124.

Hernberg, S., 2000. Lead poisoning in a historical perspective. Am. J. Ind. Med. 38, 244–254.

Holson, J.F., Stump, D.G., Ulrich, C.E., 1999. Absence of prenatal developmental toxicity from inhaled arsenic trioxide in rats. Toxicol. Sci. 51, 87–97.

Hornbuckle, K., Robertson, L., 2010. Polychlorinated biphenyls (PCBs): sources, exposures, toxicities. Environ. Sci. Technol. 44 (8), 2749–2751.

Hsu, J.M., 1981. Lead toxicity as related to glutathione metabolism. J. Nutr. 111 (1), 26–33.

Hughes, M.F., Beck, B.D., Chen, Y., Lewis, A.S., Thomas, D.J., 2011. Arsenic exposure and toxicology: a historical perspective. Toxicol. Sci. 123, 305–332.

Hutton, M., 1987. In: Hutchinson, T.C., Meema, K.M. (Eds.), Human Health Concerns of Lead, Mercury, Cadmium and Arsenic. SCOPE, John Wiley & Sons Ltd. (Chapter 6) Chichester, New York, Brisbane, Toronto.

IARC, 1983. Polynuclear aromatic compounds, Part 1, chemical, environmental and experimental data. IARC Monogr. Eval. Carcinog. Risk Chem. Hum. 32, 1–453. PMID:6586639.

Ide, C.W., Bullough, G.R., 1988. Arsenic and old glass. J. Soc. Occup. Med. 38, 85–88.

Jacobs, I.A., Taddeo, J., Kelly, K., Valenziano, C., 2002. Poisoning as a result of barium styphnate explosion. Am. J. Ind. Med. 41, 285–288.

Jaffe, K.M., Shurtleff, D.B., Robertson, W.O., 1983. Survival after acute mercury vapor poisoning—role of intensive supportive care. Am. J. Dis. Child. 137, 749–751.

Jensen, G.E., Hansen, M.L., 1998. Occupational arsenic exposure and glycosylated haemoglobin. Analyst. 123 (1), 77–80.

Kagamimori, S., Watanabe, M., Nakagawa, H., 1986. Case-control study on cardiovascular function in females with a history of heavy exposure to cadmium. Bull. Environ. Contam. Toxicol. 36, 484–490.

Kanluen, S., Gottlieb, C.A., 1991. A clinical pathologic study of four adult cases of acute mercury inhalation toxicity. Arch. Pathol. Lab. Med. 115 (1), 56–60.

Karpathios, T., Zervoudakis, A., Thodoridis, C., 1991. Mercury vapor poisoning associated with hyperthyroidism in a child. Acta Paediatr. Scand. 80 (5), 551–552.

Kellen, E., Zeegers, M.P., Hond, E.D., Buntinx, F., 2007. Blood cadmium may be associated with bladder carcinogenesis: the Belgian case-control study on bladder cancer. Cancer Detect. Prev. 31 (1), 77–82.

Kiefer, F., Cumpelik, O., Wiebel, F.J., 1988. Metabolism and cytotoxicity of benzo (a) pyrene in the human lung tumour cell line NCI-H322. Xenobiotica. 18 (6), 747–755.

Kjellstrom, T., Nordberg, G.F., 1985. Kinetic model of cadmium metabolism. In: Friberg, L., Elinder, C.G., Kjellström, T., et al.,Cadmium and Health: A Toxicological and Epidemiological Appraisal. Vol. I. Exposure, Dose and Metabolism. CRC Press, Boca Raton, FL, pp. 179–197.

Kulkarni, M., Chaudhari, A., 2007. Microbial remediation of nitro-aromatic compounds: an overview. J. Environ. Manage. 85 (2), 496–512.

Larregle, E.V., Varas, S.M., Oliveros, L.B., 2008. Lipid metabolism in liver of rat exposed to cadmium. Food Chem. Toxicol. 46 (5), 1786–1792.

Lee-Feldstein, A., 1983. Arsenic and respiratory cancer in man: follow-up of an occupational study. In: Lederer, W., Fensterheim, R. (Eds.), Arsenic: Industrial, Biomedical and Environmental Perspectives. Van Nostrand Reinhold, New York, NY, pp. 245–265.

Liu, J., Zheng, B., Aposhian, H.V., 2002. Chronic arsenic poisoning from burning high-arsenic containing coal in Guizhou, China. Environ. Health Perspect. 110 (2), 119–122.

Mackenzie, K.M., Angevine, D.M., 1981. Infertility in mice exposed in utero to benzo[a]pyrene. Biol. Reprod. 24, 183–191.

Maher, W., Butler, E., 1988. Arsenic in the marine environment. Appl. Organomet. Chem. 2, 191–214.

Maiti, A., Das, S., Bhattacharyya, N., 2012. Bioremediation of high molecular weight polycyclic aromatic hydrocarbons by *Bacillus thuringiensis* strain NA2. J. Sci. 1 (4), 72–75.

Mangwani, N., Shukla, S., Rao, T.S., Das, S., 2014. Calcium-mediated modulation of *Pseudomonas mendocina* NR802 biofilm influences the phenanthrene degradation. Colloids Surf. B. 114, 301–309.

McCormick, D.L., Burns, F.J., Alberg, R.E., 1981. Inhibition of benzo[a]pyrene-induced mammary carcinogenesis by retinyl acetate. J. Natl. Cancer Inst. 66, 559–564.

Meador, J.P., Casillas, E., Sloan, C.A., Varanasi, U., 1995. Comparative bioaccumulation of polycyclic aromatic hydrocarbons from sediment by two in faunal invertebrates. Mar. Ecol. Prog. Ser. Oldendorf. 123 (1), 107–124.

Megharaj, M., Avudainayagam, S., Naidu, R., 2003. Toxicity of hexavalent chromium and its reduction by bacteria isolated from soil contaminated with tannery waste. Curr. Sci. 47, 51–54.

Milton, A.H., Smith, W., Rahman, B., 2005. Chronic arsenic exposure and adverse pregnancy outcomes in Bangladesh. Epidemiology. 16 (1), 82–86.

Mohanty, M., Patra, H.K., 2011. Attenuation of chromium toxicity by bioremediation technology. Rev. Environ. Contam. Toxicol. 210, 1–34.

Moore, M.M., Harrington-Brock, K., Doerr, C.L., 1994b. Genotoxicity of arsenic and its methylated metabolites. Environ. Geochem. Health 16, 191–198.

Morton, W.E., Caron, G.A., 1989. Encephalopathy: an uncommon manifestation of workplace arsenic poisoning? Am. J. Ind. Med. 15, 1–5.

Mukhopadhyay, J., Gutzmer, J., Beukes, N.J., 2005. Organotemplate structures in sedimentary manganese carbonates of the Neoproterozoic Penganga Group, Adilabad, India. J. Ear. Syst. Sci. 114, 247–257.

Muzi, G., dell'Omo, M., Madeo, G., 2001. Arsenic poisoning caused by Indian ethnic remedies. J. Pediatr. 139 (1), 169.

Nordberg, G., Slorach, S., Steinstrom, T., 1973. Cadmium poisoning caused by a cooled-soft-drink machine. Lakartidingen. 70, 601–604.

Nriagu, J.O., 1979. Global inventory of natural and anthropogenic emissions of trace metals to the atmosphere. Nature. 279, 409–411.

Ohyama, S., Ishinishi, N., Hisanaga, A., 1988. Comparative chronic toxicity, including tumorigenicity, of gallium arsenide and arsenic trioxide intratracheally instilled into hamsters. Appl. Organomet. Chem. 2, 333–337.

Paçyna, J.M., Pacyna, E.G., 2001. An assessment of global and regional emissions of trace metals to the atmosphere from anthropogenic sources worldwide. Environ. Rev. 9, 269–298.

Pan, J., Plant, J.A., Voulvoulis, N., Oates, C.J., Ihlenfeld, C., 2010. Cadmium levels in Europe: implications for human health. Environ. Geochem. Health 32 (1), 1–12.

Patra, M., Bhowmik, N., Bandopadhyay, B., Sharma, A., 2004. Comparison of mercury, lead and arsenic with respect to genotoxic effects on plant systems and the development of genetic tolerance. Environ. Exp. Bot. 52 (3), 199–223.

Perelo, L.W., 2010. Review: in situ and bioremediation of organic pollutants in aquatic sediments. J. Hazard. Mater. 177 (1), 81–89.

Perfus-Barbeoch, L., Leonhardt, N., Vavasseur, A., Forestier, C., 2002. Heavy metal toxicity: cadmium permeates through calcium channels and disturbs the plant water status. Plant J. 32 (4), 539–548.

Peters, J.L., Perlstein, T.S., Perry, M.J., 2010. Cadmium exposure in association with history of stroke and heart failure. Environ. Res. 110 (2), 199–206.

Piver, A., 1989. Medical consequences of stratospheric ozone depletion. Can. Fam. Physician. 35, 2283–2284.

Purdey, M., 2004. Chronic barium intoxication disrupts sulphated proteoglycan synthesis: a hypothesis for the origins of multiple sclerosis. Med. Hypotheses. 62, 746–754.

Qiao, Y., Taylor, P.R., Yao, S., 1997. Risk factors and early detection of lung cancer in a cohort of Chinese tin miners. Ann. Epidemiol. 7, 533–541.

Ratnaike, R.N., 2003. Acute and chronic arsenic toxicity. Postgrad. Med. J. 79, 391–396.

Rautio, P., Kukkola, E., Huttunen, S., 2005. Growth alterations in Scots pine seedlings grown in metal-polluted forest soil: implications for restorative forest management. J. Appl. Bot. Food Qual. 79, 52–58.

Reimann, C., de Caritat, P., 2005. Distinguishing between natural and anthropogenic sources for elements in the environment: regional geochemical surveys versus enrichment factors. Sci. Total Environ. 337 (1), 91–107.

Rosenman, K.D., Sims, A., Luo, Z., 2003. Occurrence of lead-related symptoms below the current occupational safety and health act allowable blood lead levels. J. Occup. Environ. Med. 45 (5), 546–555.

Samanta, S.K., Singh, O.V., Jain, R.K., 2002. Polycyclic aromatic hydrocarbons: environmental pollution and bioremediation. Trends Biotechnol. 20 (6), 243–248.

Sandstrom, A.I.M., Wall, S.G.I., Taube, A., 1989. Cancer incidence and mortality among Swedish smelter workers. Br. J. Ind. Med. 46, 82–89.

Saraiya, M., Berg, C.J., Shulman, H., Green, C.A., Atrash, H.K., 1999. Estimates of the annual number of clinically recognized pregnancies in the United States, 1981–1991. Am. J. Epidemiol. 149 (11), 1025–1029.

Schwartz, J.G., Snider, T.E., Montiel, M.M., 1992. Toxicity of a family from vacuumed mercury. Am. J. Emerg. Med. 10 (3), 258–261.

Schwarz, D., Kisselev, P., Cascorbi, I., Schunck, W.H., Roots, I., 2001. Differential metabolism of benzo[a]pyrene and benzo[a]pyrene-7, 8-dihydrodiol by human CYP1A1 variants. Carcinogen. 22 (3), 453–459.

Scott, N., Hatlelid, K.M., MacKenzie, N.E., Carter, D.E., 1993. Reactions of arsenic (III) and arsenic (V) species with glutathione. Chem. Res. Toxicol. 6 (1), 102–106.

Shaikh, Z.A., Smith, J.C., 1980. Metabolism of orally ingested cadmium in humans. Dev. Toxicol. Environ. Sci. 8, 569–574.

Shea, K.M., 2003. Pediatric exposure and potential toxicity of phthalate plasticizers. Pediatrics. 111 (6), 1467–1474.

Shipman, D.L., 1986. Cadmium food poisoning in a Missouri school. J. Environ. Health 49, 89.

Shubik, P., Porta, G.D., 1957. Carcinogenesis and acute intoxication with large doses of polycyclic hydrocarbons. Am. Med. Assoc. Arch. Pathol. 64, 691–703.

Solecki, R., Hothorn, L., Holzweissig, M., et al., 1991. Computerized analysis of pathological findings in long term trials with phenylmercuric acetate in rats. Arch. Toxicol. Suppl. 14, 100–103.

Spallholz, J.E., 1997. Free radical generation by selenium compounds and their prooxidant toxicity. Biomed. Environ. Sci. 10, 260–270.

Stauber, J.L., Florence, T.M., Gulson, B.L., Dale, L.S., 1994. Percutaneous absorption of inorganic lead compounds. Sci. Total Environ. 145 (1), 55–70.

Stavenow, L., Pessah-Rasmussen, H., 1988. Effects of polycyclic aromatic hydrocarbons on proliferation, collagen secretion and viability of arterial smooth muscle cells in culture. Artery. 15, 94–108.

Stevens, J.T., DiPasquale, L.C., Farmer, J.D., 1979. The acute inhalation toxicology of the technical grade organoarsenical herbicides, cacodylic acid and disodium methanearsonic acid; a route comparison. Bull. Environ. Contam. Toxicol. 21, 304–311.

Szczeklik, A., Szczeklik, J., Galuszka, Z., 1994. Humoral immunosuppression in men exposed to polycyclic aromatic hydrocarbons and related carcinogens in polluted environments. Environ. Health Perspect. 102 (3), 302–304.

Taueg, C., Sanfilippo, D.J., Rowens, B., 1992. Acute and chronic poisoning from residential exposures to elemental mercury. J. Toxicol. Clin. Toxicol. 30 (1), 63–67.

Timmreck, C., 2012. Modeling the climatic effects of large explosive volcanic eruptions. Wiley Interdiscip. Rev. Clim. Change 3, 545–564.

Tinggi, U., 2003. Essentiality and toxicity of selenium and its status in Australia: a review. Toxicol. Lett. 137, 103–110.

Topping, D.C., Pal, B.C., Martin, D.H., 1978. Pathologic changes induced in respiratory tract mucosa by polycyclic hydrocarbons of differing carcinogenic activity. Am. J. Pathol. 93 (2), 311–324.

Townshend, R.H., 1982. Acute cadmium pneumonitis: a 17-year follow-up. Br. J. Ind. Med. 39, 411–412.

Tsai, S.M., Wang, T.N., Ko, Y.C., 1999. Mortality for certain diseases in areas with high levels of arsenic in drinking water. Arch. Environ. Health 54 (3), 186–193.

Tseng, C.H., Tai, T.Y., Chong, C.K., 2000. Long-term arsenic exposure and incidence of non-insulin dependent diabetes mellitus: a cohort study in arseniasis-hyperendemic villages in Taiwan. Environ. Health Perspect. 108 (9), 847–851.

Uede, K., Furukawa, F., 2003. Skin manifestations in acute arsenic poisoning from the Wakayama curry-poisoning incident. Br. J. Dermatol. 149 (4), 757–762.

Uherek, E., Halenka, T., Borken-Kleefeld, J., Balkanski, Y., Berntsen, T., Borrego, C., et al., 2010. Transport impacts on atmosphere and climate: land transport. Atm. Environ. 44 (37), 4772–4816.

Vantroyen, B., Heilier, J.F., Meulemans, A., 2004. Survival after a lethal dose of arsenic trioxide. J. Toxicol. Clin. Toxicol. 42 (6), 889–895.

Aposhian, H.V., Zakharyan, R.A., Avram, M.D., Sampayo-Reyes, A., Wollenberg, M.L., 2004. A review of the enzymology of arsenic metabolism and a new potential role of hydrogen peroxide in the detoxication of the trivalent arsenic species. Toxicol. Appl. Pharmacol. 198 (3), 327–335.

Vinodhini, R., Narayanan, M., 2008. Bioaccumulation of heavy metals in organs of fresh water fish *Cyprinus carpio* (Common carp). Int. J. Environ. Sci. Tech. 5 (2), 179–182.

Wang, S.L., Chiou, J.M., Chen, C.J., 2003. Prevalence of non-insulin-dependent diabetes mellitus and related vascular diseases in southwestern arseniasis-endemic and nonendemic areas in Taiwan. Environ. Health Perspect. 111 (2), 155–159.

Warfvinge, K., 1995. Mercury distribution in the mouse brain after mercury vapour exposure. Int. J. Exp. Pathol. 76, 29–35.

Webb, D.R., Wilson, S.E., Carter, D.E., 1986. Comparative pulmonary toxicity of gallium arsenide, gallium (III) oxide, or arsenic (III) oxide intratracheally instilled into rats. Toxicol. Appl. Pharmacol. 82, 405–416.

Webb, D.R., Wilson, S.E., Carter, D.E., 1987. Pulmonary clearance and toxicity of respirable gallium arsenide particulates intratracheally instilled into rats. Am. Ind. Hyg. Assoc. J. 48 (7), 660–667.

Welch, K., Higgins, I., Oh, M., 1982. Arsenic exposure, smoking and respiratory cancer in copper smelter workers. Arch. Environ. Health. 37 (6), 325–335.

Wilson, R.H., DeEds, F., Cox, A.J., 1941. Effects of continued cadmium feeding. J. Pharmacol. Exp. Ther. 71, 222–235.

Wisniewska-Knypl, J.M., Jablonska, J., Myslak, Z., 1971. Binding of cadmium on metallothionein in man: an analysis of a fatal poisoning by cadmium iodide. Arch. Toxicol. 28, 46–55.

Woods, J.S., Southern, M.R., 1989. Studies on the etiology of trace metal-induced porphyria: effects of porphyrinogenic metals on coproporphyrinogen oxidase in rat liver and kidney. Toxicol. Appl. Pharmacol. 97, 183–190.

Wright, D.A., Welbourn, P., 2002. Environmental Toxicology (Vol. 11). Cambridge University Press, Cambridge, pp. 249–348.

Xuan, X.Z., Lubin, J.H., Li, J.Y., 1993. A cohort study in southern China of tin miners exposed to radon and radon decay products. Health Phys. 64 (2), 120–131.

Yang, C.Y., Chang, C.C., Tsai, S.S., 2003. Arsenic in drinking water and adverse pregnancy outcome in an arseniasis-endemic area in northeastern Taiwan. Environ. Res. 91 (1), 29–34.

3 Nanotoxicity: Aspects and Concerns in Biological Systems

Supratim Giri

Department of Chemistry, National Institute of Technology, Rourkela, Odisha, India

3.1 Introduction

3.1.1 Perspective

Around the middle of the past decade, a nanotechnology innovation center was on the verge of being operational in a famous city in Europe. Globally, such research centers, as well as nanotechnology-based industries, were not quite uncommon at that time. Much to the surprise of the investors, media, and the researchers, there had been a serious protest against the setting up of the nanotechnology center. The protests mostly came from the locals on the basis of various ethical and socio-political issues. Activists also expressed fears about environmental contamination due to engineered nanomaterials and exposure to nanoparticles. Anti-nanotechnology slogans were spray-painted on the walls of a historic landmark situated in the city. In spite of such protests, the innovation center became functional but with some delay.

Social activism against the formation of industry-oriented technological hubs, more popularly known as technopoli, is not new and occurs around the globe. Environmental impact is definitely one of the key issues in dispute. But a new and unusual aspect that arose from the protest activities against the European nanotechnology center was the fear about nanoscale materials, or man-made engineered nanomaterials. Over the past two decades, a vast amount of research and product development based on nanotechnology had been carried out globally (Warheit, 2010). Like any other technology, the advent of nanotechnology has been fueled by the basic need for the betterment of human life and the urge to discover unknown frontiers. Day by day, the use of nanotechnology and its products are becoming an integral part of modern civilization. Naturally, just like the concerns that arose in the case of the European city, some fundamental questions arise about the effects of engineered nanomaterials, especially in the context of ecology and the world of living organisms. Because nanotechnology is a rapidly emerging subject, there has been a continuing debate over whether the progress of this technology should

Microbial Biodegradation and Bioremediation. DOI: http://dx.doi.org/10.1016/B978-0-12-800021-2.00003-0

be unlimited or restricted in order to have a minimum impact on the ecological and biological worlds. There is an urgent need for detailed assessment of any negative impact of nanomaterials on living organisms. Starting with a general overview of nanostructured materials and their biotechnological, medicinal, and industrial uses, this chapter will discuss aspects of the toxicity of engineered nanomaterials, their influence on biological systems, and their effects on the microbial world.

3.1.2 Nanotechnology and Biological Research

In current times, we come across the term "nanotechnology" quite frequently. It would not be an exaggeration to state that in the twenty-first century, nanotechnology has become one of the key technologies influencing science on a global scale. In layman's term, nanotechnology can be described simply as engineering with nanomaterials. This naturally raises the fundamental question of the definition of *nanomaterials*. One general definition could be that nanomaterials are those materials that are intentionally designed to attain at least one dimension in about the 1−100 nm range, so that the materials show some special properties that are neither present in their bulk nor intrinsic to their atomic or molecular state (Rao et al., 2005). Interestingly, this definition points out that nanomaterials are essentially manmade. Many entities existing naturally, e.g., protein, DNA, and viruses, are not considered as nanomaterials by definition although their size is in the 1−100 nm range. Therefore, nanomaterials do not appear naturally and these man-made materials exhibit unique or novel properties due to their small size and relatively huge surface area. Modern science is focused on exploiting some of the special properties expressed by nanomaterials to generate useful technologies for the benefit of humanity. Having found its application in clothing materials, icar paints, computer hard drives, cosmetics, water purification, etc., nanotechnology is prevalent as a commercial commodity in many aspects of modern life.

As mentioned earlier in the definition, engineered nanomaterials exhibit a unique property by virtue of their existing in nanodimensions. For example, cadmium selenide (CdS) is a semiconductor in its bulk state. If this bulk material were broken down into subsequent smaller and smaller particles, there would be a stage where each tiny particle of CdS would contain a few hundred to a few thousand atoms, and the respective dimension of the particles would be between 2 and 4 nm. In this situation, CdS particles show a unique photoluminescent property in the visible region that is neither present in its bulk state nor intrinsic to its atomic or ionic state (Klabunde, 2002). Nanoparticles of photoluminescent semiconductor materials are commonly termed *quantum dot* (QD). The important feature of QD is the fluorescence emission wavelength can be fine-tuned from the red to the blue end of the spectrum by controlling the material's composition as well as the dimensions of the particles within the 2−4 nm limit. Similarly, engineered gold nanoparticles having one of their dimensions within 10−100 nm absorb visible light by a special phenomenon termed *surface plasmon absorption*, which is present neither in bulk nor in the atomic gold state (Klabunde, 2002).

Many biological components, e.g., viruses, segments of a DNA double helix, and proteins, have their size dimensions within a few to a thousand nanometers. Sometimes these biomolecular entities are protected by a barrier-like nuclear membrane that has a pore diameter of 9 nm on average. Due to alterations in these biological components at their atomic or molecular levels, such as misfolding of a protein or mutation in a gene, or simply due to infection of by a virus, many diseases may occur (Kim et al., 2010). Not surprisingly, since the engineered nanomaterials have size ranges matching those of some important biological components, and since the surfaces of nanomaterials can be customized to interface with the biological matrix, there has been a wide application of nanomaterials in biological research and biomedicine. From controlled release delivery agents, cancer therapeutics, and subcellular organelle tracking to bioimaging and biosensors, various types of nanomaterials have been tailor made, modified, and functionalized to serve useful purposes (Cho et al., 2008). The unique electronic, magnetic, or optical properties shown by nanomaterials may be fine-tuned if the dimensions of the nanoparticles are changed to within 1−100 nm. In other words, manipulation of their atomic-level assembly has a direct effect on the behavior of electrons in the nanomaterial. Electrons have two important behaviors: their spin and their ability to move in a quantized manner in various energy levels of matter. If a bulk material is reduced to nanometer dimensions, the electrons feel special constraints in terms of free movement from one energy level to another. In this situation, either a small change in the sizes of the nanoscale materials or changes in the composition of the materials give rise to special behavior of electrons that is consequently manifested in the optical, electronic, or magnetic properties of the nanomaterials. For example, all electrons of a ferromagnetic bulk material do not spin in the same direction. However, if the nanoparticles of the same material having a size <20 nm in diameter are considered, the electrons are found to be spinning in the same direction. Thus, individual magnetic fields originating from each electron's spin point in the same direction, and the ferromagnetic nanoparticles behave as giant magnets with a huge net magnetic moment. This phenomenon is also known as superparamagnetism (Klabunde, 2002). From hyperthermia-based tumor therapy to magnetic resonance imaging (MRI) contrast agents, these superparamagnetic iron oxide nanoparticles (SPIONs) have been widely applied in biomedical research. One such SPION is being used as the MRI contrast agent for imaging of the liver under the trade name Feridex, approved by the US Federal Drug Administration (FDA) (Kim et al., 2010). Some SPIONs are being used in cancer therapy under the trade name of NanoTherm, which is currently in phase 3 of clinical trials (Kim et al., 2010). Some nanomaterials that are under clinical study are listed in Table 3.1.

Similar to the above example, when manipulated to attain a rod-shaped structure or when the external surface is covered with thin layer of silica, gold nanoparticles absorb near infrared light (NIR) between 700 and 800 nm. Electrons of these types of gold nanoparticles interact with surrounding water molecules and generate a local heating effect when excited by the NIR radiation. Some of these types of gold nanoparticles are currently being studied for thermal ablations of tumors in mouse models (Hirsch et al., 2003; Huang et al., 2009). Apart from these metal or metal

Table 3.1 Examples of Commercially Available Nanomaterials Used in Biomedicine

Nanomaterial	Trade Name	Application	Target	Adverse Effect	Manufacturer	Current Status
Iron oxide	Feridex	MRI contrast	Liver	Back pain, vasodilation	Bayer Schering	FDA approved
Gold	Aurimmune	Cancer therapy	Various forms	Fever	CytImmune Sciences	In phase 2 clinical trials
Gold nanoshells	Auroshell	Cancer therapy	Head and neck	Under investigation	Nanospectra Biosciences	In phase 1 clinical trials
Quantum dot	Qdots, EviTags	*In vitro* diagnostics	Tumor cells, tissues	Not applicable	Life Technologies, eBioscience	Research use only
Liposome	Abraxane	Cancer therapy	Breast	Cytopenia	Abraxis Bioscience	FDA approved
Polymer	Oncasper	Cancer therapy	Lymphoblastic leukemia	Urticaria, rash	Rhóne-Poulenc Rorer	FDA approved

Source: Kim et al., 2010.

oxide nanomaterials, several organic or polymer-based nanostructures, such as liposomes, dendrimers, micelles, carbon nanotubes (CNTs), are being used in biomedicine; quite a few among these are either FDA approved or under advanced clinical trial. In essence, physical and chemical properties of nanoparticles can be customized by atomic-level engineering to perform specific functions in biology. The advantage of size matching with biological entities and the ease of modification of nanomaterial surfaces with biocompatible layers has led to widespread use of nanotechnology in biological science. Nanotechnology-based biomedical applications are not confined only to *in vitro* regimes; however, they are being thoroughly explored in live mammalian systems as an *in vivo* platform and are also being tested clinically in humans in some cases (Jordan et al., 2001).

3.2 Entry of Nanomaterials into Living Organisms

Nanomaterials may enter animal body either by environmental exposure or by systematic administration. The former way is classified as an involuntary or unintentional mode of entry because living organisms have no control over the rate or extent of entry. Environmental pollution or accidental contamination of the immediate surroundings of the living being is usually responsible for involuntary entry. In terms of their systematic administration, however, nanoparticles are routinely being employed in ventures ranging from research involving drug delivery to bioimaging using live animals as experimental platforms. In this section, the details of the entry pathways of nanoparticles into living organisms are discussed.

3.2.1 Unintentional Entry of Nanomaterials and Routes of Entry

As engineered nanomaterials are being used in various commercial commodities like water filtration systems, cosmetics, clothing, etc., the industrial-scale production of such materials has now become commonplace. Almost unavoidably, these synthetic particles are likely to be present in the environment due to factory releases. Like any other industrial waste, nanomaterial-based factory release may take place through air, water, and soil. Airborne nanoparticles may be emitted into air through gas exhaust. Solid- or liquid-phase waste may contaminate water and soil via accidental spills or leakage while being transported. Another way that the environment can be contaminated by nanomaterials is through degradation, or via improper disposal by consumers or end-users of nanomaterial-based products and commodities. Once in the environment, nanoparticles may interact with other particles that are present, be transformed, and vary from their original composition and size depending on the lifetime of these materials. These nanoparticles then have the potential to interact with and affect the living organisms that are present in air, water, and soil. The extent to which any living organism would face exposure to such nanomaterials depends on many factors. The availability of nanoparticles, i.e., whether they are bonded to other particles, surfaces, molecules, etc., plays a vital role in determining the level of exposure. Also, the charge of the nanoparticles (positive or negative) due to ionization of the surface, as well as the quantity and speed of sedimentation in the environment, directly affect the exposure (Holden et al., 2013). Due to their small size, nanoparticles that are intact and not attached to any other particles or surfaces are extremely mobile in the environment. Therefore, means of transportation, such as rain, flooding, storms, snow, further assist in moving these materials from their point of origin or accumulation to other regions—particularly the habitats of living organisms. Once the nanomaterials affect the flora and fauna of a particular region, the contamination propagates upward following the food chain. Not only ecosystems but humans are also inadvertently exposed to nanomaterials in this manner (Holden et al., 2013).

There are three major pathways by which nanomaterials can enter into an animal body unintentionally from the environment (Arora et al., 2012). Inhalation is one of the major routes, and entry through the skin and the gastrointestinal (GI) tract are the others. Through inhalation, airborne particulates can easily enter into the respiratory systems of animals. A mammalian respiratory system can be divided into three regions: nasopharyngeal, tracheobronchial, and alveolar. The nasopharyngeal region consists of the nose and throat, and the tracheobronchial region comprises the trachea, or windpipe, and its subsequent branches, the bronchi and bronchioles. The third region contains alveoli that closely resemble tiny airbags; here, networks of blood vessels encompass the alveolar wall so that the exchange of gas can take place between the alveolar cavities and the bloodstream through the alveolar membrane. While the effect of airborne particulates (e.g., asbestos, carbon black) on the respiratory system has long been studied in environmental and occupational medicine, comprehensive data on the effect of engineered nanomaterials on the same is only now being investigated (Oberdörster et al., 2005). Research shows that nanoparticles can be deposited in all regions of the respiratory system through nasal breathing. Interestingly, 80% of nanoparticles may be deposited in the nasopharyngeal region if the particle size is as small as 1 nm, according to a mathematical model published by the International Commission on Radiological Protection (ICRP) (ICRP, 1994). As particle size increases, the probability of deposition in the alveolar region increases, with a maximum of 50% deposition predicted for the 20 nm particles. As diameter increases above 20 nm, the deposition probability subsides in the alveolar region and that in the nasopharyngeal region subsequently increases. The deposited particles in the lung alveoli are able to cross the alveolar epithelium barrier and get to the bloodstream, from where the particles may travel to the lymph nodes, spleen, heart, bone marrow, etc. (Hagens et al., 2007). In an inhalation experimental study, silver nanoparticles were found to be present inside alveolar macrophage up to 7 days after their application (Oberdörster et al., 2005).

As the largest organ and the primary outer layer of the body, skin presents vast possibilities for direct contact with nanoparticles. In this context, some considerable research data is available on the entry of engineered nanomaterials through human skin and their subsequent effects. Anatomically, human skin can be divided into the epidermis (top layer), dermis (middle layer), and hypodermis (bottom layer). The hypodermis also contains the subcutaneous fat tissues. Any invasion of nanoparticles into the dermis layer could lead to their entry into systematic vasculature by the lymphatic system present in this region. The outermost sheath of epidermis contains a keratin-based covering known as the *stratum corneum*. Usually, this layer is able to provide effective protection against any micron-sized particles. However, there have been contrasting results regarding penetration experiments using nanoparticles. For instance, entry of titanium dioxide (TiO_2) nanoparticles with ranging in size from 20 to 100 nm was confined within the uppermost *stratum corneum* layer due to the systematic administration of the materials to healthy human grafted skin samples as reported by Gontier et al. (2008). However, there are reports of localization of TiO_2 microparticles in the dermis region of skin via penetration through grafted human skin samples, as observed by Lademann et al. (1999). QDs, which are nanosized semiconductor materials with particle size below 10 nm, have been reported to be localized in the dermis layer by entry through an intact *stratum corneum* as studied by Ryman-Rasmussen et al. (2006). An increase of silver in the blood has been reported by Trop et al. (2006) from the usage of silver nanoparticle-based wound dressings that were applied topically. The results in this study indicated invasion of ultrafine silver nanoparticles through human skin, which led to the silver contamination in blood. Although silver nanoparticle-based wound dressings have been reported to offer better antimicrobial activities in wound healing and are clinically approved, skin penetration studies put their use into question (Arora et al., 2012).

The probability of nanoparticles entering living organisms through the GI tract is rather limited and detailed research data in this context are not available. However, nanoparticles enter through this pathway mainly by contaminated food, water, or nanoparticle-based oral drug delivery systems. Chen et al. (2006) reported toxicity of copper nanoparticles via oral ingestion in mouse models. Copper nanoparticles were able to cross the epithelial tissue covering the lumen of the intestines to reach the blood vessels and finally deposited in other vital organs like the liver, kidney, spleen, etc. by systematic blood circulation. This study suggests that nanoparticles can migrate from the GI tract to the blood circulation system and pose a risk for nanomaterial exposure in living organisms.

3.2.2 Systematic Administration of Nanomaterials (In Vivo)

The biomedical application of nanomaterials is an emerging field of research in the current era (Kim et al., 2010). The main purpose of such research is to develop a nanomaterial-based therapeutic system for targeted drug delivery, and for improving the bioimaging and treatment of cancers. Before carrying out research studies on the therapeutic effects of nanomaterials in human clinical trials, extensive studies were made on small animal models, from fish to mammalian systems (Davis, 1997). Small rodents like mice, rats, guinea pigs, etc. are commonly used as mammal subjects. Stable colloidal dispersion of nanoparticles can be delivered directly into the stomach of the rodents by a procedure called oral gavage (Teo et al., 2002). In this procedure a bulb-tipped gastric gavage needle is attached to a syringe and used to deliver the compound into the stomach, minimizing pain or trauma in the animals. Nanoparticles dispersed in simulated body fluids can also be delivered inside a rodent's body by various types of injection. Intravenous injection is one of the major ways by which nanoparticles can be introduced into the tail vein of the animals such that the materials can be directly channeled into the blood circulation system (Senior et al., 1985). Systematic administration can also be performed by intraperitoneal and subcutaneous injection procedures. In the intraperitoneal method, nanoparticles can be injected directly into the peritoneum or the body cavity of the rodents (Marino and Mitchell, 1972). Later the nanoparticles are absorbed into the blood vessels through the peritoneum. Subcutaneous injection, although more commonly used to administer anesthetics, is also used to deliver nanomaterials systematically in the tissue underneath the skin (Kala et al., 1998). Nanoparticles thereafter reach the blood vessels located in the subcutaneous tissue. The main purpose of systematic *in vivo* administration is to deliver the nanoparticles in the circulating bloodstreams of the animal. Researchers also use zebrafish embryos as an aquatic animal model by administering nanomaterials inside the embryos and studying the subsequent biological effect (Hu et al., 2011).

3.3 Fate of Nanoparticles Inside Living Organisms

In the previous section, it has been discussed how nanoscale particles can enter into animal body through various routes of entry. The next big question concerns the effect or outcome of these nanoparticles inside animal body in both the short and the long term. Irrespective of the routes of entry, nanoparticles ultimately reach an animal's blood circulation system. The consequence of the presence of nanoparticles in the blood circulation has been most extensively studied in the rodent-type mammalian model.

3.3.1 Accumulation and Biodistribution

Any nanosized particle entering into the bloodstream appears as a "foreign" object to the natural immune systems of mammals. This is almost the same mechanism by which a pathogen, when it enters the bloodstream, is coated with molecules like opsonin. This molecule contains specific antibodies that help the pathogens to be recognized by the receptors present in various phagocytic cells of the immune system. A receptor-mediated phagocytosis then takes place. The mononuclear phagocytes that act as scavengers destroy the pathogens by engulfing them in this manner. As a part of the natural process of immune response, various plasma proteins are also deposited on the surface of nanoparticles that get introduced into the blood circulation. The protein coverage is generically termed *opsonization* and makes the nanoparticles recognizable to the phagocytes (Parham, 2005). The mononuclear phagocyte system (MPS) is the chief component of the immune system and consists of phagocytic cells that are classified as monocytes and macrophages. These phagocytic cells are primarily present in the spleen, liver, lungs, and other connective tissues. The macrophage cells of the liver are called Kupffer cells (KCs). After being coated with plasma proteins, the nanoparticles are taken up by phagocytes present mainly in the spleen and liver (Wang et al., 2013). Therefore nanoparticles are quickly removed from the blood circulation and end up in the liver and spleen. Sadauskas et al. (2009) demonstrated that gold nanoparticles with 40 nm diameters accumulated in the KCs of mouse livers following intravenous injection and stayed in the liver up to 6 months. Similar studies were carried out by Balasubramanian et al. (2010) and nanoparticles were found to have accumulated in the liver and spleen.

The fact that nanoparticles are cleared rapidly from the bloodstream and localized in few specific organs presents a significant challenge to biomedical researchers who want to use nanoparticles as carriers of therapeutic agents and other important biomolecular cargos to various vital organs of the body like the heart, kidney, etc. through systematic administration. Studies show that longer the nanoparticles circulate in the bloodstream, the greater the probability of these particles reaching organs other than the liver and spleen (Li and Huang, 2008). The prolonged circulation time also helps the nanoparticles to reach tumor sites, where the outer endothelial layer of blood vessels are either defective or poorly aligned such that there exist gaps much bigger than that in normal tissues (Danhier et al., 2010). Also, due to the absence of any lymphatic drainage system at the tumor site, nanoparticles can easily penetrate into the tumor tissues through the "leaky junctions" present in the blood vessels at that site (Figure 3.1). Thus, this nonspecific phenomenon by which nanoparticles reach tumor tissues and become localized is known as the enhanced permeability and retention (EPR) effect.

Biomedical researchers have been utilizing EPR to establish tumor targeting through using nanoparticles as carriers (Danhier et al., 2010). Interestingly, if the

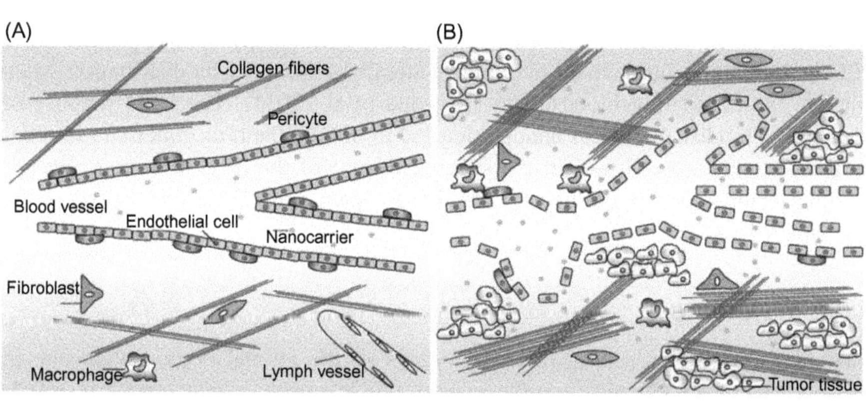

Figure 3.1 Schematic diagram of vasculature in normal and tumor tissues. Endothelial cell lines exhibit tight junctions in case of normal tissue and leaky junction in case of tumor tissue, through which nanoparticles can reach the tumor site from blood circulation (Danhier et al., 2010).
Source: Reprinted with permission from the publisher.

surface of the nanoparticles is covered with a polymer called polyethylene glycol (PEG), the absorption of plasma proteins on the surface of nanoparticles is minimized. The molecular weight of PEG varies greatly, but anywhere between 500 and 20,000 moles per gram polymer is generally used to obtain a brush layer structure on the nanoparticle surface. The PEG layer acts as a "stealth layer," preventing opsonization and allowing PEG-coated nanoparticles to circulate for a longer period of time in the blood circulation system (Gref et al., 1995). However, the existence of a PEG brush layer alone is not an indicator of a longer circulation time for nanoparticles; the overall size of the PEG-coated nanoparticles and the thickness of the brush layer are also important. If the overall hydrodynamic diameter of the nanoparticles is <5 nm, the particles are quickly removed from the blood by renal clearance (Wang et al., 2013). On the other hand, nanoparticles >200 nm size are filtered in the sinusoidal spleen, leading to fast clearance from the blood circulation (Wang et al., 2013). The study of time spent by a particulate in blood circulation is termed *pharmacokinetics* (PK). Perrault et al. (2009) conducted extensive PK studies using PEG-coated spherical gold nanoparticles in mice. They studied gold nanoparticles coated with PEG having molecular weights ranging from 2000 to 10,000 moles per gram. They also investigated the effect of gold nanoparticles' core size and hydrodynamic diameter on blood PK. Their studies confirmed that 17 nm-sized gold nanoparticles coated with a PEG layer of 10,000 molecular weight, possessed a hydrodynamic diameter of approximately 60 nm. The half-life value in the blood circulation of such nanoparticles was found to be around 51 h. Perrault et al. also reported localization of such gold nanoparticles in tumor tissue by the EPR effect. As longer circulating nanoparticles have a tendency to reach all the major organs of the animal body, including tumor tissue, PEG- and protein-coated gold nanoparticles have been reported to be allocated in the lungs, kidney, spleen, liver, heart, and tumors (Choi et al., 2010a). Such allocation is termed *biodistribution* (Figure 3.2). Although a significant amount of work has been done in the case of PEG-coated gold nanoparticles, any nanomaterials with a PEG-based stealth layer show a similar generalized trend of biodistribution when put into the circulating bloodstream (Almeida et al., 2011). The main purpose of obtaining a biodistribution of nanoparticles is to deliver them to tumor site in order to achieve a therapy. However, the striking feature of the biodistribution effect is that the majority of nanoparticles, even those coated with PEG, still end up in the liver and spleen.

3.3.2 Clearance

Nanoparticles, after reaching different organs of the animal body, either stay in those organs or are cleared from the system; this is a primary concern for researchers. Various *in vivo* studies have shown that a large fraction of biodistributed nanoparticles are captured in the liver (Wang et al., 2013). The liver consists mainly of two types of cells: structural cells, known as hepatocytes, and phagocytic macrophage cells, or KCs. Nanoparticles that are larger than 100 nm are essentially taken up by KCs. However, nanoparticles with a size range of around 20–50 nm can

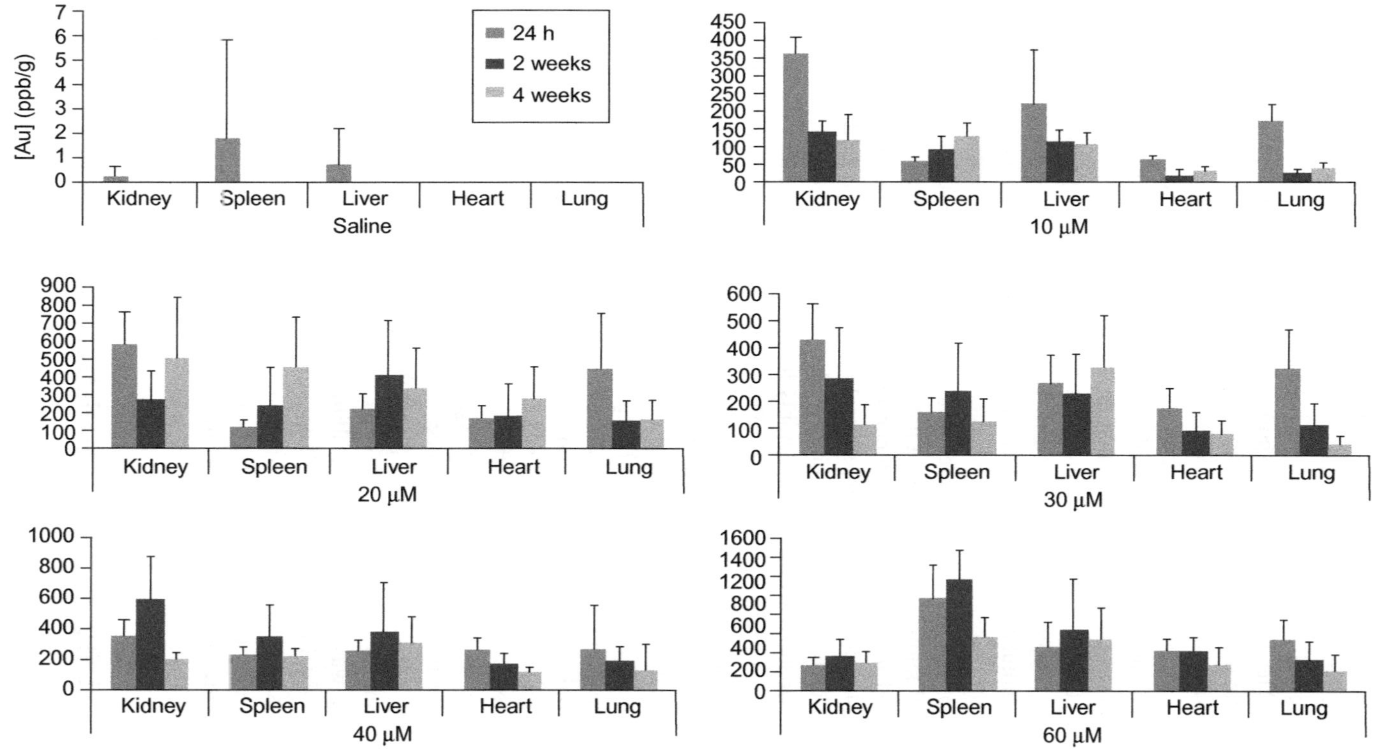

Figure 3.2 Biodistribution plot showing the accumulation of glutathione-coated gold nanoparticles in different organs of a rodent upon intravenous injection. With time, relative accumulation in kidney increases. The distribution pattern was not affected by increase of doses (Simpson et al., 2013).

Source: Reprinted with permission from the publisher.

evade the KCs and penetrate inside the liver sinusoids (Johnston et al., 2010). At that point, these particles are either internalized by hepatocytes or put into the lymphatic circulation. Once in the lymphatic circulation, nanoparticles migrate to the lymph nodes. Internalized nanoparticles in hepatocytes have been observed to be cleared via bile. The hepatobiliary transport mechanism is expected to carry the nanoparticles to the digestive system, from which clearance takes place in the form of fecal matter (Johnston et al., 2010). Nanoparticles that are engulfed by KCs are generally trapped in the cellular lysosome, where the particles are exposed to an acidic environment and various acid hydrolase enzymes (Schluep, 2010). Some types of nanomaterials are degraded in that enzymatic or acidic environment. For example, Lunov et al. (2010) demonstrated that dextran-coated magnetic iron oxide nanoparticles are subjected to biodegradation in phagocytic lysosomes. Single-walled carbon nanotubes (SWCNTs) have been reported to be susceptible with respect to enzymatic biodegradation (Liu et al., 2008). The nanomaterials that are degraded in Kupffer-type hepatic macrophage cells undergo clearance easily. Although it is well known that there are a number of degrading enzymes present in the liver, consorted *in vivo* data on nanoparticle degradation in the hepatic region are still unavailable. However, there are several nanomaterials, e.g., gold nanoparticles, CdSe-based QDs, that cannot be biodegraded in the liver and instead remain deposited in the hepatic cells for a considerably longer period of time.

In comparison with hepatic clearance, renal clearance is not a major mode of clearance for nanoparticles but plays an active role in clearing smaller sized particles. Glomerular filtration is the first step in the renal clearance of nanoparticles. In the nephron of the kidney, the glomerulus is a network of blood vessels residing inside the Bowman's capsule. The glomerulus has a specialized vascular bed known as the glomerular-basal membrane, which has filtration slits with an effective physiological opening of 4–5 nm (Ota et al., 1979). Therefore, nanoparticles having dimensions smaller than 5 nm can be filtered out of the blood circulation and excreted from the body in the form of urine. In mice models, Choi et al. (2011) have demonstrated that QD nanoparticles having a diameter <5.5 nm were cleared via renal clearance, while the same nanoparticles with a greater size were not cleared. Interestingly, SWCNTs with an average length of 340 nm were found to be eliminated through the kidney (Choi et al., 2007). The apparent anomaly was explained by the ability of SWCNTs to orient themselves longitudinally in the blood flow as well as during the passage through the glomerular filter slits. Not only the dimensions but also the shape and orientation of the nanoparticles play a role in renal clearance. Nanoparticles that are not cleared through any hepatic or renal routes, or are not degraded in phagocytes, remain within the animal tissues.

3.4 Nanotoxicity, *In Vivo* Degradation, and Effects

There has long been a debate as to whether nanomaterials are harmful to biological systems. By virtue of possessing a higher surface-to-volume ratio, a nanoparticle's surface is more accessible to the surrounding environment compared to the bulk phase of the same material having the same mass. Such a situation sometimes increases the chemical response due to the exposure of a number of active sites in the nanoparticles. For example, in some cases reactive oxygen species (ROS) are formed upon interaction between molecular oxygen and the nanoparticle's surface (Carlson et al., 2008). In addition, the increased surface area of the nanomaterials sometimes leads to the leaching out of metal ions due to the dissolution of surface layers. If the dissolved ions are of heavy metals like Cd^{2+} or Ag^{+}, which are potentially hazardous to biological systems, the toxic effects of nanoparticles become a huge cause of concern.

Historically, toxicity studies of nanoparticles were first carried out *in vivo* (Oberdörster et al., 2005). Those studies involved the effect of airborne nanoscale particles entering in animal body mostly due to inhalation. At the same time, researchers required a lot of toxicological data involving *in vitro* studies of nanoparticles. Consequently, a huge information database regarding *in vitro* toxicity of nanoparticles became available (Arora et al., 2012). In an *in vivo* environment, a nanoparticle is able to interact with proteins, cells, biological membranes, and tissues. Thus *in vitro* studies of toxicity provide a platform to gain systematic knowledge about the interaction of nanoparticles with animal cells, proteins, etc. Important information, such as the mechanism of nanoparticle uptake, cell viability data, range of permissible doses, obtained from *in vitro* studies serves various "proof-of-principle" understandings of nanoparticles' toxicological behavior. For a long time, these *in vitro* data were believed to provide an extrapolating pathway to predict the *in vivo* toxicity of nanomaterials.

For example, in several toxicity assays involving CdSe-based QDs, it has been found that the Cd^{2+} ions dissolved from the ionic lattice are responsible for toxicity (Hardman, 2006). Derfus et al. (2004) demonstrated that due to those leached-out Cd^{2+} ions, CdSe-based QDs appeared to be toxic towards the model hepatocytes that were cultured *in vitro*. Additionally, reacting oxygen species (ROS) can appear via the reaction of molecular oxygen on the surface of the QDs induced by UV light. These ROS are also toxic to animal cells (Carlson et al., 2008). The toxicity of the QDs can be minimized by rendering a stable protective layer of polymers, proteins, etc. on the particle surface; the surface covering can also play an important role in controlling the particle aggregation behavior of these nanomaterials.

However, there have been quite a few comprehensive *in vivo* toxicity studies of CdSe-based QD using rodents as animal models. Hauck et al. (2010) showed that administration of ZnS-capped spherical QDs (with diameter between 13 and 21 nm) to rats by intravenous injection did not incur any toxicity over either a short (<7 days) or long (>80 days) period. When analyzing body mass, animal survival, hematology, and histology data and comparing these with a control group of animals that did not receive the QD doses, the researchers did not observe any appreciable toxicity. Interestingly, a considerable amount of Cd was found in the kidney as a result of biodistribution after 30 days of QD administration. Since any intact spherical nanoparticle with a diameter >5.5 nm cannot be accumulated in the kidney through filtration, Hauck et al. justified that some portion of QDs were disintegrated to constituent Cd^{2+} ions, yet did not incur any observable toxicity. Their study also revealed that a level of 3.0 μg Cd per gram of animal body weight present in the kidney was not sufficient to cause toxicity.

In a separate study, Fischer et al. (2010) focused on the short-term (12 h) *in vivo* toxicity effect as a result of uptake of QDs by liver macrophage cells or KCs of rodents. In this study, a set of bovine serum albumin (BSA)-coated QDs and another set of PEG-coated QDs were intravenously injected into rats; the researchers analyzed the KCs 12 h postinjection. The QDs used here contained a CdSe core capped with a ZnS layer and stabilized by a carboxylic acid-based polymeric ligand. As albumin is one of the major component that usually gets deposited on any bare nanoparticle surface due to the opsonization process, BSA-coated QDs were chosen to mimic naturally occurring nanoparticles entering the animal body. Quite expectedly, the BSA-coated QDs were found to have accumulated in liver KCs at a greater extent than the PEG-coated ones 6 h postinjection. Even 12 h postinjection, photoluminescence was observed from the KCs entrapping the QDs, which suggested the intact structure of QDs. However, they recorded the exocytosis process of BSA-coated QDs from the KCs as well as breakdown of the same type of QDs. Approximately 28% of the BSA-coated QDs were found to be intact within the KCs. A portion of QDs were expected to be biodegraded within the KC, since hypochlorous acid present in these type of cells is capable of releasing Cd^{2+} ions from QDs. The released Cd^{2+} was expected either to reach the circulation or to be present in the hepatic tissue, creating a major concern about long-term toxicity. The KC cells, after being extracted from the sacrificed rat livers, did not show any cytotoxicity and appeared normal in a histology study. Those extracted KC cells also expressed normal cellular biochemical function. Interestingly, liver macrophage cells like KCs release cytokines when confronted with pathogens or foreign material to induce an additional immune response. Cytokines are small cell-signaling protein molecules that are secreted by a few specific types of cells of the immune and nervous systems. The increased level of cytokine release represents an inflammatory condition in animal body. Fischer et al. observed the presence of some inflammatory cytokines from the KC cells that could potentially increase antibody production in the immune system. The extent of inflammatory response depends on the type and the doses of the nanoparticles. Since the aspects of inflammation and immune response are inseparably related to toxicity, the *in vitro* results of nanotoxicity studies may not provide an actual road map in predicting the *in vivo* toxicity scenario. But this fact does not underestimate the necessity of *in vitro* studies, as the detailed *in vitro* data help immensely to understand the mechanism of interaction between nanoparticles, cells, and biomolecules *in vivo*. Although biodistributed nanoparticles are entrapped in KC cells and any immediate cytotoxicity may not be observed, KC cells migrate to lymph nodes and exhibit an average life span of about 14 months (Bouwens et al., 1986). Whether such retention of nanoparticles *in vivo* for a considerably longer period induces any genotoxicity or unwanted mutation of DNA is a matter of current research in nanotoxicology.

3.5 Ecology, Environment, and Nanomaterials

In comparison with the recent emergence and progress of nanotechnology-based industry, relevant research assessing its impact on the environment has lagged behind. Although humans use chemicals on a daily basis, the problem of the resulting chemical pollution should encourage environmentalists to obtain a systematic study on nanomaterials' impact on ecology and the environment.

As discussed earlier, undesired emission of nanomaterials in the environment leads to the contamination of soil, water, and air. Once engineered nanomaterials enter the environment, ecosystem processes are at risk (Batley et al., 2013). Human society as a whole depends on collective ecosystem components such as nutrient cycling, crop pollination, food production, and biodiversity conservation. Such components originate from complex ecological interactions among different living organisms placed at different levels in ecosystems. Nutrient cycling is one of the major aspects of ecosystem and takes place through various reactions catalyzed by organisms capable of performing certain biochemical reactions (Holden et al., 2013). In this way, and over a large spatial scale, microbial communities may interact with other ecological communities present in soil or on the ocean bed and finally with the human population. Since the environment depends on complex ecological processes, any impact on the environment by engineered nanomaterials will affect ecological processes. Therefore, ecological nanotoxicity is a great cause for concern in the life sciences. For example, soil bacteria are abundant on the Earth's crust and can absorb and disperse nanoparticles effectively (Horst et al., 2010). If the soil becomes contaminated with CdSe-based QDs, however, bacteria will be associated with these. As a result, QDs can either enter into the bacteria, causing further toxicity, or can generate ROS by reacting with environmental oxygen in the presence of UV radiation. These contaminated bacteria may then be consumed by protozoan predators. In a similar manner, higher living organisms would be affected by nanotoxicity moving higher up the food chain. In an experiment conducted by Werlin et al. (2011), a *Pseudomonas aeruginosa* strain of Gram-negative bacteria was treated with CdSe-based QDs. Those QDs, once internalized by the bacterial population, transferred in intact form to a protozoa (*Tetrahymena thermophilia*) that feeds on the *P. aeruginosa* bacteria in the food chain. As the protozoa consumed multiple bacteria, the total number of QD nanoparticles became amplified in those protozoa. This effect demonstrated the possibility of gradual amplification and accumulation of nanoparticles in the organisms existing at the next-highest trophic level of the food chain (Figure 3.3).

In some well-characterized studies, nanoparticles of zinc oxide (ZnO) and TiO_2 have been found to affect the community of soil bacteria (Ge et al., 2011). Although the effects were dose dependent, alteration of natural biological processes, such as nitrogen fixation, decomposition of complex organic molecules, and methane oxidation, has been observed. Similarly, engineered nanoparticles affect aquatic organisms, including zooplanktons and phytoplanktons. In a study performed by King-Heiden et al. (2009), zebrafish embryos were exposed to

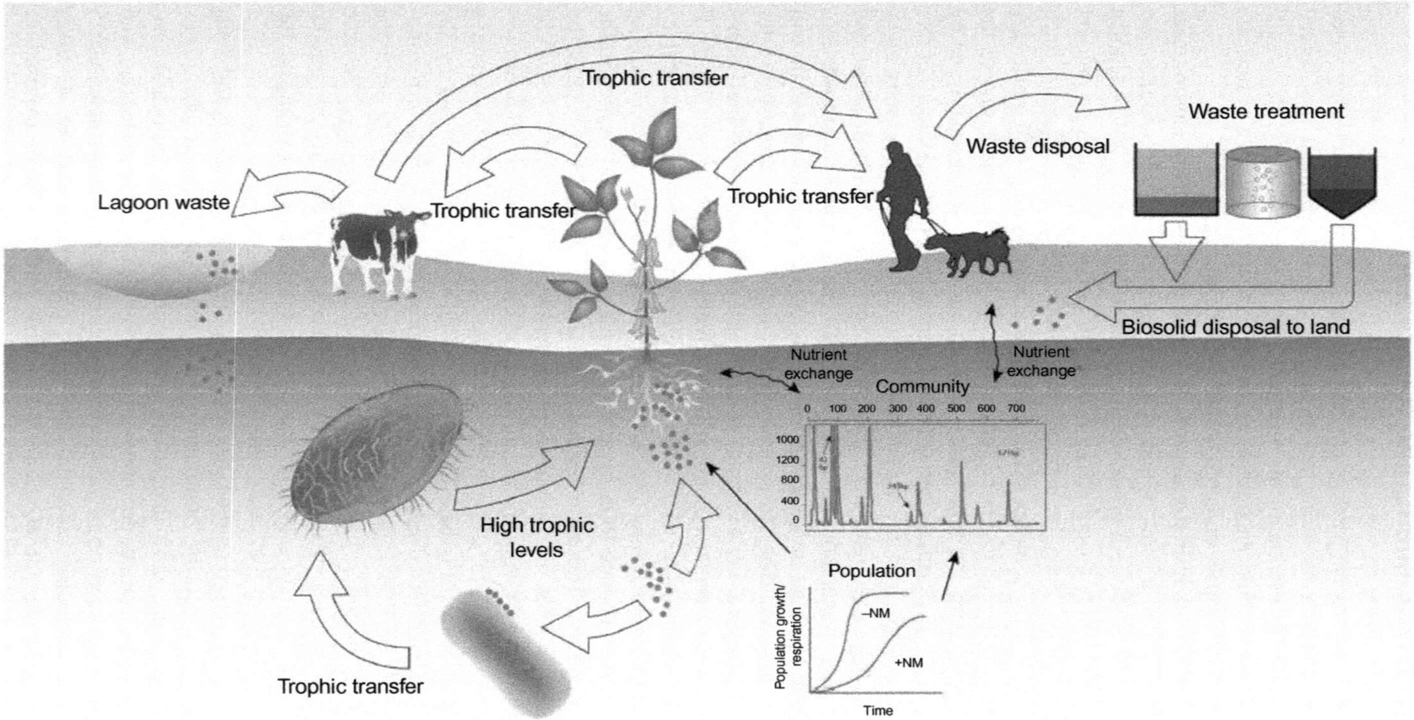

Figure 3.3 Schematic diagram showing the concept of ecological nanotoxicity. Engineered nanomaterials (red dots) entering bacteria (purple) existing in soil below ground can pass to protozoa (green). These nanomaterials can also enter plant systems via the roots. In the food chain, engineered nanomaterials propagate upwards in the ecosystem (Holden et al., 2013). (For interpretation of the references to color in this figure legend, the reader is referred to the web version of this book.)
Source: Reprinted with permission from the publisher.

CdSe-based QDs. The developing embryos were incubated in an aqueous suspension of QDs coated with hydrophilic polymers. The results indicated that QDs disintegrated into Cd^{2+} ions after being absorbed by the embryos and caused toxicity at sublethal concentrations. This type of toxicity affects propagation in fishes and other aquatic organisms, endangering the food chain and aquatic ecosystems.

In an environmental study, Das et al. (2012) attempted to assess the effect of silver nanoparticles on bacterioplankton communities that are present in natural water resources. Their result showed that although the bacterial community tolerated single low-level silver nanoparticle doses, the composition of the bacterial community was different than that of the control cases where no nanoparticles were applied. Due to the antimicrobial properties of silver ions, silver nanoparticle-based water purification systems are widely available in the commercial market. Engineered silver nanoparticles are one of the leading types of nanomaterials produced annually on a mass industrial scale. Any contamination of such silver nanoparticles in natural water resources could interfere with the microbial ecological balance.

Application of nanotechnology or nanomaterials in the plant system has not been thoroughly investigated yet. There have been a few important studies available in the literature (Gonzalez-Melendi et al., 2007). However, achieving a comprehensive analysis based on the research outcome will take several years.

3.6 The Microbial World and Engineered Nanomaterials

Microorganisms play a vital role in the biological and ecological balance of the environment. Various engineered nanomaterials, on the other hand, have been debated in terms of their role in the microbial world, which includes prokaryotic bacteria, bacteriophages, fungi, algae, and protozoa. While much of the literature indicates a negative role for nanomaterials, as they may disrupt the health of the microbial population, experimental proof of engineered nanomaterials killing or harming microbial communities largely depends on the extent of exposure and the condition of the local environment (Ripp, 2011). The presence of any organic matter, local pH, the salinity of the local environment, etc. often become the deciding factors for microbial toxicity, as reported by a wide range of literature published during the last decade in this field. Thus two contradictory opinions, one supporting and the other countering microbial nanotoxicology, exist. This calls for scientific investigation in assessing the risk possessed by nanomaterials toward the microbial world. The influence of a few important classes of nanomaterials, such as QD, nanoparticles of metals and metal oxides, CNTs, and fullerene, on the global microbial biosphere will be discussed briefly in the following section.

3.6.1 Effect of Nanotoxicity in the Microbial Domain

While assessing the toxicity effects of CdSe-based QDs on animal cells in the previous sections, it has been discussed that the surface capping, functionalization, and dose concentrations of QDs are important parameters of toxicity. A similar trend is also observed in microbial applications. Aruguete et al. (2010) observed no toxicity when nonfunctionalized and stable CdSe-based QDs were applied to a *P. aeruginosa* strain up to a QD concentration of 675 nM over a period of 24 h. The same type of QD was also found to be nontoxic when incubated with *P. aeruginosa*, *Escherichia coli*, and *Bacillus subtilis* strains for 48 h at a lower concentration (Mahendra et al., 2008). Interestingly, zero bacterial viability was observed when the pH of the QD solution was changed to a higher or lower value. This result was rationalized in terms of the dissolution of QDs and the subsequent release of toxic Cd^{2+} and Se^{2-} ions (Mahendra et al., 2008). Moreover, dissolved QDs were observed to be more bactericidal than equivalent concentrations of Cd^{2+} and Se^{2-} salts. *P. aeruginosa* cells exhibited intracellular incorporation of CdSe QD due to the cell membrane damage and generation of ROS, whereas no such effects were observed upon incubation with Cd^{2+} ions as control (Metz et al., 2009). The QD uptake of bacteria has been reported to increase in the presence of light due to probable generation of ROS (Metz et al., 2009). In *E. coli*, observed bacterial cell death induced by QDs was mainly attributed to the direct binding with DNA, causing genotoxicity (Zhang et al., 2011). Similar trends of bacterial toxicity are shown by CdTe-based QDs, but to a different extent (Dumas et al., 2009).

Among a wide range of metal oxide nanoparticles, TiO_2 and ZnO are the most common. They are found in cosmetic formulations as well as in industrial products like paints, etc. Both of these oxide show microbial toxicity toward Gram-positive as well as Gram-negative bacteria (Jiang et al., 2009). By virtue of their photocatalytic activity, TiO_2 nanoparticles produce efficient ROS when excited by UV light. Wei et al. (1994) reported almost 100% mortality of *E. coli* under UV radiation within several minutes. Even under dark conditions, TiO_2 nanoparticles are able to induce bacterial toxicity. Nanoparticles of ZnO are effective in halting bacterial population growth but not effective against *E. coli* (Adams et al., 2006). The nano-ZnO-induced toxicity mechanism includes Zn^{2+} ion-mediated oxidative stress, direct cell binding, and subsequent cell wall and cell membrane damage (Raghupati et al., 2011). Smaller nano-TiO_2 or -ZnO exhibit greater toxicity as these are able to penetrate the bacteria to a greater extent. These TiO_2 and ZnO nanoparticles are also harmful to microbes like algae, fungi, bacteriophages, etc. (Ripp, 2011).

Silver nanoparticles demonstrate a wide spectrum of antimicrobial activity. These particles are toxic to Gram-positive and -negative bacteria, algae, fungi, bacteriophages, etc. (Marambio-Jones and Hoek, 2010). At a concentration of 1 μg/mL, silver nanoparticles were observed to arrest growth of *P. aeruginosa* in both planktonic and biofilm form (Kora and Arunachalam, 2011). Generation of ROS by silver ions released from nanoparticles and interference with DNA and ATP production by those ions are believed to be the prime reasons for toxicity (Chudasama et al., 2010). Silver nanoparticles were also found to reduce microbial biomass when added to soil ecosystems within a period of for months in a dose-dependent manner (Hansch and Emmerling, 2010).

CNTs are rolled-up graphene sheets of hexagonally connected carbon atoms through covalent bonding. If a CNT is encased in a single layer of carbon sheet (graphene), it is known as a SWCNT; if there are multiple concentric, encasing graphene layers it is known as a multiple-walled carbon nanotube (MWCNT). Although a large volume of literature is available evaluating the effect of CNT on animal cells, the same on microbes is somewhat limited in comparison. In a study reported by Kang et al. (2007), a SWCNT concentration of 5 μg/mL induced an 80% reduction in *E. coli* viability after an exposure of 1 h. Compromise of cell wall integrity by direct contact of the SWCNT with the bacterial cell was the damaging factor, as evident from the study. Interestingly, CNTs with a smaller diameter had been found to be more efficient in piercing through the bacterial cell membrane (Kang et al., 2008). Generally, SWCNTs are more effective in killing Gram-positive than Gram-negative bacteria, owing to the fact that the outer bacterial membrane structure is much more complex in the latter type. Research on the effect of MWCNTs on bacteria has revealed a toxicity trend similar to that of SWCNTs; however, not much work has been reported involving MWCNTs. Another carbon-based nanostructure that exhibits antimicrobial behavior is fullerene, or bucky balls, which consists of 60 carbon atoms arranged to form a soccer ball-like structure. There is considerable literature describing the antimicrobial effects of these fullerenes in a detailed manner (Lyon et al., 2006).

Nanostructures formed from polymeric macromolecules are an important and exciting class of materials showing antimicrobial properties. A somewhat detailed discussion is provided in the next sections.

3.6.2 Nanomaterials and Microbial Drug Resistance

One of the most recent applications of nanotechnology in the microbial domain is in the development of antibiotics. Since their discovery, antibiotics have helped to prevent serious diseases like tuberculosis, pneumonia, etc. However, this revolutionary treatment paradigm is facing an increasing threat from a number of microorganisms known as multiple drug resistant (MDR) or, more commonly, "superbugs" (Aruguete et al., 2013). Widespread production and indiscriminate use as well as misuse of antibiotics is believed to be responsible for the origin of these MDR strains and microorganisms. In recent times, engineered nanomaterials have been put forward as effective agents in treating MDR microbes. Nanomaterial-based antibacterial therapeutics generally contain two basic parts: a nanosized carrier and an organic molecule acting as an antibiotic drug. Polymeric carbohydrate nanoparticles (e.g., chitosans), nanoliposomes (lipid vesicles), dendrimers (branched organic polymers), inorganic nanoparticles (e.g., metal, metal oxides), etc. typically serve as nanocarriers. The drug molecules are either encapsulated within liposomes or functionalized on the outer surface of dendrimers or inorganic nanoparticles. The assembled design of drug delivery shows effectiveness in killing pathogenic MDR bacterial strains as described by Huh and Kwon (2011). In a study reported by Chakraborty et al. (2010), vancomycin-resistant *Staphylococcus aureus* was treated with vancomycin conjugated with chitosan nanoparticles tagged with folic acid and unconjugated free vancomycin alone as a control *in vitro*. Interestingly, the chitosan nanoparticle–mediated delivery system exhibited a much lower minimum inhibitory concentration (MIC) of drugs as compared to the free vancomycin. Chitosan–antibiotic conjugates that were not tagged with folic acid were not effective against *S. aureus*. This experimental finding implies that drug-resistant bacteria mistakenly consumed the nanochitosan–drug conjugate as food. This deceptive "Trojan horse" approach is one of the mechanisms responsible for antimicrobial nanomaterial-mediated defense against MDR strains. In addition, metal nanoparticles conjugated with conventional antibiotic drugs sometimes show synergistic effects. For example, silver nanoparticles conjugated with ampicillin were more effective in killing MDR strains of *P. aeruginosa* and *Enterobacter aerogenes* compared to either silver nanoparticles or the ampicillin alone *in vitro* (Brown et al., 2012). The detailed pathways and rationalization behind this type of synergistic effect are yet to be investigated.

Traditional antibiotics show their efficacy by interfering in many biological processes of bacteria, including cell wall synthesis, ribosomal function, DNA replications, etc. Antibiotics can also cause cell wall damage in bacteria (Aruguete et al., 2013). There are some defense mechanisms offered by bacteria to resist the actions of antibiotics. Sometimes bacteria modify their own biochemical processes to prevent the antibiotics from hitting a specific biochemical target. Drugs may also be degraded, altered, or expelled from the bacterial system (Aruguete et al., 2013). The major problem stems from the fact that all these drug resistance mechanisms are encoded by DNA sequences commonly known as antibiotic resistance genes (ARGs). Upon activation, these ARGs may pass from one generation to another either vertically or horizontally between neighboring populations of the same generation. As discussed previously, there are several types of nanomaterials, such as fullerene, CNTs, metal or metal oxide nanoparticles, polymeric nanoparticles, that demonstrate intrinsic antimicrobial properties. One such material is macromolecular antimicrobial polymer (MAP), which exists in nanoscale dimensions (Hancock and Sahl, 2006). These polymeric nanomaterials carry out antibacterial activity by selectively lysing the bacterial cell membrane without interfering in any complex metabolic pathway. Action against the bacterial membrane is nonspecific and does not involve any participation of membrane-specific receptor components. Thus the probability of mutation or any other alteration causing drug resistance is low for such materials. Furthermore, appropriate chemical modification can make these materials biodegradable, which in turn makes them a useful therapeutic option for human health care (Nederberg et al., 2011). On the other hand, some materials like silver nanoparticles tend to latch specifically onto membranes and selectively inactivate certain biomolecules. The thiol groups present in the membrane receptor proteins may interact with silver nanoparticles. Triggered by ARG response, drug resistance may be set off in such situations.

Many surface active nanomaterials are capable of generating ROS that can damage biomolecules, including DNA, RNA, and proteins. Silver ions leaching out from silver nanoparticles can produce such ROS, through which antimicrobial activities are often manifested. In response to the ROS, bacteria are able to generate enzymes like superoxide dismutase that can negate the reactive oxygen radicals. In *E. coli*, the genetically induced oxidative stress response is well studied

(Adams et al., 2006). Maintaining the reducing environment of the bacteria is the main purpose of the oxidative stress response. As this response pathway is genetically regulated, it is suspected that a mechanism resistant to ROS-generating nanoparticles would prevail through vertical and horizontal propagation of genes in bacteria.

Bacteria may also defend the antimicrobial behavior of nanomaterials by formation of biofilms (Choi et al., 2010b). Biofilms are an extracellular polymeric matrix in which microorganisms are embedded to form a two-dimensional coverage. They are composed of polysaccharides, lipids, proteins, etc. Typically, when bacteria are within a biofilm environment, they exhibit stronger antibiotic resistance compared to their free-floating planktonic form. Studies involving *E. coli* have shown that silver nanoparticles are ineffective in eradicating *E. coli* biofilm compared to the planktons (Choi et al., 2010b). It has been postulated that silver nanoparticles are unable to diffuse through biofilm and are aggregated locally on certain areas of biofilm surfaces.

Some types of engineered nanomaterials show their antibiotic effect by releasing toxic metal ions from the respective nanoparticles. These metal ions interfere with enzyme systems or DNA to express damaging effects. However, it is reported that if the metal ion is the sole cause of toxicity in bacteria, nanomaterial resistance can be developed through an adaptation pathway, triggered by a genetic response (Silver, 2003). Bacteria may either eject or sequester the metal ions as a part of the resistance process. Chronic exposure of these types of metal or metal oxide nanoparticles to the microbial world may initiate activation of ARGs, mutation, or genetic propagation that may cause MDR effects among microbes (Silver, 2003). Qiu et al. (2012) reported a horizontal gene transfer between *E. coli* and *Salmonella* in a drinking water treatment plant where nanoalumina was used as a part of the treatment process. The widespread use of silver nanoparticle-based commercial commodities, such as nano-TiO_2- or nano-ZnO-based cosmetic products, raises the same concern about the antibacterial effect of such nanomaterials in the long term. Any engineered nanomaterials that put a stress on bacteria are potentially responsible for creating an adaptive genetic response due to which bacteria may develop resistance to traditional antibiotics as well. At the end, the microbes may become resistant to both the antibiotics and the nanomaterials. Although, engineered nanomaterials show antimicrobial activity and are effective against MDR strain of bacteria to some extent; a controlled and well planned use of the nanomaterials is desired in the microbial domain.

3.6.3 Biodegradable Nanomaterials and Microbes

Most of the traditional antibiotics do not damage bacterial cell walls; rather, they penetrate the cell via a specific interaction with the membrane component (Brogden, 2005). After going inside the cell they block important cellular biochemistry, for example, by blocking cell division, inhibiting DNA replication, or causing the breakdown of double-stranded DNA, etc. Because the bacterial morphology largely remains intact in this case, bacteria can develop drug resistance in a relatively simpler way. On the other hand, there are some types of cationic peptide-based antibiotics that enter the bacteria by making ruptures in the cell wall and thereby disintegrating the bacterial structures. Since the antibiotic actions of such drugs are nonspecific and physical in nature, bacterial development of drug resistance becomes difficult (Brogden, 2005). Such cationic peptide-based antibiotics (e.g., alamethicin, protegrins, etc.) are chemically amphiphilic in nature and attain α-helical or β-sheet-like secondary structures upon interacting with bacterial cell walls that are negatively charged. After being diffused through the cell walls, multiple molecules of these cationic peptides become aggregated and form a molecular assembly in the lipophilic layer of the bacterial cell membrane. This phenomenon leads to the rupture of the cell membrane and subsequent death of the microorganism (Oren and Shai, 1999). Despite the fact that cationic peptide-based antibiotics are effective against MDR bacteria, clinical trials of such drugs have yielded limited success (Hancock and Sahl, 2006). Cytotoxicity of cationic peptides towards animal cells, low *in vivo* lifetimes, and high manufacturing costs are the main reasons for this failure. As an alternative, less toxic antimicrobial polymeric compounds, such as poly (acrylamide) and poly(β-lactam), have been developed as drugs. However, these polymeric compounds suffer from low biodegradation behavior.

In this context, Nederberg et al. (2011) have recently developed a polycarbonate-based cationic polymer that is biodegradable as well as less toxic. These polymers did not form any secondary structures before interacting with cell membranes, as in the case of cationic peptide-based polymers. Furthermore, in an aqueous medium the monomers of these polycarbonate polymers self-assembled to form spherical nanoparticles with an approximately 200 nm average diameter. Nederberg et al. hypothesized that the positively charged cationic nanoparticles of polycarbonate polymers had prominent interactions with bacterial cell membranes and expressed a bactericidal effect via a locally accumulated cationic charge, followed by disintegration of the cell membrane. Also, larger size of those nanoparticles compared to other polymeric macromolecules helped in the cell membrane damage (Figure 3.4). In a series of experiments, Nederberg et al. demonstrated successful inhibition of microbes including a range of bacteria and fungi. In a dose-dependent growth inhibition study, the polycarbonate polymeric material inhibited the growth of *B. subtilis*, *S. aureus*, Methicillin-resistant *S. aureus*, *E. faecalis*, and *C. neoformans* with corresponding MIC of 4.3, 6.5, 7.0, 10.8, and 10.8 μM, respectively. These MIC values were found to be comparable to the MIC values of conventional antimicrobial drugs in clinical settings.

The biocompatibility of polycarbonate nanostructures was tested in mouse models. Hemolysis, or the breakdown of red blood cells, is a major problem pertaining to cationic polymeric drugs. Nederberg et al. reported no appreciable hemolysis up to a concentration of 500 μg/mL. Additional *in vivo* toxicity studies revealed unchanged functional parameters of the livers and kidneys of mouse models upon intravenous injection of polycarbonate nanostructures with a concentration (39 μM) much above the highest MIC measured. These results indicate that in future, nanostructure-based antimicrobial agents could effectively be applied in clinical

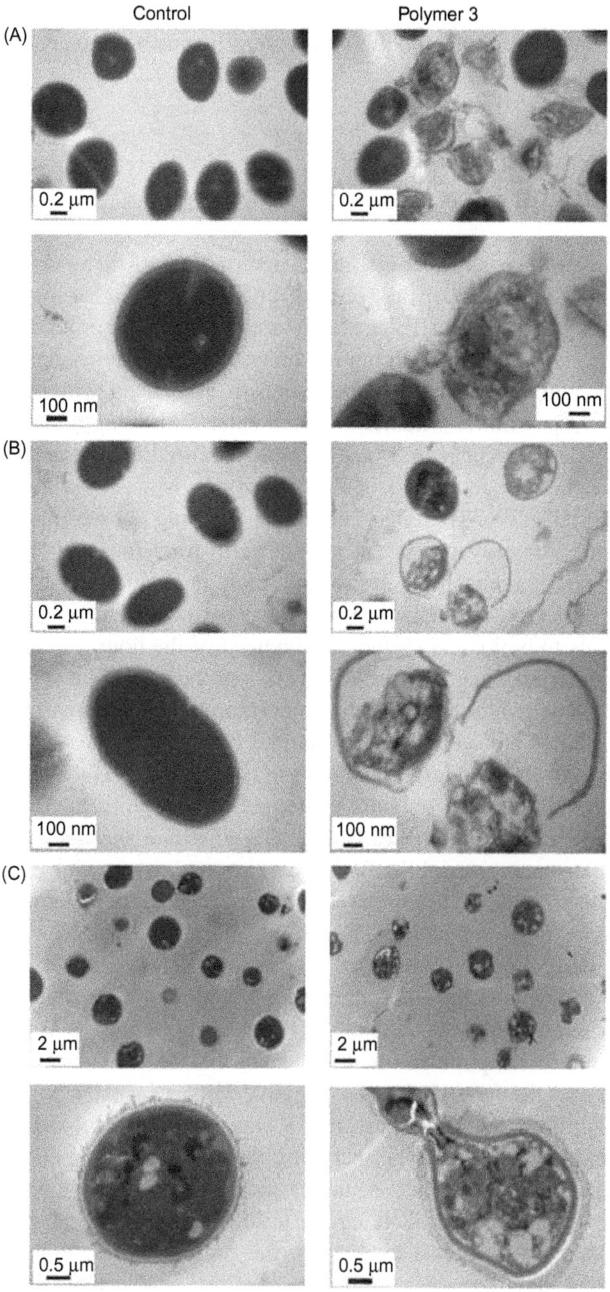

Figure 3.4 Comparative TEM images of microbes in the absence and presence of polycarbonate-based antimicrobial polymer (polymer 3). Methicillin-resistant *Staphylococcus aureus* (A), *Enterococcus faecalis* (B), and *Cryptococcus neoformans* (C) are shown before (left) and after (right) incubation with the polymer for 8 h at lethal doses. After treatment, cell walls and membranes were found to be damaged and cell death was observed (Nederberg et al., 2011). *Source*: Reprinted with permission from the publisher.

medicine with significant success against multi-drug-resistant bacterial strains. Such biodegradable and biocompatible polymeric nanostructures could open a new horizon in antibiotic research.

3.7 Conclusion

Nanotechnology, which has become an indispensable part of modern life, is still a new discipline. While most of the research utilizing nanotechnology and nanomaterials has been oriented chiefly towards application and product development, impact of such materials on the living organisms and the environment has not been studied to an equal extent. The greatest concern associated with nanomaterials is their size, which determines whether they can enter into living organisms and interact with cells, tissues, and genes. As a result, continuing research and industrial applications involving nanomaterials require the provision of assurances for health and environmental safety. There has been much ongoing debate among governmental agencies and advocacy groups over whether to implement any specific regulations on the use of nanomaterials in medical, biological, and industrial research and development. Therefore, systematic knowledge of the extent of toxicity imparted by nanomaterials to biological systems is the need of the hour.

How nanomaterials enter a living organism and the fate of an organism once the materials are inside it are the two most important questions in assessing nanotoxicity. *In vitro* studies are carried out on animal cells and tissues in order to understand the potential harmfulness of nanoparticles. Later, *in vivo* studies reveal the short- and long-term effects of those materials in live animals. However, studies show that a predictive relationship between *in vivo* and *in vitro* studies does not exist: a particular nanomaterial that shows *in vitro* toxicity may or may not be toxic *in vivo*. Parameters that influence the expression of toxicity in animal systems are difficult to understand and predict. Therefore, more systematic knowledge of and data on short- and long-term *in vivo* toxicity is still needed.

As microorganisms easily come into contact with nanomaterials existing in soil and water via unplanned or accidental exposure, nanotoxicity plays a vital role in the health of microbial populations. Since microbes are important components of the ecosystem and the food chain, any negative impact on their population is most likely to be amplified and affect humans. However, microbial toxicity due to nanoparticles depends on many factors, including their concentration and the local conditions. Another important fact is that nanomaterials are not entirely new to microbes. Bacteria have been evolving for millions of years in the presence of natural antibiotics; they have also evolved in the presence of nanomaterials such as metal oxides and clay particles. From an adaptation perspective, bacteria have the potential to develop defense mechanisms against engineered nanomaterials as well. However, the intelligent design of engineered nanomaterials may produce effective antimicrobial agents working against harmful multi-drug-resistant bacteria that are otherwise difficult to treat with conventional molecular antibiotics.

Due to the potentially hazardous effects of a few types of nanomaterials, there has been a fear among the lay population about nanomaterials and their detrimental effects. Research has shown that careful design and wise application of such materials will lead to a better future in the fields of science and technology. However, with the growth of nanotechnology-based industries, it is equally important to keep the environment free from unwanted exposure to nanomaterials as much as possible, because a complete assessment of the effect of nanotoxicity on the biological world has not been carried out yet. There is a gap of knowledge about nanomaterials that may appear nontoxic under short-term study, but whose long-term effects are not known. The fact that we are still far from attaining a conclusive picture makes nanotoxicity an exciting area of research in biology.

Reference

Adams, L.K., Lyon, D.Y., Alvarez, P.J.J., 2006. Comparative eco-toxicity of nanoscale TiO_2, SiO_2, and ZnO water suspensions. Water Res. 40 (19), 3527–3532.

Almeida, J.P.M., Chen, A.L., Foster, A., Drezek, R., 2011. *In vivo* biodistribution of nanoparticles. Nanomedicine. 6 (5), 815–835.

Arora, S., Rajwade, J.M., Paknikar, M.K., 2012. Nanotoxicology and *in vitro* studies: the need of the hour. Toxicol. Appl. Pharmacol. 258, 151–165.

Aruguete, D.M., Guest, J.S., Yu, W.W., Love, N.G., Hochella, M.F., 2010. Interaction of CdSe/CdS core–shell quantum dots and *Pseudomonas aeruginosa*. Environ. Chem. 7 (1), 28–35.

Aruguete, D.M., Kim, B., Hochella, M.F., Yanjun, M., Cheng, Y., Hoegh, A., et al., 2013. Antimicrobial nanotechnology: its potential for the effective management of microbial drug resistance and implications for research needs in microbial nanotoxicology. Environ. Sci. Processes Impacts. 15, 93–102.

Balasubramanian, S.K., Jittiwat, J., Manikandan, J., Ong, C., Yu, L.E., Ong, W., 2010. Biodistribution of gold nanoparticles and gene expression changes in the liver and spleen after intravenous administration in rats. Biomaterials. 31 (8), 2034–2042.

Batley, G.E., Kirby, J.K., McLaughlin, M.J., 2013. Fate and risks of nanomaterials in aquatic and terrestrial environments. Acc. Chem. Res. 46 (3), 854–862.

Bouwens, L., Baekeland, M., De Zanger, R., Wisse, E., 1986. Quantitation, tissue distribution and proliferation kinetics of Kupffer cells in normal rat liver. Hepatology. 6 (4), 718–722.

Brogden, K.A., 2005. Antimicrobial peptides: pore formers or metabolic inhibitors in bacteria. Nat. Rev. Microbiol. 3, 238–250.

Brown, A.N., Smith, K., Samuels, T.A., Lu, J., Obare, S.O., Scott, M.E., 2012. Nanoparticles functionalized with ampicillin destroy multiple-antibiotic-resistant isolates of *Pseudomonas aeruginosa* and *Enterobacter aerogenes* and methicillin resistant *Staphylococcus aureus*. Appl. Environ. Microbiol. 78, 2768–2774.

Carlson, C., Hussain, S.M., Schrand, A.M., Braydich-Stolle, L.K., Hess, K.L., Jones, R.L., et al., 2008. Unique cellular interaction of silver nanoparticles: size-dependent generation of reactive oxygen species. J. Phys. Chem. B. 112 (43), 13608–13619.

Chakraborty, S.P., Sahu, S.K., Mahapatra, S.K., Santra, S., Bal, M., Roy, S., et al., 2010. Nanoconjugated vancomycin: new opportunities for the development of anti-VRSA agents. Nanotechnology. 21 (10), 105103.

Chen, Z., Meng, H., Zing, G., Chen, C., Zhao, Y., Jia, G., et al., 2006. Acute toxicological effects of copper nanoparticles *in vivo*. Toxicol. Lett. 163, 109–120.
Cho, K., Wang, X., Nie, S., Chen, Z., Shin, D.M., 2008. Therapeutic nanoparticles for drug delivery in cancer. Clin. Cancer Res. 14 (5), 1310–1316.
Choi, C.H.J., Alabi, C.A., Webster, P., Davis, M.E., 2010a. Mechanism of active targeting in solid tumors with transferrin-containing gold nanoparticles. PNAS USA. 107 (3), 1235–1240.
Choi, C.H.J., Zuckerman, J.E., Webster, P., Davis, M.E., 2011. Targeting kidney mesangium by nanoparticles of defined size. PNAS USA. 108 (16), 6656–6661.
Choi, H., Liu, W., Misra, P., Tanaka, E., Zimmer, J., Itty Ipe, B., et al., 2007. Renal clearance of quantum dots. Nat. Biotechnol. 25, 1165–1170.
Choi, O.Y., Yu, C.P., Fernandez, G.E., Hu, Z., 2010b. Interactions of nanosilver with *Escherichia coli* cells in planktonic and biofilm cultures. Water Res. 44, 6095–6103.
Chudasama, B., Vala, A.K., Andhariya, N., Mehta, R.V., Upadhyay, R.V., 2010. Highly bacterial resistant silver nanoparticles: synthesis and antibacterial activities. J. Nanopart. Res. 12 (5), 1677–1685.
Danhier, F., Feron, O., Preat, V., 2010. To exploit the tumor microenvironment: passive and active tumor targeting of nanocarriers for anti-cancer drug delivery. J. Controlled Release. 148, 135–146.
Das, P., Williams, C.J., Fulthorpe, R.R., Hoque, M.E., Metcalfe, C.D., Xenopoulos, M.A., 2012. Changes in bacterial community structure after exposure to silver nanoparticles in natural waters. Environ. Sci. Technol. 46, 9120–9128.
Davis, S.S., 1997. Biomedical applications of nanotechnology—implications for drug targeting and gene therapy. Trends Biotechnol. 15 (6), 217–224.
Derfus, A.M., Chan, W.C.W., Bhatia, S.N., 2004. Probing the cytotoxicity of semiconductor quantum dots. Nano Lett. 4 (1), 11–18.
Dumas, E.M., Ozenne, V., Mielke, R.E., Nadeau, J.L., 2009. Toxicity of CdTe quantum dots in bacterial strains. IEEE Trans. Nano Biosci. 8 (1), 58–64.
Fischer, H.C., Hauck, S.T., Gomez-Aristizabal, A., Chan, W.C.W., 2010. Exploring primary liver macrophages for studying quantum dot interactions with biological systems. Adv. Mater. 22, 2520–2524.
Ge, Y., Schimel, J.P., Holden, P.A., 2011. Evidence for negative effects of TiO_2 and ZnO nanoparticles on soil bacterial communities. Environ. Sci. Technol. 45, 1659–1664.
Gontier, E., Ynsa, M.-D., Biro, T., Hunyadi, J., Kiss, B., Gaspar, K., et al., 2008. Is there penetration of titania nanoparticles in sunscreens through skins? A comparative electron and ion microscopy study. Nanotoxicology. 2, 218–231.
Gonzalez-Melendi, P., Fernandez-Pacheco, R., Coronado, M.J., Coredor, E., Testillano, P.S., Risueno, M.C., et al., 2007. Nanoparticles as smart treatment-delivery systems in plants: assessment of different techniques of microscopy for their visualization in pant tissues. Ann. Bot. 101 (1), 187–195.
Gref, R., Domb, A., Quellec, P., Blunk, T., Müller, R.H., Verbavatz, J.M., et al., 1995. The controlled intravenous delivery of drugs using PEG-coated sterically stabilized nanospheres. Adv. Drug Delivery Rev. 16 (2–3), 215–233.
Hagens, W.I., Oomen, A.G., de Jong, W.H., Cassee, W.H., Sips, A.J., 2007. What do we (need to) know about the kinetic properties of nanoparticles in the body? Regul. Toxicol. Pharmacol. 49, 217–219.
Hancock, R.E.W., Sahl, H.G., 2006. Antimicrobial and host–defence peptides as new anti-infective therapeutic strategies. Nat. Biotechnol. 24, 1551–1557.

Hansch, M., Emmerling, C., 2010. Effects of silver nanoparticles on the microbiota and enzyme activity in soil. J. Plant Nutr. Soil Sci. 173 (4), 554–558.

Hardman, R., 2006. A toxicological review of quantum dots: toxicity depends on physicochemical and environmental factors. Environ. Health Perspect. 114 (2), 165–172.

Hauck, T.S., Anderson, R.E., Fischer, H.C., Newbigging, S., Chan, W.C.W., 2010. *In vivo* quantum-dot toxicity assessment. Small. 6 (1), 138–144.

Hirsch, L.R., Stafford, R.J., Bankson, J.A., Sershen, S.R., Rivera, B., Price, R.E., et al., 2003. Nanoshell-mediated near-infrared thermal therapy of tumors under magnetic resonance guidance. PNAS USA. 100, 13549–13554.

Holden, P.A., Nisbet, M.R., Lenihan, H.S., Miller, R.J., Cherr, G.N., Schimel, P.J., et al., 2013. Ecological nanotoxicity: integrating nanomaterial hazard considerations across the subcellular, population, community, and ecosystems levels. Acc. Chem. Res. 46 (3), 813–822.

Horst, A.M., Neal, A.C., Mielke, R.E., Sislian, P.R., Suh, W.H., Madler, L., et al., 2010. Dispersion of TiO_2 nanoparticle agglomerates by *Pseudomonas aeruginosa*. Appl. Environ. Microbiol. 76, 7292–7298.

Hu, Y.L., Qi, W., Han, F., Shao, J.Z., Gao, J.Q., 2011. Toxicity evaluation of biodegradable chitosan nanoparticles using a zebra fish embryo model. Int. J. Nanomed. 6, 3351–3359.

Huang, X., Neretina, S., El-Sayed, M.A., 2009. Gold nanorods: from synthesis and properties to biological and biomedical applications. Adv. Mater. 21, 4880–4910.

Huh, A.J., Kwon, Y.J., 2011. "Nanoantibiotics": a new paradigm for treating infectious diseases using nanomaterials in the antibiotics resistant era. J. Controlled Release. 156, 128–145.

ICRP, 1994. Human respiratory tract model for radiological protection. ICRP Publication 66. Ann. ICRP. 24 (1–3).

Jiang, W., Mashayekhi, H., Xing, B.S., 2009. Bacterial toxicity comparison between nano- and micro-scaled oxide particles. Environ. Pollut. 157 (5), 1619–1625.

Johnston, H., Semmler-Behnke, M., Brown, D., Kreyling, W., Tran, L., Stone, V., 2010. Evaluating the uptake and intracellular fate of polystyrene nanoparticles by primary and hepatocyte cell lines *in vitro*. Toxicol. Appl. Pharmacol. 242, 66–78.

Jordan, A., Scholz, R., Maier-Hauff, K., Johannsen, M., Wust, P., Nadobny, J., et al., 2001. Presentation of a new magnetic field therapy system for the treatment of human solid tumors with magnetic fluid hyperthermia. J. Magn. Magn. Mater. 225, 118–126.

Kala, S.V., Lykissa, E.D., Neely, M.W., Lieberman, M.W., 1998. Low molecular weight silicones are widely distributed after a single subcutaneous injection in mice. Am. J. Pathol. 152 (3), 645–649.

Kang, S., Pinault, M., Pfefferle, L.D., Elimelech, M., 2007. Single-walled carbon nanotubes exhibit strong antimicrobial activity. Langmuir. 23 (17), 8670–8673.

Kang, S., Herzberg, M., Rodrigues, D.F., Elimelech, M., 2008. Antibacterial effects of carbon nanotubes: size does matter. Langmuir. 24 (13), 6409–6413.

Kim, B.Y.S., Rutka, J.T., Chan, W.C.W., 2010. Nanomedicine. N. Engl. J. Med. 363 (25), 2434–2443.

King-Heiden, T.C., Wiecinski, P.N., Mangham, A.N., Metz, K.M., Nesbit, D., Pedersen, J.A., et al., 2009. Quantum dot nanotoxicity assessment using the zebrafish embryo. Environ. Sci. Technol. 43, 1605–1611.

Klabunde, K.J. (Ed.), 2002. *Nanoscale Materials in Chemistry*. John Wiley & Sons, Inc., USA.

Kora, A.J., Arunachalam, J., 2011. Assessment of antibacterial activity of silver nanoparticles on *Pseudomonas aeruginosa* and its mechanism of action. World J. Microbiol. Biotechnol. 27 (5), 1209–1216.

Lademann, J., Weighmann, H.-J., Rickmeyer, C., Barthelmes, H., Schaefer, H., Mueller, G., et al., 1999. Penetration of titanium dioxide microparticles in a sunscreen formulation into the horny layer and the follicular orifice. Skin Pharmacol. Appl. Skin Physiol. 12, 247–256.

Li, S., Huang, L., 2008. Pharmacokinetics and biodistribution of nanoparticles. Mol. Pharm. 5 (4), 496–504.

Liu, Z., Davis, C., Cai, W., He, L., Chen, X., Dai, H., 2008. Circulation and long-term fate of functionalized, biocompatible single-walled carbon nanotubes in mice probed by Raman spectroscopy. PNAS USA. 105, 1410–1415.

Lunov, O., Syrovets, T., Rocker, C., Tron, K., Nienhaus, G., Rasche, V., et al., 2010. Lysosomal degradation of the carboxydextran shell of coated superparamagnetic iron oxide nanoparticles and the fate of professional phagocytes. Biomaterials. 31, 9015–9022.

Lyon, D.Y., Adams, L.K., Falkner, J.C., Alvarez, P.J.J., 2006. Antibacterial activity of fullerene water suspensions: effects of preparation method and particle size. Environ. Sci. Technol. 40 (14), 4360–4366.

Mahendra, S., Zhu, H.G., Colvin, V.L., Alvarez, P.J., 2008. Quantum dot weathering results in microbial toxicity. Environ. Sci. Technol. 42 (24), 9424–9430.

Marambio-Jones, C., Hoek, E.M.V., 2010. A review of the antibacterial effects of silver nanomaterials and potential implications for human health and the environment. J. Nanopart. Res. 12 (5), 1531–1551.

Marino, A.A., Mitchell, J.T., 1972. Lung damage in mice following intraperitoneal injection of butylated hydroxytoluene. Exp. Biol. Med. (Maywood). 140 (1), 122–125.

Metz, K.M., Mangham, A.N., Bierman, M.J., Jin, S., Hamers, R.J., Pedersen, J.A., 2009. Engineered nanomaterial transformation under oxidative environmental conditions: development of an *in vitro* biomimetic assay. Environ. Sci. Technol. 43 (5), 1598–1604.

Nederberg, F., Zhang, Y., Tan, J.P.K., Xu, K., Wang, H., Yang, C., et al., 2011. Biodegradable nanostructures with selective lysis of microbial membranes. Nat. Chem. 3, 409–414.

Oberdörster, G., Oberdörster, E., Oberdorster, J., 2005. Nanotoxicology: an emerging discipline evolving from studies of ultrafine particles. Environ. Health Perspect. 113 (7), 823–839.

Oren, Z., Shai, Y., 1999. Cyclization of a cytolytic amphipathic α-helical peptide and its diastereomer: effect on structure, interaction with model membranes, and biological function. Biochemistry. 39, 6103–6114.

Ota, Z., Makino, H., Miyoshi, A., Hiramatsu, M., Takahashi, K., Ofuji, T., 1979. Molecular-sieve in glomerular basement-membrane as revealed by electron microscopy. J. Electron Microsc. 28, 20–28.

Parham, P., 2005. Elements of immune system and their roles in defense, The Immune System. second ed. Garland Science Publishing, New York, NY.

Perrault, S.D., Walkey, C., Jennings, T., Fischer, H.C., Chan, W.C., 2009. Mediating tumor targeting efficiency of nanoparticles through design. Nano Lett. 9, 1909–1915.

Qiu, Z., Yu, Y., Chen, Z., Jin, M., Yang, D., Zhao, Z., et al., 2012. Nanoalumina promotes the horizontal transfer of multiresistance genes mediated by plasmids across genera. PNAS USA. 109 (13), 4944–4949.

Raghupathi, K.R., Koodali, R.T., Manna, A.C., 2011. Size-dependent bacterial growth inhibition and mechanism of antibacterial activity of zinc oxide nanoparticles. Langmuir. 27 (7), 4020–4028.

Rao, C.N.R., Muller, A., Cheetham, A.K. (Eds.), 2005. The Chemistry of Nanomaterials: Synthesis, Properties and Application. John Wiley & Sons, Inc., USA.

Ripp, S., 2011. Nanotoxicology in the microbial world. ACS Symposium Series. 1079, 121–140 (Chapter 6).

Ryman-Rasmussen, J.P., Riviere, J.E., Monteiro-Riviere, N.A., 2006. Penetration of intact skin by quantum dots with diverse physicochemical properties. Toxicol. Sci. 91, 159–165.

Sadauskas, E., Danscher, G., Stoltenberg, M., Vogel, U., Larsen, A., Wallin, H., 2009. Protracted elimination of gold nanoparticles from mouse liver. Nanomedicine. 5 (2), 162–169.

Senior, J., Crawley, J.C.W., Gregoriadis, G., 1985. Tissue distribution of liposomes exhibiting long half-lives in the circulation after intravenous injection. Biochim. Biophys. Acta. 839 (1), 1–8.

Silver, S., 2003. Bacterial silver resistance: molecular biology and uses and misuses of silver compounds. FEMS Microbiol. Rev. 27 (2–3), 341–353.

Simpson, C.A., Salleng, K.J., Cliffel, D.E., Feldheim, D.L., 2013. *In vivo* toxicity, biodistribution, and clearance of glutathione-coated gold nanoparticles. Nanomed. Nanotechnol. Biol. Med. 9 (2), 257–263.

Teo, S., Stirling, D., Thomas, S., Hoberman, A., Kiorpes, A., Vikram, K., 2002. A 90-day oral gavage toxicity study of D-methylphenidate and D,L-methylphenidate in Sprague-Dawley rats. Toxicology. 179 (3), 183–196.

Trop, M., Novak, M., Rodl, S., Hellbom, B., Kroell, W., Goessler, W., 2006. Silver-coated dressing Acticoat caused raised liver enzymes and argyria-like symptoms in burn patient. J. Trauma. 60, 648–652.

Wang, B., He, X., Zhang, Z., Zhao, Y., Feng, W., 2013. Metabolism of nanomaterials *in vivo*: blood circulation and organ clearance. Acc. Chem. Res. 46 (3), 761–769.

Warheit, D.B., 2010. Debunking some misconceptions about nanotoxicology. Nano Lett. 10, 4777–4782.

Wei, C., Lin, W.Y., Zainal, Z., Williams, N.E., Zhu, K., Kruzic, A.P., et al., 1994. Bactericidal activity of TiO_2 photocatalyst in aqueous media—toward a solar-assisted water disinfection system. Environ. Sci. Technol. 28 (5), 934–938.

Werlin, R., Priester, J.H., Mielke, R.E., Kramer, S., Jackson, S., Stoimenov, P.K., et al., 2011. Biomagnification of cadmium selenide quantum dots in a simple experimental microbial food chain. Nat. Nanotechnol. 6 (1), 65–71.

Zhang, W., Yao, Y., Chen, Y.S., 2011. Imaging and quantifying the morphology and nanoelectrical properties of quantum dot nanoparticles interacting with DNA. J. Phys. Chem. C. 115 (3), 599–606.

4 Application of Molecular Techniques for the Assessment of Microbial Communities in Contaminated Sites

Anirban Chakraborty[a], *Chanchal K. DasGupta*[a] *and Punyasloke Bhadury*[b]

[a]Department of Life Science and Biotechnology, Jadavpur University, Kolkata, India, [b]Integrative Taxonomy and Microbial Ecology Research Group, Department of Biological Sciences, Indian Institute of Science Education and Research Kolkata, Nadia, West Bengal, India

4.1 Introduction

Invisible to the normal eye they may be, but microorganisms outnumber the combined total plant and animal cells in the biosphere by 2–3 orders of magnitude and constitute ca. 60% of the Earth's biomass (Whitman et al., 1998). In addition to existing in such enormous numbers, microorganisms are ubiquitous in all spheres of the environment and represent a large, unexplored reservoir of biodiversity. Almost every ecological process is associated, either directly or indirectly, with microbial activities. In addition, microbes play crucial roles in global biogeochemical cycling of essential nutrients (e.g., carbon, nitrogen, phosphorous, and sulfur) (Falkowski et al., 2008) and degradation of persistent environmental pollutants (e.g., biotransformation and detoxification of xenobiotics) (Peng et al., 2008; Haritash and Kaushik, 2009; Fuchs et al., 2011). In fact, the removal of natural or anthropogenic environmental contaminants by microorganisms, known as bioremediation, has emerged and is consistently developing in recent years as a powerful and cost-effective green technology for a wide variety of contaminated environments (Juwarkar et al., 2010; Chakraborty et al., 2012). Since green technologies are considered flagship agents for environmental management and sustainable development nowadays, considerable research efforts are increasingly devoted by the scientific community towards successful application of bioremediation across a variety of contaminated environments.

Since most bioremediation strategies, be they natural attenuation (Mulligan and Yong, 2004), bioaugmentation (Vogel, 1996), biotransformation (Vahter, 2002), biomineralization (Benzerara et al., 2011), or their combinations (Bento et al., 2005),

Microbial Biodegradation and Bioremediation. DOI: http://dx.doi.org/10.1016/B978-0-12-800021-2.00004-2

rely on stimulation of the activities of *in situ* microbial populations (McCarty, 1993; Farhadian et al., 2008), they hold numerous advantages over alternative physico-chemical and mechanical remediation strategies with respect to the required logistics, maintenance, and eco-friendliness. However, successful application of microbial remediation of a particular contaminated site requires extensive reconnaissance of the indigenous microbial communities, including their diversity, physiology, and functional dynamics (Paul et al., 2005). Thus, a plethora of laboratory-scale studies are usually carried out before a field application strategy is designed. Microbial communities are thoroughly scrutinized in order to determine the dominant taxa inhabiting the contaminated environment followed by rigorous experimentation on stimulating the predominant microbes, often by external amendments, to mediate the natural attenuation of the contaminant(s). In a nutshell, a critical understanding of microbial communities and their interactions with different biotic and abiotic factors in contaminated sites is a mandatory prerequisite for successful bioremediation.

Till the early 1980s, characterization of microbial community composition was limited to culturing microorganisms from environmental samples. Standard culture techniques involving enrichment and isolation of pure cultures in defined growth media followed by phylogenetic, physiological, and functional characterization used to be the only tools available. Over the course of time, however, numerous modifications taking into account specific growth conditions such as oxygen concentration gradient (microaerophilic organisms) (Emerson and Merrill Floyd, 2005) have been incorporated into culture-based techniques to enrich microbes from specific habitats. Consequently, the significance of continuous utilization of traditional culture-based techniques in order to isolate the dominant members of microbial communities at contaminated sites cannot be denied. Notwithstanding the above utilities, $<1\%$ of the total number of microorganisms can currently be cultured in laboratories (Rossello-Mora and Amann, 2001; Hugenholtz, 2002). Moreover, fast-growing microbes or ones able to quickly adapt to laboratory culture conditions usually outgrow other members of the community. As a result, culture-dependent processes introduce considerable statistical bias when used for microbial diversity analysis as they do not accurately represent authentic community composition (Rappe and Giovannoni, 2003), thereby overlooking the ecological importance of unculturable organisms at contaminated sites (Widada et al., 2002).

Research in microbial ecology took an amazing turn in the early 1990s when a series of contemporary discoveries laid the foundation for modern-day molecular microbial tools. To begin with, Norman Pace and his group established the concept of using molecular markers (e.g., ribosomal RNAs) for phylogenetic characterization of microorganisms (Lane et al., 1985; Olsen et al., 1986). Karry Mullis, on the other hand, made his Nobel-winning invention of polymerase chain reaction (PCR) which enabled scientists to amplify DNA *in vitro* (Saiki et al., 1988). As a result of these findings, along with other protocols such as extraction of total nucleic acids from different environments (Tsai and Olson, 1991; Griffiths et al., 2000), microbial ecologists found themselves equipped with a number of culture-independent tools to study the diversity and dynamics of microbial communities in a much greater resolution. These modern molecular techniques provided exciting opportunities to address the central questions in the microbial ecology of contaminated environments, i.e., (i) what type of microorganisms are present? (ii) what do these microorganisms do? and (iii) what are the impacts of the activities of these microorganisms? Instead of culturing a particular organism, these methods rely on characterization of cellular constituents such as nucleic acids, proteins, and fatty acids, which, upon analysis, reflect not just the composition of the microbial community, but their functional roles in the environment as well (Amann et al., 1995). Unlike culture-dependent processes, molecular techniques are better equipped to sustain the *in situ* composition and metabolic function of microbial communities by immediate preservation or direct extraction of molecules of interest from environmental samples. These new approaches allow linkage between ecological processes in the environment and specific microbial populations and help microbial ecologists to better understand the interrelationships of environmental factors with an enormous genetic and metabolic diversity of microorganisms.

An overview of the current molecular technologies that are integrated into bioremediation approaches in order to conduct a comprehensive census of intrinsic microbial populations of contaminated environments is depicted in Figure 4.1. This chapter intends to present an in-depth discussion of the advantages and limitations of such culture-independent molecular techniques with specific emphasis on high-throughput "-omics" approaches in the context of qualitative and quantitative evaluation of the composition and activity of microbial communities at contaminated sites.

4.2 Microbial Community Profiling

Cultivation-independent community profiling based on molecular diversity of phylogenetic markers has become a traditional approach in microbial ecology in the last couple of decades and is therefore widely utilized to survey indigenous microbial populations in contaminated environments (Nocker et al., 2007). The small-subunit ribosomal RNA markers, harboring evolutionarily conserved as well as highly diverse regions, are conventionally considered as "gold standards" for phylogenetic analyses and thus, are most widely exploited in microbial community analysis (Pace, 1997). Nevertheless, several catabolic genes have also been successfully used as molecular markers to detect specific subpopulations within microbial communities with certain physiological capabilities, thus reflecting their biodegradation potential (Torsvik and Ovreas, 2002). Therefore, microbial communities can be studied at both their compositional and functional levels. A number of traditional techniques, such as clone libraries and genetic fingerprinting, along with more recent metagenomics methods are routinely used by microbial ecologists to evaluate microbial diversity and community compositions.

4.2.1 Clone Libraries and Sequencing

The clone library method is one of the pioneer diversity analysis techniques and is still among the most widely used methods (Rastogi and Sani, 2011). This process

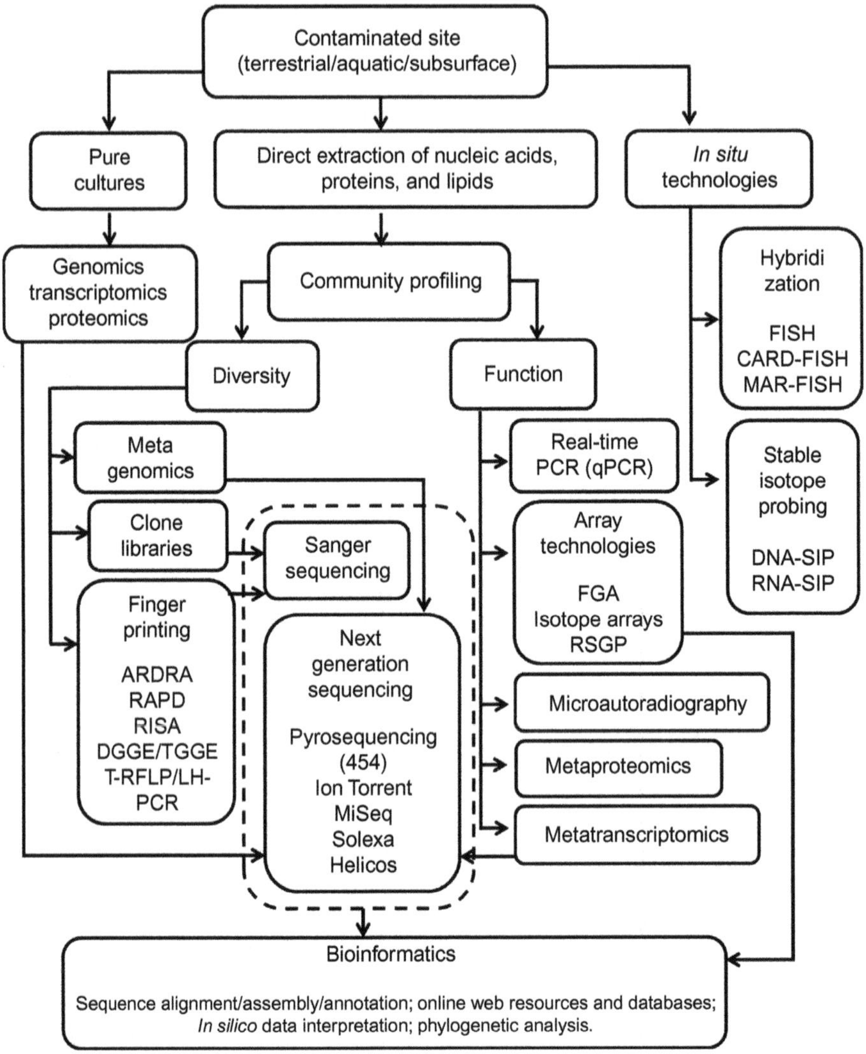

Figure 4.1 Schematic illustration of traditional and novel molecular techniques for surveying the diversity and activity of intrinsic microbial communities observed in contaminated sites.

relies on cloning of PCR-amplified phylogenetic marker sequences from DNA collected from a particular environment followed by subsequent sequencing of gene fragments. Some cloning strategies make use of an overhanging 3′-A (known as the sticky end) added to PCR products by the Taq polymerase. This allows efficient ligation into plasmid vectors with an overhanging 3′-T (Marchuk et al., 1991). Sequencing of inserts allows either species identification or determination of similarity by comparing with already known sequences in extensive and rapidly growing sequence databases, e.g., GenBank and Ribosomal Database Project (RDP). The RDP currently has more than 2.7 million sequences (as of May 2013) and is frequently updated (http://rdp.cme.msu.edu). Sequencing of clone libraries from environmental samples has led to a wealth of information about prokaryotic diversity (Singleton et al., 2001). While clone libraries of 16S rRNA markers permit an initial survey of diversity and identification of novel taxa, a large number of clones must be sequenced to detect rare organisms against a background of a few dominant species (Tiedje et al., 1999). However, typical clone libraries of 16S rRNA contain fewer than 1000 sequences and therefore reveal only a small portion of the microbial diversity present in a sample. However, despite their limitations (e.g., labor intensive, time-consuming, and expensive), clone libraries are still frequently used for preliminary microbial diversity surveys (DeSantis et al., 2007). With decreasing costs of capillary sequencing and an increasing trend of automation at different procedural levels such as automated colony picking, more progress is expected in this method in the context of accurate diversity analysis.

The clone library method has been used extensively in community profiling of microbes in polluted environments (Torsvik et al., 1998; Malik et al., 2008). Dojka et al. (1998) studied the microbial diversity in aquifers contaminated with hydrocarbons (jet fuel and chlorinated solvents) using this approach. Among a total of 812 clones screened, 104 unique sequence types without any significant similarities with known taxonomic divisions were obtained. A more recent study by Lin et al. (2012) examined vertical stratification of microbial diversity in the subsurface sediments of the infamous Hanford Site, the most nuclear waste-polluted site in the United States. The researchers used a combination of quantitative, real-time PCR and clone library methods to analyze 8000 sequences from 21 different samples collected from sediments of different geological strata including natural redox transition zones. Their study revealed 13 novel phylogenetic orders within Deltaproteobacteria, a class rich in microbes involved in redox transformations of heavy metals and radionuclides. Another combined approach of denaturing gradient gel electrophoresis (DGGE) and clone library methods was used by Sutton et al. (2009) to characterize bacterial populations associated with arsenic mobilization in the contaminated aquifers of Bangladesh. The authors observed that microbial communities in the deep-water Pleistocene aquifers were dominated by arsenic-tolerant bacteria, rather than arsenate- or Fe(III)-reducing bacteria, suggesting that arsenic mobilization may not occur at depth. Another study by Ghosh et al. (2013) investigated the temporal dynamics and community structure of arsenite-oxidizing bacteria from groundwater aquifers in Bengal Delta Plain (BDP), which were contaminated with a high level of arsenic, based on clone library and sequencing of arsenite oxidase subunit A (*aioA*) gene and 16S rRNA marker. Based on the library result, comprising 460 clones, the authors found that BDP aquifers harbor indigenous novel arsenite-oxidizing bacteria, as evident from deep phylogeny, and that their temporal dynamics in BDP aquifers were strongly influenced by increasing concentrations of arsenic and other heavy metals. In general, clone libraries provide essential information about the dominant microbial groups in contaminated environments and thus, are crucial for the purpose of designing appropriate bioremediation strategies.

4.2.2 Genetic Fingerprinting Techniques

Genetic fingerprinting techniques share a common trait of generating profiles of microbial communities based on direct analysis of PCR products amplified from environmental DNA (Muyzer, 1999). Over the years, microbial ecologists have been using a number of techniques that produce community fingerprints based on either sequence polymorphism or length polymorphism. The community profiles from different samples can also be compared using computer-assisted cluster analysis by specific software packages. In general, genetic fingerprinting techniques are rapid and allow simultaneous analyses of multiple samples. Since fingerprinting approaches have been devised to demonstrate effects on or differences between microbial communities and do not necessarily provide direct taxonomic identities, these are specifically significant in the context of assessment of impacts of abiotic and biotic factors on microbial communities in contaminated sites.

4.2.2.1 Denaturing- and Temperature-Gradient Gel Electrophoresis (DGGE/TGGE)

DGGE was developed in the 1980s to detect point mutations in defined genes (Fischer and Lerman, 1979; Myers et al., 1987). However, this technique was subsequently introduced in microbial ecology research as a novel fingerprinting technique in 1993 (Muyzer et al., 1993). This method is based on separation of PCR-amplified specific molecular markers, such as 16S rRNA fragments, of the same length but with separate base pair compositions. The separation of the fragments is dependent on the decreased electrophoretic mobility of partially melted double-stranded DNA molecules in two-dimensional polyacrylamide gels containing a linear gradient of DNA denaturants (DGGE) or a linear temperature gradient (TGGE) (Muyzer and Smalla, 1998). During PCR amplification prior to electrophoresis, a 30–50 nucleotide long GC-rich sequence (known as GC clamp) is generally attached to the 5′-end of one of the primers to prevent complete dissociation of the DNA fragments and generation of single strands. The amplified DNA fragments are usually limited to 500 bp in size and are separated on the basis of melting behaviors governed by the sequence differences. The number of bands produced during DGGE or TGGE generally correlates with the number of dominant microbial taxa in the sample. Thus, electrophoresis of mixed amplicons from a complex community through denaturing gradients results in a fingerprint consisting of bands at different migration distances in the gel. The individual bands can be excised from the gel, re-amplified, and ultimately sequenced or hybridized to molecular probes in order to identify specific taxonomic groups. This particular feature is immensely significant in the context of environmental samples where microbial diversity is largely unknown.

Since the community profiles generated by DGGE/TGGE are not as phylogenetically exhaustive as those provided by 16S rRNA clone libraries, this method is generally chosen in order to determine the dominant members of microbial communities with average phylogenetic resolution. DGGE in particular has been widely used for the assessment of microbial community structure in contaminated soil and water. A study utilized DGGE to evaluate the impacts of multiple heavy metal contaminations along a concentration gradient on the soil microbial diversity (Li et al., 2006). The authors observed that the number of DGGE bands increased in the samples collected from sites distant from the heavily contaminated zones, suggesting lower abundance and diversity of bacterial communities in contaminated sites. In another study, a combination of DGGE and fluorescent *in situ* hybridization (FISH) was used to examine the diversity of sulfate-reducing bacteria (SRB) in an aquifer contaminated with petroleum hydrocarbons (Kleikemper et al., 2002). The authors observed a certain degree of variability in the rate coefficients of sulfate reduction indicating activities of specific genera of SRB, whereas culture-independent analyses revealed a high SRB diversity. A very recent study used this technique to assess the effects of biostimulants on microbial community composition in enrichment cultures obtained from a 2,4,6-trinitritoluene (TNT)-contaminated aquifer (Fahrenfeld et al., 2013). The researchers established anoxic microcosms amended with different organic substrates. Subsequent DGGE analysis showed the dominance of Gamma- and Betaproteobacteria, along with Gram-positive Negativicutes and *Clostridia*, suggesting their possible involvement in TNT biodegradation. DGGE analysis has also been recently used by Coulon et al. (2012) to explore the reverse-transcribed bacterial 16S rRNA from the upper 1.5 cm of a hydrocarbon-polluted sediment in coastal mudflats in order to study the functional significance of dynamic tidal biofilms dominated by aerobic hydrocarbonoclastic bacteria and diatoms in biodegradation of hydrocarbons. Apart from microbial community profiling, the DGGE technique has also been used to examine functional gene clusters such as dissimilatory sulfite reductase beta-subunit (*dsrB*) genes in sulfate-reducing bacterial communities (Geets et al., 2006) and BTEX-monooxygenase genes from bacterial strains obtained from hydrocarbon-polluted aquifers (Hendrickx et al., 2006).

There are many advantages of DGGE/TGGE methods, including monitoring of spatiotemporal changes in microbial community structure with simultaneous understanding of the dominant members of a specific community. On the other hand, this technique is limited in various aspects. For example, sequence information derived from microbial populations is limited to 500 bp fragments of 16S rRNA sequences which may lack the specificity required for phylogenetic identification of some organisms. Sequence heterogeneities of multiple copies of rRNA in one organism can lead to multiple bands in DGGE, resulting in overestimation of diversity. Contrarily, different 16S rRNA fragments may have identical melting points, thus culminating into the same band. An additional shortcoming of DGGE is that the band intensity may not be the actual representation of the abundance of microbial population as multiple copies of a particular gene from the same organism would lead to the formation of a strong band, thus leading to underestimation of diversity. Some of these limitations can be addressed by designing improved group-specific primers (Muhling et al., 2008) or by using a variant technique known as denaturing high-performance liquid chromatography (dHPLC), which utilizes chromatographic separation instead of electrophoresis for finer separation of gene fragments (Wagner et al., 2009). In principle, an ion-pair reversed phase HPLC is used to separate a mixture of GC-clamped rRNA amplicons under partly denaturing conditions. The dHPLC technique has recently been successfully applied to monitor oilfield sulfate-reducing bacterial communities, which play a crucial role in oilfield souring and microbially induced corrosion (Priha et al., 2013). Targeting the beta subunit of the dissimilatory sulfite reductase gene (*dsrB*), this study monitored the usefulness of dHPLC compared to DGGE in terms of reproducibility, accuracy, and labor intensity. The authors found out that not only are the results of dHPLC reproducible but the process is also less labor intensive compared to DGGE.

4.2.2.2 Amplified Ribosomal DNA Restriction Analysis

Amplified ribosomal DNA restriction analysis (ARDRA) is one of the earlier methods designed for community profiling. In this simple and cost-effective method, PCR-amplified 16S rRNA fragments are digested or cut at specific sites with restriction enzymes and the resulting digests are subsequently separated by gel electrophoresis. Different DNA sequences are cut in different locations, generating a fingerprint unique to the community being analyzed. Divergence of the community rRNA restriction pattern on a gel is highly influenced by the type of restriction enzyme used. Banding patterns in ARDRA can be used either to screen clones or to measure bacterial community structure. Microbial community composition and succession in an aquifer exposed to phenol, toluene, and chlorinated aliphatic hydrocarbons were assessed by ARDRA with the aim of identifying the dominant microbial taxa involved in biodegradation of trichloroethene (TCE) following biostimulation (Fries et al., 1997). The authors observed that although dichloroethene treatment severely affected the species richness, microbial communities returned to their original levels upon subsequent treatments with phenol and TCE. Another study used ARDRA to observe shifts in a sulfur-transforming microbial community towards a highly diverse community dominated by *Dehalococcoides*-type microorganisms in well waters contaminated with TCE (Hohnstock-Ashe et al., 2001). While ARDRA is useful for detecting structural changes in microbial communities, a major limitation of ARDRA is that fingerprints obtained from complex microbial communities are difficult to interpret accurately due to the generation of complicated banding patterns. Additionally, optimization with restriction enzymes is often difficult if sequences are unknown. Nevertheless, ARDRA has been combined with other molecular techniques such as terminal restriction fragment length polymorphism (T-RFLP) and DGGE to characterize microbial communities from contaminated sources (Watts et al., 2001).

4.2.2.3 Terminal Restriction Fragment Length Polymorphism

T-RFLP is another widely popular molecular tool used to study microbial communities. This technique can be considered an improved ARDRA method. The major difference between T-RFLP and ARDRA is the use of one fluorescently labeled primer at the 5′-terminus in T-RFLP. The resultant PCR products are subsequently digested with restriction enzymes to produce multiple terminal restriction fragments (T-RFs). The T-RFs are separated by high-resolution gel electrophoresis on an automated genetic analyzer along with simultaneous visualization and quantification (Liu et al., 1997). The electrophoretogram represents the profile of a microbial community as a series of peaks varying in migration distance. Since the use of fluorescently tagged primers limits the analysis to only terminal fragments of the digestion, the banding pattern is simplified. This enables the analysis of complex communities as well as providing information on diversity as each visible band represents a single operational taxonomic unit (OTU) or ribotype.

Microbial diversity in a hydrocarbon- and chlorinated solvent-contaminated aquifer undergoing intrinsic bioremediation was monitored by Dojka et al. (1998) using T-RFLP and sequence types characteristic of *Syntrophus* spp. and *Methanosaeta* spp. were observed. These findings led the authors to infer that the terminal step of hydrocarbon degradation in the methanogenic zone of the aquifer was aceticlastic methanogenesis. Another study utilized T-RFLP to study bacterial community changes in microbial mats following exposure to crude oil (Bordenave et al., 2004a). The authors observed that OTUs related to *Chloroflexus*, *Burkholderia*, *Desulfovibrio*, and *Cytophaga* were dominant in the communities, suggesting their adaptive capability to crude oil exposure. The same group of researchers used T-RFLP in another related study to assess the impact of the Erika oil spill on microbial communities in the Northern French Atlantic coast (Bordenave et al., 2004b). They targeted the 16S rRNA marker for overall bacterial diversity as well as the *pufM* gene for phototrophic purple sulfur bacteria. Their analysis indicated that communities from the contaminated and uncontaminated sites evolved differently. In another contemporary study, Turpeinen et al. (2004) examined microbial community structure and activities in arsenic-, copper-, and chromium-contaminated soils of abandoned wood impregnating plants. They generated community profiles using a combination of T-RFLP and phospholipid fatty acid (PLFA) methods. This study revealed that the communities were dominated by As(III)-resistant bacteria and their proportions varied according to the varying concentrations of contaminant. Another study used a combination of amplicon library and RFLP approaches to investigate the interdependence between geoelectrical signatures at underground petroleum plumes and the structures of subsurface microbial communities (Allen et al., 2007). Their results revealed that the zone contaminated with residual hydrocarbons above the free-phase petroleum contained aromatic hydrocarbon degraders and large populations of methylotrophs and methanotrophs. Among comparatively recent studies, Anneser et al. (2010) utilized T-RFLP to generate depth-resolved community profiles for a tar oil-contaminated aquifer. These community profiles were used in conjunction with geochemical parameters to scrutinize hot spots of biodegradation in the homogeneous aquifer. A similar study by Larentis et al. (2013) investigated the composition and activity of toluene-degrading organisms from another tar oil-contaminated aquifer using T-RFLP and quantitative PCR (Q-PCR). The results of this study pointed towards a novel, competitive niche partitioning behavior by aerobic and anaerobic hydrocarbon-degrading communities. Tipayno et al. (2012) also used T-RFLP to determine the structure and diversity of bacterial populations from metal-contaminated soils undergoing phytoremediation.

Automated detection systems and capillary electrophoresis in T-RFLP analysis allow high-throughput quantitative analysis of microbial community samples in deeper resolution than any other genetic fingerprinting methods. Despite the higher accuracy and sensitivity, T-RFLP is highly dependent on DNA extraction followed by PCR amplification of 16S rRNA marker, which can be affected by PCR biases and the choice of universal primers (Kirk et al., 2004). Additionally, the choice of restriction enzymes is another critical factor as different enzymes produce different community fingerprints and incomplete digestion by the restriction enzymes may lead to an overestimation of diversity (Osborn et al., 2000). Another major pitfall of T-RFLP in comparison with DGGE/TGGE is the difficulty of retrieving suitable phylogenetic information from the T-RFs generated since fragments are difficult to isolate and are usually too short to be sequenced properly.

4.2.2.4 Length Heterogeneity Polymerase Chain Reaction

Length heterogeneity polymerase chain reaction (LH-PCR) analysis is quite similar to T-RFLP, except different microorganisms are discriminated based on natural length polymorphisms that occur due to mutation within genes (Mills et al., 2007). This method targets the hypervariable regions present in 16S rRNA markers and produces a characteristic profile. During LH-PCR, DNA is amplified using a forward primer labeled with a fluorescent dye while a fluorescent internal size standard is run with each sample in order to measure the amplicon lengths in base pairs. The peak intensity is proportional to the relative abundance of that particular amplicon. One advantage of using LH-PCR over the T-RFLP is that the former does not require any restriction digestion and therefore PCR products can be directly analyzed by a fluorescent detector. The drawbacks of the LH-PCR technique include inability to resolve complex amplicon peaks and underestimation of diversity, as phylogenetically distinct taxa may produce same length amplicons.

LH-PCR was used in a study by Connon et al. (2005) to determine shifts in bacterial communities during an ongoing propane-stimulated bioremediation strategy of trichloroethene-contaminated groundwater samples. A 385 bp fragment accounted for 83% of the total fragments analyzed during peak propane removal rates and it decreased when the nitrate level in the aquifer decreased. The 16S amplicon libraries confirmed a TM7 division bacterium harboring that 385 bp LH-PCR fragment. Recently, Robertson et al. (2011) used this technique to identify changes in the root-associated bacterial communities in the shared rhizosphere of pine and lingonberry as a result of hydrocarbon contamination.

4.2.2.5 Ribosomal Intergenic Spacer Analysis

Ribosomal intergenic spacer analysis (RISA) relies on PCR amplification of a portion of the intergenic spacer (IGS) region between the small (16S) and the large

(23S) rRNA subunits (Fisher and Triplett, 1999). The IGS region, depending on the species, has significant sequence and length (50–1500 bp) variability, facilitating taxonomic identification of organisms. Application of RISA in microbial community analysis from contaminated sources is somewhat limited due to the limited database of ribosomal intergenic spacer sequences. Furthermore, RISA sequence variability may be too great for environmental applications. Its level of taxonomic resolution is greater than 16S rRNA and hence may lead to very complex community profiles. An automated version of this technique (ARISA) is also available which detects the abundance and size of PCR amplicons by measuring the fluorescence emission of labeled primers (Ranjard et al., 2001). In environmental studies, RISA has been used to detect microbial populations involved in the degradation of polyaromatic hydrocarbons at low temperature under aerobic and nitrate-reducing enriched soil conditions (Eriksson et al., 2003). Their study showed dominance of bacteria belonging to Alpha-, Beta-, and Gammaproteobacteria in the enrichments. A recent study utilized ARISA to investigate the diversity of hydrocarbon-degrading bacteria and bacterial community response in oil spill-contaminated beach sands in the Gulf of Mexico (Kostka et al., 2011). The authors observed increased abundance of *Alcanivorax* sequences in the oil-contaminated sand.

4.3 Functional Analysis of Microbial Communities

As phylogenetic markers have paved the way for obtaining detailed information on the composition of microbial communities, other catabolite-specific, functional genes have also been used to detect the functional capabilities of indigenous microbes in different environments including contaminated sites. Functional profiling of microbial communities has enabled researchers not only to link microbial phylogeny to function but also to gain insights about specific microbial groups that are sensitive or most affected, resilient, and predominant, or actively involved in biotransformation of contaminants. A variety of culture-independent approaches, comprising PCR-based, array-based, mass-spectrometry-based (metaproteomics), and high-throughput sequencing-based techniques (metatranscriptomics), are currently being used for community functional analysis and they routinely yield immensely significant information regarding the metabolic versatility of microorganisms.

4.3.1 Quantitative Polymerase Chain Reaction

Real-time Q-PCR has emerged as a promising new technology for rapid and accurate measurement of the abundance and expression of taxonomic and functional gene markers (Smith and Osborn, 2009). This novel tool has enabled careful monitoring of the functional dynamics of microbial communities and assessment of the catabolic activity during bioremediation of contaminated environments. During real-time PCR, the amplification can be detected and the amplicons can be quantified while the reaction is occurring, more specifically, during the early exponential phase of the reaction (Heid et al., 1996). It involves the use of either fluorescent dyes (such as SYBR Green) (Ponchel et al., 2003) or fluorogenic probes (such as TaqMan probes) (McGoldrick et al., 1998), and the amount of fluorescence measured at the end of each cycle is directly related to the amount of product in the PCR reaction. Initially, fluorescence remains at background levels, and increases in fluorescence are not detectable even though product accumulates exponentially. Eventually, enough amplified product accumulates to yield a detectable fluorescent signal. The cycle number at which this occurs is called the threshold cycle or Ct. Since the Ct value is measured in the exponential phase when reagents are not limited, real-time PCR can be used to reliably and accurately calculate the initial amount of template present in the reaction, forming the basis for the quantitative aspect of this technique.

Real-time PCR has been used in several environmental studies such as the identification and quantification of the arsenate reductase gene (*arsC*) in soil (Sun et al., 2004) and aromatic oxygenase genes (Baldwin et al., 2003). Sun and co-workers developed a Q-PCR assay to detect the abundance of *arsC* genes in arsenic-contaminated soil samples (Heid et al., 1996). This assay detected $0.8 \times 10^4 - 1.56 \times 10^5$ *arsC* gene copies per nanogram of soil DNA. Baldwin and co-workers, on the other hand, optimized this technique for the detection of aromatic oxygenase genes (Ponchel et al., 2003). The authors observed that using a polymerization temperature 4–5°C below the melting temperature reduced background fluorescence signals by a noticeable margin, allowing higher detection limits. qPCR has also been used in quantifying the proportion of microorganisms containing alkane monooxygenase and the subsequent assessment of microbial community changes in hydrocarbon-contaminated Antarctic soil (Powell et al., 2006). Their study indicated that the abundance of the *alkB* gene was correlated with the concentration of n-alkanes in soil samples. More recently, a study by Alonso-Gutierrez et al. (2011) utilized a qPCR approach to analyze alkane degrading properties of *Dietzia* sp., a key contributor in the Prestige oil spill biodegradation. Their study suggested two previously unknown alkane degradation pathways in *Dietzia* sp. instigated by two novel, putative oil-degrading enzymes. Another study examined microbial activity during bioaugmentation of atrazine-contaminated soils (Wang et al., 2013). The researchers performed qPCR assays targeting three functional genes in the atrazine degradation pathway (*trzN*, *atzB*, and *atzC*) and observed a continuous increase in the abundance of these genes during bioaugmentation. The qPCR technique was also used by Holmes et al. (2013) in another recent study which quantified bacterial activity during *in situ* bioremediation of a uranium-contaminated aquifer.

The advantages that real-time PCR offers include speed, sensitivity, accuracy, and the possibility of robotic automation. Although real-time PCR can measure gene quantity, the results obtained do not necessarily link gene expression with a specific measurable microbial activity or population. RNA samples extracted from soil and water samples are low in yield and often do not represent soil microbial population. Also, specific probes used in the amplification reactions may fail to capture the sequence diversity that is present within environmental samples. PCR molecular techniques have completely revolutionized the detection of DNA/RNA, especially in microbial ecological studies. However, differential amplification of target markers such as 16S rRNA can incorporate a considerable amount of bias into PCR-based diversity studies.

4.3.2 Microarray Technologies

Microarrays ("chips") containing nucleic acids as probes represent a major advancement in molecular detection technology. They are ideal for simultaneous, high-throughput analysis of multiple genes in environmental samples (Cho and Tiedje, 2002; Zhou, 2003). A microarray is a miniaturized array of complementary DNA probes ($\sim$500–5000 nucleotides in length) or oligonucleotides (10–75 bp) attached directly to a solid support (usually a glass slide) using metal pins (contact printing) or by inkjets (noncontact printing), which permits simultaneous hybridization of a large set of probes complementary to their corresponding DNA/RNA targets in a sample (Schena et al., 1995). Compared to traditional nucleic acid membrane hybridization, microarrays present the advantages of miniaturization, high sensitivity, and rapid detection. Moreover, probe-target specificity is ensured by the presence of single mismatch probes on the array, which allows distinguishing of sequence-specific signals from nonspecific ones. Fluorescent dyes can be enzymatically or chemically incorporated in the sample to be hybridized, rendering the readout of microarray based on detection of a fluorescent signal, usually by confocal laser scanning microscopy. Microarrays differ according to the immobilization technology used to attach the probes, the length and nature of the probes, and the synthesis and labeling of the targets. The choice between the different technologies is based on parameters such as cost, probe density, specificity, sensitivity, and quantification. For example, probes can be directly synthesized *in situ* using photolithographic masks (e.g., GeneChip arrays) or electrochemical reactions, or spotted onto the solid support after *ex situ* synthesis. The latter microarrays are of lower cost and they offer a much greater flexibility with respect to the nature of the probes, which can be synthetic oligonucleotide sequences of 10- to 75-*mers*, PCR products, cDNAs, or clone libraries. Longer probes offer the advantage of lower production cost and higher signal intensities. However, spotted microarrays are of lower density than *in situ* synthesized microarrays.

Three major classes of environmental microarray formats are usually employed in microbial ecology research, namely, phylogenetic oligonucleotide arrays (POAs) or phylochips, functional gene arrays (FGAs), and community genome arrays (CGAs) (Wu et al., 2004). Phylochips rely on the use of phylogenetic markers (16S rRNA) for the identification of microorganisms present in an environment. Due to the high-throughput capacity of microarrays and the availability of extensive rRNA sequence databases, phylochips provide a very convenient means of simultaneously identifying many microorganisms from a sample. Several studies have employed POA in environmental investigations of microbial populations in environmental samples (Small et al., 2001). FGAs, on the other hand, identify or measure genes encoding key enzymes in a metabolic process (Wu et al., 2001). Such an approach provides vital information about the presence of important genes as well as the expression of the genes in the environment by measuring the mRNA. FGA techniques have been optimized and successfully implemented by Denef et al. (2003) for the detection of specific aromatic oxygenase genes in a soil community responsible for degrading polychlorinated biphenyls. Rhee et al. (2004) developed and applied a 50-*mer* oligonucleotide FGA in a contemporary study to detect the presence and expression of naphthalene-degrading genes in soil contaminated with polyaromatic hydrocarbons. This array contained 1662 unique and group-specific probes with <85% similarity with their nontarget sequences and was applied for the analyses of naphthalene-amended enrichment cultures as well as soil microcosms. The authors observed that the naphthalene-degrading genes from *Rhodococcus*-type organisms were predominant in the enrichment cultures, while soil microcosms were dominated by naphthalene- and hydrocarbon degradation genes from *Ralstonia*, *Comamonas*, and *Burkholderia*. Finally, CGA uses nonporous hybridization surfaces and fluorescence-based detection systems for high-throughput analysis. Wu et al. (2004) initiated the development and testing of CGA as a tool to detect specific microorganisms within a natural microbial community. CGA has been shown to achieve species-to-strain level differentiation depending on hybridization temperature and has an added potential for the determination of genomic relatedness of isolated bacteria. The major disadvantage of CGA is that culturable organisms are needed in the array preparation thus making the application of CGA for environmental community analysis quite difficult.

The application of microarrays in environmental microbiology, specifically in the examination of microbial populations engaged in biodegradation, has the potential for organism identification as well as identification of their ecological role. However, more rigorous and systematic assessment and development are needed to realize the full potential of microarrays for microbial detection and community analysis. Microarrays detect only the dominant populations in many environments. In addition, probes designed to be specific to known sequences can cross-hybridize to similar or unknown sequences and may produce misleading signals. Moreover, soil, water, and sediments often contain humic acids and other organic materials that may inhibit DNA hybridization onto microarrays. Finally, limitations in quality RNA extraction from many environmental samples imply that advances in RNA extraction and purification and amplification methods are needed to make microarray gene expression analysis possible for a broader range of samples.

The concept of microarray technology has been modified by incorporating other techniques in order to increase the breadth of the data generated. Such modifications include isotope arrays and reverse sample genome probing (RSGP). Isotope arrays can be used for simultaneous monitoring of substrate uptake profiles along with taxonomic identities of active microbial communities (Adamczyk et al., 2003). In this technique, samples are incubated with a radioactively labeled substrate, which during the course of growth becomes incorporated into a microbial biomass. The labeled rRNA is separated from unlabeled rRNA and then further labeled with fluorochromes. Fluorescent labeled rRNA is hybridized to a phylogenetic microarray, followed by scanning for radioactive and fluorescent signals. The technique thus allows parallel study of microbial community composition and specific substrate consumption by metabolically active microorganisms of complex microbial communities. The major strengths of the technique lie in the fact that it does not involve an amplification step and is hence free of biases associated with PCR. The limitations of the technique include difficulties in obtaining high-quality rRNA and detecting low abundance but active microbial populations from environmental samples (Wagner et al., 2006).

Another variation of genome array is known as the RSGP (Voordouw et al., 1991). This approach involves the isolation of chromosomal DNA in pure culture from standard microbial species followed by cross-hybridization to an array developed using pure cultures. Genomes that exhibit more than 70% cross-hybridization are often regarded as the same species (Greene and Voordouw, 2003). A genome array or master filter is then prepared from the genome with less cross-hybridization. Community DNA probes from an area of interest are prepared and subsequently used to hybridize with the genome array. Since its introduction, RSGP has been extensively used in the characterization of SRB from oilfields (Voordouw et al., 1991) and in the identification of hydrocarbon-degrading bacteria in soil (Shen et al., 1998). RSGP has also been used in the characterization of the dominant members of soil microbial communities enriched with a mixture of aromatic hydrocarbons (Greene et al., 2000). In a more recent RSGP study investigating TCE contaminated soils, it was observed that TCE affected the microbial community composition when TCE was actively metabolized in soils (Hubert et al., 2005). The application of RSGP analysis to environmental samples is limited to the development of suitable master filters or genome arrays representative of the microbial community. The major drawback of this technique is the choice of organism to be used as the standard in the preparation of the genome arrays. In contaminated soil and water, where the identity, physiology, and biochemistry of the degrading microbial populations are unknown or nonculturable, the RSGP application is very limited.

4.3.3 Stable Isotope Probing

Stable isotope probing (SIP) involves offering a stable isotope, e.g., ^{13}C-labeled substrate to microbial communities whose utilization is of interest to decipher a key biogeochemical process (Wellington et al., 2003; Dumont and Murrell, 2005). Active microbial communities that utilize the labeled substrate during growth incorporate the isotopes within their biomass. The labeled biological macromolecules are then separated from the biomass and the phylogenetic identity of microorganisms metabolizing the substrate is established using molecular techniques. SIP relying on DNA biomarkers involves labeling of DNA with ^{13}C that could be separated from ^{12}C by cesium chloride (CsCl) equilibrium density gradient centrifugation (Uhlik et al., 2013). The ^{13}C-labeled DNA could be analyzed by genetic fingerprinting or clone library techniques, leading to the identification of microorganisms. In recent years, with advances in imaging and spectroscopic techniques,

SIP has been combined with other techniques such as FISH and Raman microscopy to simultaneously investigate the taxonomic identities and activity of microbial communities at single-cell resolution (Huang et al., 2007). In the Raman–FISH method, environmental samples are incubated with a substrate labeled with ^{13}C stable isotope. After incorporation, the spectral profiles of uncultured microbial cells at single-cell resolution are generated using Raman microscopy, which measures the laser light scattered by chemical bonds of different cell biomarkers. The proportion of stable isotope incorporation in cells affects the amount of light scattered, resulting in measurable peak shifts for labeled cellular components. The Raman–FISH provides much higher resolution and overcomes many of the limitations associated with conventional SIP/microautoradiography (MAR)-FISH techniques.

SIP was applied to characterize the herbicide 2,4-dichlorophenoxyacetic acid (2,4-D)-degrading soil microbial communities (Cupples and Sims, 2007). Soil samples were spiked with ^{13}C-labeled 2,4-D and 16S sequencing of the heavy, ^{13}C-labeled fraction indicated predominance of Betaproteobacteria, suggesting their involvement in 2,4-D biodegradation. An RNA-based SIP method was used by Kasai et al. (2006) to isolate anaerobic benzene-degrading bacteria from gasoline-contaminated groundwater samples. Anaerobic microcosms were prepared using groundwater samples spiked with ^{13}C-labeled benzene and a range of electron acceptors such as nitrate, sulfate, and oxygen. A phylotype related to *Azoarcus* was observed to be dominant in the nitrate-amended microcosms. Another study used DNA-SIP coupled with DGGE and clone libraries to characterize salicylate, naphthalene, and phenanthrene-degrading bacterial communities from a soil bioreactor (Singleton et al., 2005). Akob et al. (2011) used DNA-SIP to glean an understanding of the shifts in terminal electron-accepting processes during biostimulation of heterotrophic bacteria in subsurface sediments contaminated with uranium and nitrate. Using ^{13}C ethanol-amended microcosms, the researchers observed that nitrate reduction was preferred by the microbial communities (dominated by members of Betaproteobacteria) as the principal electron-accepting process compared to U(VI) and Fe(III) reduction. The Raman–FISH method was used to investigate the importance of naphthalene-degrading *Pseudomonas* communities in groundwater microcosms (Huang et al., 2007).

4.4 Determination of *In Situ* Abundance of Microorganisms

Quantification of specific bacterial populations *in situ* is key to understanding the mechanisms that underlie many biologically mediated processes, particularly microbial processes at contaminated sites. Conventional cultivation-based methods to measure microbial abundance are unsuitable for quantifying uncultured microorganisms that constitute the majority of microbial life in most environmental samples. The use of hybridization-based, cultivation-independent techniques provides potentially useful means to overcome this limitation as well as insights into the native physiological state, cellular rRNA content, and structural dynamics of indigenous microbial communities under specific environmental conditions. Numerous rRNA-based techniques are now commonly employed to quantify microbial communities in environmental samples. Direct *in situ* probing using rRNA or functional gene targets in combination with autoradiography and cell quantification techniques have become useful in monitoring the impacts of environmental conditions on the ecophysiology of the microbial communities in their native environments.

4.4.1 *Fluorescence* In Situ *Hybridization*

In order to quantify the presence and relative abundance of microbial populations in a community sample, FISH uses the sequence information of cloned, rRNA markers from environmental habitats to develop specific phylogenetic oligonucleotide probes that allow selective hybridization to the target region of the ribosomal RNA in fixed cells (Poulsen et al., 1993). Microbial cells are treated with fixatives, then hybridized with fluorescently labeled, oligonucleotide probes (usually 15–25 bp) on a glass slide, followed by visualization with either epifluorescence or confocal laser microscopy. Hybridization with rRNA-targeted probes enhances the characterization of uncultured microorganisms and also facilitates the description of complex microbial communities. Two types of FISH probes based on conserved or unique regions of 16S rRNA genes can be developed: domain- or group-specific probe and strain-specific probes. Domain- or group-specific probes discriminate or identify members of larger phylogenetic groups, while strain-specific probes quantify or assess the abundance of a specific species or strain within a microbial community (Dubey et al., 2006).

A major drawback of FISH is that a limited number of probes can be used in a single hybridization experiment and background fluorescence can be problematic in some environmental samples. A prior knowledge of the sample and the microorganisms most likely to be detected is necessary for the design of specific probes. Another obstacle of the standard FISH technique is its limited sensitivity because bacterial cells with reduced ribosome contents, a condition that often occurs in oligotrophic environments, are not satisfactorily stained for microscopic analysis. Additionally, FISH alone cannot provide any insight into metabolic activities of microorganisms. Thus, it can be modified by combining it with other techniques such as MAR to describe the functional properties of microorganisms in their natural environment (Wagner et al., 2006). MAR-FISH allows for the identification of bacteria and concomitantly gives an indication of their specific *in situ* activity using suitable isotope-labeled substrates. The substrate uptake patterns in FISH-labeled bacteria can be investigated *in situ* in mixed natural communities at a single-cell level, even for unculturable bacteria. Another variation of this process is known as STAR (substrate tracking autoradiography)-FISH. However, STAR-FISH differs from MAR-FISH only in methodological details, and the basic principle of the technique remains the same. Catalyzed reporter deposition-FISH (CARD-FISH) is another version of FISH with increased sensitivity and reliability. Signal intensities of hybridized cells are increased by enzymatic signal amplification using horseradish peroxidase (HRP)-labeled oligonucleotide probes in combination with tyramide signal amplification (TSA). TSA is based upon the patented CARD technique using derivatized tyramide.

FISH, along with its several modifications, is used extensively in studying *in situ* microbial community dynamics in contaminated environments (Malik et al., 2008; Rastogi and Sani, 2011). Richardson et al. (2002) combined group-specific FISH and T-RFLP in the characterization of microbial communities engaged in TCE biodegradation in an anoxic microbial consortium. Their study indicated *in situ* clustering among the members of *Cytophaga*, *Flavobacterium*, and *Bacteroides* in the consortium. Recently, the MAR-FISH technique was used by Ariesyady et al. (2007) to investigate the functional bacteria and archaea community structures responsible for the decomposition process in a full-scale anaerobic sludge digester. MAR-FISH results suggested that glucose-degrading microbial communities had higher abundance and diversity compared to the limited fatty acid-degrading communities.

4.5 Application of "-omics" Technologies

The "-omics"-based technologies are empowered with novel sequencing techniques, often referred to as next-generation sequencing (NGS) or deep sequencing. Compared to traditional Sanger sequencing, these newer sequencing formats enable much faster production of gigabases of sequence information in a matter of few days. A number of NGS platforms, e.g., 454 pyrosequencing (Roche), MiSeq, Solexa (Illumina), and Ion Torrent are widely used by biologists (Mardis, 2013). Needless to say, the application of these novel, high-resolution sequencing methods has immense potential in microbial community analysis of contaminated environments, not only at the composition level, but at the functional level as well.

4.5.1 Metagenomics

Metagenomics, developed in the first decade of the twenty-first century, laid the foundation of the "-omics" techniques and revolutionized research in microbial ecology (Handelsman, 2004). Metagenomics is the reconfiguration of collective microbial genomes available in DNA collected from environmental samples and does not depend on prior knowledge of the communities (Riesenfeld et al., 2004). Essentially, genetic fingerprinting techniques do not provide information on diversity beyond the selected genes that are being amplified and consequently, are useful only for partial community analysis. Metagenomic techniques, however, are based on the concept that the DNA retrieved directly from environmental samples theoretically represents entire community genomes which could be sequenced and analyzed in the same way as that of a whole genome of a pure bacterial culture, thus enabling microbial ecologists to glean a comprehensive understanding of whole communities. Additionally, metagenomics is crucial for understanding the biochemical roles of uncultured microorganisms and their interaction with other biotic and abiotic factors. Environmental metagenomic libraries have proved to be great resources for new microbial enzymes and antibiotics with potential applications in biotechnology, medicine, and industry (Riesenfeld et al., 2004). Metagenomic libraries are constructed by isolating total DNA from environmental samples followed by shotgun cloning of random DNA fragments into suitable vectors (e.g., plasmid, cosmid, and bacterial artificial chromosome). The vectors are then transformed into a host bacterium and the clones are subsequently screened either by sequencing or testing for certain physiological functions. Metagenomic libraries containing small DNA fragments in the range of 2–3 kb provide better coverage of the metagenome of an environment than those with larger fragments. It has been estimated that to retrieve the genomes from rare members of microbial communities, at least 1011 genomic clones would be required (Rastogi and Sani, 2011). Small-insert DNA libraries are also useful to screen for phenotypes that are encoded by single genes and to reconstruct the metagenomes for genotypic analysis. Large-fragment metagenomic libraries (100–200 kb) are desirable when investigating multigene biochemical pathways. Sequence-driven massive whole-genome metagenomic sequencing provides insights on many important genomic features such as redundancy of functions in a community, genomic organizations, and traits that are acquired from distinctly related taxa through horizontal gene transfers (Handelsman, 2004). In function-driven metagenomic analysis (functional metagenomics), libraries are screened based on the expression of a selected phenotype on a specific medium. In a metagenomic library, the frequency of active gene clones expressing a phenotype is typically very low, necessitating improved high-throughput screening and detection assays.

Metagenomic analysis is frequently applied in order to investigate complex microbial communities from different polluted sites (Stenuit et al., 2008; Desai et al., 2010). Hemme et al. (2010) used metagenomic analysis to understand adaptive bacterial communities in heavy metal-contaminated groundwater. Their study indicated that prolonged exposure (more than 50 years) to heavy metals resulted in significant loss of species and metabolic diversity in groundwater microbial communities. The authors inferred that lateral gene transfer could be a crucial adaptive mechanism, furnishing the surviving microbial communities with genes conferring resistance to the stress incurred from different contaminants. Another study used pyrosequencing-based metagenomics to characterize the metagenome of a hydrocarbon-contaminated site, in terms of its composition and metabolic potentials (Abbai et al., 2012). Analyses of the metagenome by the same group of researchers revealed adaptation of complex microbial communities to hydrocarbon-contaminated geochemical conditions (Abbai and Pillay, 2013).

4.5.2 Metatranscriptomics

Transcriptomic or metatranscriptomic tools are used to gain functional insights into the activities of environmental microbial communities by studying their mRNA transcriptional profiles (McGrath et al., 2008). Transcriptomic analyses consist of extraction and enrichment of the total mRNA, then subsequent cDNA synthesis via reverse transcription, followed by sequencing of the complete cDNA transcriptome. Jennings et al. (2009) performed transcriptomics analysis on a *cis*-dichloroethene (cDCE)-assimilating *Polaromonas* strain in order to identify the upregulated genes using DNA microarrays. Their study revealed that the genes encoding for glutathione S-transferase, cyclohexanone monooxygenase, and haloacid dehalogenase were highly upregulated during cDCE biodegradation. In a similar study, comparative transcriptomics was used by Holmes et al. (2009) to decipher the metabolic potentials of a *Geobacter uraniireducens* strain growing in uranium-contaminated subsurface sediments. The authors observed higher transcripts of a total of 1084 genes including 34 genes encoding for *c*-type cytochromes that are homologous to the well-known *c*-type cytochromes required for Fe(III) and U(VI) reduction by *Geobacter sulfurreducens*. The above-mentioned studies demonstrate that it is feasible to monitor gene expression of microorganisms growing in the presence of anthropogenic contaminants. Recent advances in direct extraction of mRNA from archaeal, bacterial, and eukaryotic microbial cells have enabled researchers to obtain gene expression profiles of the entire microbial community, also known as "metatranscriptome." Extraction of metatranscriptomes coupled with pyrosequencing or construction of cDNA microarrays provides a useful tool to monitor the transcriptional activities of entire microbial communities. However, obtaining prokaryotic metatranscriptomes is often trickier than its eukaryotic counterpart due to a number of reasons such as absence of poly-A tails in prokaryotic mRNAs and the usually high rRNA to mRNA ratio in prokaryotic transcriptomes. Urich et al. (2008) employed a metatranscriptomic approach to simultaneously obtain information on both the structure and function of soil microbial communities. Among a total of 258,411 RNA tags ($\sim$98 bp length) generated from a soil sample in this study, 193, 219 and 21,133 were of rRNAs and mRNAs, respectively. Another recent study used a comparative metatranscriptomic strategy to analyze bacterial community responses during phenanthrene degradation (de Menezes et al., 2012). Phenanthrene addition caused up to a 33-fold increase in the transcripts of dioxygenase and other genes involved in stress response and detoxification. The transcriptomics approaches discussed above have a wide applicability in linking the structure and function of environmental microbial communities by performing a single experiment.

4.5.3 Metaproteomics

Proteomics-based investigations have been useful in determining changes in the composition and abundance of proteins, as well as in identification of key proteins involved in the physiological response of microorganisms when exposed to anthropogenic pollutants (Maron et al., 2007). Recently, traditional proteomic approaches are being applied to detect protein expression profiles directly from mixed microbial communities of an environmental sample reflecting their actual functional activities in a given ecosystem. These "community proteomics" or "metaproteomics" approaches have advanced significantly because of technological developments in two-dimensional gel electrophoresis (2-DE) coupled with mass spectrometry (MS) techniques, along with upgrades in protein sequence and structure databases (Wilmes et al., 2008). Matrix-assisted laser desorption ionization time-of-flight (MALDI-ToF) MS is the most commonly used MS technique for identifying proteins of interest excised from 2-DE gel spots, based on the principle of peptide mass fingerprinting. Likewise, liquid chromatography linked to MS via electrospray ionization source (LC-ESI-MS) is also used to separate and identify peptide fragments from different microorganisms. Identification of peptide mass fingerprints generated by MALDI-ToF MS or LC-ESI-MS in a metaproteomic analysis is achieved by their comparisons with previously known peptide fingerprints from the protein data banks (Benndorf et al., 2007).

Benndorf et al. (2007) conducted a functional metaproteomic analysis on microbial communities within 2,4-D-contaminated soil and chlorobenzene-contaminated ground waters of an aquifer. This study was instrumental in formulating and further optimizing the protocol for metaproteome analysis of soil and groundwater samples. In a similar study, 2-DE in combination with MALDI-ToF and quadrupole time-of-flight (Q-ToF) MS/MS were utilized to identify highly expressed proteins during microbial transformations in mixed culture activated sludge of an enhanced biological phosphorous removal (EBPR) system (Wilmes et al., 2008). In a very recent study, Chourey et al. (2013) applied high-performance MS instrumentation along with improved protein extraction processes to characterize the metaproteomes of microbial communities during biostimulation with emulsified vegetable oil (EVO) in uranium- and nitrate-contaminated sites. Proteome characterization revealed distinct differences in the microbial biomass collected from the EVO-biostimulated groundwater compared to the groundwater upgradient to the EVO injection points.

4.6 Conclusion

Developing efficient and consistent bioremediation strategies for contaminated environments requires an in-depth understanding of the community structures and metabolic functioning of indigenous microbial communities which are undoubtedly the key driving force behind the biological treatment of pollutants. This is particularly challenging for microbiologists, as most environmental microbes are not readily culturable and their biology is not fully characterized. A few decades ago, the emergence of molecular tools revolutionized the field of microbial ecology, as they provided direct access to the phylogeny and physiology of environmental microbes independent of their cultivability. Nowadays, high-throughput technologies represent the next technological leap towards massive and faster characterization of microbial communities. In the field of bioremediation, they are expected to boost the discovery of new catabolic activities and provide valuable information for the management and cleanup of contaminated sites and effluents in the context of sustainable development. In the modern era, the emergence of specialized techniques for monitoring the genome, transcriptome, and proteome of *in situ* microbial communities, regardless of cultivation, has paved the way towards development and successful execution of efficient and innovative bioremediation strategies. However, appropriate interpretation and management of the massive data generated specifically by the NGS approaches still requires development of efficient statistical algorithms and bioinformatics tools. In other words, rapid advances in computation and modeling are of immediate necessity in order for microbial ecologists to fully exploit the potential of high-throughput technologies.

References

Abbai, N., Pillay, B., 2013. Analysis of hydrocarbon-contaminated groundwater metagenomes as revealed by high-throughput sequencing. Mol. Biotechnol. 54, 900–912.

Abbai, N., Govender, A., Shaik, R., Pillay, B., 2012. Pyrosequence analysis of unamplified and whole genome amplified DNA from hydrocarbon-contaminated groundwater. Mol. Biotechnol. 50, 39–48.

Adamczyk, J., Hesselsoe, M., Iversen, N., Horn, M., Lehner, A., Nielsen, P.H., et al., 2003. The isotope array, a new tool that employs substrate-mediated labeling of rRNA for determination of microbial community structure and function. Appl. Environ. Microbiol. 69, 6875–6887.

Akob, D.M., Kerkhof, L., Kusel, K., Watson, D.B., Palumbo, A.V., Kostka, J.E., 2011. Linking specific heterotrophic bacterial populations to bioreduction of uranium and nitrate in contaminated subsurface sediments by using stable isotope probing. Appl. Environ. Microbiol. 77, 8197–8200.

Allen, J.P., Atekwana, E.A., Duris, J.W., Werkema, D.D., Rossbach, S., 2007. The microbial community structure in petroleum-contaminated sediments corresponds to geophysical signatures. Appl. Environ. Microbiol. 73, 2860–2870.

Alonso-Gutierrez, J., Teramoto, M., Yamazoe, A., Harayama, S., Figueras, A., Novoa, B., 2011. Alkane-degrading properties of *Dietzia* sp. H0B, a key player in the Prestige oil spill biodegradation (NW Spain). J. Appl. Microbiol. 111, 800–810.

Amann, R.I., Ludwig, W., Schleifer, K.H., 1995. Phylogenetic identification and *in situ* detection of individual microbial cells without cultivation. Microbiol. Rev. 59, 143–169.

Anneser, B., Pilloni, G., Bayer, A., Lueders, T., Griebler, C., Einsiedl, F., et al., 2010. High resolution analysis of contaminated aquifer sediments and groundwater—what can be learned in terms of natural attenuation? Geomicrobiol. J. 27, 130–142.

Ariesyady, H.D., Ito, T., Okabe, S., 2007. Functional bacterial and archaeal community structures of major trophic groups in a full-scale anaerobic sludge digester. Water Res. 41, 1554–1568.

Baldwin, B.R., Nakatsu, C.H., Nies, L., 2003. Detection and enumeration of aromatic oxygenase genes by multiplex and real-time PCR. Appl. Environ. Microbiol. 69, 3350–3358.

Benndorf, D., Balcke, G.U., Harms, H., von Bergen, M., 2007. Functional metaproteome analysis of protein extracts from contaminated soil and groundwater. ISME J. 1, 224–234.

Bento, F.M., Camargo, F.A.O., Okeke, B.C., Frankenberger, W.T., 2005. Comparative bioremediation of soils contaminated with diesel oil by natural attenuation, biostimulation and bioaugmentation. Bioresour. Technol. 96, 1049–1055.

Benzerara, K., Miot, J., Morin, G., Ona-Nguema, G., Skouri-Panet, F., Ferard, C., 2011. Significance, mechanisms and environmental implications of microbial biomineralization. C.R. Geosci. 343, 160–167.

Bordenave, S., Fourcans, A., Blanchard, S., Goni, M.S., Caumette, P., Duran, R., 2004a. Structure and functional analyses of bacterial communities changes in microbial mats following petroleum exposure. Ophelia 58, 195–203.

Bordenave, S., Jezequel, R., Fourcans, A., Budzinski, H., Merlin, F.X., Fourel, T., et al., 2004b. Degradation of the "Erika" oil. Aquat. Living Res. 17, 261–267.

Chakraborty, R., Wu, C.H., Hazen, T.C., 2012. Systems biology approach to bioremediation. Curr. Opin. Biotechnol. 23, 483–490.

Cho, J.C., Tiedje, J.M., 2002. Quantitative detection of microbial genes by using DNA microarrays. Appl. Environ. Microbiol. 68, 1425–1430.

Chourey, K., Nissen, S., Vishnivetskaya, T., Shah, M., Pfiffner, S., Hettich, R.L., et al., 2013. Environmental proteomics reveals early microbial community responses to biostimulation at a uranium- and nitrate-contaminated site. Proteomics. Available from: http://dx.doi.org/doi:10.1002/pmic.201300155.

Connon, S.A., Tovanabootr, A., Dolan, M., Vergin, K., Giovannoni, S.J., Semprini, L., 2005. Bacterial community composition determined by culture-independent and -dependent methods during propane-stimulated bioremediation in trichloroethene-contaminated groundwater. Environ. Microbiol. 7, 165–178.

Coulon, F., Chronopoulou, P.M., Fahy, A., Païsse, S., Goni-Urriza, M., Peperzak, L., et al., 2012. Central role of dynamic tidal biofilms dominated by aerobic hydrocarbonoclastic bacteria and diatoms in the biodegradation of hydrocarbons in coastal mudflats. Appl. Environ. Microbiol. 78, 3638–3648.

Cupples, A.M., Sims, G.K., 2007. Identification of *in situ* 2,4-dichlorophenoxyacetic acid-degrading soil microorganisms using DNA-stable isotope probing. Soil Biol. Biochem. 39, 232–238.

de Menezes, A., Clipson, N., Doyle, E., 2012. Comparative metatranscriptomics reveals widespread community responses during phenanthrene degradation in soil. Environ. Microbiol. 14, 2577–2588.

Denef, V.J., Park, J., Rodrigues, J.L.M., Tsoi, T.V., Hashsham, S.A., Tiedje, J.M., 2003. Validation of a more sensitive method for using spotted oligonucleotide DNA microarrays for functional genomics studies on bacterial communities. Environ. Microbiol. 5, 933–943.

Desai, C., Pathak, H., Madamwar, D., 2010. Advances in molecular and "-omics" technologies to gauge microbial communities and bioremediation at xenobiotic/anthropogen contaminated sites. Bioresour. Technol. 101, 1558–1569.

DeSantis, T.Z., Brodie, E.L., Moberg, J.P., Zubieta, I.X., Piceno, Y.M., Andersen, G.L., 2007. High-density universal 16S rRNA microarray analysis reveals broader diversity than typical clone library when sampling the environment. Microbial. Ecol. 53, 371–383.

Dojka, M.A., Hugenholtz, P., Haack, S.K., Pace, N.R., 1998. Microbial diversity in a hydrocarbon- and chlorinated-solvent-contaminated aquifer undergoing intrinsic bioremediation. Appl. Environ. Microbiol. 64, 3869–3877.

Dubey, S.K., Tripathi, A.K., Upadhyay, S.N., 2006. Exploration of soil bacterial communities for their potential as bioresource. Bioresour. Technol. 97, 2217–2224.

Dumont, M.G., Murrell, J.C., 2005. Stable isotope probing [mdash] linking microbial identity to function. Nat. Rev. Microbiol. 3, 499–504.

Emerson, D., Merrill Floyd, M., 2005. Enrichment and isolation of iron-oxidizing bacteria at neutral pH. In: Jared, R.L. (Ed.), Methods in Enzymology, vol. 397. Academic Press, Waltham, Massachusetts, USA, pp. 112–123.

Eriksson, M., Sodersten, E., Yu, Z., Dalhammar, G., Mohn, W.W., 2003. Degradation of polycyclic aromatic hydrocarbons at low temperature under aerobic and nitrate-reducing conditions in enrichment cultures from northern soils. Appl. Environ. Microbiol. 69, 275–284.

Fahrenfeld, N., Zoeckler, J., Widdowson, M., Pruden, A., 2013. Effect of biostimulants on 2,4,6-trinitrotoluene (TNT) degradation and bacterial community composition in contaminated aquifer sediment enrichments. Biodegradation. 24, 179–190.

Falkowski, P.G., Fenchel, T., Delong, E.F., 2008. The microbial engines that drive Earth's biogeochemical cycles. Science. 320, 1034–1039.

Farhadian, M., Vachelard, C., Duchez, D., Larroche, C., 2008. *In situ* bioremediation of monoaromatic pollutants in groundwater: a review. Bioresour. Technol. 99, 5296–5308.

Fischer, S.G., Lerman, L.S., 1979. Length-independent separation of DNA restriction fragments in two-dimensional gel electrophoresis. Cell. 16, 191–200.

Fisher, M.M., Triplett, E.W., 1999. Automated approach for ribosomal intergenic spacer analysis of microbial diversity and its application to bacterial communities. Appl. Environ. Microbiol. 65, 4630–4636.

Fries, M.R., Hopkins, G.D., McCarty, P.L., Forney, L.J., Tiedje, J.M., 1997. Microbial succession during a field evaluation of phenol and toluene as the primary substrates for trichloroethene co-metabolism. Appl. Environ. Microbiol. 63, 1515–1522.

Fuchs, G., Boll, M., Heider, J., 2011. Microbial degradation of aromatic compounds—from one strategy to four. Nat. Rev. Microbiol. 9, 803–816.

Geets, J., Borremans, B., Diels, L., Springael, D., Vangronsveld, J., van der Lelie, D., et al., 2006. DsrB gene-based DGGE for community and diversity surveys of sulfate-reducing bacteria. J. Microbiol. Methods 66, 194–205.

Ghosh, D., Routh, J., Bhadury, P., 2013. Temporal trends of arsenite oxidizing indigenous bacterial communities in arsenic-rich deltaic aquifers in West Bengal, India (in review).

Greene, E.A., Voordouw, G., 2003. Analysis of environmental microbial communities by reverse sample genome probing. J. Microbiol. Methods 53, 211–219.

Greene, E.A., Kay, J.G., Jaber, K., Stehmeier, L.G., Voordouw, G., 2000. Composition of soil microbial communities enriched on a mixture of aromatic hydrocarbons. Appl. Environ. Microbiol. 66, 5282–5289.

Griffiths, R.I., Whiteley, A.S., O'Donnell, A.G., Bailey, M.J., 2000. Rapid method for coextraction of DNA and RNA from natural environments for analysis of ribosomal DNA- and rRNA-based microbial community composition. Appl. Environ. Microbiol. 66, 5488–5491.

Handelsman, J., 2004. Metagenomics: application of genomics to uncultured microorganisms. Microbiol. Mol. Biol. Rev. 68, 669–685.

Haritash, A.K., Kaushik, C.P., 2009. Biodegradation aspects of polycyclic aromatic hydrocarbons (PAHs): a review. J. Hazard Mater. 169, 1–15.

Heid, C.A., Stevens, J., Livak, K.J., Williams, P.M., 1996. Real time quantitative PCR. Genome Res. 6, 986–994.

Hemme, C.L., Deng, Y., Gentry, T.J., Fields, M.W., Wu, L., Barua, S., et al., 2010. Metagenomic insights into evolution of a heavy metal-contaminated groundwater microbial community. ISME J. 4, 660–672.

Hendrickx, B., Dejonghe, W., Faber, F., Boenne, W., Bastiaens, L., Verstraete, W., et al., 2006. PCR-DGGE method to assess the diversity of BTEX mono-oxygenase genes at contaminated sites. FEMS Microbiol. Ecol. 55, 262–273.

Hohnstock-Ashe, A.M., Plummer, S.M., Yager, R.M., Baveye, P., Madsen, E.L., 2001. Further biogeochemical characterization of a trichloroethene-contaminated fractured dolomite aquifer: electron source and microbial communities involved in reductive dechlorination. Environ. Sci. Technol. 35, 4449–4456.

Holmes, D.E., O'Neil, R.A., Chavan, M.A., N'Guessan, L.A., Vrionis, H.A., Perpetua, L.A., et al., 2009. Transcriptome of *Geobacter uraniireducens* growing in uranium-contaminated subsurface sediments. ISME J. 3, 216–230.

Holmes, D.E., Giloteaux, L., Williams, K.H., Wrighton, K.C., Wilkins, M.J., Thompson, C. A., et al., 2013. Enrichment of specific protozoan populations during *in situ* bioremediation of uranium-contaminated groundwater. ISME J. 7, 1286–1298.

Huang, W.E., Stoecker, K., Griffiths, R., Newbold, L., Daims, H., Whiteley, A.S., et al., 2007. Raman–FISH: combining stable-isotope Raman spectroscopy and fluorescence *in situ* hybridization for the single cell analysis of identity and function. Environ. Microbiol. 9, 1878–1889.

Hubert, C., Shen, Y., Voordouw, G., 2005. Changes in soil microbial community composition induced by cometabolism of toluene and trichloroethylene. Biodegradation. 16, 11–22.

Hugenholtz, P., 2002. Exploring prokaryotic diversity in the genomic era. Gen. Biol. 3, Reviews 0003.1–Reviews 0003.8.

Jennings, L.K., Chartrand, M.M., Lacrampe-Couloume, G., Lollar, B.S., Spain, J.C., Gossett, J. M., 2009. Proteomic and transcriptomic analyses reveal genes upregulated by *cis*-dichloroethene in *Polaromonas* sp. strain JS666. Appl. Environ. Microbiol. 11, 3733–3744.

Juwarkar, A.A., Singh, S.K., Mudhoo, A., 2010. A comprehensive overview of elements in bioremediation. Rev. Environ. Sci. Biotechnol. 9, 215–288.

Kasai, Y., Takahata, Y., Manefield, M., Watanabe, K., 2006. RNA-based stable isotope probing and isolation of anaerobic benzene-degrading bacteria from gasoline-contaminated groundwater. Appl. Environ. Microbiol. 72, 3586–3592.

Kirk, J.L., Beaudette, L.A., Hart, M., Moutoglis, P., Klironomos, J.N., Lee, H., et al., 2004. Methods of studying soil microbial diversity. J. Microbiol. Methods 58, 169–188.

Kleikemper, J., Schroth, M.H., Sigler, W.V., Schmucki, M., Bernasconi, S.M., Zeyer, J., 2002. Activity and diversity of sulfate-reducing bacteria in a petroleum hydrocarbon-contaminated aquifer. Appl. Environ. Microbiol. 68, 1516–1523.

Kostka, J.E., Prakash, O., Overholt, W.A., Green, S.J., Freyer, G., Canion, A., et al., 2011. Hydrocarbon-degrading bacteria and the bacterial community response in Gulf of Mexico beach sands impacted by the deep horizon oil spill. Appl. Environ. Microbiol. 77, 7962–7974.

Lane, D.J., Pace, B., Olsen, G.J., Stahl, D.A., Sogin, M.L., Pace, N.R., 1985. Rapid determination of 16S ribosomal RNA sequences for phylogenetic analyses. Proc. Natl. Acad. Sci. USA 82, 6955–6959.

Larentis, M., Hoermann, K., Lueders, T., 2013. Fine-scale degrader community profiling over an aerobic/anaerobic redox gradient in a toluene-contaminated aquifer. Environ. Microbiol. Rep. 5, 225–234.

Li, Z., Xu, J., Tang, C., Wu, J., Muhammad, A., Wang, H., 2006. Application of 16S rDNA-PCR amplification and DGGE fingerprinting for detection of shift in microbial community diversity in Cu-, Zn-, and Cd-contaminated paddy soils. Chemosphere 62, 1374–1380.

Lin, X., Kennedy, D., Fredrickson, J., Bjornstad, B., Konopka, A., 2012. Vertical stratification of subsurface microbial community composition across geological formations at the Hanford Site. Environ. Microbiol. 14, 414–425.

Liu, W.T., Marsh, T.L., Cheng, H., Forney, L.J., 1997. Characterization of microbial diversity by determining terminal restriction fragment length polymorphisms of genes encoding 16S rRNA. Appl. Environ. Microbiol. 3, 4516–4522.

Malik, S., Beer, M., Megharaj, M., Naidu, R., 2008. The use of molecular techniques to characterize the microbial communities in contaminated soil and water. Environ. Int. 34, 265–276.

Marchuk, D., Drumm, M., Saulino, A., Collins, F.S., 1991. Construction of T-vectors, a rapid and general system for direct cloning of unmodified PCR products. Nucleic Acids Res. 19, 1154.

Mardis, E.R., 2013. Next-generation sequencing platforms. Annu. Rev. Anal. Chem. 6, 287–303.

Maron, P.A., Ranjard, L., Mougel, C., Lemanceau, P., 2007. Metaproteomics: a new approach for studying functional microbial ecology. Microb. Ecol. 53, 486–493.

McCarty, P.L., 1993. *In situ* bioremediation of chlorinated solvents. Curr. Opin. Biotechnol. 4, 323–330.

McGoldrick, A., Lowings, J.P., Ibata, G., Sands, J.J., Belak, S., Paton, D.J., 1998. A novel approach to the detection of classical swine fever virus by RT-PCR with a fluorogenic probe (TaqMan). J. Virol. Methods 72, 125–135.

McGrath, K.C., Thomas-Hall, S.R., Cheng, C.T., Leo, L., Alexa, A., Schmidt, S., et al., 2008. Isolation and analysis of mRNA from environmental microbial communities. J. Microbiol. Methods. 75, 172–176.

Mills, D.K., Entry, J.A., Gillevet, P.M., Mathee, K., 2007. Assessing microbial community diversity using amplicon length heterogeneity polymerase chain reaction. Soil Sci. Soc. Am. J. 71, 572–578.

Muhling, M., Woolven-Allen, J., Murrell, J.C., Joint, I., 2008. Improved group-specific PCR primers for denaturing gradient gel electrophoresis analysis of the genetic diversity of complex microbial communities. ISME J. 2, 379–392.

Mulligan, C.N., Yong, R.N., 2004. Natural attenuation of contaminated soils. Environ. Int. 30, 587–601.

Muyzer, G., 1999. DGGE/TGGE a method for identifying genes from natural ecosystems. Curr. Opin. Microbiol. 2, 317–322.

Muyzer, G., Smalla, K., 1998. Application of denaturing gradient gel electrophoresis (DGGE) and temperature gradient gel electrophoresis (TGGE) in microbial ecology. Antonie Van Leeuwenhoek. 73, 127–141.

Muyzer, G., de Waal, D.C., Uitterlinden, A.G., 1993. Profiling of complex microbial populations by denaturing gradient gel electrophoresis analysis of polymerase chain reaction-amplified genes coding for 16S rRNA. Appl. Environ. Microbiol. 59, 695–700.

Myers, R.M., Maniatis, T., Lerman, L.S., 1987. Detection and localization of single base changes by denaturing gradient gel electrophoresis. Methods Enzymol. 155, 501–527.

Nocker, A., Burr, M., Camper, A., 2007. Genotypic microbial community profiling: a critical technical review. Microb. Ecol. 54, 276–289.

Olsen, G.J., Lane, D.J., Giovannoni, S.J., Pace, N.R., Stahl, D.A., 1986. Microbial ecology and evolution: a ribosomal RNA approach. Annu. Rev. Microbiol. 40, 337–365.

Osborn, A.M., Moore, E.R.B., Timmis, K.N., 2000. An evaluation of terminal-restriction fragment length polymorphism (T-RFLP) analysis for the study of microbial community structure and dynamics. Environ. Microbiol. 2, 39–50.

Pace, N.R., 1997. A molecular view of microbial diversity and the biosphere. Science 276, 734–740.

Paul, D., Pandey, G., Pandey, J., Jain, R.K., 2005. Accessing microbial diversity for bioremediation and environmental restoration. Trends Biotechnol. 23, 135–142.

Peng, R.H., Xiong, A.S., Xue, Y., Fu, X.Y., Gao, F., Zhao, W., et al., 2008. Microbial biodegradation of polyaromatic hydrocarbons. FEMS Microbiol. Rev. 32, 927–955.

Ponchel, F., Toomes, C., Bransfield, K., Leong, F., Douglas, S., Field, S., et al., 2003. Real-time PCR based on SYBR-Green I fluorescence: an alternative to the TaqMan assay for a relative quantification of gene rearrangements, gene amplifications and micro gene deletions. BMC Biotechnol. 3, 18.

Poulsen, L.K., Ballard, G., Stahl, D.A., 1993. Use of rRNA fluorescence *in situ* hybridization for measuring the activity of single cells in young and established biofilms. Appl. Environ. Microbiol. 59, 1354–1360.

Powell, S., Ferguson, S., Bowman, J., Snape, I., 2006. Using real-time PCR to assess changes in the hydrocarbon-degrading microbial community in Antarctic Soil during bioremediation. Microb. Ecol. 52, 523–532.

Priha, O., Nyyssonen, M., Bomberg, M., Laitila, A., Simell, J., Kapanen, A., et al., 2013. Application of denaturing high-performance liquid chromatography for monitoring sulfate-reducing bacteria in oil fields. Appl. Environ. Microbiol. 79, 5186–5196.

Ranjard, L., Poly, F., Lata, J.C., Mougel, C., Thioulouse, J., Nazaret, S., 2001. Characterization of bacterial and fungal soil communities by automated ribosomal intergenic spacer analysis fingerprints: biological and methodological variability. Appl. Environ. Microbiol. 67, 4479–4487.

Rappe, M.S., Giovannoni, S.J., 2003. The uncultured microbial majority. Annu. Rev. Microbiol. 57, 369–394.

Rastogi, G., Sani, R.K., 2011. Molecular techniques to assess microbial community structure, function, and dynamics in the environment. In: Ahmad, I., Ahmad, F., Pichtel, J. (Eds.), Microbes and Microbial Technology. Springer, New York, NY, pp. 29–57.

Rhee, S.K., Liu, X., Wu, L., Chong, S.C., Wan, X., Zhou, J., 2004. Detection of genes involved in biodegradation and biotransformation in microbial communities by using 50-mer oligonucleotide microarrays. Appl. Environ. Microbiol. 70, 4303–4317.

Richardson, R.E., Bhupathiraju, V.K., Song, D.L., Goulet, T.A., Alvarez-Cohen, L., 2002. Phylogenetic characterization of microbial communities that reductively dechlorinate TCE based upon a combination of molecular techniques. Environ. Sci. Technol. 36, 2652–2662.

Riesenfeld, C.S., Schloss, P.D., Handelsman, J., 2004. Metagenomics: genomic analysis of microbial communities. Annu. Rev. Genet. 38, 525–552.

Robertson, S., Rutherford, P.M., Massicotte, H., 2011. Plant and soil properties determine microbial community structure of shared Pinus-Vaccinium rhizospheres in petroleum hydrocarbon contaminated forest soils. Plant Soil. 346, 121–132.

Rossello-Mora, R., Amann, R., 2001. The species concept for prokaryotes. FEMS Microbiol. Rev. 25, 39–67.

Saiki, R.K., Gelfand, D.H., Stoffel, S., Scharf, S.J., Higuchi, R., Horn, G.T., et al., 1988. Primer-directed enzymatic amplification of DNA with a thermostable DNA polymerase. Science. 239, 487–491.

Schena, M., Shalon, D., Davis, R.W., Brown, P.O., 1995. Quantitative monitoring of gene expression patterns with a complementary DNA microarray. Science. 270, 467–470.

Shen, Y., Stehmeier, L.G., Voordouw, G., 1998. Identification of hydrocarbon-degrading bacteria in soil by reverse sample genome probing. Appl. Environ. Microbiol. 64, 637–645.

Singleton, D.R., Furlong, M.A., Rathbun, S.L., Whitman, W.B., 2001. Quantitative comparisons of 16S rRNA gene sequence libraries from environmental samples. Appl. Environ. Microbiol. 67, 4374–4376.

Singleton, D.R., Powell, S.N., Sangaiah, R., Gold, A., Ball, L.M., Aitken, M.D., 2005. Stable-isotope probing of bacteria capable of degrading salicylate, naphthalene, or phenanthrene in a bioreactor treating contaminated soil. Appl. Environ. Microbiol. 71, 1202–1209.

Small, J., Call, D.R., Brockman, F.J., Straub, T.M., Chandler, D.B., 2001. Direct detection of 16S rRNA in soil extracts by using oligonucleotide microarrays. Appl. Environ. Microbiol. 67, 4708–4716.

Smith, C.J., Osborn, A.M., 2009. Advantages and limitations of quantitative PCR (Q-PCR)-based approaches in microbial ecology. FEMS Microbiol. Ecol. 67, 6–20.

Stenuit, B., Eyers, L., Schuler, L., Agathos, S.N., George, I., 2008. Emerging high-throughput approaches to analyze bioremediation of sites contaminated with hazardous and/or recalcitrant wastes. Biotechnol. Adv. 26, 561–575.

Sun, Y., Polishchuk, E.A., Radoja, U., Cullen, W.R., 2004. Identification and quantification of arsC genes in environmental samples by using real-time PCR. J. Microbiol. Methods. 58, 335–349.

Sutton, N.B., van der Kraan, G.M., van Loosdrecht, M.C.M., Muyzer, G., Bruining, J., Schotting, R.J., 2009. Characterization of geochemical constituents and bacterial populations associated with As mobilization in deep and shallow tube wells in Bangladesh. Water Res. 43, 1720–1730.

Tiedje, J.M., Asuming-Brempong, S., Nusslein, K., Marsh, T.L., Flynn, S.J., 1999. Opening the black box of soil microbial diversity. Appl. Soil Ecol. 13, 109–122.

Tipayno, S., Kim, C.G., Sa, T., 2012. T-RFLP analysis of structural changes in soil bacterial communities in response to metal and metalloid contamination and initial phytoremediation. Appl. Soil Ecol. 61, 137–146.

Torsvik, V., Ovreas, L., 2002. Microbial diversity and function in soil: from genes to ecosystems. Curr. Opin. Microbiol. 5, 240–245.

Torsvik, V., Daae, F.L., Sandaa, R.A., Ovreas, L., 1998. Novel techniques for analysing microbial diversity in natural and perturbed environments. J. Biotechnol. 64, 53–62.

Tsai, Y.L., Olson, B.H., 1991. Rapid method for direct extraction of DNA from soil and sediments. Appl. Environ. Microbiol. 57, 1070–1074.

Turpeinen, R., Kairesalo, T., Haggblom, M.M., 2004. Microbial community structure and activity in arsenic-, chromium- and copper-contaminated soils. FEMS Microbiol. Ecol. 47, 39–50.

Uhlik, O., Leewis, M.C., Strejcek, M., Musilova, L., Mackova, M., Leigh, M.B., et al., 2013. Stable isotope probing in the metagenomics era: a bridge towards improved bioremediation. Biotechnol. Adv. 31, 154–165.

Urich, T., Lanzen, A., Qi, J., Huson, D.H., Schleper, C., Schuster, S.C., 2008. Simultaneous assessment of soil microbial community structure and function through analysis of the meta-transcriptome. PLoS ONE. 3, e2527.

Vahter, M., 2002. Mechanisms of arsenic biotransformation. Toxicology. 181–182, 211–217.

Vogel, T.M., 1996. Bioaugmentation as a soil bioremediation approach. Curr. Opin. Biotechnol. 7, 311–316.

Voordouw, G., Voordouw, J.K., Karkhoff-Schweizer, R.R., Fedorak, P.M., Westlake, D.W., 1991. Reverse sample genome probing, a new technique for identification of bacteria in environmental samples by DNA hybridization, and its application to the identification of sulfate-reducing bacteria in oil field samples. Appl. Environ. Microbiol. 57, 3070–3078.

Wagner, A.O., Malin, C., Illmer, P., 2009. Application of denaturing high-performance liquid chromatography in microbial ecology: fermentor sludge, compost, and soil community profiling. Appl. Environ. Microbiol. 75, 956–964.

Wagner, M., Nielsen, P.H., Loy, A., Nielsen, J.L., Daims, H., 2006. Linking microbial community structure with function: fluorescence *in situ* hybridization-microautoradiography and isotope arrays. Curr. Opin. Biotechnol. 17, 83–91.

Wang, Q., Xie, S., Hu, R., 2013. Bioaugmentation with *Arthrobacter* sp. strain DAT1 for remediation of heavily atrazine-contaminated soil. Int. Biodeter. Biodegr. 77, 63–67.

Watts, J.E., Wu, Q., Schreier, S.B., May, H.D., Sowers, K.R., 2001. Comparative analysis of polychlorinated biphenyl-dechlorinating communities in enrichment cultures using three different molecular screening techniques. Environ. Microbiol. 3, 710–719.

Wellington, E.M.H., Berry, A., Krsek, M., 2003. Resolving functional diversity in relation to microbial community structure in soil: exploiting genomics and stable isotope probing. Curr. Opin. Microbiol. 6, 295–301.

Whitman, W.B., Coleman, D.C., Wiebe, W.J., 1998. Prokaryotes: the unseen majority. Proc. Natl. Acad. Sci. USA 95, 6578–6583.

Widada, J., Nojiri, H., Omori, T., 2002. Recent developments in molecular techniques for identification and monitoring of xenobiotic-degrading bacteria and their catabolic genes in bioremediation. Appl. Microbiol. Biotechnol. 60, 45–59.

Wilmes, P., Wexler, M., Bond, P.L., 2008. Metaproteomics provides functional insight into activated sludge wastewater treatment. PLoS ONE 3, e1778.

Wu, L., Thompson, D.K., Li, G., Hurt, R.A., Tiedje, J.M., Zhou, J., 2001. Development and evaluation of functional gene arrays for detection of selected genes in the environment. Appl. Environ. Microbiol. 67, 5780–5790.

Wu, L., Thompson, D.K., Liu, X., Fields, M.W., Bagwell, C.E., Tiedje, J.M., et al., 2004. Development and evaluation of microarray-based whole-genome hybridization for detection of microorganisms within the context of environmental applications. Environ. Sci. Technol. 38, 6775–6782.

Zhou, J., 2003. Microarrays for bacterial detection and microbial community analysis. Curr. Opin. Microbiol. 6, 288–294.

5 Microbial Indicators for Monitoring Pollution and Bioremediation

Tingting Xu[a], *Nicole Perry*[b], *Archana Chuahan*[a], *Gary Sayler*[a] *and Steven Ripp*[a]

[a]The University of Tennessee, The Center for Environmental Biotechnology, The Joint Institute for Biological Sciences, Knoxville, TN, [b]Department of Biology, Wittenberg University, Springfield, OH

5.1 Introduction

Microorganisms metabolically rely upon chemicals scavenged from their environment and correspondingly defend themselves against toxic chemicals to ensure their growth and survival. These essential needs can be genetically exploited to create bioreporter organisms that can rapidly and specifically indicate the presence and bioavailability of target environmental contaminants. Using the toolbox of genetic engineering, genes or their operon units that are activated upon exposure to a target chemical agent or chemical class can be isolated and linked to reporter genes to serve as operational switches that bioindicate chemical presence with sensitivity and reproducibility that oftentimes rivals conventional analytical methods. Reporter genes most often consist of light signaling elements such as the bioluminescent bacterial (*lux*) and firefly (*luc*) luciferases, green fluorescent protein (*gfp* as well as its palette of multicolored fluorescent derivatives), and colorimetric indicators like β-galactosidase (*lacZ*) (Close et al., 2009). These reporter genes are widely available within numerous types of cloning vectors for straightforward manipulation and have been successfully applied toward the creation of a large number of bioreporter organisms. The challenge in bioreporter development, however, lies in discovering and isolating the appropriate promoter gene switch that activates in response to the desired chemical target or targets. Although often provided via the evolutionary adaptation of microbial populations to new chemical bioavailability, particular promoter elements still must be isolated and characterized, which requires significant time, effort, and cost. Additionally, what nature supplies is not always optimal: sensitivity and specificity may be lacking. Synthetic biology addresses some of these issues by providing the tools necessary to modify and strengthen promoters for improved and more tailored sensing capacities and effectively complements directed evolution methods to create wholly new chemical-responsive genetic architectures

Microbial Biodegradation and Bioremediation. DOI: http://dx.doi.org/10.1016/B978-0-12-800021-2.00005-4

that nature has not yet designed (van der Meer and Belkin, 2010; Checa et al., 2012; Cobb et al., 2013). Notwithstanding this ability to create a new generation of upgraded bioreporters, decades of past bioreporter development and unique applications testify to their utility as successful monitors for numerous environmental contaminants (Trogl et al., 2012; Xu et al., 2013).

5.2 Choosing a Whole Cell Bioreporter

Bioreporters are typically designed with either a "lights-off" or "lights-on" signaling mechanism. A lights-off bioreporter is built by linking the reporter gene to a constitutive genetic element that remains activated as long as the cell is healthy, and thus this type of bioreporter emits its bioluminescent, fluorescent, or colorimetric light signal continuously. Upon exposure to a chemical or chemical mixture that harms the cell or disrupts cellular metabolism, an analogous reduction in reporter signal intensity, or a lights-off response, occurs that correlates to the degree of toxic interaction. Although these bioreporters do not specifically identify the chemical(s) affecting the cell, they clearly indicate that there is a toxic, biologically relevant interaction occurring, and pre-alarm to the need for further downstream analyses to identify the chemical(s) of concern. The classic lights-off bioreporter is exemplified in the Microtox® assay, where bioluminescence from *Vibrio fischeri* now reclassified as *Aliivibrio fischeri* (Urbanczyk et al., 2007) bacteria is quantified after exposure to a dilution series of an allegedly toxic sample (Hermens et al., 1985). A resulting decrease in bioluminescence denotes toxicity, with subsequent bioluminescent measurements being correlated to relative levels of toxicity expressed as an EC_{50} value (a concentration at which a 50% reduction in light is produced).

In a lights-on bioreporter, the microorganism is designed to remain "dark" until it encounters its designated target(s), upon which it emits a light signal. These bioreporters rely upon a fusion between the reporter gene and an inducible promoter that activates after interacting with the designated target(s). Activation can be designed to occur either specifically or nonspecifically. Under specific activation, the promoter/reporter gene fusion directly associates with a particular chemical or class of chemicals, with the intensity of subsequent signal generation exhibiting proportionality to the bioavailable portion of the chemical. This ability to measure bioavailability rather than strictly the total amount of chemical present in the sample is a key characteristic of bioreporters. Being living entities, bioreporters are capable of describing that portion of the chemical that is freely available to cross the cellular membrane—i.e., the bioavailable portion—and therefore denote the biological effect of a chemical rather than the mere presence or absence of a chemical as is defined using analytical measurement techniques like gas chromatography/mass spectrometry (GC/MS). However, when using bioreporters, one must be careful in the interpretation of bioavailability since its measurement is heavily influenced by the bioreporter's physiological status, growth rate, membrane composition, membrane transport mechanisms, and many other factors that shift in tandem with myriad environmental influences (Harms et al., 2006).

A lights-on bioreporter designed to be nonspecific uses a promoter/reporter gene complex whose promoter is activated in the presence of a general class of chemicals that can be identified as affecting certain cytotoxic, genotoxic, mutagenic, or stress-related cellular pathways. Thus, although the chemical itself cannot be specifically identified, the chemical's effect on a living system can be described within certain specifications that assist in delineating and defining exposure risk profiles.

The conventional inventory of reporter genes and their advantages and disadvantages are discussed below. The reader is directed to several detailed reviews for more in-depth information (Girotti et al., 2008; Diplock et al., 2010; Ripp et al., 2011; Shin, 2011; Struss et al., 2010; Xu et al., 2013).

5.2.1 *Bacterial Luciferase (*lux*)*

Bioluminescence is defined as light generated and emitted by a living organism using luciferin and a luciferase enzyme. Luciferase has been found in a number of species including bacteria, jellyfish, fungi, algae, insects, shrimp, and squid. The bacterial luciferase (*lux)* gene cassette contains a series of five genes—*luxA*, *luxB*, *luxC*, *luxD*, and *luxE*—that bioluminesce by producing a blue-green 490 nm light signal when expressed (Meighen, 1994). The gene system is commonly referred to as *luxCDABE*, which corresponds to the order of the genes in the operon. Expression of *luxA* and *luxB* results in a heterodimeric luciferase enzyme that catalyzes the light-emitting reaction by converting the long-chain aldehyde synthesized by the *luxC*, *luxD*, and *luxE* gene products to a carboxylic acid. The other substrate for this reaction is a reduced flavin mononucleotide ($FMNH_2$), which becomes oxidized to flavin mononucleotide (FMN) by the luciferase enzyme as the reaction proceeds. The bioluminescent reaction cannot progress in the absence of oxygen and, therefore, bioreporters that incorporate the gene cassette *luxCDABE* are to date restricted to aerobic conditions. Depending on the desired application, it is possible to engineer a bioreporter to contain either the *luxAB* or the *luxCDABE* gene cassettes. The *lux* genes have additionally undergone a codon optimization process that allows for their expression under eukaryotic genetic controls. This permits broader application of bioluminescent bioreporters beyond bacteria and more toward yeast and human cells that ultimately serve as better proxies for describing a chemical's true effect on higher animal life forms.

5.2.1.1 *luxAB*

A *luxAB* bioreporter harbors only the *luxA* and *luxB* genes from the *luxCDABE* gene cassette. This allows the bioreporter to generate the luciferase enzyme but, due to lack of *luxCDE*, requires that the aldehyde, commonly *n*-decanal, be added exogenously. One advantage of this method of integration is that it allows the saturation of the reaction with substrate, which results in a stronger, brighter bioluminescent signal. However, this makes time point data contingent upon the addition of aldehyde substrate, which could cause bioluminescent signals to be missed due to early or late addition of the aldehyde. Some studies have incorporated the aldehyde substrate into a growth medium to allow for continuous availability. Working with the fairly small 2.1 kb *luxAB* genes also simplifies the cloning process and numerous such bioreporters have been created in bacterial, yeast, and mammalian cells and used for various bioassay applications in the environment such as the detection of inorganic and organic compounds (Ripp et al., 2011; Xu et al., 2013).

5.2.1.2 luxCDABE

The *luxCDABE* bioreporter contains all elements of the *luxCDABE* gene cassette, thus permitting the organism to bioluminesce autonomously, independent of substrate or cofactor addition. This type of monitoring allows for real-time detection and has become popular in water quality and other environmental applications. *luxCDABE*-based bioreporters are particularly useful in biosensor applications where they are mated to transducers designed to monitor the light signal (Su et al., 2011). With no user interaction required, remote and deployable long-term biosensor surveillance of desired environments becomes feasible. Synthetic optimization of the *luxCDABE* gene cassette has permitted it to move away from its native AT-rich requirements towards GC-rich microorganisms, creating a wider availability of microbial hosts and allowing for a better complement of bioreporter hosts to test environments (Craney et al., 2007). At 7 kb, the size of the *luxCDABE* cassette does impede some cloning procedures but commercial cloning kits designed for larger size gene integrations have for the most part alleviated these problems. A unique application of *luxCDABE* biosensing has been its assimilation into the DuPont LuxArray, which represents a genome-wide collection of 689 *Escherichia coli* bioreporters that contain unique promoter regions linked to the *luxCDABE* (Van Dyk et al., 2001). The LuxArray allows for analysis of the transcriptional response of a cell to its environment and has the ability to collect kinetic data by measuring bioluminescence continuously over time.

5.2.1.3 Eukaryotic Optimized luxCDABE

The *luxCDABE* reporter gene cassette, although it originated from and has been used predominately in bacterial hosts, has also been optimized for expression in eukaryotic organisms to provide a bioreporter system that more accurately mimics human bioavailability with respect to human toxicity concerns. The major challenges for functional bioluminescent expression in mammalian cells were the low translational efficiency of bacterial gene sequences and the inability to polycistronically express multiple genes from a single promoter in the eukaryotic background. Thanks to the process of codon optimization and the utilization of unique genetic elements such as the internal ribosomal entry site (IRES) element (Baird et al., 2006) and the viral 2A element (Szymczak and Vignali, 2005), the *luxCDABE* cassette has been successfully expressed in the lower eukaryotic organism *Saccharomyces cerevisiae* (Gupta et al., 2003) and human cell lines (Close et al., 2010) for autonomous bioluminescent emission. It has also been discovered during the development process that co-expression of a sixth gene, *frp*, which encodes a flavin mononucleotide oxidoreductase, is required for optimized light production in eukaryotic host cells (Gupta et al., 2003; Close et al., 2010). The Frp protein recycles and regenerates $FMNH_2$, which appears to be limited in the eukaryotic cellular background. Autonomously bioluminescent yeast bioreporters for estrogenic and androgenic compounds have been developed using the *luxCDABEfrp* genes with demonstrated dose-responsive bioluminescent signaling occurring within 3–4 h (Sanseverino et al., 2005, 2009; Eldridge et al., 2007; Eldridge et al., 2011).

5.2.2 Firefly Luciferase (luc)

The *luc* genes were originally identified and isolated from the firefly *Photinus pyralis*. This reporter gene has become popularized due to the high output of yellow/green light (550–575 nm) upon oxidation of the reduced luciferin substrate in the presence of ATP-Mg^{2+} and oxygen. The *luc* genes also harbor the greatest quantum yield of current bioluminescent systems (Fraga, 2008). Since posttranslational modifications are not needed for the Luc protein, it is made accessible immediately following translation; however, exogenous addition of the luciferin substrate is required for the ensuing light reaction. Thus, similar to the *luxAB* reporter system, *luc*-based bioreporters emit light discontinuously and are impeded from functioning autonomously. Nonetheless, with their robust signal strength, small size (1.7 kb), and wide availability of cloning vectors, a large variety of *luc*-based bioreporters have been constructed for the detection of environmentally relevant organic compounds, heavy metals, and estrogenic and endocrine disrupter agents (Xu et al., 2013).

5.2.3 Green Fluorescent Protein

Green fluorescent protein (GFP) is part of a family of natural and recombinant photoproteins isolated and cloned from the jellyfish *Aequorea victoria*. The activation of GFP does not require a substrate; instead, bioreporters that contain GFP emit a fluorescent signal at 508 nm upon exposure to ultraviolet or blue light. Consequently, GFP-based bioreporters must always be tied to instrumentation capable of delivering the excitation wavelength and measuring the resulting emission wavelength. Nonetheless, the ability to autofluoresce allows for near real-time sensing, which makes GFP bioreporters an effective method of assessment for monitoring environmental conditions. The small size of GFP ($\sim$750 bp) also facilitates straightforward cloning. Besides GFP, there are other fluorescent variants that emit in the cyan, red, yellow, and orange wavelengths (Shaner et al., 2005). With this selectable palette of colors to choose from, bioreporters capable of sensing multiple analytes can be developed by linking individual promoters to differently colored fluorescent reporter gene outputs. For example, Hever and Belkin (2006) created fusions of the stress-responsive *recA* and *grpE* promoters to green and red fluorescent reporter protein genes, respectively, to create a dual reporter for simultaneous monitoring of both toxic and genotoxic events. Dual bioreporter systems that combine fluorescence with bioluminescence can also be created, as demonstrated by Mitchell and Gu (2004) in their similar work linking the stress-responsive *recA* and *katG* promoters to GFP and *luxCDABE*, respectively.

5.2.4 *lacZ*

The *lacZ* gene originates from *E. coli* and encodes β-galactosidase (β-gal), which hydrolyzes β-galactoside disaccharides into monosaccharides. In the presence of o-nitrophenyl-β-D-galactoside (ONPG), *lacZ* bioreporters yield a yellow by-product that can be measured colorimetrically. Color intensity corresponds to the β-gal activity and provides an estimate of the chemical concentration. For example, the SOS Chromotest is a commercially available kit that utilizes *lacZ*-based fusions to DNA damage-sensitive reporters to detect genotoxic compounds (Quillardet et al., 1982). One disadvantage to colorimetric analysis is low sensitivity, which has resulted in the development of new detection methods, such as the incorporation of *lacZ* bioreporters with different β-galactoside substrates that allow for fluorescent, luminescent, or chemiluminescent endpoints. Another drawback to *lacZ* bioreporters is the required lysis of reporter cells to allow quantification of β-gal activity, which causes delayed and discontinuous accumulation of results. This issue has been overcome with the introduction of electrochemical and amperometric interfaces to bypass permeabilization steps (Ron and Rishpon, 2010). For environmental analyses, it is important to account for the endogenous presence of β-gal in sample matrices and its high background activity potentially interfering with endpoint measurements.

5.3 Applying the Bioreporter as a Pollution Monitoring and Bioremediation Tool

The environmental application of bioreporters is a challenging prospect due to their genetically engineered nature and the consequent regulatory boundaries that limit their use outside the confines of a laboratory. Most applications of bioreporters, therefore, rely upon transport of the sample to the laboratory where they can be appropriately processed and then integrated into the bioreporter assay. However, bioreporters can be environmentally applied if properly safeguarded against uncontrolled release. This typically requires encapsulation or immobilization of the bioreporter into an interface device that receives the sample *ex situ* or, alternatively, *in situ* if used in a flow-through manner directly from the source sample. These biosensor interfaces ultimately form an all-in-one sensing unit that measures and monitors for signal emission by the integrated bioreporter organisms.

5.3.1 *Keeping the Bioreporters Alive and Healthy*

For sustained, long-term environmental biosensing, the bioreporters must be packaged in a format that maintains their viability such that they are continuously available to report on their surroundings or can do so as rapidly as possible following a triggered resuscitation event. Encapsulants designed to maintain cells in a viable state for extended durations typically consist of liquid matrices to which the bioreporter culture is added; these then solidify to form sheets, layers, blocks, or discrete beads of cells. Agar, agarose, polydimethylsiloxane (PDMS), polyvinyl alcohol/polyvinyl pyridine copolymer, latex copolymer, carrageenan, polyacrylamide, alginate, polyurethane/polycarbonyl sulfonate, polyvinyl alcohol, and sol–gel silica glass are only a few of the matrices applied as bioreporter encapsulants (Date et al., 2010; Michelini and Roda, 2012). Encapsulation can also be as simple as drying the bioreporters onto a filter surface (Toba and Hay, 2005) or packaging lyophilized (freeze-dried) bioreporters into single-use test vials (Siegfried et al., 2012). The challenge is to retain the bioreporter cells within a microenvironment that contributes to long-term survival while maintaining compatibility with the measuring scheme in terms of optical clarity and minimizing signal interference, since these parameters ultimately affect assay sensitivity. Considering the vast number of biotic and abiotic factors that can impinge upon cellular bioreporter growth, survival, and longevity, effective encapsulation is a daunting challenge that has yet to be adequately solved; there likely will never be a universal matrix for doing so. Nonetheless, many research groups have demonstrated reasonably adequate microorganismal stability from days to months but, unfortunately, seldom under the authentic extreme conditions inherent to real-world environmental exposures. Other researchers have alternatively exploited what nature has already provided and used the innate ability of certain microorganisms to form spores as vehicles for long-term bioreporter storage (Knecht et al., 2011), or cultivated bioreporters in robust biofilm assemblages to enable resilient surface attachment on biocompatible materials followed by encapsulation (Ben-Yoav et al., 2011). Another well-applied option is to couple the bioreporter cells with surface-attached linkers, which typically entails chemical methods to facilitate covalent binding between the surface and functional groups on the cell (Fleming, 2010); nanomaterials have found a unique niche in this process that allows for high cell loading capacities with improved stability and viability (Dai and Choi, 2013). Surface-to-cell linkages can also be fabricated using antibodies targeted to cell surface antigens (Premkumar et al., 2001) as well as bacteriophages (bacterial viruses) that bind to specific cell surface receptors (Tolba et al., 2010).

5.3.2 Integrating Bioreporter Organisms with Biosensor Devices

The biosensor device consists of a merger between the bioreporter organisms and a transducer capable of measuring the endpoint signal emanating from the bioreporters. A simple embodiment of such a biosensor device is a fiber optic cable terminated on one end with encapsulated bioreporters and on the other end with an appropriate signal measurement instrument (Eltzov and Marks, 2010). For example, Ivask et al. (2007) combined *luxCDABE*-based heavy metal-specific bioluminescent bioreporters with a sodium alginate encapsulant that was solidified onto the tip of a fiber optic cable and measured resulting bioluminescent responses with a photomultiplier interface. The primary benefit of fiber optic interfaces is that they can terminate far from the measurement device, thus allowing infiltration of bioreporters into remote areas without requiring human presence. However, being tied to such an interface raises logistical concerns dealing with power requirements and traversal and subsequent degradation of the signal over extended distances. More compact, all-in-one biosensors can more ideally be fabricated using lab-on-chip technologies where the reporter organisms are directly interfaced with miniaturized transducer units. The bioluminescent bioreporter integrated circuit (BBIC) was an early example wherein *luxCDABE*-based bioreporters were adhered to a 1.5 mm × 1.5 mm complementary metal oxide semiconductor (CMOS) microluminometer for remote sensing of water and airborne chemical contaminants (Ripp et al., 2003; Nivens et al., 2004). The more recent addition of more sensitive silicon photomultipliers and avalanche photodiodes as sensing platforms has increased the utility of these chip-based biosensors (Daniel et al., 2008; Lopes et al., 2012). When combined with microfluidics, true lab-on-chip biosensors can be fabricated to meet a wide variety of sensing needs. This typically entails the formation of on-chip channels fabricated from PDMS with a combination of pumps and valves directing the sample to the bioreporter cells and the transducer interface. Elad et al. (2011), for example, positioned single-photon avalanche diodes (SPADs) paired with *luxCDABE*-based bioluminescent bioreporters above a network of PDMS channels to enable on-line, flow-through surveillance and early warning of water contamination events. A similar on-line system referred to as Lumisens was developed by the Thouand research group wherein a charge-coupled device (CCD) camera was used to capture bioluminescent signals emanating from immobilized or lyophilized *luxCDABE*-based bioreporters localized in multiwell cards or plates after exposure to contaminated water delivered via pumps or microinjectors (Charrier et al., 2011a,b; Jouanneau et al., 2012). With a large number of wells being available within the Lumisens biosensor, multiple bioreporters could be accommodated for replicate and multiplexed sensing in a fairly easy-to-use plug-and-play format. Moving toward even higher level multiplexing are the bacterial sensor arrays consisting of whole genome bioreporters that incorporate hundreds to thousands of reporter gene fusions to globally indicate chem/bio effects across the entire living cell (Elad et al., 2010). Elad and Belkin (2013), for example, linked approximately 2000 *E. coli* K12 promoters to GFP to create a library of reporter cells that, when spotted into microtiter plate wells, were capable of collectively bioindicating a specific "fingerprint" based on each type of chemical exposure event. The previously mentioned LuxArray developed by Van Dyk et al. (2001) provides a similar array interface based on *luxCDABE*-based bioluminescent fingerprint endpoints. The challenge remains in accommodating these thousands of bioreporters on a single miniaturized detector interface, although the proof-of-concept for smaller numbers of bioreporters has been demonstrated using a variety of optical and electrical biochip platforms (Shacham-Diamand et al., 2010).

5.4 Examples of *In Situ* Field Applications

Despite the fact that prokaryotic and eukaryotic whole cell bioreporters for a wide range of pollutants are being continuously developed and refined in the laboratory, very few bioreporters actually achieve practical application toward conventional on-site monitoring and testing due to logistical challenges of long-term maintenance and viability and mandatory restrictions on the release of genetically engineered organisms. Although advances in cell immobilization and preservation techniques as well as progress made in biosensor interface and device design (see discussions in the previous section) have significantly accelerated the technical development of field-operable biosensor systems, the lack of patent protection of the bioreporters is still a significant obstacle for biosensor devices to reach the market. Nevertheless, biosensor devices intended for on-site applications have been developed in the laboratory as proof-of-concept prototypes, some of which have been tested with real environmental samples either in the field or in laboratory-simulated settings that mimic potential filed applications (Table 5.1). Such devices incorporate various separation or immobilization technologies to comply with mandatory regulations in order to avoid uncontrolled release of recombinant bioreporter cells into the environment. Most on-site surveillance devices are designed for water quality monitoring due to the relative simplicity of prebioassay sample manipulations, while applications of soil testing are limited largely because soil samples often require complicated pretreatments or extractions prior to the bioassays, which have yet to be effectively integrated into the on-site biosensing pipeline.

Some of the early efforts to realize on-site monitoring have utilized bioreporter cell suspensions trapped in sealed containers to ensure that no recombinant microorganisms are freely discharged to the environment. Gu et al. (1996) originally designed a single miniature bioreactor with a 58 mL working volume to house *luxCDABE*-expressing recombinant bioluminescent *E. coli* bioreporter cells for on-line toxicity monitoring, which was later upgraded to a two-stage bioreactor system in which the first mini-bioreactor functioned to provide a stable and continuous supply of the bioreporter cells for the second mini-bioreactor where exposure to test sample and collection of signal occurred (Gu et al., 1999). Bioluminescent signals emitted from the bioreporter cells were transmitted via a fiber optic probe to a luminometer interface for data acquisition and analysis. Gu and Gil (2001) further exploited the modular feature of this design and developed a multichannel monitoring system for continuous multiplexed toxicant profiling to satisfy the demand for real-world water quality monitoring since environmental water samples are rarely contaminated by a single (type of) pollutant. The parallel setup of four two-stage mini-bioreactor channels with each channel employing *E. coli* bioreporter cells whose bioluminescent emission was linked to a target-specific promoter allowed for simultaneous detection of toxic contaminants causing DNA (strain DPD2794 with the *recA* promoter), membrane (DPD2540 with the *fabA* promoter), and protein (TV1061 with the *grpE* promoter) damage, as well as general cellular toxicity (GC2 with a constitutive *lac* promoter), in a water sample artificially contaminated with 50 ppb mitomycin C, 5 ppm cerulenin, and 100 ppm phenol. A perhaps more interesting aspect of the multichannel system for field applications is that it provides a means of reporting unexpected additive, antagonistic, or synergistic effects of mixture samples to produce a unique toxicity signature of the test sample (Gu and Gil, 2001). In addition, since the test sample can be pumped or injected into

Table 5.1 Examples of *In Situ* Field Applications of Biosensor Devices

Biosensor Design		Sample Matrix	Bioreporter Strain	Target Analyte
Bioreporter Phase	Signal Detection			
Continuous culture in bioreactor (Gu and Gil, 2001; Gu et al., 2001; Kim and Gu, 2005)	Fiber optic cable and luminometer	Surface water and wastewater	*E. coli recA-luxCDABE*	DNA damage
			E. coli fabA-luxCDABE	Membrane damage
			E. coli grpE-luxCDABE	Protein damage
			E. coli katG-luxCDABE	Oxidative stress
			E. coli lac-luxCDABE	General toxicity
Immobilized on a transducer (Ivask et al., 2007)	Fiber optic cable and luminometer	Soil and sediment suspension	*E. coli merR-luxCDABE*	Mercury
			E. coli ars-luxCDABE	Arsenic
Immobilized and freely released *in situ* (Ripp et al., 2000)	Fiber optic cable and photomultiplier	Soil	*P. fluorescens HK44 nahG-luxCDABE*	PAH
Immobilized on a transducer cable (Eltzov and Marks, 2010)	Fiber optic cable and photomultiplier tube	Tap water and wastewater	*E. coli recA-luxCDABE*	DNA damage
			E. coli grpE-luxCDABE	Protein damage
Immobilized on a multiwell cartridge (Roda et al., 2011)	Fiber optic taper and CCD camera	Chemical mixture	*S. cerevisiae* human estrogen receptor regulated wild-type firefly luciferase	Estrogenic compounds
			S. cerevisiae human androgen receptor regulated red-shifted firefly luciferase	Androgenic compounds

Immobilized on a PDMS chip (Elad et al., 2011)	Single-photon avalanche diode	Wastewater	*E. coli recA-luxCDABE*	DNA damage
			E. coli micF-luxCDABE	Oxidative stress
			E. coli arsR-luxCDABE	Arsenic
			E. coli CP38-luxCDABE	General toxicity
Immobilized on a multiwell card (Jouanneau et al., 2012)	CCD camera	Wastewater	*E. coli* pBMerlux	Mercury
Lyophilized in a 96-well plate (Siegfried et al., 2012)	CCD camera	Wastewater	*E. coli* pBMerlux	Mercury
			E. coli pBZntlux	Zinc
			E. coli pBArslux	Arsenic
			E. coli pBCoplux	Copper
Lyophilized in a sealed vial (Siegfried et al., 2012)	Luminometer	Groundwater	*E. coli arsR-luxCDABE*	Arsenic
Lyophilized on a paper strip (Stocker et al., 2003)	Color change	Wastewater	*E. coli arsR-lacZ*	Arsenic
Lyophilized in a sealed vial (Shin et al., 2005)	Color change	Wastewater	*E. coli capR-lacZ*	Phenolic compounds

the bioreactor in an adjustable fashion (i.e., flow rate, step-wise, or continuous flow) and with the bioluminescent signal being continuously transmitted to and processed on the luminometer/computer interface, such a system can be potentially placed at sites near drinking water facilities or wastewater treatment plants for on-line monitoring to serve as an early warning checkpoint. The feasibility of this concept has been tested with surface water (Gu et al., 2001) and wastewater discharged from power plants (Kim and Gu, 2005). In the surface water test, samples collected at different times and locations displayed distinct toxicity signatures, highlighting the importance of integrating multiple channels into the water quality monitoring scheme. Kim and Gu (2005) also demonstrated in their test with power plant wastewater that a feedback-dependent dilution loop might be essential for samples with high toxicity to minimize false negative results.

While the use of continuous bioreporter culture in bioreactors most closely mimics the optimal laboratory bioassay conditions, it is somehow less ideally conceptualized as a field-deployable strategy due to its lack of long-term self-sustainability. The relatively large size of the system, the external input required to maintain continuous bioreactors for stable performance, as well as appropriate handling of the collected waste containing genetically engineered microorganisms prior to disposal represent some of the major hurdles for its implementation in the field. Alternatively, biosensor devices exploiting encapsulated bioreporter organisms immobilized on an easily sterilizable and disposable interface offer the advantages of being compact, low maintenance, and providing a simple postassay procedure. A common design of such a device is to utilize a fiber optic transducer cable coated with layers of encapsulated bioreporter cells on one end with the other end connected to signal detection and processing device (recently reviewed in Eltzov and Marks, 2010). As an example, Ivask et al. (2007) developed fiber optic cables layered with alginate-encapsulated *luxCDABE*-based *E. coli* bioreporter cells to detect mercury and arsenic. Although not as sensitive as nonimmobilized bioreporters, these fiber optic biosensors were shown to be able to detect as little as 0.0026 mg Hg/L, 0.012 mg As(III)/L, and 0.14 mg As(V)/L within a 3 h incubation time and successfully detect the presence of heavy metals in known contaminated soil and sediment samples. However, since multiple steps of pretreatment (drying, homogenization, resuspension, etc.) were necessary for the effective sensing-and-reporting reaction chain, such a biosensor is still restrained to on-site applications for solid matrices such as soil and sediment. On the other hand, samples already in an aqueous phase are more or less immediately ready for analysis as they can be directed to the fiber optic biosensor in a flow-through mode, as demonstrated in the flow-through on-line water toxicity biosensor developed by Eltzov et al. (2009). They immobilized the *E. coli* DPD2794 (*recA-luxCDABE*) genotoxicity bioreporter and the *E. coli* TV1061 (*grpE-luxCDABE*) protein-damaging agent bioreporter on a fiber optic transducer using alginate as an encapsulant and tested their ability to detect toxicant in flowing tap water and wastewater continuously over 24 h. Using mitomycin C (100 μg/L) and *p*-chlorophenol (50 μg/L) as respective positive controls for strains DPD2794 and TV1061, this flow-through system displayed a 1 h response time and an up to 10 h recovery time. The dose-dependent detection range for the DPD2794 and TV1061 biosensor was determined to be 32–2000 μg mitomycin C/L and 10–5000 μg *p*-chlorophenol/L, respectively. Also addressed in this study was the importance of nutrient supplementation in the maintenance and viability of the immobilized bioreporter cells. It was found that a minimum Luria-Bertani media concentration of 7.5% was necessary to maintain the system while still minimizing the potential of biofouling. Although the system maintained functionality during the 24 h tap water test, its long-term use for more polluted wastewater remained challenging due to the loss of cell viability and sensitivity after 20 h.

Instead of directly immobilizing the bioreporter organisms on the surface of signal transducers, they can be encapsulated and incorporated into disposable chips or cartridges that are not in direct physical contact with the optical transducer or other signal capturing device, thereby improving field deployability. For example, Elad et al. (2011) applied a panel of recombinant *luxCDABE*-expressing *E. coli* bioreporters on a multiwell PDMS biochip coupled with SPAD detectors for simultaneous detection of arsenic and toxicants causing genotoxicity and oxidative stress. The PDMS chip featured multiple wells in which the bioreporter culture was immobilized using agar as an encapsulant and prefabricated channels to facilitate on-chip water sample movement. The multiwell platform not only allowed concurrent analysis of various contaminants but also enabled on-chip replications for robust statistical evaluation. In this study, the multichannel biochip was challenged with a 10-day continuous flow of tap water with five simulations of distinct contamination events which consisted of a 2 h pulse spiking of nalidixic acid (genotoxicity), paraquat (oxidative stress), arsenic, or a mixture of the three model compounds. With a surprisingly small volume of bioreporter cells (60 μL/well), this system remained functional during the 10-day exposure period and was able to detect each simulated contamination within 0.5–2.5 h. Similarly, Charrier et al. (2011a,b) developed the Lumisens 3 biosensor system, which consisted of a removable multichannel card housing the immobilized bioluminescent bioreporter cells, fluidics for sample transportation, and a CCD camera placed directly above the card for photon collection. Unlike the SPAD detectors that collect signal on a well-by-well basis (Elad et al., 2011), the CCD camera used in Lumisens 3 and other similar devices collects optical signal on a per-card (or cartridge) basis, thus allowing high-throughput analysis of all channels on the entire card surface. Jouanneau et al. (2012) recently tested the feasibility of the Lumisens 3 system housing a mercury bioreporter for environmental applications. It was demonstrated that the system was capable of responding to daily 100 min spikes of 500 nM mercury with a 6 h response cycle over a 10-day continuous flow setting. Unfortunately, contradictory to the Elad study (Elad et al., 2011), the Lumisens 3 produced unstable bioluminescence levels with an overall variation of approximately 40% during the 10-day period and also suffered from biofouling in long-term analysis.

An interesting conclusion from the Jouanneau et al. (2012) study was that the freeze-drying preservation approach was shown to improve the stability and reproducibility of the system. A Lumisens 4 biosensor device was subsequently developed, similar to that of Lumisens 3, with the exception of the use of lyophilized bioreporter cells entrapped in an easily exchangeable 96-well plate. Even with an initial 30 min reconstitution step for the freeze-dried bioreporters, Lumisens 4 was able to respond to the target toxicant within 1 h as compared to the 6 h for Lumisens 3. In the same 10-day simulated mercury exposure experiment described above, Lumisens 4 exhibited consistent responses to each spike, displaying only a 3% variation in bioluminescence. To further evaluate its potential for field applications, Lumisens 4 with a panel of heavy metal bioreporters was applied to monitor a wastewater effluent spiked with several heavy metal contaminants over a 1-week period. With assistance from the data analysis software Metalsoft (Jouanneau et al., 2011), Lumisens 4 successfully detected each pollution event at the level of specific heavy metals and their concentrations, showing potential for future field applications. Another successful example of field application of lyophilized bioreporters is the portable ARSOlux biosensor arsenic test kit developed by Siegfried et al. (2012). Instead of using a flow-through system, this group packaged freeze-dried bioluminescent arsenic-sensitive bioreporters into single-use vials, which were subsequently transported to Bangladesh for a test campaign of local tube well water arsenic contamination. The full test kit consisted of 160 sealed vials of lyophilized *E. coli arsR-luxCDABE* bioreporter cells (enough for a 10-point calibration curve and 150 tests), a miniature battery-powered luminometer, and other supplies such as syringes and racks. Field testing was performed by injecting water samples into the bioreporter vials, followed by 2 h incubation before reading of the bioluminescent signal using the luminometer. Bioluminescent data were immediately analyzed using a prepared datasheet on a laptop computer and the results were communicated to local residents within the same day. Used vials were then heat inactivated in boiling water to ensure no genetically engineered microorganisms were released into the environment. It was shown that the bioreporter kit produced results comparable to other analytical chemical analyses, resulting in only two false positives and no false negatives with respect to the 10 μg/L total arsenic World Health Organization standard. This campaign highlighted some practical advantages of the bioreporter kit: compact and portable, fast and high-throughput (estimated ~150 tests including sample collection, incubation, and data analysis during an 8 h work day), minimal training for operators, easy waste handling, and low costs for manufacturing, storage, and transportation.

In addition to the use of bioluminescent signal for detection, colorimetric-based biosensors are particularly attractive for deployable field applications since no specialized instrument is required to detect the color change in response to target contaminant exposure. However, because color comparison with the naked eye is somewhat arbitrary, such biosensors better serve as qualitative rather than quantitative sensing tools. Stocker et al. (2003) developed an arsenic test paper strip coated with freeze-dried *E. coli arsR-lacZ* bioreporters which were able to rapidly detect arsenic present at >10 μg/L following a 30-min incubation with the test water sample. Although not capable of distinguishing arsenite concentrations above 0.3 or 0.4 μM, the test trip correctly identified arsenic contamination at >10 μg/L when tested with known contaminated Bangladesh water samples. Similarly, Shin et al. (2005) packaged lyophilized *lacZ*-based phenolic compound-sensitive bioreporters in sealed vials which were directly tested with untreated wastewater samples. Color development was triggered by incubation with 3 mM β-D-galactopyranoside for 5–7 h. To improve measurement precision, a regression curve using absorbance readings previously obtained in the laboratory was also introduced into the on-site testing scheme to enable semiquantitative determination of the measured concentration. Although not capable of producing the same concentration measurements compared to analytical methods, the biosensors successfully identified the presence of phenolic compounds in the wastewater samples, presenting a viable preliminary detection method.

5.5 Field Release of *Pseudomonas fluorescens* HK44 for Monitoring PAH Bioremediation in Subsurface Soils

Pseudomonas fluorescens strain HK44 is a genetically engineered whole cell bioreporter that emits visible light upon exposure to naphthalene, salicylate, or 4-methyl salicylate when physiologically active (King et al., 1990). This property is attributed to the presence of an engineered plasmid, pUTK21, which harbors a transposon-based bioluminescent *luxCDABE* gene cassette derived from *A. fischeri* inserted within its naphthalene degradation pathway (Figure 5.1). Successful applications of strain HK44 as a real-time bioreporter have been clearly established and documented in both on-line and *in situ* studies (King et al., 1990; Burlage et al., 1994; Heitzer et al., 1994; Trogl et al., 2007). To test its biosensing capabilities under actual field conditions, strain HK44 was released in 1996 into semi-contained lysimeter structures (4 m deep × 2.5 m diameter) containing PAH-contaminated soil to serve as a bioremediation process monitoring and control tool

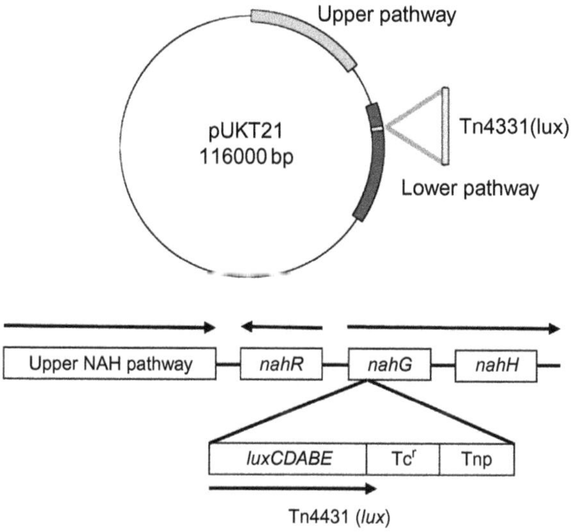

Figure 5.1 The microbial bioreporter *P. fluorescens* HK44 carries plasmid pUTK21 which contains a transposon-based *luxCDABE* gene cassette inserted into the degradation pathway for naphthalene. *Source*: Adapted with permission from Trogl et al., 2012.

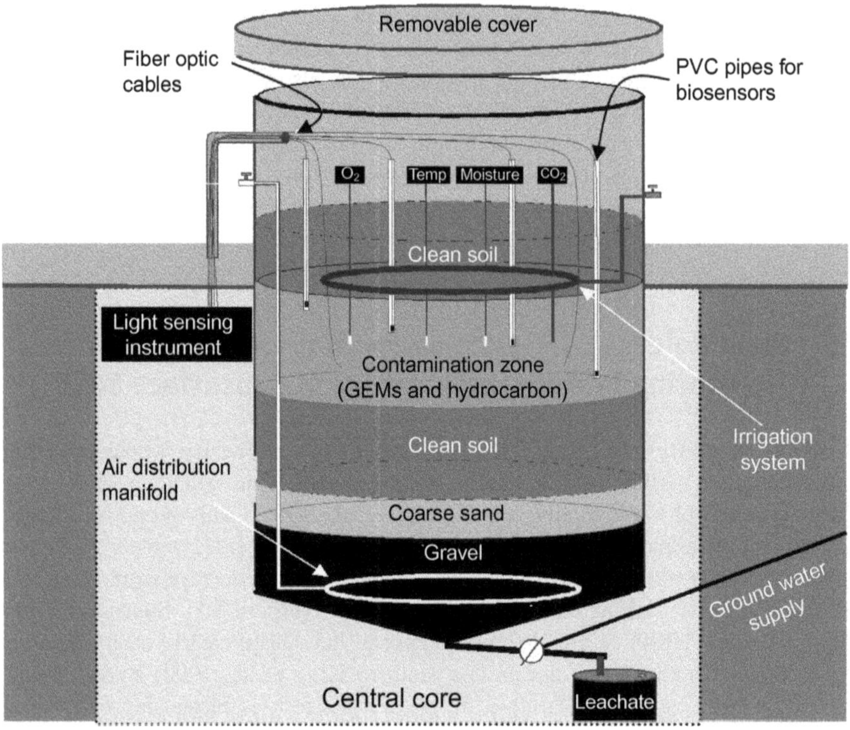

Figure 5.2 Schematic of the lysimeter structure used to contain the *P. fluorescens* HK44 microorganisms during their *in situ* testing. Fiber optic cables terminating at a photomultiplier tube penetrated the soil matrix to record HK44-derived bioluminescence in real time directly from the PAH-contaminated treatment zone of the soil.

(Ripp et al., 2000) (Figure 5.2). A secondary objective was to test the hypothesis that a genetically engineered microorganism could be successfully introduced and maintained for a significant time span under field release exposures. Thus, both the dynamics of bioluminescent light emission and the population status of the HK44 microorganisms were monitored over the initial 2 years of the field release. Bioluminescence from active HK44 cells at different depths of soil lysimeters was measured using a portable multiplex light detection system involving fiber optics and photomultiplier tubes, whereas biosensors, utilizing alginate-encapsulated HK44 cells, monitored for the presence of naphthalene in the vapor phase. Results showed that the field study successfully provided real-time data that reflected naphthalene bioavailability, degradative activity, and optimal degradation conditions for *P. fluorescens* HK44 (Ripp et al., 2000). Thus, the bioremediation field practitioner could potentially use the bioluminescent or other preferred signaling response from an appropriately engineered bioreporter to progressively track the effectiveness of a bioremediation event *in situ*.

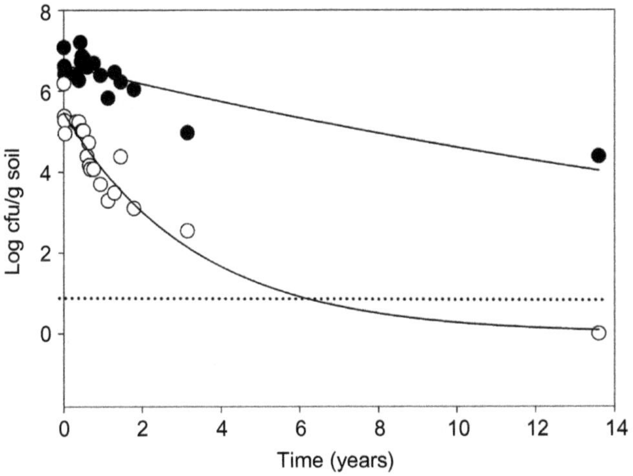

Figure 5.3 Population decay rates of tetracycline resistant microorganisms (○), inclusive of *P. fluorescens* HK44, and total heterotrophic microbial populations (●) over a 14-year duration of sampling of the lysimeter soils. The dotted line indicates the approximate culturable detection limit of 10 cfu/g soil.
Source: Adapted with permission from Layton et al., 2012.

A perhaps more important aspect of this field release study was gaining an enhanced understanding of the long-term survivability of the engineered HK44 microorganisms. It is often assumed that an organism created in a laboratory environment and designed to carry, express, and maintain nonnative genes has reduced fitness, and will therefore quickly perish when released into a highly competitive microbial ecosystem. Indeed, over the first 2 years of this study, HK44 populations were reduced from an initial inoculum of $\sim 1 \times 10^6$ colony-forming units (cfu)/g soil to a final concentration of $\sim 1 \times 10^3$ cfu/g soil. The lysimeter soils were again extensively sampled in 2000, approximately 4 years postrelease and, surprisingly, HK44 cells were successfully revived by selective plate counting even though the selective pressure of the hydrocarbon contaminants that it degraded were depleted at least 2 years prior. Bioluminescence was also detected, indicating that HK44 and the recombinant *luxCDABE* genes were still functionally active. Recovery of *lux* mRNA from the soil complemented these findings. In 2010, approximately 14 years postrelease, the lysimeter soils were again sampled (Figure 5.3). Interestingly, HK44 microorganisms could not be revived using traditional plate counting methods but *luxA*, HK44-specific tetracycline resistant (*tetA*), and naphthalene dioxygenase (*nahA*) genes were detected at low concentrations using more sensitive quantitative polymerase chain reaction (qPCR) methods ($\sim$100–2000 copies/g soil). Thus, the inability to culture viable HK44 cells did not imply their absence from the microbial community, but rather pointed towards the limitation of culture-based techniques to capture less abundant microbial populations. Therefore, more sensitive and robust culture-independent approaches, such as metagenomics, need to be considered and applied to better delineate microbial community structures. Metagenomics can additionally provide information on how the native microbial community changes over time after introduction of non-native engineered microbial populations as well as on the risks associated with horizontal gene transfer of the introduced recombinant DNA. In this work, 16S rDNA and shotgun libraries were made from the pooled DNA of one of the lysimeters and sequenced using a next-generation Roche454 platform (Layton et al., 2012). Although the results obtained from the metagenomic analysis did not confirm the presence of HK44-derived genes after >14 years residency in the lysimeter soils, genetic markers related to HK44 such as the *nah* and *tet* (tetracycline) genes were identified but were more closely related to other strains because of their widespread distribution across different genera. These findings clearly point towards the scientific need for the development of more robust strain-specific molecular markers and/or techniques and development of new methods for confident analysis of data generated from the community analysis. However, accumulation of such information from field studies is necessary and provides a better understanding of the impact of recombinant gene introduction into environmental ecosystems.

Acknowledgments

Portions of this review reflecting work by the authors were supported by the National Science Foundation under Grant Numbers CBET-1159344, CBET-0853780, and DBI-0963854. Ms. Perry was supported by an NSF REU under Grant Number DBI-1156644. Any opinions, findings, and conclusions or recommendations expressed in this material are those of the authors and do not necessarily reflect the views of the National Science Foundation.

References

Baird, S.D., Turcotte, M., Korneluk, R.G., Holcik, M., 2006. Searching for IRES. RNA. 12, 1755–1785.

Ben-Yoav, H., Melamed, S., Freeman, A., Shacham-Diamand, Y., Belkin, S., 2011. Whole-cell biochips for bio-sensing: integration of live cells and inanimate surfaces. Crit. Rev. Biotechnol. 31, 337–353.

Burlage, R.S., Palumbo, A.V., Heitzer, A., Sayler, G., 1994. Bioluminescent reporter bacteria detect contaminants in soil samples. Appl. Biochem. Biotechnol. 45, 731–740.

Charrier, T., Chapeau, C., Bendria, L., Picart, P., Daniel, P., Thouand, G., 2011a. A multi-channel bioluminescent bacterial biosensor for the on-line detection of metals and toxicity. Part II: Technical development and proof of concept of the biosensor. Anal. Bioanal. Chem. 400, 1061–1070.

Charrier, T., Durand, M.J., Jouanneau, S., Dion, M., Pernetti, M., Poncelet, D., et al., 2011b. A multi-channel bioluminescent bacterial biosensor for the on-line detection of metals and toxicity. Part I: Design and optimization of bioluminescent bacterial strains. Anal. Bioanal. Chem. 400, 1051–1060.

Checa, S.K., Zurbriggen, M.D., Soncini, F.C., 2012. Bacterial signaling systems as platforms for rational design of new generations of biosensors. Curr. Opin. Biotechnol. 23, 766–772.

Close, D.M., Ripp, S.A., Sayler, G.S., 2009. Reporter proteins in whole-cell optical bioreporter detection systems, biosensor integrations, and biosensing applications. Sensors. 9, 9147–9174.

Close, D.M., Patterson, S.S., Ripp, S.A., Baek, S.J., Sanseverino, J., Sayler, G.S., 2010. Autonomous bioluminescent expression of the bacterial luciferase gene cassette (*lux*) in a mammalian cell line. PLoS One. 5, e12441.

Cobb, R.E., Chao, R., Zhao, H., 2013. Directed evolution: past, present, and future. AICHE J. 59, 1432–1440.

Craney, A., Hohenauer, T., Xu, Y., Navani, N.K., Li, Y.F., Nodwell, J., 2007. A synthetic *luxCDABE* gene cluster optimized for expression in high-GC bacteria. Nucleic Acids Res. 35, e46.

Dai, C., Choi, S., 2013. Technology and applications of microbial biosensor. Open J. Appl. Biosens. 2, 83–93.

Daniel, R., Almog, R., Ron, A., Belkin, S., Diamand, Y.S., 2008. Modeling and measurement of a whole-cell bioluminescent biosensor based on a single photon avalanche diode. Biosens. Bioelectron. 24, 882–887.

Date, A., Pasini, P., Daunert, S., 2010. Fluorescent and bioluminescent cell-based sensors: strategies for their preservation. In: Belkin, S., Gu, M.B. (Eds.), Whole Cell Sensing Systems I: Reporter Cells and Devices. Springer-Verlag, Berlin, Germany, pp. 57–75.

Diplock, E.E., Alhadrami, H.A., Paton, G.I., 2010. Application of microbial bioreporters in environmental microbiology and bioremediation. In: Belkin, S., Gu, M.B. (Eds.), Whole Cell Sensing Systems II. Springer-Verlag, Berlin, Germany, pp. 189–209.

Elad, T., Belkin, S., 2013. Broad spectrum detection and "barcoding" of water pollutants by a genome-wide bacterial sensor array. Water Res. 47, 3782–3790.

Elad, T., Lee, J.H., Gu, M.B., Belkin, S., 2010. Microbial cell arrays. In: Whole Cell Sensing Systems I: Reporter Cells and Devices, pp. 85–108.

Elad, T., Almog, R., Yagur-Kroll, S., Levkov, K., Melamed, S., Shacham-Diamand, Y., et al., 2011. Online monitoring of water toxicity by use of bioluminescent reporter bacterial biochips. Environ. Sci. Technol. 45, 8536–8544.

Eldridge, M.L., Sanseverino, J., Layton, A.C., Easter, J.P., Schultz, T.W., Sayler, G.S., 2007. *Saccharomyces cerevisiae* BLYAS, a new bioluminescent bioreporter for detection of androgenic compounds. Appl. Environ. Microbiol. 73, 6012–6018.

Eldridge, M.L., Sanseverino, J., de Aragao Umbuzeiro, G., Sayler, G.S., 2011. Analysis of environmental samples with yeast-based bioluminescent bioreporters. In: Ekundayo, E. (Ed.), Environmental Monitoring. Intech Publishers, Rijeka, Croatia.

Eltzov, E., Marks, R.S., 2010. Fiber-optic based cell sensors. In: Whole Cell Sensing Systems I: Reporter Cells and Devices, pp. 131–154.

Eltzov, E., Marks, R.S., Voost, S., Wullings, B.A., Heringa, M.B., 2009. Flow-through real time bacterial biosensor for toxic compounds in water. Sens. Actuators, B. 142, 11–18.

Fleming, J.T., 2010. Electronic interfacing with living cells. In: Belkin, S., Gu, M.B. (Eds.), Whole Cell Sensing Systems I: Reporter Cells and Devices. Springer-Verlag, Berlin, Germany, pp. 155–178.

Fraga, H., 2008. Firefly luminescence: a historical perspective and recent developments. Photochem. Photobiol. Sci. 7, 146–158.

Girotti, S., Ferri, E.N., Fumo, M.G., Maiolini, E., 2008. Monitoring of environmental pollutants by bioluminescent bacteria. Anal. Chim. Acta. 608, 2–29.

Gu, M.B., Gil, G.C., 2001. A multi-channel continuous toxicity monitoring system using recombinant bioluminescent bacteria for classification of toxicity. Biosens. Bioelectron. 16, 661–666.

Gu, M.B., Dhurjati, P.S., VanDyk, T.K., LaRossa, R.A., 1996. A miniature bioreactor for sensing toxicity using recombinant bioluminescent *Escherichia coli* cells. Biotechnol. Prog. 12, 393–397.

Gu, M.B., Gil, G.C., Kim, J.H., 1999. A two-stage minibioreactor system for continuous toxicity monitoring. Biosens. Bioelectron. 14, 355–361.

Gu, M.B., Kim, B.C., Cho, J., Hansen, P.D., 2001. The continuous monitoring of field water samples with a novel multi-channel two-stage mini-bioreactor system. Environ. Monit. Assess. 70, 71–81.

Gupta, R.K., Patterson, S.S., Ripp, S.A., Sayler, G.S., 2003. Expression of the *Photorhabdus luminescens lux* genes (*luxA*, *B*, *C*, *D*, and *E*) in *Saccharomyces cerevisiae*. FEMS Yeast Res. 4, 305–313.

Harms, H., Wells, M.C., van der Meer, J.R., 2006. Whole-cell living biosensors—are they ready for environmental application? Appl. Microbiol. Biotechnol. 70, 273–280.

Heitzer, A., Malachowsky, K., Thonnard, J.E., Bienkowski, P.R., White, D.C., Sayler, G.S., 1994. Optical biosensors for environmental online monitoring of naphthalene and salicylate bioavailability with an immobilized bioluminescent catrabolic. Appl. Environ. Microbiol. 60, 1487–1494.

Hermens, J., Busser, F., Leeuwangh, P., Musch, A., 1985. Quantitative structure–activity relationships and mixture toxicity of organic chemicals in *Photobacterium phosphoreum*: the microtox test. Ecotoxicol. Environ. Saf. 9, 17–25.

Hever, N., Belkin, S., 2006. A dual-color bacterial reporter strain for the detection of toxic and genotoxic effects. Eng. Life Sci. 6, 319–323.

Ivask, A., Green, T., Polyak, B., Mor, A., Kahru, A., Virta, M., et al., 2007. Fibre-optic bacterial biosensors and their application for the analysis of bioavailable Hg and As in soils and sediments from Aznalcollar mining area in Spain. Biosens. Bioelectron. 22, 1396–1402.

Jouanneau, S., Durand, M.J., Courcoux, P., Blusseau, T., Thouand, G., 2011. Improvement of the identification of four heavy metals in environmental samples by using predictive decision tree models coupled with a set of five bioluminescent bacteria. Environ. Sci. Technol. 45, 2925–2931.

Jouanneau, S., Durand, M.J., Thouand, G., 2012. Online detection of metals in environmental samples: comparing two concepts of bioluminescent bacterial biosensors. Environ. Sci. Technol. 46, 11979–11987.

Kim, B.C., Gu, M.B., 2005. A multi-channel continuous water toxicity monitoring system: its evaluation and application to water discharged from a power plant. Environ. Monit. Assess. 109, 123–133.

King, J.M.H., Digrazia, P.M., Applegate, B.M., Burlage, R., Sanseverino, J., Dunbar, P., et al., 1990. Rapid, sensitive bioluminescent reporter technology for naphthalene exposure and biodegradation. Science. 249, 778–781.

Knecht, L.D., Pasini, P., Daunert, S., 2011. Bacterial spores as platforms for bioanalytical and biomedical applications. Anal. Bioanal. Chem. 400, 977–989.

Layton, A.C., Smartt, A.E., Chauhan, A., Ripp, S., Williams, D.E., Burton, W., et al., 2012. Ameliorating risk: culturable and metagenomic monitoring of the 14 year decline of a genetically engineered microorganism at a bioremediation field site. J. Bioremed. Biodegrad.S1:009.

Lopes, N., Hawkins, S.A., Jegier, P., Menn, F.-M., Sayler, G.S., Ripp, S., 2012. Detection of dichloromethane with a bioluminescent (*lux*) bacterial bioreporter. J. Ind. Microbiol. Biotechnol. 39, 45–53.

Meighen, E.A., 1994. Genetics of bacterial bioluminescence. Annu. Rev. Genet. 28, 117–139.

Michelini, E., Roda, A., 2012. Staying alive: new perspectives on cell immobilization for biosensing purposes. Anal. Bioanal. Chem. 402, 1785–1797.

Mitchell, R.J., Gu, M.B., 2004. Construction and characterization of novel dual stress-responsive bacterial biosensors. Biosens. Bioelectron. 19, 977–985.

Nivens, D.E., McKnight, T.E., Moser, S.A., Osbourn, S.J., Simpson, M.L., Sayler, G.S., 2004. Bioluminescent bioreporter integrated circuits: potentially small, rugged and inexpensive whole-cell biosensors for remote environmental monitoring. J. Appl. Microbiol. 96, 33–46.

Premkumar, J.R., Lev, O., Marks, R.S., Polyak, B., Rosen, R., Belkin, S., 2001. Antibody-based immobilization of bioluminescent bacterial sensor cells. Talanta. 55, 1029–1038.

Quillardet, P., Huisman, O., Dari, R., Hofnung, M., 1982. SOS Chromotest, a direct assay of induction of an SOS function in *Escherichia coli* K-12 to measure genotoxicity. PNAS USA. 79, 5971–5975.

Ripp, S., Nivens, D.E., Ahn, Y., Werner, C., Jarrell, J., Easter, J.P., et al., 2000. Controlled field release of a bioluminescent genetically engineered microorganism for bioremediation process monitoring and control. Environ. Sci. Technol. 34, 846–853.

Ripp, S., Daumer, K.A., McKnight, T., Levine, L.H., Garland, J.L., Simpson, M.L., et al., 2003. Bioluminescent bioreporter integrated circuit sensing of microbial volatile organic compounds. J. Ind. Microbiol. Biotechnol. 30, 636–642.

Ripp, S.A., Layton, A.C., Sayler, G.S., 2011. The microbe as a reporter: microbial bioreporter sensing technologies for chemical and biological detection. In: Sen, K., Ashbolt, N.J. (Eds.), Environmental Microbiology: Current Technology and Water Applications. Caister Academic Press, Norfolk, UK.

Roda, A., Cevenini, L., Michelini, E., Branchini, B.R., 2011. A portable bioluminescence engineered cell-based biosensor for on-site applications. Biosens. Bioelectron. 26, 3647–3653.

Ron, E.Z., Rishpon, J., 2010. Electrochemical cell-based sensors. In: Belkin, S., Gu, M.B. (Eds.), Whole Cell Sensing Systems I: Reporter Cells and Devices. Spring-Verlag, Berlin, Germany, pp. 77–84.

Sanseverino, J., Gupta, R.K., Layton, A.C., Patterson, S.S., Ripp, S.A., Saidak, L., et al., 2005. Use of *Saccharomyces cerevisiae* BLYES expressing bacterial bioluminescence for rapid, sensitive detection of estrogenic compounds. Appl. Environ. Microbiol. 71, 4455–4460.

Sanseverino, J., Eldridge, M.L., Layton, A.C., Easter, J.P., Yarbrough, J., Schultz, T.W., et al., 2009. Screening of potentially hormonally active chemicals using bioluminescent yeast bioreporters. Toxicol. Sci. 107, 122–134.

Shacham-Diamand, Y., Belkin, S., Rishpon, J., Elad, T., Melamed, S., Biran, A., et al., 2010. Optical and electrical interfacing technologies for living cell bio-chips. Curr. Pharm. Biotechnol. 11, 376–383.

Shaner, N.C., Steinbach, P.A., Tsien, R.Y., 2005. A guide to choosing fluorescent proteins. Nat. Methods. 2, 905–909.

Shin, H.J., 2011. Genetically engineered microbial biosensors for *in situ* monitoring of environmental pollution. Appl. Microbiol. Biotechnol. 89, 867–877.

Shin, H.J., Park, H.H., Lim, W.K., 2005. Freeze-dried recombinant bacteria for on-site detection of phenolic compounds by color change. J. Biotechnol. 119, 36–43.

Siegfried, K., Endes, C., Bhuiyan, A.F.M.K., Kuppardt, A., Mattusch, J., van der Meer, J.R., et al., 2012. Field testing of arsenic in groundwater samples of Bangladesh using a test kit based on lyophilized bioreporter bacteria. Environ. Sci. Technol. 46, 3281–3287.

Stocker, J., Balluch, D., Gsell, M., Harms, H., Feliciano, J., Daunert, S., et al., 2003. Development of a set of simple bacterial biosensors for quantitative and rapid measurements of arsenite and arsenate in potable water. Environ. Sci. Technol. 37, 4743–4750.

Struss, A.K., Pasini, P., Daunert, S., 2010. Biosensing systems based on genetically engineered whole cells. In: Zourob, M. (Ed.), Recognition Receptors in Biosensors. Springer, New York, NY, pp. 565–598.

Su, L.A., Jia, W.Z., Hou, C.J., Lei, Y., 2011. Microbial biosensors: a review. Biosens. Bioelectron. 26, 1788–1799.

Szymczak, A.L., Vignali, D.A.A., 2005. Development of 2A peptide-based strategies in the design of multicistronic vectors. Expert Opin. Biol. Ther. 5, 627–638.

Toba, F.A., Hay, A.G., 2005. A simple solid phase assay for the detection of 2,4-D in soil. J. Microbiol. Methods. 62, 135–143.

Tolba, M., Minikh, O., Brovko, L.Y., Evoy, S., Griffiths, M.W., 2010. Oriented immobilization of bacteriophages for biosensor applications. Appl. Environ. Microbiol. 76, 528–535.

Trogl, J., Kuncova, G., Kubicova, L., Parik, P., Halova, J., Demnerova, K., et al., 2007. Response of the bioluminescent bioreporter *Pseudomonas fluorescens* HK44 to analogs of naphthalene and salicylic acid. Folia Microbiol. (Praha). 52, 3–14.

Trogl, J., Chauhan, A., Ripp, S.A., Layton, A.C., Kuncova, G., Sayler, G.S., 2012. *Pseudomonas fluorescens* HK44: lessons learned from a model whole-cell bioreporter with a broad application history. Sensors. 12, 1544–1571.

Urbanczyk, H., Ast, J.C., Higgins, M.J., Carson, J., Dunlap, P.V., 2007. Reclassification of *Vibrio fischeri*, *Vibrio logei*, *Vibrio salmonicida* and *Vibrio wodanis* as *Aliivibrio fischeri* gen. nov., comb. nov., *Aliivibrio logei* comb. nov., *Aliivibrio salmonicida* comb. nov and *Aliivibrio wodanis* comb. nov. Int. J. Syst. Evol. Microbiol. 57, 2823–2829.

van der Meer, J.R., Belkin, S., 2010. Where microbiology meets microengineering: design and applications of reporter bacteria. Nat. Rev. Microbiol. 8, 511–522.

Van Dyk, T.K., DeRose, E.J., Gonye, G.E., 2001. LuxArray, a high-density, genomewide transcription analysis of *Escherichia coli* using bioluminescent reporter strains. J. Bacteriol. 183, 5496–5505.

Xu, T., Close, D.M., Sayler, G.S., Ripp, S.A., 2013. Genetically modified whole-cell bioreporters for environmental assessment. Ecol. Indic. 28, 125–141.

6 Mercury Pollution and Bioremediation—A Case Study on Biosorption by a Mercury-Resistant Marine Bacterium

Jaysankar De[a], Hirak R. Dash[b] and Surajit Das[b]

[a]Department of Biotechnology, UNESCO Chair—Life Sciences International Postgraduate Educational Center, Yerevan, Armenia, [b]Laboratory of Environmental Microbiology and Ecology (LEnME), Department of Life Science, National Institute of Technology, Rourkela, Odisha, India

6.1 Introduction

Mercury is the most toxic heavy metal and is unique in character, being liquid at room temperature and highly volatile owing to its low vapor pressure (De et al., 2008). Acute poisoning from mercury is rare, though chronic poisoning is of great concern in many parts of the world. Threats from mercury raise an alarm because it becomes biomagnified, as much as seven times, along the food chain (Hintelmann, 2010). Bioremediation includes technologies that accelerate natural processes for degrading or reducing toxic effects of harmful chemicals and thereby provides a good cleanup strategy for many, if not all, types of pollution (De et al., 2007, 2008). The use of microorganisms has provided a safer and more economic alternative to conventional physicochemical practices. Though indigenous microflora present in the natural environmental conditions carry out natural remediation of contaminants in the environments, the process is very slow and often results in accumulation of toxic substances in the environment, leading to more hazardous effects (Monteiro et al., 1996).

Metabolic processes of the organisms mostly use the contaminants as energy sources resulting in nontoxic or less toxic by-products. Bioremediation technology exploits naturally occurring mitigation processes such as natural attenuation, biostimulation, and bioaugmentation (Rajesh and Bhateria, 2011). This is because microorganisms interact with metals and minerals in both natural and synthetic environments, altering their physical and chemical states and affecting their growth, activity, and survival. It is worth considering whether a microorganism is native or

Microbial Biodegradation and Bioremediation. DOI: http://dx.doi.org/10.1016/B978-0-12-800021-2.00006-6

newly introduced to the site, in understanding and implementing the method of bioremediation. Microorganisms can gain energy by catalyzing chemical reactions, which involves breakdown of chemical bonds and transfer of electrons from the contaminants. Most of the microorganisms are of potential use for the treatment of environmental pollution and many such processes are of commercial value (Lloyd et al., 2005; Gadd, 2010).

Mercury, a naturally occurring element in the environment, is found in air, water, and soil, and can exist in many forms including elemental or metallic mercury, organic and inorganic mercury compounds, etc. It is the only metal to be present in liquid form at normal temperature and pressure. This metal is known to exert greater toxicity at trace concentrations by affecting the nervous system, brain (Azevedo et al., 2012), heart (Frustaci et al., 1999), kidneys (Sommar et al., 2013), lungs, and immune system. Mercury is a neurotoxin and it is clear that adults, children, and developing fetuses are at risk from ingestion of mercury (Zahir et al., 2005). However, the most common route of exposure to mercury is by eating fish containing methylmercury (http://www.who.int/phe/news/Mercury-flyer.pdf). Other methods of exposure include improper handling of mercury containing products and compounds. As mercury has a high residence time, it can cross any geographical boundary causing havoc around the globe, (Marusczak et al., 2011). The major problem arises only when the natural transformation of methylmercury from inorganic mercury occurs more often than the reverse reaction.

Mercury pollution is a major, persistent problem these days and many conventional practices have been employed for its cleanup (Lone et al., 2008; Wang and Chen, 2009; Henneberry et al., 2011). However, the major challenges of these techniques are their costliness, release of harmful by-products, etc. (Stelting et al., 2010). Hence, there has been a shift of attention towards microorganism-mediated mechanisms of mercury bioremediation. Many bacteria have been reported to have novel genetic mechanisms to transform the toxic form of mercury to less toxic forms for cleanup of contaminated environments (Barkay et al., 2003; Barkay and Wagner-Döbler, 2005; De et al., 2008; Chien et al., 2010; Pepi et al., 2013). One of the best-studied bacterial detoxification mechanism is the *mer* operon-mediated mechanism. The bacterial isolates harboring *mer* operons contain many functional genes like *merA*, *merB* (optional in broad-spectrum mercury-resistant bacteria), *merT*, *merP*, and *merF* as well as operator and promoter regions. The *mer* operon is a positively inducible operon and is responsible for the detoxification of mercury by converting the organic form (C−Hg) to the inorganic form (Hg^{2+}, ${Hg_2}^{2+}$), mediated by MerB (organomercurial lyase) followed by conversion of inorganic mercury to elemental/volatile mercury by MerA (mercuric ion reductase) (Osborn et al., 1997; Barkay et al., 2003; De et al., 2008; Dash and Das, 2012). The volatile mercury is relatively less toxic in nature and goes to the atmosphere, causing less damage to the environment. As this can cause secondary air pollution, however, it should be limited to very low mercury contamination and be case specific. However, in several laboratory (Canstein et al., 1999; De et al., in press) and field (Leonhäuser et al., 2013) scale packed bed bioreactors this volatile mercury has been converted back to the metallic form and reused. Thus, microbe-mediated mechanisms of bioremediation of mercury hold more promise and potential than all other contemporary conventional remediation practices.

6.2 The Mercury Cycle in the Environment

The pathway of mercury cycle in the environment is very complex. All the forms of mercury are interconvertable in nature out of which formation of methylmercury draws huge attention due to its high toxicity nature. Finally, mercury ends up in fish, animals, and sediments; in addition, it can go back into the atmosphere by means of volatilization.

As depicted in Figure 6.1, mercury in the aquatic ecosystem comes from the known point sources, with rainfall being the most primary source. Once mercury enters an aquatic system, it can be converted from one form to another inside the ecosystem to form a complex cycle. It may reach the sediments by particle settling

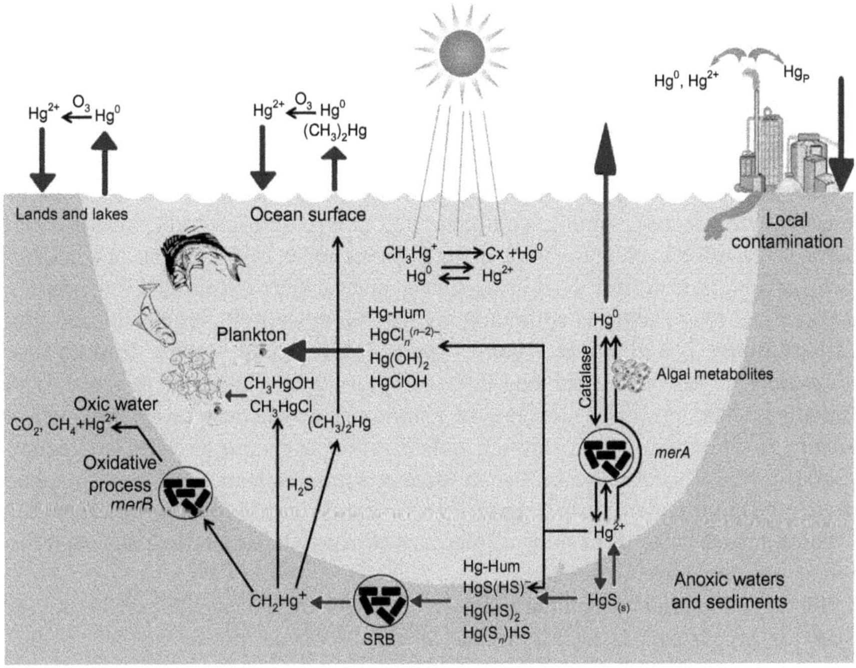

Figure 6.1 Fate of mercury in the environment.
Source: Modified from Barkay et al., 2003.

and later be released by simple diffusion or resuspension. It may enter the food chain or go back to the atmosphere by volatilization. Dissolved organic carbon (DOC) and pH are among the limiting factors to decide the fate of mercury in an ecosystem (Ravichandran, 2004; Gu et al., 2011). Higher acidity and DOC levels enhance the mobility of mercury in the ecosystem, thus making it more likely to enter the food chain.

Though the exact mechanism of mercury entering the food chain is still unknown, most probably it varies from one ecosystem to another. It has been reported that bacteria processing sulfate (SO_4^{2-}), i.e., sulfate-reducing bacteria, take up mercury in its inorganic form and can convert it to the organic form through their metabolic processes. These bacteria are either consumed by higher organisms in the food chain or they may release the methylmercury in the surrounding environment, which is absorbed quickly by planktons. It subsequently accumulates in the next trophic level of the food chain (Barkay and Wagner-Döbler, 2005).

There are various sources of mercury in the environment, both natural and anthropogenic. Natural sources may include volcanoes, natural mercury deposits, and volatilization from the ocean. As mercury is a natural element occurring throughout the solar system, it is found in small concentrations in rocks and it is the main component of mineral cinnabar. Forest fires and burning of fossil fuels such as coal and petroleum also contribute to the natural sources of mercury pollution in the environment. It has been estimated that natural emission from continental sources is approximately 1000 tons. In the preindustrial era, the natural emission of mercury was estimated to be 600 tons, whereas, due to anthropogenic load, the deposited amount has increased to around 2000 tons in the postindustrial era (Wangberg et al., 2010).

Currently, the level of mercury is increasing at a greater pace due to discharges from the hydroelectric, mining, and paper and pulp industries. Other major contributors to the global mercury pool include incineration of municipal and medical wastes and emissions from coal-using power plants. Mercury emitted into water or air can travel long distances without being degraded due to its high residence time, and hence there is a global reservoir of mercury circulating worldwide at any particular time. It has been estimated that the annual global input of mercury to the reservoir is 4900 tons. The recycling of anthropogenic mercury raises the levels of mercury in the environment through volatilization from the terrestrial or aquatic ecosystem (Selin, 2009). According to some recent models on the flow of mercury through the environment, it is suggested that natural sources account for about 10% of the estimated 5500–8900 tons of mercury currently being emitted and reemitted to the atmosphere from all sources.

Mercury is a priority pollutant and is of global importance due to its adverse effects on ecosystems as well as human health. Concentrations of global mercury have increased approximately threefold due to various anthropogenic activities (Figure 6.2) and the world's oceans are the major reservoirs for its deposition (Mason et al., 2012).

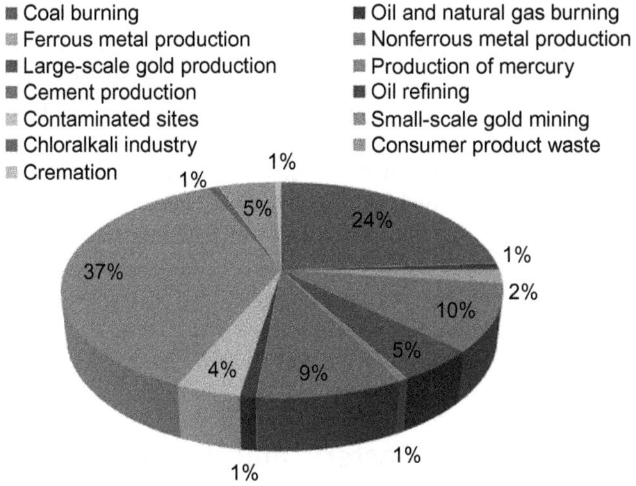

Figure 6.2 Relative contribution of anthropogenic sources to the global mercury content (UNEP, 2013).

6.3 Health Effects Associated with Mercury Contamination

Although acute poisoning by mercury is rare, chronic mercury poisoning poses a severe threat to large areas of the world today. Chronic mercury poisoning causes neurological symptoms, initially known as Minamata disease, and defects in the fetus (Bernhoft, 2012). The toxic effect of mercury is mostly dependent upon its route of exposure and its chemical form. Methylmercury is the most toxic among all the forms of mercury, affecting the immune system, altering the genetic and genetic system, and causing damage to the nervous system including coordination and the senses of touch, taste, and sight (Hutton, 1987; Young et al., 2001). It can affect the developing embryo at a higher rate i.e. 5-10 fold higher to that of the adults. Elemental mercury, when inhaled for a long period, causes tremors, gingivitis, and excitability. Ingestion of other common forms of mercury such as Hg^{2+} damages the gastrointestinal tract. Human beings suffer the most harm from methylmercury contamination by consuming contaminated fishes and other animals present in the top level of the aquatic food chain. It has been estimated that more than 60,000 children born each year are at risk for adverse neurodevelopmental effects due to *in utero* exposure to methylmercury. Hence, the population at highest risk is the offspring of women who consume large amounts of fish and seafood.

Mercury is of environmental concern because it biomagnifies in the food chain by up to 7 orders of magnitude, resulting in high concentrations in top predators such as fish and polar bears (Hintelmann, 2010; Sonne, 2010). Mercury affects not only humans but also other aquatic wildlife in a similar fashion. It has been shown that concentrations of mercury that are toxic in the lab and controlled lab studies are common in certain environments. In a study from Wisconsin, reductions in loon chick production have been found in lakes where mercury concentrations in eggs exceeded concentrations that are toxic in laboratory studies (Meyer, 2005). At dietary mercury concentrations that are typical of parts of the Everglades, the behavior of juvenile great egrets is affected. Studies with mallards, great egrets, and other aquatic birds have shown that protective enzymes are less effective following exposure to mercury. Analyses of such biochemical indicators indicate that mercury is adversely affecting diving ducks from San Francisco Bay, herons and egrets from the Carson River in Nevada, and heron embryos from colonies along the Mississippi River. Finally, other contaminants also affect the toxicity of mercury. Methylmercury can be more harmful to bird embryos when selenium, another potentially toxic element, is present in the diet (USGS, 2000).

6.4 Mercury-Resistant Bacteria and Mechanisms of Resistance

As a response to toxic mercury compounds globally distributed by geological and anthropogenic activities, microbes have developed a surprising array of resistance mechanisms to overcome Hg toxicity. An extensively studied resistance system based on clustered genes in an operon (i.e., *mer*) allows bacteria to detoxify Hg^{2+} into volatile mercury by enzymatic reduction (Komura et al., 1971; Summers, 1986; Misra, 1992; Silver, 1996; Osborn et al., 1997; Barkay et al., 2003). It appears that bacterial resistance to mercury is an ancient mechanism, probably acquired even before anthropogenic usage of mercury. Since the same biotransformations that constitute the Hg biogeochemical cycle can take place inside the human body, understanding its external transformations and transport processes will help in figuring out which of these processes can exacerbate or ameliorate Hg toxicity in humans (Barkay et al., 2003).

Bacteria carry out chemical transformations of mercury by oxidation, reduction, methylation, and demethylation, and develop resistance determined by the extra chromosomal genetic material, including plasmids. (Silver and Misra, 1984). Earlier, there were many reports on the mercury-resistant isolates from industrial wastes, effluents, oceans, dental amalgams, etc. Though there are many proposed mechanisms of mercury resistance, *mer* operon-mediated mechanism has been well studied and elucidated since the discovery of bacterial mercury resistance phenomenon. (Summers and Lewis, 1973) to date. Five different types of resistance mechanisms have been reported in bacteria for mercury which include (i) reduced uptake of mercuric ions, (ii) demethylation of methylmercury followed by conversion to mercuric sulfide compounds, (iii) sequestration of methylmercury, (iv) methylation of mercury, and (v) enzymatic reduction of Hg^{2+} to Hg^0 (Barkay et al., 2003).

6.4.1 Mer *Operon-Mediated Mercury Resistance*

Mercury-resistant determinants have been found in a wide range of Gram-negative and Gram-positive bacteria isolated from different environments. They vary in number and identity of genes involved, and are encoded by the *mer* operon located on plasmids (Summers and Silver, 1978; Brown et al., 1986; Griffin et al., 1987; Radstrom et al., 1994; Osborn et al., 1997), chromosomes (Wang et al., 1987, 1989; Inoue et al., 1989, 1991; Iohara et al., 2001), transposons (Misra et al., 1984; Kholodi et al., 1993; Liebert et al., 1997, 1999; Hobman et al., 1994), and integrons (Liebert et al., 1999). Gram-negative bacterial *mer* operons have been studied more extensively than those of Gram-positive bacteria although Gram-positive bacteria have the same set of genes arranged in a similar order. The most widely studied eubacterial mercury resistance is the reduction of the highly reactive cationic form of mercury to the volatile form, which is a relatively inert form of monatomic mercury vapor. MerA, the mercuric ion reductase, is the central protein for this reduction, which is encoded by the *mer* loci. This is a cytosolic flavin disulfide oxidoreductase (a homodimer of around 120 kDa) and uses NADPH as its reductant. Even though the exact complement of genes/proteins varies in these loci, their expression is always regulated by a Hg^{2+}-sensing transcriptional regulator located in the cytosol, and all of them utilize an integral membrane protein (e.g., MerT, MerP) to facilitate uptake of Hg^{2+} into the cytosol where MerA acquires and reduces it to elemental mercury (Hg^0). MerA has a flexible amino-terminal domain and is homologous to the small periplasmic mercury-binding protein MerP. MerP uses its two cysteine residues to displace nucleophiles like Cl^- to which Hg^{2+} is coordinated. Subsequently MerP exchanges Hg^{2+} with two cysteines of the first three transmembrane helices of MerT, while the transfer of Hg^{2+} from MerT to cytosol may involve the second cysteine pair lying on the cytosolic side of the inner membrane (Osborn et al., 1997). Products of *mer*C and *mer*F, both membrane proteins, act as agents in the mercury transport system (Kusano et al., 1990; Wilson et al., 2000). In some cases, the *mer* operon contains other functional genes. The *mer*G product provides phenylmercury resistance, presumably by reducing the in-cell permeability to phenylmercury (Kiyono and Pan-Hou, 1999).

The heart of *mer*-mediated mercury resistance is mercuric ion reductase (MerA), which catalyzes the conversion of thiol-avid Hg^{2+} to volatile Hg^0, which lacks significant affinity with any other ligand of functional groups. This enzyme is energy dependent in the form of NADPH and is found in the cytoplasm. However, in some bacterial isolates they have been reported for their resistance towards organic mercury compounds and, hence, they are known as broad-spectrum mercury-resistant bacteria. In addition to the other structural and functional genes, these bacteria harbor an additional gene known as *merB*, which codes for organomercurial lyase (MerB) (Figure 6.3). Characterization of MerB encoded by R831 revealed it to have a molecular mass of 22.4 kDa, while it lacks any cofactor and is a monomer. This enzyme is responsible for simply cleaving the C—Hg bond to release Hg^{2+} to the cytoplasm, which is further taken care of by the protein coded by *merA* to form the Hg^0 (Melnick and Parkin, 2007). Since the transposons carrying *mer* operons have been identified from both clinical and environmental bacteria, it is considered that the horizontal transfer of the mobile genetic elements may contribute to the worldwide distribution of *mer* operons (Chien et al., 2010).

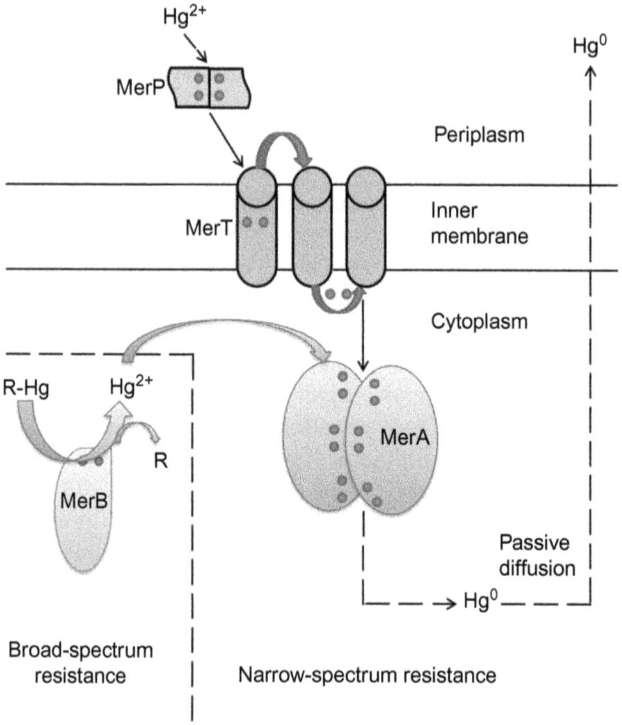

Figure 6.3 A schematic presentation of narrow- and broad-spectrum mercury resistance in bacteria.
Source: Modified from Dash and Das, 2012.

6.4.2 Regulation of mer Operon

The bacterial metalloregulator MerR is the directory of an eponymous family of regulatory proteins that controls the transcription of a set of genes (*mer* operon) conferring mercury resistance in many bacteria. Expression of the *mer* operon is regulated by the products of *merR* and *merD*, and is inducible by Hg^{2+}. The homodimeric product of *merR* represses operon expression in the absence of an inducer and activates transcription in the presence of an inducer. The product of *merD* coregulates expression of the operon (Misra, 1992; Silver and Phung, 1996). MerR, the metal responsive regulator protein, is a transcriptional activator of the *mer* operon. This apparently homodimer protein binds to a region of dyad symmetry called MerO (operator) which is located just upstream of the *merT* gene. MerO lies between -10 and -35 RNA polymerase recognition sites for the promoters of the functional genes and MerR serves as a negative autoregulator. However, in the presence of Hg^{2+}, it binds to MerR, provokes an allosteric change in the protein, and propagates to the operator region of DNA, which in turn results in the improved access of RNA polymerase to the transcription start site. Another negative regulator protein present

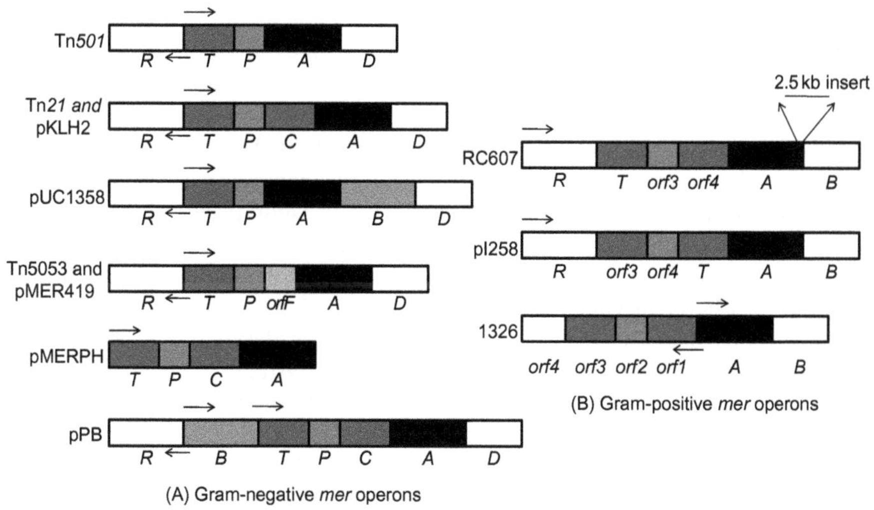

Figure 6.4 Schematic presentation of *mer* operons in both Gram-negative and Gram-positive bacteria.
Source: Modified from Osborn et al., 1997.

in the bacterial system includes MerD. When the intracellular Hg^{2+} concentration decreases due to continuous volatilization by MerA, *merD* is expressed to negatively regulate the *mer* operon system (Brown et al., 2003).

6.4.3 Genetic Diversity of mer Genes Within an Operon

Most of the studies on genetic diversity of mercury-resistant genotypes are based on the DNA–DNA hybridization analysis. The pioneer study in this regard was conducted by Barkay et al. (1985), who demonstrated a 2.6 kb Tn*21 mer* operon in Gram-negative mercury-resistant bacteria that is absent in Gram-positive mercury-resistant bacteria. Thus, there exists a significant variation in DNA sequences between *mer* genes of Gram-negative and Gram-positive bacteria.

Tn*21* and Tn*501* were reported to be the most abundant mercury-resistant operon in Gram-negative bacteria. The *merC* gene also plays an important role in the *mer* operon; in the case of Gram-negative bacteria, *mer* operons can be classified on the basis of the presence or absence of the *merC* gene (Figure 6.4). In the case of Gram-negative bacteria, *merC* genes are found more commonly in isolates from mercury-polluted environments than those of nonpolluted environments, thus confirming the enhanced resistance associated with the harboring of a *merC* gene. However, in many cases mercury-resistant genotypes, most of the isolates do not hybridize with the *mer* probes that are significantly different from the normal characterized *mer* sequences, and they harbor certain alternative mechanisms of resistance to Hg^{2+} ions. As 0.001–3% of the total bacteria can be cultured in laboratory conditions, the unculturable approach reports that the Tn*501* homologous sequences are more common in fresh water communities than in marine communities. Restriction fragment length polymorphism (RFLP) analysis of the *mer* sequences revealed the same result, indicating that the Tn*21* and Tn*501* are the most commonly occurring *mer* operons in Gram-negative mercury-resistant bacteria.

Although the presence of mercuric reductase is essential for enzymatic detoxification—and hence resistance to inorganic mercury—expression of the *merA* gene (and *merR*) has been reported in a high proportion of Gram-positive environmental strains sensitive to mercury, suggesting the presence of nonfunctional *mer* operons in which the mercury transport genes are either absent or nonfunctional (Bogdanova et al., 1992). Microbial volatilization mechanisms without the involvement of *mer* operon-encoded mercuric reductase enzyme have been discovered (Takeuchi and Sugio, 2006; Wiatrowski et al., 2006; Al Mailem et al., 2011). The most remarkable has been the discovery of a constitutive mechanism for volatilization of mercury in dissimilatory metal-reducing bacteria (e.g., *Shewanella oneidensis* MR-1, *Geobacter sulfurreducens* PCA, and *Geobacter metallireducens* GS-15), which is not linked to reduction by the MerA and is dependent on iron (Wiatrowski et al., 2006). Although the physical arrangement of the *mer* operons may vary, all contain the essential genes but surprisingly, only limited studies have attempted to characterize mercury resistance at the molecular level in marine bacterial isolates (Barkay et al., 1989; Rasmussen and Sorensen, 1998). Barkay et al. (1989) found that only 12% of culturable MRB from estuarine environments hybridized to the *mer* (Tn*21*) probe, suggesting that such MRB from the marine environment encodes novel *mer* genes or other mechanism(s) that provide Hg resistance (Reyes et al., 1999; De et al., 2008). It has been proposed that *mer* is an ancient system, which evolved at a time when levels of available mercury kept rising in natural environments, possibly because of increased volcanic activity (Osborn et al., 1997). Barkay et al.(2010) suggested a thermophilic origin of the mercury resistance operon.

6.4.4 Tolerance to Mercury by Biosorption

Mercuric mercury complexes with organic and inorganic ligands and is easily adsorbed to surfaces of particulates, owing to its high reactivity and affinity to thiol groups (Barkay, 2000) as well as adsorption of Hg(II) to cell surface components (Francois et al., 2012). Mercury bioavailability therefore plays a crucial role in the evaluation of microbial resistance levels. There is some tolerance towards mercury owing solely to unspecific sequestration by cell walls and lipopolysaccharide (LPS) layers (Langley and Beveridge, 1999). All types of biological materials, including many bacterial cells (Noghabi et al., 2007; Cain et al., 2008), have been reported to sequester mercury by live or dead biomass consisting of passive immobilization procedures. These practices rely on different physicochemical mechanisms such as adsorption, surface complexation, ion exchange, or surface precipitation (Francois et al., 2012). This procedure appears to be fast and is a metabolism-independent process that allows the use of dead biomass. Both cell walls as well as secreted extracellular polymeric substances (EPS) participate in this process. Bacteria have a greater capacity to adsorb mercury from solution than any other form of life due to their high surface-to-volume ratio (Beveridge, 1989). However, many bacterial isolates have been characterized for mercury biosorptions, although the mechanism is yet to be understood clearly.

6.5 Mercury-Resistant Bacteria in Bioremediation

Exploitation of *mer* operon-mediated mechanisms for the bioremediation of contaminated wastes dates back many decades. Recently, several approaches using engineered and naturally occurring *mer* operons have been widely applied for treatment of contaminated wastewaters and environments (reviewed in Dash and Das, 2012). Packed bed bioreactors inoculated with several types of naturally isolated mercury-resistant bacteria have been reported to be used widely for treatment of wastewater from chloralkali industries. On-site mercury cleanup by technical-scale bioreactors has been optimized to decrease the mercury load of wastewater by up to 90% (Canstein et al., 2002a). Additionally, mixed culture biofilms have been reported to have higher mercury retention efficiency than industrial chloralkali wastewater (Canstein et al., 2002b). Though many conventional practices are followed for the removal of mercury from contaminated wastes, the use of mercury-resistant bacteria has many major advantages. Hence, the use of bacteria for metal removal from contaminated sites is a promising technology despite certain limitations like production of large volumes of mercury-loaded biomass.

Mercury-resistant bacteria used in bioremediation acquire mercury-resistant mechanisms naturally (Cursino et al., 2000). The bacteria in use for bioremediation should possess metal processing mechanisms like biosorption, intracellular assimilation, uptake and reflux, immobilization, complexation, and precipitation (Stelting et al., 2010). The biochemical and molecular basis of plasmid- or transposon-borne mercurial and organomercurial resistance are of utmost importance, as these substances constitute a natural strategy for the detoxification of mercury-contaminated sites (Silver et al., 1994). Because of its high levels of efficacy and specificity, the microbial pathway encoded by the *mer* operon is well recognized for the enzymatic reduction of mercury ions to a water-insoluble metallic form. Inside the cell, Hg^{2+} is reduced to Hg^0, which passively diffuses out of the cell without any energy expenditure (Canstein et al., 1999). Therefore, the bacterial biomass acts as a catalyst, without the accumulation of a large volume of contaminated biomass. Despite the great promise of mercury-resistant bacteria in bioremediation of mercury-contaminated sites, use of the *mer* operon for the treatment of industrial waste is scarce (Canstein et al., 1999). Most reports on this technique are restricted to laboratory studies. *Deinococcus geothermalis* was examined for the microbial bioremediation of mercury in radioactive wastes (Brim et al., 2000). Synergistic use of combined bacterial-plant processes has also been reported for the cleaning of mercury from the environment. A metal-resistant *Cupriavidus metallidurans* strain was examined for its potential to bioremediate mercury (Rojas et al., 2011), while hydrocarbon-utilizing haloarchaea was reported to remediate mercury under hypersaline conditions. Microbial surfactant has also been reported to be useful in bioremediation of mercury. The mercury bioremediation potential of engineered bacteria expressing metallothionein and polyphosphate kinase has been evaluated (Ruiz et al., 2011). Tributyltin-resistant isolates of *Enterobacter cloacae* and *Alkaligenes faecalis* have been found promising in bioremediation of high concentrations of methylmercury even in the absence of the primary nutrient glycerol (Adelaja and Keenan, 2012). A recent report on the biofilm-forming, nonpathogenic marine bacterial isolate *Pseudomonas putida* SP-1, capable of volatilizing 89% of mercury, has confirmed the efficiency of mercury-resistant marine bacteria in bioremediation (Zhang et al., 2012). However, the gap between laboratory research and field application is bridged to some extent by using case studies (Gluszcz et al., 2008; Randall et al., 2013; Pepi et al., 2013). Thus, mercury-resistant bacteria can be used for enhanced bioremediation of highly contaminated sites.

Anaerobic microbes, such as sulfate-reducing bacteria with their capacity to produce sulfide, could potentially be used to precipitate Hg(II) as insoluble cinnabar (HgS). Unfortunately, these bacteria have dual and opposite roles, as they methylate mercury in anoxic environments (Wagner-Döbler, 2013). For example, the *Geobacter* species, which volatilize mercury, could also produce methylmercury (Kerin et al., 2006). It has recently been shown that elemental mercury can be both oxidized and reduced by natural organic matter in anoxic environments (Zheng et al., 2012), which further makes the option of using anaerobes in mercury bioremediation a risky approach. Thus, the associated danger with mercury transformations under anoxic conditions needs careful inspection in any intended bioremediation application.

6.6 Bioaccumulating Mercury-Resistant Marine Bacteria as Potential Candidates for Bioremediation of Mercury: Case Study

6.6.1 Background Knowledge

Bacteria are able to adapt to heavy metals in contaminated environments by developing specific mechanisms of resistance. They develop the endogenous capability to regulate their physiological functions to overcome the toxic effects. Thus, in order to use them for bioremediation purposes, these microbes should have the potential to convert the toxic form of the metal to its nontoxic form(s) or to remove the toxic form from the contaminated environment. Mercury is a heavy metal with extreme toxicity, and, owing to its ability to biomagnify and its high residence time in the environment, it causes havoc for other living organisms as well as the ecosystem. Technologies for treating mercury-polluted environments have been a major concern over the last couple of decades (Williams and Silver, 1984; Nakamura et al., 1990; Brunke et al., 1993; Canstein et al., 1999; Canstein, 2000; Chang et al., 1997, Chen et al., 1998; Essa et al., 2002; Wagner-Döbler, 2003; Deckwer et al., 2004). Common methods to remove Hg^{2+} from contaminated water are mostly based on sorption to materials such as ion exchange resins (Osteen and Bibler, 1991; Ritter and Bibler, 1992). Removal of mercury in a laboratory test reactor using mercury-resistant bacteria was first reported in 1984 (Williams and Silver, 1984). Since then, various attempts have been made to improve this technology. However, attention to bioremediation began in earnest in the 1990s (Summers, 1992). One of the initial efforts to retain mercury in bacterial bioreactors was made by Brunke et al. (1993). Reduction of Hg by MRB as the best such mechanism for its removal from chloralkali waste has been demonstrated (Canstein et al., 1999; Wagner-Döbler, 2003). Biosorption, another biological method involving adsorption of metals into the algal or bacterial biomass (either dead or alive), has been inexpensive and also promising (Chang and Hong, 1994; Mulligan et al., 2001), but is applicable only to low concentrations of metals in water (Chen et al., 1998). Saouter et al. (1994, 1995) reported their preliminary investigation on using Hg^{2+}-reducing strains to decontaminate a polluted pond in Tennessee. Diels et al. (1995) reported bioprecipitation of metal ions on the cell surface as a removal mechanism. Biosorption using natural (Volesky and Holan, 1995) or recombinant microbial biomass (Pazirandeh et al., 1995) has been also tried successfully. Chen and Wilson (1997a,b) used a genetically engineered *Escherichia coli*-expressing mercury transport system and metallothionein for removing Hg. Chang and Law (1998) developed a detoxification process using *Pseudomonas aeruginosa* PU21 in batch, fed batch, and continuous bioreactor systems. Canstein et al. (1999) demonstrated the removal of mercury from chloralkali electrolysis wastewater by a mercury-resistant *P. putida* strain. These laboratory-scale reactor results formed the basis for the development of a technical-scale bioreactor that decontaminated mercury-polluted chloralkali wastewater *in situ* (Wagner-Döbler et al., 2000, 2003). Deckwer et al. (2004) have described a three-phase fluidized bioreactor system. Sorption by biomass as well as precipitation as HgS are the main other biological methods beside the biofilm based bioreactor system for detoxification and/or removal of toxic mercury. In spite of large initial interest in using sorption as a removal technology, no initiative has been taken for its commercial application (Gadd, 2010). Among many efforts have been undertaken for the use of engineered microorganisms for mercury accumulation in the cytoplasm, the most noteworthy is the development of a transgenic bacterium expressing metallothionein and polyphosphate kinase genes (*pkk*) for accumulation of Hg^{2+} (Ruiz et al., 2011). Kiyono and Pan-Hou (2006) have also reported genetically engineered *E. coli* for simultaneous expression of mercury transport proteins and mercury lyase enzymes for intracellular expression of mercury. In an earlier effort, a genetically engineered *E. coli* strain with a Hg^{2+} transport system and metallothionein has been used to bioaccumulate mercury from wastewater (Deng and Wilson, 2001). High concentrations of NaCl, which cause less bioavailability of metal, usually inhibit the volatilization of mercury by bacteria (Selifonova and Barkay, 1994; Barkay et al., 1997). As most mercury-containing wastewater contains salt, screening of salt-tolerant bacteria with inherent properties of mercury bioaccumulation in its native form will be of great importance. Therefore, the current work is aimed at screening and isolating mercury-resistant marine bacteria by bioaccumulation processes for efficient and sustainable bioremediation of mercury from contaminated environments.

6.6.2 Experimental Procedures

6.6.2.1 Sampling, Isolation, and Selection of Bacteria

A water sample was collected from the Odisha coast of the Bay of Bengal along four study sites, i.e., Chilika (19°44.582′N and 85°12.768′E), Bhitarakanika (20°44.33′N and 86°52.06′E), Gopalpur (19°19.218′N and 84°57.730′E), and Rushikulya (19°22.647′N and 85°03.165′E). The water was serially diluted and plated on to a seawater nutrient agar (SWNA) medium (peptone 5 g, yeast extract 3.0 g, agar 15 g, aged seawater 500 mL, and deionized water 500 mL; pH 7.5 ± 0.1). The SWNA was supplemented with 10 ppm of Hg as $HgCl_2$ and incubated at 30°C for 48 h. Enumeration of mercury-resistant marine bacteria was conducted, followed by the isolation of pure cultures from the individual colonies with visible differences in colony morphology. The isolates were subjected to determination of minimum inhibitory concentration (MIC) using a microbroth dilution technique (CLSI, 2006). The isolate designated as BW-03 showed the highest resistance to mercury and was selected for further study.

6.6.2.2 Molecular Identification of the Isolate

The mercury-resistant marine bacterium BW-03 selected for further study was identified by sequencing of the 16S rRNA gene fragment. DNA was extracted from the cells grown in SWN broth medium using a bacterial genomic DNA extraction kit (QiaGen), and the amplification reaction was performed using the universal 16S rDNA primers (Sigma-Aldrich) of 27F (5′-AGAGTTTGATCMTGGCTCAG-3′) and 1492R (5′-ACGGCTACCTTGTTACGA-3′) in a thermal cycler (BioRad). A polymerase chain reaction was performed in 50 μL volumes containing 2 mM $MgCl_2$, 2.5 U Taq polymerase (Sigma-Aldrich), 100 μM of each dNTP, 0.2 μM of each primer, and 3 μL template DNA. The PCR program consisted of denaturation at 96°C for 5 min followed by 30 cycles of 95°C for 15 s, 49°C for 30 s, and 72°C for 1 min and a final extension at 72°C for 10 min. Amplified DNA were purified by a PCR purification kit (Sigma-Aldrich) following manufacturer's instructions and finally quantified by Nano drop (Eppendorf). Sequencing of the cleaned PCR product was carried out by Xcelris Genomics, India, using the Sanger method. Sequence data were compiled, and a consensus sequence was obtained by using a BioEdit (7.0.5.3) program and examined for sequence homology with the archived 16S rDNA sequences from GenBank at www.ncbi.nlm.nih.gov/nucleotide, employing BLASTN. Multiple alignments of sequences were performed with the ClustalX (1.83) (Thompson et al. 1997). A phylogenetic tree was constructed using the neighbor-joining DNA distance algorithm (Saitou and Nei 1987) using Mega 5. The resultant tree topologies were evaluated by bootstrap analysis (Felsenstein 1985) of neighbor-joining data sets based on 1000 resamplings.

6.6.2.3 GenBank Submission

The partial sequence of the 16S rRNA gene of BW-03 was submitted at NCBI GenBank, and the assigned accession number is KF241550.

6.6.2.4 *Amplification of* merA *Gene*

The conserved region of *merA* gene was amplified using the lysates as a template. PCR was carried out using the primer set of F1 merA-5′TCGTGATGTTCGACCGCT3′ and F2 merA-5′TACTCCCGCCGTTTCCAAT3′ (Sotero-Martins et al., 2008). The amplification reactions were performed in a total volume of 20 μL by using a thermal cycler (BioRad). The PCR mixture contained 1 U/μL Taq polymerase (Sigma-Aldrich), 1 × Enzyme buffer, 200 μM of each dNTP (Sigma-Aldrich), 1.25 mM $MgCl_2$, 0.5 μM of each primer. The optimized amplification conditions included a predenaturation step at 94°C for 1 min followed by 30 cycles of 94°C for 1 min, 55°C for 1 min, an extension step at 72°C for 1 min, and final extension at 72°C for 7 min. The PCR products were analyzed using agarose gel electrophoresis (1.5%) and visualized in a gel documentation system (BioRad). A sensitive strain of *E. coli* (DH5α) was used as the negative control while *Bacillus thuringiensis* PW-05 was taken as a positive control for this study (Dash et al., 2013).

6.6.2.5 *Mercury Biosorption Study*

Taking advantage of the higher affinity of mercury towards sulfur molecules (Flora et al., 2008), H_2S assay was performed to determine the biosorption of mercury by the isolated bacterial cell mass. Briefly, the pure culture of the isolate was grown for 48 h in SWN broth supplemented with 50 ppm of $HgCl_2$ at 35°C. The cell mass was harvested by centrifugation at 6000 rpm and was washed twice with 0.07 M phosphate buffer saline (PBS) (pH 7.0) containing 0.5 mM EDTA, 0.2 mM magnesium acetate, and 5 mM sodium thioglycolate, and the cell mass was subjected to H_2S gas for the development of black color.

6.6.2.6 *Biofilm Development*

The ability of BW-03 to form biofilm was studied by using a microtiter plate assay and by a tube assay with slight modifications (Jain et al., 2013) both in the presence and absence of mercury. Isolates from fresh SWN agar plates were inoculated in Luria Bertani (LB) broth and incubated for 18 h at 37°C in static conditions. This overnight culture was diluted 100 times, and individual wells of sterile, polystyrene, 96-well flat bottom tissue culture plates were filled with 200 μL aliquots of the above diluted culture and broth without culture (control). The plates were incubated for 72 h at 37°C to observe biofilm formation. After incubation, the content of each well was gently removed by slightly tapping the plates, and the wells were washed with PBS (pH 7.3) to remove the free-floating planktonic bacteria. The plates were then stained with a 0.1% (w/v) crystal violet solution. Excess stain was washed off thoroughly with 95% ethanol, and the plates were kept for drying. Optical density (OD_{595}) of the wells was determined with a microplate reader (Victor X3 2030 multilabel reader, Perkin-Elmer).

6.6.2.7 Modification of Functional Groups

The function of the chemical groups present in bacterial cells was identified by FTIR spectroscopy. Bacteria cell samples grown with and without $HgCl_2$ (50 ppm) were harvested by centrifugation at 8000 *g* for 10 min and mixed with 2% KBr. The mixture was compressed into translucent sample disks and fixed in the FTIR spectrometer (Perkin-Elmer RX I) for analysis.

6.6.2.8 Determination of Mercury Removal Potential

BW-03 was inoculated in SWN medium containing 50 ppm of Hg as $HgCl_2$ and grown at 30°C for 48 h. Both the cell pellet and the supernatant were collected by centrifugation, and the cell pellet was resuspended in a lysis buffer (100 mM NaH_2PO_4, 10 mM Tris–Cl, 8 M urea (pH 8.0) for 1 h) and centrifuged at 13,000 rpm for 10 min to collect cell-associated mercury. The level of mercury in both supernatant and cell pellet was determined by cold vapor atomic absorption spectrophotometer (Perkin-Elmer AAnalyst™ 200). A standard solution of $HgCl_2$ (Merck) was used in this experiment.

6.6.2.9 Metal Resistance Pattern

The four metals used in this study were mercury as $HgCl_2$, cadmium as $CdCl_2$, zinc as $ZnSO_4$, lead as $PbNO_3$, and arsenic as Na_2HAsO_4. The MIC for different metals was determined by a microbroth dilution technique following CLSI guidelines (CLSI, 2006). Briefly, bacterial isolate was inoculated into an LB broth medium (casein enzymic hydrolysate 10.0 g, yeast extract 0.5 g, NaCl 10.0 g, and distilled water 1000 mL, pH 7.5 ± 0.2). A stock solution of the metal was prepared by dissolving the metal-containing salt in autoclaved LB broth. The MIC was determined by using twofold serial dilutions in a Muller Hinton Broth (MHB) medium containing various dilutions of the metal as metal salts in a microtiter plate. To each dilution of the compound, 10 μL of 0.5 McFarland culture was added and the plate was incubated at 37°C for 24 h. Un-inoculated MHB was taken as the negative control and MHB without mercury inoculated with the isolate was taken as the positive control. After incubation, the bacterial growth was monitored by measuring the turbidity of the culture with a microtiter plate reader (OD_{630}). The MIC was determined as the lowest concentration of compound at which the visible growth of the organisms was completely inhibited.

6.6.3 Results

6.6.3.1 Sampling, Screening, and Molecular Identification of Bacteria

Collected seawater samples were plated onto an SWN agar medium supplemented with 10 ppm of Hg as $HgCl_2$. After incubation at 30°C, 10 colonies appeared with distinguished colony morphology. All of these 10 isolates were subjected to an Hg tolerance test and BW-03 was found to tolerate the highest concentration of Hg as

$HgCl_2$, i.e., 50 ppm. In order to verify the identity of the isolate at the molecular level, the 16S rRNA gene was amplified, purified, and sequenced. Comparison of known 16S rRNA sequence data for NCBI database indicates the isolate as *Bacillus cereus*. The obtained sequence was submitted to the GenBank, and the obtained accession number was KF241550. The phylogenetic tree (Figure 6.5) based on the 16S rRNA gene sequence was constructed to determine the relationship of BW-03 with other *Bacillus* sp.

6.6.3.2 Amplification of the merA *Gene*

Though the isolate was resistant to a higher amount of mercury, amplification of the *merA* gene gave a negative result. A volatilization assay by X-ray film study also confirmed the non-*merA*, nonvolatilizing mercury-resistant nature of the isolate (data not shown here). The amplification result showed the presence of a *merA* gene fragment only in the positive control. The isolate BW-03 as well as the negative control *E. coli* DH5α showed a negative result for the amplification of the *merA* gene.

6.6.3.3 Mercury Biosorption Study

Accumulation of inorganic mercury by the bacterial cell was confirmed by the development of black color precipitate of the bacterial cell mass within 5 min of exposure to H_2S gas (Figure 6.6).

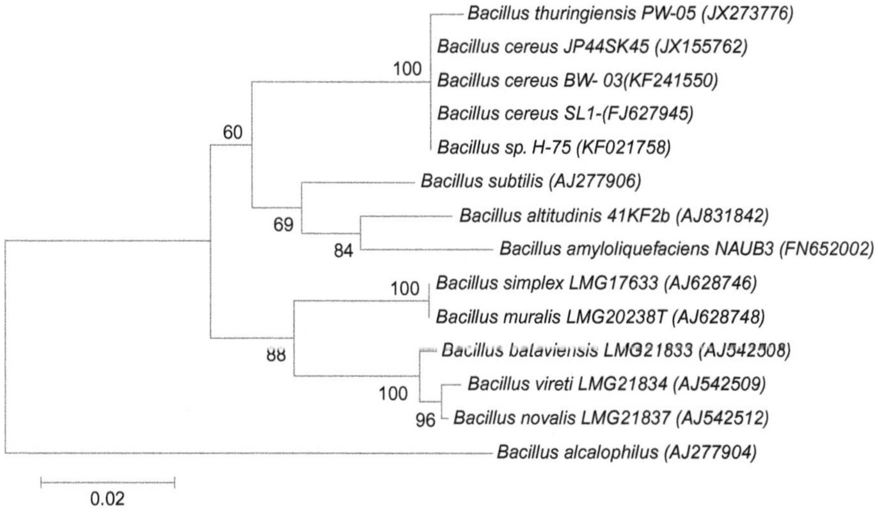

Figure 6.5 Phylogenetic tree for 16S rRNA gene sequences of BW-03 which has been identified as *B. cereus*.

6.6.3.4 Biofilm Development

The isolate BW-03 showed potential biofilm formation after 48 and 72 h both in a glass tube assay and a microtiter plate assay in LB broth medium (Figure 6.7). The bacterium was found to be viable up to 7 days in experimental conditions both in the presence and absence of $HgCl_2$, which has been confirmed by measuring the cell concentration by spectroscopy and by viable cell count. It has been observed that >50% of the residual mercury was entrapped in the bacterial EPS which has been estimated by Atomic absorption spectrophotometer (AAS).

6.6.3.5 Modification of Functional Groups

A sharp shift in wave numbers of some groups has been found in the range of 400–4000 cm^{-1} of BW-03 with and without supplementation of Hg^{2+} (Table 6.1). The peaks at 3283.16, 1650.14, 1385.63, 1095.78, and 668.53 cm^{-1} in the absence of Hg^{2+} were attributed to bonded hydroxyl groups (–OH), amides (C=O), nitrile

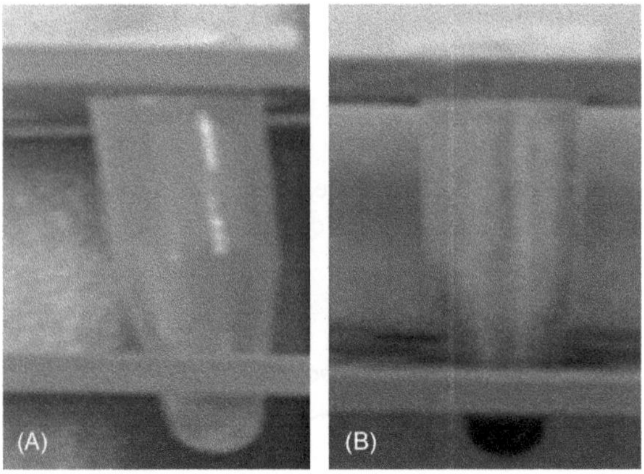

Figure 6.6 Development of black color precipitate of the bacterial cell mass of *B. cereus* BW-03 after exposure to H_2S gas, confirming biosorption of Hg^{2+} by bacterial cell. (A) Cell mass of *B. cereus* BW-03 before H_2S exposure and (B) cell mass of *B. cereus* BW-03 after H_2S exposure.

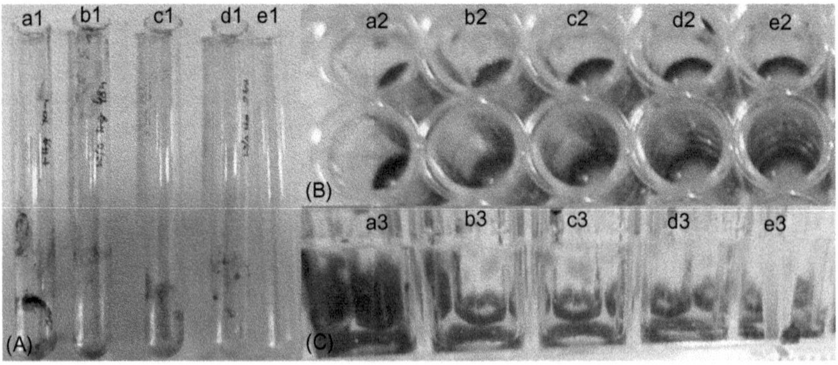

Figure 6.7 Biofilm formation of *B. cereus* BW-03 (A) by glass tube assay. (B) Aerial view of microtiter plate assay. (C) Transverse view of microtiter plate assay.

Table 6.1 Shift of Wave Numbers (cm^{-1}) of the Functional Groups in IR Spectra of Biomass of *B. cereus* BW-03 Grown with (50 ppm Hg as $HgCl_2$) and without Hg^{2+} Supplementation

Functional Groups	Without Hg^{2+}	With Hg^{2+}
Bonded –OH	3432.41	3423.65
S–H stretch	2349.26	2322.01
Amide C=O stretch	1721.56	1694.31
Nitrile C–N stretch	1275.84	1285.57
Phosphate P=O stretch	1093.85	1085.09
Alkyl halide	821.36	739.61

Table 6.2 Tolerance of *B. cereus* BW-03 Towards Various Metals

Metal	Tolerance
$HgCl_2$	50 ppm
$CdCl_2$	458.3 ppm
$ZnSO_4$	1614 ppm
$PbNO_3$	2500 ppm
Na_2HAsO_4	464.8 ppm

(C–N), phosphate groups (P=O stretch), and alkyl halide groups, respectively. These peaks of PW-05 in the presence of Hg^{2+} shifted a certain amount when compared with the control in experimental conditions, which indicates the chemical modification of these functional groups as well as their possible role in mercury bioremediation by this bacterium.

When the bacteria was grown without Hg^{2+} supplement, the peak corresponding to this group (S–H stretch) (2349.68 cm^{-1}) was negligible; however, the peak height increased and the wave number shifted (2339.90 cm^{-1}) when the bacteria was grown with Hg^{2+}, suggesting that this functional group plays a major role in the bioremediation of mercury.

6.6.3.6 Determination of Mercury Removal Potential

The accumulated mercury concentration was quantified by Atomic Absorption Spectrophotometer and was found to be 40.06% of the initial concentrations of 50 ppm. The scanning electron microscopy image shows no structural deformities, confirming the nontoxic nature of the accumulated mercury to the bacterial isolate.

6.6.3.7 Metal Tolerance by the Isolate

The growth of *B. cereus* BW-03 in MHB medium amended with different concentrations of metals was tested, and the MIC was calculated by comparing the growth with a negative control. BW-05 was found to tolerate higher concentrations of all the metals tested (Table 6.2). The isolate was found to tolerate 50 and 2500 ppm of metals like mercury and lead, respectively, to their corresponding metal salts. It could also tolerate higher concentrations of other heavy metals and their salts like cadmium (458.3 ppm), zinc (1614 ppm), and arsenic (464.8 ppm).

6.7 Discussion

Mercury is a global threat to humans as well as other living organisms due to its detrimental effects on every life form. Though mercury is a naturally occurring element present throughout the world, human activity—especially mining and the burning of fossil fuels—has increased the mobilization of mercury into the environment. Most of the emissions and releases of mercury into the environment have occurred since 1800 and are associated with the industrial revolution, based on coal burning, metal ore smelting, and gold mining. The gradual increase in industry, specifically in Asia and South America, has worsened the situation, leading to on the need for regulation, strict legislation, and determination of remedial measures (Dash and Das, 2012).

As elemental mercury is highly volatile at ambient conditions, the most suggested and practiced remediation strategies for soil and water involve transforming highly toxic mercury compounds, such as ionic mercury and organomercurials, into less toxic elemental mercury and then volatilizing this from the system. In such a way, local cleanups can be achieved and food chain accumulation can be prevented. However, the lesser toxic volatile mercury can return to the atmosphere to travel longer distance and can be reconverted to the toxic form of mercury i.e. Hg^{2+}, to create further panic. This has led researchers to search for alternative methods for the bioremediation of mercury. Cinnabar is the most stable form of mercury in the environment and therefore precipitation as HgS emerged as a potential remediation strategy (Lefebvre et al., 2007; Abdrashitova, 2011). However, the fact that HgS is unstable under anoxic conditions, i.e., when it is taken up by sulfate-reducing bacteria and methylated, which is not a much effective solution. Essa et al. (2006) suggested another novel mechanism for mercury sequestering with potential applicability for bioremediation. They used a bacterium, *Klebsiella pneumoniae*, which produced a volatile organothiol, identified as dimethyldisulfide (DMDS), that could precipitate mercury efficiently. However, the chemical nature of the precipitate was not elucidated and there has not been any follow-up study. Moreover, DMDS is a hazardous substance and hydrogen sulfide is toxic, which caused failure of this approach. Biotechnology, using a genetic engineering approach, also opened up new avenues in mercury bioremediation research. One of the novel approaches was the transfer of the *mer* operon into a highly radiation-resistant strain of *Deinococcus radiodurans* (Brim et al., 2000) that considered the nature of mixed contaminants and extreme environmental conditions. Bae et al. (2000, 2001, 2002) engineered the regulator MerR onto the surface of *E. coli*, resulting in highly specific binding of Hg(II), which was six times higher than in the wild type. Bioaugmentation (inoculation of contaminated sites) with selected or engineered bacteria has often been unsatisfactory, partly because the introduced metabolic potential was not the limiting factor for pollutant degradation (Cases and De Lorenzo, 2005).

Currently, research into mercury bioremediation has shifted to take advantage of the biosorption capabilities of mercury-resistant bacteria for entrapping contaminated mercury; these bacteria can be harvested again and again to be reused for many commercial practices. Marine bacteria are better than terrestrial bacteria for use in bioremediation practices due to their rapid rate of acclimatization to environmental conditions, greater bioremediation capability, and high buffering capacity. In this regard, the current study involving screening of potential mercury bioabsorbing marine bacteria holds promise. There are many reports of mercury-resistant bacteria having biosorption capabilities. Many species of *Pseudomonas*, lactic acid bacteria (LAB), *Bacillus*, and *Lysinibacillus* have been reported from environmental conditions to produce EPS for biosorption of mercury (Noghabi et al., 2007; De et al., 2008; Francois et al., 2012; Kinoshita et al., 2013). Microbes for biosorption of mercury were constructed using binding and transport proteins from the *mer* operon, e.g., the periplasmic binding protein MerP, and the membrane-bound transport protein MerT, which have a very high affinity for Hg(II). *Rhodopseudomonas palustris*, a photosynthetic bacterium, was engineered to accumulate Hg(II) using MerP and MerT, and to sequester it by chelating with phytochelatin (Deng and Jia, 2011). Pan-Hou et al. (2002) engineered the bacterial ppk gene for polyphosphate kinase into *E. coli*, while a mouse metallothionein was utilized by Ruiz et al. (2011).

Most of these approaches are based on nonenzymatic reactions and are characterized by a limited loading efficiency. Moreover, biosorption approaches do need to compete with established nonbiological solutions for adsorbing mercury. Examples of such competitors are activated carbon, hydrogen sulfide gas, and sulfur-containing compounds, which are used for adsorbing mercury. The applicability of biosorption therefore depends strongly on process parameters like stability of the material, loading capacity for mercury, ability for regeneration, production costs, effect of co-contaminants, etc. (Wagner-Döbler, 2013).

However, *B. cereus* BW-03 showed its ability to grow in the presence of high concentrations of salt, which is one of the few reports of its kind. This isolate also exhibited resistance to higher concentrations of other toxic metals like Cd, Pb, As, and Zn, which increases the ability of the isolate to be applied in various contaminated environmental conditions with varying environmental parameters. The isolate showed greater tolerance levels than many other reports of *Bacillus* species of marine origin (Ivanova et al., 1999; Oguntoyinbo 2007; Ki et al., 2009), suggesting the higher potential of the isolate.

The isolate BW-03 showed resistance towards only two antibiotics, namely amoxicillin and vancomycin, which is of great importance as there is less chance of transmitting the resistant genotypes to other environmental strains by horizontal gene transfer. In this regard, the isolate could be of great use in comparison to other reported strains (Noghabi et al., 2007; Francois et al., 2012; Kinoshita et al., 2013) possessing mercury biosorption capacity.

The isolate does not harbor the mercury resistance genotype, i.e., the *mer* operon, and it cannot volatilize Hg^{2+}. Therefore, no chance exists for generation of Hg^0 in the environment that could further contaminate other areas. There are few reports so far regarding non-*mer*-mediated mercury resistance in bacteria (Rasmussen and Sorensen, 1998; Reyes et al., 1999; De et al., 2008). Production of EPS and H_2S

assays followed by mercury removal assay confirms the mercury biosorption capability of the isolate, which is the sole cause of mercury resistance by the isolate. Due to greater potential for EPS secretion and biofilm formation, the Hg^{2+} could be entrapped in the bacterial exopolymeric substances (Braissant et al., 2007; Panwichian et al., 2011), thus not allowing Hg^{2+} to enter into the bacterial system and thereby conferring resistance to the bacterial isolate. Furthermore, the bacterial isolates, after their use in the contaminated environments, may be harvested and the adsorbed mercury may be extracted from them to be used as a raw material for further processing. Considering the fact that the isolate BW-03 could also tolerate toxic metals other than mercury, it holds promise in the bioremediation of waste that is of a complex nature and which most likely contains many of those toxic metals.

6.8 Conclusion

Marine bacteria provide a useful source for bioremediation and have an advantage over terrestrial bacteria because they can be used in extreme environmental conditions. As the bacteria form biofilms, there is a greater increase in their bioremediation capability due to the presence of EPS like polysaccharides, amyloids, extracellular enzymes, and biosurfactants. In this regard, the isolated marine biofilm-forming mercury-resistant bacterium *B. cereus* BW-03 holds promise as a suitable candidate species and as an alternative for current remediation of inorganic mercury and mixed waste. It can be suggested that the biosorption aspect of mercury bioremediation should be given priority over the volatilization aspect in future.

Acknowledgments

H.R.D. acknowledges a research fellowship from the Ministry of Human Resource Development, Government of India. S.D. thanks the Department of Biotechnology, Government of India for research grants.

References

Abdrashitova, S.A., 2011. Application of native bacteria for in situ bioremediation of mercury contaminated groundwater occurring in Northern Kazakhstan as a result of operation of the former PO "Khimprom" chemical plant in Pavlodar. ISTC Project No K-1477p Unrestricted Summary of Technical Report.

Adelaja, O.A., Keenan, H.E., 2012. Tolerance of TBT resistance bacteria to methylmercury. Res. J. Environ. Sci. Available from: http://dx.doi.org/doi:10.3923/rjes.2012.

Al Mailem, D.M., Al Awadh, H., Sorkhoh, N.A., Eliyas, M., Radwan, S.S., 2011. Mercury resistance and volatilization by oil utilizing haloarchaea under hypersaline conditions. Extremophiles. 15, 39–44.

Azevedo, B.F., Furieri, L.B., Pecanha, F.M., Wiggers, G.A., Vassallo, P.F., Simoes, M.R., et al., 2012. Toxic effects of mercury on the cardiovascular and central nervous systems. J. Biomed. Biotechnol. Available from: http://dx.doi.org/doi:10.1155/2012/949048.

Bae, W., Chen, W., Mulchandani, A., Mehra, R.K., 2000. Enhanced bioaccumulation of heavy metals by bacterial cells displaying synthetic phytochelatins. Biotechnol. Bioeng. 70, 518–524.

Bae, W., Mehra, R.K., Mulchandani, A., Chen, W., 2001. Genetic engineering of *Escherichia coli* for enhanced uptake and bioaccumulation of mercury. Appl. Environ. Microbiol. 67, 5335–5338.

Bae, W., Mulchandani, A., Chen, W., 2002. Cell surface display of synthetic phytochelatins using ice nucleation protein for enhanced heavy metal bioaccumulation. J. Inorg. Biochem. 88, 223–227.

Barkay, T., 2000. The Mercury Cycle in Encyclopedia of Microbiology. second ed. Academic Press, Inc., San Diego, CA, pp. 171–181.

Barkay, T., Wagner-Döbler, I., 2005. Microbial transformations of mercury: potentials, challenges, and achievements in controlling mercury toxicity in the environment. Adv. Appl. Microbiol. 57, 1–52.

Barkay, T., Fouts, D.L., Olson, B.H., 1985. The preparation of a DNA gene probe for the detection of mercury resistance genes in Gram-negative communities. Appl. Environ. Microbiol. 49, 686–692.

Barkay, T., Liebert, C., Gillman, M., 1989. Environmental significance of the potential for mer(Tn21)-mediated reduction of Hg^{2+} to Hg^0 in natural waters. Appl. Environ. Microbiol. 55, 1196–1202.

Barkay, T., Gillman, M., Turner, R.R., 1997. Effects of dissolved carbon and salinity on bioavailability of mercury. Appl. Environ. Microbiol. 63, 4267–4271.

Barkay, T., Miller, S.M., Summers, A.O., 2003. Bacterial mercury resistance from atoms to ecosystem. FEMS Microbiol. Rev. 27, 355–384.

Barkay, T., Kritee, K., Boyd, E., Geesey, G., 2010. A thermophilic bacterial origin and subsequent constraints by redox, light and salinity on the evolution of the microbial mercuric reductase. Environ. Microbiol. 12, 2904–2917.

Bernhoft, R.A., 2012. Mercury toxicity and treatment: a review of the literature. J. Environ. Public Health. http://dx.doi.org/doi:10.1155/2012/460508 (Epub December 22, 2011).

Beveridge, T.J., 1989. Role of cellular design in bacterial metal accumulation and mineralization. Annu. Rev. Microbiol. 43, 147–171.

Bogdanova, E.S., Mindlin, S.Z., Pakrova, E., Kocur, M., Rouch, D.A., 1992. Mercuric reductase in environmental Gram-positive bacteria sensitive to mercury. FEMS Microbiol. Lett. 97, 95–100.

Braissant, O., Decho, A.W., Dupraz, C., Glunk, C., Przekop, K.M., Visscher, P.T., 2007. Exopolymeric substances of sulfate-reducing bacteria: interactions with calcium at alkaline pH and implication for formation of carbonate minerals. Geobiology. Available from: http://dx.doi.org/doi:10.1111/j.1472-4669.2007.00117.x.

Brim, H., McFarlan, S.C., Fredrickson, J.K., Minton, K., Zhai, M., Wackett, L.P., et al., 2000. Engineering *Deinococcus radiodurans* for metal remediation in radioactive mixed waste environments. Nat. Biotechnol. 18, 85–90.

Brown, N.L., Misra, T., Winnie, J.N., Schmidt, A., Seiff, M., 1986. The nucleotide sequence of the mercuric resistance operons of plasmid R100 and transposon Tn501: further evidence for mer genes, which enhance the activity of the mercuric ion detoxification system. Mol. Gen. Genet. 202, 143–151.

Brown, N.L., Stoyanov, J.V., Kidd, S.P., Hobman, J.L., 2003. The MerR family of transcriptional regulators. FEMS Microbiol. Rev. 27, 145–163.

Brunke, M., Deckwer, W.D., Fritschmuth, J.M., Horn, H., Lunsdorf, M., Rhode, M., et al., 1993. Microbial retention of mercury from waste systems in a laboratory column containing *merA* gene bacteria. FEMS Microbiol. Rev. 11, 45–52.

Cain, A., Vannela, R., Woo, L.K., 2008. Cyanobacteria as a biosorbent for mercuric ion. Bioresour. Technol. 99, 6578–6586.

Canstein, H.V.F., 2000. Mercury-Reducing Biofilms: Insights in their Activity and Composition and Application for Bioremediation of Mercury-Polluted Wastewaters (Ph.D. thesis). Technical University of Braunschweig, Germany.

Canstein, V.H., Li, Y., Timmis, K.N., Deckwer, W.D., Wagner-Döbler, I., 1999. Removal of mercury from chloralkali electrolysis wastewater by a mercury-resistant *Pseudomonas putida* strain. Appl. Environ. Microbiol. 65, 5279–5284.

Canstein, V.H., Li, Y., Leonhäuser, J., Haase, E., Felske, A., Deckwer, W.D., et al., 2002a. Spatially oscillating activity and microbial succession of mercury-reducing biofilms in a technical-scale bioremediation system. Appl. Environ. Microbiol. 68, 1938–1946.

Canstein, V.H., Kelly, S., Li, Y., Wagner-Döbler, I., 2002b. Species diversity improves the efficiency of mercury-reducing biofilms under changing environmental conditions. Appl. Environ. Microbiol. 68, 2829–2837.

Cases, I., De Lorenzo, V., 2005. Genetically modified organisms for the environment: Stories of success and failure and what we have learned from them. Int. Microbiol. 8, 213–222.

Chang, J.S., Hong, J., 1994. Biosorption of mercury by the inactivated cells of *Pseudomonas aeruginosa* PU21 (Rip64). Biotechnol. Bioeng. 44, 999–1006.

Chang, J.S., Law, W.S., 1998. Development of microbial mercury detoxification process using a mercury-hyperesistant strain of *Pseudomonas aeruginosa* PU21. Biotechnol. Bioeng. 57, 462–470.

Chang, J.S., Law, R., Chang, C.C., 1997. Biosorption of lead, copper and mercury by biomass of *Pseudomonas aeruginosa* PU21. Water Res. 31, 1651–1658.

Chen, S., Wilson, D.B., 1997a. Construction and characterization of *Escherichia coli* genetically engineered for bioremediation of Hg^{2+} contaminated environments. Appl. Environ. Microbiol. 63, 2442–2445.

Chen, S., Wilson, D.B., 1997b. Genetic engineering of bacteria and their potential for Hg^{2+} bioremediation. Biodegradation. 8, 97–103.

Chen, S., Kim, E., Shuler, M.L., Wilson, D.B., 1998. Hg^{2+} removal by genetically engineered *Escherichia coli* in a hollow fiber bioreactor. Biotechnol. Prog. 14, 667–671.

Chien, M.F., Lin, K.H., Chang, J.E., Huang, C.C., Endo, G., Suzuki, S., 2010. Interdisciplinary studies on environmental chemistry—biological responses to contaminants. In: Hamamura, N., Suzuki, S., Mendo, S., Barroso, C.M., Iwata, H., Tanabe, S. (Eds.), Distribution of Mercury Resistance Determinants in a Highly Mercury Polluted Area in Taiwan. TERRAPUB, pp. 31–36.

Clinical and Laboratory Standards, 2006. Institute Methods for Dilution Antimicrobial Susceptibility Tests for Bacteria That Grow Aerobically, seventh ed. Approved Standard M7-A7, CLSI, Wayne, PA.

Cursino, L., Mattos, S.V.M., Azevedo, V., Galarza, F., Bucker, D.H., Chartone-Souza, E., et al., 2000. Capacity of mercury volatilization by *mer* (from *Escherichia coli*) and glutathione S-transferase (from *Schistosoma mansoni*) genes cloned in *Escherichia coli.* Sci. Total Environ.(261), 109–113.

Dash, H.R., Das, S., 2012. Bioremediation of mercury and importance of bacterial *mer* genes. Int. Biodeter. Biodegrad. 75, 207–213.

Dash, H.R., Mangwani, N., Das, S., 2013. Characterization and potential application in mercury bioremediation of highly mercury resistant marine bacterium *Bacillus thuringiensis* PW-05. Environ. Sci. Pollut. Res. Available from: http://dx.doi.org/doi:10.1007/s11356-013-2206-8.

De, J., Leonhäuser, J., Vardanyan, L., in press. Removal of mercury in fixed-bed continuous upflow reactors by mercury-resistant bacteria and effect of sodium chloride on their performance. QScience Connect.

De, J., Ramaiah, N., Bhosle, N.B., Garg, A., Vardanyan, L., Nagle, V.L., et al., 2007. Potential of mercury-resistant marine bacteria for detoxification of chemicals of environmental concern. Microbes Environ. 22 (4), 336–345.

De, J., Ramaiah, N., Vardanyan, L., 2008. Detoxification of toxic heavy metals by marine bacteria highly resistant to mercury. Mar. Biotechnol. 10, 471–477.

Deckwer, W.D., Becker, F.U., Ledakowicz, S., Wagner-Döbler, I., 2004. Microbial removal of ionic mercury in a three-phase fluidized bioreactor. Environ. Sci. Technol. 38, 1858–1865.

Deng, X., Jia, P., 2011. Construction and characterization of a photosynthetic bacterium genetically engineered for Hg^{2+} uptake. Bioresour. Technol. 102, 3083–3088.

Deng, X., Wilson, D.B., 2001. Bioaccumulation of mercury from wastewater by genetically engineered *Escherichia coli*. Appl. Microbiol. Biotechnol. 56, 276–279.

Diels, L., Dong, Q., Van der Lelie, D., Baeyens, W., Mergeay, A., 1995. The *czc* operon of *Alcaligenes eutrophus* CH34: from resistance mechanism to the removal of heavy metals. J. Ind. Microbiol. 14, 142–153.

Essa, A.M., Macaskie, L.E., Brown, N.L., 2002. Mechanisms of mercury bioremediation. Biochem. Soc. Trans. 30, 672–674.

Essa, A.M.M., Creamer, N.J., Brown, N.L., Macaskie, L.E., 2006. A new approach to the remediation of heavy metal liquid wastes via off-gases produced by *Klebsiella pneumoniae* M426. Biotechnol. Bioeng. 95, 576–583.

Felsenstein, J., 1985. Confidence limits on phylogenies: an approach using the bootstrap. Evolution. 39, 783–791.

Flora, S.J.S., Mittal, M., Mehta, A., 2008. Heavy metal induced oxidative stress & its possible reversal by chelation therapy. Indian J. Med. Res. 128, 501–523.

Francois, F., Lombard, C., Guigner, J.M., Soreau, P., Jaisson, F.B., Martino, G., et al., 2012. Isolation and characterization of environmental bacteria capable of extracellular biosorption of mercury. Appl. Environ. Microbiol. Available from: http://dx.doi.org/doi:10.1128/AEM.06522-11.

Frustaci, A., Magnavita, N., Chimenti, C., Caldarulo, M., Sabbioni, E., Pietra, R., et al., 1999. Marked elevation of myocardial trace elements in idiopathic dilated cardiomyopathy compared with secondary dysfunction. J. Am. Coll. Cardiol. 33, 1578–1583.

Gadd, G.M., 2010. Metals, minerals and microbes: geomicrobiology and bioremediation. Microbiology. 156, 609–643.

Gluszcz, P., Zakrzewska, K., Wagner-Döbler, I., Ledakowicz, S., 2008. Bioreduction of ionic mercury from wastewater in a fixed-bed bioreactor with activated carbon. Chem. Pap. 623, 232–238.

Griffin, H.G., Foster, T.J., Silver, S., Mishra, T.K., 1987. Cloning and DNA sequence of mercuric reductase and organomercurial resistance determinants of plasmids pDU 1358. PNAS USA. 84, 3112–3116.

Gu, B., Biana, Y., Miller, C.L., Dong, W., Jiang, X., Liang, L., 2011. Mercury reduction and complexation by natural organic matter in anoxic environments. PNAS USA. 108, 1479–1483.

Henneberry, Y.K., Kraus, T.E.C., Fleck, J.A., Krabbenhoft, D.P., Bachand, P.M., Horwath, W.R., 2011. Removal of inorganic mercury and methyl-mercury from surface waters following coagulation of dissolved organic matter with metal based salts. Sci. Total Environ. 409, 631–637.

Hintelmann, H., 2010. Organomercurials: their formation and pathways in the environment. Met. Ions Life Sci. 7, 365–401.

Hobman, J., Kholodii, G., Nikiforov, V., Ritchie, D.A., Strike, P., Yurieva, O., 1994. The sequence of the *mer* operon of p327/419, 330 and 05. Gene. 277, 73–78.

Hutton, M., 1987. Human health concerns of lead, mercury, cadmium and arsenic. In: Hutchinson, T.C., Meema, K.M. (Eds.), Lead, Mercury, Cadmium and Arsenic in the Environment. SCOPE, John Wiley & Sons Ltd, Chichester, New York.

Inoue, C., Sugawara, K., Shiratori, T., Kusano, T., Kitagawa, Y., 1989. Nucleotide sequence of the *Thiobacillus ferroxidans* chromosomal gene encoding mercuric reductase. Gene. 84, 47–54.

Inoue, C., Sugawara, K., Kusano, T., 1991. The *merR* regulatory gene in *Thiobacillus ferrooxidans* is spaced apart from the *mer* structural genes. Mol. Microbiol. 5, 2707–2718.

Iohara, K., Iiyama, R., Nakamura, K., Silver, S., Sakai, M., Takeshita, M., et al., 2001. The *mer* operon of a mercury-resistant *Pseudoalteromonas haloplanktis* strain isolated from Minamata Bay, Japan. Appl. Microbiol. Technol. 56, 736–741.

Ivanova, E.P., Vysotskii, M.V., Svetashev, V.I., Nedashkovskaya, O.I., Gorshkova, N.M., Mikhailov, V.V., et al., 1999. Characterization of *Bacillus* strains of marine origin. Int. Microbiol. 2, 267–271.

Jain, K., Parida, S., Mangwani, N., Dash, H.R., Das, S., 2013. Isolation and characterization of biofilm forming bacteria and associated extracellular polymeric substances from oral cavity. Ann. Microbiol. Available from: http://dx.doi.org/doi:10.1007/s13213-013-0618-9.

Kerin, E.J., Gilmour, C.C., Roden, E., Suzuki, M.T., Coates, J.D., Mason, R.P., 2006. Mercury methylation by dissimilatory iron-reducing bacteria. Appl. Environ. Microbiol. 72, 7919–7921.

Kholodi, G.Y., Yurieva, O.V., Lomovskaya, O.L., Gorlenko, Z.M., Mindlin, S.Z., Nikiforov, V.G., 1993. Tn5053, a mercury resistance transposon with integron ends. J. Mol. Biol. 230, 1103–1107.

Ki, J.S., Zhang, W., Qian, P.Y., 2009. Discovery of marine *Bacillus* species by 16S rRNA and *rpoB* comparisons and their usefulness for species identification. J. Microbiol. Methods. 77, 48–57.

Kinoshita, H., Sohma, Y., Ohtake, F., Ishida, M., Kawai, Y., Kitazawa, H., et al., 2013. Biosorption of heavy metals by lactic acid bacteria and identification of mercury binding protein. Res. Microbiol. 164, 701–709.

Kiyono, M., Pan-Hou, H., 1999. The *merG* gene product is involved in phenylmercury resistance in *Pseudomonas* strain K-62. J. Bacteriol. 181, 726–730.

Kiyono, M., Pan-Hou, H., 2006. Genetic engineering of bacteria for environmental remediation of mercury. J. Health Sci. 52, 199–204.

Komura, I., Funaba, T., Izaki, K., 1971. Mechanism of mercuric chloride resistance in microorganisms. II. NADPH-dependent reduction of mercuric chloride and vaporization of mercury from mercuric chloride by a multiple drug resistant strain of *Escherichia coli.* J. Biochem. 70, 895–901.

Kusano, T., Ji, G., Inoue, C., Silver, S., 1990. Constitutive synthesis of a transport function encoded by the *Thiobacillus ferrooxidans merC* gene cloned in *Escherichia coli.* J. Bacteriol. 172, 2688–2692.

Langley, S., Beveridge, T.J., 1999. Effect of o-side-chain lipopolysaccharide chemistry on metal binding. Appl. Environ. Microbiol. 65, 489–498.

Lefebvre, D.D., Kelly, D., Budd, K., 2007. Biotransformation of Hg(II) by cyanobacteria. Appl. Environ. Microbiol. 73, 243–249.

Leonhäuser, J., Canstein, V.H., Deckwer, W.D., Wagner-Döbler, I., 2013. Long-term operation of a microbiological pilot plant for clean-up of mercury-contaminated wastewater at electrolysis factories in Europe. In: Wagner- Döbler, I. (Ed.), Bioremediation of Mercury: Current Research and Industrial Applications. Caister Academic Press, Germany, pp. 119–132.

Liebert, C.A., Wireman, J., Smith, T., Summers, A.O., 1997. Phylogeny of mercury resistance (*mer*) operons from Gram-negative bacteria isolated from the fecal flora of primates. Appl. Environ. Micobiol. 63, 1066–1076.

Liebert, C.A., Hall, R.M., Summers, A.O., 1999. Transposon Tn21, flagship of the floating genome. Microbiol. Mol. Biol. Rev. 63, 507–522.

Lloyd, J.R., Renshaw, J.C., May, I., Livens, F.R., Burke, I.T., Mortimer, R.G.T., et al., 2005. Biotransformation of actinides: microbial reduction of actinides and fission products. J. Nucl. Radiochem. Res. 6, 17–20.

Lone, M.I., He, Z.L., Stoffella, P.J., Yang, X., 2008. Phytoremediation of heavy metal polluted soils and water: progresses and perspectives. J. Zhejiang Univ. Sci. B. 9, 210–220.

Marusczak, N., Larose, C., Dommergue, A., Paquet, S., Beaulne, J.S., Brachet, R.M., et al., 2011. Mercury and methylmercury concentrations in high altitude lakes and fish (Arctic charr) from the French Alps related to watershed characteristics. Sci. Total Environ. 409, 1909–1915.

Mason, R.P., Choi, A.L., Fitzgerald, W.F., Hammerschmidt, C.R., Lamborg, C.H., Soerensen, A.L., et al., 2012. Mercury biogeochemical cycling in the ocean and policy implications. Environ. Res.<http://dx.doi.org/10.1016/j.envres.2012.03.013>.

Melnick, J.G., Parkin, G., 2007. Cleaving mercury-alkyl bonds: a functional model for mercury detoxification by MerB. Science. 317, 225–227.

Meyer, M.W., 2005. The Wisconsin Loon Population Project: insuring loons will be here for the grand kids. Available from: <http://dnr.wi.gov/topic/ShorelandZoning/documents/Loon-Project.pdf>.

Misra, T.K., 1992. Bacterial resistances to inorganic mercury salts and organomercurials. Plasmid. 27, 4–16.

Misra, T.K., Brown, N.L., Fritzinger, D.C., Pridmore, R.D., Barnes, W.M., Haberstroh, L., et al., 1984. The mercuric-ion resistance operon of plasmid R100 and transposon Tn501: the beginning of the operon including the regulatory region and the first two structural genes. PNAS USA. 81, 5975–5979.

Monteiro, L.R., Costa, V., Furness, R.W., Santos, R.S., 1996. Mercury concentrations in prey fish indicate enhanced bioaccumulation in mesopelagic environments. Mar. Ecol. Prog. Ser. 141, 21–25.

Mulligan, C.N., Yong, R.N., Gibbs, B.F., 2001. Remediation technologies for metal-contaminated soils and groundwater: an evaluation. Eng. Geol. 60, 193–207.

Nakamura, K., Mineshi, S., Uchiyama, H., Yagi, O., 1990. Organomercurial volatilizing bacteria in the mercury polluted sediment of Minamata Bay, Japan. Appl. Environ. Microbiol. 56, 304–305.

Noghabi, K.A., Zahiri, H.S., Yoon, S.C., 2007. The production of a cold-induced extracellular biopolymer by *Pseudomonas fluorescens* BM07 under various growth conditions and its role in heavy metals absorption. Process Biochem. 42, 847–855.

Oguntoyinbo, F.A., 2007. Monitoring of marine *Bacillus* diversity among the bacteria community of sea water. Afr. J. Biotechnol. 6, 163–166.

Osborn, A.M., Bruce, K.D., Strike, P., Ritchie, D.A., 1997. Distribution, diversity and evolution of the bacterial mercury resistance (*mer*) operon. FEMS Microbiol. Rev. 19, 239–262.

Osteen, A.B., Bibler, J.P., 1991. Treatment of radioactive laboratory waste for mercury removal. Water Air Soil Pollut. 56, 63–74.

Pan-Hou, H., Kiyono, M., Omura, H., Omura, T., Endo, G., 2002. Polyphosphate produced in recombinant *Escherichia coli* confers mercury resistance. FEMS Microbiol. Lett. 10325, 159–164.

Panwichian, S., Kantachote, D., Wittayaweerasak, B., Mallavarapu, M., 2011. Removal of heavy metals by exopolymeric substances produced by resistant purple nonsulfur bacteria isolated from contaminated shrimp ponds. Electron. J. Biotechnol. Available from: http://dx.doi.org/doi:10.2225/vol14-issue4-fulltext-2.

Pazirandeh, M., Chrisey, L.A., Mauro, J.M., Campbell, J.R., Gaber, B.P., 1995. Expression of the *Neurospora crassa* metallothionein gene in *Escherichia coli* and its effect on heavy-metal uptake. Appl. Microbiol. Biotechnol. 43, 1112–1117.

Pepi, M., Focardi, S., Tarabelli, A., Volterrani, M., Focardi, S.E., 2013. Bacterial strains resistant to inorganic and organic forms of mercury isolated from polluted sediments of the Orbetello Lagoon, Italy, and their possible use in bioremediation processes. E3S Web of Conferences 1, 31002 (2013). <http://dx.doi.org/10.1051/e3sconf/20130131002>.

Radstrom, P., Skold, O., Swedberg, G., Flensburg, J., Roy, P.H., Sundstrom, L., 1994. Transposon Tn5090 of the plasmid R751, which carries integron, is related to Tn7, Mu, and the retroelements. J. Bacteriol. 176, 3257–3268.

Rajesh, D., Bhateria, G.R., 2011. Strategies for management of metal contaminated soil. Int. J. Environ. Sci. 1, 1884–1898.

Randall, P., Ilyushchenko, M.A., Panichkin, V.Y., Kamberov, R.I., 2013. Former Chlor-alkali factory in Pavlodar, Kazakhstan: mercury pollution, treatment options, and results of post-demercurization monitoring. Bioremediation of Mercury: Current Research and Industrial Applications. Horizon Press, January 1, 2013.

Rasmussen, L.D., Sorensen, S.J., 1998. The effect of long term exposure of mercury in the bacterial community in marine sediment. Curr. Microbiol. 36, 291–297.

Ravichandran, M., 2004. Interactions between mercury and dissolved organic matter—a review. Chemosphere. 55, 319–331.

Reyes, N.S., Frischer, M.E., Sobecky, P.A., 1999. Characterization of mercury resistance mechanisms in marine sediment microbial communities. FEMS Microbiol. Ecol. 30, 273–284.

Ritter, J.A., Bibler, J.P., 1992. Removal of mercury from wastewater: large scale performance of an ion exchange process. Water Sci. Technol. 25, 165–172.

Rojas, L.A., Yanez, C., Gonzalez, M., Lobos, S., Smalla, K., Seeger, M., 2011. Characterization of the metabolically modified heavy metal resistant *Cupriavidus metallidurans* strain MSR33 gene rated for mercury bioremediation. PLoS ONE. 6, 10.

Ruiz, O.N., Alvarez, D., Gonzalez-Ruiz, G., Torres, C., 2011. Characterization of mercury bioremediation by transgenic bacteria expressing metallothionein and polyphosphate kinase. BMC Biotechnol. 11, 2–8.

Saitou, N., Nei, M., 1987. The neighbour-joining method: a new method for reconstructing phylogenetic trees. Mol. Biol. Evol. 4, 406–425.

Saouter, E., Turner, R., Barkay, T., 1994. Microbial reduction of ionic mercury for the removal of mercury from contaminated environments. Ann. N. Y. Acad. Sci. 721, 423–427.

Saouter, E., Gillman, M., Barkay, T., 1995. An evolution of mer-specified reduction of ionic mercury as a remedial tool of a mercury-contaminated freshwater pond. J. Ind. Microbiol. 14, 343–348.

Selifonova, O.V., Barkay, T., 1994. Role of Na^+ in transport of Hg^{2+} and induction of the Tn21 *mer* operon. Appl. Environ. Microbiol. 60, 3502–3507.

Selin, N.E., 2009. Global biogeochemical cycling of mercury: a review. Annu. Rev. Environ. Resour. 34, 43–63.

Silver, S., 1996. Bacterial resistances to toxic metals—a review. Gene. 179, 9–19.

Silver, S., Misra, T.K., 1984. Bacterial transformations of and resistances to heavy metals. Basic Life Sci. 28, 23–46.

Silver, S., Phung, L.T., 1996. Bacterial heavy metal resistance: new surprise. Annu. Rev. Microbiol. 50, 753–789.

Silver, S., Endo, G., Nakamura, K., 1994. Mercury in the environment and laboratory. J. Jpn. Soc. Water Environ. 17, 235–244.

Sommar, J.N., Svensson, M.K., Bjor, B.M., Elmstahl, S.I., Hallmans, G., Lundh, T., et al., 2013. End-stage renal disease and low level exposure to lead, cadmium and mercury; a population-based, prospective nested case-referent study in Sweden. Environ. Health. Available from: http://dx.doi.org/doi:10.1186/1476-069X-12-9.

Sonne, C., 2010. Health effects from long-range transported contaminants in Arctic top predators: an integrated review based on studies of polar bears and relevant model species. Environ. Int. 36, 461–491.

Sotero-Martins, A., Jesus, M.S., Lacerda, M., Moreira, J.C., Filgueiras, A.L.L., Barrocas, P. R.G., 2008. A conservative region of the mercuric reductase gene (*merA*) as a molecular marker of bacterial mercury resistance. Braz. J. Microbiol. 39, 307–310.

Stelting, S., Burns, R.G., Sunna, A., Visnovsky, G., Bunt, C., 2010. Immobilization of *Pseudomonas* sp. strain ADP: a stable inoculant for the bioremediation of atrazine. In: 19th World Congress of Soil Science, Soil Solutions for a Changing World, Brisbane, Australia, August 1–6, 2010.

Summers, A.O., 1986. Organization, expression and evolution of genes for mercury resistance. Annu. Rev. Microbiol. 40, 607–634.

Summers, A.O., 1992. The hard stuff: metals in bioremediation. Curr. Opin. Biotechnol. 3, 271–276.

Summers, A.O., Lewis, E., 1973. Volatilization of mercuric chloride by mercury resistant plasmid bearing strain of *Escherichia coli*, *Staphylococcus aureus* and *Pseudomonas aeruginosa*. J. Bacteriol. 113, 1070–1072.

Summers, A.O., Silver, S., 1978. Microbial transformations of metals. Annu. Rev. Microbiol. 32, 637–672.

Takeuchi, F., Sugio, T., 2006. Volatilization and recovery of mercury from mercury-polluted soils and wastewaters using mercury-resistant *Acidithiobacillus ferrooxidans* strains SUG 2–2 and MON-1. Environ. Sci. 13, 305–316.

Thompson, J.D., Gibson, T.J., Plewniak, F., Jeanmougin, F., Higgins, D.G., 1997. CLUSTAL X windows interface: flexible strategies for multiple sequence alignment aided by quality analysis tools. Nucleic Acids Res. 25, 4876–4882.

UNEP, 2013. Global mercury assessment 2013. Sources, Emissions, Releases And Environmental Transport. UNEP Chemicals Branch, Geneva, Switzerland.

U.S. Geological Survey, 2000. Mercury in the Environment. Fact sheet 146-00. Available from: <http://www.usgs.gov/themes/factsheet/146-00/>.

Volesky, B., Holan, Z.R., 1995. Biosorption of heavy metals. Biotechnol. Prog. 11, 235–250.

Wagner-Döbler, I., 2003. Pilot plant for bioremediation of mercury containing industrial wastewater. Appl. Microbiol. Biotechnol. 62, 124–133.

Wagner-Döbler, I., 2013. Current research for bioremediation of mercury. In: Wagner-Döbler, I. (Ed.), Bioremediation of Mercury: Current Research and Industrial Applications, 2013. Caister Academic Press, Germany, pp. 1–16.

Wagner-Döbler, I., Ltinsdorf, H., Ltibbehilsen, T., von Canstein, H.F., Li, Y., 2000. Structure and species composition of mercury-reducing biofilms. Appl. Environ. Microbiol. 66, 4559–4563.

Wang, J., Chen, C., 2009. Biosorbents for heavy metals removal and their future. Biotechnol. Adv. 27, 195–226.

Wang, Y., Mahler, I., Levinson, H.S., Halvorson, H.O., 1987. Cloning and expression in *Escherichia coli* of chromosomal mercury resistance genes from a *Bacillus* sp. J. Bacteriol. 169, 4848–4851.

Wang, Y., Moore, M., Levinson, H.S., Silver, S., Walsh, C., Mahler, I., 1989. Nucleotide sequence of a chromosomal mercury resistance determinant from a *Bacillus* sp. with broad-spectrum mercury resistance. J. Bacteriol. 171, 83–92.

Wangberg, I., Moldanova, J., Munthe, J., 2010. Mercury cycling in the environment effects of climate change. Report by Swedish Environmental Research Institute. Report for EUROLIMPAC. Available from: <http://www.ivl.se/webdav/files/B-rapporter/B1921.pdf>.

Wiatrowski, H.A., Ward, P.M., Barkay, T., 2006. Novel reduction of mercury (II) by mercury-sensitive dissimilatory metal reducing bacteria. Environ. Sci. Technol. 40, 6690–6696.

Williams, J.W., Silver, S., 1984. Bacterial resistance and detoxification of heavy metals. Enzyme Microb. Biotechnol. 6, 530–537.

Wilson, J.R., Leang, C., Morby, A.P., Hobman, J.L., Brown, N.L., 2000. MerF is a mercury transport protein: different structures but a common mechanism for mercuric ion transporters? FEBS Lett. 472, 78–82.

Young, J.F., Wosilait, W.D., Luecke, R.H., 2001. Analysis of methylmercury disposition in humans utilizing a pbpk model and animal pharmacokinetic data. J. Toxicol. Environ. Health A. 63, 19–52.

Zahir, F., Rizwi, S.J., Haq, S.K., Khan, R.H., 2005. Low dose mercury toxicity and human health. Environ. Toxicol. Pharmacol. 20, 351–360.

Zhang, W., Chen, L., Liu, D., 2012. Characterization of a marine-isolated mercury resistant *Pseudomonas putida* strain SP1 and its potential application in marine mercury reduction. Appl. Microbiol. Biotechnol. 93, 1305–1314.

Zheng, W., Liang, L., Gu, B., 2012. Mercury reduction and oxidation by reduced natural organic matter in anoxic environments. Environ. Sci. Technol. 46, 292–299.

7 Biosurfactant-Based Bioremediation of Toxic Metals

Jaya Chakraborty and Surajit Das

Laboratory of Environmental Microbiology and Ecology (LEnME), Department of Life Science, National Institute of Technology, Rourkela, Odisha, India

7.1 Introduction

Rising industrialization and activities of mankind in the present era have resulted in heavy metal contamination of soil and water bodies. Heavy metals are a group of metals and metalloids with a density greater than 6 g/cm^3 (Alloway and Ayres, 1997). Heavy metals such as lead, cadmium, chromium, copper, arsenic, zinc, mercury, and nickel have been reported as major environmental pollutants and are included in the Environmental Protection Agency's list of priority pollutants (Cameron, 1992). Accretion of these toxic metals in soil and water signify a potential health hazard for human beings as well as animals. These metals at trace concentrations are nonbiodegradable and persistent; thus, they are a potential threat to mankind and the environment. Ingression of toxic metals into the environment occurs through different industries such as mines, tanneries, electroplating industries, and the manufacture of paints, metal pipes, batteries, and ammunition. The sulfide and oxidized ores produced during mining operations potentially produce leachate solutions that contain high concentrations of dissolved metals. This leachate solution adversely affects the quality of plant and animal health as contaminated surface water is used for agriculture, recreation, and consumption purposes.

Heavy metal contamination has been associated with birth defects, liver and kidney damage, and certain learning disabilities (Singh and Cameotra, 2004). Various chemical conventional remediation techniques used for the removal of heavy metal contamination include treatment of contaminated soil with water, inorganic and organic acids, chemical surfactants, and metal-chelating agents (Dahrazma and Mulligan, 2007). But these methods render toxic effects on aquatic and terrestrial habitats and also do not ensure proper removal of the contaminating metal ions from the soil. Techniques of remediation for contaminated soils like thermal treatment, stabilization, excavation, and landfill are infrequently used (Wang and Mulligan, 2009). For suitable removal of the metals from contaminated environments, a good metal-complexing agent possessing the properties of solubility and environmental stability is required (Juwarkar et al., 2007). One major agent helpful in remediation

Microbial Biodegradation and Bioremediation. DOI: http://dx.doi.org/10.1016/B978-0-12-800021-2.00007-8

is the surfactants, which are amphiphilic molecules consisting of strong hydrophilic and hydrophobic groups. The hydrophilic portion of surfactants renders them highly soluble in water, while the hydrophobic portion helps in residing in a hydrophobic phase thus reducing surface tension (Mulligan et al., 2001b). Critical micelle concentration (CMC) is defined as the concentration of surfactant where they arrange themselves into organized molecular assemblies known as micelles (Mattei et al., 2013). CMC is the point at which surfactant results in variation in detergency, molar conductivity, solubilization, and osmotic pressure (Shaw, 1996). Due to high toxic manifestations, a biobased alternative known as biosurfactants have come to the forefront. Microbial surface-active metabolites, called biosurfactants, are metal-complexing agents that have been reported to be effective in the remediation of heavy metal-contaminated environments (Mulligan et al., 2001; Singh and Cameotra, 2004). The special attributes of biosurfactants make them a potential substitute for remediation purposes. These are less toxic in nature (Poremba et al., 1991) and have better environmental compatibility and biodegradability (Georgiou et al., 1992). Other advantages include their production from inexpensive agro-based raw materials and organic wastes (Mukherjee et al., 2006) and retention of their activity even at extremes of temperature, pH, and salt concentration. Most biosurfactants reported to date are obtained from microorganisms of terrestrial origin. However, the marine environment that covers the majority of the earth's surface serves as a vast repertoire of microbial population, producing a variety of active metabolites and molecules. Hence, the least explored marine environment should be explored more to find some potential microorganisms for enhanced synthesis of biosurfactants.

7.2 Microbial Surface-Active Compounds: Biosurfactants

Biosurfactants are an extensive group of structurally diverse surface-active compounds produced by a variety of microorganisms which are primarily classified by their chemical structure and their microbial origin. They are commonly composed of a hydrophilic part, consisting of amino acid or peptide anions or cations, mono- or polysaccharides, and a hydrophobic part consisting of saturated or unsaturated fatty acids that preferentially partition at the interface between fluid phases having different degrees of polarity and hydrogen bonding, e.g., oil and water, or air and water interfaces (Evans et al., 1990). This property enables them to locate at the cell envelope and make nonsoluble substrates bioavailable to the cell. Therefore, they have a strong influence over interfacial rheological behavior and mass transfer (Lin, 1996). Surfactants of microbial origin reduce the free energy of the solvent system by replacing the bulk molecules of higher energy at an interface (Rosen, 1978). According to a classification proposed by Neu (1996), the term "biosurfactants" represents low-molecular-weight microbial surfactants, whereas high-molecular-weight polymers can be collectively defined as bioemulsifiers (Rosenberg and Ron, 1997). Biosurfactant molecules proficiently lower surface and interfacial tension, while bioemulsifiers composed of amphiphilic and polyphilic polymers are more effective in stabilizing oil-in-water emulsions but do not lower the surface tension to a great extent. The low-molecular-weight biosurfactants are generally glycolipids, such as rhamnolipids, trehalose lipids, sophorolipids, and fructose lipids, or lipopeptides, such as surfactin, gramicidin S, and polymixin. The high-molecular-weight bioemulsifiers are amphiphilic or polyphilic polysaccharides, proteins, lipopolysaccharides (LPS), and lipoproteins. Biosurfactants increase the surface area of hydrophobic water-insoluble substrates, and it also reduces metal toxicity by complexation of metal with the biosurfactant interacted with the cell surface to alter metal uptake (Sandrin et al., 2000).

7.2.1 Chemistry and Types

Classification of biosurfactants usually occurs by different chemical composition and microbial origin. Ron and Rosenberg (2001) classified biosurfactants into low-molecular-mass molecules, which efficiently lower surface and interfacial tension, and high-molecular-mass polymers, which are more effective as emulsion-stabilizing agents (Kappeli and Finnerty, 1979). The major classes of low-mass biosurfactants include glycolipids, lipopeptides, and phospholipids; high-mass surfactants include polymeric and particulate surfactants. Microbial biosurfactants are derived mostly from diverse sources and are either anionic or neutral, and the hydrophobic moiety is based on long-chain fatty acids or fatty acid derivatives; the hydrophilic portion can be a carbohydrate, amino acid, phosphate, or cyclic peptide (Nitschke and Coast, 2007). Different types of biosurfactants based on their nature and chemical composition are represented in the following sections.

7.2.1.1 Glycolipids

One of the best studied microbial surfactants is the glycolipids. The constituent mono-, di-, tri, and tetrasaccharides include glucose, mannose, galactose, glucuronic acid, rhamnose, and galactose sulfate. The fatty acid component usually has a composition similar to that of the phospholipids of the source microorganism secreting it (Veenanadig et al., 2000; Chen et al., 2007). Among these glycolipids, the most-known compounds are rhamnolipids, trehalolipids, and sophorolipids, which are disaccharides, combined with long-chain aliphatic acids or hydroxyaliphatic acids (Desai and Banat, 1997; Karanth et al., 1999). At *in situ* conditions, glycolipid production by microorganisms is modified by the carbon source, pH, temperature, nitrogen content, oxygen availability, and concentration of salts (Gautam and Tyagi, 2006). There are several importances of glycolipids. They help in mobilization of microorganisms by reducing the interfacial tension of the environment helping in search of better location for their growth, reproduction and colonization (Kearns and Losick, 2003). Glycolipids are further divided into three types based on the attached carbohydrate group.

1. *Rhamnolipids*: Rhamnolipids are composed of one or two molecules of rhamnose linked to one or two molecules of β-hydroxydecanoic acid. There is a glycosidic linkage between the hydroxyl groups in one of the acids with the reducing end of the rhamnose disaccharide,

whereas the hydroxyl group of the second acid is involved in ester formation. As one carboxylic group is free, the rhamnolipids are anionic in nature, having pH 4. When the lipid moiety of rhamnolipid is bonded with one or two rhamnose group it is known as monorhamnolipid (Type I) or dirhamnolipid (Type II) respectively (Figure 7.1).

Rhamnolipid production was first reported in *Pseudomonas aeruginosa* and also studied in other *Pseudomonas* species. Rhamnolipids can reduce the surface tension of water to 25–30 mN/m and also the interfacial tension against n-hexadecane to 1 mN/m (Qazi et al., 2013). CMC of rhamnolipid ranges from 10 to 30 mg/L. Rhamnolipid type 1 (L-hamnosyl-L-rhamnosyl-β-hydroxydecanoyl-β hydroxydecanoate) and rhamnolipid type 2 (L-rhamnosyl-β-hydroxydecanoyl-β-hydroxydecanoate) are the main glycolipids produced by *P. aeruginosa*. These rhamnolipids produced by the microbes are fully commercialized for bioremediation purposes.

2. *Trehalolipids*: Trehalolipids are an extensive group of glycolipids, consisting of disaccharide trehalose linked at C-6 and C-6′ position to mycolic acids, which are long-chain α-branched-β-hydroxy fatty acids (Figure 7.2). The values of interfacial tension of water against n-hexadecane achieved with the trehalolipids range between 1 and 17 mN/m and the surface tension is lowered in the range of 25–40 mN/m (Shao, 2011). The CMC for trehalolipids is in the range of about 2 mg/L. Besides having a significant role in the industrial and environmental arenas, emerging applications are also exhibited as therapeutic agents (Lang and Philp, 1998; Banat et al., 2000; Rodrigues et al., 2006). Microorganisms like

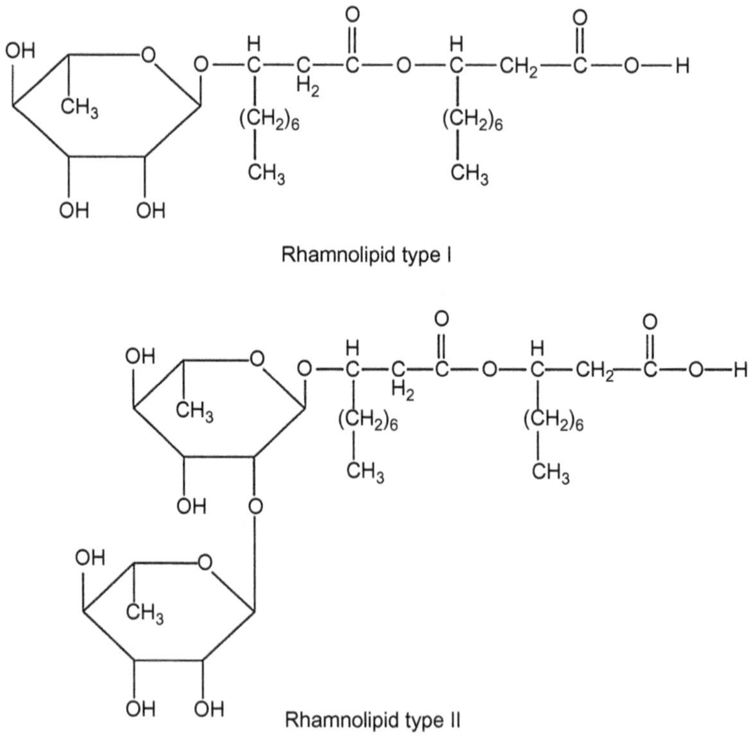

Figure 7.1 Chemical structure of monorhamnolipid and dirhamnolipid.

Mycobacterium, Nocardia, Corynebacterium and Rhodococcus produce trehalolipids, of which Rhodococcus erythropolis produce trehalose dimycolates. Trehalolipids produced by different microorganisms differ in their structure, size, and degree of saturation.

3. *Sophorolipids*: Sophorolipids are surface-active compounds secreted from microbial origin. They consist of a hydrophilic part having disaccharide sophorose, which consists of two glucose molecules linked with an unusual β-1,2 bond (Figure 7.3). The hydrophobic part of the amphiphilic molecule is made up of a terminal or subterminal hydroxylated fatty acid,

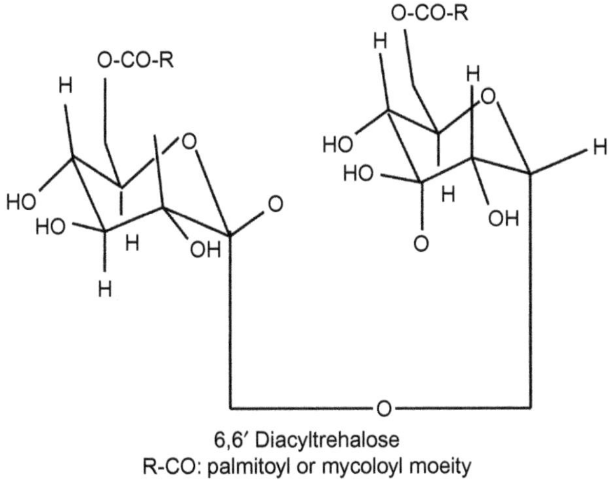

Figure 7.2 Chemical structure of trehalolipid.

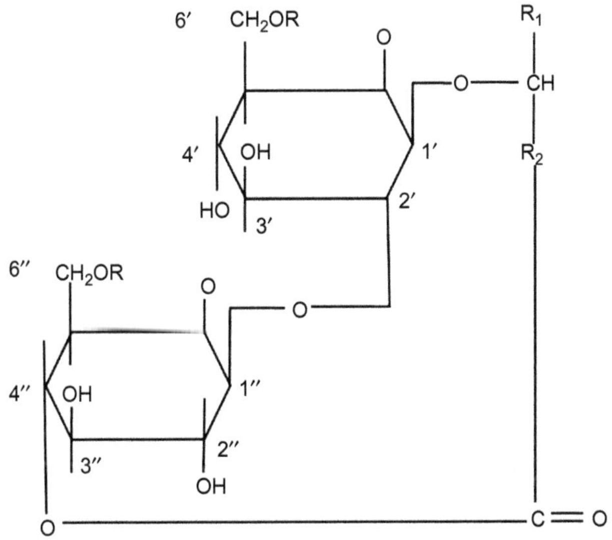

Figure 7.3 Chemical structure of sophorolipid.

glycosidically linked to the sophorose molecule. The synthesis of sophorolipids occurs as minor different molecules at two major points. The terminal carboxyl group can be acetylated or lactonized by containing acetyl groups at the 6 or 6′ position and the sophorolipid can occur in the acidic form with a free fatty acid tail or in the lactonic form with an internal esterification between the carboxylic end of the fatty acid and the 4′ position of the sophorose head molecule. The only minor difference is that lactonization occurs at the 6 or 6′ position. The physicochemical properties are changed based on the structural variation. As compared to the acidic form, lactonic sophorolipids have reduced surface tension, having lower CMC and good antimicrobial properties (Garcia-Ochoa and Casas, 1999; Lang et al., 2000). Acetylated sophorolipids are less water-soluble, having enhanced antiviral and cytokine-stimulating activity (Shah et al., 2005). Lactonic and anionic sophorolipids have been reported to lower the interfacial tension of water against n-hexadecane or vegetable oils to 1–5 mN/m over a wide range of pH, temperature, and salt concentration. *Torulopsis bombicola*, *Torulopsis petrophilum*, *Torulopsis apicola*, and *Candida bogoriensis* are potential candidates for sophorolipid production.

7.2.1.2 Lipopeptides

Lipopeptides are short linear chains or cyclic structures of amino acids, linked to a fatty acid via ester or amide bond or both. The configuration of amino acids is usually in D-form rather than the usual L-configuration to resist enzymatic action of protease. The peptide portion may contain either cationic or anionic residues, and sometimes also contains nonproteinaceous amino acids (Jerala, 2007; Strieker and Marahiel, 2009). Lipopeptide biosurfactants are generally synthesized by different fungal species like *Aspergillus* spp., and numerous bacterial genera such as *Streptomyces*, *Pseudomonas*, and *Bacillus*. Based on these structural characteristics, *Pseudomonas* sp. initially classified their cyclic lipopeptides into four major groups: viscosin, amphisin, tolaasin, and syringomycin (Nybroe and Sorensen, 2004).

1. *Viscosin*: Viscosin group has a peptide moiety consisting of 9 amino acids.
2. *Amphisin*: Amphisin group harbors 11 amino acids in the peptide moiety.
3. *Tolaasin*: Tolaasin group has a length of 19–25 amino acids containing unusual 2,3-dehydro-2-aminobutyric acid and homoserine consisting of the lipid tail of 3-hydroxydecanoic acid or 3-hydroxyoctanoic acid.
4. *Syringomycin*: Syringomycin group has 9 amino acids and contains 2,4-diaminobutyric acid and a C-terminal chlorinated threonine residue (Grgurina et al., 1994; Bender et al., 1999; Bender and Scholz-Schroeder, 2004; Gross and Loper, 2009). The lipid tail consists of 3-hydroxydecanoic acid in both cases.

   Recently, putisolvins I and II from *Pseudomonas putida* have been discovered to have a hexanoic lipid tail (Kuiper et al., 2004), also orfamide from *Pseudomonas fluorescens* (Paulsen et al., 2005; Gross et al., 2007), Pseudomensins A and B from *Pseudomonas* sp. (Sinnaeve et al., 2009), and syringofactin from *P. aeruginosa* have also been studied (Berti et al., 2007).

   Lipopeptides from *Bacillus* sp. are classified into three families of cyclic compounds, i.e., surfactin, iturin, and fengycin. The three families contain variants with the same peptides but heterogeneity in residues at specific positions (Ongena and Jacques, 2008).
5. *Surfactin*: The surfactin family encompasses the heptapeptide variants of the esperin, lichenysin, pumilacidin, and surfactin groups. The peptide moiety is linked to a β-hydroxyl

fatty acid ($C_{12}-C_{16}$) with linear, iso, or anteiso branches. Surfactins are macrolactone rings catalyzed by the β-hydroxyl fatty acid and the C-terminal peptide (Figure 7.4). The surface tension reduction by this lipopeptide biosurfactant can range from 72 to 27 mN/m at concentrations with less than 0.005%, making it one of the most powerful biosurfactants having wide applications in medical and environmental aspects. This is the most relevant cyclic lipopeptide produced by *Bacillus subtilis*, because of its very high surface activity. Surfactin has a CMC of 25−50 mg/L, while the lowest interfacial tension against n-hexadecane is 1 mN/m.

6. *Iturin*: This family consists of six variants: iturin A and C, bacillomycin D, F, L, and mycosubtilin. The structural linkage consists of a heptapeptide attached to a β-amino fatty acid chain of variable length ($C_{14}-C_{17}$).
7. *Fengycins (Plisplastins—when Tyr9 is D configured)*: These are decapeptides linked to β-hydroxyl fatty acid chain ($C_{14}-C_{18}$) in linear, iso, or anteiso form in saturated or unsaturated form.
8. *Lichenysin*: *Bacillus licheniformis* synthesizes lichenysin which acts synergistically and exhibits excellent stability at unfavorable temperature, pH, and salt conditions. They have high structural and physicochemical similarities with surfactin. The main difference is the presence of a glutaminyl residue in position 1 of the peptide sequence in place of glutamic acid in surfactin (Figure 7.5). This minor variation causes significant changes in the properties of the molecule compared to surfactin. Lichenysins are capable of lowering the surface

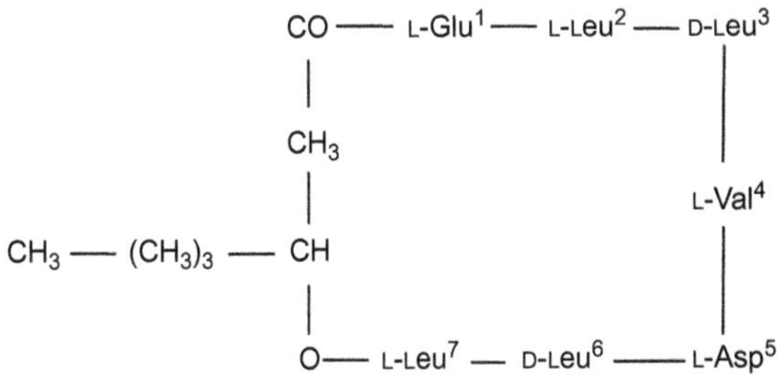

Figure 7.4 Chemical structure of surfactin.

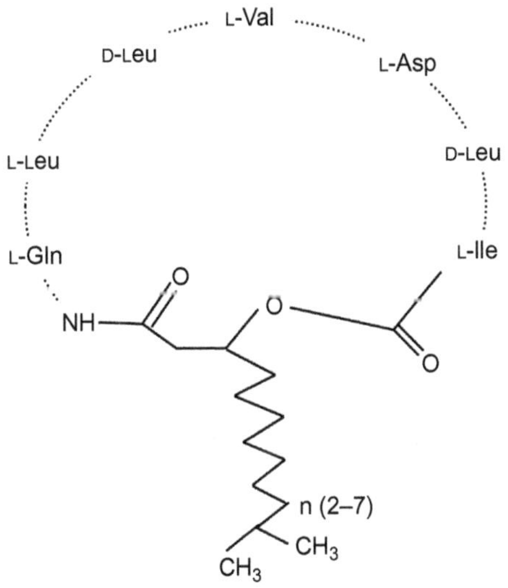

Figure 7.5 Chemical structure of lichenysin.

tension of water to 27 mN/m and the interfacial tension between water and hexadecane to 0.36 mN/m. The CMC is strongly reduced from 220 to 22 μM and they have a much higher hemolytic activity. Lichenysin is a better chelating agent having high association constants with Ca^{2+} and Mg^{2+} increased by a factor of 4 and 16, respectively. This change is attributed to an increase in the accessibility of the carboxyl group to cations due to a change in the side chain topology induced by the Glu/Gln exchange (Grangemard et al., 2001).

Bacillus spp. produce a number of cyclic lipopeptide antibiotics during the early stages of sporulation. One of the examples, *Bacillus polymyxa* produces polymixin, a decapeptide in which amino acids 3–10 form a ring structure, linked to a branched fatty acid, while *Bacillus brevis* produces gramicidin S, a cyclic decapeptide consisting of a rigid ring with two positively charged ornithine side chains on one side and the hydrophobic side chains of the other residues on the other side. Apart from the major lipopeptides discussed above, various lipopeptides have been studied in other *Bacillus* spp. like kurstakin from *Bacillus thuringiensis* (a heptapeptide with the residues Thr–Gly–Ala–Ser–His–Gln–Gln) (Hathout et al., 2000), the 12 amino acid-containing maltacines from *B. subtilis* (Hagelin et al., 2007), polymyxins from *B. polymyxa* with high diaminobutyric acid content (Storm et al., 1977), and the surfactin-like bamylocin A from *Bacillus amyloliquefaciens* (Lee et al., 2007). Lipopeptides have many functions, e.g., they have role in motility and attachment to surfaces and antimicrobial activity against other pathogenic microorganisms (Nybroe and Sorensen, 2004; Raaijmakers et al., 2006). These also show functioning as signal molecules for coordinated growth and differentiation (Lopez et al., 2009a,b).

7.2.1.3 Fatty Acids, Phospholipids, and Neutral Lipids

Fatty acids having surfactant properties are a result of microbial oxidation of alkanes (Rehn and Reiff, 1981). The length of the hydrocarbon chain in their structures determines the hydrophilic and lipophilic balance of the molecule. Saturated fatty acid should be in the range of C_{12}–C_{14} for effective lowering of surface and interfacial tension (Rosenberg and Ron, 1999). One primary example is corynomucolic acid, which acts as surfactant (Kretschmer et al., 1982). In *Acinetobacter* sp., phosphatidyl ethanolamine-rich vesicles are produced, which form optically clear microemulsions of alkanes in water. Phosphatidyl ethanolamine produced by *R. erythropolis* grown on alkane causes a lowering of interfacial tension between water and hexadecane to less than 1 mN/m at CMC 30 mg/L (Kretschmer et al., 1982).

7.2.1.4 Polymeric Biosurfactants

Emulsan, liposan, alasan, lipomanan, and other polysaccharide–protein complexes are good examples of polymeric biosurfactants. *Acinetobacter calcoaceticus* RAG-1 produces an extracellular potent polyanionic amphipathic heteropolysaccharide bioemulsifier (Rosenberg et al., 1979). Emulsan is an effective emisifying agent for hydrocarbons in water, even at concentrations as low as 0.001–0.01%. Liposan is an extracellular water-soluble emulsifier synthesized by *Candida lipolytica* and is composed of 83% carbohydrate and 17% protein.

7.2.1.5 Particulate Biosurfactants

Extracellular vesicles that play a role in hydrocarbon uptake by cells and microbial cells with surface-active properties are referred to as “particulate” biosurfactants

(Rosenberg, 1986). Extracellular membrane vesicles partition hydrocarbons to form a microemulsion, which plays an important role in alkane uptake by microbial cells. Vesicles of *Acinetobacter* sp. strain HO1-N with a diameter of 20–50 nm and a buoyant density of 1.158 g/cm^3 are composed of protein, phospholipids, and LPS (Kappeli and Finnerty, 1980).

7.2.2 Microorganisms Producing Biosurfactants

Diverse assemblages of microorganisms have been discovered that reduce surface tension and show a high emulsification index. Bacteria-producing biosurfactants include *Ochrobactrum* sp. *and Brevibacterium* sp. from crude oil-contaminated regions (Ferhat et al., 2011). *Brevibacillus* sp., *Dietzia* sp., *Pusillimonas* sp., *Sphingopyxis* sp., and *Achromobacter* sp. were reported as emulsifying agents (Tambekar and Gadakh, 2013). *Bacillus* sp. is a potential biosurfactant-producing strain having wide application in enhanced oil recovery (Amin, 2010). Different microbial genera, namely *Acinetobacter*, *Arthrobacter*, *Pseudomonas*, *Halomonas*, *Bacillus*, *Rhodococcus*, *Enterobacter*, and yeasts have been reported to produce biosurfactants (Schulz et al., 1991; Passeri et al., 1992; Banat, 1993; Abraham et al., 1998; Maneerat, 2005; Perfumo et al., 2006; Das et al., 2008a–c). Most of the microorganisms solubilize hydrophobic compounds in their environment by producing biosurfactants to utilize them as substrates in case of nutrition stress conditions (Floodgate, 1978; Margesin and Schinner, 2001a; Olivera et al., 2003). *A. calcoaceticus* isolated from the Mediterranean Sea produced a biosurfactant which was commercially known as Emulsan (Reisfeld et al., 1972; Rosenberg et al., 1979). *Bacillus cereus*, *Bacillus sphaericus*, *B. fusiformis fusiformis*, *Acinetobacter junii*, *Pseudomonas* sp., and *Bacillus pumilus* are good biosurfactant producers as proven with the various screening techniques and have profound surface tension reduction ability with diesel oil as the sole carbon source (Bento et al., 2005). Bioremediation of these hydrocarbons is dependent upon the microbial ability to degrade these complex mixtures and their rate-limiting kinetics (Margesin and Schinner, 2001b). *Aneurinibacillus migulanus*, *P. aeruginosa*, *Achromobacter insolitus*, *Bacillus circulans*, *Nocardia farcinica*, *Ochrobactrum pseudintermedium*, *O. intermedium*, *Stenotrophomonas maltophilia*, *Halobacterium salinarum*, and *Ochrobactrum oryzae* have been shown to produce biosurfactant from petroleum-contaminated sites (Tambekar and Gadakh, 2013). A thermophilic bacterial strain with biosurfactant-producing capability, *A. calcoaceticus* BU03, was isolated from petroleum-contaminated soil with an increased solubilization of polycyclic aromatic hydrocarbons (PAHs), i.e., naphthalene and phenanthrene (PHE) (Zhao and Wong, 2009). The bacterium *Myroides odoratus* was isolated from lepidopteran gut and was hypothesized to produce some biosurfactants (Spiteller et al., 2000). *Stenotrophomonas koreensis* is a novel biosurfactant-producing bacterium isolated from petroleum-contaminated sites and is also highly efficient in heavy metal removal from the environment (Patil et al., 2012). Various applications of biosurfactants with their microbial origins are listed in Table 7.1.

Table 7.1 Classification, Microbial Origin, and Application of Biosurfactants

Type	Class	Microorganisms Producing	Applications	References
Glycolipid	Rhamnolipid	*P. aeruginosa*, *Pseudomomas* sp., *Serratia rubidea*, *Tetragenococcus koreensis*, *Pseudomonas chlororaphis*, *Acinetobacter*, *Enterobacter*	Enhancement of the degradation and dispersion of different classes of hydrocarbons; emulsification of hydrocarbons and vegetable oils; removal of metals from soil	Herman et al. (1995), Maier and Soberon-Chavez et al. (2000), Lee et al. (2005), Gunther et al. (2005), Sifour et al. (2007), Whang et al. (2008), Hoskova et al. (2013)
	Trehalolipid	*Arthrobacter paraffineus*, *Corynebacterium* sp., *R. erythropolis*, *Nocardia* sp., *Mycobacterium tuberculosis*, *Micrococcus*, *Gordonia*, *Brevibacteria*	Enhancement of the bioavailability of hydrocarbons	Franzetti et al. (2010)
	Sophorolipid	*T. bombicola*, *T. petrophilum*, *T. apicola*, *C. apicola*, *C. bombicola*, *C. lipolytica*, *C. bogoriensis*	Recovery of hydrocarbons from dregs and muds; removal of heavy metals from sediments; enhancement of oil recovery	Baviere et al. (1994), Pesce (2002), Whang et al. (2008)
Lipopeptide	Surfactin	*B. subtilis*, *B. pumilus*, *B. velezensis*	Enhancement of the biodegradation of hydrocarbons and chlorinated pesticides; removal of heavy metals from a contaminated soil, sediment and water	Arima et al. (1968), Awashti et al. (1999), Ruiz-Garcia et al. (2005)

	Lichenysin	*B. licheniformis*, *Arthrobacter* sp.	Enhancement of oil recovery	Thomas et al. (1993)
	Viscosin	*P. fluorescens*	Pathogenicity factor of pectolytic strain causes decay of the difficult-to-wet waxy surfaces of broccoli heads	Hildebrand et al. (1998)
	Amphisin	*Pseudomonas* sp.	Inhibits the growth of plant pathogenic fungi	Koch et al. (2002)
	Tolaasin	*Pseudomonas tolaasi*	Helps in binding with metal ions	Lee et al. (2011)
	Syringomycin	*P. syringae*	Function as detergents to dissolve plant membranes at high concentrations	Gross (1985)
	Iturin	*Bacillus* sp., *B. amyloliquefacians*, *B. pumilus*	Inhibits growth of aflatoxin-producing fungi	Phister et al. (2004)
	Fengycin	*B. subtilis*, *B. amyloliquefaciens*, *B. thuringiensis*	Acts as a fungicide	Vanittanakom et al. (1986), Hu et al. (2007)
	Lichenysin	*B. licheniformis*	Boar sperm motility inhibition, hemolytic, significant for oil recovery	Yakimov et al. (1995), Grangemard et al. (2001), Hoornstra et al. (2003), Nerurkar (2010)
Fatty acids and neutral lipids		*Capnocytophaga* sp., *Penicillium spiculisporum*, *Corynebacterium lepus*, *Arthrobacter paraffineus*, *Talaramyces trachyspermus*, *Nocardia erythropolis*, *Acinetobacter*, *Pseudomonas* sp., *Micrococcus* sp., *Mycococcus* sp., *Candida* sp., *Aspergillus* sp.	Potent surfactant	Peypoux et al. (1999), Quentin et al. (1982)

(Continued)

Table 7.1 (Continued)

Type	Class	Microorganisms Producing	Applications	References
	Corynomycolic acid	*Corynebacterium lepus*	Enhancement of bitumen recovery	Gerson and Zajic (1978)
	Spiculisporic acid	*P. spiculisporum*	Removal of metal ions from aqueous solution; dispersion action for hydrophilic pigments; preparation of new emulsion-type organogels, superfine microcapsules (vesicles or liposomes), heavy metal sequestrants	Ishigami et al. (1983, 2000), Hong et al. (1998)
	Phosphatidylethanolamine	*Acinetobacter* sp., *R. erythropolis*	Increasing the tolerance of bacteria to heavy metals	Appanna et al. (1995)
Phospholipids		*Acinetobacter* sp., *Candida* sp., *Corynebacterium* sp., *Micrococcus* sp., *Thiobacillus* sp.	Helps in wetting elemental sulfur, which is necessary for growth	Hasumi et al. (1995)
Polymeric biosurfactants	Alasan Emulsan	*A. radioresistens* *A. calcoaceticus* RAG-1	Stabilization of the hydrocarbon-in-water emulsions	Zosim et al. (1982), Toren et al. (2001)
	Biodispersan	*A. calcoaceticus* A2	Dispersion of limestone in water	Rosenberg et al. (1988)
	Liposan Mannoprotein	*C. lipolytica* *Saccharomyces cerevisiae*	Stabilization of hydrocarbon-in-water emulsions	Cirigliano and Carman (1984)

7.3 Biosurfactant-Based Toxic Metal Remediation

Remediation of soil contaminated with toxic metals is the need of the hour to endeavor a greener approach toward a clean environment. Metals like cobalt, copper, iron, molybdenum, manganese and zinc play vital role in various biological pathways. At excessive levels they can be damaging to organisms. The other heavy metals like cadmium, lead, mercury, and plutonium are toxic metals and their accumulation over time in the bodies of animals can cause serious illness. Metals are cationic species; therefore, there is a difference between organic compound removal and metal removal. Contaminant sorption in soil by the biosurfactants depends on the chemical and physiochemical properties of the contaminant in complexation with the biosurfactant.

The mechanism of biosurfactant-mediated toxic metal sorption is promoted from solid phases in basic two ways. Usually the first way is mediated through complexation of the free form of metal present in the solution. This diminishes the solution-phase activity of the metal and promotes desorption based on Le Chatelier's principle. The second way is accumulation of metal with the biosurfactant under conditions of reduced interfacial tension at the solid–solution interface. This allows direct contact between the biosurfactant and the sorbed metal. The mechanisms of heavy metal extraction by microbial biosurfactants are ion exchange, precipitation–dissolution, and counter binding. In the case of the ion exchange mechanism, the anionic biosurfactant carrying a negative charge binds with the cationic metal ion carrying a positive charge, forming a stronger bond than the soil–metal bond. The polar head groups of the micelles can bind metals, making them more soluble in water (Figure 7.6). Biosurfactant complexed with metals are nontoxic to bacterial cells. This interaction can reduce metal toxicity, but the detailed mechanism is still unclear. One study was carried out on cadmium toxicity reduction during naphthalene biodegradation by *P. putida* strain (Sandrin et al., 2000). The concept here applies that metal chelators, such as EDTA, alter cell surface properties through the release of LPS (Gray and Wilkinson, 1965; Leive, 1965; Gilleland et al., 1973; Goldberg, et al., 1983). As LPS confers a considerable negative charge upon the cell surface (Remacle, 1990), it favors electrostatic interactions with cations, especially the metals. Removal of LPS may decrease the degree of the negativity of the cell surface charge, thus reducing interactions with cations, such as cadmium. The detailed mechanism by which metal-chelating agents reduce toxicity clearly needs further investigation.

The major efficacy of metal-activated biosurfactant action depends on several factors like soil composition, soil pH, cation-exchange capacity (CEC), particle size, time and type of contamination, and geological layout. The long duration of contaminance gives the metal ample time to stabilize and hence removal tends to be more complicated. Research has shown that metals such as lead and cadmium have stronger affinities for rhamnolipid than for many of the soil components to which they are bound in contaminated soils (Ochoa-Loza et al., 2001). Significant work has been carried out on rhamnolipid biosurfactant produced by various *P. aeruginosa* strains capable of selectively complexing cationic metal species such as Cd, Pb, and Zn. A study showed that a 5 mM solution of rhamnolipid produced

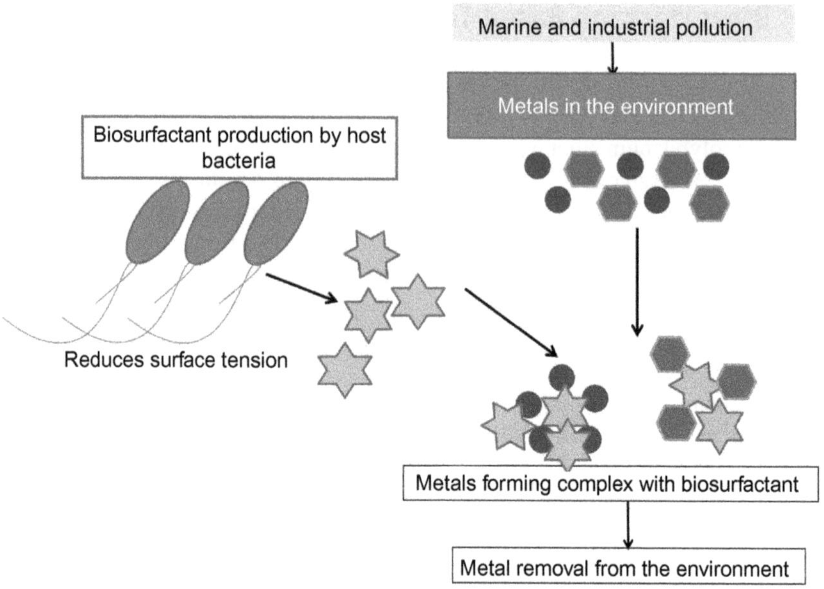

Figure 7.6 The mechanism of metal–biosurfactant complex formation.

by *P. aeruginosa* complexed 92% of cadmium, a complexation of 22 L g/mg rhamnolipid (Wang et al., 1997). Mulligan et al. (2001a) estimated the practicability of using surfactin, rhamnolipid, and sophorolipid for the removal of heavy metals like Cu and Zn from sediments. This showed that a single washing with 0.5% rhamnolipid removed 65% of copper and 18% of zinc, whereas 4% sophorolipid removed 25% of copper and 60% of zinc (Mulligan et al., 2001b). The effectiveness of surfactin was not as much as other biosurfactants, removing 15% of copper and 6% of zinc. Therefore, extraction of sediments after washing with the various surfactants indicated that the biosurfactants, rhamnolipid and surfactin, could efficiently remove the organically bound copper, and sophorolipid could remove the carbonate and oxide-bound zinc. It has been postulated that metal removal by the biosurfactants occurs through sorption of the surfactant onto the soil surface and complexation with the metal, separation of the metal from the soil into solution, and hence association with surfactant micelles. One industrial application of biosurfactants for metal removal from soil is the treatment of contaminated soil in a huge cement mixer followed by treatment with biosurfactants, which results in biosurfactant-metal complex removal leaving the soil behind. The interaction between the positively charged metal and the negatively charged surfactant is so strong that flushing water through soil removes the surfactant metal complex from the soil matrix (Mulligan, 2005). The metal further can be extracted from the biosurfactant-metal complex and used for different applications. A similar study by Jeong-Jin et al. (1998) illustrated that biosurfactant-based ultrafiltration was used to remove the divalent metal ions like copper, zinc, cadmium, and nickel from an aqueous solution containing either a single metal species or a mixture of metal ions. The vast applications of biosurfactant in organic- as well as metal-contaminated sites are attributed to its small size, lower toxicity, high specificity, biodegradability, and high cost-effectiveness.

7.4 Genetic Basis of Biosurfactant Production

Genetic basis of biosurfactant production and consideration of the genetic regulatory mechanisms facilitates in developing metabolically engineered strains with superior product characteristics and acquired potential of utilizing inexpensive agro-industrial wastes as substrates.

7.4.1 Surfactin Production

Most *Bacillus* species synthesize surfactin, a potent biosurfactant whose biosynthesis is catalyzed nonribosomally by a large multienzyme peptide synthetase complex, surfactin synthetase, consisting of three protein subunits—*SrfA*, *ComA* (earlier known as *SrfB*), and *SrfC* (Figure 7.7A). The function of amino acid moiety

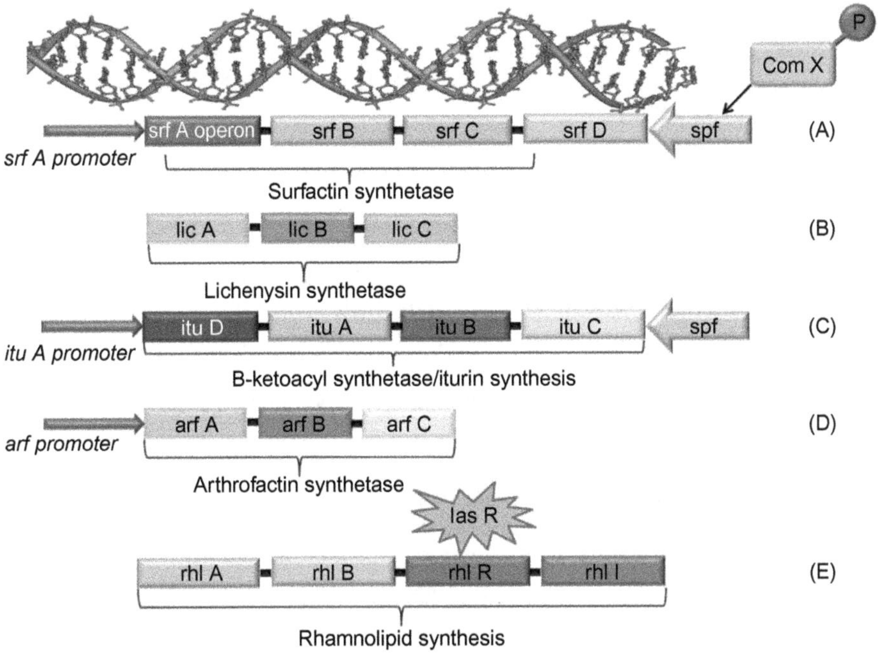

Figure 7.7 Structural organizations of the genes encoding various biosurfactant synthetases: (A) surfactin, (B) lichenysin, (C) iturin, (D) arthrofactin, and (E) rhamnolipid.

encoded by the peptide synthetase enzyme of surfactin is encoded by four open reading frames (ORFs) in the *srfA* operon, namely *srfAA*, *srfAB*, *srfAC*, and *srfAD* or *srfA*. This operon also contains a *comS* gene lying within and out-of-frame with the *srfB*. Deletion mutant studies have indicated that *srfAD* is not essential for surfactin biosynthesis, whereas the other three ORFs are enormously essential for this process. *sfp* is another gene-encoding phosphopantetheinyl transferase required for activation of surfactin synthetase by post-translational modification. It is absolutely essential for surfactin production because few mutants have been found that have all the genes required for surfactin biosynthesis, with the exception of *sfp*. Acyl transferase is one more gene that is responsible for the transfer of hydroxy fatty acid moiety to SrfAA but it is yet to be characterized (Peypoux et al., 1999). Menkhaus et al. (1993) has reported one species of *B. subtilis* that could regulate surfactin production by cell-density responsive mechanism utilizing a peptide pheromone, ComX. On an increase in cell density of the bacteria, ComX, a signal peptide, accumulates in the growth medium. The gene product ComQ activates ComX, assembling it to form a signal peptide. *srfA* gene expression is controlled by quorum sensing by ComX, which, when it interacts with ComP and ComA, activates the signal transduction system. The histidine protein kinase ComP donates a phosphate to the response regulator ComA, which gets activated and stimulates the transcription of the *srf* operon. ComA-phosphate phosphatase RapC is inhibited by the pheromone CSF which results in activation of *srf* transcription. The introduction of CSF inside the cell, which is an extracellular peptide, is done by the oligopeptide permease SpoOK. ComR and SinR also influence *srfA* expression of which ComR post-transcriptionally enhances *srfA* expression and SinR negatively controls *srfA* by regulating *comR* (Liu et al., 1996; Luttinger et al., 1996; Cosby et al., 1998).

7.4.2 Lichenysin Biosurfactant

A biosurfactant produced by *B. licheniformis* under growth in aerobic and anaerobic conditions is lichenysin (Yakimov et al., 1995). Structural genes required for lichenysin synthesis have been isolated and they show high sequence homology with those of surfactin. Therefore, the biosynthesis of both of these substances follows similar pathways. Lichenysin-like surfactin is synthesized nonribosomally by a multienzyme peptide synthetase complex. Identification of the cloned putative lichenysinA synthetase operon revealed that it contains seven amino acid activation–thiolation, two epimerization, and one thioesterase domain similar to that of surfactin (Yakimov et al., 1998). The lichenysin biosynthesis operon from *B. licheniformis* ATCC 10716 has been cloned and sequenced. The lichenysin operon consists of three peptide synthetase genes *licA*, *licB*, and *licC* (Figure 7.7B) and they are transcribed in the same direction (Marahiel et al., 1999). The *lic* operon of *B. licheniformis* is 26.6 kb long and consists of genes *licA* (three modules), *licB* (three modules), and *licC* (one module). The domain structures of these seven modules of lichenysin synthetase resemble those of surfactin synthetases SrfA-C. The modular organization of lichenysin synthetases LicA to LicC was also found to be exactly identical with that of surfactin synthetases. There is another gene called *licTE* which codes for a thioesterase-like protein (Yakimov et al., 1998).

7.4.3 Iturin Lipopeptide

IturinA is an antifungal lipopeptide biosurfactant produced by certain *B. subtilis* strains such as *B. subtilis* RB14. IturinA operon is composed of four ORFs, *ituD*, *ituA*, *ituB*, and *ituC* (Figure 7.7C). The *ituD* gene encodes a putative malonyl coenzymeA transacylase, whose disruption results in a specific deficiency in iturinA production. The striking feature of ItuA is that the three functional domains homologous to β-ketoacyl synthetase, amino transferase, and amino acid adenylation are combined. The *ituB* gene encodes a peptide synthetase consisting of four amino acid adenylation domains, two of which are flanked by an epimerization domain. The *ituC* gene encodes another peptide synthetase that has two adenylation domains, one epimerization domain and a thioesterase domain which probably helps in peptide cyclization. On replacement of the promoter of the iturin operon by the *repU* promoter of the plasmid pUB110 replication protein, a threefold increase in the production of iturinA was observed (Tsuge et al., 2001).

7.4.4 Arthrofactin Lipopeptide

The genes responsible for lipopeptide biosurfactant biosynthesis code for nonribosomal peptide synthetases (NRPSs) which are multimodular enzyme complexes. They display a high degree of structural similarity among themselves of which the lipopeptide-producing genes from *Bacillus* and *Pseudomonas* species are highly similar in structural organization. Three genes of arthrofactin operon of *Pseudomonas* spp. are *arfA*, *arfB*, and *arfC* (Figure 7.7D) which encode ArfA, ArfB, and ArfC respectively and contain two, four, and five functional modules. Each module has condensation, adenylation, and thiolation domains, but there is no epimerization domain (Roongsawang et al., 2003). Arthrofactin produced by *Pseudomonas* sp. MIS38 is the most potent cyclic lipopeptide type biosurfactant ever reported. Three genes termed *arfA*, *arfB*, and *arfC* form the arthrofactin synthetase gene cluster and encode ArfA, ArfB, and ArfC which assemble to form a unique structure. ArfA, ArfB, and ArfC contain two, four, and five functional modules, respectively. A module is defined as the unit that catalyzes the incorporation of a specific amino acid into the peptide product. The arrangement of the modules of a peptide synthetase is usually collinear with the amino acid sequence of the peptide. The modules can be further subdivided into different domains that are characterized by a set of short conserved sequence motifs. Each module bears a condensation domain [C] (responsible for formation of a peptide bond between two consecutively bound amino acids), an adenylation domain [A] (responsible for amino acid recognition and adenylation at the expense of ATP), and a thiolation domain [T] (serves as an attachment site of 4-phosphopantetheine cofactor and a carrier of thioesterified amino acid intermediates). However, none of the 11 modules possesses the epimerization domain [E] responsible for the conversion of amino acid residues from L to D form. Moreover, two thioesterase domains are tandemly located at the C-terminal end of ArfC. *arfB* is the gene absolutely essential for arthrofactin production as its disruption impaired this act (Roongsawang et al., 2003).

7.4.5 Rhamnolipid Biosurfactant

Rhamnolipid, a glycolipid biosurfactant produced by *Pseudomonas* spp., has many structural and regulatory genes encoding the rhamnolipid synthesis pathway (Figure 7.7E). The complex mechanisms involved in rhamnolipid synthesis facilitates the overproduction of these extracellular compounds, allowing their production in heterologous hosts under controlled conditions. High yield has been obtained in an industrial scale by continuous cultivation under optimized media and growth conditions by using refined methods of cell recycling, gas exchange, and downstream processing (Daniels et al., 2004). The *rhl* quorum-sensing system in *P. aeruginosa* regulates the production of the rhamnolipid type of biosurfactants. Rhamnolipid 1 was obtained from *P. aeruginosa* KY 4025 when the culture was grown on 10% alkane (Ochsner et al., 1996). *P. aeruginosa* S7B1 formed rhamnolipid 2 while growing on n-hexadecane and n-paraffin. It was also the first rhamnolipid to be identified (Itoh, 1971). Rhamnolipids 3 and 4 were synthesized by resting cells only (Hisatsuka et al., 1971; Syldatk et al., 1985a,b). Genetic details of rhamnolipid biosynthesis were obtained from genetic complementation of a mutant strain of *P. aeruginosa* PG 201 with the wild type. Genes involved in rhamnolipid biosynthesis are plasmid encoded. *rhlA*, *rhlB*, *rhlR*, and *rhlI* genes are required for production of rhamnolipids in a heterologous host (Ochsner et al., 1995) and they are transcribed in 5′-*rhlABRI*-3′ direction. According to a proposed biosynthetic pathway, rhamnolipid synthesis proceeds by two sequential glycosyl transfer reactions, each catalyzed by a different rhamnosyl transferase (Burger et al., 1963). Rhamnolipid 1 synthesis is catalyzed by the enzyme rhamnosyl transferase 1, an *rhlAB* gene product, organized in one operon. Both the genes are co-expressed from the same promoter and are essential for rhamnolipid synthesis. RhlA is presumably involved in the synthesis or transport of rhamnosyl transferase precursor substrates or in the stabilization of the RhlB protein (Ochsner et al., 1994). The second rhamnosyl transferase, encoded by *rhlC*, has been characterized and its expression has been shown to be coordinately regulated with *rhlAB* by the same quorum-sensing system (Rahim et al., 2001). The *rhlR* and *rhlI* act as regulators of the *rhlAB* gene expression. RhlI protein forms *N*-acylhomoserine lactones, which act as autoinducers and influence RhlR regulator protein. Induction of *rhlAB* depends on quorum-sensing transcription activator RhlR complexed with the autoinducer *N*-butyryl homoserine lactone (C4-HSL). Another quorum-sensing system encoded by *lasR* and *lasI* has an influence on rhamnolipid biosynthesis. The *las* system is both a positive and a negative regulator of the *rhl* system (Pesci et al., 1997). The *lasI* and *rhlI* products are *N*-oxododecanoyl homoserine lactone (OdDHL, 3OC12HSL or PAI-1) (Pearson et al., 1994) and *N*-butyryl homoserine lactone (BHL, C4-HSL, or PAI-2), respectively (Winson et al., 1995). The *las* system regulates the *rhl* system which in turn regulates rhamnolipid synthesis. It has been found that transcription of *rhlAB* genes involves σ54 and this is overexpressed under nitrogen-limiting conditions.

7.4.6 Viscosin

Viscosin is produced by *P. fluorescens* PfA7B. It acts as a wetting agent and thus the bacterium becomes able to adhere to broccoli heads and cause decay of the wounded as well as unwounded florets of broccoli. Triparental matings of mutants with their corresponding wild-type clones and the helper *Escherichia coli* HB101 (with the mobilizable plasmid pPK2013) yielded transconjugants. Their linkage maps indicated that a 25 kb chromosomal DNA after transcription and translation forms three proteins which forms a synthetase complex and is required for viscosin production.

7.4.7 Amphisin

Amphisin is produced by *Pseudomonas* sp. DSS73. It has both biosurfactant and antifungal properties which brings about the inhibition of plant pathogenic fungi. The two-component regulatory system GacA/GacS (GacA is a response regulator and GacS is a sensor kinase) controls the amphisin synthetase gene (*amsY*) (Koch et al., 2002). The surface motility of this bacterium requires the production of this biosurfactant as is indicated by the mutants defective in the genes *gacS* and *amsY*. Amphisin synthesis is regulated by *gacS* gene as the *gacS* mutant regains the property of surface motility upon the introduction of a plasmid encoding the heterologous wild-type *gacS* gene from *Pseudomonas syringae* (Andersen et al., 2003).

7.4.8 Putisolvin

P. putida PCL1445 produces two surface-active cyclic lipopeptides designated as putisolvins I and II. The ORF encoding the synthesis of the putisolvins bears amino acid homology to various lipopeptide synthetases (Kuiper et al., 2004). Putisolvins are produced by a putisolvin synthetase designated as *psoA*. Three heat shock genes, *dnaK*, *dnaJ*, and *grpE*, positively regulate the biosynthesis of putisolvin (Dubern et al., 2005). The *ppuI*–*rsaL*–*ppuR* quorum-sensing system controls putisolvin biosynthesis. *ppuI* and *ppuR* mutants exhibit decreased putisolvin production, whereas *rsaL* mutants show enhanced putisolvin production (Dubern et al., 2006).

7.4.9 Emulsan and Alasan

Acinetobacter spp. are known to produce high-molecular-weight biosurfactants—emulsan and alasan (Navon-Venezia, et al., 1995; Bach et al., 2003). The RAG-1 emulsan of *Acinetobacter* is a noncovalently linked complex of a lipoheteropolysaccharide and a protein. The polysaccharide part, called apoemulsan, consists of various sugar components such as D-galactosamine, D-galactosaminuronic acid, and diaminodideoxy glucosamine. The fatty acids make 12% of this biopolymer and make it amphipathic in nature. The BD4 emulsan of *A. calcoaceticus* BD4 consists of a repeating heptasaccharide unit comprising L-rhamnose, D-glucose, D-glucuronic acid, and D-mannose in molar ratios of 4:1:1:1 (Kaplan et al., 1987). On the other hand, alasan produced by *Acinetobacter radioresistens* is an anionic, high-molecular-weight, alanine-containing heteropolysaccharide and protein. *Acinetobacter lwoffii* RAG-1 produces a potent bioemulsifier, emulsan. The logarithmic phase cells of this bacterium secrete this compound as a mini-capsule on the cell surface which is, however, released into the medium as a protein–polysaccharide complex when the cells reach the stationary state. This release is caused by an esterase which, if removed, forms a polymer called apoemulsan which cannot bring about the emulsification of nonpolar, hydrophobic, aliphatic materials (Zosim et al., 1986). A 27 kb gene cluster termed *wee* encodes the genes (*wza, wzb, wzc, wzx, wzy*) required for emulsan biosynthesis (Nakar and Gutnick, 2001). It was later demonstrated that Wzc and Wzb are protein tyrosine kinase and protein tyrosine phosphatase respectively, and deletion in either of the two genes gives rise to an emulsan-defective phenotype (Nakar and Gutnick, 2003). *Acinetobacter venetianus* RAG-1 also forms emulsan. Removal of the protein fraction yields apoemulsan, which exhibits much lower emulsifying activity on hydrophobic substrates such as n-hexadecane. The genes encoding the biosynthetic enzymes required for the synthesis of apoemulsan has been cloned and sequenced. One key protein associated with the emulsan complex is a cell surface esterase. The esterase has been cloned and overexpressed in *E. coli* BL21 (DE3) behind the phage T7 promoter with the His tag system. After overexpression, most of the proteins were found in inclusion bodies (Bach et al., 2003).

7.4.10 Serrawettin

Serratia, a group of Gram-negative bacteria, produce surface-active cyclodepsipeptides known as serrawettin W1, W2, and W3 (Matsuyama et al., 1986, 1989). Different strains of *Serratia marcescens* produce these different serrawettins, e.g., serrawettin W1 is produced by strains 274 and ATCC 13880 or NS 38, W2 is produced by strain NS 25, and W3 is produced by strain NS 45. Besides this, *Serratia liquefaciens* produces serrawettin W2. Temperature-dependent synthesis of two novel lipids—rubiwettin R1 and RG1—is observed in *Serratia rubidaea* (Matsuyama et al., 1990). *S. marcescens* forms a biosurfactant serrawettin W1. A single gene *pswP* is responsible for the production of this biosurfactant. This gene has a high homology with gene of the NRPSs family. Another serrawettin W1 synthetase putative gene *swrW* was identified through genetic analysis of serrawettin-less mutants of *S. marcescens* 274 (Das et al., 2008c). Homology analysis of this gene demonstrated the presence of condensation, adenylation, thiolation, and thioesterase domains characteristic of NRPS. This putative serrawettin synthetase gene was uni-modular in contrast to the multimodular nature of NRPS. This presumed that SwrW may be the simplest enzyme in the NRPS family (Li et al., 2005). *S. liquefaciens* MG1 forms a biosurfactant, serrawettin W2. Its synthesis is catalyzed by a peptide synthetase which is encoded by the *swrA* gene. The *swrI* gene product catalyzes the formation of *N*-butanoyl-L-homoserine lactone (BHL) and *N*-hexanoyl-L-homoserine lactone (HHL).

7.4.11 Fungal Surfactants

Various fungi secrete glycolipid types of surface-active agents but the genetic basis of their production is largely unknown. Mannosylerythritol lipids (MEL) were first isolated from the dimorphic fungus *Ustilago maydis* and were also detected later in *Candida antarctica*, *Schizonella melanogramma*, and *Geotrichum candidum*. Sophorose lipids are secreted by *Candida bombicola*. *U. maydis* produces two kinds of glycolipid biosurfactants, mannosylerythritol lipid (MEL), referred to as ustilipids (Uchida et al., 1989), and ustilagic acid, that are cellobiose lipids. These compounds are secondary metabolites as indicated by the fact that the mutants generated by deletion of the genes involved in their production are not lethal. Two genes, *emt1* and *cyp1*, are involved in the synthesis of these glycolipids. *emt1* is for MEL synthesis and *cyp1* is for ustilagic acid production. It is assumed that Cyp1 is involved in terminal and/or subterminal hydroxylation of an unusual fatty acid, 15, 16-dihydroxyhexadecanoic acid, which is present in ustilagic acid. *Trichoderma reesei* forms hydrophobins, which are low-molecular-weight proteins having high cysteine content and high surface and amphiphilic properties (Linder et al., 2001). *hfb1* and *hfb2* are the genes regulating the synthesis of hydrophobins.

7.5 Application in Metal Remediation

Biosurfactants have become an important product of biotechnology for industrial and environmental applications. They are high-value microbial products having low toxicity, relative ease of preparation, and widespread applicability. They are used as emulsifiers, de-emulsifiers, wetting agents, spreading agents, foaming agents, functional food ingredients, and detergents in various industrial sectors such as petroleum and petrochemicals, organic chemicals, foods and beverages, cosmetics and pharmaceuticals, mining and metallurgy, agrochemicals and fertilizers, environmental control and management, and many others.

Biosurfactants and biosurfactant-producing bacterial strains are highly capable of enhancing organic contaminants' availability and biodegradation rates. Apart from metal complexation and environmental cleanup, biosurfactants have manifold applications: they enhance oil recovery, make hydrocarbon bioavailable, enhance biodegradation, increase emulsification, help in microcapsule synthesis, enhance bitumen recovery, and increase phytoextraction (Figure 7.8) Obayori et al. (2009) investigated the biodegradative properties of biosurfactant produced by *Pseudomonas* sp. LP1 strain on crude oil and diesel. They reported 92.34% degradation of crude oil and 95.29% removal of diesel oil. The results obtained confirmed the ability of strain LP1 to metabolize the hydrocarbon components of crude and diesel oil. Reddy et al. (2010) studied the biodegradative properties of biosurfactant-producing *Brevibacterium* sp. and reported that this strain could degrade 93.92% of the PHE and also had the ability to degrade other polyaromatic hydrocarbons (PAHs) such as anthracene and fluorene. Kang et al. (2010) used sophorolipid on biodegradation of aliphatic and aromatic hydrocarbons under laboratory conditions. The addition of this

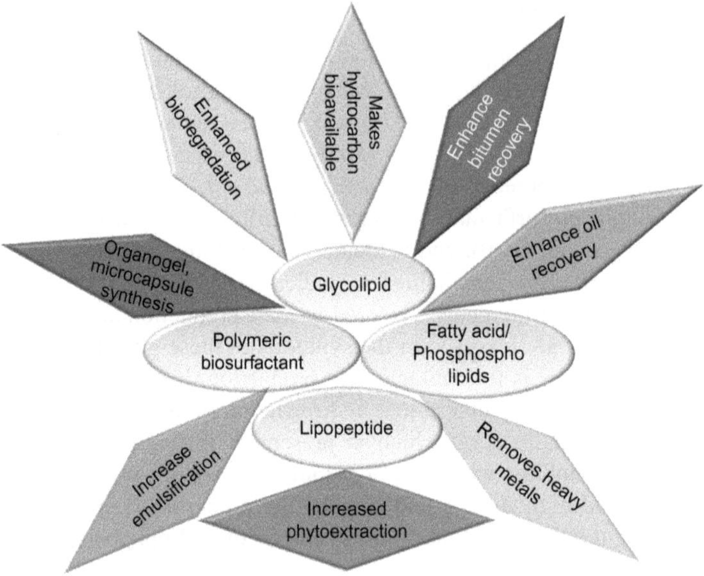

Figure 7.8 Application of biosurfactants in bioremediation.

biosurfactant to soil also increased biodegradation of tested hydrocarbons, with the rate of degradation ranging from 85% to 97% of the total amount of hydrocarbons. Their results indicated that sophorolipid may have the potential to facilitate the bioremediation of sites contaminated with hydrocarbons having limited water solubility and increasing the bioavailability of microbial consortia for biodegradation.

The cell-free culture broth containing the biosurfactants can be applied directly or in diluted form to the contaminated site. The other benefit of this approach is that the biosurfactants are very stable and effective in the culture medium that is used for their synthesis. The usefulness of biosurfactant-producing strains in bioremediation of sites highly contaminated with crude petroleum–oil hydrocarbons was confirmed by Das and Mukherjee (2007). The ability of three biosurfactant-producing strains, *B. subtilis* DM-04, *P. aeruginosa M*, and *P. aeruginosa NM*, to remediate petroleum crude oil-contaminated soil samples was investigated by treating the soil samples with aqueous solutions of biosurfactants obtained from the respective bacteria strains. Bioaugmentation with *P. aeruginosa M* and *NM* consortium and *B. subtilis* strain showed that TPH (total petroleum hydrocarbons) levels were reduced from 84 to 21 and 39 g/kg of soil, respectively. In contrast, the TPH level was decreased to 83 g/kg in control soil. Joseph and Joseph (2009) separated the oil from the petroleum sludge by induced biosurfactant produced by bacteria by directly inoculating it with *Bacillus* sp. strains and by addition of the cell-free supernatant. The biosurfactant displayed the ability to reduce surface and interfacial tensions in both aqueous and hydrocarbon mixtures and hence had potential for oil recovery.

Biosurfactants have often been used to enhance bioavailability and biodegradation of hydrophobic compounds, but there is little knowledge available about the effect of simultaneous emulsifier production on biodegradation of complex hydrocarbon mixtures. Nievas et al. (2008) studied the biodegradation of a bilge waste, which is a fuel oil–type complex residue produced in normal ship operations. Bilge waste is a hazardous waste composed of a mixture of seawater and hydrocarbon residue; n-alkanes, resolvent total hydrocarbons, and unsolvent complex mixture are its main constituents. Insolvent complex mixture is principally composed of branched and cyclic aliphatic hydrocarbons and aromatic hydrocarbons, which usually show the greatest resistance to biodegradation. In their studies, they investigated the biodegradation of an oily bilge waste by an emulsifier-producing microbial consortium. As the result for both levels of oily wastes, 136 g/kg of resolving hydrocarbons and 406 g/kg of unsolvent mixture, they found that all of the hydrocarbon types showed an important concentration reduction from their initial values. They observed that the extent of biodegradation followed the order: n-alkanes > resolved total hydrocarbon > unsolvent complex mixture.

An emulsifier-producing microbial consortium used for biodegradation of bilge wastes showed reduction of n-alkanes, resolvent hydrocarbons, and unsolvent mixture of around 85%, 75%, and 58%, respectively. Barkay et al. (1999) tested the effect of a bioemulsifier alasan produced by *A. radioresistens* KA53 on the solubilization of PAHs, PHE, and fluoranthene (FLA). They also studied the influence of alasan on mineralization of PHE and FLA by *Sphingomonas paucimobilis* EPA505. They indicated that aqueous solubility of PHE and FLA increased linearly in the presence of increasing concentrations of bioemulsifier (50–500 μg/mL) and that mineralization of PAHs by *S. paucimobilis* EPA505 was stimulated by appearance of alasan. The presence of alasan at concentrations of up to 300 μg/mL more than doubled the degradation rate of FLA and significantly increased the degradation rate of PHE. Increasing the alasan concentration over 300 μg/mL had no further effect on PAHs mineralization.

It was estimated by the US Environmental Protection Agency that 40% of sites are co-contaminated with organic and metal pollutants (Sandrin and Maier, 2003). The presence of toxic metals (lead, cadmium, arsenic) in some cases causes inhibition of organic compound biodegradation. However, a number of possible approaches that can lower metal bioavailability and/or increase microbial tolerance to metals have been studied. These include inoculation with metal-resistant microorganisms and addition of materials like clay minerals—kaolinite and montmorillonite, calcium carbonate, phosphate, chelating agents (EDTA), and surfactants. Biosurfactants produced by microorganisms show good results for enhancing organic compound biodegradation in the presence of metals. Application of biosurfactants or microorganism-produced biosurfactants in *in situ* co-contaminated sites for bioremediation seems to be more environmentally compatible and more economical than using modified clay complexes or metal chelators. Sandrin et al. (2000) showed that metal-complexing rhamnolipids reduced metal toxicity to allow enhanced organic biodegradation by *Burkholderia* sp. This suggested that rhamnolipid was able to reduce metal toxicity to microbial consortia in co-contaminated soils through a combination of metal complexation and the alteration of cell surface properties through the release of lipopolysaccharide (LPS), resulting in an enhanced bioremediation effect. Maslin and Maier (2000) studied the effect of rhamnolipids produced by various *P. aeruginosa* strains on PHE degradation by indigenous populations in two soils co-contaminated with PHE and cadmium. The authors showed that rhamnolipids had the ability to complex cationic metals, increasing the PHE bioavailability. The biodegradation of PHE was increased from 7.5% to 35% in one soil, and from 10% to 58% in another, in response to rhamnolipid application.

7.6 Conclusion

With scientific upliftment, ongoing research on mechanisms of interaction among hydrocarbons, surfactants and cells have revealed the potential application of biosurfactant in bioremediation. The complex mixture of different components produced by various organisms impedes applications, and further research is required to resolve specific issues. Biosurfactant-assisted removal of heavy metal ions by complex formation and succeeding mobilization has received much interest. This method is highly efficient and less environmentally toxic as compared with synthetic surfactants. A study carried out by Gao et al. (2012) regarding potential recovery of heavy metal ions in sludge from an industry water treatment plant by application of biosurfactants confirms that biobased surface-active compounds exhibit high selectivity toward certain heavy metal ions. The type of biosurfactant may also impact the removal efficiency, as seen in the case of the effect of saponins, which were found to be superior to sophorolipids. The results obtained by Lima et al. (2011a–c) imply that biosurfactants may be successfully used for simultaneous removal of heavy metal ions and organic pollutants. It was reported that the application of lipopeptides obtained from different bacterial strains notably enhanced the removal rate of cadmium (99%) as well as PHE (80–88%). The primary reason may be the inconsistency between the anticipated role of biosurfactants in contaminant treatment processes and their actual role in the ecology of microorganisms, which by far exceeds the boundaries of bioremediation (Tremblay et al., 2007; Glick et al., 2010; Chrzanowski et al., 2012a,b).

Future studies should be designed not only on an efficiency-focused approach, but also on expanding this challenging problem by elucidating the complex interactions of biosurfactants, microorganisms, and pollutants. Biosurfactant production from extremophilic microbes can be aimed at isolating new surface-active compounds with novel properties. Biosurfactants have antimicrobial, antiadhesive, immunomodulating properties which have been successfully applied in gene therapy, immunotherapy, and medical insertion. Advances in the area of biomedical application are probably going to take the lead due to higher potential economic returns. Moreover, due to their self-assembly properties, new and fascinating applications in nanotechnology are predicted for biosurfactants. The commercial success of microbial surfactants is currently limited by the high cost of production. Cheaper renewable substrates are optimized for efficient multistep downstream processing methods that can make biosurfactant production more profitable and economically feasible. Additionally, recombinant and mutant hyperproducer microbial strains, if produced, will be able to grow on a wide range of cheaper substrates. Although the biosurfactants are thought to be ecofriendly, some experiments indicated that under certain circumstances they can be toxic to the environment. Nevertheless, careful and controlled use of these interesting surface-active molecules will surely help in the enhanced cleanup of toxic environmental pollutants and provide us with a clean environment.

The role of biosurfactants in facilitating systems as biocontrol agents is yet not much explored and warrants investigation. Such studies will help in replacing harsh chemical surfactants with green ones. There is also more work required on the production cost of green surfactants to achieve net economic gain from application of biosurfactants in agriculture as well as in other sectors. The use of agricultural waste for overproduction of biosurfactants also requires more serious thought and application. The chemical compositions of biosurfactants' reported potent biocontrol agents can be transformed by changing the production scheme, leading to biosynthesis of highly target-specific green surfactants. The rhizosphere region has a high prevalence of biosurfactants and biosurfactant-producing bacteria, which acts as a positive indicator for its potent role in sustainable agriculture. Mainly species of *Pseudomonas* and *Bacillus* appear in literature as producers of biosurfactants, indicating that only limited genera have been studied to date. A metagenomics-based functional screening modern approach can also lead to the discovery of novel green surfactants. Intense work on green surfactants is a priority to prevent the adverse effects of synthetic surfactants largely employed in many commercial sectors, including agrochemical industries. Therefore, a combined effort from researchers of various attributes and fields need to design and engineer bacteria as well as their products like biosurfactants for proper utilization in bioremediation of toxic metals.

References

Abraham, W.R., Meyer, H., Yakimov, M., 1998. Novel glycine containing glucolipids from the alkane using bacterium *Alcanivorax borkumensis*. Biochem. Biophys. Acta. 1393, 57–62.

Alloway, B.J., Ayres, D.C., 1997. Chemical Principles of Environmental Pollution. CRC Press, 2nd edn, Blackie Academic and Professional, London.

Amin, G.A., 2010. A potent biosurfactant producing bacterial strain for application in enhanced oil recovery applications. J. Petrol. Environ. Biotechnol. 1, 104. Available from: http://dx.doi.org/doi:10.4172/2157-7463.1000104.

Andersen, J.B., Koch, B., Nielsen, T.H., Sørensen, D., Hansen, M., 2003. Surface motility in *Pseudomonas* sp. DSS73 is required for efficient biological containment of the root-pathogenic microfungi *Rhizoctonia solani* and *Pythium ultimum*. Microbiology. 149, 37–46.

Appanna, V.D., Finn, H., Pierre, M.S.t., 1995. Exocellular phosphatidylethanolamine production and multiple-metal tolerance in *Pseudomonas fluorescens*. FEMS Microbiol. Lett. 131, 53–56.

Arima, K., Kakinuma, A., Tamura, G., 1968. Surfactin, a crystalline peptide lipid surfactant produced by *Bacillus subtilis*: isolation, characterization and its inhibition of fibrin clot formation. Biochem. Biophys. Res. Commun. 31, 488–494.

Awashti, N., Kumar, A., Makkar, R., Cameotra, S., 1999. Enhanced biodegradation of endosulfan, a chlorinated pesticide in presence of a biosurfactant. J. Environ. Sci. Health B. 34, 793–803.

Bach, H., Berdichevsky, Y., Gutnick, D., 2003. An exocellular protein from the oil-degrading microbe *Acinetobacter venetianus* RAG-1 enhances the emulsifying activity of the polymeric bioemulsifier emulsan. Appl. Environ. Microbiol. 69 (5), 2608–2615.

Banat, I.M., 1993. The isolation of a thermophilic biosurfactant producing *Bacillus* sp. Biotechnol. Lett. 15, 591–594.

Banat, I.M., Makkar, R.S., Cameotra, S.S., 2000. Potential commercial applications of microbial surfactants. Appl. Microbiol. Biotechnol. 53 (5), 495–508.

Barkay, T., Navon-Venezia, S., Ron, E.Z., Rosenberg, E., 1999. Enhancement of solubilization and biodegradation of polyaromatic hydrocarbons by the bioemulsifier alasan. Appl. Environ. Microbiol. 65, 2697–2702.

Baviere, M., Degouy, D., Lecourtier, J., 1994. Process for washing solid particles comprising a sophoroside solution. US Patent 5,326,407.

Bender, C.L., Scholz-Schroeder, B.K., 2004. New insights into the biosynthesis, mode of action, and regulation of syringomycin, syringopeptin and coronatine. In: Ramos, J.L. (Ed.), The Pseudomonads, vol. 2. Kluwer Academic Press, Dordrecht, The Netherlands, pp. 125–158.

Bender, C.L., Alarcon-Chaidez, F., Gross, D.C., 1999. *Pseudomonas syringae* phytotoxins: mode of action, regulation, and biosynthesis by peptide and polyketide synthetases. Microbiol. Mol. Biol. Rev. 63, 266–292.

Bento, F.M., de Oliveira Camargo, F.A., Okeke, B.C., Frankenberger Jr., W.T., 2005. Diversity of biosurfactant producing microorganisms isolated from soils contaminated with diesel oil. Microbiol. Res. 160 (3), 249–255.

Berti, A.D., Greve, N.J., Christensen, Q.H., Thomas, M.G., 2007. Identification of a biosynthetic gene cluster and the six associated lipopeptides involved in swarming motility of *Pseudomonas syringae* pv. tomato DC3000. J. Bacteriol. 189, 6312–6323.

Burger, M.M., Glaser, L., Burton, R.M., 1963. The enzymatic synthesis of a rhamnose-containing glycolipid by extracts of *Pseudomonas aeruginosa*. J. Biol. Chem. 238, 2595–2601.

Cameron, R.E., 1992. Guide to Site and Soil Description for Hazardous Waste Site Characterization Metals, vol. 1. Environmental Protection Agency, Las Vegas, EPA/600/4-91/029.

Chen, S.Y., Wei, Y.H., Chang, J.S., 2007. Repeated pH-stat fed batch fermentation for rhamnolipid production with indigenous *Pseudomonas aeruginosa* S2. Appl. Microbiol. Biotechnol. 76 (1), 67–74.

Chrzanowski, L., Dziadas, M., Ławniczak, L., Cyplik, P., Bialas, W., Szulc, A., et al., 2012a. Biodegradation of rhamnolipids in liquid cultures: effect of biosurfactant dissipation on diesel fuel/B20 blend biodegradation efficiency and bacterial community composition. Bioresour. Technol. 111, 328–335.

Chrzanowski, L., Lawniczak, L., Czaczyk, K., 2012b. Why do microorganisms produce rhamnolipids? World J. Microbiol. Biotechnol. 28, 401–419.

Cirigliano, M.C., Carman, G.M., 1984. Purification and characterization of liposan, a bioemulsifier from *Candida lipolytica*. Appl. Environ. Microbiol. 50, 846–850.

Cosby, W.M., Vollenbroich, D., Lee, O.H., Zuber, P., 1998. Altered *srf* expression in *Bacillus subtilis* resulting from changes in culture pH is dependent on the Spo0K oligopeptide permease and the ComQX system of extracellular control. J. Bacteriol. 180, 1438–1445.

Dahrazma, B., Mulligan, C.N., 2007. Investigation of the removal of heavy metals from sediments using rhamnolipid in a continuous flow configuration. Chemosphere. 69 (5), 705–711.

Daniels, R., Vanderleyden, J., Michiels, J., 2004. Quorum sensing and swarming migration in bacteria. FEMS Microbiol. Rev. 28, 261–289.

Das, K., Mukherjee, A.K., 2007. Crude petroleum-oil biodegradation efficiency of *Bacillus subtilis* and *Pseudomonas aeruginosa* strains isolated from a petroleum-oil contaminated soil from North-East India. Bioresour. Technol. 98, 1339–1345.

Das, P., Mukherjee, S., Sen, R., 2008a. Antimicrobial potential of a lipopeptide biosurfactant derived from a marine *Bacillus circulans*. J. Appl. Microbiol. 104, 1675–1684.

Das, P., Mukherjee, S., Sen, S., 2008b. Improved bioavailability and biodegradation of a model polyaromatic hydrocarbon by a biosurfactant producing bacterium of marine origin. Chemosphere. 72 (9), 1229–1234.

Das, P., Mukherjee, S., Sen, R., 2008c. Genetic regulations of the biosynthesis of microbial surfactants: an overview. Biotechnol. Genet. Eng. Rev. 25 (1), 165–186.

Desai, J.D., Banat, I.M., 1997. Microbial production of surfactant and their commercial potential. Microbiol. Mol. Biol. Rev. 61 (1), 47–64.

Dubern, J.F., Lagendijk, E.L., Lugtenberg, B.J., Bloemberg, G.V., 2005. The heat shock genes dnaK, dnaJ, and grpE are involved in regulation of putisolvin biosynthesis in *Pseudomonas putida* PCL1445. J. Bacteriol. 187, 5967–5976.

Dubern, J.F., Lugtenberg, B.J., Bloemberg, G.V., 2006. The ppuI–rsaL–ppuR quorum sensing system regulates biofilm formation of *Pseudomonas putida* PCL 1445 by controlling biosynthesis of the cyclic lipopeptides putisolvins I and II. J. Bacteriol. 188 (8), 2898–2906.

Evans, K.M., Gill, R.A., Robotham, P.W.J., 1990. The source, composition and flux of polycyclic aromatic hydrocarbons in sediments of the river Derwent, Derbyshire, U.K. Water Air Soil Pollut. 51, 1–12.

Ferhat, S., Mnif, S., Badis, A., Kamel, E., Redha, A., Boucherit, A., et al., 2011. Screening and preliminary characterization of biosurfactants produced by *Ochrobactrum* sp. 1C and *Brevibacterium* sp. 7G isolated from hydrocarbon-contaminated soils. Int. Biodeterior. Biodegrad. 65, 1182–1188.

Floodgate, D., 1978. The formation of oil emulsifying agents in hydrocarbonclastic bacteria. In: Loutit, M.W., Miles, J.A.R. (Eds.), Microbial Ecology. Springer-Verlag, New York, NY, pp. 82–85.

Franzetti, A., Gandolfi, I., Bestetti, G., Smyth, T.J., Banat, I.M., 2010. Production and applications of trehalose lipid biosurfactants. Eur. J. Lipid Sci. Technol. 112, 617–627.

Gao, L., Kano, N., Sato, Y., Li, C., Zhang, S., Imaizumi, H., 2012. Behavior and distribution of heavy metals including rare earth elements, thorium, and uranium in sludge from industry water treatment plant and recovery method of metals by biosurfactants application. Bioinorg. Chem. Appl. Available from: http://dx.doi.org/doi:10.1155/2012/173819.

Garcia-Ochoa, F., Casas, J.A., 1999. Unstructured kinetic model for sophorolipid production by *Candida bombicola*. Enzyme Microb. Technol. 25, 613–621.

Gautam, K.K., Tyagi, V.K., 2006. Microbial surfactants: a review. J. Oleo Sci. 55, 155–166.

Georgiou, G., Lin, S.C., Sharma, M., 1992. Surface active compounds from microorganisms. Biotechnology. 10, 60–65.

Gerson, O.F., Zajic, J.E., 1978. Surfactant production from hydrocarbons by *Corynebacterium lepus*, sp. nov. and *Pseudomonas asphaltenicus*, sp. nov. Dev. Ind. Microbiol. 19, 577–599.

Gilleland, H.E., Stinnett Jr., J.D., Roth, I.L., Eagon, R.G., 1973. Freeze-etch study of *Pseudomonas aeruginosa*—localization within the cell wall of an ethylenediaminetetraacetate-extractable component. J. Bacteriol. 113, 417–432.

Glick, R., Gilmour, C., Tremblay, J., Satanower, S., Avidan, O., Deziel, E., et al., 2010. Increase in rhamnolipid synthesis under iron-limiting conditions influences surface motility and biofilm formation in *Pseudomonas aeruginosa*. J. Bacteriol. 192, 2973–2980.

Goldberg, S.S., Cordeiro, M.N., Silva Pereira, A.A., Mares-Guia, M.L., 1983. Release of lipopolysaccharide (LPS) from cell surface of *Trypanosoma cruzi* by EDTA. Int. J. Parasitol. 13, 11–18.

Grangemard, I., Wallach, J., Maget-Dana, R., Peypoux, F., 2001. Lichenysin: a more efficient cation chelator than surfactin. Appl. Biochem. Biotechnol. 90 (3), 199–210.

Gray, G.W., Wilkinson, G., 1965. The action of ethylene diaminotetraacetic acid on *Pseudomonas aeruginosa*. J. Appl. Bacteriol. 28, 153–164.

Grgurina, I., Barca, A., Cervigni, S., Gallo, M., Scaloni, A., Pucci, P., 1994. Relevance of chlorine-substituent for the antifungal activity of syringomycin and syringotoxin, metabolites of the phytopathogenic bacterium *Pseudomonas syringae* pv. syringae. Experientia. 50, 130–133.

Gross, D.C., 1985. Regulation of syringomycin synthesis in *Pseudomonas syringae* pv. syringae and defined conditions for its production. J. Appl. Microbiol. 58 (2), 167–174.

Gross, H., Loper, J.E., 2009. Genomics of secondary metabolite production by *Pseudomonas* spp. Nat. Prod. Rep. 26, 1408–1446.

Gross, H., Stockwell, V.O., Henkels, M.D., Nowak-Thompson, B., Loper, J.E., Gerwick, W.H., 2007. The genomisotopic approach: a systematic method to isolate products of orphan biosynthetic gene clusters. Chem. Biol. 14, 53–63.

Gunther, N.W., Nunez, A., Fett, W., Daniel, K., Solaiman, Y., 2005. Production of rhamnolipids by *Pseudomonas chlororaphis*, a nonpathogenic bacterium. Appl. Environ. Microbiol. 71, 2288–2293.

Hagelin, G., Indrevoll, B., Hoeg-Jensen, T., 2007. Use of synthetic analogues in confirmation of structure of the peptide antibiotics maltacines. Int. J. Mass Spectrom. 268, 254–264.

Hasumi, K., Takizawa, K., Takahashi, F., 1995. Inhibition of acylcoA: cholesterol acyltransferase by isohalobacillin, a complex of novel cyclic acylpeptides produced by *Bacillus* sp. A1238. J. Antibiot. 48, 1419–1424.

Hathout, Y., Ho, Y.P., Ryzhov, V., Demirev, P., Fenselau, C., 2000. Kurstakins: a new class of lipopeptides isolated from *Bacillus thuringiensis*. J. Nat. Prod. 63, 1492–1496.

Herman, D.C., Artiola, J.F., Miller, R.M., 1995. Removal of cadmium, lead, and zinc from soil by a rhamnolipid biosurfactant. Environ. Sci. Technol. 29, 2280–2285.

Hildebrand, P.D., Braun, P.G., McRae, K.B., Lu, X., 1998. Role of the biosurfactant viscosin in broccoli head rot caused by a pectolytic strain of *Pseudomonas fluorescens*. Can. J. Plant Pathol. 20 (3), 296–303.

Hisatsuka, K., Nakahara, T., Sano, N., Yamada, K., 1971. Formation of rhamnolipid by *Pseudomonas aeruginosa* and its function in hydrocarbon fermentation. Agric. Biol. Chem. 35, 686–692.

Hong, J.J., Yang, S.M., Lee, C.H., Choi, Y.K., Kajiuchi, T., 1998. Ultrafiltration of divalent metal cations from aqueous solution using polycarboxylic acid type biosurfactants. J. Colloid Interface Sci. 202, 63–73.

Hoornstra, D., Andersson, M.A., Mikkola, R., Salkinoja-Salonen, M.S., 2003. A new method for *in vitro* detection of microbially produced mitochondrial toxins. Toxicol. In Vitro. 17 (5), 745–751.

Hoskova, M., Schreiberova, O., Jezdík, R., Chudoba, J., Masak, J., Sigler, K., et al., 2013. Characterization of rhamnolipids produced by non-pathogenic *Acinetobacter* and *Enterobacter* bacteria. Bioresour. Technol. 130, 510–516.

Hu, L.B., Shi, Z.Q., Zhang, T., Yang, Z.M., 2007. Fengycin antibiotics isolated from B-FS01 culture inhibit the growth of *Fusarium moniliforme* Sheldon ATCC 38932. FEMS Microbiol. Lett. 272 (1), 91–98.

Ishigami, Y., Yamazaki, S., Gama, Y., 1983. Surface active properties of biosoap from spiculisporic acid. J. Colloid Interface Sci. 94, 131–139.

Ishigami, Y., Zhang, Y., Ji, F., 2000. Spiculisporic acid. Functional development of biosurfactants. Chim. Oggi. 18, 32–34.

Itoh, S., 1971. Rhamnolipids produced by *Pseudomonas aeruginosa* grown on n-paraffin. J. Antibiot. 24, 855–859.

Jeong-Jin, H., Yang, S., Lee, C., Choi, Y., Kajuichi, T., 1998. Ultrafiltration of divalent metal cations from aqueous solution using polycarboxylic acid type biosurfactant. J. Colloid Interface Sci. 202, 63–73.

Jerala, R., 2007. Synthetic lipopeptides: a novel class of anti-infectives. Expert Opin. Investig. Drugs. 16, 1159–1169.

Joseph, P.J., Joseph, A., 2009. Microbial enhanced separation of oil from a petroleum refinery sludge. J. Hazard. Mater. 161, 522–525.

Juwarkar, A.A., Nair, A., Dubey, K.V., Singh, S.K., Devotta, S., 2007. Biosurfactant technology for remediation of cadmium and lead contaminated soils. Chemosphere. 68, 1996–2002.

Kappeli, O., Finnerty, W.R., 1980. Characteristics of hexadecane partition by the growth medium of *Acinetobacte*r sp. Biotechnol. Bioeng. 22 (3), 495–503.

Kang, S.W., Kim, Y.B., Shin, J.D., Kim, E.K., 2010. Enhanced biodegradation of hydrocarbons in soil by microbial biosurfactant, sophorolipid. Appl. Biochem. Biotechnol. 160, 780–790.

Kaplan, N., Zosim, Z., Rosenberg, E., 1987. Reconstitution of emulsifying activity of *Acinetobacter calcoaceticus* BD4 emulsan by using pure polysaccharide and protein. Appl. Environ. Microbiol. 53 (2), 440–446.

Kappeli, O., Finnerty, W.R., 1979. Partition of alkane by an extracellular vesicle derived from hexadecane-grown *Acinetobacter*. J. Bacteriol. 140, 707–712.

Karanth, N.G.K., Deo, P.G., Veenanadig, N.K., 1999. Microbial production of biosurfactants and their importance. Curr. Sci. 77, 116–123.

Kearns, D.B., Losick, R., 2003. Swarming motility in undomesticated *Bacillus subtilis*. Mol. Microbiol. 49, 581–590.

Koch, B., Nielsen, T.H., Sorensen, D., Andersen, J.B., Christophersen, C., Molin, S., et al., 2002. Lipopeptide production in *Pseudomonas* sp. strain DSS73 is regulated by components of sugar beet seed exudate via the Gac two-component regulatory system. Appl. Environ. Microbiol. 68 (9), 4509–4516.

Kretschmer, A., Bock, H., Wagner, F., 1982. Chemical and physical characterization of interfacial-active lipids from *Rhodococcus erythropolis* grown on n-alkane. Appl. Environ. Microbiol. 44, 864–870.

Kuiper, I., Lagendijk, E.L., Pickford, R., Derrick, J.P., Lamers, G.E., Thomas-Oates, J.E., et al., 2004. Characterization of two *Pseudomonas putida* lipopeptide biosurfactants, putisolvin I and II, which inhibit biofilm formation and break down existing biofilms. Mol. Microbiol. 51 (1), 97–113.

Lang, S., Philp, J.C., 1998. Surface-active compounds in rhodococci. Antonie Van Leeuwenhoek. 74, 59–70.

Lang, S., Brakemeier, A., Heckmann, R., Spockner, S., Rau, U., 2000. Production of native and modified sophorose lipids. Chim. Oggi-Chem. Today. 18, 76–79.

Lee, M., Kim, M.K., Kwon, M.J., Park, B.D., Kim, M.H., Goodfellow, M., et al., 2005. Effect of the synthesized mycolic acid on the biodegradation of diesel oil by *Gordonia nitida* strain LE31. J. Biosci. Bioeng. 100 (4), 429–436.

Lee, S., Jo, G., Hwang, D., Woo, Y., Lee, Y., Yong, Y., et al., 2011. A peptide produced by *Pseudomonas tolaasi*, tolaasin binds to metal ions. J. Korean Soc. Appl. Biol. Chem. 54 (4), 633–636.

Lee, S.C., Kim, S.H., Park, I.H., Chung, S.Y., Choi, Y.L., 2007. Isolation and structural analysis of bamylocin A, novel lipopeptide from *Bacillus amyloliquefaciens* LP03 having antagonistic and crude oil-emulsifying activity. Arch. Microbiol. 188, 307–312.

Leive, L., 1965. Release of lipopolysaccharide by EDTA treatment of *E. coli*. Biochem. Biophys. Res. Commun. 21, 290–296.

Li, H., Tanikawa, T., Sato, Y., Nakagawa, Y., Matsuyama, T., 2005. *Serratia marcescens* gene required for surfactant serrawettin W1 production encodes putative aminolipid synthetase belonging to nonribosomal peptide synthetase family. Microbiol. Immunol. 49, 303–310.

Lima, T.M.S., Procopio, L.C., Brandao, F.D., Carvalho, A.M.X., Totola, M.R., Borges, A.C., 2011a. Biodegradability of bacterial surfactants. Biodegradation. 22, 585–592.

Lima, T.M.S., Procopio, L.C., Brandao, F.D., Carvalho, A.M.X., Totola, M.R., Borges, A.C., 2011b. Simultaneous phenanthrene and cadmium removal from contaminated soil by a ligand/biosurfactant solution. Biodegradation. 22, 1007–1015.

Lima, T.M.S., Procopio, L.C., Brandao, F.D., Leao, B.A., Totola, M.R., Borges, A.C., 2011c. Evaluation of bacterial surfactant toxicity towards petroleum degrading microorganisms. Bioresour. Technol. 102, 2957–2964.

Lin, S.C., 1996. Biosurfactants: recent advances. J. Chem. Technol. Biotechnol. 66, 109–120.

Linder, M., Selber, K., Nakari-Setala, T., Qiao, M., Kula, M.R., Penttila, M., 2001. The hydrophobins HFBI and HFBII from *Trichoderma reesei* showing efficient interactions with nonionic surfactants in aqueous two-phase systems. Biomacromolecules. 2 (2), 511–517.

Lopez, D., Fischbach, M.A., Chu, F., Losick, R., Kolter, R., 2009a. Structurally diverse natural products that cause potassium leakage trigger multicellularity in *Bacillus subtilis*. P. Natl. Acad. Sci. USA. 106, 280–285.

Lopez, D., Vlamakis, H., Losick, R., Kolter, R., 2009b. Cannibalism enhances biofilm development in *Bacillus subtilis*. Mol. Microbiol. 74, 609–618.

Liu, L., Nakano, M., Lee, O.H., Zuber, P., 1996. Plasmid-amplified *comS* enhances genetic competence and suppresses *sinR* in *Bacillus subtilis*. J. Bacteriol. 178, 5144–5152.

Luttinger, A., Hahn, J., Dubnau, D., 1996. Polynucleotide phosphorylase is necessary for competence development in *Bacillus subtilis*. Mol. Microbiol. 19, 343–356.

Maier, R.M., Soberon-Chavez, G., 2000. *Pseudomonas aeruginosa* rhamnolipids: biosynthesis and potential applications. Appl. Microbiol. Biotechnol. 54, 625–633.

Maneerat, S., 2005. Biosurfactants from marine microorganisms. Songklanakarin J. Sci. Technol. 27, 1263–1272.

Marahiel, M.A., Konz, D., Dokel, S., 1999. Molecular and biochemical characterization of the protein template controlling biosynthesis of the lipopeptide lichenysin of *Bacillus licheniformis*. J. Bacteriol. 181, 133–140.

Margesin, R., Schinner, F., 2001a. Biodegradation and bioremediation of hydrocarbons in extreme environments. Appl. Microbiol. Biotechnol. 56, 650–663.

Margesin, R., Schinner, F., 2001b. Bioremediation (natural attenuation and biostimulation) of diesel oil contaminated soil in an alpine glacier skiing area. Appl. Environ. Microbiol. 67, 3127–3133.

Maslin, P.M., Maier, R.M., 2000. Rhamnolipid enhanced mineralization of phenanthrene in organic metal co-contaminated soils. Bioremediat. J. 4, 295–308.

Matsuyama, T., Kaneda, K., Yano, I., 1986. Two kinds of bacterial wetting agents: aminolipid and glycolipid. Proc. Jpn. Soc. Mass Spectrom. 11, 125–128.

Matsuyama, T., Sogawa, M., Nakagawa, Y., 1989. Fractal spreading growth of *Serratia marcescens* which produces surface-active exolipids. FEMS Microbiol. Lett. 61, 243–246.

Matsuyama, T., Keneda, K., Ishizuka, I., Toida, T., Yano, I., 1990. Surface-active novel glycolipid and linked 3-hydroxy fatty acids produced by *Serratia rubidaea*. J. Bacteriol. 172 (6), 3015–3022.

Mattei, M., Kontogeorgis, G.M., Gani, R., 2013. Modeling of the critical micelle concentration (CMC) of nonionic surfactants with an extended Group-Contribution Method. Ind. Eng. Chem. Res. 52 (34), 12236–12246.

Menkhaus, M., Ullrich, C., Kluge, B., Vater, J., Vollenbroich, D., 1993. Structural and functional organization of the surfactin synthetase multienzyme system. J. Biol. Chem. 268, 7678–7684.

Mukherjee, S., Das, P., Sen, R., 2006. Towards commercial production of microbial surfactants. Trends Biotechnol. 24 (11), 509–515.

Mulligan, C.N., 2005. Environmental applications for biosurfactants. Environ. Pollut. 133, 183–198.

Mulligan, C.N., Yong, R.N., Gibbs, B.F., 2001a. Heavy metal removal from sediments by biosurfactants. J. Hazard. Mater. 85 (1), 111–125.

Mulligan, C.N., Yong, R.N., Gibbs, B.F., 2001b. Surfactant-enhanced remediation of contaminated soil: a review. Eng. Geol. 60 (1), 371–380.

Nakar, D., Gutnick, D.L., 2001. Analysis of the wee gene cluster responsible for the biosynthesis of the polymeric bioemulsifier from the oil-degrading strain *Acinetobacter lwoffii* RAG-1. Microbiology. 147, 1937–1946.

Nakar, D., Gutnick, D.L., 2003. Involvement of a protein tyrosine kinase in production of the polymeric bioemulsifier emulsan from the oil-degrading strain *Acinetobacter lwoffii* RAG-1. J. Bacteriol. 185, 1001–1009.

Navon-Venezia, S., Zosim, Z., Gottlieb, A., Legmann, R., Carmeli, S., Ron, E.Z., et al., 1995. Alasan, a new bioemulsifier from *Acinetobacter radioresistens*. Appl. Environ. Microbiol. 61 (9), 3240–3244.

Nerurkar, A.S., 2010. Structural and molecular characteristics of lichenysin and its relationship with surface activity. Adv. Exp. Med. Biol. 672, 304–315.

Neu, T.R., 1996. Significance of bacterial surface-active compounds in interaction of bacteria with interfaces. Microbiol. Rev. 60, 151–166.

Nievas, M.L., Commendatore, M.G., Estevas, J.L., Bucala, V., 2008. Biodegradation pattern of hydrocarbons from a fuel oil-type complex residue by an emulsifier-producing microbial consortium. J. Hazard. Mater. 154, 96–104.

Nitschke, M., Coast, S.G., 2007. Biosurfactants in food industry. Trends Food Sci. Technol. 18, 252–259.

Nybroe, O., Sorensen, J., 2004. Production of cyclic lipopeptides by fluorescent pseudomonads. In: Ramos, J.-L. (Ed.), Pseudomonas, Biosynthesis of Macromolecules and Molecular Metabolism. Kluwer Academic/Plenum Publishers, New York, NY, pp. 147–172.

Obayori, O.S., Ilori, M.O., Adebusoye, S.A., Oyetibo, G.O., Omotayo, A.E., Amund, O.O., 2009. Degradation of hydrocarbons and biosurfactant production by *Pseudomonas* sp. strain LP1. World J. Microbiol. Biotechnol. 25 (9), 1615–1623.

Ochoa-Loza, F.J., Artiola, J.F., Maier, R.M., 2001. Stability constants for the complexation of various metals with a rhamnolipid biosurfactant. J. Environ. Qual. 30 (2), 479–485.

Ochsner, U.A., Fiechter, A., Reiser, J., 1994. Isolation, characterization, and expression in *Escherichia coli* of the *Pseudomonas aeruginosa rhlAB* genes encoding a rhamnosyltransferase involved in rhamnolipid biosurfactant synthesis. J. Biol. Chem. 269, 19787–19795.

Ochsner, U.A., Fiechter, A., Reiser, J., Witholt, B., 1995. Production of *Pseudomonas aeruginosa* rhamnolipid biosurfactants in heterologous hosts. Appl. Environ. Microbiol. 61, 3503–3506.

Ochsner, U.A., Hembach, T., Fiechter, A., 1996. Production of rhamnolipid biosurfactants. Adv. Biochem. Eng. Biotechnol. 53, 89–118.

Olivera, N.L., Commendatore, M.G., Delgado, O., Esteves, J.L., 2003. Microbial characterization and hydrocarbon biodegradation potential of natural bilge waste microflora. J. Ind. Microbiol. Biotechnol. 30, 542–548.

Ongena, M., Jacques, P., 2008. *Bacillus* lipopeptides: versatile weapons for plant disease biocontrol. Trends Microbiol. 16, 115–125.

Passeri, S., Schmidt, M., Haffner, T., Wray, V., Lang, S., Wagner, F., 1992. Marine biosurfactants. IV. Production, characterization and biosynthesis of an anionic glucose lipid from the marine bacterial strain MM1. Appl. Microbiol. Biotechnol. 37, 281–286.

Patil, S.N., Aglave, B.A., Pethkar, A.V., Gaikwad, V.B., 2012. *Stenotrophomonas koreensis* a novel biosurfactant producer for abatement of heavy metals from the environment. Afr. J. Microbiol. Res. 6 (24), 5173–5178.

Paulsen, I.T., Press, C.M., Ravel, J., 2005. Complete genome sequence of the plant commensal *Pseudomonas fluorescens* Pf-5. Nat. Biotechnol. 23, 873–878.

Pearson, J.P., Gray, K.M., Passador, L., Tucker, K.D., Eberhard, A., 1994. Structure of the autoinducer required for the expression of *Pseudomonas aeruginosa* virulence genes. PNAS USA. 91, 197–201.

Perfumo, A., Banat, I.M., Canganella, F., Marchant, R., 2006. Rhamnolipid production by a novel thermotolerant hydrocarbon-degrading *Pseudomonas aeruginosa* AP02-1. J. Appl. Microbiol. Biotechnol. 72, 132–138.

Pesce, L., 2002. A biotechnological method for the regeneration of hydrocarbons from dregs and muds, on the base of biosurfactants. World Patent 02/062,495.

Pesci, E.C., Pearson, J.P., Seed, P.C., Iglewski, B.H., 1997. Regulation of las and rhl quorum sensing in *Pseudomonas aeruginosa*. J. Bacteriol. 179, 3127–3132.

Peypoux, F., Bonmatin, J.M., Wallach, J., 1999. Recent trends in the biochemistry of surfactin. Appl. Microbiol. Biotechnol. 51 (5), 553–563.

Phister, T.G., O'Sullivan, D.J., McKay, L.L., 2004. Identification of bacilysin, chlorotetaine, and iturin A produced by *Bacillus* sp. strain CS93 isolated from pozol, a Mexican fermented maize dough. Appl. Environ. Microbiol. 70 (1), 631–634.

Poremba, K., Gunkel, W., Lang, S., Wagner, F., 1991. Marine biosurfactants. III. Toxicity testing with marine microorganisms and comparison with synthetic detergents. Z. Naturforsch. 46, 210–216.

Qazi, M.A., Malik, Z.A., Qureshi, G.D., Hameed, A., Ahmed, S., 2013. Yeast extract as the most preferable substrate for optimized biosurfactant production by *rhlB* gene positive *Pseudomonas putida* SOL-10 isolate. J. Bioremed. Biodeg. 4 (204), p. 2.

Quentin, M.J., Besson, F., Peypoux, F., Michel, G., 1982. Action of peptidolipidic antibiotics of the iturin group on erythrocytes. Effect of some lipids on haemolysis. Biochim. Biophys. Acta. 684, 207–211.

Raaijmakers, J.M., de Bruijn, I., de Kock, M.J., 2006. Cyclic lipopeptide production by plant-associated *Pseudomonas* spp. diversity, activity, biosynthesis, and regulation. Mol. Plant Microbe Int. 19, 699–710.

Rahim, R., Ochsner, U.A., Oliveira, C., Graninger, M., Messner, P., 2001. Cloning and functional characterization of the *Pseudomonas aeruginosa* rhlC gene that encodes rhamnosyltransferase 2, an enzyme responsible for di-rhamnolipid biosynthesis. Mol. Microbiol. 40, 708–718.

Reddy, M.S., Naresh, B., Leela, T., Prashanthi, M., Madhusudhan, N.C., Dhanasri, G., et al., 2010. Biodegradation of phenanthrene with biosurfactant production by a new strain of *Brevibacillus* sp. Bioresour. Technol. 101, 7980–7983.

Rehn, H.J., Reiff, I., 1981. Mechanisms and occurrence of microbial oxidation of long chain alkanes. Adv. Biochem. Eng. 19, 175–216.

Reisfeld, A., Rosenberg, E., Gutnick, D., 1972. Microbial degradation of crude oil: factors affecting the dispersion in sea water by mixed and pure cultures. Appl. Environ. Microbiol. 24, 363–368.

Remacle, J., 1990. The cell wall and metal binding. In: Volesky, B. (Ed.), Biosorption of Heavy Metals. CRC Press, Inc., Boca Raton, FL, pp. 83–92.

Rodrigues, L., Banat, I.M., Teixeira, J., Oliveira, R., 2006. Biosurfactants: potential applications in medicine. J. Antimicrob. Chemother. 57, 609–618.

Ron, E.Z., Rosenberg, E., 2001. Natural roles of biosurfactants. Environ. Microbiol. 3 (4), 229–236.

Roongsawang, N., Hase, K., Haruki, M., Imanaka, T., Morikawa, M., 2003. Cloning and characterization of the gene cluster encoding arthrofactin synthetase from *Pseudomonas* sp. MIS38. Chem. Biol. 10, 869–880.

Rosen, M., 1978. Surfactants and Interfacial Phenomena. Wiley and Sons. Pub. Wiley Interscience, New York, NY, pp. 149–159.

Rosenberg, E., Ron, E.Z., 1997. Bioemulsans: microbial polymeric emulsifiers. Curr. Opin. Biotechnol. 8 (3), 313–316.

Rosenberg, E., Ron, E.Z., 1999. High- and low-molecular-mass microbial surfactants. Appl. Microbiol. Biotechnol. 52 (2), 154–162.

Rosenberg, E., Zuckerberg, A., Rubinovitz, C., Gutnick, D.L., 1979. Emulsifier Arthrobacter RAG-1: isolation and emulsifying properties. Appl. Environ. Microbiol. 37, 402–408.

Rosenberg, E., Rubinovitz, C., Legmann, R., Ron, E.Z., 1988. Purification and chemical properties of *Acinetobacter calcoaceticus* A2 Biodispersan. Appl. Environ. Microbiol. 54, 323–326.

Ruiz-Garcia, C., Bejar, V., Martinez-Checa, F., Llamas, I., Quesada, E., 2005. *Bacillus velezensis* sp. nov., a surfactant-producing bacterium isolated from the river Vélez in Málaga, southern Spain. Int. J. Syst. Evol. Microbiol. 55 (1), 191–195.

Sandrin, T.R., Maier, R.M., 2003. Impact of metals on the biodegradation of organic pollutants. Environ. Health Perspect. 111, 1093–1100.

Sandrin, T.R., Chech, A.M., Maier, R.M., 2000. A rhamnolipid biosurfactant reduces cadmium toxicity during naphthalene biodegradation. Appl. Environ. Microbiol. 66 (10), 4585–4588.

Schulz, D., Passeri, A., Schmidt, M., Lang, S., Wagner, F., Wray, V., 1991. Marine biosurfactants. I. Screening for biosurfactants among crude oil degrading marine microorganisms from the North Sea. Z. Naturforsch. C. 46 (3–4), 197–203.

Shah, V., Doncel, G.F., Seyoum, T., Eaton, K.M., Zalenskaya, I., Hagver, R., 2005. Sophorolipids, microbial glycolipids with anti-human immunodeficiency virus and sperm-immobilizing activities. Antimicrob. Agents Chemother. 49, 4093–4100.

Shao, Z., 2011. Trehalolipids. Microbiol. Monogr. 20, 121–143.

Shaw, D.J., 1996. Introduction to Colloid and Surface Chemistry. second ed. Butterworth, London.

Sifour, M., Al-Jilawi, M.H., Aziz, G.M., 2007. Emulsification properties of biosurfactant produced from *Pseudomonas aeruginosa* RB 28. Pak. J. Biol. Sci. 10, 1331–1335.

Singh, P., Cameotra, S.S., 2004. Potential applications of microbial surfactants in biomedical sciences. Trends Biotechnol. 22 (3), 142–146.

Sinnaeve, D., Michaux, C., Van Hemel, J., 2009. Structure and X-ray conformation of pseudodesmins A and B, two new cyclic lipodepsipeptides from *Pseudomonas* bacteria. Tetrahedron. 65, 4173–4181.

Spiteller, D., Dettner, K., Bolan, W., 2000. Gut bacteria may be involved in interactions between plants, herbivores and their predators: microbial biosynthesis of *N*-acylglutamine surfactants as elicitors of plant volatiles. Biol. Chem. 381 (8), 755–762.

Storm, D.R., Rosenthal, K.S., Swanson, P.E., 1977. Polymyxin and related peptide antibiotics. Annu. Rev. Biochem. 46, 723–763.

Strieker, M., Marahiel, M.A., 2009. The structural diversity of acidic lipopeptide antibiotics. Chembiochem. 10, 607–616.

Syldatk, C., Lang, S., Matulovic, U., Wagner, F., 1985a. Chemical and physical characterization of four interfacial-active rhamnolipids from *Pseudomonas* sp. DSM 2874 grown on n-alkanes. Z. Naturforsch. 40, 51–60.

Syldatk, C., Lang, S., Wagner, F., Wray, V., Witte, L., 1985b. Production of four interfacial active rhamnolipids from n-alkanes or glycerol by resting cells of *Pseudomonas* sp. DSM 2874. Z. Naturforsch. 40, 61–67.

Tambekar, D.H., Gadakh, P.V., 2013. Biochemical and molecular detection of biosurfactant producing bacteria from soil. Int. J. Life Sci. Biotechnol. Pharm. Res. 2 (1), 2250–3137.

Thomas, C.P., Duvall, M.L., Robertson, E.P., Barrett, K.B., Bala, G.A., 1993. Surfactant-based EOR mediated by naturally occurring microorganisms. SPE Res. Eng. 8 (4), 285–291.

Toren, A., Navon-Venezia, S., Ron, E.Z., Rosenberg, E., 2001. Emulsifying activity of purified alasan proteins from *Acinetobacter radioresistens*. Appl. Environ. Microbiol. 67, 1102–1106.

Tremblay, J., Richardson, A.P., Lepine, F., Deziel, E., 2007. Self-produced extracellular stimuli modulate the *Pseudomonas aeruginosa* swarming motility behavior. Environ. Microbiol. 9, 2622–2630.

Tsuge, K., Akiyama, T., Shoda, M., 2001. Cloning, sequencing and characterization of the iturin A operon. J. Bacteriol. 183, 6265–6273.

Uchida, Y., Tsuchiya, R., Chino, M., Hirano, J., Tabuchi, T., 1989. Extracellular accumulation of mono- and di-succinoyl trehalose lipids by a strain of *Rhodococcus erythropolis* grown on n-alkanes. Agric. Biol. Chem. 53, 757–763.

Vanittanakom, N., Loefer, W., Koch, U., Jung, G., 1986. Fengycin a novel antifungal lipopeptide antibiotic produced by *Bacillus subtilis* F-29-3. J. Antibiot. 39, 888–901.

Veenanadig, N.K., Gowthaman, M.K., Karanth, N.G.K., 2000. Scale up studies for the production of biosurfactant in packed column bioreactor. Bioprocess Biosyst. Eng. 22 (2), 95–99.

Wang, C.L., Michels, P.C., Dawson, S.C., Kitisakkul, S., Baross, J.A., Keasling, J.D., et al., 1997. Cadmium removal by a new strain of *Pseudomonas aeruginosa* in aerobic culture. Appl. Environ. Microbiol. 63 (10), 4075–4078.

Wang, S., Mulligan, C.N., 2009. Rhamnolipid biosurfactant-enhanced soil flushing for the removal of arsenic and heavy metals from mine tailings. Process Biochem. 44 (3), 296–301.

Whang, L.M., Liu, P.W.G., Ma, C.C., Cheng, S.S., 2008. Application of biosurfactant, rhamnolipid, and surfactin, for enhanced biodegradation of diesel-contaminated water and soil. J. Hazard. Mater. 151, 155–163.

Winson, M.K., Camara, M., Latifi, A., Foglino, M., Chhabra, S.R., 1995. Multiple *N*-acyl-L-homoserine lactone signal molecules regulate production of virulence determinants and secondary metabolites in *Pseudomonas aeruginosa*. PNAS USA. 92, 9427–9431.

Yakimov, M.M., Timmis, K.N., Wray, V., Fredrickson, H.L., 1995. Characterization of a new lipopeptide surfactant produced by thermotolerant and halotolerant subsurface *Bacillus lechiniformis* BAS50. Appl. Environ. Microbiol. 61, 1706–1713.

Yakimov, M.M., Kröger, A., Slepak, T.N., Giuliano, L., Timmis, K.N., Golyshin, P.N., 1998. A putative lichenysin A synthetase operon in *Bacillus licheniformis*: initial characterization. Biochim. Biophys. Acta. 1399, 141–153.

Zhao, Z., Wong, J.W., 2009. Biosurfactants from *Acinetobacter calcoaceticus* BU03 enhance the solubility and biodegradation of phenanthrene. Environ. Technol. 30 (3), 291–299.

Zosim, Z., Gutnick, D.L., Rosenberg, E., 1982. Properties of hydrocarbon-in-water emulsions stabilized by *Acinetobacter* RAG-1 emulsan. Biotechnol. Bioeng. 24, 281–292.

Zosim, Z., Rosenberg, E., Gutnick, D.L., 1986. Changes in emulsification specificity of the polymeric bioemulsifier emulsan: effects of alkanols. Colloid Polym. Sci. 264, 218–223.

8 Biofilm-Mediated Bioremediation of Polycyclic Aromatic Hydrocarbons

Sudhir K. Shukla[a], Neelam Mangwani[b], T.Subba Rao[a] and Surajit Das[b]

[a]Biofouling and Biofilm Processes Section, Water and Steam Chemistry Division, BARC Facilities, Kalpakkam, India, [b]Laboratory of Environmental Microbiology and Ecology (LEnME), Department of Life Science, National Institute of Technology, Rourkela, Odisha, India

8.1 Introduction

The most important environmental issue that planet Earth has encountered over the past few decades is contamination with toxic waste effluents and recalcitrant compounds due to anthropogenic activities. Among the diversity of pollutants available, one of the lethal contaminants of soil and marine sediment is polycyclic aromatic hydrocarbons (PAHs). These carbon compounds are widespread in nature, contain two or more aromatic ring structures, and cause mutagenic and carcinogenic effects in humans (Seo et al., 2009). Exposure to PAHs can occur in several ways: through air, soil, food, water, and as an occupational hazard. PAHs are widely distributed and relocated in the environment as a result of incomplete combustion of organic matter, electricity-generating power plants, automobile exhausts, coal combustion, etc. (Samanta et al., 2002). The US Environmental Protection Agency (US EPA) has listed 16 PAHs as priority pollutants that need immediate remediation. The ubiquitous occurrence of PAHs represents an obvious health risk and the public are concerned regarding their removal from the environment. Many abiotic and biotic factors like volatilization, bioaccumulation, photooxidation, and microbial transformation are associated with PAHs' toxicity to the environment. Of late, biological methods are proving to be economical and safe in the efficient removal of PAHs from the environment (Singh et al., 2006). When it comes to biological treatment of pollutants, bacteria are of particular interest, since diverse genera of bacteria capable of degrading PAHs have been isolated from different environments (Chauhan et al., 2008). In nature, most of the microbiota survive as biofilms and are enmeshed in a

Microbial Biodegradation and Bioremediation. DOI: http://dx.doi.org/10.1016/B978-0-12-800021-2.00008-X

matrix of extracellular polymeric substances (EPS). The term *biofilms* was coined by Costerton et al. (1978), who described them as "a microcosm of bacterial consortium enmeshed with abiotic constituents in an exopolymeric matrix at interfacial regions." Biofilms are very common in environments like natural waters and soil, living tissues, human prostheses (medical biofilms), and industrial or potable water distribution systems (Chandki et al., 2011). EPSs are rich in water, nucleic acids, proteins, carbohydrates, and uronic acids (Jain et al., 2013), which play a significant role in biofilm development. There are many physiological and physicochemical conditions in biofilm mode that make the bacterial cells robust and have impact on bioremediation; for instance, competition among bacteria nutrients and oxygen, pH, O_2 and temperature conditions. Catabolite repression by other organic metabolites can decrease the expression of the pollutant degradation genes. Biofilm-associated cells exhibit specific gene expression, often controlled by quorum-sensing systems, or dormancy. This can increase tolerance to contaminants (Shimada et al., 2012). When compared with artificially immobilized and encapsulated cells, the advantage of biofilms is that the high density and tolerance of the constituent cell is naturally achieved. Thus, forming biofilms is considered a natural strategy of microorganisms to construct and maintain a favorable niche in stressful environments (Shemesh et al., 2010). Rapid degradation activity of biofilm cultures could be attributed to their producing detached or dispersal cells. It is suggested that biofilms are a bacterial cell breeding place that releases superactivated detached cells for effective survival in stringent environments. According to Morikawa (2010), application of biofilms to bioproduction and bioaugmentation processes is a simple and rational approach for a successful bioremediation technology. Thus in recent times, there have been several reports that demonstrated the efficacy of biofilms in bioremediation of xenobiotic compounds, toxic metals, and biotransformation technology (Singh et al., 2006; Yamaga et al., 2010; Fida et al., 2012).

8.2 Environmental Pollutants and Bioremediation

Release of unwanted substances into the environment beyond stipulated limits, where deleterious effects begin to manifest, is termed *pollution*. Pollutants have been present in the environment since the preindustrial era. However, rapid industrialization and unscrupulous use of synthetic and innate chemicals like heavy metals, hydrocarbons, chlorinated hydrocarbons, nitro-aromatic compounds, organo-phosphorus, and pesticides has resulted in serious environmental pollution (Seo et al., 2009). To prevent the spread of chemical pollution, biological or a preferential combinational of both remediation methods are to be adapted for successful cleanup. Biological remediation, i.e., *bioremediation*, is the process that uses the metabolic potential of microbes and plants to eliminate environmental pollutants (Glazer and Nikaido, 1995) (Figure 8.1). The demand for and complexity of bioremediation strategy predominantly depends upon understanding of the pollutants' nature and the metabolic pathway the microbe adopts for degradation.

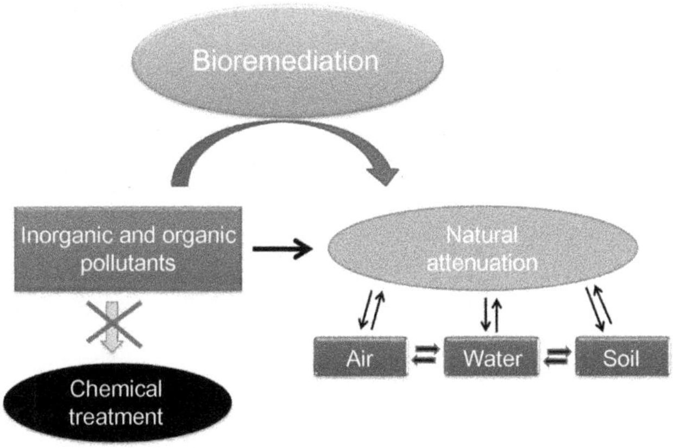

Figure 8.1 Natural attenuation and bioremediation are widely accepted environment cleanup procedures.

8.2.1 Organic Compounds

A number of new hydrocarbon molecules have been discovered, synthesized, and utilized by chemical and pharmaceutical industries. The persistence of these organic pollutants and emerging contamination is of key concern. These chemicals dominate the human habitat owing to their application in industry, agriculture, therapeutic products, and households (Doll and Frimmel, 2004). Integers of natural or anthropogenic organic complexes are there in our biosphere: ~16 million molecular species are known of which roughly 40,000 are under scrutiny due to their daily use. The detail structure, application, and toxicity of some PAHs are illustrated in Table 8.1.

8.2.1.1 Polycyclic Aromatic Hydrocarbons

PAHs are ubiquitous and are natural pollutants of air, water, and soil. PAHs are mainly formed due to incomplete combustion of fossil fuels. They are a usual component of dyes, plastics, and pesticides. Crude oil and coal deposits are natural repositories of PAHs. PAHs are composed of two or more fused aromatic rings. They are a class of multiphase compounds that are released into the environment naturally and by anthropogenic means. PAHs are one of the most significant classes of organic compounds that have caused rising concern regarding their harmful effects to humans and other living organisms. A total of 16 PAHs have been listed as toxic pollutants by the US EPA (Perelo, 2010) (Figure 8.2). They are highly hydrophobic in nature and tend to adsorb onto the surface of soil (or sediments in a marine environment).

8.2.1.2 Nitro-Aromatic Compounds

Nitro-aromatic compounds consist of a minimum of one nitro group ($-NO_2$) attached to an aromatic ring. These compounds consist of an imperative group

Table 8.1 Organic Pollutants: Structure, Toxicity, and Source

	Pollutants	Structure	Toxicity/Health Effect	Use/Source
PAHs	Pyrene	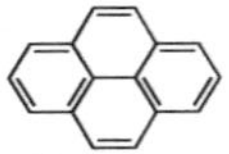	Toxic to kidneys and liver, carcinogenic	Dye and dye precursor, combustion
	Napthalene	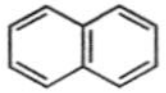	Carcinogenic, hemolytic anemia	Industrial application
	Phenanthrene	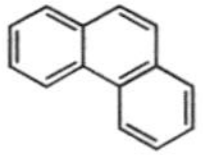	Carcinogenic, dye	Coal tar, pesticides
	Benzo(a)pyrene	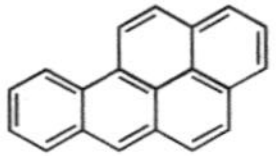	Mutagenic, carcinogen	Industries, coal tar, automobile exhaust fumes
Nitroaromatic compounds	2,4,6-Trinitrotoluene (TNT)	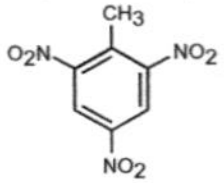	Skin irritation, carcinogenic, immune modulator	Explosive, pesticides
	Hexanitrobenzene	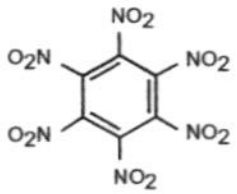	Skin irritation, carcinogenic, Fertility	Highly explosive
Organochlorines	DDT (dichlorodiphenyltrichloroethane)	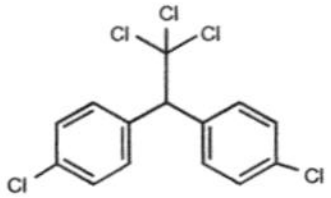	Xenoestrogenic, endocrine disruptor	Insecticide
	2,4-D (2,4-Dichlorophenoxyacetic acid)	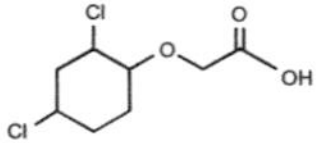	Neurotoxic, carcinogenic	Pesticide
Phthalates	Di(2-ethylhexyl) phthalate	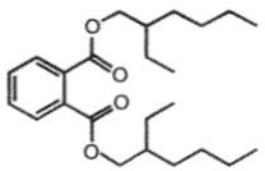	Endocrine disruptor	Plasticizer, dielectric fluid
Azo dye	Azobenzene	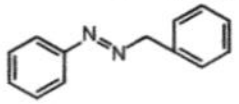	Carcinogenic	Dyeing

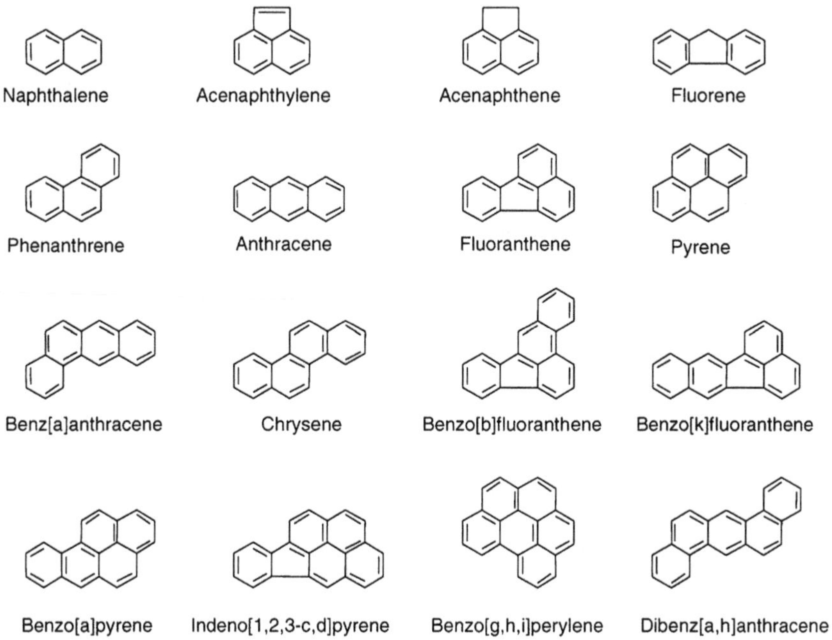

Figure 8.2 Molecular structure of the 16 PAHs considered priority pollutants by the United States Environmental Protection Agency (US EPA).

of chemicals which are widely used at present in various industries such as agrochemicals, textile, and chemicals in pharmaceutical industry (Su et al., 2012). Nitro-aromatic compounds are released to the environment by both synthetic and natural means; however, the major contribution is through synthetic means. They are highly resistant to degradation because of their recalcitrant character. They are toxic to both humans and animals. The reported manifestations in humans are sensitization of skin, immuno-toxicity, and methaemoglobinemia. Nitro-aromatic compounds are also a common toxicant observed in many aquatic ecosystems (Katritzky et al., 2003).

8.2.1.3 Organo-Chlorine Compounds

Organo-chlorines include a class of organic compounds containing chlorine (Cl^-) atoms bounded by covalent attachment to the carbon structure. Organo-chlorines have diverse structures and applications. They are mainly synthetic compounds that have both aliphatic chains and aromatic rings. Aliphatic organo-chlorides such as vinyl chloride are used to make polymers, whereas compounds like chloromethane are common solvents used in dry cleaning and the textile industry. Aromatic organo-chlorines like DDT, endosulfan, chlordane, aldrin, dieldrin, and endrin, mirex are used as pesticides. They are designated in the list of persistent organic pollutants (POPs) due to their toxicity and environmental persistence. Organo-chlorines like polychlorinated biphenyls (PCBs) are used as coolants and insulators in transformers. Tons of commercial PCBs (e.g., Aroclors) are present in the terrestrial and aquatic environments (Salvo et al., 2012). Commercially, PCBs are frequently used as plasticizers, adhesives, lubricants, and hydraulic fluids (Perelo, 2010).

8.2.1.4 Phthalates

Phthalates are esters of phthalic acid. They are broadly used in the manufacture of plastics, insect repellents, synthetic fibers, lubricants, and cosmetics (Liang et al., 2008). They are also used in the pharmaceutical industry as enteric coatings of pills and nutritional supplements as viscosity control and gelling agents. The paper and plastic industries are major manufacturers of phthalates, often discharging them into the environment at relatively high levels, which causes environmental pollution. Other applications of phthalates include their use in blood transfusion bags and intravenous fluid bags and tubes. Phthalates are classified as endocrine disruptors or hormonal active agents (HAAs). Phthalates like butyl benzyl phthalate are reported to cause eczema and rhinitis in children. Butyl benzyl phthalate is also classified as a possible class-C human carcinogen (under the 1986 US EPA guidelines) (Matsumoto et al., 2008).

8.2.1.5 Azo Dyes

The effluents from textile industries have been designated as major pollutants of groundwater resources and it has become an area for scientific and legislative inspection. The contaminated area is enriched with huge amount of organic and inorganic components. Among the textile dyes, azo dyes are foremost. The dyes are extremely recalcitrant and associated with toxicity to aquatic flora and fauna. Azo dyes have the functional group R–N=N–R′ (R and R′ can be either aryl or alkyl) (Puvaneswari et al., 2006). A survey by the National Institute for Occupational Safety and Health reported the carcinogenicity of azo dyes such as benzidine based on animal experiments (Mittal et al., 2013). Around 2000 different azo dyes are presently used and annually over 7×10^5 tons of these compounds are produced globally (Zollinger, 2003).

8.2.2 Heavy Metals

Metals with a density above 5 g/cm^3 are referred to as heavy metals. Most heavy metals are included in the transition elements, i.e., d block element (Duruibe et al., 2007). They have an incomplete filled d orbital; thus their cations form complex compounds. Many heavy metal cations are also known as *trace elements* as they are part of biochemical reactions. Metals like Co, Cu, Cr, Fe, Mg, Mn Mo, Ni, Si, and Zn are vital micronutrients that are necessary for various biochemical and physiological functions (Ernst, 1996). However, at higher concentrations, they form nonspecific complexes with cell components, which may have toxic effects. Trace elements like Zn^{2+}, Ni^{2+}, and Cu^{2+} are toxic at higher concentrations. Metals like

Al, Sb, As, Ti, V Ag, Pb, Hg, and Cd are not biologically important. However, metals like Hg^{2+} and Cd^{2+} form strong complexes with cell organelles, which make them physiologically nonfunctional. Table 8.2 illustrates the source and toxic effects of heavy metal. Similarly, Pb and As are of concern because of their toxicity to human and aquatic animals (Buschmann et al., 2008).

8.2.3 Bioremediation

Microbial degradation is one of the major strategies used for bioremediation of organic compounds and heavy metals. The sustainability of the bioremediation process depends on the degradation and transformation potential of microorganisms. The uniqueness of the bioremediation method is due to the fact that it removes pollutants from the natural environment or mitigates the pollutants using the indigenous available site-specific microbial community. Broadly, *in situ* bioremediation can be divided into (i) bioattenuation, which depends on the natural process of biodegradation, (ii) biostimulation, where stimulation of degradation of pollutants is accomplished by the addition of electron donors or acceptors, and (iii) bioaugmentation, where the microbes with inherent capabilities of degrading or transformation of the pollutants are added to the site (Madsen, 1991). The resident microbiota at a contaminated site develop tolerance to the pollutant and also modify their genomes or changes their genetic machinery to detoxify or degrade chemicals that persist in the environment (Grace et al., 2011). Thus, the assets in bioremediation technology are metabolic potential of a bacteria and the genetic machinery of the microbes present in the environment. Many researchers have studied the metabolic potential of microbes (Juhasz and Naidu, 2000; Johnsen et al., 2005). Depending upon the metabolic potential of the microorganism, the bioremediation process can be designed and implemented. Thus, the process of bioremediation is also designed based upon the microbial potential to detoxify the pollutants either by a mineralization process or by immobilization

Table 8.2 Toxicity of Heavy Metals

Metal	Toxicity	Source
Pb	Neurotoxic	Battery, architectural metal, oil paints, gasoline
Hg	Poisonous	Electrical and electronic applications
Cd	Carcinogenic, poisonous, neumonitis, pulmonary edema	Batteries and electroplating.
Co	Contact dermatitis	Batteries, catalysis, electroplating
Ni	Alloy metal, rechargeable batteries	Carcinogenic, dermatitis
Cu	Electrical motors, metal alloy, architecture and industries	Wilson's disease, liver cirrhosis, DNA damage
As (metalloid)	Arsenicosis	Wood preservative, military, car batteries

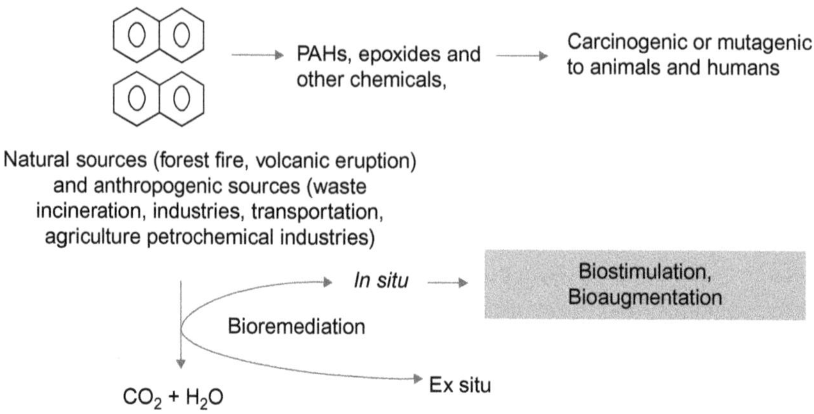

Figure 8.3 Schematic illustration showing the source, distribution, toxicity, and bioremediation of PAHs.

(Parales and Haddock, 2004). In general, microbes adapt to the toxicity of the pollutant by genetic modification (Singh et al., 2006). Microbes detoxify metals by adsorption, or by metallothionein production to trap metal ions (Ryvolova et al., 2011). However, immeasurable intracellular and extracellular enzymes degrade the organic compounds. A number of genes (like *pah*, *nid*, *nah*, and *bph*) involved in the degradation of organic aromatic pollutants (PAHs, PCB) have been studied in detail (Furukawa and Fujihara, 2008; Seo et al., 2009). Bioremediation strategies are either *ex situ* or *in situ* based on the specific process. *In situ* bioremediation is the removal or attenuation under natural environment, whereas an *ex situ* bioremediation method involves degradation of chemicals from the excavated samples (Pandey et al., 2009). The targets for *in situ* bioremediation are biostimulation, which exploits active indigenous microbial population, and bioaugmentation, which incorporates specific proficient seeding of bacteria at the contaminated site (Thomassin-Lacroix et al., 2002). Several factors, such as site conditions, indigenous populations of microorganisms, and the type, quantity, and toxicity of pollutant chemical species present, determine the suitability of a particular bioremediation technology (Figure 8.3).

8.3 Bioremediation of PAHs

Due to the hydrophobic nature and low solubility of PAHs, they are resistant to biodegradation and can bioaccumulate in the environment through the food chain. Therefore, PAHs in the environment represent a long-term threat to human health (Figure 8.3). Several PAH compounds contain A, B, Bay, K, and L regions. These regions, when undergoing metabolism, produce highly reactive epoxides (Figure 8.4). The persistence of PAHs in the environment depends on the physical

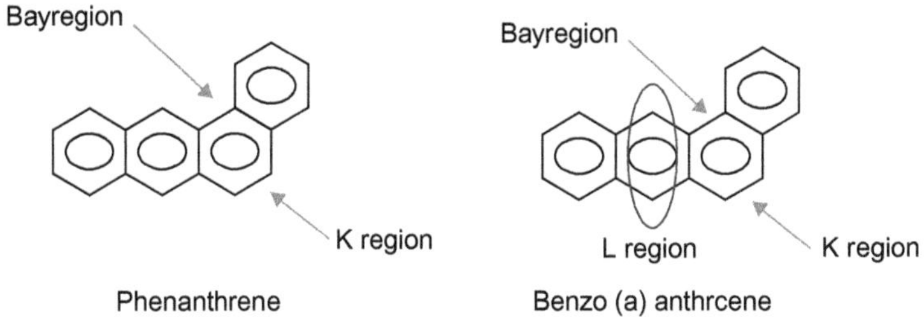

Figure 8.4 Metabolically active regions of PAH degradation.

and chemical characteristics of the PAHs, their concentration, and their bioavailability. Higher-molecular-weight PAHs have higher toxicity and longer environmental persistence (Chauhan et al., 2008).

8.3.1 Source and Distribution

The release of PAHs (or intermediates) in the environment is from both natural (e.g., forest fire, volcanic eruption) and anthropogenic activity (e.g., waste incineration, industries, transportation, agriculture, petrochemical industries). The foremost source of PAHs is coal, which is rich in a variety of PAHs and other organic pollutants. PAHs are released from coal during combustion, cooking, pyrolysis, and coal preparation processes (Guo et al., 2011; Wang et al., 2007). Another major discharge of PAHs to the environment is through motor vehicle exhausts. PAHs are present in the form of particulates or vapor and distributed in air, water, sediments, and soil. The persistence and migration of specific PAHs in the environment depends on their structure, molecular weight, solubility, and evaporation capability. PAHs can easily persist in any environment, particularly in the atmosphere. Low-molecular-weight PAHs remain within the vapor phase of the atmosphere, whereas high-molecular-weight PAHs are found in aggregates and form particulates. Low-molecular-weight PAHs with four or less aromatic rings (e.g., naphthalene, phenanthrene, anthracene, pyrene) exist both in the vapor phase and as adsorbed particles. Those with five or six rings (e.g., benzo(α)pyrene, coronene) are present principally as adsorbed airborne particulate matter, such as fly ash and soot. The presence of atmospheric PAHs varies significantly with temperature, relative humidity, and solar radiation. The other major source of PAHs' presence is the soil, where incomplete combustion of organic matter and atmospheric precipitation results in their accumulation (Chetwittayachan et al., 2002; Nam et al., 2008). PAHs remain in soil for quite a long period due to their hydrophobicity and aggregation property. The concentrations of PAHs in soil reflect pollution to some ambient degree (Guo et al., 2011). PAHs' distribution in soil also related to microbial degradation, photooxidation, and the volatilization pattern of the area (or site); such a process is observed frequently in tropical soils. This could be the reason why the PAH concentrations vary strikingly between temperate and tropical soils. The persistence of PAHs is greater in marine sediments, since they are less subject to photochemical or biological oxidation. PAHs also have the tendency to bioaccumulate in vegetative and aquatic habitats. With the characteristics of environmental persistence and bioaccumulation, PAHs are harmful to human health; however, very little information is available regarding exposure to PAHs and the levels found in the human body (Samanta et al., 2002).

8.3.2 Toxicity

Many PAHs and their compounds are mutagenic and carcinogenic. They are highly lipophilic and are readily absorbed in body tissues of mammals. The carcinogenic and mutagenic potency of a PAH is related to the structural complexity: usually a more complex compound is more toxic. Several epidemiological studies have verified that contact with environmental PAHs is associated with elevated risk of cancers in humans. The carcinogenicity of PAHs is mainly due to some of their metabolic intermediates (such as epoxide and phthalates) or interaction with other chemicals (Guo et al., 2011). Naphthalene is the most widely studied PAH for toxic effects. It is easily absorbed into the tissues of the liver, kidney, and lung, thereby manifesting its toxicity. Naphthalene can also lead to hemolytic anemia. It is also an inhibitor of mitochondrial respiration (Samanta et al., 2002). Many PAHs are genotoxic and cause cytogenetic damages by intercalating DNA and forming DNA adduct.

8.3.3 Bacterial Metabolism of PAHs

Microorganisms have been found to degrade an assortment of PAHs. Metabolism of PAHs is largely determined by genetic aspects. For most of the low-molecular-weight PAHs, both aerobic and anaerobic pathways are well documented. But the mainstream attention has been paid to aerobic metabolism. The aerobic degradation of PAHs begins with the action of dioxygenase with the addition of an oxygen atom between two carbon atoms of the benzene ring. The outcome is a *cis*-dihydrodiol and opening of aromatic ring. The next step is formation of dihydroxylated intermediate by the action of enzyme dehydrogenase. Subsequently, a ring cleavage reaction produces intermediates of a tri-carboxylic acid cycle (Chauhan et al., 2008). The general mechanism of aerobic degradation of PAHs is summarized in Figure 8.5 by taking naphthalene, phenanthrene, and pyrene as representative models. Electron acceptors in the absence of molecular oxygen are usually nitrate, ferrous, and sulfate ions. A plethora of genes encoding PAH catabolic enzymes have been characterized and listed (Table 8.3). The investigation of the PAHs' catabolic genes is of high significance from both a fundamental and an applied research point of view. It will be useful to evaluate their evolution, structural diversity, mutations, gene transfer, and relationship over time. Moreover, the overall acquaintance will facilitate the molecular mechanisms of bacterial adaptation to PAHs and development of bioremediation strategies. A large number of

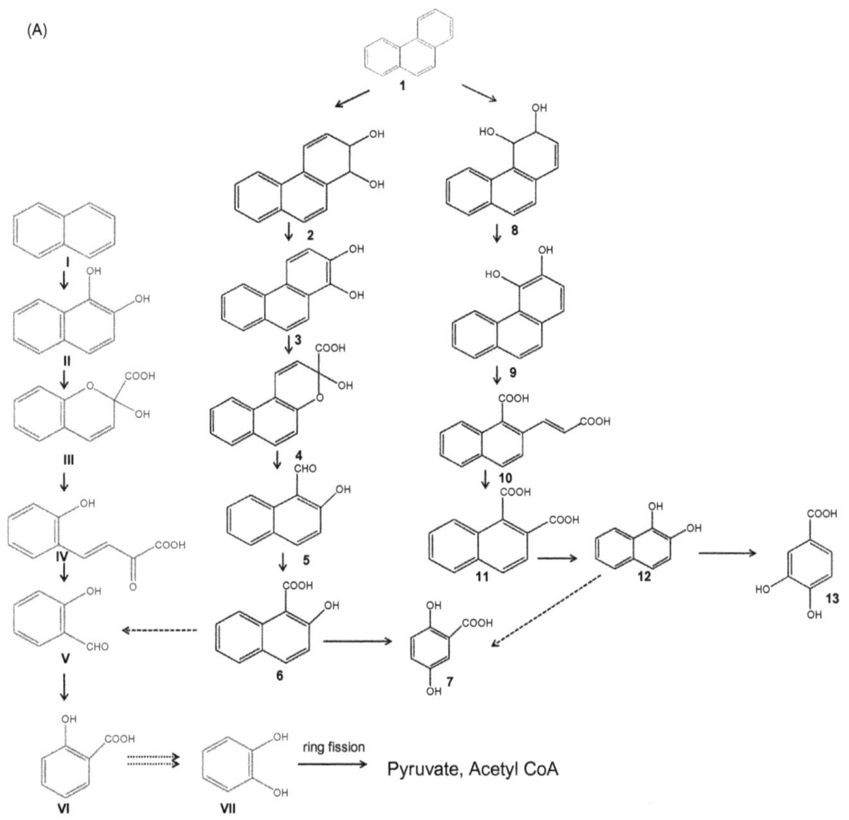

Figure 8.5 (A) Schematic pathway of biodegradation of naphthalene (in blue) and phenanthrene. *Compound designations:* (**1**) Phenanthrene; (**2**) *cis*-1,2-dihydroxy-1,2-dihydrophenanthrene; (**3**) 1,2-dihydroxyphenanthrene; (**4**) 3-hydroxy-3H-benzo[f]chromene-3-carboxylic acid; (**5**) 2-hydroxy-naphthalene-1-carbaldehyde; (**6**) 2-hydroxy-1-naphthoic acid; (**7**) gentisic acid; (**8**) *cis*-3,4-dihydroxy-3,4-dihydrophenanthrene; (**9**) 3,4-dihydroxyphenanthrene; (**10**) 2-(2-carboxy-vinyl)-naphthalene-1-carboxylic acid; (**11**) naphthalene-1,2-dicarboxylic acid; (**12**) naphthalene-1,2-diol; (**13**) protocatechuic acid; (**I**) naphthalene; (**II**) 1,2-dihydroxynaphthalene; (**III**) 2-hydroxychromene-2-carboxylate; (**IV**) cis-*O*-hydroxybenzalpyruvate; (**V**) 2-hydroxybenzaldehyde; (**VI**) salicylate; (**VII**) catechol. (**B**) Biodegradation of pyrene. *Compound designations***:** (**1**) pyrene; (**2**) pyrene-*cis*-1,2-dihydrodiol; (**3**) pyrene-1,2-diol; (**4**) 2-hydroxy-3-(perinaphthenone-9-yl)-propenic acid; (**5**) 2-hydroxy-2H-1-oxa-pyrene-2-carboxylic acid; (**6**) 4-hydroxyperinaphthenone; (**7**) 1,2-dimethoxypyrene; (**8**) pyrene-trans-4,5-dihydrodiol; (**9**) pyrene-*cis*-4,5-dihydrodiol; (**10**) pyrene-4,5-diol; (**11**) phenanthrene-4,5-dicarboxylic acid; (**12**) 4-carboxyphenanthrene-5-ol; (**13**) 4-carboxy-5-hydroxy-phenanthrene-9,10-dihydrodiol; (**14**) 4-carboxyphenanthrene-5,9,10-triol; (**15**) 2,6,6′-tricarboxy-2′-hydroxybiphenyl; (**16**) 2,2′-dicarboxy-6,6′-dihydroxybiphenyl; (**17**) phthalic acid; (**18**) 4-phenantroic acid; (**19**) 3,4-dihydroxy-3,4-dihydro-phenanthrene-4-carboxylic acid; (**20**) phenanthrene-3,4-diol; (**21**) 4-phenanthroic acid methyl ester; (**22**) 4-hydroxyphenanthrene; (**23**) 7,8-benzocoumarin; (**24**) 2-hydroxy-2-(phenanthrene-5-one-4-enyl)-acetic acid; (**25**) 5-hydroxy-5H-4-oxapyrene-5-carboxylic acid; (**26**) pyrene-4,5-ione; (**27**) 4-oxa-pyrene-5-one. (For interpretation of the references to color in this figure legend, the reader is referred to the web version of this book.)
Source: Adapted from Seo et al., 2009.

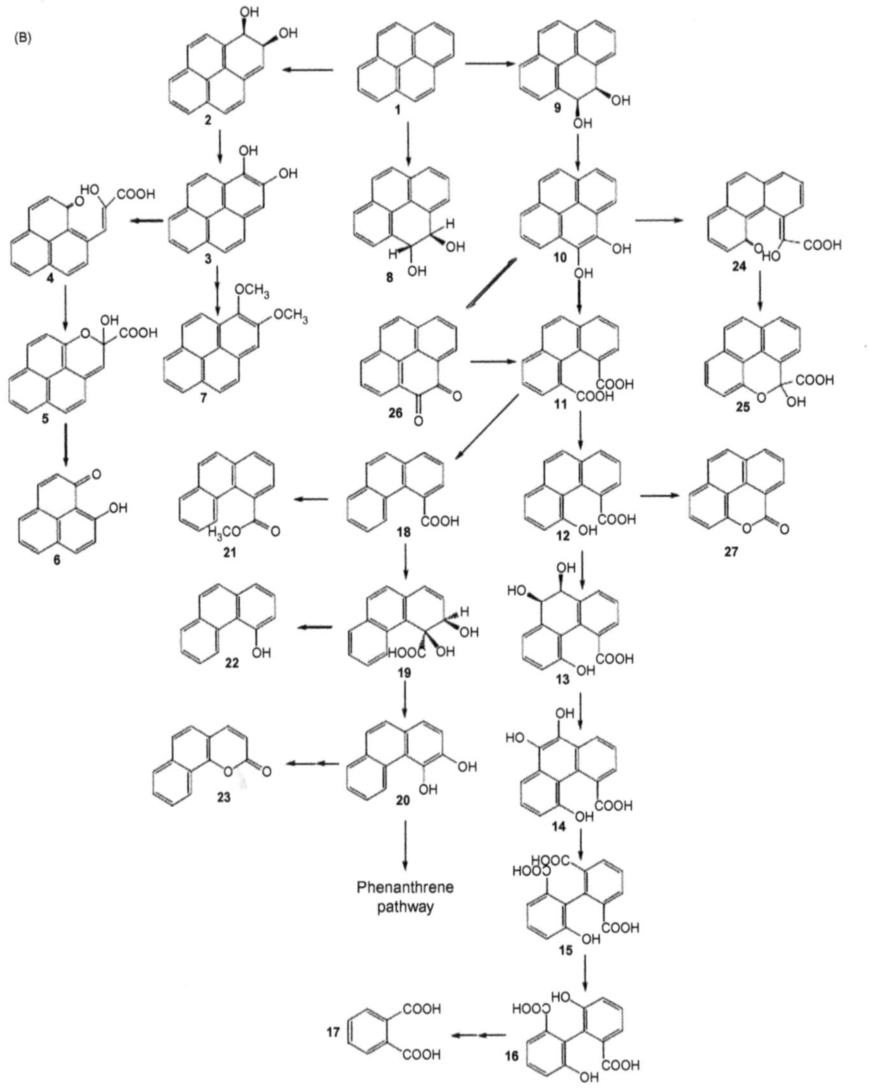

Figure 8.5 (Continued)

PAH-degrading bacteria are reported, namely, *Pseudomonas putida*, *Pseudomonas fluorescens*, *Pseudomonas paucimobilis*, *Pseudomonas mendocina*, *Pseudomonas vesicularis*, *Pseudomonas cepacia*, *Alcaligenes denitrificans*, *Mycobacterium* sp, *Aeromonas* sp., *Alcaligenes faecalis*, *Bacillus cereus*, *Vibrio* sp., *Cyclotrophicus* sp., *Stenotrophomonas maltophilia*, *Beijerinckia* sp., *Micrococcus* sp., *Nocardia* sp., and *Flavobacterium* sp. have been studied (Samanta et al., 2002).

Table 8.3 PAHs-Degrading Genes (Chauhan et al., 2008)

Strain	Gene	PAHs Degradation	Enzyme/Function
P. putida	*nahAa*	Naphthalene	Reductase
	nahH	Naphthalene	Catechol oxygenase
	ndoA	Naphthalene	Dioxygenase
P. putida OUS82	*pahC*	Phenanthrene	Dioxygenase
Pseudomonas stutzeri AN10	nahG	Naphthalene	Naphthalene-salicylate 1-hydroxylase
A. faecalis AKF2	*phnC*	Phenanthrene	Dihydroxyphenanthrene dioxygenase
Nocardiodes sp. KP7	*phdA*	Phenanthrene	Alpha subunit of dioxygenase
Mycobacterium sp. PYR-1	*nidA, nidB*	Pyrene	Dioxygenase
Sphingomonas paucimobilis var. EPA505	*pbhA*	Phenanthrene, anthracene, benzo [b]fluoranthene	Ring fission dioxygenase
Burkholderia sp. RP007	*phnAc*	Naphthalene, phenanthrene, and anthracene	Dioxygenase
Rhodococcus sp. NCIMB12038	*narB*	Naphthalene	*cis*-naphthalene dihydrodiol dehydrogenase

8.4 Bacterial Biofilms and Bioremediation

8.4.1 Biofilms

The term "biofilm" stands for the surface-associated microbial communities wherein the cells are embedded in self-produced EPSs. Microorganisms residing within the biofilm live in a cooperative manner and benefit each other by forming different ecological niches within the biofilm. The properties of these microbial communities are governed partially by the structure, diffusion of nutrients, and physiological activity of the cells (Costerton et al., 1995). In nature, the biofilm phase is the predominant lifestyle of microbes. Biofilms possess a physical barrier and cells remain in the physiological dormant state; therefore, biofilms show a high tolerance to physical, chemical, and biological stresses. Biofilms are resilient to a wide variety of environmental stresses and have thus transformed into the most resistant microbial life forms on planet Earth.

8.4.2 Biofilm Development

Biofilm development is a sequential process governed by cellular, surface-related, and environmental factors. Biofilm formation involves different physiological

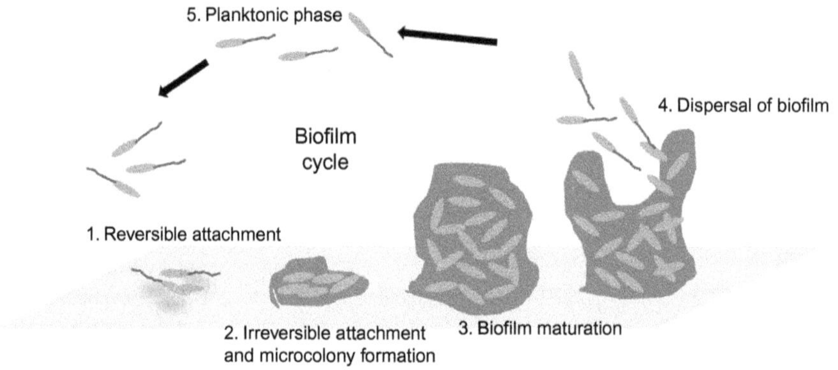

Figure 8.6 Illustration showing different developmental stages of bacterium biofilm cycle.

states of the respective organisms (Costerton et al., 1995) (Figure 8.6). In the first step, the substratum is conditioned by many inorganic and organic molecules secreted by microbes themselves, thereby providing a favorable place for cell settlement. Conditioning of the surface is followed by the interaction of the planktonic cells. In the second step, microbial cells attach to the surface, first in a reversible phase and then irreversibly. In the third step, irreversible attachment is followed by multiplication of cells. In addition, they secrete EPS, which attaches the cells to the surface and to each other. The EPS matrix consists of polysaccharides, proteins, lipids, and nucleic acids; these provide the mechanical stability of biofilms. EPS also mediates bacterial adhesion to surfaces and forms a cohesive, three-dimensional polymeric network that interconnects and transiently immobilizes biofilm cells (Flemming and Wingender, 2010).

8.4.3 Biofilm Components

Three microbial products, EPSs, soluble microbial products, and inert biomass, form the structure of the biofilm. The chemical composition of biofilms markedly influences the metabolism of the biofilm microbiota and their growth substrates. The biomolecules present in the biofilm milieu can be differentiated between utilization-associated products by the resident bacteria and biomass-associated products which are biodegradable. The autotrophs are the significant microorganisms which release a variety of soluble microbial products or soluble organic compounds during their normal biomass metabolism. The soluble organic compounds control and support a substantial heterotrophic population of the biofilm. Major components of the biofilm matrix and their respective function have been summarized in Table 8.4. Major components of the biofilm matrix are detailed in the following sections.

8.4.3.1 Exopolysaccharides

EPSs are also referred to as the dark matter of biofilms because of their large range of matrix polymers and the difficulty in analyzing them (Wingender et al., 1999). Earlier EPSs were meant for the "extracellular" polysaccharides alone but over

Table 8.4 Major Components of EPSs and Their Function in Bacterial Biofilms (Flemming and Wingender, 2010)

EPS Components	Functions in the Context of Biofilms
Polysaccharides	Adhesion, aggregation of bacterial cells, cohesion of biofilms, protective barrier, retention of water (by hydrophilic polysaccharide), sorption of organic and inorganic compounds, sink for excess energy
Proteins	Enzymatic activity, electron donor or acceptor, adhesion, aggregation of bacterial cells, cohesion of biofilms, protective barrier, sorption of organic and inorganic compounds, sink for excess energy, export of cell components
DNA	Exchange of genetic Information, adhesion, aggregation of bacterial cells, cohesion of biofilms
Divalent ions	Mechanical stability, regulation of bap, regulation of EPS production
Surfactants and lipids	Bacterial attachment and detachment, hydrophobicity
Water	Provides hydrated environment, medium for movements for nutrients

time, when it became evident from several studies that biofilm matrix also contains proteins, nucleic acids, lipids, etc., it was renamed "EPSs" (Flemming et al., 2007). EPS content varies greatly depending upon various biological and physical factors. EPS also determines the biofilm structure and stability, which also depends on the presence of multivalent cations such as Ca^{2+} (Chen and Stewart, 2002; Shukla and Rao, 2013a). The major component of the EPS matrix is exopolysaccharides (Wingender et al., 2001). They are long, linear, or branched molecules with high molecular weight ranging from 500 to 2000 kDa. Exopolysaccharides have been isolated and extensively characterized from various bacterial species and diverse environments. Most exopolysaccharides are hetero-polysaccharides that consist of neutral as well as charged sugar residues. On the other hand, some exopolysaccharides are homo-polysaccharides, including the sucrose-derived glucans, fructans, and cellulose (Wingender et al., 2001). Exopolysaccharides contain organic or inorganic constituents that greatly affect their physical (polyanionic and polycationic) properties and determine their biological function. Exopolysaccharides can be highly diverse, even among the bacterial strains of a single species (Table 8.5). For instance, *Pseudomonas aeruginosa* produces at least three kinds of exopolysaccharide: *alginate*, *Pel*, and *Psl37*, of which alginate is the most studied exopolysaccharide and is an unbranched heteropolymer with high molecular mass. Alginate is important at the beginning of biofilm formation, i.e., in the establishment of microcolonies, and aids mechanical stability in mature biofilms.

8.4.3.2 Extracellular Proteins

Surface proteins are found to be regularly present in the biofilm matrix, but their role in intercellular adhesion and integrity of biofilm matrix remained unnoticed.

Table 8.5 Nature of Exopolysaccharides and Their Examples

Nature of Exopolysaccharides	Examples	Responsible Residues	Organism
Polyanionic	Alginate, xanthan and colonic acid	Uronic acids, ketal-linked pyruvate, or sulfate	*P. aeruginosa*
Polycationic	Polymeric intercellular adhesion (PIA), PNAG	β-1,6-linked *N*-acetylglucosamine with partly deacetylated residues	*Staphylococcus aureus* and *Staphylococcus epidermidis*

In the last decade, several studies established that biofilm matrix consists of significant amounts of proteins and plays a critical role in biofilm architecture and stability (Cucarella et al., 2001; Shukla and Rao, 2013a,b). Some of these proteins are able to establish biofilm formation even in the absence of exopolysaccharides. This fact shows that exopolysaccharides are not essential for the development of mature biofilms. Broadly, they can be divided into two subcategories: enzymes and structural proteins.

8.4.3.2.1 Enzymes

Many extracellular enzymes have been reported in biofilms. In general, these enzymes participate in the degradation of EPS and are known as EPS-modifying enzymes. The substrates of these enzymes are biopolymers such as polysaccharides, cellulose, proteins, and nucleic acids. EPS constituents act as a nutrient store; when required, this store is digested with the help of these extracellular enzymes. Apart from this, they are also useful in the dispersal of biofilm and hence the release of microbial cells and resettlement on a new substratum occurs.

8.4.3.2.2 Structural Proteins

Another group of extracellular proteins are biofilm-associated proteins (bap) and homologous bap-like proteins (Latasa et al., 2006). This family of proteins shares some common features like very high molecular mass, a core domain of tandem repeats, and attachment with the cell surface that promotes biofilm formation. Apart from these, amyloids are the second most found proteinaceous components of the biofilm matrix. The amyloids have been detected ubiquitously in various types of environmental biofilms such as freshwater lakes, brackish water, drinking-water reservoirs, and wastewater treatment plants (Otzen and Nielsen, 2008).

8.4.3.3 Extracellular DNA

Earlier extracellular DNA (eDNA) was regarded as the product of lysed cells within the biofilm; however, it has now become clear that it is in fact an essential part of the biofilm matrix (Whitchurch et al., 2002; Molin and Tolker-Nielsen, 2003). eDNA comprises fragments of high molecular weight, up to 30 Kb

(Tetz et al., 2009). Some reports, using exogenous DNA and DNase effect on biofilm cells, have shown that there is an association between eDNA and higher antibiotic resistance in biofilms (Mulcahy et al., 2008; Tetz et al., 2009). eDNA aids in exchange of genetic information and aggregation of bacterial cells and determines the mechanical stability of biofilms. A recent study has shown that eDNA also facilitates the self-organization of cells inside bacterial biofilms (Gloag et al., 2013).

8.4.3.4 Lipids and Biosurfactants

EPSs also have hydrophobic property due to the substantial presence of hydrophobic molecules apart from highly hydrated hydrophilic molecules like polysaccharides, proteins, and DNA. The hydrophobic character of the EPS is attributed to lipids and alkyl group–linked polysaccharides, such as methyl and acetyl groups (Neu et al., 1992; Busalmen et al., 2002). Many microbes produce hydrophobic molecules that are very critical for adhesion phenomenon, e.g., secretion of lipopolysaccharides in *Thiobacillus ferrooxidans* (Sand and Gehrke, 2006) and surface-active extracellular lipids in *Serratia marcescens* (Matsuyama and Nakagawa, 1996). These surface-active molecules, which influence the surface tension at the air–water interface, are generally called biosurfactants. They play an important role by altering the surface tension and in turn aid in the exchange of gases (Leck and Bigg, 2005).

8.4.4 Physiological State of Cells in a Biofilm

Bacterial cells switch over to the sessile form of growth from a free-swimming planktonic phase; they express phenotypic traits that are often distinct from those that are expressed during planktonic growth (Stoodley et al., 2004). These distinct phenotypes are with respect to gene transcription and growth rate as compared to planktonic counterparts. Apparently, biofilm cells are physiologically distinct from their planktonic state. The metabolic requirements in a biofilm mode and planktonic lifestyle are entirely different. The metabolic activities of biofilm cells are governed by their local variations in the biofilm milieu. Different concentrations of nutrients, availability of oxygen, and signaling molecules inside the matrix result in a gradient of various niches. The sessile bacterial cells react to these changes by varying either gene expression patterns or physiological function, which results in their adapting to their local surroundings. In general, bacteria that are embedded in the biofilm matrix exhibit decreased respiration and metabolic activity and remain in a dormant state, which is also the basis of physiological resistance of biofilms toward antimicrobials (Cogan et al., 2005).

8.4.5 Quorum Sensing

Quorum sensing (QS) is a cell density–dependent phenomenon mediated by signaling molecules known as *autoinducers* (AIs). In bacteria, three types of AIs have been widely studied, namely *N*-acyl homo-serine lactones (AHLs), auto-inducing peptides (AIP), and AI-2. The elementary phase of QS is the binding of AIs to transcriptional factors or cell surface receptors and establishment of the relevant gene expression. AIs are very specific for bacterial groups. Gram-positive bacteria employ AHLs for QS-based gene regulation, whereas Gram-negative bacteria have AIP as a QS AI (Mangwani et al., 2012). QS regulates a number of vital functions in bacteria such as virulence, biofilm formation, and sporulation via AIs. Biofilm formation and various phases in the biofilm development are tightly regulated by QS-related mechanisms (Li and Tian, 2012). Studies on different QS mutant strains have revealed the involvement of an array of quorum regulatory units in biofilm formation and maturation (Kjelleberg and Molin, 2002).

8.5 Application of Biofilms in Bioremediation Technology

Bioremediation is the use of the biological process for mitigation/elimination of the toxic environmental pollutant. The bioremediation approach has, for the most part, been found to be better than chemical and physical methods in terms of its efficiency and techno-economic viability (Paul et al., 2005). It is an emerging *in situ* technology for the cleanup of environmental pollutants using microorganisms. Most of the bioremediation processes use biocatalysts, i.e., enzymes or whole living cells for the metabolism of the pollutant. Application of whole cells as biocatalysts is often preferred over isolated enzymes because of their catalytic stability, economic aspects (no enzyme purification step), self-generation of expensive cofactors, and their ability to carry out reactions that involve multicomponent assembly (Woodley, 2006). To deploy a whole-cell process at higher scale, cells need to be retained and reused, which can be accomplished by cell immobilization (Avnir et al., 2006). Cell immobilization is performed either by chemically binding cells to a carrier material or physical entrapment or encapsulation. However, cell immobilization is associated with some severe disadvantages, such as (i) reduced activity and viability of the biocatalyst, (ii) diffusion mass-transfer limitations of substrates and oxygen lowering reaction rates, (iii) additional costs, and (iv) lack of universally applicable immobilization concepts (Buchholz et al., 2012; Hekmat et al., 2007).

The potential of biofilm communities in bioremediation processes has recently been realized, as it has many advantages over whole cells which are used as biocatalysts. Naturally immobilized microbial cells in biofilms exclude the necessity of cell immobilization, as biofilm cells are already embedded in self-produced EPS. Therefore, using biofilm for any bioprocess could be an elegant and economical method. Moreover, biofilm-mediated bioremediation offers a promising alternative to planktonic cell—mediated bioremediation. Cells in a biofilm mode are robust and show tolerance to toxic materials present in the waste as cells are embedded in biofilm matrix, which acts as a physical barrier for cells. Their overall robustness makes them attractive living catalysts for organic syntheses, which can be harnessed for technological applications (Halan et al., 2012).

Before deploying any microbe for bioremediation purposes, many questions and parameters should be investigated and optimized. For instance, which microbe forms biofilms or novel isolates among the environmental bacteria? Which surfaces are suitable for biofilm growth and which environmental conditions suit biofilm development? Do the conditions that favor the biofilm development also support the biological process of our interest? These are some ideal questions that arise when designing a biofilm-based process and need to be worked out in detail.

8.5.1 Biofilms for PAH Remediation

Over the past few decades, rapid growth of chemical industries and persistence of many toxic chemical pollutants (e.g., PAHs), and the consequent environmental problems, have raised concerns about long-term environmental disasters with a public conscience (Pandey and Jain, 2002). Therefore, various physical, chemical, and biological research projects are being carried out currently to develop a plethora of processes for a sustaining environment. Our recent report has showed the importance of EPS in biofilm-mediated bioremediation that increased the amount of exopolysaccharides due to the presence of high concentrations of calcium, thereby significantly enhancing the degradation of PAHs by a marine strain of *P. mendocina* NR802 (Mangwani et al., 2013).

8.5.2 Factors Influencing the Bioremediation of PAHs

PAHs are highly hydrophobic compounds. Their poor solubility in water makes it difficult for a living system to utilize them as a source of energy. However, several bacterial genera can efficiently degrade PAHs. A number of physical factors, such as temperature and pH, together with genetic assets like QS, plasmid, and chemotaxis can influence their degradation. Understanding of these factors will be useful in designing a suitable bioremediation strategy.

8.5.2.1 Bioavailability

Bioavailability is defined as the effect of physicochemical and microbiological factors on the rate of biodegradation of a pollutant. It is one of the most critical factors in bioremediation of PAHs (Mueller et al., 1996). Since PAH compounds are hydrophobic, organic molecules with low water solubility are thus resistant to biological, chemical, and photolytic breakdown, and hence have a very low bioavailability (Semple et al., 2003). Solubility decreases with increase in size and molecular weight PAHs and in turn its availability for the microbial metabolism diminishes (Fewson, 1988). Apart from the solubility in water, persistence of PAHs also determines the bioavailability of contaminants. PAHs undergo rapid sorption to soil constituents such as mineral surfaces (e.g., clay) and organic matter (e.g., humic acids). Longer contact time of PAHs with soil results in more irreversible sorption, which leads to lower chemical and biological extractability of the PAHs. This phenomenon is called "ageing" of the contaminant (Hatzinger and Alexander, 1995).

PAHs can be released from the surface of minerals and organic matter by the use of surface-active agents or surfactants. Surfactants contain both a hydrophobic and hydrophilic moiety. Thus by solubilizing the hydrophobic PAH, it can gain access to the hydrophilic microbial cell. Some microorganisms also produce surfactants known as biosurfactants during the dispersion phase of the biofilms. These biosurfactants can assist in desorption of PAHs from the soil (Makkar and Rockne, 2003). One major advantage of using biosurfactants is that they are less toxic to indigenous microbes and do not produce micelles, which can encapsulate contaminant PAHs and prevent microbial access (Makkar and Rockne, 2003). Rhamnolipids are biosurfactants that have been found in *P. aeruginosa* biofilm matrix. They display a variety of surface activities and have been reported to assist in microcolony formation, formation of water channels in mature biofilm, and biofilm dispersion (Boles et al., 2004). Therefore, biosurfactants may be highly useful for bioremediation of oil recovery and oil spills, as they can successfully disperse hydrophobic substances and make them more bioavailable.

8.5.2.2 Temperature

Among various physical factors, temperature is a major governing factor in determination of microbial degradation ability of PAHs *in situ*. In general, contaminated sites may not be at the optimum temperature for bioremediation throughout the year. The solubility of PAHs increases with an increase in temperature, which in turn increases the bioavailability of the PAH molecules (Margesin and Schinner, 2001). On the other hand, oxygen concentration in water goes down at higher temperatures, which in turn reduces the metabolic activity of aerobic microorganisms. Therefore, most studies tend to use mesophilic temperatures to achieve higher bioavailability with substantial microbial metabolic activity (Lau et al., 2003).

8.5.2.3 pH

The hydrogen ion concentration (pH) influences the growth and metabolism of the microorganisms. It is very critical to monitor the pH of a contaminated site and regulate it to accomplish optimum growth of *in situ* PAH-degrading microbes. Since the pollutant may also influence the pH of contaminated sites, the indigenous microorganisms at the sites that do have PAH-degrading ability might not transform PAHs under these altered acidic or alkaline conditions. Moreover, many sites contaminated with PAHs are not at the optimal pH for bioremediation. Therefore, the pH of such sites should be adjusted, for instance, by the addition of lime, etc. Sometimes, *in situ* microorganisms at a contaminated site may be tolerant to the site conditions, but may have the potential to metabolize PAHs in suboptimal conditions. Therefore, it is also necessary to isolate and characterize the indigenous PAH-degrading microorganisms and accordingly adjust the optimum pH at the site.

8.5.2.4 Oxygen

Depending on the indigenous microbes, *in situ* bioremediation of PAHs can proceed under both aerobic and anaerobic conditions. Most of the reports on work PAHs degradations showed the involvement of aerobic metabolism of PAHs. This is in part due to the ease of study and culture of aerobic microorganisms relative to anaerobic microorganisms. Moreover, it is easier to stimulate aerobic *in situ* microbial communities by using hydrogen peroxide (Pardieck et al., 1992), sodium nitrate (Bewley and Webb, 2001), and perchlorate (Coates et al., 1999). The aerobic biodegradation of hydrocarbons is up to an order of magnitude higher when compared to anaerobic biodegradation (Rockne and Strand, 1998). However, some reports showed that rates of anaerobic PAHs degradation could be comparable to those under aerobic and denitrifying conditions (McNally et al., 1998).

Though the future of anaerobic bioremediation is promising, there are several drawbacks to anaerobic bioremediation. In general, the shortcomings of anaerobic bioremediation dominate the benefits of the process. For example, (1) all environments do not contain anaerobes with PAH-degrading capability (Coates et al., 1997). (2) Imposing the reducing conditions may significantly change the geochemistry of the soil surface. In an anaerobic condition, other electron acceptors such as nitrate, ferric iron, and sulfate are reduced during respiration (Stumm and Morgan, 1981). The reduction to ferrous iron and the release of phosphate from iron–phosphate complexes are toxic processes to the environment. (3) Reducing conditions also enhance the pH which in turn results in the solubilization of carbonate minerals and the release of trace metals (Ponnamperuma, 1972). Therefore, a thorough understanding of the implications of anaerobic bioremediation is needed before it can be used for *in situ* bioremediation. The discovery of a wide diversity of PAH-degrading anaerobes is a significant step toward the process of bioremediation.

8.5.2.5 Cell–Cell Signaling

Biofilm architecture and its composition significantly affect the PAHs' degradation ability (Mangwani et al., 2013). QS is known to be one of the major factors influencing biofilm architecture (Nadell et al., 2008); therefore, it is proposed that a QS-mediated bioremediation process of organic pollutants be employed and developed. For instance, in a study by Yong and Zhong (2010), an improved biodegradation of phenol by *P. aeruginosa* on exogenous addition of AHLs was reported. The contribution of QS to hexadecane degradation by *Acinetobacter* sp. was also described (Kang and Park, 2010).

8.5.2.6 Chemotaxis

Chemotaxis is the movement of organisms in response to a nutrient source or chemical gradient (Paul et al., 2005). Cells with chemotactic capabilities can sense xenobiotic chemicals adsorbed to soil particles and swim toward them, thereby overcoming the mass-transfer limitations in the bioremediation process (Figure 8.7). Under conditions of limited carbon or energy sources, chemotaxis

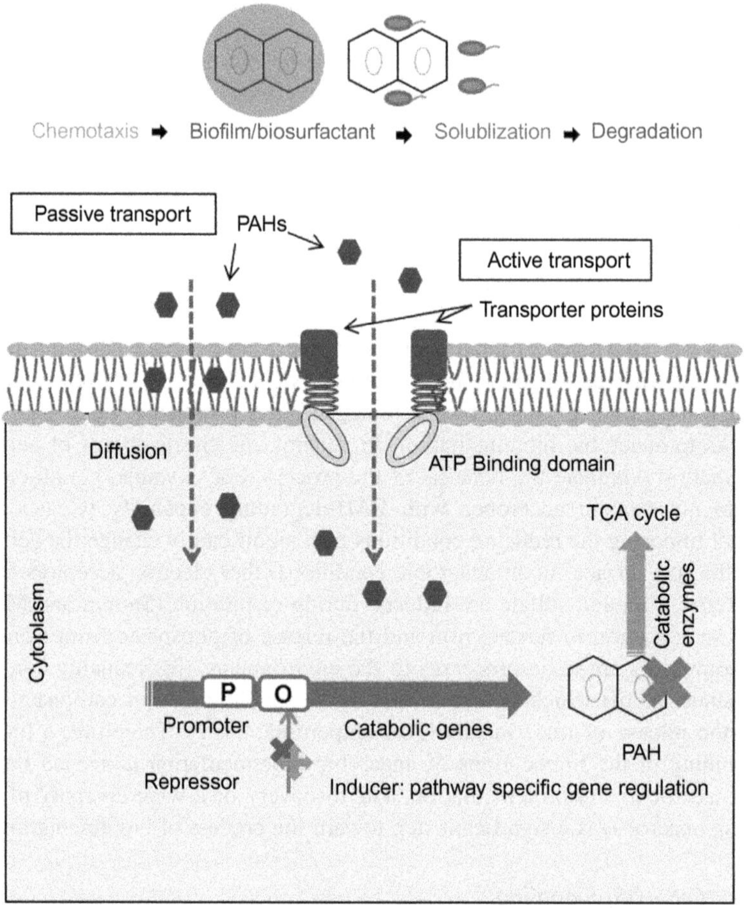

Figure 8.7 Metabolism of a PAH by bacteria: role of biofilm and chemotaxis.

helps bacteria to find the optimum conditions for growth after exposure to such chemicals. Chemotaxis also governs biofilm formation in several microorganisms (Pratt and Kolter, 1999). It guides bacteria to travel toward hydrophobic pollutants which act as a carbon source and is followed by surface attachment. The flagellum has been reported to be a critical locomotory organ for adhesion to surfaces and also facilitates the initiation of biofilm formation (Pratt and Kolter, 1999). Chemotaxis and motility are required for a bacterium to develop a biofilm and move along the surface, grow, and form microcolonies (Nicolella et al., 2000). Thus bioavailability and biodegradation of organic contaminants can be significantly improved by exploiting the chemotactic response of bacteria. PAHs (naphthalene) are reported to be chemoattractants for bacteria (Law and Aitken, 2003).

8.5.2.7 Horizontal Gene Transfer

Horizontal gene transfer (HGT), or lateral transfer of genetic material between existing bacteria, is a very common process in a biofilm community. This process is helpful for bioremediation of an assortment of PAHs. The HGT processes occur in biofilm generally by conjugation and transformation. Moreover, DNA release and transformation processes are very common in a biofilm, which adds new promises for augmentation of metabolic potential of a strain with an adaptable catabolic potential. The catabolic gene encoding for degradation of PAHs (or xenobiotic) is often encoded on plasmids or transposons (Springael et al., 2002). Plasmids are an important structural and functional component of biofilms, stabilized by plasmid-coded mechanisms. They are often lost by a rapidly growing planktonic culture. In terms of that, plasmid immovability is high in biofilms because of the quiescent nature of cells in the biofilm (Madsen et al., 2012). Consequently, a biofilm community is often a good reservoir of active plasmid. The natural transformation potential of biofilm is 600-fold more as compared to planktonic cells (Molin and Tolker-Nielsen, 2003). Therefore, horizontal transfers of those catabolic plasmids in biofilms provide a transformed population that has the potential to degrade several compounds. The auto-gene relocation among cells is promoted by the dense and packed structure of the biofilms. Owing to this, HGT is more helpful as compared to traditional bioaugmentation (*in situ* method) for bioremediation.

8.5.3 Bioremediation Strategies for PAHs Degradation

Hydrocarbon compound structure is important in biodegradability. PAHs with more than three rings poses challenges to bacterial degradation. This is due to their low solubility in water and high affinity to organic liquids, which generally limits their bioavailability to microorganisms (Krell et al., 2013). Surfactant-enhanced bioremediation (SEB) of less soluble hydrocarbons such as PAHs is a potential biodegradation technique (Bustamante et al., 2012). SEB is defined as the application and regulation of certain physicochemical properties that aid the bioremediation of contaminants. This includes nonionic surfactant mixtures adjusting physicochemical parameters such as oxygen, phosphorus, nitrogen, pH, and temperature. The total PAH concentration is also an important determinant to develop a successful bioremediation process. Higher concentrations of PAHs can be toxic to microbes; therefore, the recommended concentration is around $\leq 5\%$.

In a natural environment, PAH-degrading microbes are found ubiquitously. PAH-contaminated soils and sediments should be screened for the bacteria/microorganism having PAH degradation potential. For instance, many reports showed that phenanthrene-degrading bacteria were isolated from PAH-contaminated sites (Tam et al., 2002). On the other hand, bacterial isolates optimized for PAH degradation with an inherent property of PAH degradation (Mangwani et al., 2013), as well as genetically modified microorganisms, can also be introduced in a bioremediation process. In addition to aerobic environments, many potential PAH-degrading bacteria have been isolated from various anaerobic environments such as marine sediments (Chang et al., 2003) and municipal sewage sludges (Coates et al., 1997).

8.6 Conclusion

Elaborate studies on PAHs have provided the necessary information on and understanding of their persistence and toxicity in nature. Many bioremediation strategies have been developed and adopted effectively. However, each strategy has certain specific advantages and disadvantages in cleanup processes. The biofilm-mediated bioremediation process has the advantage of reusability of bacterial biomass and a low cost as compared to cell immobilization. However, the very low bioavailability and dormant state of the cells in biofilms is the disadvantage. Therefore, besides the bioavailability of the pollutant itself, low temperature, anaerobic conditions, low levels of nutrients and co-substrates, the presence of toxic substances, and the physiological potential of microorganisms are important parameters to be considered for cleanups of polluted sites. The biological response of biofilm cells to environmental pollutants by chemotaxis or QS can elicit variable physiological responses in bacteria. The choice of methods in each technology requires careful consideration of all these factors. Therefore, it is important to gain a better understanding of the metabolic cooperation among the microbiota within a biofilm community. The structural and functional *in situ* studies of microbial communities contaminated with PAHs using community fingerprinting and environmental genomic techniques are valuable. Despite the known inherent limitations of the bioremediation of PAH-contaminated sites, further research is required to test these limitations and exploit the potential of the *in situ* microbial communities to metabolize PAHs in suboptimal conditions. In addition, further research is required to develop potential anaerobic technologies that can be deployed to remediate subsurface sites such as marine sediments.

Acknowledgments

The authors would like to acknowledge the authorities of BARC, Kalpakkam, and NIT, Rourkela for their encouragement and support. N.M. gratefully acknowledges the receipt of a fellowship from the Ministry of Human Resource Development, Government of India, for the doctoral research. S.D. thanks the Department of Biotechnology, Government of India, for a research grant on biofilm-based-enhanced bioremediation.

References

Avnir, D., Coradin, T., Lev, O., Livage, J., 2006. Recent bio-applications of sol–gel materials. J. Mater. Chem. 16, 1013–1030.

Bewley, R.J., Webb, G., 2001. *In situ* bioremediation of groundwater contaminated with phenols, BTEX and PAHs using nitrate as electron acceptor. Land Contam. Reclam. 9, 335–347.

Boles, B.R., Thoendel, M., Singh, P.K., 2004. Self-generated diversity produces "insurance effects" in biofilm communities. PNAS USA. 101, 16630–16635.

Buchholz, K., Kasche, V., Bornscheuer, U.T., 2012. Biocatalysts and Enzyme Technology. second ed. Wiley-VCH, Weinheim.

Busalmen, J., Vazquez, M., De Sanchez, S., 2002. New evidences on the catalase mechanism of microbial corrosion. Electrochim. Acta. 47, 1857–1865.

Buschmann, J., Berg, M., Stengel, C., Winkel, L., Sampson, M.L., Trang, P.T.K., et al., 2008. Contamination of drinking water resources in the Mekong delta floodplains: arsenic and other trace metals pose serious health risks to population. Environ. Int. 34, 756–764.

Bustamante, M., Duran, N., Diez, M., 2012. Biosurfactants are useful tools for the bioremediation of contaminated soil: a review. J. Soil Sci. Plant Nutr. 12, 667–687.

Chandki, R., Banthia, P., Banthia, R., 2011. Biofilms: a microbial home. J. Indian Soc. Periodontol. 15, 111–114.

Chang, B.V., Chang, W., Yuan, S.Y., 2003. Anaerobic degradation of polycyclic aromatic hydrocarbons in sludge. Adv. Environ. Res. 7, 623–628.

Chauhan, A., Oakeshott, J.G., Jain, R.K., 2008. Bacterial metabolism of polycyclic aromatic hydrocarbons: strategies for bioremediation. Indian J. Microbiol. 48, 95–113.

Chen, X., Stewart, P.S., 2002. Role of electrostatic interactions in cohesion of bacterial biofilms. Appl. Environ. Microbiol. 59, 718–720.

Chetwittayachan, T., Shimazaki, D., Yamamoto, K., 2002. A comparison of temporal variation of particle-bound polycyclic aromatic hydrocarbons (pPAHs) concentration in different urban environments: Tokyo, Japan, and Bangkok, Thailand. Atmos. Environ. 36, 2027–2037.

Coates, J.D., Woodward, J., Allen, J., Philp, P., Lovley, D.R., 1997. Anaerobic degradation of polycyclic aromatic hydrocarbons and alkanes in petroleum-contaminated marine harbour sediments. Appl. Environ. Microbiol. 63, 3589–3593.

Coates, J.D., Michaelidou, U., Bruce, R.A., O'Connor, S.M., Crespi, J.N., Achenbach, L.A., 1999. Ubiquity and diversity of dissimilatory (per) chlorate-reducing bacteria. Appl. Environ. Microbiol. 65, 5234–5241.

Cogan, N.G., Cortez, R., Fauci, L., 2005. Modeling physiological resistance in bacterial biofilms. Bull. Math. Biol. 67, 831–853.

Costerton, J.W., Geesey, G., Cheng, K., 1978. How bacteria stick. Sci. Am. 238, 86–95.

Costerton, J.W., Lewandowski, Z., Caldwell, D.E., Korber, D.R., Lappin-Scott, H.M., 1995. Microbial biofilms. Annu. Rev. Microbiol. 49, 711–745.

Cucarella, C., Solano, C., Valle, J., Amorena, B., Lasa, I., Penades, J.R., 2001. Bap, a *Staphylococcus aureus* surface protein involved in biofilm formation. J. Bacteriol. 183, 2888–2896.

Doll, T.E., Frimmel, F.H., 2004. Kinetic study of photocatalytic degradation of carbamazepine, clofibric acid, iomeprol and iopromide assisted by different TiO_2 materials determination of intermediates and reaction pathways. Water Res. 38, 955–964.

Duruibe, J.O., Ogwuegbu, M.O.C., Egwurugwu, J.N., 2007. Heavy metal pollution and human biotoxic effects. Int. J. Phys. Sci. 2, 112–118.

Ernst, W.H.O., 1996. Bioavailability of heavy metals and decontamination of soils by plants. Appl. Geochem. 11, 163–167.

Fewson, C.A., 1988. Biodegradation of xenobiotic and other persistent compounds: the causes of recalcitrance. Trends Biotechnol. 6, 148–153.

Fida, T.T., Breugelmans, P., Lavigne, R., Coronado, E., Johnson, D.R., van der Meer, J.R., et al., 2012. Exposure to solute stress affects genome-wide expression but not the polycyclic aromatic hydrocarbon-degrading activity of *Sphingomonas* sp. strain LH128 in biofilms. Appl. Environ. Microbiol. 78, 8311–8320.

Flemming, H.C., Wingender, J., 2010. The biofilm matrix. Nat. Rev. Microbiol. 8, 623–633.

Flemming, H.C., Neu, T.R., Wozniak, D.J., 2007. The EPS matrix: the "house of biofilm cells". J. Bacteriol. 189, 7945–7947.

Furukawa, K., Fujihara, H., 2008. Microbial degradation of polychlorinated biphenyls: biochemical and molecular features. J. Biosci. Bioeng. 105, 433–449.

Glazer, A.N., Nikaido, H., 1995. Application of biotechnology for mineral processing. Microbial Biotechnology: Fundamentals of Applied Microbiology. Freeman, New York, NY, pp. 268–287.

Gloag, E.S., Turnbull, L., Huang, A., Vallotton, P., Wang, H., Nolan, L.M., et al., 2013. Self-organization of bacterial biofilms is facilitated by extracellular DNA. PNAS USA. 110, 11541–11546.

Grace, L.P.W., Chang, T.C., Whang, L.M., Kao, C.H., Pan, P.T., Cheng, S.S., 2011. Bioremediation of petroleum hydrocarbon contaminated soil: effects of strategies and microbial community shift. Int. Biodeter. Biodegrad. 65, 1119–1127.

Guo, Y., Wu, K., Huo, X., Xu, X., 2011. Sources, distribution, and toxicity of polycyclic aromatic hydrocarbons. J. Environ. Health. 73, 22–29.

Halan, B., Buehler, K., Schmid, A., 2012. Biofilms as living catalysts in continuous chemical syntheses. Trends Biotechnol. 30, 453–465.

Hatzinger, P.B., Alexander, M., 1995. Effect of aging of chemicals in soil on their biodegradability and extractability. Environ. Sci. Technol. 29, 537–545.

Hekmat, D., Bauer, R., Neff, V., 2007. Optimization of the microbial synthesis of dihydroxyacetone in a semi-continuous repeated-fed-batch process by *in situ* immobilization of *Gluconobacter oxydans*. Process Biochem. 42, 71–76.

Jain, K., Parida, S., Mangwani, N., Dash, H.R., Das, S., 2013. Isolation and characterization of biofilm-forming bacteria and associated extracellular polymeric substances from oral cavity. Ann. Microbiol. 63, 1553–1562. Available from: http://dx.doi.org/doi:10.1007/s13213-013-0618-9.

Johnsen, A.R., Wick, L.Y., Harms, H., 2005. Principles of microbial PAH-degradation in soil. Environ. Pollut. 133, 71–84.

Juhasz, A.L., Naidu, R., 2000. Bioremediation of high molecular weight polycyclic aromatic hydrocarbons: a review of the microbial degradation of benzo(a)pyrene. Int. Biodeter. Biodegrad. 45, 57–88.

Kang, Y.S., Park, W., 2010. Contribution of quorum-sensing system to hexadecane degradation and biofilm formation in *Acinetobacter* sp. strain DR1. J. Appl. Microbiol. 109, 1650–1659.

Katritzky, A.R., Oliferenko, P., Oliferenko, A., Lomaka, A., Karelson, M., 2003. Nitrobenzene toxicity: QSAR correlations and mechanistic interpretations. J. Phys. Org. Chem. 16, 811–817.

Kjelleberg, S., Molin, S., 2002. Is there a role for quorum sensing signals in bacterial biofilms? Curr. Opin. Microbiol. 5, 254–258.

Krell, T., Lacal, J., Reyes-Darias, J.A., Jimenez-Sanchez, C., Sungthong, R., Ortega-Calvo, J.J., 2013. Bioavailability of pollutants and chemotaxis. Curr. Opin. Biotechnol. 24, 451–456.

Latasa, C., Solano, C., Penades, J.R., Lasa, I., 2006. Biofilm-associated proteins. C. R. Biol. 329, 849–857.

Lau, K., Tsang, Y., Chiu, S.W., 2003. Use of spent mushroom compost to bioremediate PAH-contaminated samples. Chemosphere. 52, 1539–1546.

Law, A.M., Aitken, M.D., 2003. Bacterial chemotaxis to naphthalene desorbing from a non-aqueous liquid. Appl. Environ. Microbiol. 69, 5968–5973.

Leck, C., Bigg, E.K., 2005. Biogenic particles in the surface microlayer and overlaying atmosphere in the central Arctic Ocean during summer. Tellus B. 57, 305–316.

Li, Y.H., Tian, X., 2012. Quorum sensing and bacterial social interactions in biofilms. Sensors. 12, 2519–2538.

Liang, P., Xu, J., Li, Q., 2008. Application of dispersive liquid–liquid microextraction and high-performance liquid chromatography for the determination of three phthalate esters in water samples. Anal. Chim. Acta. 609, 53–58.

Madsen, E.L., 1991. Determining *in situ* biodegradation. Environ. Sci. Technol. 25, 1662–1673.

Madsen, J.S., Burmolle, M., Hansen, L.H., Sorensen, S.J., 2012. The interconnection between biofilm formation and horizontal gene transfer. FEMS Immunol. Med. Microbiol. 65, 183–195.

Makkar, R.S., Rockne, K.J., 2003. Comparison of synthetic surfactants and biosurfactants in enhancing biodegradation of polycyclic aromatic hydrocarbons. Environ. Toxicol. Chem. 22, 2280–2292.

Mangwani, N., Dash, H.R., Chauhan, A., Das, S., 2012. Bacterial quorum sensing: functional features and potential applications in biotechnology. J. Mol. Microbiol. Biotechnol. 22, 215–227. Available from: http://dx.doi.org/doi:10.1159/000341847.

Mangwani, N., Shukla, S.K., Rao, T.S., Das, S., 2013. Calcium-mediated modulation of *Pseudomonas mendocina* NR802 biofilm influences the phenanthrene degradation. Colloids Surf. B. 114, 301–309. Available from: http://dx.doi.org/doi:10.1016/j.colsurfb.2013.10.003.

Margesin, R., Schinner, F., 2001. Biodegradation and bioremediation of hydrocarbons in extreme environments. Appl. Microbiol. Biotechnol. 56, 650–663.

Matsumoto, M., Hirata-Koizumi, M., Ema, M., 2008. Potential adverse effects of phthalic acid esters on human health: a review of recent studies on reproduction. Regul. Toxicol. Pharmacol. 50, 37–49.

Matsuyama, T., Nakagawa, Y., 1996. Surface-active exolipids: analysis of absolute chemical structures and biological functions. J. Microbiol. Methods. 25, 165–175.

McNally, D.L., Mihelcic, J.R., Lueking, D.R., 1998. Biodegradation of three-and four-ring polycyclic aromatic hydrocarbons under aerobic and denitrifying conditions. Environ. Sci. Technol. 32, 2633–2639.

Mittal, A., Thakur, V., Gajbe, V., 2013. Adsorptive removal of toxic azo dye Amido Black 10B by hen feather. Environ. Sci. Pollut. Res. Int. 20, 260–269.

Molin, S., Tolker-Nielsen, T., 2003. Gene transfer occurs with enhanced efficiency in biofilms and induces enhanced stabilisation of the biofilm structure. Curr. Opin. Biotechnol. 14, 255–261.

Morikawa, M., 2010. Dioxygen activation responsible for oxidation of aliphatic and aromatic hydrocarbon compounds: current state and variants. Appl. Microbiol. Biotechnol. 87, 1596–1603.

Mueller, J.G., Cerniglia, C., Pritchard, P.H., 1996. Bioremediation of environments contaminated by polycyclic aromatic hydrocarbons. Biotechnol. Res. Ser. 6, 125–194.

Mulcahy, H., Charron-Mazenod, L., Lewenza, S., 2008. Extracellular DNA chelates cations and induces antibiotic resistance in *Pseudomonas aeruginosa* biofilms. PLoS Pathog. 4, e1000213.

Nadell, C.D., Xavier, J.B., Levin, S.A., Foster, K.R., 2008. The evolution of quorum sensing in bacterial biofilms. PLoS Biol. 6, e14.

Nam, J.J., Thomas, G.O., Jaward, F.M., Steinnes, E., Gustafsson, O., Jones, K.C., 2008. PAHs in background soils from Western Europe: influence of atmospheric deposition and soil organic matter. Chemosphere. 70, 1596–1602.

Neu, T.R., Dengler, T., Jann, B., Poralla, K., 1992. Structural studies of an emulsion-stabilizing exopolysaccharide produced by an adhesive, hydrophobic *Rhodococcus* strain. J. Gen. Microbiol. 138, 2531–2537.

Nicolella, C., van Loosdrecht, M., Heijnen, S.J., 2000. Particle-based biofilm reactor technology. Trends Biotechnol. 18, 312–320.

Otzen, D., Nielsen, P.H., 2008. We find them here, we find them there: functional bacterial amyloid. Cell. Mol. Life Sci. 65, 910–927.

Pandey, G., Jain, R.K., 2002. Bacterial chemotaxis toward environmental pollutants: role in bioremediation. Appl. Environ. Microbiol. 68, 5789–5795.

Pandey, J., Chauhan, A., Jain, R.K., 2009. Integrative approaches for assessing the ecological sustainability of *in situ* bioremediation. FEMS Microbiol. Rev. 33, 324–375.

Parales, R.E., Haddock, J.D., 2004. Biocatalytic degradation of pollutants. Curr. Opin. Biotechnol. 15, 374–379.

Pardieck, D.L., Bouwer, E.J., Stone, A.T., 1992. Hydrogen peroxide use to increase oxidant capacity for in situ bioremediation of contaminated soils and aquifers: a review. J. Contam. Hydrol. 9, 221–242.

Paul, D., Pandey, G., Pandey, J., Jain, R.K., 2005. Accessing microbial diversity for bioremediation and environmental restoration. Trends Biotechnol. 23, 135–142.

Perelo, L.W., 2010. Review: *in situ* and bioremediation of organic pollutants in aquatic sediments. J. Hazard. Mater. 177, 81–89.

Ponnamperuma, F., 1972. The Chemistry of Submerged Soils. Academic Press, NY and London.

Pratt, L.A., Kolter, R., 1999. Genetic analyses of bacterial biofilm formation. Curr. Opin. Microbiol. 2, 598–603.

Puvaneswari, N., Muthukrishnan, J., Gunasekaran, P., 2006. Toxicity assessment and microbial degradation of azo dyes. Indian J. Exp. Biol. 44, 618.

Rockne, K.J., Strand, S.E., 1998. Biodegradation of bicyclic and polycyclic aromatic hydrocarbons in anaerobic enrichments. Environ. Sci. Technol. 32, 3962–3967.

Ryvolova, M., Krizkova, S., Adam, V., Beklova, M., Trnkova, L., Hubalek, J., et al., 2011. Analytical methods for metallothionein detection. Curr. Anal. Chem. 7, 243–261.

Salvo, L.M., Bainy, A.C., Ventura, E.C., Marques, M.R., Silva, J.R.M., Klemz, C., et al., 2012. Assessment of the sublethal toxicity of organochlorine pesticide endosulfan in juvenile common carp (*Cyprinus carpio*). J. Environ. Sci. Health Part A. 47, 1652–1658.

Samanta, S.K., Singh, O.V., Jain, R.K., 2002. Polycyclic aromatic hydrocarbons: environmental pollution and bioremediation. Trends Biotechnol. 20, 243–248.

Sand, W., Gehrke, T., 2006. Extracellular polymeric substances mediate bioleaching/biocorrosion via interfacial processes involving iron (III) ions and acidophilic bacteria. Res. Microbiol. 157, 49–56.

Semple, K.T., Morriss, A., Paton, G., 2003. Bioavailability of hydrophobic organic contaminants in soils: fundamental concepts and techniques for analysis. Eur. J. Soil Sci. 54, 809–818.

Seo, J.S., Keum, Y.S., Li, Q.X., 2009. Bacterial degradation of aromatic compounds. Int. J. Environ. Res. Public Health. 6, 278–309.

Shemesh, M., Kolter, R., Losick, R., 2010. The biocide chlorine dioxide stimulates biofilm formation in *Bacillus subtilis* by activation of the histidine kinase KinC. J. Bacteriol. 192, 6352–6356.

Shimada, K., Itoh, Y., Washio, K., Morikawa, M., 2012. Efficacy of forming biofilms bynaphthalene degrading *Pseudomonas stutzeri* TiO2 toward bioremediation technology and its molecular mechanisms. Chemosphere. 87, 226–233.

Shukla, S.K., Rao, T.S., 2013a. Effect of calcium on *Staphylococcus aureus* biofilm architecture: a confocal laser scanning microscopic study. Colloids Surf. B. 103, 448–454.

Shukla, S.K., Rao, T.S., 2013b. Dispersal of Bap-mediated *Staphylococcus aureus* biofilm by proteinase K. J. Antibiot. 66, 55–60.

Singh, R., Paul, D., Jain, R.K., 2006. Biofilms: implications in bioremediation. Trends Microbiol. 14, 389–397.

Springael, D., Peys, K., Ryngaert, A., Roy, S.V., Hooyberghs, L., Ravatn, R., et al., 2002. Community shifts in a seeded 3-chlorobenzoate degrading membrane biofilm reactor: indications for involvement of in situ horizontal transfer of the clc-element from inoculum to contaminant bacteria. Environ. Microbial. 4, 70–80.

Stoodley, L.H., Costerton, J.W., Stoodley, P., 2004. Bacterial biofilms: from the natural environment to infectious diseases. Nat. Rev. Microbiol. 2, 95–108.

Stumm, W., Morgan, J.J., 1981. Aquatic Chemistry. Wiley-Interscience, New York, NY.

Su, L., Zhang, X., Yuan, X., Zhao, Y., Zhang, D., Qin, W., 2012. Evaluation of joint toxicity of nitroaromatic compounds and copper to *Photobacterium phosphoreum* and QSAR analysis. J. Hazard. Mater. 241–242, 450–455.

Tam, N.F.Y., Guo, C.L., Yau, W.Y., Wong, Y.S., 2002. Preliminary study on biodegradation of phenanthrene by bacteria isolated from mangrove sediments in Hong Kong. Mar. Pollut. Bull. 45, 316–324.

Tetz, G.V., Artemenko, N.K., Tetz, V.V., 2009. Effect of DNase and antibiotics on biofilm characteristics. Antimicrob. Agents Chemother. 53, 1204–1209.

Thomassin-Lacroix, E., Eriksson, M., Reimer, K., Mohn, W., 2002. Biostimulation and bioaugmentation for on-site treatment of weathered diesel fuel in Arctic soil. Appl. Microbiol. Biotechnol. 59, 551–556.

Wang, Z., Chen, J., Qiao, X., Yang, P., Tian, F., Huang, L., 2007. Distribution and sources of polycyclic aromatic hydrocarbons from urban to rural soils: a case study in Dalian, China. Chemosphere. 68, 965–971.

Whitchurch, C.B., Tolker-Nielsen, T., Ragas, P.C., Mattick, J.S., 2002. Extracellular DNA required for bacterial biofilm formation. Science. 295, 1487.

Wingender, J., Neu, T., Flemming, H.C., 1999. In: Wingender, J., Neu, T., Flemming, H.-C. (Eds.), Microbial Extracellular Polymeric Substances. Springer, Heidelberg, pp. 1–19.

Wingender, J., Strathmann, M., Rode, A., Leis, A., Flemming, H.C., 2001. Isolation and biochemical characterization of extracellular polymeric substances from *Pseudomonas aeruginosa*. Methods Enzymol. 336, 302–314.

Woodley, J.M., 2006. Microbial biocatalytic processes and their development. Adv. Appl. Microbiol. 60, 1–15.

Yamaga, F., Washio, K., Morikawa, M., 2010. Sustainable biodegradation of phenol by *Acinetobacter calcoaceticus* P23 isolated from the rhizosphere of duckweed *Lemna aoukikusa*. Environ. Sci. Technol. 44, 6470–6474.

Yong, Y.C., Zhong, J.J., 2010. *N*-acylated homoserine lactone production and involvement in the biodegradation of aromatics by an environmental isolate of *Pseudomonas aeruginosa*. Process Biochem. 45, 1944–1948.

Zollinger, H., 2003. Color Chemistry: Syntheses, Properties, and Applications of Organic Dyes and Pigments. Wiley-VCH, Weinheim.

9 Nanoremediation: A New and Emerging Technology for the Removal of Toxic Contaminant from Environment

Avinash P. Ingle[a], Amedea B. Seabra[b], Nelson Duran[c,d] and Mahendra Rai[a,d]

[a]Department of Biotechnology, Sant Gadge Baba Amravati University, Amravati, Maharashtra, India, [b]Universidade Federal de São Paulo—Unifesp, Diadema, SP, Brazil, [c]Center of Natural and Human Sciences, Universidade Federal do ABC, SP, Brazil, [d]Institute of Chemistry, Biological Chemistry Laboratory, Universidade Estadual de Campinas, Campinas, SP, Brazil

9.1 Introduction

Remediation is mainly associated with the environment. Environmental remediation means the removal of pollutants or contaminants from environment such as soil, air, groundwater, sediments, or water surfaces for the general protection of human health and the environment. Remediation is generally subjected to an array of regulatory requirements and also based on assessments of human health and ecological risks where no authorized standards exist.

Air and water resources contaminated by any means are the major sources responsible for environmental pollution. In industrial areas, the air is filled with numerous pollutants due to industrial pollution. This air may contain different types of pollutants such as carbon monoxide (CO), chlorofluorocarbons (CFCs), different heavy metals (namely, arsenic, chromium, lead, cadmium, mercury, zinc) (Table 9.1), hydrocarbons, nitrogen oxides, organic chemicals (volatile organic compounds, known as VOCs, and dioxins), sulfur dioxide, and particulate materials. Different kinds of pollutants cause different hazards. For example, nitrogen and sulfur-dioxide degenerated from industries are responsible for the generation of acid rain which infiltrates and ultimately causes soil pollution. Other sources responsible for the release of nitrogen and sulfur dioxide in high concentrations in the atmosphere are the burning of oil, coal, and gas. Moreover, water pollution is also caused by numerous factors like industrial effluent; sewage; oil spills; leaking

Microbial Biodegradation and Bioremediation. DOI: http://dx.doi.org/10.1016/B978-0-12-800021-2.00009-1

Table 9.1 Heavy Metals and Their Source of Contamination in Soil (Malik and Biswasa, 2012)

Heavy Metal	Sources of Contamination in Soil
Lead (Pb)	Batteries, metal products
Cadmium (Cd)	Electroplating, batteries, and fertilizers
Arsenic (As)	Timber treatment, paints, and pesticides
Chromium (Cr)	Timber treatment, leather tanning, pesticides, and dyes
Copper (Cu)	Timber treatment, fertilizers, fungicides, electrical, and pigments
Manganese (Mn)	Fertilizer
Zinc (Zn)	Dyes, paints, timber treatment, fertilizers, and mine tailings
Mercury (Hg)	Instruments, fumigants, and fertilizers
Nickel (Ni)	Alloys, batteries, and mine tailings
Molybdenum (Mo)	Fertilizer

of fertilizers, herbicides, and pesticides from land; by-products from manufacturing; and extracted or burned fossil fuels (Filipponi and Sutherland, 2010).

Contaminants are most often measured in parts per million (ppm) or parts per billion (ppb) and their toxicity is defined by a "toxic level." The toxic level varies for different kinds of contaminants according to the sources, namely, water, soil, air. For example, the toxic level for arsenic is 10 ppm in soil and for mercury it is 0.002 ppm in water. Therefore, even very low concentrations of a specific contaminant can be toxic. In addition, contaminants are mostly found as mixtures.

Consequently, there is a need for technologies that are capable of monitoring, recognizing, and, ideally, treating such small amounts of contaminants in air, water, and soil. In this context, nanotechnologies offer numerous opportunities to prevent, reduce, sense, and treat environment contamination. Nanotechnologies can enhance and enable preexisting technologies and develop new ones (Filipponi and Sutherland, 2010). The application of nanotechnology for the remediation of contaminants may give promising results in future. The search for new and advanced materials is a need of hour for the environmental protection. In recent years, a great deal of attention have been focused on the application of nanostructured materials as adsorbents or catalysts to remove toxic and harmful substances from the environment, i.e., from wastewater and air (Agarwal and Joshi, 2010). This chapter is mainly focused on different traditional methods for remediation of contaminants from the environment and their limitations, and on nanoremediation as an alternative approach for cleaning the polluted environment.

9.2 Different Kinds of Remediation

Environmental remediation can be achieved using three kinds of approaches: physical, chemical, and biological.

9.2.1 Physical Remediation

These are the oldest remediation methods for soil. Different techniques like capping, soil mixing, soil washing, soil vapor extraction, land-farming, soil flushing, solidification, and excavation processes can be used to clean soil. Contaminated sites can be cleaned up rapidly through this method, but due to the high cost and risk of contaminant shifting, it is not a good remedial technique for the removal of heavy metals from a large area. Some of the above-mentioned techniques are discussed here.

9.2.1.1 Soil Washing

Soil washing uses liquids, usually water, occasionally combined with solvents, and mechanical processes to scrub soils. Solvents are selected on the basis of their ability to solubilize specific contaminants and on their environmental and health effects (Chu and Chan, 2003; Urum et al., 2003). The soil washing process separates fine soil (clay and silt) from coarse soil (sand and gravel). Since hydrocarbon contaminants generally bind to smaller soil particles (clay and silt), separating them from the larger ones reduces the volume of contaminated soil (Riser-Roberts, 1998). The smaller volume of soil, which contains the majority of clay and silt particles, can be further treated by other methods (such as incineration or bioremediation).

9.2.1.2 Soil Vapor Extraction

Soil vapor extraction, also known as soil venting or vacuum extraction, is an accepted, recognized, and cost-effective technology for remediating unsaturated soils contaminated with different contaminants (Zhan and Park, 2002). Soil vapor extraction involves the installation of vertical and/or horizontal wells in the area of soil contamination. Air "blowers" are often used to aid the evaporation process. Vacuums are applied through the wells near the source of contamination to evaporate the volatile constituents of the contaminated mass which are subsequently withdrawn through an extraction well. Extracted vapors are then treated (commonly with carbon adsorption) before being released into the atmosphere. This procedure is also used with groundwater pumping and air stripping for treating contaminated groundwater (Khan et al., 2004).

9.2.1.3 Land-Farming

Land-farming is an above-ground remediation technology that reduces the concentration of pollutants like petroleum constituents present in soils through processes associated with bioremediation. This technology usually involves the spreading of excavated contaminated soils in a thin layer (no more than 1.5 m) on the ground surface of a treatment site and stimulating aerobic microbial activity within the soils through aeration and/or the addition of nutrients, minerals, and water/moisture (Hejazi et al., 2003).

9.2.1.4 Soil Flushing

Soil flushing is a pioneering remediation technology that "floods" contaminated soils with a solution that moves the contaminants to an area where they can be removed (Logsdon et al., 2002; Di-Palma et al., 2003). Soil flushing is generally carried out by passing an extraction fluid through in-place soils using an injection or infiltration process. Contaminated groundwater and extraction fluids are captured and pumped to the surface using standard groundwater extraction wells. Recovered groundwater and extraction fluids with the adsorbed contaminants are again processed to meet the appropriate discharge standards before being recycled or released to local, publicly owned, wastewater treatment works or receiving streams (Son et al., 2003). Soil flushing applies to all types of soil contaminants and is generally used in conjunction with other remediation technologies such as activated carbon, biodegradation, and pump-and-treat (Boulding, 1996).

Many other physical remediation techniques have been reviewed in detail with specific discussion in an excellent review by Khan et al. (2004).

9.2.2 Chemical Remediation

In chemical remediation, heavy metals in contaminated soil are transformed with added chemicals to a less toxic form, which is not easily absorbed by plants. So, in the stabilization process, heavy metals remain in the soil but in a less harmful form. Due to the high rate of success, this method is becoming popular. It is a periodical treatment method that requires special equipment and operators and can affect the physical structure and biological activity of the treated medium at low levels. Therefore, it is not applicable on large scales for heavy metal remediation. Many chemical remediation methods have been proposed and used for remediation of toxic substances from the environment. These methods include remediation using actinide chelators, chemical immobilization, critical fluid extraction, oxidation, *in situ* catalyzed peroxide remediation, and photodegradation with uranium recovery. (Czupyrna et al., 1989; Gopalan et al., 1993; Gates and Siegrist, 1994; Ho et al., 1995).

Young and Jordan (1995) demonstrated that cyanide contamination from soil and water can be minimized by addition of oxidants. Oxidants have high electron affinity and therefore strip cyanide of electrons predominantly yielding cyanate as a reaction product. Oxygen, ozone, hydrogen peroxide, chlorine, hypochlorite, and sulfur dioxide are the most common oxidants which can be used.

Maier et al. (2001) developed an environment friendly approach for the remediation of metal contaminants from metal-contaminated soils and sewage sludge. Instead of using old techniques like landfilling or metal extraction techniques which involved caustic or toxic agents, they used a metal-chelating biosurfactant, rhamnolipid, for removal of metals or metal-associated toxicity from metal-contaminated waste.

Chromium is a common water and soil contaminant because of wide applications in metallurgy, staining glass, anodizing aluminum, organic synthesis, leather tanning, and wood preserving industries. Cr(VI) affects human physiology, accumulates in food chain, and causes severe health problems ranging from simple skin irritation to lung carcinoma. Contrarily, Cr(III) is nontoxic and an essential human nutrient, which does not readily migrate in groundwater since it usually precipitates as hydroxides, oxides, etc. Therefore, reduction of Cr(VI) to Cr(III) is necessary so as to minimize its effects from the environment.

Chemical precipitation is effective and most widely used process in the industries because of its relatively simple nature and inexpensiveness to operate. The removal of Cr(VI) by chemical reduction from industrial wastewater involves a two-step process: reduction of Cr(VI) under acidic conditions (usually pH 2–3) and the precipitation of trivalent chromium as hydroxyl species. The most commonly used reducing agents are gaseous sulfur dioxide, sodium sulfite, sodium meta-bisulfite, ferrous sulfate, and barium sulfite. Madhavi et al. (2013) proposed hydroxide precipitation using lime and limestone. These are the commonly employed precipitant agents due to their availability and low cost in most countries. Lime precipitation can be employed to effectively treat inorganic effluent with a metal concentration of higher than 1000 mg/l. Similarly, they also proposed sulfide precipitation for the removal of Cr(VI) from contaminated soils and groundwater. In a similar way, other heavy metals and various kinds of contaminants can be removed or their concentration minimized to safe level using different chemical methods.

9.2.3 Biological Remediation

Bioremediation is the use of any organism metabolism to remove pollutants. Technologies can be generally classified as *in situ* or *ex situ*. *In situ* bioremediation involves treating the contaminated material at the site, while *ex situ* involves the removal of the contaminated material to be treated elsewhere. Some examples of bioremediation-related technologies are phytoremediation, bioventing, bioleaching, land-farming, bioreactor, composting, bioaugmentation, rhizo-filtration, and biostimulation. Bioremediation can occur on its own (natural attenuation or intrinsic bioremediation) or can be spurred on via the addition of fertilizers to increase the bioavailability within the medium (biostimulation). Recent advancements have also proven successful via the addition of matched microbe strains to the medium to enhance the resident microbe population's ability to break down contaminants. Microorganisms used to perform the function of bioremediation are known as bioremediators. Some microorganisms can decompose or transform the chemical substances present in petroleum and petroleum derivatives. Hydrocarbons from crude oil represent substrates for microorganisms when an accidental oil spill occurs (http://en.wikipedia.org/wiki/Bioremediation#cite_note-1). Biological remediation can be categorized into two types: microbial remediation and phytoremediation.

9.2.3.1 Microbial Remediation

Different microbial systems like bacteria, fungi, yeasts, and actinomycetes can be used for removal of toxic and other contaminants from the environment. Although many microorganisms are capable of degrading crude oil present in soil, it has been found beneficial to employ mix culture approach then the pure cultures in bioremediation as it shows the synergistic interactions. Thapa et al. (2012) reviewed that different bacteria can be used for the removal of petroleum hydrocarbon contaminants from soil. According to them, the bacteria that can degrade petroleum products include *Pseudomonas*, *Aeromonas*, *Moraxella*, *Beijerinckia*, *F. lavobacteria*, *chrobacteria*, *Nocardia*, *Corynebacteria*, *Atinetobacter*, *Mycobactena*, *Modococci*, *Streptomyces*, *Bacilli*, *Arthrobacter*, *Aeromonas*, and *Cyanobacteria*.

Sediments in the Grand Calumet River in northwestern Indiana are heavily contaminated with polycyclic aromatic hydrocarbons (PAHs). Dean-Ross (2003) used enriched culture of some bacteria for the degradation of multiple PAHs. One strain of *Mycobacterium flavescens* was isolated using pyrene as the sole source of carbon and energy, while a second, *Rhodococcus* species, was isolated on anthracene. They developed a sediment assay system to optimize conditions for bioremediation and reported that both the strains significantly removed the PAHs from the sediments. A comparative study was carried out to investigate the complexation and competition effects on Cd and Pb uptake using metal-resistant bacterium *Cupriavidus metallidurans*. The results supported that accumulation of Cd was more affected by competition with Ca, Mg, and Zn, whereas Pb accumulation was more influenced by complexation with humic acids. Further study concluded that there is a need to consider chemical site-specificity in the removal of metals from contaminated environments (Hajdu and Slaveykova, 2012). Seo et al. (2009) reviewed the degradation pathways of selected aromatic compounds by different strains of bacteria (Table 9.2).

The remediation carried out using fungi is commonly known as "mycoremediation." The term *mycoremediation* refers specifically to the use of fungal mycelia in bioremediation. One of the primary roles of fungi in the ecosystem is decomposition, which is performed by the mycelia. The mycelium secretes extracellular enzymes and acids that break down lignin and cellulose, the two main building blocks of plant fiber. These are organic compounds composed of long chains of carbon and hydrogen, structurally similar to many organic pollutants. The key to mycoremediation is determining the right fungal species to target a specific pollutant (http://en.wikipedia.org/wiki/Bioremediation#cite_note-1).

In an experiment conducted for bioremediation, a plot of soil contaminated with diesel oil was inoculated with mycelia of oyster mushrooms; traditional bioremediation techniques (bacteria) were used on control plots. After 4 weeks, more than 95% of many of the PAH had been reduced to nontoxic components in the mycelial-inoculated plots. It appears that the natural microbial community participates with the fungi to break down contaminants, eventually into carbon dioxide and water. Wood-degrading fungi are particularly effective in breaking down aromatic pollutants (toxic components of petroleum), as well as chlorinated compounds (http://en.wikipedia.org/wiki/Bioremediation#cite_note-1).

Table 9.2 List of Different Strains of Bacteria Capable of Degrading Various Aromatic Compounds (Seo et al., 2009)

Bacterial Species	Aromatics Compounds
Achromobacter sp. (NCW)	CBZ
Alcaligenes denitrificans	FLA
Arthrobacter sp. (F101)	FLE
Arthrobacter sp. (P1-1)	DBT, CBZ, PHE
Arthrobacter sulphureus(RKJ4)	PHE
Acidovorax delafieldii (P4-1)	PHE
Bacillus cereus (P21)	PYR
Brevibacterium sp. (HL4)	PHE
Burkholderia sp. (S3702, RP007, 2A-12TNFYE-5, BS3770)	PHE
Burkholderia sp. (C3)	PHE
Burkholderia cepacia (BU-3)	NAP, PHE, PYR
Burkholderia cocovenenans	PHE
Burkholderia xenovorans(LB400)	BZ, BP
Chryseobacterium sp. (NCY)	CBZ
Cycloclasticus sp. (P-1)	PYR
Janibacter sp. (YY-1)	DBF, FLE, DBT, PHE, ANT, DD
Marinobacter (NCE312)	NAP
Mycobacterium sp.	PYR, BaP
Mycobacterium sp. (JS14)	FLA
Mycobacterium sp. (6PY1, KR2, AP1)	PYR
Mycobacterium sp. (RJGII-135)	PYR, BaA, BaP
Mycobacterium sp. (PYR-1, LB501T)	FLA, PYR, PHE, ANT
Mycobacterium sp. (CH1, BG1, BB1, KR20)	PHE, FLE, FLA, PYR
Mycobacterium flavescens	PYR, FLA
Mycobacterium vanbaalenii (PYR-1)	PHE, PYR, dMBaA
Mycobacterium sp. (KMS)	PYR
Nocardioides aromaticivorans (IC177)	CBZ
Pasteurella sp. (IFA)	FLA
Polaromonas naphthalenivorans (CJ2)	NAP
Pseudomonas sp. (C18, PP2, DLC-P11)	NAP, PHE
Pseudomonas sp. (BT1d)	HFBT
Pseudomonas sp. (B4)	BP, CBP
Pseudomonas sp. (HH69)	DBF
Pseudomonas sp. (CA10)	CBZ, CDD
Pseudomonas sp. (NCIB 9816-4)	FLE, DBF, DBT
Pseudomonas sp. (F274)	FLE
Pseudomonas paucimobilis	PHE
Pseudomonas vesicularis (OUS82)	FLE
Pseudomonas putida (P16, BS3701, BS3750, BS590-P, BS202-P1)	NAP, PHE

(*Continued*)

Table 9.2 (Continued)

Bacterial Species	**Aromatics Compounds**
Pseudomonas putida (CSV86)	MNAP
Pseudomonas fluorescens (BS3760)	PHE, CHR, BaA
Pseudomonas stutzeri (P15)	PYR
Pseudomonas saccharophilia	PYR
Pseudomonas aeruginosa	PHE
Ralstonia sp. (SBUG 290, U2)	DBF, NAP
Rhodanobacter sp. (BPC-1)	BaP
Rhodococcus sp.	PYR, FLA
Rhodococcus sp. (WU-K2R)	NAT, BT
Rhodococcus erythropolis (I-19)	ADBT
Rhodococcus erythropolis (D-1)	DBT
Staphylococcus sp. (PN/Y)	PHE
Stenotrophomonas maltophilia (VUN 10,010)	PYR, FLA, BaP
Stenotrophomonas maltophilia (VUN 10,003)	PYR, FLA, BaA, BaP, DBA, COR
Sphingomonas yanoikuyae (R1)	PYR
Sphingomonas yanoikuyae (JAR02)	BaP
Sphingomonas sp. (P2, LB126)	FLE, PHE, FLA, ANT
Sphingomonas sp.	DBF, DBT, CBZ
Sphingomonas paucimobilis (EPA505)	FLA, NAP,
Sphingomonas wittichii (RW1)	ANT, PHE, CDD
Terrabacter sp. (DBF63)	DBF, CDBF, CDD, FLE
Xanthamonas sp.	PYR, BaP, CBZ

Abbreviations: PYR, pyrene; BaP, benzo[*a*]pyrene; PHE, phenanthrene; FLA, fluoranthene; FLE, fluorene; ANT, anthracene; NAP, naphthalene; BaA, benz[*a*]anthracene; dMBaA, dimethylbenz[*a*]anthracene; DBA, dibenz[*a,h*] anthracene; COR, coronene; CHR, chrysene; DBF, bibenzofuran; CDBF, chlorinated dibenzothophene; HFBT, 3-hydroxy-2-formylbenzothiophene; BP, biphenyl; CBP, Chlorobiphenyl; NAT, naphthothiophene; BT, benzothiophene; BZ, benzoate; ADBT, alkylated dibenzothiophene; CBZ, carbazole; DD, dibenzo-*p*-dioxin; CDD, chlorinated dibenzo-*p*-dioxin; MNAP, methyl naphthalene.

A basidiomycete *Phlebia* sp. strain MG-60 isolated from mangrove stands was reported as a hypersaline-tolerant lignin-degrading fungus that participated in biobleaching of pulp and decolorization of dyes (Li et al., 2002) in the presence of different concentrations of sea salts. Treatment with basidiomycetous fungus or its lignin-degrading enzymes, lignin peroxidase, manganese-dependent peroxidase, and laccase has been widely reported (Blanquez et al., 2008; Raghukumar et al., 2008). These act on a broad range of substrates and hence are able to degrade several xenobiotics (Kim and Nicell, 2006) including synthetic dyes (Wesenberg et al., 2003). Several ascomycetous and hyphomycetous fungi also produce laccase (Baldrian, 2006). Laccases (EC 1.10.3.2) have lately been reported to be produced by several marine and marine-derived fungi (Raghukumar et al., 2008; D'Souza et al., 2009).

In another study, *Cladosporium resinae* was found to degrade polyurethane (Gautam et al., 2006). Hong et al. (2009) surveyed a gas station soil and observed that some strains of *Fusarium* and *Hypocrea* could degrade one carcinogenic high-weight PAH, pyrene, as well as uptake copper and zinc. This may be because these strains were able to use the pyrene as their sole carbon source. Jecu et al. (2010) examined polyvinyl alcohol films under a scanning electron microscope and reported substantial degradation by fungi like *Aspergillus niger*. Verma et al. (2010) reported the ability of four marine-derived fungi namely, *Diaporthe* sp., *Pestalotiopsis* sp., *Coriolopsis byrsina*, and *Cerrena unicolor* for their bioremediation of raw textile mill effluents. Several other studies have been successfully carried out which proved that microbial systems are efficient candidates for the removal of toxic chemicals and substances from the environment.

Bioremediation is emerging as a potential tool to address the problem of Cr(VI) pollution. The interaction of chromium with various microbial/bacterial strains isolated and their reduction capacity toward Cr(VI) were recently discussed (Dhal et al., 2013).

9.2.3.2 Phytoremediation

Phytoremediation denotes the treatment of pollutants through the use of plants that mitigate the environmental problem without the need to excavate the contaminant material and dispose of it elsewhere (US-EPA, 1998). Phytoremediation consists of mitigating pollutant concentrations in contaminated soils, water, or air, with plants able to contain, degrade, or eliminate metals, pesticides, solvents, explosives, crude oil and its derivatives, and various other contaminants from the media that contain them. The use of metal-accumulating plants to clean soil and water contaminated with toxic metals is the most rapidly developing component of this environmentally friendly and cost-effective technology (Raskin et al., 1997; Ali et al., 2013).

Raskin et al. (1997) reviewed the metal-accumulating capacities of different plants like *Brassica juncea* (Indian mustard), *Zea mays* (corn), and *Ambrosia artemisiifilia* (ragweed) for different metals. The results suggested that certain varieties of *B. juncea* showed enhanced ability to accumulate metals like Pb, Cu, and Ni from a hydroponic solution into their above-ground (harvestable) parts. These plants concentrated toxic heavy metals to a level up to several percent of their dried shoot biomass. Similarly, *Z. mays* and *A. artemisiifilia* were also found to accumulate some metals at varied concentrations.

In another study, Lone et al. (2008) reviewed that about 400 plant species have potential to remove toxic substances from soil and water. Among them, *Thlaspi*, *Brassica*, *Sedum alfredii*, and *Arabidopsis* were the species mostly studied. They also suggested that recent advances in biotechnology will play a promising role in the development of new hyper-accumulators by transferring metal hyper-accumulating genes from low biomass wild species to the higher biomass producing cultivated species in the future.

From the past literature, it can be stated that the most frequently cited species in phytoremediation studies was *Brassica juncea* (L.) Czern. (148 citations), followed by *Helianthus annuus* L. (57), *Brassica napus* L. and *Zea mays* L. (both 39 citations). The greater interest in the *Brassicaceae* family is due to their high potential for heavy metal accumulation. Among the plants of the *Brassica* species, the *Brassica juneca* deserves special attention because of its relevance to the process of phytoextraction of heavy metals from soil (Szczyglowska et al., 2011).

Recently, the potentials of aquatic plant species such as *Salvinia molesta* and the terrestrial plant species spinach (*Spinaciao leracea*) and somato (*Solanum lycopersicum*) were investigated in terms of their ability to remove the persistent organochlorine pesticide endosulfan from contaminated water and soil respectively. Endosulfan is a persistent, toxic broad-spectrum organochlorine insecticide and acaricide used on food and nonfood crops. It was also reported that *Salvinia molesta* is more efficient in the removal of endosulfan from an aquatic environment. It reduces the percentage of endosulfan from an initial concentration of 123 μg/L to 97.94 μg/L ± 0.33% in 21 days. Both the terrestrial plant species, spinach and tomato, also showed complete removal of this pesticide from soil in 21 and 28 days respectively, from an initial concentration of 140 μg/kg (Harikumar et al. 2013). Apart from the above-mentioned plants, many other plants have been successfully used for the process of phytoremediation (Table 9.3).

Table 9.3 Metals and Their Hyper-Accumulators Plant Species (Malik and Biswasa, 2012)

Metal	Plants
Lead (Pb)	*Brassica juncea*, Water hyacinth (*Eichhornia crassipes*), Hydrilla (*Hydrilla verticillata*), Sunflower (*Helianthus annuus*), *Lemna minor, Salvinia molesta, Spirodela polyrhiza*
Cadmium (Cd)	Alpine pennycress (*Thlaspica crulescens*), *Cardaminopsis halleri*, Eel grass (*Vallisneria spiralis*), Water hyssop (*Bacopa monnieri*), Water hyacinth (*Eichhornia crassipes*), Hydrilla (*Hydrilla verticillata*), Duckweed (*Lemna minor*), Giant duckweed (*Spirodela palyrhiza*)
Arsenic (As)	Chinese brake fern (*Pteris vittata*), Fern (*Pteris cretica*)
Chromium (Cr)	Duckweed (*Lemna minor*), *Ceratophyllum demersum*, Giant reed (*Arundo donax*), Cattail (*Typhaan gustifolia*), Alfoalbo (*Medicago sativa*), Water hyssop (*Bacopa monnieri*), Pistastratiotes, Water fern (*Salvinia molesta*), *Spirodela polyrhiza*
Copper (Cu)	*Aeolanthus bioformifollus, Lemna minor, Vigna radiata*, Creosote bush (*Larrea tridantata*), Water hyssop (*Bacopa monnieri*), Indian mustard (*Brassica juncea*)
Manganese (Mn)	*Alyxia rubricaulis, Macademia neurophylla*
Zinc (Zn)	Alpine pennycress (*Thlaspica erulescens*), *Brassica juncea*
Mercury (Hg)	*Lemna minor*, Water lettuce (*Pistia stratiotes*), Water hyacinth (*Eichhornia crassipes*), Hydrilla (*Hydrilla verticillata*)
Nickel (Ni)	*Phyllanthus serpentines, Lemna minor, Salvinia molesta, Brassica juncea, Spirodela polyrhiza*
Molybdenum (Mo)	Eel grass of Africa (*Haumania strumrobertii*)

9.3 Limitations of Traditional Remediation Methods

Soil and groundwater contamination/pollution occurs due to different human activities and also arises from manufacturing processes. It is a matter of great complexity and concern. Contamination may come from industrial sites (including lakes and rivers in their vicinity), underground storage tank leakages, landfills, and abandoned mines. Pollutants in these areas include heavy metals (e.g., mercury, lead, cadmium) and organic compounds (e.g., benzene, chlorinated solvents, creosote). Current available approaches, including physical and chemical methods, employed for remediation are generally laborious, time consuming, and significantly expensive. Similarly, bioremediation has some disadvantages. Since bioremediation needs organisms to successfully reduce pollutant levels, it is not instantaneous and often need to test and optimize conditions empirically for organism's growth. Also, there may be inhibitors present; for example, compounds may not be in a biodegradable form (polymers, plastics) or may be recalcitrant (Lu et al., 2011).

A pretreatment process and removal of the contaminated area is often required, with a consequent disturbance of the ecosystem. Nanotechnology allows developing technologies that can perform remediation and reach inaccessible areas such as crevices and aquifers, thus eliminating the necessity for costly traditional operations. Moreover, nanotechnology can have ability to manipulate matter at its molecular levels; it can be used to develop remediation tools that are specific for a certain pollutant (e.g., metal), therefore increasing the affinity, selectivity, and sensitivity of the technique. Hence, nanotechnology can help to develop techniques that will allow for more specific and cost-effective remediation tools (Filipponi and Sutherland, 2010).

9.4 Nanoremediation: An Alternative for Traditional Remediation Processes

Nanotechnology is an emerging branch of science. The production of different kinds of nanomaterials and their products are a rapidly developing field which provides many opportunities for innovation. Nanotechnology has already proved its effective bioapplications in various fields including pharmaceuticals, medicines, and agriculture (Hasnain et al., 2013). Nanoremediation methods entail the application of reactive nanomaterials for transformation and detoxification of pollutants. These nanomaterials have properties that enable both chemical reduction and catalysis to mitigate the pollutants of concern (Karn et al., 2009).

For the abatement of pollution, application of nanotechnology is just beginning to be explored which could dramatically catalyze the most radical changes in the field of environment. One of the previous studies carried out has shown that nanoscale iron particles are very effective for the transformation and detoxification of a wide variety of common environmental contaminants, such as chlorinated organic solvents, organochlorine pesticides, and polychlorinated biphenyls (PCBs).

Therefore, modified iron nanoparticles, such as catalyzed and supported nanoparticles, have been synthesized to further enhance the speed and efficiency of remediation (Zhang, 2003). Moreover, highly reactive iron oxide nanoparticles have the potential to increase the efficacy of the remediating the contaminant. However, if used with a microorganism, they could give a better result. In this context, *Geobacter metallireducens*, when bound to macroparticulate ferrous oxide, were reported to reduce 4-nitroacetophane, a highly contaminating organic compound (Braunschweig et al., 2013).

With the advancement of industrial, agricultural, and urban activities, the levels of groundwater pollution have increased enormously in the last few decades (Mehndiratta et al., 2013; Mura et al., 2013). In India, this has led to an increase in the concentration of many organic and inorganic pollutants far above the permissible limits of drinking water standards. The majority of groundwater quality problems are caused by contamination, overexploitation, or a combination of the two. Soil and groundwater quality are slowly but surely declining everywhere. Direct dumping of untreated effluents into wells and other water sources are the major causes of groundwater pollution. Moreover, extensive use of pesticides has also led to the increase in pollutant concentrations in groundwater. The most harmful compounds that are finding their way into groundwaters through various illegal industrial and agricultural activities include various organochlorines, organophosphorous, pesticides, heavy metals, and other carcinogenic compounds (Agarwal and Joshi, 2010).

Some organic pollutants are also known to persist in the environment for a longer period. These chemicals usually possess the potential for becoming bioaccumulated or biomagnified. In both cases, such chemicals can have a tremendously hazardous effect on humans and the environment. Lindane, or γ-HCH (hexachlorocyclohexane), an organochlorine pesticide (gamma isomer of 1,2,3,4,5,6-HCH), is used all over the world for controlling various agricultural pests. However, this compound is known to be bioaccumulated, causing cancer and disturbing the homeostasis of the endocrinal system (Vijgen et al., 2011). Therefore, it affects both human and environmental health. Recently, Singh et al. (2013) investigated an integrated nano-biotechnique for degrading Lindane. The study concluded that Pd/Fe^0 bimetallic nanoparticles (CMC-Pd/nFe^0) as well as *Sphingomonas* sp. strain NM05 are effective in degrading γ-HCH. But, interestingly, the synergistic effect of both of them showed a degradation efficiency of $\sim$1.7–2.1 times greater as compared to the system containing either *Sphingomonas* sp. strain NM05 or CMC-Pd/nFe^0 alone.

The application of nanotechnology for the remediation of contaminants has shown promising results. Rajan (2011) reported that remediation of groundwater can be achieved by using nanotechnology. Similar points were also discussed recently (Kemp et al., 2013). According to Rajan (2011), nanomaterials such as zero-valent iron (nZVI) and carbon nanotubes (CNTs) can be used in environmental cleanups such as groundwater remediation for drinking and reuse. However, there may be some concerns remaining regarding the potential risks to the environment and human health associated with the use of nanomaterials (Kumar et al., 2013;

Taghizadeh et al., 2013). There are many other nanoparticles and nanomaterials that have been used for the remediation of different kinds of pollutants/contaminants present in water (Table 9.4).

Soil contamination is also a matter of great concern. Hydrophobic organic groundwater contaminants such as polynuclear aromatic hydrocarbons (PAHs) are generally found to be associated with soil and are very difficult to remove from it. Tungittiplakorn et al. (2004) demonstrated that amphiphilic polyurethane (APU)

Table 9.4 Examples of Nanoparticles and Nanomaterials for Use in Water Remediation (Theron et al., 2008)

Nanoparticle/Nanomaterials	Pollutant
Nanocrystalline zeolites	Toluene, nitrogen dioxide
Carbonaceous nanomaterials	
Activated carbon fibers (ACFs)	Benzene, toluene, xylene, ethylbenzene
CeO_2 carbon nanotubes (CNTs)	Heavy metal ions
CNTs functionalized with polymers	*p*-Nitrophenols, benzene, toluene, dimethylbenzene, heavy metal ions
CNTs functionalized with Fe	
Single-walled CNTs	Trihalomethanes (THMs)
Multi-walled CNTs	Heavy metal ions, THMs, chlorophenols, herbicides, microcystin toxins
Self-assembled monolayer on mesoporous support (SAMMS)	Inorganic ions
	Heavy metal ions
Anion-SAMMS	Actinides and lanthanides
Thiol-SAMMS	
HOPO-SAMMS	
Biopolymers	Heavy metal ions
Single-enzyme nanoparticles (SENs)	Not tested
Zero-valent iron nanoparticles (nZVI)	Polychlorinated biphenyls (PCBs)
	Inorganic ion, chlorinated organic compounds, heavy metals
Bimetallic nanoparticles	
Pd/Fe nanoparticles	PCBs, chlorinated ethane, chlorinated methanes
Ni/Fe nanoparticles and Pd/Au nanoparticles	Trichloroethylene (TCE) and PCBs
	Dichlorophenol, trichlorobenzene, chlorinated ethane, brominated organic compounds (BOCs)
TiO_2 photocatalyst	
Nanocrystalline TiO_2	Heavy metal ions
Nitrogen (N)-doped TiO_2	Azo dyes
Fe (III)-doped TiO_2	Phenols
Supported TiO_2 nanoparticles	Aromatic pollutants
TiO_2-based p–n junction nanotubes	Toluene

nanoparticles made up of polyurethane acrylate anionomer (UAA) or poly (ethylene glycol)-modified urethane acrylate (PMUA) precursor chains emulsified and cross-linked in water have the ability to remove PAH from contaminated soil. Similarly, hexavalent chromium [Cr(VI)] is a potential carcinogen, teratogen, and mutagen. It is on the top priority list of toxic soil and environmental pollutants defined by the US EPA. Singh et al. (2011) demonstrated that use of zero-valent iron nanoparticles (nZVI) can be useful for the remediation of Cr(VI) from soil. In their experiment they reported that 1.5 g of nZVI entrapped in alginate beads removes 98% Cr(VI) from spiked soil within a contact time of 60 min. Pavuıa-Sanders et al. (2013) constructed magnetic shell cross-linked Knedel-like nanoparticles (MSCKs) from amphiphilic block copolymers of PAA20-b-PS280 and oleic acid-stabilized magnetic iron oxide nanoparticles using tetrahydrofuran, *N*, *N*-dimethylformamide, and water for the isolation and removal of hydrocarbons present in crude oil by providing an external magnetic field. The results obtained are significant and it was found that the ratio of removal of hydrophobic contaminants (hydrocarbons) from crude oil is 10 mg of oil per 1 mg of MSCKs.

Titanium dioxide (TiO_2) and zinc oxides (ZnO) are on the list of most popular materials used in various applications because of their semiconducting, photocatalytic, energy converting, and electronic- and gas-sensing properties. Considering these properties, many researchers have focused research on TiO_2 and ZnO nanoparticles for their application as photocatalysts in water treatment. TiO_2 and ZnO nanoparticles are frequently studied for their ability to remove organic contaminants from various media. These nanoparticles have the advantages of being readily available, inexpensive, and of low toxicity (Mansoori et al., 2008).

Among the various nanomaterials, the magnetic nanoparticles (MNPs) have been attracting particular attention because of their convenient magnetic field–assisted separation which is quality step for nanoremediation. Farrukh et al. (2013) prepared polymer brush–grafted MNPs for highly efficient water remediation. They reported the highly efficient removal of mercury (II) ions from water by employing polymer brush–functionalized MNPs. In another study on MNPs, biologically synthesized magnetite (Fe_3O_4) nanoparticles were studied using X-ray absorption and X-ray magnetic circular dichroism for the remediation of Cr(VI). The possibility of reducing toxic Cr(VI) into a stable, nontoxic form, such as a Cr^{3+} spinel layer, makes these biogenic magnetite nanoparticles an attractive candidate for Cr remediation (Telling et al., 2009).

9.5 Conclusion

From all the above facts it can be concluded that air, water, and soil are the main sources that can be contaminated with different kinds of pollutants or contaminants generated by human activity. Heavy metal ions, carbon monoxide, CFCs, hydrocarbons, nitrogen oxides, organic chemicals, sulfur dioxide, etc. are major pollutants that are hazardous to human beings, aquatic life, plants, etc. Although there are traditional remediation methods—physical, chemical, and biological—available for the removal of the above-mentioned pollutants, due to some limitations, they are unable to attain safety levels These methods are generally laborious, time consuming, and significantly expensive.

Apart from these techniques, remediation of toxic compounds or substances from the environment can be achieved by emerging technology, i.e., nanotechnology. Different nanoparticles or nanomaterials have been found to be very effective for the removal of a wide range of contaminants from the environment as compared to traditional methods. Moreover, nanotechnology-based approaches are comparatively cheaper and more efficient. Therefore, nanotechnology could be a better alternative for remediation and for sustaining our environment. Considering these facts, this technology is assumed to be the next generation of remediation technology.

References

Agarwal, A., Joshi, H., 2010. Application of nanotechnology in the remediation of contaminated groundwater: a short review. Recent Res. Sci. Technol. 2, 51−57.

Ali, H., Khan, E., Sajad, M.A., 2013. Phytoremediation of heavy metals: concepts and applications. Chemosphere. 91, 869−881.

Baldrian, P., 2006. Fungal laccases—occurrence and properties. FEMS Microbiol. Rev. 30, 215−242.

Blanquez, P., Sarra, M., Vicent, T., 2008. Development of a continuous process to adapt the textile wastewater treatment by fungi to industrial conditions. Process Biochem. 43, 1−7.

Boulding, J.R., 1996. USEPA Environmental Engineering Sourcebook. Ann Arbor Press, Chelsea, MI.

Braunschweig, J., Bosch, J., Meckenstock, R.U., 2013. Iron oxide nanoparticles in geomicrobiology: from biogeochemistry to bioremediation. N. Biotechnol. 30, 793−802.

Chu, W., Chan, K.H., 2003. The mechanism of the surfactant-aided soil washing system for hydrophobic and partial hydrophobic organics. Sci. Total Environ. 307, 83−92.

Czupyrna, G., Levy, R.D., MacLean, A.I., Gold, H., 1989. In-situ Immobilization of Heavy-Metal Contaminated Soils. Noyes Data Corporation, Park Ridge, NJ.

Dean-Ross, D., 2003. Use of PAH-degrading bacteria in bioremediation of PAH-contaminated sediments. In: Proceedings of Second International Symposium on Contaminated Sediments, held at Quebec City, Canada during 26−28 May, 2003, pp. 252−257.

Dhal, B., Thatoi, H.N., Das, N.N., Pandey, B.D., 2013. Chemical and microbial remediation of hexavalent chromium from contaminated soil and mining/metallurgical solid waste: a review. J. Hazard. Mater. 250−251, 272−291.

Di-Palma, L., Ferrantelli, P., Merli, C., Biancifiori, F., 2003. Recovery of EDTA and metal precipitation from soil flushing solutions. J. Hazard. Mater. 103, 153−168.

D'Souza, D.T., Sharma, D., Raghukumar, C., 2009. A thermostable metal-tolerant laccase with bioremediation potential from a marine-derived fungus. Mar. Biotechnol. 11, 725−737.

Farrukh, A., Akram, A., Ghaffar, A., Hanif, S., Hamid, A., Duran, H., et al., 2013. Design of polymer-brush-grafted magnetic nanoparticles for highly efficient water remediation. ACS Appl. Mater. Interfaces. 5, 3784−3793.

Filipponi, L., Sutherland, D., 2010. Environment: application of nanotechnologies. In: Nanoyou Teachers Training Kit in Nanotechnologies. pp. 1–26.

Gates, D.D., Siegrist, R.L., 1994. In-situ chemical oxidation of trichloroethylene using hydrogen peroxide. J. Environ. Eng. 121, 639–644.

Gautam, R., Bassi, A.S., Yanful, E.K., 2006. A review of biodegradation of synthetic plastic and foams. Appl. Biochem. Biotechnol. 141, 85–109.

Gopalan, A., Zincircioglu, O., Smith, P., 1993. Minimization and remediation of DOE nuclear waste problems using high selectivity actinide chelators. Radioactive Waste Manage. Nucl. Fuel Cycle. 17, 161–175.

Hajdu, R., Slaveykova, V.I., 2012. Cd and Pb removal from contaminated environment by metal resistant bacterium *Cupriavidus metallidurans* CH34: importance of the complexation and competition effects. Environ. Chem. 9, 389–398.

Harikumar, P.S.P., Jesitha, K., Sreechithra, M., 2013. Remediation of endosulfan by biotic and abiotic methods. J. Environ. Protect. 4, 418–425.

Hasnain, S., Ali, A.S., Uddin, Z., Zafar, R., 2013. Application of nanotechnology in health and environmental research: a review. Res. J. Environ. Earth Sci. 5, 160–166.

Hejazi, R., Husain, T., Khan, F.I., 2003. Land farming operation in arid region-human health risk assessment. J. Hazard. Mater. 99, 287–302.

Ho, C.L., Shebl, M.A.A., Watts, R.J., 1995. Development of an injection system for in situ catalyzed peroxide remediation of contaminated soil. Hazard. Waste Hazard. Mater. 12, 15–25.

Hong, J.W., Park, J.Y., Gadd, G.M., 2009. Pyrene degradation and copper and zinc uptake by *Fusarium solani* and *Hypocrea lixii* isolated from petrol station soil. J. Appl. Microbiol. 108, 2030–2040.

<http://en.wikipedia.org/wiki/Bioremediation#cite_note-1>. [Retrieved on 05.10.13].

Jecu, L., Gheorghe, A., Rosu, A., Raut, I., Grosu, E., Ghuirea, M., 2010. Ability of fungal strains to degrade PVA based materials. J. Polym. Environ. 18, 284–290.

Karn, B., Kuiken, T., Otto, M., 2009. Nanotechnology and in situ remediation: a review of the benefits and potential risks. Environ. Health Perspect. 117, 1823–1831.

Kemp, K.C., Seema, H., Saleh, M., Le, N.H., Mahesh, K., Chandra, V., et al., 2013. Environmental applications using graphene composites: water remediation and gas adsorption. Nanoscale. 5, 3149–3171.

Khan, F.I., Husain, T., Hejazi, R., 2004. An overview and analysis of site remediation technologies. J. Environ. Manage. 71, 95–122.

Kim, Y.J., Nicell, J.A., 2006. Laccase-catalyzed oxidation of bisphenol A with the aid of additives. Process Biochem. 41, 1029–1037.

Kumar, R., Khan, M.A., Haq, N., 2013. Application of carbon nanotubes in heavy metals remediation. Crit. Rev. Environ. Sci. Technol. Available from: http://dx.doi.org/doi:10.1080/10643389.2012.741314.

Li, X., Kondo, R., Sakai, K., 2002. Studies on hypersaline-tolerant white-rot fungi. II: Biodegradation of sugarcane bagasse with marine fungus *Phlebia* sp. MG-60. J. Wood Sci. 48, 159–162.

Logsdon, S.D., Keller, K.E., Moorman, T.B., 2002. Measured and predicted solute leaching from multiple undisturbed soil columns. Soil Sci. Soc. Am. J. 66 (3), 686–695.

Lone, M.I., He, Z.L., Stoffella, P.J., Yang, X., 2008. Phytoremediation of heavy metal polluted soils and water: progresses and perspectives. J. Zhejiang Univ. Sci. B. 9 (3), 210–220.

Lu, J.C., Li, Z.T., Hussain, K., Yang, G.K., 2011. Bioremediation: the new directions of oil spill cleanup. Middle East J. Sci. Res. 7 (5), 738–740.

Madhavi, V., Reddy, A.V.B., Reddy, K.G., Madhavi, G., Prasad, T.N.V.K.V., 2013. An overview on research trends in remediation of chromium. Res. J. Recent Sci. 2 (1), 71–83.

Maier, R.M., Neilson, J.W., Artiola, J.F., Jordan, F.L., Glenn, E.P., Descher, S.M., 2001. Remediation of metal-contaminated soil and sludge using biosurfactant technology. Int. J. Occup. Med. Environ. Health. 14 (3), 241–248.

Malik, N., Biswasa, A.K., 2012. Role of higher plants in remediation of metal contaminated sites. Sci. Rev. Chem. Commun. 2 (2), 141–146.

Mansoori, G.A., Rohani, T., Bastami, A., Ahmadpour, Z., Eshaghi, 2008. Environmental application of nanotechnology. Annu. Rev. Nano Res. 2 (2), 1–73.

Mehndiratta, P., Jain, A., Srivastava, S., Gupta, N., 2013. Environmental pollution and nanotechnology. Environ. Pollut. 2, 49–58.

Mura, S., Seddaiu, G., Bacchini, F., Roggero, P.P., Greppi, G.F., 2013. Advances of nanotechnology in agro-environmental studies. Ital. J. Agron. 8, e18.

Pavıa-Sanders, P., Zhang, S., Flores, J.A., Sanders, J.E., Raymond, J.E., Wooley, K.L., 2013. Robust magnetic/polymer hybrid nanoparticles designed for crude oil entrapment and recovery in aqueous environments. ACS Nano. 7 (9), 7552–7561.

Raghukumar, C., D'Souza-Ticlo, D., Verma, A.K., 2008. Treatment of colored effluents with lignin-degrading enzymes: an emerging role of marine-derived fungi. Crit. Rev. Microbiol. 34, 189–206.

Rajan, C.S., 2011. Nanotechnology in groundwater remediation. Int. J. Environ. Sci. Dev. 2, 182–187.

Raskin, I., Smith, R.D., Salt, D.E., 1997. Phytoremediation of metals: using plants to remove pollutants from the environment. Curr. Opin. Biotechnol. 8, 221–226.

Riser-Roberts, E., 1998. Remediation of Petroleum Contaminated Soils. Lewis Publishers, Boca Raton, pp. 542.

Seo, J.S., Keum, Y.S., Li, Q.X., 2009. Bacterial degradation of aromatic compounds. Int. J. Environ. Res. Public Health. 6, 278–309.

Singh, R., Misra, V., Singh, R.P., 2011. Remediation of Cr(VI) contaminated soil by zero-valent iron nanoparticles (nZVI) entrapped in calcium alginate beads. Second International Conference on Environmental Science and Development, IPCBEE 4. IACSIT Press, Singapore.

Singh, R., Manickam, N., Mudiam, M.K.R., Murthy, R.C., Misra, V., 2013. An integrated (nano-bio) technique for degradation of γ-HCH contaminated soil. J. Hazard. Mater. 258–259, 35–41.

Son, A.J., Shin, K.H., Lee, J.U., Kim, K.W., 2003. Chemical and ecotoxicity assessment of PAH-contaminated soils remediated by enhanced soil flushing. Environ. Eng. Sci. 20, 197–206.

Szczyglowska, M., Piekarska, A., Konieczka, P., Namiesnik, J., 2011. Use of Brassica plants in the phytoremediation and biofumigation processes. Int. J. Mol. Sci. 12, 7760–7771.

Taghizadeh, M., Kebria, D.Y., Darvishi, G., Kootenaei, F.G., 2013. The use of nano zero valent iron in remediation of contaminated soil and groundwater. Int. J. Sci. Res. Environ. Sci. 1, 152–157.

Telling, N.D., Coker, V.S., Cutting, R.S., van der Laan, G., Pearce, C.I., Pattrick, R.A.D., et al., 2009. Remediation of Cr(VI) by biogenic magnetic nanoparticles: an X-ray magnetic circular dichroism study. Appl. Phys. Lett. 95, 10.

Thapa, B., Ajay-Kumar, K.C., Ghimire, A., 2012. A review on bioremediation of petroleum hydrocarbon contaminants in soil. J. Sci. Eng. Technol. 8, 164–170.

Theron, J., Walker, J.A., Cloete, T.E., 2008. Nanotechnology and water treatment: applications and emerging opportunities. Crit. Rev. Microbiol. 34, 43–69.

Tungittiplakorn, W., Lion, L.W., Cohen, C., Kim, J.Y., 2004. Engineered polymeric nanoparticles for soil remediation. Environ. Sci. Technol. 38, 1605–1610.

U.S. Environmental Protection Agency (US EPA), 1998. A citizen's guide to phytoremediation, office of solid waste and emergency response (5102 G) EPA 542-F-98-001.

Urum, K., Pekdemir, T., Gopur, M., 2003. Optimum conditions for washing of crude oil-contaminated soil with biosurfactant solutions. Proc. Safety Environ. Prot. Trans. Inst. Chem. Eng. Part B. 81, 203–209.

Verma, A.K., Raghukumar, C., Verma, P., Shouche, Y.S., Naik, C.G., 2010. Four marine-derived fungi for bioremediation of raw textile mill effluents. Biodegradation. 21, 217–233.

Vijgen, J., Abhilash, P.C., Li, Y.F., Lal, R., Forter, M., Torres, J., et al., 2011. Hexachlorocyclohexane (HCH) as new Stockholm Convention POPs—a global perspective on the management of Lindane and its waste isomers. Environ. Sci. Pollut. Res. 18, 152–162.

Wesenberg, D., Kyriakides, I., Agathos, S.N., 2003. White-rot fungi and their enzymes for the treatment of industrial dye effluents. Biotechnol. Adv. 22, 161–187.

Young, C.A., Jordan, T.S., 1995. Cyanide remediation: current and past technologies. In: Erickson, L.E., Tillison, D.L., Grant, S.C., McDonald, J.P. (Eds.), Proceedings of the 10th Annual Conference on Hazardous Waste Research. Held at Kansas State University Manhattan, Kansas during May 23–24, 1995.

Zhan, H., Park, E., 2002. Vapor flow to horizontal wells in unsaturated zones. Soil Sci. Soc. Am. J. 66, 710–721.

Zhang, W.X., 2003. Nanoscale iron particles for environmental remediation: an overview. J. Nanopart. Res. 5, 323–332.

10 Bioremediation Using Extremophiles

Tonya L. Peeples

Department of Chemical and Biochemical Engineering, Seamans Center, The University of Iowa, Iowa City, IA

10.1 Bioremediation Using Extremophiles

Microorganisms from extreme environments provide robust enzymatic and whole-cell biocatalytic systems that are attractive under conditions that limit the effectiveness of typical bioconversions. In the path to applying extreme microbes, scientists and engineers have deepened their understanding of molecular systems that govern extremozyme action, stability, and expression while at the same time gaining insight into metabolic strategies for whole-cell catalysis under extreme conditions. Advances in the application of extremophilic systems for remediation operations are made through examining selective conditions that dominate contaminated environments, identifying extremophilic systems that can persist and adapt in contaminated environments, and engineering remediation capacity into organisms that can survive extreme selective pressures.

10.2 Identifying Extremophiles for Remediation Applications

The advent of industrial chemical processes has created environments where pollution is prevalent. The organisms that inhabit these altered ecosystems have the ability to adapt and develop metabolic characteristics that may be useful in a variety of new processes. Naturally occurring microbes can be used to actively degrade pollutants and convert them into less harmful substances. Advances in molecular biology have enabled scientists, engineers, and entrepreneurs to probe the biosphere for remediation catalysts, to identify biological indicators of ecosystem vitality and to search for sources of new commercially valuable products. The utility of extremophilic systems in environmental remediation is related to the nature of the environment where the system is likely to be applied. Selective conditions for extremophilic bacteria, archaea, and higher life forms include extremes of

Microbial Biodegradation and Bioremediation. DOI: http://dx.doi.org/10.1016/B978-0-12-800021-2.00010-8

temperature, pH, radiation exposure, high salinity, hydrophobicity, and combinations of these various conditions. Each descriptor presents challenges and advantages for bioremediation.

10.2.1 Extremes of Temperature

Biodiversity has provided organisms across numerous temperature niches. Hence, temperature has been a defining characteristic in categorizing extremophilic organisms and their impact on the surrounding environments. The temperature descriptor for an organism is typically related to the temperature at which the organism achieves its maximum specific growth rate. This temperature optimum (T_{opt}) is used to distinguish groups of extreme organisms. The range of temperature over which an organism has been demonstrated to grow also helps to describe the adaptability of the organism and may result in additional extremophilic designations (Figure 10.1).

Organisms that thrive in cold temperatures are classified as psychrophilic, psychrotrophic, or psychrotolerant species. Extreme psychrophiles require temperatures from 0°C to 15°C for growth. Some psychrophiles have T_{opt} values greater than 15°C but exhibit no growth above 20°C (Paustian, 2012). Psychrotrophic microbes can grow at low temperatures but typically have T_{opt} values between 20°C and 30°C (Willey et al., 2011). Organisms that grow optimally at temperatures greater than 20°C, but which are not able to grow well below 5°C, are designated as psychrotolerant (Canganella and Wiegel, 2011). Experts estimate that over 80% of the total biosphere of the Earth is at temperatures below 5°C (Cavicchioli et al., 2000). As such there are a multitude of natural sites where these cold

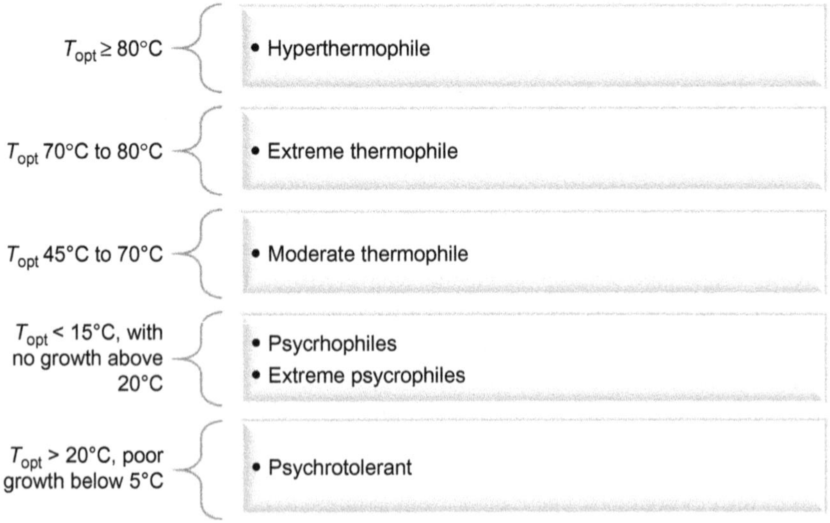

Figure 10.1 Temperature classifications for prokaryote.

temperatures are permanent conditions. There are also a few man-made low-temperature environments including sites of refrigeration and freezing. Enzymes from psychrophiles and psychrotolerant organisms have been applied in sludge treatment (Lettinga et al., 1999) and bioremediation of hydrocarbons (Margesin and Schinner, 1997, 1998, 2001).

Natural habitats for thermophilic organisms include areas where geothermal activity has made an impact on the ecosystem. "Human-made" environments include industrial processes, mining piles, compost piles, and water heaters. Canganella divides organisms that thrive at high temperatures into three categories: hyperthermophiles ($T_{opt} \geq 80°C$); extreme thermophiles (T_{opt} 70–80°C); and moderate thermophiles (T_{opt} 45–70°C) (Canganella and Wiegel, 2011).

Hyperthermophiles are located on the least evolved branches of the phylogenetic tree and are associated with substrates and products that exist in abundance in chemical settings. As such these organisms are studied extensively with an eye toward understanding mechanisms for survival and adaptation. The synergy of increased solubility of pollutants and increasing metabolic activity of thermophiles with temperature makes thermophilic systems attractive for conversion of recalcitrant pollutants. Thermophilic microbes and their enzymes have been applied in processing industrial effluents (Goh et al., 2013), as well as for *in situ* remediation of deep subsurface sites where temperatures are elevated (Tor and Lovley, 2001; Richardson et al., 2002; Lovley et al., 2004; Cason et al., 2012; Zhang et al., 2012a,b).

Thermophilic systems used in remediation have been isolated from both natural high-temperature sites, such as hot springs, and human-created sites, such as dairy processing streams (Burgess et al., 2010). Such organisms persist due to spore formation, biofilm formation, and up-regulation of molecular chaperonin proteins during higher-temperature processing.

10.2.2 Extremes of pH

Microbes can tolerate larger gradients in pH across the cell membrane than higher life forms. Many bacteria that have been applied in fermentations for food production have demonstrated tolerance to acidic environments (Burgess et al., 2010). Beyond compatibility with traditional fermentations and tolerance of extreme pH, bacteria and archaea that require more extreme conditions have been identified. Acidophiles that grow at cold or moderate temperatures have been applied in bioleaching applications in tanks or in ore piles. Thermophilic acidophiles have been applied to oxidize mineral sulfide inclusions or concentrates, thereby liberating metals of interest (Peeples et al., 1991; Peeples and Kelly, 1993, 1995; Gemmell and Knowles, 2000; Hallberg, 2010; Johnson, 2012).

Although acidophilic organisms are generally defined as microbes growing at pH values lower than 5 and having pH optima between 2 and 4, microbes that can thrive at pH values near 0 have been identified. Prokaryotic acidophiles can grow heterotrophically (using organic compounds for carbon and energy), autotrophically (using carbon dioxide and other C-1 compounds for carbon and minerals or light for energy), or mixotrophically (using a combination of heterotrophic and autotrophic modes). Some eukaryotic organisms including yeasts and filamentous fungi can grow at low pH values. In addition, obligatory acidophilic protozoa have been isolated (Stierle and Stierle, 2005).

Alkaliphilic microbes thrive at pH values above 9 and grow much more slowly or not at all at pH values near neutrality. Alkali-tolerant organisms can grow at a pH of 9 or 10 but have pH optima near neutrality. Industrial enzymes from alkaliphiles that carry out enzymatic conversions under extremes of pH are primarily secreted proteins. Intracellular spaces are kept closer to neutrality and thus produce less extreme enzymes. Because of their persistence in alkaline environments, these organisms have produced a variety of useful enzymes, particularly hydrolases for use in pulp and paper processing and detergent formulations and for bioconversion of fats, proteins, and carbohydrates (Karadzic et al., 2006; Mesbah and Wiegel, 2008; Antranikian et al., 2009; Goh et al., 2013).

Members of the recently proposed *Anoxybacillus* genus are alkali-tolerant thermophiles (Goh et al., 2013). These microbes may have utility for the conversion of starch and lignocellulosic feedstocks, bioremediation, waste treatment, and bioenergy production. Despite the name *Anoxybacillus*, bacteria of this group can grow aerobically, or function as aerotolerant or facultative anaerobes. Most species are primarily moderate thermophiles, growing at from 50°C to 65°C. At least one species grows optimally at 37°C but still is extremophilic based on growth in alkali.

10.2.3 Extremes of Radiation

Organisms that can survive under high levels of radiation may be useful for the cleanup of a variety of severely contaminated sites associated with weapons proliferation and the energy industry. Sources for useful microbes include human-generated sites of higher, persistent radiation exposure (Zahradka et al., 2006; Dhaker et al., 2011; Lage et al., 2012; Miralles et al., 2012) and natural environments. Natural sites where microbes may have been exposed to high radiation levels include deep subsurface regions associated with radioactive ore deposits as well as areas of high ultraviolet (UV) radiation (Zahradka et al., 2006; Chen et al., 2012). Organisms have been identified based on efforts to find places on Earth that resemble space environments, which mimic the chemical environments of other planetary environments including hyper-arid deserts, high elevation, or high sun exposure environments (Zahradka et al., 2006; Dhaker et al., 2011; Lage et al., 2012; Miralles et al., 2012).

Deinococcus radiodurans has been extensively studied because it can survive high levels of ionizing radiation. Many years of research have led to the discovery of several antioxidant activities that convey radiation resistance, including repair of DNA. *D. radiouridans* has been explored as a candidate for bioremediation of sites contaminated with radioactive materials. Clones have been developed that can degrade toluene and reduce toxic metals such as chromium and mercury in the presence of cesium-137 (Brim et al., 2006).

Advanced tools (such as synchrotron accelerators) have been used to simulate interplanetary solar radiation in vacuum chambers loaded with dehydrated cell powders (Lage et al., 2012). Further, low-energy particle radiation shows no effect on dehydrated cells and organisms protected by dust grains. Salt-tolerant microorganisms (*Haloferax vulcanii* and *Natrialba magadii*) show similar stability against vacuum UV radiation to that of the radio-resistant organisms (Lage et al., 2012). Silicate nanoparticles had a protective effect for cellular survival under radiation exposure. Antarctic bacteria have been shown to survive high ionizing radiation doses. These organisms are also of great interest in the production of industrial enzymes.

High-altitude Andean lakes have been explored to identify organisms that survive and to further elucidate mechanisms of stability. The extreme nature of these environments includes high levels of UV, hypersalinity, low nutrients, high concentrations of heavy metals, wide temperature fluctuations, and periodic desiccation (Bequer Urbano et al., 2013).

10.2.4 Extremes of Salinity

Organisms that thrive in high salt environments are termed halophiles. These organisms are isolated from solar salterns, saline river beds, sediments, and hypersaline waters such as the Great Salt Lake and the Dead Sea (Sei and Fathepure, 2009; Al-Mailem, D.M. et al., 2010; Rhodes et al., 2011; Luque et al., 2012; Erdogmus et al., 2013). Classifications include aerobic halophilic archaea, anaerobic methanogenic halophilic archaea, and halophilic bacteria (Kamekura, 1998). Hypersaline habitats may include other adverse conditions including high concentrations of divalent cations, low pH, elevated temperatures, and high levels of UV exposure. Demonstrating the highest specific growth rates in media saturated with salt (2.5–5.2 M NaCl), some of these prokaryotes have the ability to maintain osmotic balance by accumulating KCl. Because of high salt concentrations inside and out, enzymes produced by these organisms are adapted to function under extremes of salinity. For example, lipases and esterases have been isolated that can withstand up to 5 M NaCl (Fucinos et al., 2012). Many of these extremozymes and the organisms that produce them have been used for remediation in brine and hydrocarbon-rich environments (Sei and Fathepure, 2009; Al-Mailem et al., 2010, 2012, 2013; Najera-Fernandez et al., 2012; Erdogmus et al., 2013; Harada et al., 2013).

10.2.5 Extreme Concentration of Hydrocarbons

Hydrocarbon-rich environments provide selective pressures for the persistence of extremophiles. Natural sites associated with submarine or subsurface petroleum fields as well as areas of high polyaromatic hydrocarbon (PAH) contamination serve as habitats for organisms adapted to organics (Margesin and Schinner, 2001; Kleikemper et al., 2005; Kim and Crowley, 2007; Rocchetti et al., 2012; Zhang et al., 2012a,b; Bartolucci et al., 2013). Solvent-tolerant systems and mixed communities of organisms that thrive in hydrocarbon extremes have been augmented for the remediation of contaminated sites and have been the templates for engineering of novel biotransformation systems.

Methanogenic archaea are often associated with areas of high hydrocarbon concentration, suggesting that hydrocarbon degradation in subsurface environments is mediated through the conversion of low-molecular-weight organics to methane (Rocchetti et al., 2012). This conversion of hydrocarbons involves syntrophic species to carry out acetogenic fermentation. The interaction of mixed cultures of microbes facilitates complete mineralization of hydrocarbons, changes the mobility of metals in contaminated sites, and may include species that are mesophilic, thermophilic, and/or halophilic in nature (Dojka et al., 1998; Lin et al., 2012; Wan et al., 2012; Zhang et al., 2012a,b; Al-Mailem et al., 2013).

Oxidative products of biological activity on organic hydrocarbons can cause damage to microbial cells. Organic hydrocarbon solvents impair vital functions by facilitating loss of ions, metabolites, lipids, and proteins; by inhibiting membrane protein functions that have evolved to mitigate the negative impact of solvents by modulating membrane permeability (Chiu et al., 2007); and by pumping organics out of the intracellular space (Kieboom et al., 1998; Sardessai and Bhosle, 2002; Fernandes et al., 2003). These strategies for survival have been the subject of metabolic engineering in high-performing hosts for specialty chemicals synthesis, biofuel generation, and remediation catalysts (Srikumar et al., 1998; Dunlop et al., 2011; Rojas et al., 2001).

10.2.6 Extremes of Pressure

Deep oceanic and lithospheric sites have been the source for identifying organisms adapted to high pressures. Archaea and yeasts tolerating high metal content and hydrocarbons have been identified (Tor and Lovley, 2001; Jiao et al., 2003; Mosher et al., 2012). The activity of these organisms toward oceanic pollution leads to the conclusion that they play a role in the biogeochemical cycling of carbon and metals in the biosphere. While there are efforts to characterize adaptive features that enable cellular functions under high-pressure conditions (Bartlett and Chi, 1994; Simonato et al., 2006; Bartlett, 2009), most efforts regarding remediation with high-pressure systems are related to additional extremophilic characteristics, such as solvent tolerance, radiation tolerance, or thermophily.

10.3 Enzyme Catalysis for Remediation

The application of extremely stable enzymes for environmental remediation is an attractive alternative to whole-cell bioconversions, which are perceived to take more time to implement. For these applications, robust enzymatic catalysts are required. Efforts to find enzymes that remain stable under extreme conditions and to protect enzymes in harsh bioremediation solutions have resulted in some interesting novel catalysts (Burton, 2001; Gianfreda and Rao, 2004; Kumar, 2010;

Peixoto et al., 2011). Further, for enzymatic conversion, it is helpful to have enzymes that operate without the need for cofactor regeneration and that can be produced through large-scale fermentation with minimal purification needs. In most cases, cell lysates can also be applied to minimize biocatalyst production costs (Peixoto et al., 2011).

While the application of free enzymes for remediation is limited, a few examples for the transformation of metals, organophosphates, phenolics, and hydrocarbons exist (Table 10.1). Metal remediation in industrial effluents can be performed in a cost-effective manner with the use of enzymatic precipitation (Chaudhuri et al., 2013). Polyphenol oxidase (PPO) is a good biocatalyst for the transformation of phenolics, which abound in a variety of industrial processes.

Organophosphates are used as insecticides and herbicides and have been applied as chemical warfare agents. Detoxification of these compounds is mediated by enzymes of the aminohydrolase family. A variety of extremophilic phophotriesterase enzymes have been identified that have the potential for remediation of organophosphate compounds (Hawwa et al., 2009; Mandrich et al., 2010; Hiblot et al., 2012). The extremozymes have the ability to hydrolyze lactones as well as phosphates and have thus been named phosphotriesterase-like lactonases (PLLs) (Hawwa et al., 2009). Cell-free applications for enzymatic conversion of organopesticides include the development of catalytic biomaterials (Lu et al., 2010). These materials have been used in cleaning up munitions waste and contaminated environmental sites. Further materials with organopesticide hydrolase enzymes have been incorporated into firefighting foams and paint coatings to facilitate decontamination activities. The act of immobilization increases stability of enzymes applied against thermal denaturation and can be achieved through novel magnetosomes or nanoscaffolds to facilitate enzyme stability and accessibility to substrates during biocatalysis (Lu et al., 2010; Ginet et al., 2011; Raynes et al., 2011).

As petroleum-derived pollutants accumulate in the environment from human activity, sites high in n-alkanes and aromatic and PAHs have increased

Table 10.1 Extremozymes Used in Bioremediation

Categories and Conditions	Organism	Enzyme	Target Pollutant
Thermophiles and hyperthermophiles (T_{opt} 55–105°C)	*Sulfolobus sulfataricus*	Phosphotriesterase-like lactonase	Organophosphates
	Thermus thermophilus *HB27*	Laccase	Polyaromatic hydrocarbons
	Thermoascus aurantiacus	Phenol oxidase	Phenolic hydrocarbons
	Geobacillus stearothermophilus	Phosphotriesterase-like lactonase	Organophosphates
Acidophiles ($pH_{opt} < 3$)	*Sulfolobus sulfataricus*	Phosphotriesterase-like lactonase	Organophosphates

(Ginet et al., 2011). PAHs are prevalent in the environment and pose threats because of their mutagenic and cytotoxic properties. Enzymatic remediation of such molecules requires aromatic ring oxidation and biotransformation. Laccase enzymes show activity in PAH degradation, but for process development the natural enzymes need to be made more stable in organic solvents (Peixoto et al., 2011) or extremophilic analogues must be applied (Machuca et al., 1998; Miyazaki, 2005; Ramirez et al., 2012).

10.4 Whole-Cell Catalysis for Remediation Under Extreme Conditions

Biotransformation of environmental pollutants often requires multiple catalytic steps and enzymes that need cofactors. This multistep conversion of molecules means that most remediation applications require whole cells. A variety of extremophilic systems adapted to specific niche environments have been applied in the field and in bioreactors. These bioconversion operations attempt to augment naturally occurring remediation capacity and accelerate reactive processes by removing limitations on microbial action. Many environmental sites are contaminated with multiple pollutants and support multiple modes of extremophily that operate in concert to facilitate chemical transformation. Microbial species that function at extremes have been studied in microcosms and as single strains to increase the range of applications for remediation. Several examples of remediation technologies are discussed here.

10.4.1 Temperature, Pressure, and Whole-Cell Bioremediation

In higher-temperature (and high-pressure) environments, solubilities and mass transfer rates for organic contaminants are increased while oxygen solubility is decreased (Cui et al., 2012). These conditions present challenges and opportunities for remediation operations. Whole-cell remediation in high-temperature environments includes a wide range of processes. Applications include detoxification of process effluents, remediation of petroleum hydrocarbons in warm climates, and thermally enhanced remediation operations with patented technologies.

Pseudomonas spp. in mixed consortia have been applied in the detoxification of process effluent at moderately thermophilic temperatures, and this type of consortium has been further developed as a patented process (Lugowski et al., 1997). This process is useful for transformation of hydrocarbon mixtures including sulfur-bearing organic compounds and aromatic hydrocarbons such as benzothiazoles, hexachlorocyclohexane, and toluene. Higher-temperature processes (60–70°C) have been developed using more extreme thermophiles. Specifically, benzene, toluene, ethylbenzene, and xylene (BTEX) degradation has been performed with *Thermus aquaticus* and *Thermus* sp. (Margesin and Schinner, 2001). The significance of BTEX in contamination from underground storage of petroleum has resulted in interesting technology for stripping and steam injection of subsurface environments. One patent uses steam injection to enhance biotransformation environments for thermal bacteria (Taylor et al., 1988). This technology purports enhanced conversion of BTEX as well as remediation of chlorinated solvent contamination from industrial activity. This process likely involves more complex mixtures of microbes within the temperature and nutrient gradients of the injection and recirculation sites (Margesin and Schinner, 2001; Richardson et al., 2002). Deep sites below rocks tend to have synergistic impacts of pressure and temperature, where native extremophiles are augmented by elevated temperatures.

Composting has been explored as a remediation technology for sediments contaminated with explosives. Thermophilic conditions (55°C) enhance the degradation of trinitrotoluene (TNT). High soil loadings could be used without negatively impacting the process (Margesin and Schinner, 2001). Despite the apparent success of thermal composting, most patented technologies for this type of remediation are based on mesophilic microbes.

A recent patent for remediation of hydrocarbon waste has been issued based on application of the hyperthermophile *Archaeglobus fulgidus* (Fardeau et al., 2008). The sulfate-reducing characteristics of *A. fulgidus* have been applied for degrading linear or branched, saturated or unsaturated, aromatic hydrocarbons under anaerobic conditions.

The thermophilic *Anoxybacillus* spp. have been evaluated for the removal of colors from dye-contaminated industrial textile effluents (Goh et al., 2013). Some strains have been demonstrated to reduce to mercury (Hg^{2+}) and chromium (Cr^{4+}) and have been applied in biosorption systems to remove heavy metals from contaminated water. Strains have also been used in other biofilters to treat gaseous toluene, in composting processes, and to enhance the settleability of sewage sludge.

Despite the activity of research with thermophilic species and remediation, nature has generated more low-temperature environments where bioconversions might occur. These sites range from cold low-pressure environments to the cold high-pressure environments that dominate the deep sea. Psychrophilic communities thrive in lake sediments and other perpetually low-temperature environments. Anaerobic methanogens thrive in these systems and degrade a variety of pollutants while living in community with syntrophic strains (Parrilli et al., 2010). Analysis of hydrocarbon remediation by cold marine microbes revealed a capacity to degrade n-alkanes (C-10–C-36), naphthalenes, and PAH compounds (Brakstad et al., 2004).

Organisms in cold environments adapt rapidly to contamination in that the distribution of species changes based on the dominating contaminant. For example, significant numbers of hydrocarbon degraders are found after a petroleum spill (Margesin and Schinner, 1997; Zhang et al., 2012a,b). One technology to enhance remediation in cold sites is augmentation with additional hydrocarbon degraders; however, the results show that the native species are more efficient than the added microbes. The benefit for bioaugmentation through addition of nutrients and organisms could be in shortening the adaptation time (Margesin and Schinner, 2001).

The density of the pollutants plays into availability as depth within cold oceanic environments increases. Compounds with higher density will sink and be in contact with microbes. Still organisms may be associated with deep petroleum fields (Hallmann et al., 2008). Studies of microbial isolates from deep sites (1500 m) reveal hydrocarbon degraders such as *Microbacterium* sp., *Sphingomonas* sp., and organisms related to proteobacteria, but often with these isolations the cultivated organisms are evaluated at moderate temperatures and pressures for hydrocarbon transformation. High pressure and low temperature result in lower rates of hydrocarbon transformation (Hallmann et al., 2008; Johnson and Hill, 2003). Nonetheless, solvent-tolerant hydrocarbon degraders have been isolated from deep-sea muds (Margesin and Schinner, 2001; Brakstad et al., 2004; Singh et al., 2013).

10.4.2 Whole-Cell Remediation at Extremes of pH and Salinity

In both acidic and alkaline environments, single cells and mixed consortia have been found to metabolize organic pollutants (Gemmell and Knowles, 2000; Kletzin et al., 2004). The complex chemistry of the ions in the environmental system also impacts solubilities and availabilities of inorganic and organic pollutants (Johnson et al., 1985; Kishino and Kobayashi, 1995). Many of the acidophilic systems can grow mixotrophically, oxidizing metals for energy while utilizing organic carbon for biomass generation (Kletzin et al., 2004). Soil isolates from coal piles have yielded microbes that degrade PAHs such as naphthalene and toluene (Margesin and Schinner, 2001). These complex transformations may be done in concert with indigenous fungal systems, where the fungus carries out early steps followed by prokaryotic action.

At high pH, whole cells have been used for remediation of phenolic waste streams. Mixed cultures containing organisms such as *Arthrobacter* sp., *Bacillus cereus*, *Citrobacter fruendii*, *Pseudomonas putida* biovar *B*, and *Micrococcus agilus* have been applied for the cleanup of industrial wastes containing by-products from processes that utilize phenol as a major feedstock (Sarnaik and Kanekar, 1995). Other microbes adapted to high pH (including *Arthrobacter atrocyaneus*, *Bacillus megaterium*, and *Pseudomonas mendocina*) have been applied for the remediation of organophosphorous pesticides (Bhadbhade et al., 2002).

Salty conditions impact the bioavailability of certain contaminants by modulating solubility (Maltseva et al., 1996; Kamekura, 1998; Margesin and Schinner, 2001; Mesbah and Wiegel, 2008; Mesbah et al., 2009; Al-Mailem et al., 2010; de Lourdes Moreno et al., 2011; Al-Mailem et al., 2013). Still, halophilic microbes are capable of degrading organics such as petroleum hydrocarbons found in crude oil sites and chlorinated compounds found in other contaminated sites (Maltseva et al., 1996; Kamekura, 1998; Margesin and Schinner, 2001; Mesbah and Wiegel, 2008; Mesbah et al., 2009; Al-Mailem et al., 2010, 2013; de Lourdes Moreno et al., 2011). Alkalophilic and halophilic microbes such as *Norcardiodes* sp. have been applied to transform chlorinated phenols (Margesin and Schinner, 2001). Organisms such as *Streptomyces albaxialis and Halobacterium* spp. have been applied for the transformation of petroleum hydrocarbons. In hypersaline wastewaters, biofilm and biofilter reactors have been applied for remediation of phenols, PAHs, and halogenated hydrocarbons (Lefebvre et al., 2005; Bassin et al., 2011). Additional species that have been relevant in breaking down halogenated hydrocarbons such as trichloroethylene include *Halobacterium* sp. and *Methylmicrobium* sp. (Margesin and Schinner, 2001). Regarding soils, halophiles have a niche in the remediation of salt marshes contaminated with crude oil (Shin et al., 2001; Garcia-Blanco et al., 2007).

10.5 Evolution and Engineering of Extremophilic Character in Remediation Systems

Organisms evolve to thrive in extreme environments through persistence, adaptation, and specialization. Species that are tolerant to a wide variety of conditions are well distributed in the biosphere and serve as a reservoir for evolution and adaptation to more specialized genotypes. Species that persist show "phenotypic plasticity" and survive under a broader range of conditions. When a persistent species is exposed to a stable environment at the periphery of its growth window, mutations and adaptations can facilitate the fixation of traits that are beneficial in a particular niche. Further selective pressures can shift the optimal conditions for growth toward specialized extremes.

Wolfe-Simon et al. (2011) reported on the ability of gamma-proteobacterium strain GFAJ-1 to grow in the presence of arsenate. This organism was isolated from Mono Lake, California, which is known to be contaminated with arsenic. Based on batch growth data and XAFS and XANES data, this group concluded that the strain is arseno-tolerant and can incorporate arsenic (As) into biomolecules as a substitute for phosphorus (P). These results are controversial and have been refuted based on the redox chemistry of arsenate. Schoepp-Cothenet et al. suggest that the organism's incorporation of As into biologically active molecules is not a proved phenomenon. Arsenate resembles phosphate but is not redox active in the cytoplasm of prokaryotes (Schoepp-Cothenet et al., 2011). The authors agree that the organisms may adapt to benefit from the arsenite/arsenate redox couple for energy, but not for incorporation into biomass.

One synergy between microbial existence and the presence of petroleum in deep rock formations is the potential to stimulate the indigenous species to produce metabolites that enhance oil recovery. Analyses of cultivable microorganisms from oil reservoirs reveal that extremophilic species are present. Along with these anaerobic, thermophilic, and fermentative bacteria are metabolites including CO_2, CH_4, ethanol, acetone, acetate, and biosurfactants. Sampling of heavy oil producing reservoirs composed of carbonate rocks led to the identification of *Thermoanaerobacter*. At the depth of oil recovery, 1250 m, the temperature and pressure are 70°C and 14.2 MPa, respectively. Heavy oil properties and characteristics of the reservoir carbonate hinder oil recovery from the Cordoba platform (Castorena-Cortes et al., 2011). The authors posit that metabolic products from fermentative organisms can improve residual oil recovery due to better rock–oil interactions and potentially modified oil rheology.

Many efforts to enhance biocatalytic capacity and stability through manipulation of microbial systems have been undertaken (Ang et al., 2005; Le Borgne et al.,

2008; Kumar et al., 2008; Singh et al., 2008; Kim et al., 2009; Littlechild, 2011). These efforts go beyond probing and enrichment of existing organisms to the manipulation of metabolic pathways of mesophilic organisms in the direction of extremophily (Bizukojc et al., 2010; Dunlop et al., 2011; Foo and Leong, 2013) or the manipulation of extremophilic organisms to enhance remediation capacity (Tang et al., 2009; Ding and Lai, 2010; Parrilli et al., 2010; Bressan et al., 2013). Efforts to develop microbial systems that are more robust in challenging environments may create novel catalysts for the remediation of industrial process wastes and the generation of biofuels from waste materials. Continued advancement of molecular tools will facilitate transformation enhancements for bioremediation under extreme conditions.

References

Al-Mailem, D.M., Sorkhoh, N.A., Al-Awadhi, H., Eliyas, M., Radwan, S.S., 2010. Biodegradation of crude oil and pure hydrocarbons by extreme halophilic archaea from hypersaline coasts of the Arabian gulf. Extremophiles. 14 (3), 321–328.

Al-Mailem, D.M., Eliyas, M., Radwan, S.S., 2012. Enhanced haloarchaeal oil removal in hypersaline environments via organic nitrogen fertilization and illumination. Extremophiles. 16 (5), 751–758.

Al-Mailem, D.M., Eliyas, M., Radwan, S.S., 2013. Oil-bioremediation potential of two hydrocarbonoclastic, diazotrophic marinobacter strains from hypersaline areas along the Arabian gulf coasts. Extremophiles. 17 (3), 463–470.

Ang, E.L., Zhao, H.M., Obbard, J.P., 2005. Recent advances in the bioremediation of persistent organic pollutants via biomolecular engineering. Enzyme Microb. Technol. 37 (5), 487–496.

Antranikian, G., Ruepp, A., Gordon, P.M.K., Ballschmiter, M., Zibat, A., Stark, M., et al., 2009. Rapid access to genes of biotechnologically useful enzymes by partial genome sequencing: The thermoalkaliphile *Anaerobranca gottschalkii*. J. Mol. Microbiol. Biotechnol. 16, 81–90.

Bartlett, D., Chi, E., 1994. Genetic-characterization of omph mutants in the deep-sea bacterium *Photobacterium* sp. strain Ss9. Arch. Microbiol. 162 (5), 323–328.

Bartlett, D.H., 2009. Microbial life in the trenches of deep-trench microbiology and the characteristics of microbial life in the trenches. Mar. Technol. Soc. J. 43 (5), 128–131.

Bartolucci, S., Contursi, P., Fiorentino, G., Limauro, D., Pedone, E., 2013. Responding to toxic compounds: a genomic and functional overview of archaea. Front. Biosci. Landmark. 18, 165–189.

Bassin, J.P., Dezotti, M., Sant'Anna Jr., G.L., 2011. Nitrification of industrial and domestic saline wastewaters in moving bed biofilm reactor and sequencing batch reactor. J. Hazard. Mater. 185 (1), 242–248.

Bequer Urbano, S., Albarracin, V.H., Ordonez, O.F., Farias, M.E., Alvarez, H.M., 2013. Lipid storage in high-altitude Andean lakes extremophiles and its mobilization under stress conditions in rhodococcus sp A5, a UV-resistant actinobacterium. Extremophiles. 17 (2), 217–227.

Bhadbhade, B.J., Sarnaik, S.S., Kanekar, P.P., 2002. Bioremediation of an industrial effluent containing monocrotophos. Curr. Microbiol. 45 (5), 346–349.

Bizukojc, M., Dietz, D., Sun, J., Zeng, A., 2010. Metabolic modelling of syntrophic-like growth of a 1,3-propanediol producer, *Clostridium butyricum*, and a methanogenic archeon, *Methanosarcina mazei*, under anaerobic conditions. Bioprocess Biosyst. Eng. 33 (4), 507–523.

Brakstad, O.G., Bonaunet, K., Nordtug, T., Johansen, O., 2004. Biotransformation and dissolution of petroleum hydrocarbons in natural flowing seawater at low temperature. Biodegradation. 15 (5), 337–346.

Bressan, R.A., Park, H.C., Orsini, F., Oh, D., Dassanayake, M., Inan, G., 2013. Biotechnology for mechanisms that counteract salt stress in extremophile species: a genome-based view. Plant Biotechnol. Rep. 7 (1), 27–37.

Brim, H., Osborne, J.P., Kostandarithes, H.M., Fredrickson, J.K., Wackett, L.P., Daly, M.J., 2006. *Deinococcus radiodurans* engineered for complete toluene degradation facilitates cr(VI) reduction. Microbiology-Sgm. 152, 2469–2477.

Burgess, S.A., Lindsay, D., Flint, S.H., 2010. Thermophilic bacilli and their importance in dairy processing. Int. J. Food Microbiol. 144 (2), 215–225.

Burton, S.G., 2001. Development of bioreactors for application of biocatalysts in biotransformations and bioremediation. Pure Appl. Chem. 73 (1), 77–83.

Canganella, F., Wiegel, J., 2011. Extremophiles: from abyssal to terrestrial ecosystems and possibly beyond. Naturwissenschaften. 98 (4), 253–279.

Cason, E.D., Piater, L.A., van Heerden, E., 2012. Reduction of U(VI) by the deep subsurface bacterium, *Thermus scotoductus* SA-01, and the involvement of the ABC transporter protein. Chemosphere. 86 (6), 572–577.

Castorena-Cortes, G., Zapata-Penasco, I., Roldan-Carrillo, T., Reyes-Avila, J., Mayol-Castillo, M., Roman-Vargas, S., et al., 2011. Evaluation of indigenous anaerobic microorganisms from Mexican carbonate reservoirs with potential MEOR application. J. Petrol. Sci. Eng. 81, 86–93.

Cavicchioli, R., Thomas, T., Curmi, P.M.G., 2000. Cold stress response in archaea. Extremophiles. 4 (6), 321–331.

Chaudhuri, G., Dey, P., Dalal, D., Venu-Babu, P., Thilagaraj, W.R., 2013. A novel approach to precipitation of heavy metals from industrial effluents and single-ion solutions using bacterial alkaline phosphatase. Water Air Soil Pollut. 224 (7), 1625.

Chen, C., Lin, C., Chuankhayan, P., Huang, Y., Hsieh, Y., Huang, T., 2012. Crystal structures of complexes of the branched-chain aminotransferase from *Deinococcus radiodurans* with alpha-ketoisocaproate and L-glutamate suggest the radiation resistance of this enzyme for catalysis. J. Bacteriol. 194 (22), 6206–6216.

Chiu, H.C., Lin, T.L., Wang, J.T., 2007. Identification and characterization of an organic solvent tolerance gene in helicobacter pylori. Helicobacter. 12 (1), 74–81.

Cui, J., Chen, C., Qin, Z., Yu, C., Shen, H., Shen, C., 2012. Biodegradation of organic pollutants by thermophiles and their applications: a review. Ying Yong Sheng Tai Xue Bao = the Journal of Applied Ecology/Zhongguo Sheng Tai Xue Xue Hui, Zhongguo Ke Xue Yuan Shenyang Ying Yong Sheng Tai Yan Jiu Suo Zhu Ban. 23 (11), 3218–3226.

de Lourdes Moreno, M., Sanchez-Porro, C., Piubeli, F., Frias, L., Teresa Garcia, M., Mellado, E., 2011. Cloning, characterization and analysis of cat and ben genes from the phenol degrading halophilic bacterium *Halomonas organivorans*. Plos One. 6 (6), e21049.

Dhaker, A.S., Marwah, R., Damodar, R., Gupta, D., Gautam, H.K., Sultana, S., et al., 2011. *In vitro* evaluation of antioxidant and radioprotective properties of a novel extremophile from mud volcano: Implications for management of radiation emergencies. Mol. Cell. Biochem. 353, 243–250.

Ding, J., Lai, M., 2010. The biotechnological potential of the extreme halophilic archaea Haloterrigena sp H13 in xenobiotic metabolism using a comparative genomics approach. Environ. Technol. 31, 905–914.

Dojka, M.A., Hugenholtz, P., Haack, S.K., Pace, N.R., 1998. Microbial diversity in a hydrocarbon- and chlorinated-solvent-contaminated aquifer undergoing intrinsic bioremediation. Appl. Environ. Microbiol. 64 (10), 3869–3877.

Dunlop, M.J., Dossani, Z.Y., Szmidt, H.L., Chu, H.C., Lee, T.S., Keasling, J.D., 2011. Engineering microbial biofuel tolerance and export using efflux pumps. Mol. Syst. Biol. 7, 487.

Erdogmus, S.F., Mutlu, B., Korcan, S.E., Guven, K., Konuk, M., 2013. Aromatic hydrocarbon degradation by halophilic archaea isolated from Camalti Saltern, Turkey. Water Air Soil Pollut. 224 (3), 1449.

Fardeau, M., Ollivier, B., Hirschler-Rea, A., Khelifi, N., Assignee: Paris Cedex, F.R. (Ed.), Institut de Recherche pour le Development (I.R.D.). 2008. Use of thermophilic sulphate-reducing archaea for the implementation of a process for the degradation of hydrocarbons. Unites States Patent No. 8,455,240. Washington, DC: U.S. Patent and Trademark Office.

Fernandes, P., Ferreira, B.S., Cabral, J.M.S., 2003. Solvent tolerance in bacteria: role of efflux pumps and cross-resistance with antibiotics. Int. J. Antimicrob. Agents. 22 (3), 211–216.

Foo, J.L., Leong, S.S.J., 2013. Directed evolution of an *E. coli* inner membrane transporter for improved efflux of biofuel molecules. Biotechnol. Biofuels. 6, 81.

Fucinos, P., Gonzalez, R., Atanes, E., Fernandez Sestelo, A.B., Perez-Guerra, N., Pastrana, L., 2012. Lipases and esterases from extremophiles: overview and case example of the production and purification of an esterase from *Thermus thermophilus* HB27. Lipases Phospholipases: Methods Protoc. 861, 239–266.

Garcia-Blanco, S., Venosa, A.D., Suidan, M.T., Lee, K., Cobanli, S., Haines, J.R., 2007. Biostimulation for the treatment of an oil-contaminated coastal salt marsh. Biodegradation. 18 (1), 1–15.

Gemmell, R.T., Knowles, C.J., 2000. Utilisation of aliphatic compounds by acidophilic heterotrophic bacteria the potential for bioremediation of acidic wastewaters contaminated with toxic organic compounds and heavy metals. FEMS Microbiol. Lett. 192 (2), 185–190.

Gianfreda, L., Rao, M.A., 2004. Potential of extra cellular enzymes in remediation of polluted soils: a review. Enzyme Microb. Technol. 35 (4), 339–354.

Ginet, N., Pardoux, R., Adryanczyk, G., Garcia, D., Brutesco, C., Pignol, D., 2011. Single-step production of a recyclable nanobiocatalyst for organophosphate pesticides biodegradation using functionalized bacterial magnetosomes. Plos One. 6 (6), e21442.

Goh, K.M., Kahar, U.M., Chai, Y.Y., Chong, C.S., Chai, K.P., Ranjani, V., 2013. Recent discoveries and applications of anoxybacillus. Appl. Microbiol. Biotechnol. 97 (4), 1475–1488.

Hallberg, K.B., 2010. New perspectives in acid mine drainage microbiology. Hydrometallurgy. 104 (3–4), 448–453.

Hallmann, C., Schwark, L., Grice, K., 2008. Community dynamics of anaerobic bacteria in deep petroleum reservoirs. Nat. Geosci. 1 (9), 588–591.

Harada, R.M., Yoza, B.A., Masutani, S.M., Li, Q.X., 2013. Diversity of archaea communities within contaminated sand samples from Johnston atoll. Bioremed. J. 17 (3), 182–189.

Hawwa, R., Aikens, J., Turner, R.J., Santarsiero, B.D., Mesecar, A.D., 2009. Structural basis for thermo-stability revealed through the identification and characterization of a highly thermostable phosphotriesterase-like lactonase from *Geobacillus stearothermophilus*. Arch. Biochem. Biophy. 488 (2), 109–120.

Hiblot, J., Gotthard, G., Chabriere, E., Elias, M., 2012. Characterisation of the organophosphate hydrolase catalytic activity of SsoPox. Sci. Rep. 2, 779.

Jiao, N.Z., Sieracki, M.E., Zhang, Y., Du, H.L., 2003. Aerobic anoxygenic phototrophic bacteria and their roles in marine ecosystems. Chinese Sci. Bull. 48 (11), 1064–1068.

Johnson, D.B., 2012. Reductive dissolution of minerals and selective recovery of metals using acidophilic iron- and sulfate-reducing acidophiles. Hydrometallurgy. 127, 172–177.

Johnson, J.E., Hill, R.T., 2003. Sediment microbes of deep-sea bioherms on the northwest shelf of Australia. Microb. Ecol. 46 (1), 55–61.

Johnson, R.L., Brillante, S.M., Isabelle, L.M., Houck, J.E., Pankow, J.F., 1985. Migration of chlorophenolic compounds at the chemical waste-disposal site at alkali lake, oregon 2 contaminant distributions, transport, and retardation. Ground Water. 23 (5), 652–666.

Kamekura, M., 1998. Diversity of extremely halophilic bacteria. Extremophiles. 2 (3), 289–295.

Karadzic, I., Masui, A., Zivkovic, L.I., Fujiwara, N., 2006. Purification and characterization of an alkaline lipase from *Pseudomonas aeruginosa* isolated from putrid mineral cutting oil as component of metalworking fluid. J. Biosci. Bioeng. 102 (2), 82–89.

Kieboom, J., Dennis, J.J., de Bont, J.A.M., Zylstra, G.J., 1998. Identification and molecular characterization of an efflux pump involved in *Pseudomonas putida* S12 solvent tolerance. J. Biol. Chem. 273 (1), 85–91.

Kim, J., Crowley, D.E., 2007. Microbial diversity in natural asphalts of the rancho la brea tar pits. Appl. Environ. Microbiol. 73 (14), 4579–4591.

Kim, S., Kweon, O., Cerniglia, C.E., 2009. Proteomic applications to elucidate bacterial aromatic hydrocarbon metabolic pathways. Curr. Opin. Microbiol. 12 (3), 301–309.

Kishino, T., Kobayashi, K., 1995. Relation between toxicity and accumulation of chlorophenols at various pH, and their absorption mechanism in fish. Water Res. 29 (2), 431–442.

Kleikemper, J., Pombo, S.A., Schroth, M.H., Sigler, W.V., Pesaro, M., Zeyer, J., 2005. Activity and diversity of methanogens in a petroleum hydrocarbon-contaminated aquifer. Appl. Environ. Microbiol. 71 (1), 149–158.

Kletzin, A., Urich, T., Muller, F., Bandeiras, T.M., Gomes, C.M., 2004. Dissimilatory oxidation and reduction of elemental sulfur in thermophilic archaea. J. Bioenerg. Biomembr. 36 (1), 77–91.

Kumar, M., Leon, V., Materano, A.D.S., Ilzins, O.A., Luis, L., 2008. Biosurfactant production and hydrocarbon-degradation by halotolerant and thermotolerant *Pseudomonas* sp. World J. Microbiol. Biotechnol. 24 (7), 1047–1057.

Kumar, S., 2010. Engineering cytochrome P450 biocatalysts for biotechnology, medicine and bioremediation. Expert Opin. Drug Metab. Toxicol. 6 (2), 115–131.

Lage, C.A.S., Dalmaso, G.Z.L., Teixeira, L.C.R.S., Bendia, A.G., Paulino-Lima, I.G., Galante, D., 2012. Mini-review: probing the limits of extremophilic life in extraterrestrial environment-simulated experiments. Int. J. Astrobiol. 11 (4), 251–256.

Le Borgne, S., Paniagua, D., Vazquez-Duhalt, R., 2008. Biodegradation of organic pollutants by halophilic bacteria and archaea. J. Mol. Microbiol. Biotechnol. 15 (2–3), 74–92.

Lefebvre, O., Vasudevan, N., Torrijos, M., Thanasekaran, K., Moletta, R., 2005. Halophilic biological treatment of tannery soak liquor in a sequencing batch reactor. Water Res. 39 (8), 1471–1480.

Lettinga, G., Rebac, S., Parshina, S., Nozhevnikova, A., van Lier, J.B., Stams, A.J.M., 1999. High-rate anaerobic treatment of wastewater at low temperatures. Appl. Environ. Microbiol. 65 (4), 1696–1702.

Lin, C., Sheu, D., Lin, T., Kao, C., Grasso, D., 2012. Thermophilic biodegradation of diesel oil in food waste composting processes without bioaugmentation. Environ. Eng. Sci. 29 (2), 117–123.

Littlechild, J.A., 2011. Thermophilic archaeal enzymes and applications in biocatalysis. Biochem. Soc. Trans. 39, 155–158.

Lovley, D.R., Holmes, D.E., Nevin, K.P., 2004. Dissimilatory fe(III) and mn(IV) reduction. Adv. Microb. Physiol. 49, 219–286.

Lu, H.D., Wheeldon, I.R., Banta, S., 2010. Catalytic biomaterials: engineering organophosphate hydrolase to form self-assembling enzymatic hydrogels. Protein Eng. Des. Sel. 23 (7), 559–566.

Lugowski, A.J., Palmateer, G.A., Boose, T.R., Merriman, J.E., 1997. Assignee: Uniroyal Chemical Ltd./Ltee. In: Elmira, C.A. (Ed.), Biodegradation Process for De-Toxifying Liquid Streams. U.S. Patent No. 5,656,169 Washington, DC: U.S. Patent and Trademark Office.

Luque, R., Gonzalez-Domenech, C.M., Llamas, I., Quesada, E., Bejar, V., 2012. Diversity of culturable halophilic archaea isolated from Rambla Salada, Murcia (Spain). Extremophiles. 16 (2), 205–213.

Machuca, A., Aoyama, H., Duran, N., 1998. Production and characterization of thermostable phenol oxidases of the ascomycete *Thermoascus aurantiacus*. Biotechnol. Appl. Biochem. 27, 217–223.

Maltseva, O., McGowan, C., Fulthorpe, R., Oriel, P., 1996. Degradation of 2,4-dichlorophenoxyacetic acid by haloalkaliphilic bacteria. Microbiology-Sgm. 142, 1115–1122.

Mandrich, L., Merone, L., Manco, G., 2010. Hyperthermophilic phosphotriesterases/lactonases for the environment and human health. Environ. Technol. 31, 1115–1127.

Margesin, R., Schinner, F., 1997. Bioremediation of diesel-oil-contaminated alpine soils at low temperatures. Appl. Microbiol. Biotechnol. 47 (4), 462–468.

Margesin, R., Schinner, F., 1998. Low-temperature bioremediation of a waste water contaminated with anionic surfactants and fuel oil. Appl. Microbiol. Biotechnol. 49 (4), 482–486.

Margesin, R., Schinner, F., 2001. Biodegradation and bioremediation of hydrocarbons in extreme environments. Appl. Microbiol. Biotechnol. 56 (5–6), 650–663.

Mesbah, N., Wiegel, J., 2008. Life at Extreme Limits—The Anaerobic Halophilic Alkalithermophiles. New York Academy of Sciences, New York, NY.

Mesbah, N., Cook, G., Wiegel, J., 2009. The halophilic alkalithermophile *Natranaerobius thermophilus* adapts to multiple environmental extremes using a large repertoire of Na plus (K plus)/H plus antiporters. Mol. Microbiol. 74 (2), 270–281.

Miralles, I., Jorge-Villar, S.E., Canton, Y., Domingo, F., 2012. Using a mini-Raman spectrometer to monitor the adaptive strategies of extremophile colonizers in arid deserts: relationships between signal strength, adaptive strategies, solar radiation, and humidity. Astrobiology. 12 (8), 743–753.

Miyazaki, K., 2005. A hyperthermophilic laccase from *Thermus thermophilus* HB27. Extremophiles. 9 (6), 415–425.

Mosher, J.J., Phelps, T.J., Podar, M., Hurt Jr., R.A., Campbell, J.H., Drake, M.M., 2012. Microbial community succession during lactate amendment and electron acceptor limitation reveals a predominance of metal-reducing *Pelosinus* spp. Appl. Environ. Microbiol. 78 (7), 2082–2091.

Najera-Fernandez, C., Zafrilla, B., Jose Bonete, M., Maria Martinez-Espinosa, R., 2012. Role of the denitrifying haloarchaea in the treatment of nitrite-brines. Int. Microbiol. 15 (3), 111–119.

Parrilli, E., Papa, R., Tutino, M.L., Sannia, G., 2010. Engineering of a psychrophilic bacterium for the bioremediation of aromatic compounds. Bioeng. Bugs. 1 (3), 213–216.

Paustian, T., 2012. Through the microscope, adventures in microbiology. Retrieved 09.15.12, from: <http://www.microbiologytext.com>.

Peeples, T., Kelly, R., 1995. Bioenergetic response of the extreme thermoacidophile *Metallosphaera sedula* to thermal and nutritional stresses. Appl. Environ. Microbiol. 61 (6), 2314–2321.

Peeples, T.L., Kelly, R.M., 1993. Bioenergetics of the metal sulfur-oxidizing extreme thermoacidophile, *Metallosphaera sedula*. Fuel. 72 (12), 1619–1624.

Peeples, T.L., Hirosue, S., Olson, G.J., Kelly, R.M., 1991. Coal sulfur transformations monitored by hyperthermophilic archaebacteria. Fuel. 70 (5), 599–604.

Peixoto, R.S., Vermelho, A.B., Rosado, A.S., 2011. Petroleum-degrading enzymes: bioremediation and new prospects. Enzyme Res.475193, pp. 7.

Ramirez, M.G.C., Rivera-Rios, J.M., Tellez-Jurado, A., Maqueda Galvez, A.P., Mercado-Flores, Y., Arana-Cuenca, A., 2012. Screening for thermotolerant ligninolytic fungi with laccase, lipase, and protease activity isolated in Mexico. J. Environ. Manage. 95, S256–S259.

Raynes, J.K., Pearce, F.G., Meade, S.J., Gerrard, J.A., 2011. Immobilization of organophosphate hydrolase on an amyloid fibril nanoscaffold: towards bioremediation and chemical detoxification. Biotechnol. Prog. 27 (2), 360–367.

Rhodes, M.E., Spear, J.R., Oren, A., House, C.H., 2011. Differences in lateral gene transfer in hypersaline versus thermal environments. BMC Evol. Biol. 11, 199.

Richardson, R.E., James, C.A., Bhupathiraju, V.K., Alvarez-Cohen, L., 2002. Microbial activity in soils following steam treatment. Biodegradation. 13 (4), 285–295.

Rocchetti, L., Beolchini, F., Hallberg, K.B., Johnson, D.B., Dell'Anno, A., 2012. Effects of prokaryotic diversity changes on hydrocarbon degradation rates and metal partitioning during bioremediation of contaminated anoxic marine sediments. Mar. Pollut. Bull. 64 (8), 1688–1698.

Rojas, A., Duque, E., Mosqueda, G., Golden, G., Hurtado, A., Ramos, J.L., 2001. Three efflux pumps are required to provide efficient tolerance to toluene in *Pseudomonas putida* DOT-T1E. J. Bacteriol. 183 (13), 3967–3973.

Sardessai, Y., Bhosle, S., 2002. Tolerance of bacteria to organic solvents. Res. Microbiol. 153 (5), 263–268.

Sarnaik, S., Kanekar, P., 1995. Bioremediation of color of methyl violet and phenol from a dye-industry waste effluent using *Pseudomonas* spp. isolated from factory soil. J. Appl. Bacteriol. 79 (4), 459–469.

Schoepp-Cothenet, B., Nitschke, W., Barge, L.M., Ponce, A., Russell, M.J., Tsapin, A.I., 2011. Comment on "A bacterium that can grow by using arsenic instead of phosphorus". Science (New York, N.Y.). 332 (6034), 1149-d.

Sei, A., Fathepure, B.Z., 2009. Biodegradation of BTEX at high salinity by an enrichment culture from hypersaline sediments of rozel point at great salt lake. J. Appl. Microbiol. 107 (6), 2001–2008.

Shin, W.S., Pardue, J.H., Jackson, W.A., Choi, S.J., 2001. Nutrient enhanced biodegradation of crude oil in tropical salt marshes. Water Air Soil Pollut. 131 (1–4), 135–152.

Simonato, F., Campanaro, S., Lauro, F.M., Vezzi, A., D'Angelo, M., Vitulo, N., 2006. Piezophilic adaptation: a genomic point of view. J. Biotechnol. 126 (1), 11–25.

Singh, P., Raghukumar, C., Parvatkar, R.R., Mascarenhas-Pereira, M.B.L., 2013. Heavy metal tolerance in the psychrotolerant *Cryptococcus* sp. isolated from deep-sea sediments of the central Indian basin. Yeast. 30 (3), 93–101.

Singh, S., Kang, S.H., Mulchandani, A., Chen, W., 2008. Bioremediation: environmental clean-up through pathway engineering. Curr. Opin. Biotechnol. 19 (5), 437–444.

Srikumar, R., Kon, T., Gotoh, N., Poole, K., 1998. Expression of *Pseudomonas aeruginosa* multidrug efflux pumps MexA–MexB–OprM and MexC–MexD–OprJ in a multidrug-sensitive *Escherichia coli* strain. Antimicrob. Agents Chemother. 42 (1), 65–71.

Stierle, A.A., Stierle, D.B., 2005. Bioprospecting in the Berkeley pit: bioactive metabolites from acid mine waste extremophiles. Bioact. Nat. Prod. (Pt L). 32, 1123–1175.

Tang, Y.J., Sapra, R., Joyner, D., Hazen, T.C., Myers, S., Reichmuth, D., 2009. Analysis of metabolic pathways and fluxes in a newly discovered thermophilic and ethanol-tolerant *Geobacillus* strain. Biotechnol. Bioeng. 102 (5), 1377–1386.

Taylor, R.T., Jackson, K.J., Duba, A.G., Chen, C., 1988. Assignee: The Regents of the University of California (Ed.), In situ thermally enhanced biodegradation of petroleum fuel hydrocarbons and halogenated organic solvents. U.S. Patent No. 5,753,122 Washington, DC: U.S. Patent and Trademark Office.

Tor, J.M., Lovley, D.R., 2001. Anaerobic degradation of aromatic compounds coupled to fe (III) reduction by *Ferroglobus placidus*. Environ. Microbiol. 3 (4), 281–287.

Wan, R., Zhang, S., Xie, S., 2012. Microbial community changes in aquifer sediment microcosm for anaerobic anthracene biodegradation under methanogenic condition. J. Environ. Sci. China. 24 (8), 1498–1503.

Willey, J.M., Sherwood, L., Woolverton, C.J., Prescott, L.M., Willey, J.M., 2011. Prescott's Microbiology/Joanne M. Willey, Linda M. Sherwood, Christopher J. Woolverton. McGraw-Hill, New York, NY.

Wolfe Simon, F., 2011. Response to comments on "A bacterium that can grow using arsenic instead of phosphorus". Science. 332 (6034), 1149-j.

Zahradka, K., Slade, D., Bailone, A., Sommer, S., Averbeck, D., Petranovic, M., 2006. Reassembly of shattered chromosomes in *Deinococcus radiodurans*. Nature. 443 (7111), 569–573.

Zhang, D., Moertelmaier, C., Margesin, R., 2012a. Characterization of the bacterial archaeal diversity in hydrocarbon-contaminated soil. Sci. Total Environ. 421, 184–196.

Zhang, J., Zhang, X., Liu, J., Li, R., Shen, B., 2012b. Isolation of a thermophilic bacterium, *Geobacillus* sp. SH-1, capable of degrading aliphatic hydrocarbons and naphthalene simultaneously, and identification of its naphthalene degrading pathway. Bioresour. Technol. 124, 83–89.

11 Role of Actinobacteria in Bioremediation

Marta A. Polti[a,b], *Juan Daniel Aparicio*[a], *Claudia S. Benimeli*[c,d,e] *and María Julia Amoroso*[a,b,e]

[a]Planta Piloto de Procesos Industriales y Microbiológicos (PROIMI), CONICET. Av. Belgrano y Pasaje Caseros. 4000. Tucumán, Argentina, [b]Universidad Nacional de Tucumán (UNT). 4000. Tucumán, Argentina, [c]Centro Científico Tecnológico (CCT). Tucumán, Argentina, [d]Universidad Nacional de Catamarca (UNCA). 4700. Catamarca, Argentina, [e]Universidad del Norte Santo Tomás de Aquino (UNSTA). 4000. Tucumán, Argentina

11.1 Introduction

The class Actinobacteria represents an important component of the microbial population in soils. Their metabolic diversity and specific growth characteristics make them well suited as agents for bioremediation (Benimeli, 2004; Albarracin et al., 2005; Benimeli et al., 2007a; Polti et al., 2007; Benimeli et al., 2008; Polti et al., 2009; Albarracin et al., 2010a,b; Fuentes et al., 2011; Alvarez et al., 2012a,b; Cuozzo et al., 2012; Saez et al., 2012).

Actinobacteria are a ubiquitous group of microorganisms widely distributed in various natural ecosystems (Ghanem et al., 2000). They are Gram-positive bacteria, historically known as high guanine plus cytosine (G + C) content base in DNA (55–75 mol%); however, new species are known that do not respond to this rule (Ventura et al., 2007). This group is very heterogeneous, since it includes a broad spectrum of microorganisms that are chemically, morphologically, and physiologically very different (Goodfellow, 1989). This morphological diversity is manifested by a continuous transition from coca and bacilliform cells to hyphae that fragment and branch, forming aerial mycelium with long chains of spores (Vobis, 1997). Actinospores are formed as a result of nutrient depletion and can survive prolonged desiccation (McCarthy and Williams, 1992). This ability to sporulate is very important for their survival in the environment.

In general, optimal conditions for growth are temperatures of 25–30°C and neutral pH, but many species have been isolated from extreme environments. These bacteria are, in most cases, aerobic, although some may be microaerophilic or anaerobic. They are heterotrophic, meaning that both simple and complex carbon

Microbial Biodegradation and Bioremediation. DOI: http://dx.doi.org/10.1016/B978-0-12-800021-2.00011-X

sources can be used (Leveau et al., 2000). Physiological diversity is manifested by the production of a large number of biotechnologically important metabolites (antibiotics, enzymes, enzyme inhibitors, immunomodulators, etc.) (Goodfellow, 1989; Ensign, 1990). Also, within their particular characteristics, they have a characteristic odor of wet earth, due to the production of a metabolite called geosmin (Amoroso et al., 2013).

11.2 Actinobacteria: Growth and Reproduction

This diverse group presents a large morphological differentiation, characterized by a filamentous-type organization (Vobis, 1997), that includes septate and nonseptate multicellular filaments. In solid culture media, strains generally form compact colonies consisting of mycelium, a mass of hyphae belonging to the microorganism, and differentiating into aerial and substrate mycelium (Margulis and Chapman, 2009).

The growth rate of Actinobacteria is generally very slow. At the first stage, a branched mycelium grows on the surface of a solid culture medium, and it can be observed with a microscope after 24 h of incubation; colonies become visible after 3–4 days, but the mature aerial mycelium with actinospores appears after 7–14 days. Some slow-growing strains may require more than 1 month of incubation (Cross, 1989).

The composition of the culture media can significantly affect the growth and stability of the substrate and aerial mycelium. Actinobacteria colonies can be raised or flat. Their consistency varies from very soft to extremely hard and pasty. The range of colors includes white, yellow, orange, pink, red, purple, blue, green, brown, and black. Their surfaces may be smooth, grooved, wrinkled, granular, or flaky. Their appearance can be completely compact or have different growing areas in concentric rings of radial orientation, or a combination of both. The colony size depends on the species, age, and culture conditions and varies from one millimeter to a few centimeters in diameter (Vobis, 1997). Actinobacteria can also grow in liquid media, but need to be grown in special conditions. Liquid cultures require agitation and aeration to obtain a uniform growth, and suspension in culture medium (Cross, 1989).

Actinobacteria are an abundant microbial population in soil, which are generally present in numbers $5-6 \times 10^{10}$ CFU/g of soil. Actinobacteria are generally present in the soil as dormant spores that develop their mycelia only when certain environmental conditions, such as nutrient supply, humidity, temperature, or physiological interactions with other microorganisms, are favorable (Vobis, 1992).

Due to their metabolic diversity and association with the environment, these organisms are studied for biotechnological purposes, especially in bioremediation of toxic substances. For more than 20 years, the interaction between Actinobacteria and accumulated substances in the environment, such as oil, rubber, plastics, pesticides, and heavy metals, among others, was studied (Goodfellow et al., 1988; Vobis, 1997; Amoroso et al., 1998; Amoroso et al., 2001; Amoroso et al., 2002; Benimeli et al., 2003; Benimeli, 2004; Albarracin et al., 2005; Polti et al., 2007).

11.3 Role of Actinobacteria in the Removal of Xenobiotics

Actinobacteria possess many properties that make them good candidates for use in bioremediation of soils contaminated with organic contaminants. They produce extracellular enzymes, which metabolize a variety of compounds and spores resistant to desiccation. Furthermore, filamentous growth favors colonization of soil particles (Amoroso et al., 2013).

The systematic use of pesticides has led to a great improvement in the quality of life of humans. However, the massive and indiscriminate application of these products led to several adverse effects on human health, the environment, and even on the effectiveness of the product (Johri et al., 2000; Phillips et al., 2001). These compounds have structures that are not present (or are very rare) in nature, and so have been called xenobiotics. These are usually recalcitrants, that is, their persistence in the environment is high (Shukla and Kulshretha, 1998). However, there are microorganisms able to carry out their degradation and several metabolic pathways have been described (Table 11.1). The Food and Agriculture

Table 11.1 Reference for Biodegradation Pathways of Selected Pesticides

Pesticide	Proposed Degradation Pathway	References
Lindane	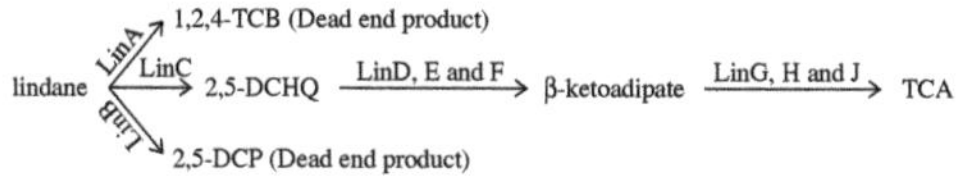	Nagata et al. (2007)
Chlordane	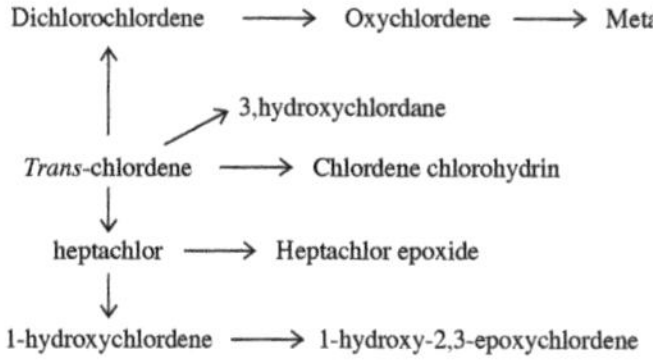	Xiao et al. (2011c)
Dieldrin, aldrin	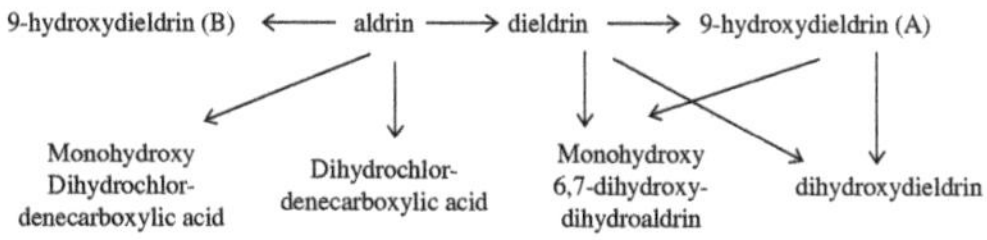	Xiao et al. (2011a)
Heptachlor	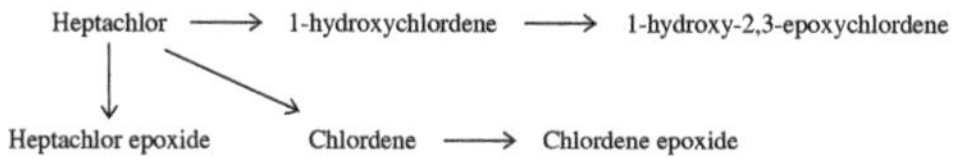	Xiao et al. (2011b)
Methoxychlor	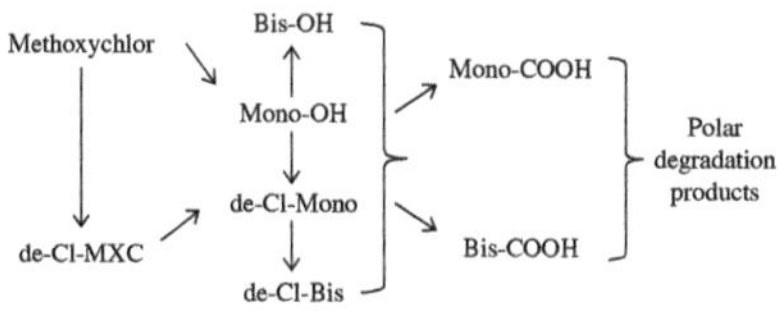	Lee et al. (2006) and Keum et al. (2009)
DDT, DDD	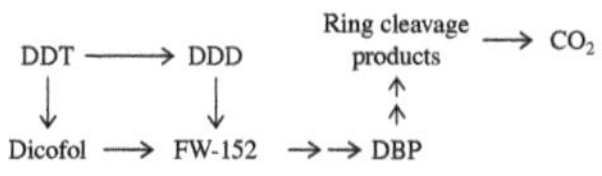	Bumpus and Aust (1987) and Purnomo et al. (2008)

Organization of the United Nations defines a pesticide as "any substance or mixture of substances intended for preventing, destroying or controlling any pest, including vectors of human or animal disease, unwanted species of plants or animals causing harm during or otherwise interfering with the production, processing, storage, transport or marketing of food, agricultural commodities, wood and wood products or animal feedstuffs, or substances that may be administered to animals for the control of insects, arachnids or other pests in or on their bodies." (FAOUN, 2002). The purpose of pesticides is to destroy certain living organisms. They constitute a particular group of biocides that can reach a wide lethality. The use of pesticide has been extensive, which explains its ubiquity (Wan et al., 2005; Amaraneni, 2006; Mansour et al., 2009; Matsumoto et al., 2009; Arias et al., 2011; Mansour, 2012; Shi et al., 2013). Agriculture consumes 85% of world production of these products to chemically control those pests which reduce the quantity and quality of crops (Al-Saleh and Obuekwe, 2005). About 10% of pesticides are used in public health to control vector-borne diseases such as malaria, dengue and Chagas disease, among other (Zaim and Guillet, 2002). They are also used to control pests in large structures such as malls, buildings, airplanes, trains, and boats. They are applied in ornamental landscaping and recreation areas such as parks and gardens, to control the proliferation of insects and fungi and the growth of grass and weeds. For the same purpose, pesticides are scattered along highways, railways, and between high-voltage power line towers (Zaim and Guillet, 2002).

Organochlorine pesticides (PO) are a group of synthetic organic compounds, produced from hydrocarbons, in which one or more hydrogen atoms are replaced by chlorine atoms, and may contain other elements such as oxygen and sulfur (Arias et al., 2011). The OPs were introduced to the market in the 1940s and 1950s. Initially they had important roles in the control of pests and vector-borne diseases. However, in the 1960s, evidence began to accumulate about their undesirable effects (Vega et al., 2007). Today, these pesticides arouse global concern and their application has been discontinued due to their toxicity, long persistence, low biodegradability, wide distribution in the environment, and chronic effects on wildlife and humans. Furthermore, the combination of physicochemical properties makes them capable of traveling long distances (O'Shea and Brownell, 1994; Shea and Brownell, 1994; Benimeli et al., 2003; Turgut, 2003; Wan et al., 2005; Tao et al., 2005; Benimeli et al., 2007a; Vega et al., 2007; Malik et al., 2009; Mansour et al., 2009; Mdegela et al., 2009; Yatawara et al., 2010; Arias et al., 2011; Aksu and Taskın, 2012; Arienzo et al., 2013; Aydin et al., 2013; Orton et al., 2013; Zhang et al., 2013a). Evidence of this is their presence in remote places like the Arctic and Antarctica where they were never used (Halsall, 2004; Bogillo and Bazylevska, 2008; Weber et al., 2010; Zhang et al., 2013b). Their persistence in soil is dependent on the nature of the soil, as well as on the physicochemical properties of the compound. Soils high in organic matter tend to retain both soluble pesticides (because of their high water storage capacity) and hydrophobic pesticides (due to their high specific surface area and other properties that favor adsorption) (Vega et al., 2007). However, their high organic content can also increase activity and improve microbial biodegradation of these compounds (Phillips et al., 2005).

Currently, it is well established that these compounds, with their high lipid solubility, have adverse effects on organisms. They accumulate mainly in the fatty tissues of those subjected to repeated or chronic exposures and pass through the food chain to higher animals, including humans, causing a variety of acute and chronic pathological effects (Vega et al., 2007). In view of the above, the United Nations Environment Programme (UNEP) considers OPs as persistent organic pollutants (POPs). In this regard, UNEP is working to implement a comprehensive strategy to eliminate the use of pesticides, due to their accumulation in the biosphere and their toxic effects on organisms other than their original target. Due to environmental and health problems caused by the OPs, the use of aldrin, chlordane, DDT, dalapon, dieldrin, endosulfan and lindane has been restricted or banned. Pesticide residues in soils and their subsequent movements in the water–soil system are key aspects of their environmental behavior. Consequently, reliable, cost-effective method(s) for remediation of OPs are needed in order to minimize contamination of sites. This approach is dependent upon inoculation with microorganisms that are capable of degrading contaminant compounds (Benimeli et al., 2003). Several studies were carried out using Actinobacteria. In this context, Benimeli et al. (2003) have isolated 93 Actinobacteria strains, and these were tested against aldrin, chlordane, DDD, DDE, DDT, dieldrin, heptachlor, heptachlor epoxide A and B, lindane, and methoxychlor. These authors reported that 62% of the strains were tolerant to pesticides tested. Particularly, four multitolerant strains, developed in a synthetic medium with aldrin, were able to remove up to 91% of aldrin (initial concentration: 36 μg/L), after 96 h at 30°C. Moreover, *Streptomyces* sp. M7 growth with 48 μg/L aldrin, removed more than 90% of aldrin after 72 h, and in the stationary growth phase less than 2.5 μg/L aldrin was detected in the medium. Continuing on these studies, Benimeli et al. (2007a) demonstrated that *Streptomyces* sp. M7 is able to grow in the presence of lindane as the only carbon source. This was the first report of lindane (γ-hexachlorocyclohexane) degradation without intracellular accumulation of the pesticide using an Actinobacteria. *Streptomyces* sp. M7 also showed an ability to remove lindane in a synthetic medium with glucose 0.6 g/L and lindane 100 μg/L, with a typical diauxic growth curve: glucose was the preferred substrate until 24 h. At 48 h, when the carbohydrate was depleted, the microorganism consumed the pesticide as a carbon source. Simultaneously, Cl^- released was detected from inoculated medium supplemented with lindane, suggesting that the pesticide was degraded by *Streptomyces* sp. M7 under aerobic conditions (Benimeli et al., 2006).

In an attempt to transfer the promising laboratory studies to the field, Benimeli et al. (2007b) carried out assays to study lindane removal, using soil extract (SE) as a culture medium, by *Streptomyces* sp. M7, and using different culture conditions: pH (5, 7, and 9) and temperature (25°C, 30°C, and 35°C). *Streptomyces* sp. M7 was cultured in SE supplemented with lindane 100 μg/L. The highest growth and pesticide removal (70.4%) was observed at pH 7 after 4 weeks of incubation. Maximum removal (approximately 70%) was observed at 30°C; although the optimal growth temperature was 25°C. Despite the fact that SE has a limited quantity of nutrients, *Streptomyces* sp. M7 was able to grow and remove lindane, showing its potential for bioremediation (Benimeli et al., 2007b).

Later, *Streptomyces* sp. M7 was used to bioremediate soil samples contaminated with lindane (Benimeli et al., 2008), and the pesticide effects on maize plants were evaluated before and after this process. Different initial pesticide concentrations (100, 150, 200, and 300 μg/kg) were added to sterile soil, microbial growth was similar to the control without lindane, and a decrease of residual lindane concentration was observed (29.1%, 78.03%, 38.81%, and 14.42%, respectively). The optimum *Streptomyces* sp M7 inoculum was 2 g/kg soil, obtaining the most efficient bioremediation process (56% removal of lindane, 100 μg/kg). On the other hand, lindane concentrations of 100 μg/kg soil affected the germination and vigor index of maize plants seeded in contaminated soils without *Streptomyces* sp. M7. When this microorganism was inoculated at the same conditions, a better vigor index was observed and lindane removal reached 68% (Benimeli et al., 2008).

Co-metabolism may be a plant-microbe interaction important to bioremediation. In this sense, maize root exudates (REs), as a primary carbon and energy source, showed potential to enhance lindane removed by *Streptomyces* sp. M7 and *Streptomyces* sp. A5, in comparison with minimal medium, using glucose as a carbon and energy source. Phytostimulation of OP-degrading Actinobacteria by maize REs are therefore likely to be a successful strategy for remediation of lindane-contaminated environments (Alvarez et al., 2012a,b).

In nature, microorganisms form microbial consortia, which carry out complex chemical processes and physiological functions in order to enable survival of the community. Microbial consortia can combine the catalytic abilities of different species to metabolize new substrates, including pesticides (Smith et al., 2005; Yang et al., 2010). Fuentes et al. (2011) compared lindane degradation by pure and mixed cultures of *Streptomyces* sp. Consortia formed by two to six microorganisms were assayed for lindane degradation and dechlorinase activity evaluation. Specific dechlorinase activity was improved by mixed cultures until 12 times, in comparison with pure cultures. Consortium formed by two, three, and four strains showed maximum lindane removal, whereas combinations of five and six strains did not efficiently remove the pesticide from the culture medium. In particular, consortium formed by *Streptomyces* sp. A2, A5, M7, and A11 presented the lowest ratio between residual lindane concentration and dechlorinase activity, indicating that it could be a promising tool for lindane biodegradation (Fuentes et al., 2011).

Immobilized cells present several advantages over free suspended cells, such as a higher retention of microorganisms in the reactor, enhanced cellular viability, and protection of cells against toxicity, among others (Poopal and Laxman, 2009). Bioremediation processes mediated by microbial and enzyme immobilization are more efficient, with a higher biodegradation rate (Saez et al., 2012). *Streptomyces* strains were immobilized, as pure culture or consortium, in four different matrices (agar, polyvinyl alcohol, silicone tubes, and cloth sachets). Lindane removal by immobilized microorganisms was significantly higher than in free cells. Moreover, the cells could be reused for two additional cycles of removal, which decreased the cost of the biotechnological process (Saez et al., 2012).

Work to date indicates that *Streptomyces* spp. have the requisite capabilities to degrade lindane, although further studies on enzyme characterization, particularly for Lin enzymes, are needed to develop effective bioremediation technologies.

11.4 Bioremediation of Heavy Metals

Heavy metals are natural components of soil and some of them are micronutrients to plants and animals, because they are used as cofactor by several enzymes. However, industrial revolution has increased pollution of the biosphere by heavy metals, which become toxic at high concentrations.

Several studies are being carried out for bioremediation of metals by Actinobacteria, on zinc (Attwa and El Awady, 2011), boron (Amoroso et al., 2013), nickel (Haferburg et al., 2008, 2009), and cadmium (Siñeriz et al., 2009), among others.

In this chapter, we will focus on the progress of studies for bioremediation of environments contaminated with chromium or copper, using Actinobacteria. Several genera of the class Actinobacteria are involved in bioremediation of these metals (Table 11.2).

11.4.1 Copper Bioremediation

Copper (Cu) is an essential heavy metal that has many known functions in biological systems. However, at elevated concentrations, Cu becomes toxic, showing a

Table 11.2 Actinobacteria Genera Involved on Chromium or Copper Bioremediation

Metal	Actinobacteria Genera	References
Chromium	*Arthrobacter*	Camargo et al. (2004)
	Streptomyces	Amoroso et al. (2001), Laxman and More (2002), Desjardin et al. (2003), Morales et al. (2007), Polti et al. (2007), Achin Vera (2008), and Poopal and Laxman (2008)
	Frankia	Richards et al. (2002)
	Amycolatopsis	Achin Vera (2008)
	Intrasporangium chromatireducens	Liu et al. (2012)
	Flexivirga alba	Sugiyama et al. (2012)
	Friedmanniella antarctica	Schumann et al. (1997)
	Halomonas sp.	Focardi et al. (2012)
	Microbacterium sp.	Pattanapipitpaisal et al. (2001)
Copper	*Amycolatopqsis*	Albarracin et al. (2005, 2008b, 2010a,b
	Streptomyces	Albarracin et al., 2008a
	Kineococcus radiotolerans	Bagwell et al., 2010

dual behavior (Benimeli et al., 2011). Industrial uses of copper have led to their wide distribution in soil, silt, waste, and wastewater with significant environmental problems that require an efficient solution. To remediate the polluted sites, many physicochemical approaches are used; these, however, are inefficient and costly. Thus, biological decontamination methods are preferable for their better efficiency and cost benefits (Benimeli et al., 2011).

Among microbes, copper-resistant Actinobacteria isolated from various polluted areas have been used as potential organisms in bioremediation technologies. Amoroso et al. (1998) have isolated 47 Cu-resistant Actinobacteria from contaminated sediments. Later, Albarracin et al. (2005) isolated 50 Actinobacteria from copper-contaminated and noncontaminated areas. A primary qualitative screening was carried out, and 100% of the isolated strains of the contaminated area were resistant up to 80 mg/L of $CuSO_4$. However, only 19.4% of isolates from the noncontaminated area grew up to that concentration. *Amycolatopsis tucumanensis* DSM 45259 was cultured in a liquid medium with 39 mg/L of Cu. The strain was able to reduce 71.2% of Cu in supernatant, after 6 days of incubation, and pellet acid digestion proved that the metal was accumulated inside the cells. Copper-resistant levels found in strains isolated from copper-contaminated area suggest that these Actinobacteria should have acquired physiological and genetic mechanisms that allow them to survive in adverse environments and that may give them competitive behavior when growing in polluted culture media (Albarracin et al., 2005).

Among the Actinobacteria isolated from the copper-polluted sediments, *Streptomyces* sp. AB2A showed an early mechanism of copper uptake/retention (80% with 0.5 mM of copper until day 3), followed by a drastic metal efflux process (days 5–7). In contrast, *Streptomyces* sp. AB3 and AB5A showed only copper retention phenotypes under the same culture conditions. Particularly, *Streptomyces* sp. AB5A showed a better efficiency in copper removal (94%). *Streptomyces* sp. AB2A showed cupric reductase activity. This was the first report of extracellular cupric reductase activity present in Actinobacteria (Albarracin et al., 2008a).

A. tucumanensis DSM 45259 demonstrated an efficient cupric reductase activity in both copper-adapted and nonadapted cells, under different temperatures of incubation (10–45°C). Preadapted cells displayed values of activity, on average, 65% higher than those obtained from nonadapted cells. In addition, preadaptation significantly improved the rate of Cu(II) reduction (Dávila Costa et al., 2011a).

A. tucumanensis DSM 45259 has shown high copper specific biosorption ability, confirmed by subcellular fractioning assays, which showed that the retained copper was associated with the extracellular fraction (exopolymer, 40%), but mainly within the cells. The copper bioaccumulation ability was also corroborated by using silver-enhanced staining of copper with the Timm's reagent technique. A fragment belonging to a copper P-type ATPase, which could be involved in copper uptake, was amplified using DNA from *A. tucumanensis* DSM 45259 (Albarracin et al., 2008b).

Taking into consideration previous findings, *A. tucumanensis* DSM 45259 was used to bioaugment soil microcosms polluted with copper in order to study the ability of this strain to diminish phytoavailable copper. The strain was capable of colonizing copper polluted soil and produced bioimmobilization of the metal (31% lower amounts of the metal were found in soil solution as compared to nonbioaugmented soil). This result was corroborated by using *Zea mays* as bioindicator: 20% and 17% lower tissue contents of copper were measured in roots and leaves, respectively, confirming the efficiency of the bioremediation process using *A. tucumanensis* DSM 45259 (Albarracin et al., 2010a,b).

Copper is a redox-active metal, which acts as a catalyst in the formation of reactive oxygen species (ROS), encouraging oxidative stress. In stress conditions, cells are forced to increase and expand their antioxidative network. Evidences of an efficient antioxidant defense system to aid microorganisms to survive in copper-stress conditions were given by Dávila Costa et al. (2011b). Biosorbed copper encouraged ROS production in a dose-dependent manner in *A. tucumanensis* DSM 45259. This increase in ROS production from baseline to stress conditions in *A. tucumanensis* DSM 45259 was lower than in a copper sensitive strain, *Amycolatopsis eurytherma*, which also suffered inexorable morphological alteration, whereas *A. tucumanensis* DSM 45259 was not affected. The levels of antioxidant enzymes and metallothioneins were all greater in *A. tucumanensis* DSM 45259 than in *A. eurytherma*. (Dávila Costa et al., 2011b).

The copper resistome of *A. tucumanensis* DSM 45259 was deeply studied. Upon copper-stress conditions, proteins of the central metabolism, energy production, transcriptional regulators, two-component system, antioxidants, and protective metabolites increased their abundance; superoxide dismutase, alkyl hydroperoxide reductase, and mycothiol reductase genes were markedly induced in expression. This work provided evidence that *A. tucumanensis* DSM 45259 has a copper resistome, a group of mechanisms that confer copper resistance to the strain (Dávila Costa et al., 2012). *A. tucumanensis* DSM 45259 could be useful to perform bioremediation process of copper-polluted soils in higher scale, as agricultural fields (Albarracin et al., 2010a,b).

11.4.2 Chromium Bioremediation

Among the heavy metals, chromium is found in nature mostly as chromite (FeO.Cr_2O_3). It exists under different oxidation states, from II to VI. Cr(II) is highly reducing, so it is easily oxidized to Cr(III) by environmental oxygen. Only Cr(III) and Cr(VI) are the stable forms in aquatic and terrestrial environments (Kamaludeen et al., 2003). Chromium's biological activity depends on its valence. Metallic chromium is inert, Cr(III) is an essential micronutrient, and Cr(VI) is a potent carcinogen (Cheung and Gu, 2003; Cefalu and Hu, 2004). The Cr(VI) compounds are approximately 1000 times more toxic and mutagenic than Cr(III) compounds (Dana Devi et al., 2001).

Industrial effluents containing chromium compounds are released directly or indirectly into natural water resources, mostly without proper effluent treatment, resulting in anthropogenic contamination of nonindustrial environments (Cheung and Gu, 2007; Benimeli et al., 2011). Biological transformation of Cr(VI) by reduction has been recognized as a means of chromium decontamination from effluents

(Laxman and More, 2002). Many bacteria can reduce Cr(VI) to Cr(III) under aerobic and anaerobic conditions, including Actinobacteria such as *Microbacterium*, *Arthrobacter*, and *Streptomyces* (Pattanapipitpaisal et al., 2001; Laxman and More, 2002) (Figure 11.1).

The first report on Cr(VI) enzymatic reduction by *Streptomyces* was given by Das and Chandra (1990), who found a threefold increase in enzyme activity after induction with Cr(VI). Laxman and More (2002) determined Cr(VI) reduction by *Streptomyces griseus*; the strain was found to grow and completely reduce Cr(VI) in semisynthetic culture media containing 5–60 ppm of Cr(VI). Reduction was faster when addition of chromium was done after 24 h growth rather than at the time of inoculation. For Cr(VI) reduction, an energy source such as glucose/sodium acetate was necessary. The metabolic inhibitor sodium azide partially inhibited the reduction. In this case, the enzyme also seemed to be of a constitutive nature. *S. griseus* immobilization on PVA-alginate increased the Cr(VI) reduction rate significantly, in comparison with free cells. Moreover, Cr(VI) reduction by immobilized cells was not inhibited by the presence of Cu^{2+}, Mg^{2+}, Mn^{2+}, and Zn^{2+} in the culture medium. This represents a biotechnological advantage for liquid effluent treatment (Poopal and Laxman, 2008).

Cr(VI)-resistant Actinobacteria were isolated from polluted and nonpolluted areas: one *Amycolatopsis* and nine *Streptomyces* strains. In general, strains from the polluted area showed higher Cr(VI)-specific removal than those from the nonpolluted area, confirming again a relationship between the environmental conditions and adaptive responses. (Polti et al., 2007). *Streptomyces* sp. MC1, a strain from a polluted area, showed an ability to reduce 100% and 75% of Cr(VI) in

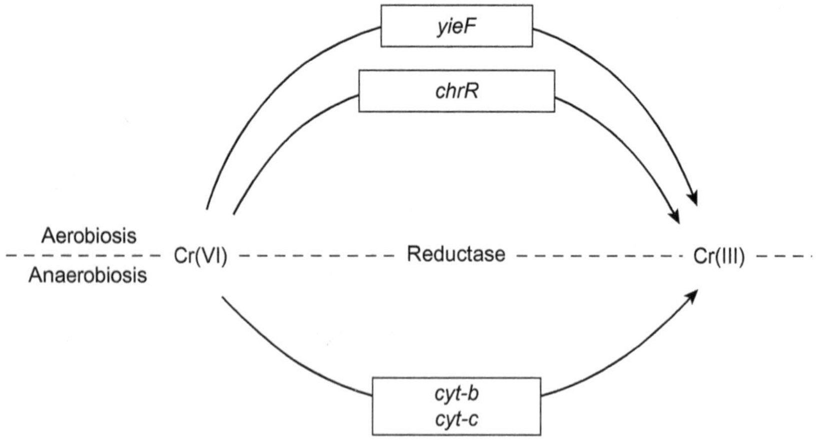

Figure 11.1 Genes involved in Cr(VI) reduction by microorganism under aerobic and anaerobic conditions. *chrR*: chromate reductase gene from *Pseudomonas. putida*; *yieF*: homologue to *chrR*, from *Escherichia coli*, YieF has high chromate reductase activity. *cyt-b* and *cyt-c*: cytochrome family genes. Cytochromes have demonstrated chromate reductase activity. *Source*: Modified from Cheung and Gu, 2007.

liquid minimal medium, with initial Cr(VI) concentrations of 5 and 50 mg/L, respectively, after 48 h of incubation (Polti et al., 2009).

Polti et al. (2010) have found chromate reductase activity in *Streptomyces* sp. MC1. Optimal conditions such as pH, temperature, growth phase, and electron donor were elucidated *in vitro* for Cr(VI) reduction by *Streptomyces* sp. MC1. *Streptomyces* sp. MC1 chromate reductase is a constitutive enzyme, mainly associated with biomass, and requires NAD(P)H as an electron donor. It showed activity over a broad temperature (19–39°C) and pH (5–8) range, with optimum activity at 30°C and pH 7. The activity was observed in supernatant, pellet, and cell-free extract; therefore, bioremediation using this enzyme could be possible in noncompatible cell reproduction systems, conditions frequently found in contaminated environments (Polti et al., 2010).

The ability of *Streptomyces* sp. MC1 to accumulate Cr(III) after the Cr(VI) reduction was also reported. This strain was able to accumulate up to 3.54 mg of Cr(III) per gram of wet biomass, reducing 98% of Cr(VI) and removing 13.9% of chromium from supernatants. Bioaccumulation ability was observed under scanning electron microscopy by using Timm's reagent technique and confirmed by energy-dispersive X-ray analysis. The general strategy for Cr(VI) control observed in *Streptomyces* sp. MC1 could include adsorption coupled with reduction to Cr(III) and, finally, Cr(III) bioaccumulation (Polti et al., 2011a).

This strain was also used in soil samples assays. *Streptomyces* sp. MC1 was able to reduce up to 94% of the bioavailable Cr(VI) from an initial concentration of 50 mg/kg, after 7 days, in comparison with noninoculated soil samples. Taking into account that soil samples were not previously sterilized, the authors inferred that bioremediation activity of *Streptomyces* sp. MC1 was not inhibited by natural soil microbial flora. It is remarkable that the soil samples did not have any previous treatment or addition of any substrate and had a normal soil humidity level. This is very important from a biotechnological perspective because it considerably reduces the process costs (Polti et al., 2009). Later, higher Cr(VI) concentrations were used. Soil samples with Cr(VI) 200 mg/kg were inoculated with *Streptomyces* sp. MC1; bioavailable chromium decreased up to 73%. *Z. mays* plantlets were negatively affected by Cr(VI); therefore, they were used as biomarkers. For that, *Z. mays* seedlings were planted in soil samples contaminated with 200 mg/kg of Cr(VI) and inoculated with *Streptomyces* sp. MC1. *Streptomyces* sp. MC1 stimulated *Z. mays* root growth and produced a decrease in chromium accumulation and an increase of *Z. mays* biomass. *Z. mays* was not only useful as a bioindicator, but it also improved soil bioremediation by *Streptomyces* sp. MC1. Both organisms could be successfully applied together in bioremediation of soils contaminated with Cr(VI) (Polti et al., 2011b).

Besides enzymes, which reduce Cr(VI), other metabolites can indirectly enhance bioremediation processes. In recent years, increasing interest has been shown in the use of bioemulsifiers as washing agents that enhance desorption of soil-bound metals. *Streptomyces* sp. MC1 has shown the ability to produce an emulsifier. Optimal production conditions were determined: culture medium with an initial pH of 8 with phosphate 2.0 g/L and Ca^{+2} 1.0 g/L, with an emulsification index about

3.5 times greater compared to the basal value. Partially purified emulsifier presented high thermostability and partial water solubility. These findings could have promising future prospects for the remediation of organic- and metal-contaminated sites (Colin et al., 2013b).

A. tucumanensis DSM 45259 showed an ability to produce emulsifiers, using different carbon and nitrogen sources. The best specific production was detected using glycerol and urea as carbon and nitrogen substrates, respectively. However, with all of the substrates used during the batch assay, the bioemulsifiers showed high levels of stability at extreme conditions of pH, temperature, and salt concentration. The bioemulsifier obtained was able to mediate Cr(VI) recovery, with the removal percentage doubled compared to that seen when using deionized water. These findings appear promising for the development of remediation technologies for hexavalent chromium compounds based upon direct use of these microbial emulsifiers (Colin et al., 2013a).

11.5 Conclusion

Environmental pollution is an increasing world-wide problem. Technologies must adapt quickly to this situation. Biological remediation, by using living organisms or their products, is a suitable alternative. Actinobacteria have proved to be a powerful tool to perform this process on different matrices, under different culture conditions, as pure cultures or consortia, or as producers of enzymes and emulsifiers. The next step involves the use of this knowledge to field scale.

References

Achin Vera, L.O., 2008. Remocion de Cu(II) y Cr(VI) por actinomycetes aislados de ambientes contaminados con metales pesados en sistemas de bimetal (*in spanish*), In: CIUNT (Ed.), Segundas Jornadas de Jóvenes Investigadores UNT AUGM. Universidad Nacional de Tucuman, San Miguel de Tucuman.

Aksu, A., Taskın, O.S., 2012. Organochlorine residue and toxic metal (Pb, Cd and Cr) levels in the surface sediments of the Marmara Sea and the coast of Istanbul, Turkey. Mar. Pollut. Bull. 64 (5), 1060–1062.

Albarracin, V.H., Amoroso, M.J., Abate, C.M., 2005. Isolation and characterization of indigenous copper-resistant actinomycete strains. Chem. Erde. 65 (S1), S145–S156.

Albarracin, V.H., Avila, A.L., Amoroso, M.J., Abate, C.M., 2008a. Copper removal ability by *Streptomyces* strains with dissimilar growth patterns and endowed with cupric reductase activity. FEMS Microbiol. Lett. 288 (2), 141–148.

Albarracin, V.H., Winik, B., Kothe, E., Amoroso, M.J., Abate, C.M., 2008b. Copper bioaccumulation by the actinobacterium *Amycolatopsis* sp. AB0. J. Basic Microbiol. 48 (5), 323–330.

Albarracin, V.H., Alonso-Vega, P., Trujillo, M.E., Amoroso, M.J., Abate, C.M., 2010a. *Amycolatopsis tucumanensis* sp. nov., a copper-resistant actinobacterium isolated from polluted sediments. Int. J. Syst. Evol. Microbiol. 60, 397–401.

Albarracin, V.H., Amoroso, M.J., Abate, C.M., 2010b. Bioaugmentation of copper polluted soil microcosms with *Amycolatopsis tucumanensis* to diminish phytoavailable copper for *Zea mays* plants. Chemosphere 79 (2), 131–137.

Al-Saleh, E.S., Obuekwe, C., 2005. Inhibition of hydrocarbon bioremediation by lead in a crude oil-contaminated soil. Int. Biodeterior. Biodegrad. 56 (1), 1–7.

Alvarez, A., Benimeli, C., Saez, J., Fuentes, M., Cuozzo, S., Polti, M., et al., 2012a. Bacterial bio-resources for remediation of hexachlorocyclohexane. Int. J. Mol. Sci. 13 (11), 15086–15106.

Alvarez, A., Yanez, M.L., Benimeli, C.S., Amoroso, M.J., 2012b. Maize plants (*Zea mays*) root exudates enhance lindane removal by native *Streptomyces* strains. Int. Biodeterior. Biodegrad. 66 (1), 14–18.

Amaraneni, S.R., 2006. Distribution of pesticides, PAHs and heavy metals in prawn ponds near Kolleru lake wetland, India. Environ. Int. 32 (3), 294–302.

Amoroso, M.J., Castro, G.R., Carlino, F.J., Romero, N.C., Hill, R.T., Oliver, G., 1998. Screening of heavy metal-tolerant actinomycetes isolated from the Sali River. J. Gen. Appl. Microbiol. 44 (2), 129–132.

Amoroso, M.J., Castro, G.R., Duran, A., Peraud, O., Oliver, G., Hill, R.T., 2001. Chromium accumulation by two *Streptomyces* spp. isolated from riverine sediments. J. Ind. Microbiol. Biotechnol. 26 (4), 210–215.

Amoroso, M.J., Oliver, G., Castro, G.R., 2002. Estimation of growth inhibition by copper and cadmium in heavy metal tolerant actinomycetes. J. Basic Microbiol. 42 (4), 231.

Amoroso, M.J., Benimeli, C.S., Cuozzo, S.A., 2013. Actinobacteria: Application in Bioremediation and Production of Industrial Enzymes. CRC Press LLC. Boca Raton, Florida.

Arias, A., Pereyra, M., Marcovecchio, J., 2011. Multi-year monitoring of estuarine sediments as ultimate sink for DDT, HCH, and other organochlorinated pesticides in Argentina. Environ. Monit. Assess. 172 (1–4), 17–32.

Arienzo, M., Masuccio, A.A., Ferrara, L., 2013. Evaluation of sediment contamination by heavy metals, organochlorinated pesticides, and polycyclic aromatic hydrocarbons in the Berre coastal lagoon (Southeast France). Arch. Environ. Contam. Toxicol. 65, 396–406.

Attwa, A.I., El Awady, M.E., 2011. Bioremediation of zinc by *Streptomyces aureofacienes*. J. Appl. Sci. 11 (5), 873–877.

Aydin, M.E., Ozcan, S., Beduk, F., Tor, A., 2013. Levels of organochlorine pesticides and heavy metals in surface waters of Konya closed basin, Turkey. Sci. World J. 2013, 849716.

Bagwell, C.E., Hixson, K.K., Milliken, C.E., Lopez-Ferrer, D., Weitz, K.K., 2010. Proteomic and physiological responses of *Kineococcus radiotolerans* to copper. PLoS One 5, e12427.

Benimeli, C.S., 2004. Biodegradación de plaguicidas organoclorados por actinomycetes acuáticos. Universidad Nacional de Tucuman., San Miguel de Tucumán.

Benimeli, C.S., Amoroso, M.J., Chaile, A.P., Castro, G.R., 2003. Isolation of four aquatic *Streptomycetes* strains capable of growth on organochlorine pesticides. Bioresour. Technol. 89 (2), 133–138.

Benimeli, C.S., Castro, G.R., Chaile, A.P., Amoroso, M.J., 2006. Lindane removal induction by *Streptomyces* sp. M7. J. Basic Microbiol. 46 (5), 348–357.

Benimeli, C.S., Castro, G.R., Chaile, A.P., Amoroso, M.J., 2007a. Lindane uptake and degradation by aquatic *Streptomyces* sp. strain M7. Int. Biodeterior. Biodegrad. 59 (2), 148–155.

Benimeli, C.S., Gonzalez, A.J., Chaile, A.P., Amoroso, M.J., 2007b. Temperature and pH effect on lindane removal by *Streptomyces* sp. M7 in soil extract. J. Basic Microbiol. 47 (6), 468–473.

Benimeli, C.S., Fuentes, M.S., Abate, C.M., Amoroso, M.J., 2008. Bioremediation of lindane-contaminated soil by *Streptomyces* sp. M7 and its effects on *Zea mays* growth. Int. Biodeterior. Biodegrad. 61 (3), 233–239.

Benimeli, C.S., Polti, M.A., Albarracín, V.H., Abate, C.M., Amoroso, M.J., 2011. Bioremediation potential of heavy metal-resistant Actinobacteria and maize plants in polluted soil biomanagement of metal-contaminated soils. In: Khan, M.S., Zaidi, A., Goel, R., Musarrat, J. (Eds.), Biomanagement of Metal-Contaminated Soils. Springer Netherlands, Gewerbestrasse, pp. 459–477.

Bogillo, V., Bazylevska, M., 2008. Variations of organochlorine contaminants in Antarctica. In: Mehmetli, E., Koumanova, B. (Eds.), The Fate of Persistent Organic Pollutants in the Environment. Springer, Netherlands, Istambul, pp. 251–267.

Bumpus, J.A., Aust, S.D., 1987. Biodegradation of DDT [1,1,1-trichloro-2,2-bis(4-chlorophenyl) ethane] by the white rot fungus *Phanerochaete chrysosporium*. Appl. Environ. Microbiol. 53 (9), 2001–2008.

Camargo, F., Bento, F., Okeke, B., Frankenberger, W., 2004. Hexavalent chromium reduction by an actinomycete, *Arthrobacter crystallopoietes* ES 32. Biol. Trace Elem. Res. 97, 183–194.

Cefalu, W.T., Hu, F.B., 2004. Role of chromium in human health and in diabetes. Diabetes Care 27 (11), 2741–2751.

Cheung, K.H., Gu, J.-D., 2003. Reduction of chromate (CrO_4^{2-}) by an enrichment consortium and an isolate of marine sulfate-reducing bacteria. Chemosphere 52 (9), 1523–1529.

Cheung, K.H., Gu, J.D., 2007. Mechanism of hexavalent chromium detoxification by microorganisms and bioremediation application potential: a review. Int. Biodeterior Biodegrad. 59 (1), 8–15.

Colin, V.L., Castro, M.F., Amoroso, M.J., Villegas, L.B., 2013a. Production of bioemulsifiers by *Amycolatopsis tucumanensis* DSM 45259 and their potential application in remediation technologies for soils contaminated with hexavalent chromium. J. Hazard. Mater. 261, 577–583.

Colin, V.L., Pereira, C.E., Villegas, L.B., Amoroso, M.J., Abate, C.M., 2013b. Production and partial characterization of bioemulsifier from a chromium-resistant Actinobacteria. Chemosphere 90 (4), 1372–1378.

Cross, T., 1989. Growth and examination of actinomycetes—some guidelines. In: Williams, S.T. (Ed.), Bergey's Manual of Systematic Bacteriology, vol. 4. Williams & Wilkins, Baltimore, MD, pp. 2340–2343.

Cuozzo, S.A., Fuentes, M.S., Bourguignon, N., Benimeli, C.S., Amoroso, M.J., 2012. Chlordane biodegradation under aerobic conditions by indigenous *Streptomyces* strains. Int. Biodeterior. Biodegrad. 66 (1), 19–24.

Dana Devi, K., Rozati, R., Saleha Banu, B., Jamil, K., Grover, P., 2001. *In vivo* genotoxic effect of potassium dichromate in mice leukocytes using comet assay. Food Chem. Toxicol. 39 (8), 859–865.

Das, S., Chandra, A., 1990. Chromate reduction in *Streptomyces*. Experientia. 46, 731–733.

Dávila Costa, J.S., Albarracín, V.H., Abate, C.M., 2011a. Cupric reductase activity in copper-resistant *Amycolatopsis tucumanensis*. Water Air Soil Pollut. 216 (1–4), 527–535.

Dávila Costa, J.S., Albarracín, V.H., Abate, C.M., 2011b. Responses of environmental *Amycolatopsis* strains to copper stress. Ecotoxicol. Environ. Saf. 74 (7), 2020–2028.

Dávila Costa, J.S., Kothe, E., Abate, C.M., Amoroso, M.J., 2012. Unraveling the *Amycolatopsis tucumanensis* copper-resistome. Biometals 25 (5), 905–917.

Desjardin, V., Bayard, R., Lejeune, P., Gourdon, R., 2003. Utilisation of supernatants of pure cultures of *Streptomyces thermocarboxydus* NH_{50} to reduce chromium toxicity and mobility in contaminated soils. Water Air Soil Pollut. Focus. 3, 153–160.

Ensign, J., 1990. Introduction to the actinomycetes. In: Balows, A., Trüper, H., Dworkin, M., Harder, W., Schleifer, K. (Eds.), The Prokaryotes. A Handbook on the Biology of Bacteria: Ecophysiology, Isolation, Identification, Applications. Springer, New York, NY, pp. 811–815.

Focardi, S., Pepi, M., Landi, G., Gasperini, S., Ruta, M., Di Biasio, P., et al., 2012. Hexavalent chromium reduction by whole cells and cell free extract of the moderate halophilic bacterial strain *Halomonas* sp. TA-04. Int. Biodeterior. Biodegrad. 66, 63–70.

Fuentes, M., Sáez, J., Benimeli, C., Amoroso, M., 2011. Lindane biodegradation by defined consortia of indigenous *Streptomyces* strains. Water Air Soil Pollut. 222 (1), 217–231.

Ghanem, N.B., Sabry, S.A., El-Sherif, Z.M., Abu El-Ela, G.A., 2000. Isolation and enumeration of marine actinomycetes from seawater and sediments in Alexandria. J. Gen. Appl. Microbiol. 46 (3), 105–111.

Goodfellow, M., 1989. Suprageneric identification of actinomycetes. In: Williams, S. (Ed.), Bergey's Manual of Systematic Bacteriology, vol. 4. Williams & Wilkins, Baltimore, MD, pp. 2333–2339.

Goodfellow, M., Williams, S., Mordarski, M., 1988. Actinomycetes in Biotechnology. Academic Press, San Diego, CA.

Haferburg, G., Kloess, G., Schmitz, W., Kothe, E., 2008. "Ni-struvite"—a new biomineral formed by a nickel resistant *Streptomyces acidiscabies*. Chemosphere 72 (3), 517–523.

Haferburg, G., Groth, I., Mollmann, U., Kothe, E., Sattler, I., 2009. Arousing sleeping genes: shifts in secondary metabolism of metal tolerant Actinobacteria under conditions of heavy metal stress. Biometals 22 (2), 225–234.

Halsall, C.J., 2004. Investigating the occurrence of persistent organic pollutants (POPs) in the arctic: their atmospheric behaviour and interaction with the seasonal snow pack. Environ. Pollut. 128 (1–2), 163–175.

Johri, A.K., Dua, M., Saxena, D.M., Sethunathan, N., 2000. Enhanced degradation of hexachlorocyclohexane isomers by *Sphingomonas paucimobilis*. Curr. Microbiol. 41 (5), 309–311.

Kamaludeen, S., Megharaj, M., Juhasz, A., Sethunathan, N., Naidu, R., 2003. Chromium–microorganism interactions in soils: remediation implications. Rev. Environ. Contam. Toxicol. 178, 93–164.

Keum, Y.S., Lee, Y.H., Kim, J.H., 2009. Metabolism of methoxychlor by *Cunninghamella elegans* ATCC36112. J. Agric. Food Chem. 57 (17), 7931–7937.

Laxman, R.S., More, S., 2002. Reduction of hexavalent chromium by *Streptomyces griseus*. Miner. Eng. 15 (11), 831–837.

Lee, S.M., Lee, J.W., Park, K.R., Hong, E.J., Jeung, E.B., Kim, M.K., et al., 2006. Biodegradation of methoxychlor and its metabolites by the white rot fungus *Stereum hirsutum* related to the inactivation of estrogenic activity. J. Environ. Sci. Health B. 41 (4), 385–397.

Leveau, J., Bouix, M., Garcia, F.J.C., 2000. Microbiologia industrial: los microorganismos de interés industrial. Acribia, Editorial Zaragoza.

Liu, H., Wang, H., Wang, G., 2012. *Intrasporangium chromatireducens* sp. nov., a chromate-reducing actinobacterium isolated from manganese mining soil, and emended description of the genus *Intrasporangium*. Int. J. Syst. Evol. Microbiol. 62, 403–408.

Malik, A., Ojha, P., Singh, K., 2009. Levels and distribution of persistent organochlorine pesticide residues in water and sediments of Gomti River (India)—a tributary of the Ganges River. Environ. Monit. Assess. 148 (1–4), 421–435.

Mansour, S., 2012. Evaluation of residual pesticides and heavy metals levels in conventionally and organically farmed potato tubers in Egypt. In: He, Z., Larkin, R., Honeycutt, W. (Eds.), Sustainable Potato Production: Global Case Studies. Springer, Netherlands, Gewerbestrasse, pp. 493–506.

Mansour, S.A., Belal, M.H., Abou-Arab, A.A.K., Gad, M.F., 2009. Monitoring of pesticides and heavy metals in cucumber fruits produced from different farming systems. Chemosphere. 75 (5), 601–609.

Margulis, L., Chapman, M.J., 2009. Kingdoms and Domains: An Illustrated Guide to the Phyla of Life on Earth. Elsevier Ltd, Oxford.

Matsumoto, E., Kawanaka, Y., Yun, S.J., Oyaizu, H., 2009. Bioremediation of the organochlorine pesticides, dieldrin and endrin, and their occurrence in the environment. Appl. Microbiol. Biotechnol. 84 (2), 205–216.

McCarthy, A.J., Williams, S.T., 1992. Actinomycetes as agents of biodegradation in the environment—a review. Gene. 115 (1–2), 189–192.

Mdegela, R., Braathen, M., Pereka, A., Mosha, R., Sandvik, M., Skaare, J., 2009. Heavy metals and organochlorine residues in water, sediments, and fish in aquatic ecosystems in urban and peri-urban areas in Tanzania. Water Air Soil Pollut. 203 (1), 369–379.

Morales, D.K., Ocampo, W., Zambrano, M.M., 2007. Efficient removal of hexavalent chromium by a tolerant *Streptomyces* sp. affected by the toxic effect of metal exposure. J. Appl. Microbiol. 103, 2704–2712.

Nagata, Y., Endo, R., Ito, M., Ohtsubo, Y., Tsuda, M., 2007. Aerobic degradation of lindane (γ-hexachlorocyclohexane) in bacteria and its biochemical and molecular basis. Appl. Microbiol. Biotechnol. 76 (4), 741–752.

Orton, T.G., Saby, N.P.A., Arrouays, D., Jolivet, C.C., Villanneau, E.J., Marchant, B.P., et al., 2013. Spatial distribution of Lindane concentration in topsoil across France. Sci. Total Environ. 443, 338–350.

O'Shea, T.J., Brownell, R.L., 1994. Organochlorine and metal contaminants in baleen whales: a review and evaluation of conservation implications. Sci. Total Environ. 154 (2), 179–200.

Pattanapipitpaisal, P., Brown, N., Macaskie, L., 2001. Chromate reduction and 16S rRNA identification of bacteria isolated from a Cr(VI)-contaminated site. Appl. Microbiol. Biotechnol. 57, 257–261.

Phillips, T.M., Seech, A.G., Lee, H., Trevors, J.T., 2001. Colorimetric assay for Lindane dechlorination by bacteria. J. Microbiol. Methods 47 (2), 181–188.

Phillips, T.M., Seech, A.G., Lee, H., Trevors, J., 2005. Biodegradation of hexachlorocyclohexane (HCH) by microorganisms. Biodegradation 16 (4), 363–392.

Polti, M.A., Amoroso, M.J., Abate, C.M., 2007. Chromium(VI) resistance and removal by actinomycete strains isolated from sediments. Chemosphere 67 (4), 660–667.

Polti, M.A., Garcia, R.O., Amoroso, M.J., Abate, C.M., 2009. Bioremediation of chromium (VI) contaminated soil by *Streptomyces* sp. MC1. J. Basic Microbiol. 49 (3), 285–292.

Polti, M.A., Amoroso, M.J., Abate, C.M., 2010. Chromate reductase activity in *Streptomyces* sp. MC1. J. Gen. Appl. Microbiol. 56 (1), 11–18.

Polti, M., Amoroso, M., Abate, C., 2011a. Intracellular chromium accumulation by *Streptomyces* sp. MC1. Water Air Soil Pollut. 214 (1), 49–57.

Polti, M.A., Atjian, M.C., Amoroso, M.J., Abate, C.M., 2011b. Soil chromium bioremediation: synergic activity of Actinobacteria and plants. Int. Biodeterior. Biodegrad. 65 (8), 1175–1181.

Poopal, A., Laxman, R., 2008. Hexavalent chromate reduction by immobilized *Streptomyces griseus*. Biotechnol. Lett. 30 (6), 1005–1010.

Poopal, A.C., Laxman, R.S., 2009. Studies on biological reduction of chromate by *Streptomyces griseus*. J. Hazard. Mater. 169 (1–3), 539–545.

Purnomo, A.S., Kamei, I., Kondo, R., 2008. Degradation of 1,1,1-trichloro-2,2-bis (4-chlorophenyl) ethane (DDT) by brown-rot fungi. J. Biosci. Bioeng. 105 (6), 614–621.

Richards, J.W., Krumholz, G.D., Chval, M.S., Tisa, L.S., 2002. Heavy metal resistance patterns of *Frankia* strains. Appl. Environ. Microbiol. 68, 923–927.

Saez, J.M., Benimeli, C.S., Amoroso, M.J., 2012. Lindane removal by pure and mixed cultures of immobilized Actinobacteria. Chemosphere 89 (8), 982–987.

Schumann, P., Prauser, H., Rainey, F.A., Stackebrandt, E., Hirsch, P., 1997. *Friedmanniella antarctica* gen. nov., sp. nov., an LL-diaminopimelic acid-containing actinomycete from Antarctic sandstone. Int. J. Syst. Bacteriol. 47, 278–283.

Shea, T.J., Brownell, R.L., 1994. Organochlorine and metal contaminants in baleen whales: a review and evaluation of conservation implications (Vol. 154).

Shi, Y., Lu, Y., Meng, F., Guo, F., Zheng, X., 2013. Occurrence of organic chlorinated pesticides and their ecological effects on soil protozoa in the agricultural soils of North Western Beijing, China. Ecotoxicol. Environ. Saf. 92, 123–128.

Shukla, O., Kulshretha, A., 1998. Pesticides, Man and Biosphere. APH Publishing, New Delhi.

Siñeriz, M.L., Kothe, E., Abate, C.M., 2009. Cadmium biosorption by *Streptomyces* sp. F4 isolated from former uranium mine. J. Basic Microbiol. 49 (S1), S55–S62.

Smith, D., Alvey, S., Crowley, D.E., 2005. Cooperative catabolic pathways within an atrazine-degrading enrichment culture isolated from soil. FEMS Microbiol. Ecol. 53 (2), 265–275.

Sugiyama, T., Sugito, H., Mamiya, K., Suzuki, Y., Ando, K., Ohnuki, T., 2012. Hexavalent chromium reduction by an actinobacterium *Flexivirga alba* ST13T in the family Dermacoccaceae. J. Biosci. Bioeng. 113, 367–371.

Tao, S., Xu, F.L., Wang, X.J., Liu, W.X., Gong, Z.M., Fang, J.Y., et al., 2005. Organochlorine pesticides in agricultural soil and vegetables from Tianjin, China. Environ. Sci. Technol. 39 (8), 2494–2499.

Turgut, C., 2003. The contamination with organochlorine pesticides and heavy metals in surface water in Kucuk Menderes River in Turkey, 2000–2002. Environ. Int. 29 (1), 29–32.

Vega, F.A., Covelo, E.F., Andrade, M.L., 2007. Accidental organochlorine pesticide contamination of soil in Porrino, Spain. J. Environ. Qual. 36 (1), 272.

Ventura, M., Canchaya, C., Tauch, A., Chandra, G., Fitzgerald, G.F., Chater, K.F., et al., 2007. Genomics of Actinobacteria: tracing the evolutionary history of an ancient phylum. Microbiol. Mol. Biol. Rev. 71 (3), 495–548.

Vobis, G., 1992. The genus *Actinoplanes* and related genera. In: Balows, A., Trüper, H.G., Dworkin, M., Harder, W., Schleifer, K. (Eds.), The Prokaryotes. A Handbook on the Biology of Bacteria: Ecophysiology, Isolation, Identification, Applications. Springer, New York, NY, pp. 1030–1060.

Vobis, G., 1997. Morphology of actinomycetes. In: Miyadoh, S. (Ed.), Atlas of Actinomycetes. Asakura Publishing Co., Japan, Tokyo, pp. 180–191.

Wan, M.T., Kuo, J.N., Pasternak, J., 2005. Residues of endosulfan and other selected organochlorine pesticides in farm areas of the lower fraser valley, British Columbia, Canada. J. Environ. Qual. 34 (4), 1186–1193.

Weber, J., Halsall, C.J., Muir, D., Teixeira, C., Small, J., Solomon, K., et al., 2010. Endosulfan, a global pesticide: a review of its fate in the environment and occurrence in the Arctic. Sci. Total Environ. 408 (15), 2966–2984.

Xiao, P., Mori, T., Kamei, I., Kiyota, H., Takagi, K., Kondo, R., 2011a. Novel metabolic pathways of organochlorine pesticides dieldrin and aldrin by the white rot fungi of the genus *Phlebia*. Chemosphere 85 (2), 218–224.

Xiao, P., Mori, T., Kamei, I., Kondo, R., 2011b. Metabolism of organochlorine pesticide heptachlor and its metabolite heptachlor epoxide by white rot fungi, belonging to genus *Phlebia*. FEMS Microbiol. Lett. 314 (2), 140–146.

Xiao, P., Mori, T., Kondo, R., 2011c. Biotransformation of the organochlorine pesticide trans-chlordane by wood-rot fungi. N. Biotechnol. 29 (1), 107–115.

Yang, C., Li, Y., Zhang, K., Wang, X., Ma, C., Tang, H., et al., 2010. Atrazine degradation by a simple consortium of *Klebsiella* sp. A1 and *Comamonas* sp. A2 in nitrogen enriched medium. Biodegradation 21 (1), 97–105.

Yatawara, M., Qi, S., Owago, O.J., Zhang, Y., Yang, D., Zhang, J., et al., 2010. Organochlorine pesticide and heavy metal residues in some edible biota collected from Quanzhou Bay and Xinghua Bay, Southeast China. J. Environ. Sci. 22 (2), 314–320.

Zaim, M., Guillet, P., 2002. Alternative insecticides: an urgent need. Trends Parasitol. 18 (4), 161–163.

Zhang, F., He, J., Yao, Y., Hou, D., Jiang, C., Zhang, X., et al., 2013a. Spatial and seasonal variations of pesticide contamination in agricultural soils and crops sample from an intensive horticulture area of Hohhot, North-West China. Environ. Monit. Assess. 185 (8), 6893–6908.

Zhang, L., Dickhut, R., DeMaster, D., Pohl, K., Lohmann, R., 2013b. Organochlorine pollutants in western antarctic peninsula sediments and benthic deposit feeders. Environ. Sci. Technol. 47 (11), 5643–5651.

12 Biology, Genetic Aspects, and Oxidative Stress Response of *Streptomyces* and Strategies for Bioremediation of Toxic Metals

Anindita Mitra

Faculty of Arts and Science (FAS), Harvard University, Cambridge, MA 02138, USA

12.1 Introduction

Solving environmental problems like toxic metal removal is an important societal issue. Bioremediation is a sustainable solution for this purpose. Because environmental systems are more complex and diverse than well-controlled laboratory conditions, the applications of biology, genetics, and oxidative stress responses of a potential microorganism offer a promising approach to understand the molecular mechanisms of bioremediation. The functional genomics provide an insight into global metabolic and regulatory networks that can enhance the understanding of gene functions and further pave the way for environmental microbiologists toward a better understanding of microbial bioremediation. To develop an efficient bioremediation strategy for toxic metals, one has to understand the physiology of the microorganisms in the contaminated environment, the physio-chemical nature of the contaminated sites (e.g., oxidation–reduction potential) and the conditions where the functional genes will be mostly expressed. Where the whole genome annotation is helpful to find the biodegrading gene/s for a specific metal, the transcriptomics provide us the information of a complete set of RNA (transcriptome) produced in one or a population of cells. Transcriptome reveals the genes that are actively expressed at any given time under specific set of conditions, thus it varies with external environmental conditions. The progress of molecular biology, system biology, and the availability of whole genome sequence data, and the techniques like genomics, transcriptomics, proteomics, and metabolomics, are potentially helpful to understand the microbial and the field based bioremediation processes (Ma and Zhai, 2012).

Bioremediation works by transforming or degrading the toxic and hazardous product into less toxic chemicals. Microorganisms cannot degrade metals, but they can interact with and bio-transform them into another chemical/s by changing their

Microbial Biodegradation and Bioremediation. DOI: http://dx.doi.org/10.1016/B978-0-12-800021-2.00012-1

oxidation state, either by adding or removing electrons. Additionally, the potential bacteria for remediation must continuously deal with stress situations *in vivo*. Such stress conditions may include changes in environmental temperature, pH, humidity, etc. along with oxidative stress induced by metal-mediated free radicals.

Metals are considered toxic at high concentrations, since they disrupt cell homeostasis by binding to enzymes, proteins, and DNA and for the production of oxygen radicals through the Fenton reaction (Schmidt et al., 2007). Therefore, to maintain a homeostasis within the cell, the potential microbes excrete metals via efflux transport systems, bind and detoxify metals inside the cells by sequestering compounds in cytosol, release chelators to bind metals, or bind metals in cell walls by sorption (Haferburg and Kothe, 2007).

In this respect, *Streptomyces* spp. has played some advantageous roles over other microorganisms for multiple reasons, which are discussed in detail here. The purpose of this review is to provide a detailed overview of the current state of knowledge of the role of *Streptomyces*, and its biology and genetics, for the purpose of toxic metals bioremediation. The chapter is organized in the following manner: First, the advantageous role of *Streptomyces* over other bacteria is mentioned in general, and then its genome-scale oxidative stress regulatory networks are discussed. Attention is paid to the genetic role of *Streptomyces* spp. for metal-specific regulatory systems and the possible mechanisms of bioremediation. Finally, we discuss how this knowledge can be used as a strategy plan for a metal bioremediation process.

12.2 Genus *Streptomyces*

Streptomyces is one of the largest genus of *Actinobacteria*. This Gram-positive soil-dwelling bacteria has three characteristic developmental phases: (1) formation of branched, filamentous vegetative mycelium, (2) formation of aerial hyphae, and (3) the production of spores. The cell wall of *Streptomyces* contains peptidoglycan and teichoic acids, which are considered the major metal-binding sites (Beveridge and Murray, 1980). The active secondary metabolisms of *Streptomyces* spp. also facilitate multiple metal-binding sites. Even some secondary metabolic products help the bacteria to cope with stressful environments including heavy-metal-contaminated sites (So et al., 2000).

Streptomyces strains resistant to metals are not confined to a single lineage but are widespread along *Streptomyces* phylogeny (Alvarez et al., 2013). It has a large linear chromosome with high G + C content (about 70%), which gives much stability. Interestingly, the genes associated with metal resistance are not only confined to the chromosome but also spread over the plasmids (Ravel et al., 1998), and thus the presence of metal resistance can be found due to the transfer of plasmids with metal resistance genes (Alvarez et al., 2013). The exchange of genetic materials and gene duplications are widespread across *Streptomyces* species (Zhou et al., 2012). Further, the complete genome sequences of five *Streptomyces* species and, additionally, the numerous sequence isolates of different *Streptomyces* species from diverse environments have strengthened the scope for using this genus for bioremediation purposes. Analysis of the "core genome" components of *Streptomyces* revealed that it contains a substantial percentage of catalytic activity, transferase activity, hydrolase activity, and metabolic processes involving genes, along with a strong oxidative stress regulatory network and metal-detoxifying network (Zhou et al., 2012). Where the "core genome" of *Streptomyces* holds most of the housekeeping genes, the "arms" of the genome consist mostly of the conditional adaptive genes, which are probably generated by lateral gene transfer and gene duplication (Hopwood, 2007).

12.3 Oxidative Stress Regulation and Metal Detoxification

Metal-contaminated sites may impose oxidative stress on the inhabitant bacteria. This oxidative stress is mainly generated due to the metal-mediated reactive oxygen species (ROS), including free radicals, oxides, and peroxide that may cause various modifications to DNA bases, lipid peroxidation, and altered calcium and sulfhydryl homeostasis. To survive in the metal-contaminated environment, the microorganism has to defend against oxidative stress via its native regulatory pathways and detoxify ROS or repair the resulting damages. In *Streptomyces*, the intracellular binding and detoxification of metal species are combined with oxidative stress defense mechanisms (Figure 12.1).

In general, iron (Fe), copper (Cu), chromium (Cr), vanadium (V), and cobalt (Co) undergo redox-cycling reactions, while for mercury (Hg), cadmium (Cd), and nickel (Ni), the primary route to interact with living cells is by depletion of glutathione or superoxide dismutase and bonding to sulfhydryl groups of proteins (Valko et al., 2005). The common mechanisms for iron, copper, chromium, vanadium, and cobalt involve Fenton reactions, and generation of superoxide radicals and hydroxyl radicals in mitochondria, microsomes, and peroxisomes (Valko et al., 2005). *Streptomyces* species exhibit different tolerance sensitivity to different metals (Abbas and Edwards, 1989). Working with 32 *Streptomyces* species, Abbas and Edwards (1989) found that these sensitivities toward metals are in order of toxicity: Hg > Cd > Co > Zn > Ni > Cu > Cr > Mn.

In order to prevent damage, *Streptomyces* has developed a number of enzymatic and nonenzymatic protection systems along with repair and detoxification processes. The enzymes named catalase-peroxidase and superoxide dismutase are widely known to detoxify intercellular ROS. It was observed that the metabolic pathways lead to precipitate metals as metal sulfides; phosphates or carbonates have a significant possible biotechnological application.

12.3.1 Thiol Systems

12.3.1.1 Thioredoxin System

In *Streptomyces*, the thiol-disulfide status of proteins is maintained by a thioredoxin system. This is one of the most important antioxidant systems in *Streptomyces* to

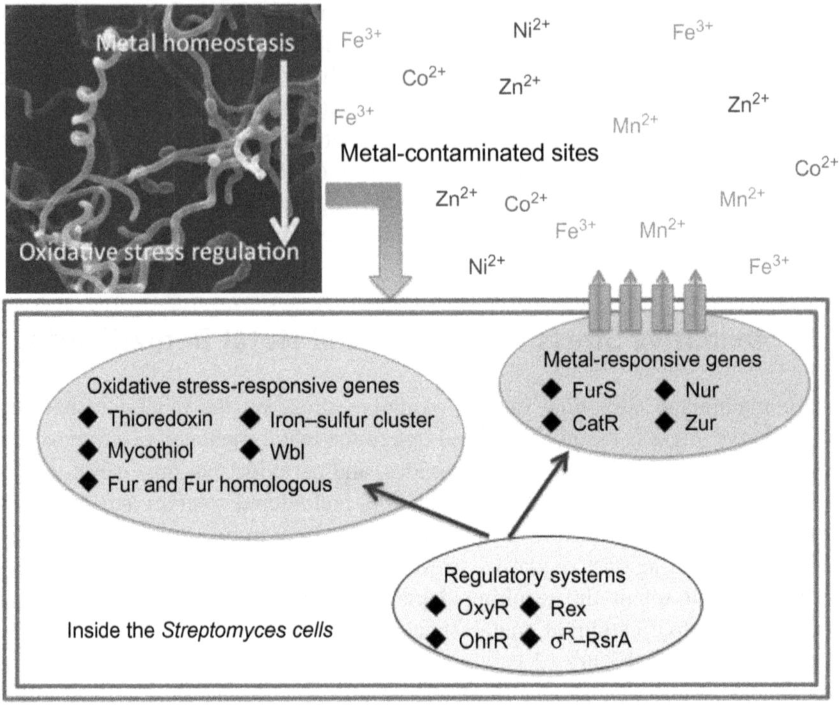

Figure 12.1 Schematic figure of the genes involved in regulation of metal homeostasis and oxidative stress in *Streptomyces* spp. in the metal-contaminated sites.
Source: Modified from Zhang et al. (2013).

protect against oxidative stress and to maintain intracellular thiol homeostasis. The system is composed of small redox-active proteins, thioredoxin (TrxA), and thioredoxin reductase (TrxB). The reduced thiol state of thioredoxins is maintained by thioredoxin reductase by transferring electrons from NADPH to two cysteine residues in the TrxB active site that, in turn, can reduce oxidised TrxA (Hengst and Buttner, 2008). The thioredoxin systems have been studied in several *Streptomyces* strains like *Streptomyces aureofaciens* (Horecka et al., 2003) and *Streptomyces coelicolor* (Stefankova et al., 2006). The model organism, *S. coelicolor*, consists of many thioredoxins (encoded by SCO3890, SCO3889, SCO5438, SCO5419, and trxC/SCO0885) (Paget et al., 2001a).

12.3.1.2 Mycothiol System

Another important thiol system in *Streptomyces* is mycothiol (MSH), which exists in reduced (MSH) and oxidized dimeric mycothiol disulfide (MSSM) states. Mycothiol is highly resistant to oxidation by metal-derived molecular oxygen, even much more resistant than cysteine or glutathione. In order to maintain the intracellular redox environment, an NADPH-dependent flavoenzyme named mycothiol-disulfide-selectivereductase (Mtr) reduces MSSM back to MSH (Hengst and Buttner, 2008). In *S. coelicolor*, MSH is under the control of sigma(R), which is regulated by a redox- sensing anti-sigma, factor, RsrA with a thiol-disulfide redox switch (discussed later). MSH can reduce RsrA to bind sigma(R), so that the RsrA-sigma(R) system senses the intracellular level of reduced MSH, and MSH serves as a natural modulator of the transcription system (Park and Roe, 2008).

12.3.2 Iron–Sulfur Clusters

Iron–sulfur [Fe–S] clusters exist as [2Fe–2S], [4Fe–4S], [3Fe–4S], or more complex forms, depending upon the cell's redox status, and controls the activities of transcriptional regulations involved in oxidative stress. These clusters of proteins are involved in electron transport and metabolic pathways across all live kingdom (Jakimowicz et al., 2005). Three transcription factors, Fumarate and Nitrate reductase Regulator (FNR), Super oxide response Regulator (SoxR), and Iron-sulphur cluster Regulator (IscR), are identified in *Streptomyces* that have Fe–S regulatory clusters. FNR regulates the expression of a number of genes involved in anaerobic respiration. In *Streptomyces lividans*, IscS-like cysteine desulfurase-DndA, which is required for the formation of an [Fe–S] cluster in apo-DndC, was identified (Chen et al., 2012). The [2Fe–2S] cluster of SoxR senses superoxide and NO stresses (Sheplock et al., 2013). SoxR, then activates other genes to remove superoxide and repair the damage that may have occurred during oxidative stress.

In *S. coelicolor*, five SoxR regulon genes (SCO2478, -4266, -7008, -1909, and -1178) were identified (Shin et al., 2011).

12.3.3 The Wbl Proteins

The *wbl* genes found in the entire *Streptomyces* genera play an important role in its biology. Its four cysteine residues, which are conserved in all members, might act as ligands for metal cofactor (Jakimowicz et al., 2005). The WhiD, a developmental protein in *Streptomyces* and a member of the WhiB-like (Wbl) family, can bind a redox-sensitive [4Fe–4S] cluster that reacts with oxygen to generate a [2Fe–2S] cluster. This WhiD is required for the late stages of sporulation in *S. coelicolor* (Jakimowicz et al., 2005). Genome sequencing of *S. coelicolor* revealed that, including *whiB* and *whiD*, there are a total of 14 *wbl* genes: 11 on the chromosome and 3 on the giant linear plasmid, SCP1 (Bentley et al., 2002; Bentley et al., 2004). However, Wbl may function as disulfide reductase, where the [4Fe–4S] cluster inactivates the enzyme until oxidative stress destroys the cluster and the enzymatic activity is released (Hengst and Buttner, 2008).

12.3.4 Fur and Fur Homologous

In *Streptomyces*, the Ferric Uptake Repressor (Fur)-like proteins regulate catalase-peroxidase (CpeB) and act as redox regulators. While catalases decompose hydrogen peroxide to water and oxygen, peroxidases use hydrogen peroxide to oxidize a number of compounds. Thus, functional catalase-peroxidases are composed of varying ratios of these two enzymatic activities. Though initially, Fur was characterized as an iron-responsive regulator, it was later revealed that in addition to iron, different *Fur* homologous can act as metal sensors of (Zn^{2+}, Ni^{2+}, Mn^{2+} and Co^{2+} (Santos et al., 2008). There are four members in the Fur family: FurS and CatR are redox-responsive regulators that control the expression of antioxidant genes, and Zur and Nur control zinc and nickel homeostasis in cell, respectively.

12.3.4.1 FurS

FurS is a zinc-containing redox regulator that regulates *cpeB* gene in *Streptomyces reticuli* in its thiol-reduced (SH) form by binding in a *furS-cpeB* operon (Lucana and Schrempf, 2000). Under oxidative stress conditions, the oxidized SH group of FurS is unable to block the transcription of *furS-cpeB*, which in turn leads to a high production of catalase peroxidase (CpeB) activity (Lucana et al., 2003). Another catalase-peroxidase–encoding *furS-cpeB* operon was isolated from *S. coelicolor* (Hahn et al., 2000a). FurA, a metal-dependent repressor, acts as a negative regulator of the *furS-cpeB* operon. The binding affinity of FurA is increased in the presence of metals and under reducing conditions, and thus decreases the production of CatC protein.

12.3.4.2 CatR

S. coelicolor produces three distinct catalases: two monofunctional catalases (CatA and CatB) and one catalase-peroxidase (CatC). CatA is induced by H_2O_2, and is required for efficient growth of mycelium, whereas CatB is induced by osmotic stress and is required for osmo-protection of the cell. CatC is transiently expressed at the late exponential to early stationary phase, but its function has not been well documented (Hahn et al., 2000b).

12.3.4.3 Nur

A nickel-responsive regulator of the Fur family named Nur, regulates superoxide dismutases (SODs) in *Streptomyces* spp. (Ahn et al., 2006). The SODs maintain the concentration of superoxide radicals in low limits through the catalysis of the dismutation of superoxide (O_2^-) into oxygen and hydrogen peroxide. The SODs are named according to the metal species attached to the redox-active site. However, the presence of cytoplasmic nickel-dependent SOD (NiSOD) is a general feature of the *Streptomyces* (Leclere et al., 1999) genus. This novel type of SOD with only nickel as the catalytic metal was first identified and characterized by Youn et al. (1996).

S. coelicolor contains two types of SODs, Ni containing SodN and Fe– containing SodF and SodF2 (Chung et al., 1999). Nur binds to the promoter region of the *sodF* and *sodF2* genes encoding Fe-containing SOD in the presence of nickel (Ahn et al., 2006).

Later, from the *Streptomyces peucetius* ATCC 27952 genome, two SODs, named sp-sod1 and sp-sod2, were identified and characterized. The sp-sod1 is an

Fe–Zn sod, while sp-sod2 is a NiSOD; they are 636 and 396 bp in length, respectively (Kanth et al., 2010). The heavy metal (Ni, Cu, Cd, Cr, Mn, Zn, Fe)–tolerant strain *Streptomyces acidiscabies* E13 isolated from a uranium mining site also showed the presence of Ni and Fe– containing superoxide dismutase in different enzymatic repression studies (Schmidt et al., 2007).

12.3.4.4 Zur

Zur is a zinc-specific regulator of the Fur family that regulates zinc transport and maintains zinc homeostasis in the cell. Zur regulates Zn mobilization with some ribosomal proteins (S14, S18, L28, L31, L32, L33, and L36) (Owen et al., 2007). The proteins, those are predicted with Zn ribbon motifs consist of two pairs of conserved cysteine residues are named C^+, that in turn bind Zn. The other proteins are C^-, as they lack cysteine containing Zn ligands. In *S. coelicolor*, the expression of four transcription units encoding C^- ribosomal proteins is elevated under conditions of zinc deprivation (Owen et al., 2007). Among these four transcriptional units, Zur controls only three; the fourth one, *rpmG3– rpmJ2*, is not controlled by Zur. In *S. coelicolor*, the *rpmG3– rpmJ2* is influenced by σ^R–RsrA redox switch, that also controls the disulfide stress condition. Depletion of zinc in *S. coelicolor* cultures leads to the release of σ^R from the σ^R–RsrA complex (Owen et al., 2007; Shin et al., 2007). Zur also regulates the zinc transport system through the *znuACB* operon (Owen et al., 2007). A putative zincophore named Coelibactin, regulated by Zur, is reported in *S. coelicolor* (Hesketh et al., 2009; Kallifidas et al., 2010). Later, a crystal structure of active Zur with three zinc binding sites (C-, M-, and D-sites) is reported from *S. coelicolor* (Shin et al., 2011). Biochemical and spectroscopic analyses revealed that while the C-site serves a structural role, the M- and D-sites regulate DNA-binding activity as an on-off switch and a fine-tuner, respectively (Shin et al., 2011).

12.3.5 The Regulatory Systems

12.3.5.1 OxyR

OxyR is a H_2O_2-sensing transcriptional regulator. In *S. coelicolor*, OxyR regulates the expression of its own structural gene and the alkyl hydroperoxide reductase system (AhpC and AhpD). In presence of H_2O_2, OxyR influences *oxyR* and *ahpCD* promoters as a positive regulator (Hahn et al., 2002). However, the mechanism by which OxyR senses peroxides is still controversial.

12.3.5.2 OhrR

Under oxidative stress conditions, lipid hydroperoxide, a nonradical product promotes further formation of reactive lipid radicals and oxidized macromolecules that has an adverse effect on membrane components (Oh et al., 2007). The organic hydroperoxide resistance (Ohr) enzymes reduces peroxides generated from lipid peroxidation in a thiol-dependent manner. The OhrR acts as an organic peroxide-sensing transcriptional repressor of *ohr* gene. Out of three paralogous *ohr* genes (*ohrA*, *ohrB*, and *ohrC*) found in *S. coelicolor*, only the expression of *ohrA* is induced by organic hydroperoxides and provides primary protection against organic hydroperoxides. In reducing conditions, OhrR represses the *ohrA* and *ohrR* genes. The oxidization by organic hydroperoxides causes de-repression of the *ohrA* gene, and the *ohrR* gene is induced through activation by OhrR (Oh et al., 2007).

12.3.5.3 Rex

The redox- sensing transcriptional repressor (Rex) homologues act as regulatory sensors of NADH/NAD^+ in most Gram-positive bacteria, including *Streptomyces*. The NADH/NAD^+ turnover in the cell is highly influenced by oxygen. Rex is a repressor that controls expression of the cytochrome bd terminal oxidase operon (*cydABCD*), which has a high affinity for oxygen and also controls its own transcription as part of the *rex–hemACD* operon. Rex binds both NADH and NAD^+, but high affinity of Rex to NADH outcompetes NAD^+. Thus, the cellular NADH/NAD^+ ratio allows Rex to repress target genes by binding to their respective Rex operator (ROP) sites and thus influence the electron transport chain (Brekasis and Paget, 2003).

12.3.5.4 σ^R–RsrA

It senses and responds to disulfide stress. In a reducing environment, σ^R is sequestered by RsrA and form 1:1 complex that prevents σ^R from interacting with RNA polymerase. But in disulfide stress, an intramolecular disulfide bond is formed in RsrA that results the removal of Zn^{2+} and loss of σ^R binding. The resultant free σ^R then binds to RNA polymerase and activates the target genes (Paget et al., 2001b). In contrast, the induced thioredoxin, in turn, reduces RsrA, which forms a negative feedback loop. It allows intracellular redox homeostatic control by RsrA.

12.3.6 Other Metal Resistance in Streptomyces

Genome sequencing of *Streptomyces xinghaiensis* NRRL $B24674^T$ revealed a number of genes related to the heavy metal resistance of mercury, copper, and nickel, indicating a potentiality of the strain for environmental bioremediation (Zhao and Yang, 2011). In heavy metal-resistant *Streptomyces acidiscabies*, a highly specific nickel transporter gene was found (Amoroso et al., 2000) that was also identified in the genome of *Streptomyces avermitilis*.

During early genetic characterization of *S. lividans*, an amplifiable sequence named AUD2, linked to the *mer* genes, was isolated from SLP3 plasmid (Cruz-Morales et al., 2013). After transcriptional analysis of *S. lividans* 66, a large genome island was identified that is rich in metal-related genes (Cruz-Morales et al., 2013).

A strong correlation was observed between production of melanin and metal resistance for phytopathogenic *Streptomyces scabies* (Beausejour and Beaulieu,

2004). Melanin can reduce the concentration of ROS and can sequestrate metals. This may be due to the presence of carboxylic groups in the melanin molecule.

It has been reported that *Streptomycete* species are able to detoxify Hg^{+2} to volatile Hg^0 by using mercuric reductase enzyme (Abbas and Edwards, 1989). Many genes are predicted for the resistance to metals and metalloids from the draft genome of *Streptomyces zinciresistens* K42, isolated from copper–zinc mine (Lin et al., 2011). The soil-borne *Streptomyces tendae* F4, produces three different hydroxamate siderophores, that reduces cadmium toxicity and increases metal uptake (Dimkpa et al., 2009).

12.4 Metal Detoxification Mechanisms and Bioremediation

Microbes can influence metals (Ledin, 2000; Barkay and Schaefer, 2001) by increasing the mobility of the contaminants or they can transform metals to precipitate out from the contaminated site. However, these mechanisms can be utilized for potential bioremediation purposes only once the microbial metabolisms, chemical reactions, and flow of metal contaminants in the environment are well understood. In Figure 12.2, the possible mechanisms for metal bioremediation by *Streptomyces* and its involved genes are summarized. Using these mechanisms, *Streptomyces* can (a) transform metals by redox or alkylation processes; (b) accumulate metals by

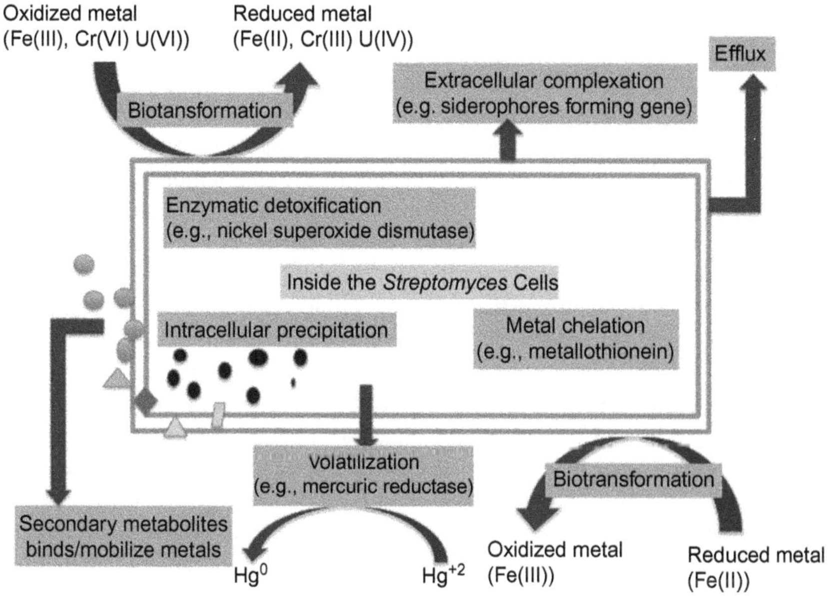

Figure 12.2 Schematic of the possible mechanisms of metal bioremediation in *Streptomyces* spp. and the involved genes.
Source: Modified from Gadd (2010).

absorption or adsorption processes. In addition, (c) the organic substances produced or released from the microorganism sometime may influence the mobility of the metals or bind the metals, (d) even microbes can influence metal mobility indirectly by changing pH and Eh of the contaminated site, and lastly, (e) organic–metal complexes degraded by microbes may alter metal specition. In order to acquire iron from the extracellular environment, *Streptomyces* produces and secretes low-molecular-weight compounds known as siderophores. These compounds chelate Fe^{3+}.

In *Streptomyces pilosusis*, iron uptake is mediated by ferrioxamines B, D1, D2, and E, while the iron transport is maintained by exogenous siderophores ferrichrome, ferrichrysin, rhodotorulic acid (RA), and synthetic enantio-RA (Muller et al., 1984). Under iron-rich conditions in a cell, Fur binds the divalent iron and inhibits DNA transcription from the genes and operons repressed by the metal. Conversely, when iron is scarce, there is de-repression of the genes that activate other iron uptake systems. In *S. coelicolor* A3(2), *des* and *cch* gene clusters are identified that direct the production of siderophores, named tris-hydroxamate ferric iron-chelators desferrioxamine E and coelichelin, respectively (Barona-Gomez et al., 2006).

12.5 Strategies, Applications, and Future Direction

The potential of microorganisms for the sustainable bioremediation of toxic metals is well established. Bioremediation strategies depend solely upon the catabolic capacities of the microbes to transform toxic metals to harmless compounds. However, there is a substantial difference between manipulated laboratory conditions and the natural environment. It is also not surprising that the genetically modified bacteria rarely function in the in situ environment.

Microbes have evolved for the last 3.8 billion years and inhabit virtually all environments, from extreme salinity and extreme pH to extreme temperature. Thus, it is important to know about their metabolic capabilities by analyzing the target gene and its potentiality before using it in the target niche. Earlier strategies to discover the potential strains were based solely on the lab-based experimental data of microbe's chemical kinetics, and on intermediate and final product identification and quantification of the metal pollutants. Now, with the advancement of genomics, it has been easier to find the specific gene responsible for a specific metal bioremediation purpose before proceeding to on-site evaluation. Selection of microbes for bioremediation purpose should be based on its functional genomics that can shed light on the understanding of the biological functions of specific sets of genes and how genes and their products work together (Zhao and Poh, 2008); and transcriptomic study that will determine when and where the genes are turned on or off in various environmental situations. Strategies for microbial bioremedation based on the understanding of the molecular mechanisms will reduce uncertainty about the functional capability of the strain in field applications. Thus, the integration of transcriptomics, proteomics, and functional genomics provide us the global metabolic and regulatory gene networks that can enhance the understanding of gene functions. This strategy will give a powerful new perspective on the holistic use of *Streptomyces* for metal bioremediation.

References

Abbas, A., Edwards, C., 1989. Effects of metals on a range of *Streptomyces* species. Appl. Environ. Microbiol. 55, 2030–2035.

Ahn, B.E., Cha, J., Lee, E.J., Han, A.R., Thompson, C.J., Roe, J.H., 2006. Nur, a nickel-responsive regulator of the Fur family, regulates superoxide dismutases and nickel transport in *Streptomyces coelicolor*. Mol. Microbiol. 59, 1848–1858.

Alvarez, A., Catalano, S.A., Amoroso, M.J., 2013. Heavy metal resistant strains are widespread along *Streptomyces* phylogeny. Mol. Phylogenet. Evol. 66, 1083–1088.

Amoroso, M.J., Schubert, D., Mitscherlich, P., Schumann, P., Kothe, E., 2000. Evidence for high affinity nickel transporter genes in heavy metal resistant *Streptomyces* species. J. Basic Microbiol. 40, 295–301.

Barkay, T., Schaefer, J., 2001. Metal and radionuclide bioremediation: issues, considerations, and potentials. Curr. Opin. Microbiol. 4, 318–323.

Barona-Gomez, F., Lautru, S., Francou, F.X., Leblond, P., Pernodet, J.L., Challis, G.L., 2006. Multiple biosynthetic and uptake systems mediate siderophore-dependent iron acquisition in *Streptomyces coelicolor* A3(2) and *Streptomyces ambofaciens* ATCC 23877. Microbiology 152, 3355–3366.

Beausejour, J., Beaulieu, C., 2004. Characterization of *Streptomyces scabies* mutants deficient in melanin biosynthesis. Can. J. Microbiol. 50, 705–709.

Bentley, S.D., Chater, K.F., Cerdeno-Tarraga, A.M., Challis, G.L., Thomson, N.R., James, K.D., et al., 2002. Complete genome sequence of the model actinomycete *Streptomyces coelicolor* A3(2). Nature 417, 141–147.

Bentley, S.D., Brown, S., Murphy, L.D., Harris, D.E., Quail, M.A., Parkhill, J., et al., 2004. SCP1, a 356,023 bp linear plasmid adapted to the ecology and developmental biology of its host, *Streptomyces coelicolor*A3(2). Mol. Microbiol. 51, 1615–1628.

Beveridge, T., Murray, R., 1980. Sites of metal deposition in the cell wall of *Bacillus subtilis*. J. Bacteriol. 141, 876–887.

Brekasis, D., Paget, M.S., 2003. A novel sensor of NADH/NAD+ redox poise in *Streptomyces coelicolor* A3(2). EMBO J. 22, 4856–4865.

Chen, F., Zhang, Z., Lin, K., Qian, T., Zhang, Y., You, D., et al., 2012. Crystal structure of the cysteine desulfurase DndA from *Streptomyces lividans* which is involved in DNA phosphorothioation. PLOS One 7, e36635.

Chung, H.J., Kim, E.J., Suh, B., Choi, J.H., Roe, J.H., 1999. Duplicate genes for Fe–containing superoxide dismutase in *Streptomyces coelicolor* A3(2). Gene 231, 87–93.

Cruz-Morales, P., Vijgenboom, E., Iruegas-Bocardo, F., Girard, G., Yanez-Guerra, L.A., Ramos-Aboites, H.E., et al., 2013. The genome sequence of *Streptomyces lividans* 66 reveals a novel tRNA-dependent peptide biosynthetic system within a metal-related genomic island. Genome Biol. Evol. 5, 1165–1175.

Dimkpa, C.O., Merten, D., Svatos, A., Buchel, G., Kothe, E., 2009. Siderophores mediate reduced and increased uptake of cadmium by *Streptomyces tendae* F4 and sunflower (*Helianthus annuus*), respectively. J. Appl. Microbiol. 107, 1687–1696.

Gadd, G.M., 2010. Metals, minerals, and microbes: geomicrobiology and bioremediation. Microbiology 156, 609–643.

Haferburg, G., Kothe, E., 2007. Microbes and metals: interactions in the environment. J. Basic Microbiol. 47, 453–467.

Hahn, J.S., Oh, S.Y., Roe, J.H., 2000a. Regulation of the *furA* and *catC* operon, encoding a ferric uptake regulator homologue and catalase-peroxidase, respectively, in *Streptomyces coelicolor*A3(2). J. Bacteriol. 182, 3767–3774.

Hahn, J.S., Oh, S.Y., Chater, K.F., Cho, Y.H., Roe, J.H., 2000b. H_2O_2-sensitive Fur-like repressor CatR regulating the major catalase gene in *Streptomyces coelicolor*. J. Biol. Chem. 275, 38254–38260.

Hahn, J.S., Oh, S.Y., Roe, J.H., 2002. Role of OxyR as a peroxide-sensing positive regulator in *Streptomyces coelicolor* A3(2). J. Bacteriol. 184, 5214–5222.

Hengst, C.D., Buttner, M.J., 2008. Redox control in actinobacteria. Biochim. Biophys. Acta. 1780, 1201–1216.

Hesketh, A., Kock, H., Mootien, S., Bibb, M., 2009. The role of *absC*, a novel regulatory gene for secondary metabolism, in zinc-dependent antibiotic production in *Streptomyces coelicolor* A3(2). Mol. Microbiol. 74, 1427–1444.

Hopwood, D.A., 2007. *Streptomyces* in Nature and Medicine: The Antibiotic Makers. Oxford University Press, Inc. New York.

Horecka, T., Perecko, D., Kutejova, E., Mikulasova, D., Kollarova, M., 2003. The activities of the two thioredoxins from *Streptomyces aureofaciens* are not interchangeable. J. Basic Microbiol. 43, 62–67.

Jakimowicz, P., Cheesman, M.R., Bishai, W.R., Chater, K.F., Thomson, A.J., Buttner, M.J., 2005. Evidence that the *Streptomyces* developmental protein WhiD, a member of the WhiB family, binds a [4Fe–4S] cluster. J. Biol. Chem. 280, 8309–8315.

Kallifidas, D., Pascoe, B., Owen, G.A., Strain-Damerell, C.M., Hong, H.J., Paget, M.S.B., 2010. The zinc-responsive regulator Zur controls expression of the Coelibactin gene cluster in *Streptomyces coelicolor*. J. Bacteriol. 192, 608–611.

Kanth, B.K., Oh, T.J., Sohng, J.K., 2010. Identification of two superoxide dismutases (FeSOD and NiSOD) from *Streptomyces peucetius* ATCC 27952. Biotechnol. Bioprocess Eng. 15, 785–792.

Leclere, V., Boiron, P., Blondeau, R., 1999. Diversity of superoxide dismutases among clinical and soil isolates of *Streptomyces* species. Curr. Microbiol. 39, 365–368.

Ledin, M., 2000. Accumulation of metals by microorganisms–processes and importance for soil systems. Earth-Sci. Rev. 51, 1–31.

Lin, Y., Hao, X., Johnstone, L., Miller, S.J., Baltrus, D.A., Rensing, C., et al., 2011. Draft genome of *Streptomyces zinciresistens* K42, a novel metal-resistant species isolated from copper–zinc mine tailings. J. Bacteriol. 193, 6408–6409.

Lucana, D.O., Schrempf, H., 2000. The DNA-binding characteristics of the *Streptomyces reticuli* regulator FurS depend on the redox state of its cysteine residues. Mol. Gen. Genet. 264, 341–353.

Lucana, D.O., Troller, M., Schrempf, H., 2003. Amino acid residues involved in reversible thiol formation and zinc ion binding in the *Streptomyces reticuli* redox regulator FurS. Mol. Genet. Genomics 268, 618–627.

Ma, J., Zhai, G., 2012. Microbial bioremediation in omics era: opportunities and challenges. J. Bioremed. Biodeg. 3, e120.

Muller, G., Matzanke, B.F., Raymond, K.N., 1984. Iron transport in *Streptomyces pilosus* mediated by ferrichromesiderophores, rhodotorulic acid, and enantio-rhodotorulic acid. J. Bacteriol. 160, 313–318.

Oh, S.Y., Shin, J.H., Roe, J.H., 2007. Dual role of OhrR as a repressor and an activator in response to organic hydroperoxides in *Streptomyces coelicolor*. J. Bacteriol. 189, 6284–6292.

Owen, G.A., Pascoe, B., Kallifidas, D., Paget, M.S., 2007. Zinc-responsive regulation of alternative ribosomal protein genes in *Streptomyces coelicolor* involves Zur and σ^R. J. Bacteriol. 189, 4078–4086.

Paget, M.S., Bae, J.B., Hahn, M.Y., Li, W., Kleanthous, C., Roe, J.H., et al., 2001a. Mutational analysis of RsrA, a zinc-binding anti-sigma factor with a thiol-disulfide redox switch. Mol. Microbiol. 39, 1036–1047.

Paget, M.S., Molle, V., Cohen, G., Aharonowitz, Y., Buttner, M.J., 2001b. Defining the disulphide stress response in *Streptomyces coelicolor* A3(2): identification of the σ^R regulon. Mol. Microbiol. 42, 1007–1020.

Park, J.H., Roe, J.H., 2008. Mycothiol regulates and is regulated by a thiol-specific antisigma factor RsrA and sigma(R) in *Streptomyces coelicolor*. Mol. Microbiol. 68, 861–870.

Ravel, J., Schrempf, H., Hill, R.T., 1998. Mercury resistance is encoded by transferable giant linear plasmids in two Chesapeake Bay *Streptomyces* strains. Appl. Environ. Microbiol. 64, 3383–3388.

Santos, C.L., Vieira, J., Tavares, F., Benson, D.R., Tisa, L.S., Berry, A.M., et al., 2008. On the nature of Fur evolution: a phylogenetic approach in actinobacteria. BMC Evol. Biol. 8, 185.

Schmidt, A., Schmidt, A., Haferburg, G., Kothe, E., 2007. Superoxide dismutases of heavy metal resistant streptomycetes. J. Basic Microbiol. 47, 56–62.

Sheplock, R., Recinos, D.A., Mackow, N., Dietrich, L.E.P., Chander, M., 2013. Species-specific residues calibrate SoxR sensitivity to redox-active molecules. Mol. Microbiol. 87, 368–381.

Shin, J.H., Oh, S.Y., Kim, S.J., Roe, J.H., 2007. The zinc-responsive regulator Zur controls a zinc uptake system and some ribosomal proteins in *Streptomyces coelicolor* A3(2). J. Bacteriol. 189, 4070–4077.

Shin, J.H., Singh, A.K., Cheon, D.J., Roe, J.H., 2011. Activation of the SoxR regulon in *Streptomyces coelicolor* by the extracellular form of the pigmented antibiotic actinorhodin. J. Bacteriol. 193, 75–81.

So, N.W., Rho, J.Y., Lee, S.Y., Hancock, I.C., Kim, J.H., 2000. A lead absorbing protein with superoxide-dismutase activity from *Streptomyces subrutilus*. FEMS Microbiol. Lett. 194, 93–98.

Stefankova, P., Perecko, D., Barak, I., Kollarova, M., 2006. The thioredoxin system from *Streptomyces coelicolor*. J. Basic Microbiol. 46, 47–55.

Valko, M., Morris, H., Cronin, M.T., 2005. Metals, toxicity, and oxidative stress. Curr. Med. Chem. 12, 1161–1208.

Youn, H.D., Kim, E.J., Roe, J.H., Hah, Y.C., Kang, S.O., 1996. A novel nickel-containing superoxide dismutase from *Streptomyces* spp. Biochem. J. 318, 889–896.

Zhang, W., Wang, Y., Lee, O.O., Tian, R., Cao, H., Gao, Z., et al., 2013. Adaptation of intertidal biofilm communities is driven by metal ion and oxidative stresses. Sci. Rep. 3, 3180. Available from: http://dx.doi.org/doi:10.1038/srep03180.

Zhao, B., Poh, C.L., 2008. Insights into environmental bioremediation by microorganisms through functional genomics and proteomics. Proteomics 8, 874–881.

Zhao, X., Yang, T., 2011. Draft genome sequence of the marine sediment-derived actinomycete *Streptomyces xinghaiensis* NRRL B24674. J. Bacteriol. 193, 5543.

Zhou, Z., Gu, J., Li, Y.Q., Wang, Y., 2012. Genome plasticity and systems evolution in *Streptomyces*. BMC Bioinformatics 13, S8.

13 Bacterial and Fungal Bioremediation Strategies

S. Gouma[a], *S. Fragoeiro*[b], *A.C. Bastos*[c] *and N. Magan*[d]

[a]School of Agricultural Technology, Technical Educational Institute of Crete, Stavromenos, Heraklion, Crete, [b]MSD Animal Health, Walton Manor, Milton Keynes, UK, [c]Department of Biology and Centre for Environmental and Marine Studies (CESAM), University of Aveiro, Aveiro, Portugal, [d]Applied Mycology Group, Cranfield Soil and AgriFood Institute, Cranfield University, Bedford, UK

13.1 Introduction

Many xenobiotic compounds in the environment have medium- to long-term stability in soil; their persistence has a significant impact on the functioning of soil ecosystems. For example, chlorinated aromatic herbicides such as triazines are heavily used worldwide for control of broad-leaved weeds in agricultural production as well as in urban and recreational areas (Gadd, 2001). Figure 13.1 provides a comparison of the relative persistence of some xenobiotic compounds in soil, including some which we will consider in more detail later in this review. Some groups, such as the triazines, are moderately persistent in soil (Radkte et al., 1994) with reported half-life values of up to 50–100 days, depending largely on soil environmental conditions (Rodriguez Couto and Toca Herrera, 2006). Microbial metabolism has long been regarded as the most important mechanism of degradation of such compounds in soil (Armstrong et al., 1967; Haggblom, 1992). Nevertheless, in conditions of low moisture and nutrient contents, microbial metabolism is compromised and the persistence of triazines and other xenobiotic compound may increase (Yadav and Loper, 2000). Thus, the relationship among soil type, moisture, pH, organic matter, and clay content will all affect the binding of individual and mixtures of pesticides in a particular soil. This will in turn influence the effectiveness of bioremediation strategies. This review will consider some of the important aspects that are relevant to bioremediation systems using microorganisms including (a) bioremediation approaches, (b) screening of bacteria and white rot fungi and the evidence for their enzyme-mediated remediation, (c) white rot fungi and

Microbial Biodegradation and Bioremediation. DOI: http://dx.doi.org/10.1016/B978-0-12-800021-2.00013-3

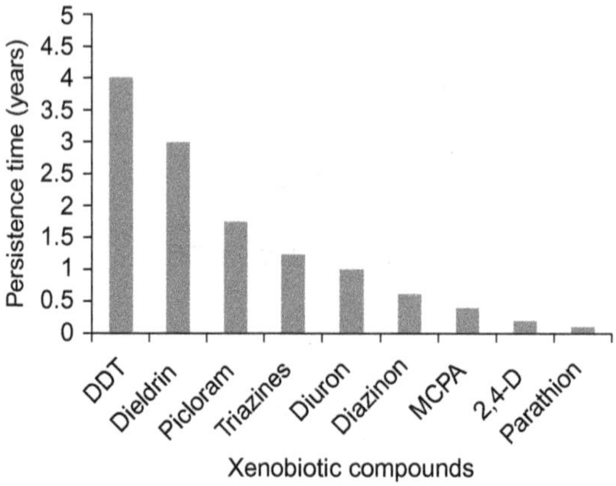

Figure 13.1 Relative persistence times in the environment of different xenobiotic compounds.

environmental screening of single and mixtures of pesticides, (d) inoculant production and delivery for soil incorporation, (e) use of spent mushroom composts (SMCs), and (f) conclusions and future strategies.

13.2 Bioremediation Considerations

Bioremediation is a process that uses mainly microorganisms, plants, or microbial or plant enzymes to detoxify contaminants in the soil and other environments. The concept includes biodegradation, which refers to the partial, and sometimes total, transformation or detoxification of contaminants by microorganisms and plants. Mineralization is a more restrictive term for the complete conversion of an organic contaminant to its inorganic constituents by a single species or a consortium of microorganisms. Co-metabolism is another more restrictive term referring to the transformation of a contaminant without the provision of carbon or energy for the degrading microorganisms (Skipper, 1998).

The process of bioremediation enhances the rate of the natural microbial degradation of contaminants by supplementing the indigenous microorganisms (bacteria or fungi) with nutrients, carbon sources, or electron donors (biostimulation, biorestoration) or by adding an enriched culture of microorganisms that have specific characteristics that allow them to degrade the desired contaminant at a quicker rate (bioaugmentation) (Mackay and Frasar, 2000). The goal of bioremediation is to at least reduce pollutant levels to undetectable, nontoxic, or acceptable levels, that is, to within limits set by regulatory agencies or, ideally, to completely mineralize organopollutants to carbon dioxide (Pointing, 2001).

13.3 Advantages and Disadvantages of Bioremediation

Bioremediation is a natural process and is therefore perceived by the public as having a reduced impact on natural ecosystems. It is typically less expensive than the equivalent physical–chemical methods. The complete destruction of target pollutants is possible on site without the need of excavation or the transport of large quantities of waste off site (Vidali, 2001). Finally, it requires little energy input and preserves the soil structure (Hohener et al., 1998). Perhaps the most attractive feature of bioremediation is the reduced impact on natural ecosystems, which should make it more acceptable to the public (Zhang and Chiao, 2002). For fungal systems this does require the soil to be aerobic, with the provision of enough oxygen to enable effective colonization to occur.

Bioremediation, however, has a number of disadvantages. This process is limited to those compounds that are biodegradable. Moreover, there are some concerns that the products of bioremediation may be more persistent or toxic than the parent compounds. Furthermore, it is difficult to extrapolate the results from bench and pilot-scale studies to full-scale field operations. All biological processes, such as this, are often highly specific and complex and take longer than other treatment options (e.g., incineration) (Vidali, 2001).

Compounds in mixtures are known to interact with biological systems in ways that can greatly alter the toxicity of individual compounds (Hernando et al., 2003). While studies on the capacity of fungi, particularly white rot fungi, and bacteria to degrade individual pesticides are extensive (Gadd, 2001), very few have examined the capacity of microorganisms to degrade mixtures of pesticides. In most pesticide-contaminated agrochemical facilities, herbicides are found in combination with other widely used agricultural chemicals and remediation strategies must take into account the presence of multiple contaminants (Grigg et al., 1997; Memic et al., 2005). Moreover, there are hardly any studies on the use of bacteria and white rot fungi to clean up mixtures of pesticides.

Grigg et al. (1997) reported the ability of an atrazine-mineralizing mixed culture isolated from soil subjected to repeated applications of atrazine to degrade other s-triazines in liquid culture. Cyanazine and simazine, either alone or combined with atrazine, were degraded in 6 days. Metribuzin was not degraded while the microbial culture completely degraded atrazine in the presence of contaminants including alachlor, metolachlor, and trifluralin. A recent study by Fragoeiro and Magan (2008) demonstrated that a mixture of dieldrin, simazine, and trifluralin is differentially degraded by *Trametes versicolor* and *Phanerochaete chrysosporium* and significant extracellular enzymes are produced in soil, even at −2.8 MPa water potential. Thus, research is needed to develop and engineer bioremediation technologies that are appropriate for sites with complex contaminants (Vidali, 2001).

For any xenobiotic compound, the threshold concentration above which remediation is necessary is referred to as the so-called "remediation trigger level." However, for many pesticides as well as for many other xenobiotic compounds, the threshold concentration has not yet been established. There is also a need to establish a target concentration when remediation is achieved. Generally, the target concentration is assumed to be in the 1 ppm (mg/L) range, but in practice it can vary from site to site and region to region (Radkte et al., 1994; Khadrani et al., 1999).

Very often, urban applications of pesticides are done at excessively high concentration, resulting in pesticide waste with prolonged persistence. When applied at the usual agricultural rates, which can be between 1–4.5 kg/ha, pesticide degradation in soil may be around 99% over the course of a growing season. However, it has been suggested that even at these concentrations, top-soil residues have been found to last for several years, ranging from 0.5 to 2.5 ppm. Unfortunately, even when present in soil at the ppb level (μg/L), many recalcitrant compounds often migrate through leaching and reach groundwater (Novotny et al., 2003).

At present, bioremediation conducted on a commercial scale predominantly utilizes prokaryotes, with comparatively few recent attempts to use white rot fungi. Bacteria are very sensitive to fluctuating environmental conditions in soil as growth requires films of water in the soil pores. While they are able to grow rapidly under controlled and optimized conditions, they have a limited range of environmental factors over which they can operate. In contrast, filamentous fungi may offer some major advantages over bacteria in the diversity of compounds they are able to oxidize (Radkte et al., 1994). In addition, they are robust organisms and are generally more tolerant to high concentrations of polluting chemicals than bacteria (Fragoeiro, 2005). They are also more tolerant of environmental stress and can produce copious amounts of extracellular enzymes during hyphal colonization of soil to enhance rates of bioremediation (Fragoeiro, 2005; Magan, 2007; Gouma, 2009). Therefore, white rot fungi probably represent a powerful prospective tool in soil bioremediation and some species have already been patented (Sasek et al., 2003). Interestingly, only a few companies have included the use of ligninolytic fungi for soil remediation into their program, for example, "EarthFax Development Corp." in the USA and "Gebruder Huber Bodenrecycling" in Germany. However, many research studies have been carried out to examine the efficacy of bioremediation fungi to degrade single xenobiotic compounds. Tables 13.1 and 13.2 give some examples of the types of bacteria and white rot fungi that have been used for bioremediation of different xenobiotic compounds.

In soil, microorganisms commonly exist in large populations. Provided with adequate supplies of carbon and energy and environmental conditions conducive to growth, microbial activity, especially the production of extracellular enzymes, can significantly assist in the amelioration of contaminated sites.

Microbial genes encode the degradative enzymes, which oxidize, reduce, dehalogenate, dealkylate, deaminate, and hydrolyze hazardous chemicals such as pesticides in the soil environment (Skipper, 1998). Once a contaminant has been enzymatically transformed to a less complex compound, it can often be metabolized using various pathways. Although a single transformation may reduce the toxicity of a contaminant, the complete mineralization of an organic compound typically requires several degradative enzymes produced by multiple genes on plasmids or chromosomes residing in a single species or among different species. The organism(s) with such enzymes may be either indigenous to the contaminated site or added as an inoculum. The enzymes can be intracellular or extracellular, and each type of enzyme has specific conditions for optimum activity (Skipper, 1998).

Table 13.1 Examples of Bacteria able to Degrade Pesticides in Pure Culture

Pesticide	Bacteria	Reference
DDT	*Alcaligenes eutrophus*	Nadeau et al. (1994)
2,4D	*A. eutrophus*, *Flavobacterium*	Pemberton and Fisher (1977)
	Arthortbacter	Sandmann and Loos (1988)
	Pseudomonas cepacia	Bhat et al. (1994)
Atrazine	*Nocardia*	Cook (1987)
	Pseudomonas	Cook (1987)
	Pseudomonas	Mandellbaum et al. (1995)
	Rhodococcus	Behki et al. (1993)
	Rhodococcus	Behki and Khan (1994)
Parathion	*Flavobacterium*	Sethunathan and Yoshida (1973)
	Pseudomonas dimuta	Serdar et al. (1982)
Diazinon	*Flavobacterium*	Sethunathan and Yoshida (1973)
Fenthion	*Bacillus*	Patel and Gopinathan (1986)
Carbofuran	*Achromobacter*	Karns et al. (1986)
	Pseudomonas	Chandhry and Ali (1988)
	Flavobacterium	Chandhry and Ali (1988)
	Flavobacterium	Head et al. (1992)
EPTC	*Arthrobacter*	Tam et al. (1987)
	Rhodococcus	Behki et al. (1993)
	Rhodococcus	Behki and Khan (1994)

EPTC, S-Ethyl Dipropylthiocarbamate.
Source: Adapted from Aislabie and Jones (1995).

13.4 Microbial Mechanisms of Transformation of Xenobiotic Compounds

These include lignin-degrading enzymes such as laccase (EC 1.10.3.2). Although known for a long time, laccases attracted considerable attention only after the beginning of studies of enzymatic degradation of wood by white rot fungi (Baldrian, 2006). Laccases are typically found in plants and fungi. Plant laccases participate in the radical-based mechanisms of lignin polymer formation, whereas in fungi, laccases probably have more roles including morphogenesis, fungal plant-pathogen host interaction, stress defense, and lignin degradation (Thurston, 1994).

Laccase (benzenediol oxygen oxidoreductase) belongs to a group of polyphenol oxidases containing copper atoms in the catalytic center and usually called multicopper oxidases. Laccases catalyze the reduction of oxygen to water accompanied by the oxidation of a broad range of aromatic compounds as hydrogen donors (Thurston, 1994), like phenols, aromatic amines, and diamines (Nyanhongo et al., 2007). Furthermore, laccase catalyzes the oxidation of non-phenolics and anilines.

Table 13.2 Degradation of Typical Environmental Pollutants by the White Rot Fungi

Fungus	Type of Pollutant	Reference
Ph. chrysosporium	Lindane, DDT, BTEX	Yadav and Reddy (1993)
	Atrazine	Hickey et al. (1994)
	Heptachlor	Arisoy (1998)
Phanerochaete eryngi	Lindane	
Pleurotus florida		
Pleurotus sajor-caju		
Ph. chrysosporium	Pentachlorophenol	
T. versicolor	Creosote, anthracene, PAHs, PCP	Alleman et al. (1992) Gianfreda and Rao (2004)
	Pesticides	Khadrani et al. (1999)
	Pesticides	Morgan et al. (1991) Bending et al. (2002)
Pl. ostreatus	Pesticides	Khadrani et al. (1999)
	PCBs	Gianfreda and Rao (2004)
	Dyes, Catechol pyrene Phenanthrene	Bezalel et al. (1996)
Pleurotus pulmonaris	Atrazine	
Pycnoporus sanguineus	Azo dyes	
Bjerkandera adusta	Pesticides	Khadrani et al. (1999)

Abbreviations: BTEX, benzene, toluene, ethylbenzene, and xylene; PAHs, polycyclic aromatic hydrocarbons; PCBs, polychlorinated biphenyls; PCP, pentachlorophenol.

This reaction is involved in transforming numerous agricultural and industrial chemicals.

Among fungi recently studied for producing extracellular laccase, *Trametes* species are probably the most actively investigated for laccase production because these fungi are commonly found in many parts of the world and apparently are excellent wood decomposers in nature. Indeed, *T. versicolor*, a representative fungus in this genus, was among the first species from which the production of large amounts of laccase was reported. It has already been marketed by several companies, although the current prices are still too high for bulk environmental application (Duran and Esposito, 2000).

The catabolic role of fungal laccase in lignin biodegradation is not well understood (Trejo-Hernandez et al., 2001), but some successful applications of this enzyme in decontamination have been reported. For example, dye decoloration by *Trametes hispida*, degradation of azo dyes by *Pyricularia oryzae* (Chivukula and Renganathan, 1995), and textile effluent degradation by *T. versicolor* have been attributed to laccase activity. Duran and Esposito (2000) also reported that laccase from *Cerrena unicolor* produced a complete transformation of 2, 4 DCP in soil colloids. In a more recent study, Zouari-Mechini et al. (2006) reported that Trametes trogii, isolated from Tunisia, decolorised Poly R 478 and several industrial dyes in agar plates. They found that the addition of CU^{2+} stimulated decolorization, suggesting that laccase could be involved in the process. The results on dye decolorization with the crude enzyme without peroxidase activity and the results obtained with the purified enzyme confirmed that *T. trogii* laccase decolorizes industrial dyes.

13.4.1 Some Enzymes Involved in Bioremediation

Lignin peroxidase: Lignin peroxidase (once called ligninase) (EC 1.11.14) was first discovered in *Ph. chrysosporium* (Glenn and Gold, 1983). Lignin biodegradation is also produced by many, but not all, white rot fungi (Hatakka, 1994). This enzyme is an extracellular hemeprotein, dependent of H_2O_2, with an unusually high redox potential and low optimum pH. It shows little substrate specificity, reacting with a wide variety of lignin model compounds and even unrelated molecules (Barr and Aust, 1994).

Manganese peroxidase: Manganese peroxidase (EC 1.11.13) is also a heme peroxidase and it forms a family of isoenzymes. Similarly to lignin peroxidase (LiP) these isoenzymes are also glycoproteins. This enzyme shows a strong preference for Mn (II) as its reducing substrate and is not able to complete its catalytic cycle without the presence of Mn (II). The redox potential of the Mn peroxidase–Mn system is lower than that lignin peroxidase and it has only shown capacity to oxidize *in vitro* phenolic substrates. Manganese peroxidase seems to be more widespread than lignin peroxidase among white rot fungi (Hatakka, 1994).

Total ligninolytic activity: The production and activity of the ligninolytic enzymes in soil (as opposed to a ligninolytic substrate) may be a prerequisite for transformation of pollutants by microorganisms (Lang et al., 2000a). Thus, quantification of the activity of these enzymes by white rot fungal inoculants is important. Historically, various ^{14}C-radiolabeled and unlabeled substrates have been used to screen for ligninolytic activity. However, these assays are relatively slow and difficult. The decolorization of dyes by white rot fungi was first reported by Glenn and Gold (1983) who developed a method to measure ligninolytic activity of *Ph. chrysosporium* based on the decolorization of a number of sulphonated polymeric dyes. Subsequently, other workers adapted the dye decolorization test for evaluating the ability of white rot fungi to degrade dyes and other xenobiotics (Nyanhongo et al., 2007). The high-molecular-weight dyes cannot be taken up by the microorganisms and thus provide a specific screen for extracellular activity (Gold et al., 1988; Field et al., 1993). The decolorization of polymeric dyes has been proposed as a useful screening method for ligninolytic activity (Lin et al., 1991). Today the polymeric dyes used are inexpensive, stable, and readily soluble and have high extinction coefficients and low toxicity toward *Ph. chrysosporium* and other white rot fungi and bacteria tested (Gold et al., 1988).

Other enzymes involved in bioremediation: The other microbial enzymes involved in the pollutant transformation are hydrolases. Several bacteria and fungi produce a group of extra- or ecto-cellular enzymes (enzymes acting outside of but still linked to their cells of origin) that include proteases, carbohydratases, esterases, phosphatases, and phytases. These enzymes are physiologically necessary to living organisms (Gianfreda and Rao, 2004). Some of them (e.g., proteases and carbohydratases) catalyze the hydrolysis of large molecules, such as proteins and carbohydrates, to smaller molecules for subsequent absorption by cells. Due to their intrinsic low substrate specificity, hydrolases may play a pivotal role in bioremediation of several pollutants (Gianfreda and Rao, 2004).

Staszczak et al. (2000) suggested that proteases are involved in the regulation of ligninolytic activities in cultures of *T. versicolor* under nutrient limitation. Margesin et al. (2000a) showed a positive influence of naphthalene on protease activity, and Baran et al. (2004) reported an increase in phosphatase, dehydrogenase, urease, and protease activities in a site in which concentrations of polycyclic aromatic hydrocarbons (PAHs) were higher than 1000 μg/kg.

Mougin et al. (1996) suggested the degradation of the pesticide lindane by *Ph. chrysosporium* via detoxification by a cytochrome P450 monooxygenase system. Cytochrome P_{450}s are hemethiolate proteins that have been characterized in animals, plants, bacteria, and filamentous fungi (Van Eerd et al., 2003). Regulation and expression of P_{450}s are not well understood in plants and microorganisms because of the very low quantities of P_{450}s enzymes usually present in these cells, particularly if the organism has not been exposed to physiochemical, physiological, or xenobiotic stress (Van Eerd et al., 2003).

Cellobiose dehydrogenase (CDH) may be an important enzyme in pollutant degradation (Cameron et al., 2000). CDH is secreted by *Ph. chrysosporium* and several other white rot fungi. It has been shown to directly reduce the munitions 2,4,6-trinitro tolouene and hexahydro-1,3,5 trinitro-1,3,5-triazine (RDX) and indirectly degrade many more chemicals (Cameron et al., 2000).

Whether the degradation of pesticides is carried out by lignin-degrading enzymes or by other enzymatic systems, or by both, the use of microorganisms, especially fungi, in bioremediation is very promising and further studies are required to understand which enzymes are involved in the process. This information is necessary to establish the best conditions for enzyme production and consequent bioremediation *in situ*.

13.5 Screening of Bacteria and White Rot Fungi for Bioremedial Applications

The approach that we have used has been the utilization of an environmental screen to help identify the best bacteria and fungi that might be appropriate for *in situ* testing for the remediation of mixtures of pesticides. This allows us to consider water, temperature, and pH effects on the capacity for tolerance and active degradation of mixtures of pesticides. In our studies, we have examined mixtures of (a) chlorpyrifos, linuron, and metribuzin and (b) simazine, trifluralin, and dieldrin. These have all been widely used as pesticides/herbicides in arable cultivation systems for many years; some of these have relative long half-lives in soil and can cause problems in terrestrial ecosystems and penetrate water courses.

In vitro modifications of soil extract–based media have been employed in which water potential was modified matrically by using PEG8000 as a solute. We have previously shown that many *Trametes* and related species are more sensitive to matric than solute stress (Mougin et al., 1996; Magan, 2007). The applicability of fungi in bioremediation of soil contaminated with pesticides depends on their capacity to grow in the presence of such compounds and their ability to produce degradative enzymes. Complementary information is required on the capacity and ability to produce the key extracellular enzymes that are required for the degradation of individual and mixtures of xenobiotic compounds. A significant amount of research on white rot fungi has been conducted in liquid and/or synthetic media, with less known about bioremediation capabilities in soil, especially under different environmental conditions. Hestbjerg et al. (2003) reported that field conditions did not always enable white rot fungi such as *Ph. chrysosporium* to achieve optimum activity and therefore, it was not a good competitor in the soil environment (Sasek, 2003). This last point was reinforced by Radkte et al. (1994) who reported that bacteria from polluted and agricultural soil antagonized the growth of *Ph. chrysosporium* on solid media. Nevertheless, some studies have described the successful application of *Ph. chrysosporium* as a bioremediation agent in soil. For example, McFarland et al. (1996) described complete alachlor transformation by this fungus, within 56 days of treatment. Reddy and Mathew (2001) also showed that this species was able to degrade DDT, lindane and atrazine.

We have screened a range of bacteria and fungi in soil-based extracts to identify those which may be tolerant and have potential for degradation of such mixtures as opposed to individual pesticides. Figure 13.2 shows an example of the comparison of five strains of *Bacillus* screened against mixtures of three xenobiotic compounds (chlorpyrifos, metribuzin, and linuron) in a soil extract broth medium. Table 13.3 shows the efficacy of different white rot fungal strains in terms of ability to grow

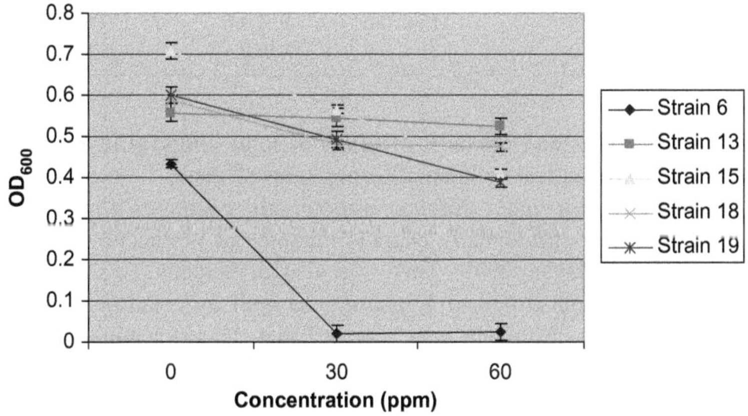

Figure 13.2 Growth (means ± standard error, $n = 3$) of the five strains of *Bacillus* spp. as optical density (OD_{600}) in a mixture of 0, 30, 60 ppm of chlorpyrifos, metribuzin, and linuron after 24 h incubation in soil extract at 25°C.

Table 13.3 Growth Rates (cm/day) in the Presence of 30 ppm Chlorpyrifos, Metribuzin, and Linuron, Individually and as a Mixture 15 and 30 ppm (Total Concentration) on a Mineral Salts Medium at 25°C

Fungi	Treatments					
	Control	Chlorpyrifos (30 ppm)	Metribuzin (30 ppm)	Linuron (30 ppm)	Mixture (15 ppm)	Mixture (30 ppm)
T. versicolor	0.79	0.36	0.79	0.36	0.72	0.72
Ph. chrysosporium	1.09	0.08	0.15	0.04	0.03	0.09
Pl. ostreatus	0.38	0.34	0.38	0.14	0.18	0.13
Pe. gigantea	0.47	0.37	0.38	0.33	0.40	0.34
Py. coccineus	0.53	0.28	0.43	0.55	0.48	0.45

They are the means of three replicates. The variances of all treatments were <5% of the means

Table 13.4 Effect of Simazine, Trifluralin, and Dieldrin (0, 5, and 10 mg/L) Individually and as a Mixture on Ligninolytic Activity by *T. versicolor* (R101) and *Pl. ostreatus* at 15°C in Relation to Solute Water Potential (Expressed as Radius of Enzymatic Clearing Zone ± Standard Deviation of the Mean, $n = 3$)

	Ψ (MPa)	*T. versicolor* (R101)		*Pl. ostreatus*	
		−0.7	−2.8	−0.7	−2.8
Simazine (mg/L)	0	21 ± 0.6	11 ± 1.0	40 ± 0.0	27 ± 0.7
	5	21 ± 0.6	10 ± 1.0	21 ± 0.4	13 ± 0.4
	10	16 ± 0.6	13 ± 1.0	21 ± 1.2	10 ± 0.1
Trifluralin (mg/L)	5	14 ± 0.6	12 ± 0.5	35 ± 0.6	25 ± 0.6
	10	11 ± 0.6	8 ± 0.6	28 ± 0.6	19 ± 0.5
Dieldrin (mg/L)	5	15 ± 1	10 ± 1.5	36 ± 1.5	13 ± 1.9
	10	11 ± 1	6 ± 1.0	25 ± 0.2	15 ± 0.6
Mixture (mg/L)	5	15 ± 0.6	8 ± 0.6	25 ± 0.6	22 ± 1.5
	10	21 ± 0.6	9 ± 0.6	17 ± 0.6	18 ± 1.0

on individual pesticides and mixtures of same. In some cases, growth is modified by the exposure to individual compounds or mixtures of them.

We demonstrated that under different osmotic stress regimes a range of white rot fungi were able to differentially degrade mixtures of pesticides in soil extract broth (Fragoeiro and Magan, 2005). There was also an increase in a range of hydrolytic enzyme production including ligninases, as well as cellulases, even under water stress conditions. Although it is accepted that the extracellular ligninolytic enzymes are at least in part responsible for the critical initial reactions of pollutant transformation, the production and activity of these enzymes in contaminated soil under different field conditions have not been examined in detail, although they are critical for successful degradation (Lang et al., 1998, 2000a,b).

The question arises as to whether environmental factors will further modify the ability of such bacteria or fungi to tolerate and utilize mixtures of pesticides. Table 13.4 shows the relative lignolytic ability of the different white rot fungi in the presence of individual and mixtures of simazine, trifluralin, and dieldrin under freely available water potential conditions (−0.7 MPa) and water stress represented by twice the wilting point of plants (−2.8 MPa). This shows the changes in activity that can occur under different environmental conditions and single versus mixtures of pesticides.

13.6 Degradation of Pesticide Mixtures by Bacteria and Fungi

When examining degradation of mixtures of pesticides, there are two key indicators of successful degradation. These are the production of enzymes by the candidate bioremedial microorganisms and measurement of the degradation rates using quantifiable analytical techniques. It is also important to ensure that the breakdown products are not also toxic in the ecosystem. When examining the remediation of mixtures of pesticides, it must also be possible to analyze for each of the component pesticides at the required threshold levels either individually or in a single extraction with the right High Pressure Liquid Chromatography (HPLC) method. For both these groups of pesticides, it was possible to develop HPLC methods to allow this to be achieved.

Studies by Fragoeiro and Magan (2005) showed that mixtures of simazine, trifluralin, and dieldrin could be degraded in soil-based broth as well as in soil microcosms at different water potentials over medium-term storage conditions. In this case, the inoculants were added as a wood chip-based inoculum. This suggested that there is a differential rate of breakdown of the component pesticides in the mixture with some being degraded faster than others after 6 and 12 weeks in microcosms (Table 13.5). Yadav and Reddy (1993) described co-mineralization of a mixture of the pesticides 2, 4-D and 2,4,5-T by a wild type strain of *Ph. chrysosporium* and a putative peroxidase mutant in nutrient-rich broth with a small amount remaining in the mycelial fractions (5%). Bending et al. (2002) showed degradation rates of metalaxyl, atrazine, terbuthylazine, and diuron by white rot fungi in nutrient solution of >86% for atrazine and terbuthylazine. The studies by Fragoeiro and Magan (2005) showed that enzymatic activity, osmotic stress, and degradation of pesticide mixtures in soil extract liquid broth inoculated with *T. versicolor* and *Ph. chrysosporium* suggested that both fungal strains had the ability to degrade the pesticides simazine, dieldrin, and trifluralin supported by the capacity for expression of a range of extracellular enzymes at −0.7 and −2.8 MPa water potential. *T. versicolor* in particular was able to produce phosphomonoesterase, protease (relevant to P and N release), β-glucosidase, cellulase (carbon cycling), and laccase activity.

Table 13.5 Comparison of Effect of Wood Chips and Fungal Inoculants on Percentage Pesticide (%) (Simazine, Trifluralin, Dieldrin, 10 mg/kg) Degraded After 6 and 12 Weeks at −7.0 and −2.8 MPa Water Potentials in Soil Microcosms at 15°C

Incubation (weeks)	Water Potential (− MPa)	Treatment	Percentage Pesticide Degraded		
			Simazine	Trifluralin	Dieldrin
6	0.7	Wood chips	41.4* *(2.5)*	56.0 *(58.4)*	71.2 *(23.7)*
		T. versicolor	89.9*	77.7	48.2*
		Ph. chrysosporium	63.8*	74.7	87.3
6	2.8	Wood chips	13.8 *(21.2)*	75.2 *(57.1)*	61.8 *(40.0)*
		T. versicolor	57.1*	81.7*	70.7*
		Ph. chrysosporium	64.4*	85.5*	69.9*
12	0.7	Wood chips	46.6* *(27.5)*	67.5 *(62.4)*	79.4 *(53.8)*
		T. versicolor	73.5*	76.5	52.7
		Ph. chrysosporium	75.6*	57.3	100*
12	2.8	Wood chips	75.7* *(29.9)*	92.1* *(64.2)*	61.6 *(40.2)*
		T. versicolor	57.3	80.9*	51.0
		Ph. chrysosporium	64.3*	93.7*	79.7*

Figures in parentheses are for comparison with degradation in natural soil.
*Significantly different from the controls based on actual concentration using HPLC ($P = 0.05$).

The results suggest that laccase is not only secreted in nutrient-rich substrates but is produced by mycelia growing in weak nutritional matrices. Of particular interest is the capacity of *T. versicolor* for laccase production in the presence of up a 50 ppm mixture of the pesticides. This presence was shown to stimulate laccase production as, in the treatment of 0 ppm, without stress laccase production was very low related to all the other concentrations over the first 4-week period. It reached the same levels at the 5th and 6 weeks of incubation. Recent studies demonstrated induction of production of ligninolytic enzymes, particularly laccase, in the presence of copper, veratryl alcohol, and a phenolic mixture, in the presence of mixtures of pesticides (Fragoeiro and Magan, 2005). Moreover, studies of some agrochemicals, industrial compounds, and their transformation products were shown to enhance laccase production in liquid cultures of *T. versicolor.* Many of them enhanced laccase activity up to 20-fold having as positive control 2, 5-xylidine (35-fold enhancement of laccase activity).

Highly significant increases in laccase produced by *Trametes hirsuta* growing in an air-lift bioreactor after the addition of glycerol has been observed. Fragoeiro and Magan (2008) studied the impact of water potential on mixtures of the pesticides simazine, dieldrin, and trifluralin, in soil inoculated with *T. versicolor* and *Ph. chrysosporium* in relation to different soil water potentials (−0.7 and −2.8 MPa). The researchers showed that in natural soil the level of laccase produced by *T. versicolor* was very low, reaching the highest levels after 6 weeks' incubation under both water regimes, whereas that amended with wood chips showed some laccase production, with the highest level after 12 weeks' incubation. Overall, the main difference between laccase production in soil microcosms and that in soil extract broth was that in soil extract–based liquid culture, laccase production was much higher, at −0.7 MPa, while in soil microcosms the optimum was at −2.8 MPa.

Total ligninolytic activity was expressed as the capacity to decolorize poly-R478. The decoloration assay of poly-R478, with a similar structure to lignin, gives information on the activity of the whole set of enzymes because the degradation of lignin is carried out by several enzymes. The results showed that decoloration of this dye occurred in all treatments, with the highest levels of decoloration after 5 weeks' incubation at −2.8 MPa water potential adjusted with KCl. In all cases, the decoloration rates were unaffected by pesticide treatment. The results suggest *T. versicolor* was tolerant to this mixture of pesticides, producing equivalent levels of decoloration in the presence and absence of the xenobiotics. The highest levels of decoloration that occurred in the treatments with KCl may be related to the production of other enzymes, as at −2.8 MPa water potential adjusted with KCl no laccase production was observed.

There was no correlation between decolorization of the dye and degradation of the pesticide mixture in this study. A similar result was described for degradation of diuron, metalaxyl atrazine, and terbuthylazine, by several fungi in liquid culture (Bending et al., 2002) as well as for the degradation of simazine, trifluralin, and dieldrin in soil (Fragoeiro and Magan, 2008), although Alcalde et al. (2002) observed correlation with oxidation of PAHs mediated by laccases.

This study on the degradation of the mixture of pesticides showed good capacity of *T. versicolor* to degrade linuron at all tested water regimes, regardless of the initial concentrations of mixture between 0 and 50 ppm. Metribuzin degradation occurred only in the treatments at −2.8 MPa solute potential adjusted with glycerol. A possible explanation, maybe, is that *T. versicolor* is unable to utilize this pesticide as a carbon source for its growth, and, as soil extract broth is a weak nutrient, medium glycerol was probably used by the fungus as an additional carbon source.

Interestingly, chlorpyrifos and its main metabolite, TCP, were not detected. The environmental fate of this pesticide has been studied extensively. The manufacturer reports that chlorpyrifos is a degradable compound, and a number of environmental forces may be active in its breakdown. In all systems (soil, water, plants, and animals), the major pathway of degradation begins with cleavage of the phosphorus ester bond to yield 3,5,6-trichloro-2-pyridinol (TCP). In soil and water, TCP is further degraded via microbial activity and photolysis to carbon dioxide and organic matter. Hydrolytic and photolytic half-lives are both around a month, at neutral pH 25°C. Under more alkaline conditions, hydrolysis proceeds more rapidly. In natural water samples, however, degradation often proceeds significantly faster: a 16-fold enhancement of hydrolysis rate has been observed in pond and canal water samples. Half-lives in the water column of less than one day are typical, due to a combination of degradation, volatilization, and partitioning into sediments.

Several published studies on the biodegradation of chlorpyrifos in liquid media have reported no significant rates of abiotic degradation of chlorpyrifos. Particularly, no abiotic degradation of chlorpyrifos in uninoculated media over a 4-day period was reported by Singh et al. (2006) over a 48 and 15 h period (Xu et al., 2008), respectively. However, the question arises of how to utilize these inoculants in terrestrial ecosystems to degrade mixtures of pesticides.

13.7 Inoculant Production for Soil Incorporation of Bioremedial Fungi

There are many studies regarding how to optimize the biodegradation potential of white rot microbial inoculants in contaminated soil (Ryan et al., 1989; Morgan et al., 1991; McFarland et al., 1996; Meysami and Baheri, 2003). If it is accepted that the extracellular ligninolytic enzymes are at least in part responsible for the critical initial reactions of pollutant transformation, the production and activity of these enzymes in contaminated soil under field conditions are two prerequisites for successful application of white rot fungi in soil bioremediation (Lang et al., 2000b).

In natural soil, a wide range of saprophytic microorganisms exist. Introduction of bacteria or white rot fungi requires effective growth and competition with these native populations. Additionally, the activity of bacteria or fungi must be stimulated to secrete the necessary enzymes into the soil matrix to enhance degradation of pesticide molecules that they would otherwise be unable to incorporate across cell walls (Canet et al., 2001).

Most of the protocols for delivering inoculum of either bacteria or wood rot fungi for soil bioremediation have adopted approaches used in the mushroom production industry. They have optimized the production of fungal spawn on lignocellulosic waste. Fungal species used in mushroom production have been formulated on inexpensive substrates such as corn cobs, sawdust, wood chips, peat, and wheat straw. When used in bioremediation, these substrates are impregnated with cells or mycelium and mixed with contaminated soil (Paszczynski and Crawford, 2000; Reddy et al., 2001). While bacteria can be used as a soil drench or on carriers such as immobilized substrates, e.g., alginate beads, from which they can grow and colonize the water films in the soil, they require enough water to allow rapid growth to occur balanced with the soil remaining aerobic. There is little information available on survival of white rot fungi in soil, especially fungi that are not used for human consumption. A number of studies have investigated ways to improve the survival of wood rot fungi in polluted soils (Ryan et al., 1989; Morgan et al., 1991; Gold et al., 2001). Certainly, better fungal growth could help introduced fungi to overcome competition from indigenous microorganisms and enhance bioremediation. This is critical as native soil microorganisms may occupy the lignocellulosic substrate and restrain growth and activity of fungal inoculants, inhibiting fungal lignino-cellulose decomposition and reducing the ability for enzyme release (Lang et al., 1998).

The introduction ratio will have an important impact on the economics of practical applications. For example, in studies by Fragoeiro and Magan (2008) a ratio of 5 g inoculant to 95 g soil was used. The effect of using this approach on differential breakdown of mixtures of xenobiotic compounds under different environmental regimes is shown in Table 13.4. Other authors have used very different ratios. For example, Novotny et al. (2003) described dye degradation in soil using a 50:50 soil:straw-based inoculant of the fungus *Irpex lacteus*; Canet et al. (2001) used a 40% incorporation rate with straw-based inoculum; Ryan and Bumpus (1989) used a 25% straw-based inoculum; Elyassi (1998) used a 10% straw inoculum; and Morgan et al. (1991) used 4 g ground maize cobs to 1 g soil. We believe that some of these are very unrealistic from a practical and economic point of view for bioremediation of single or mixtures of xenobiotics in contaminated soils. Furthermore, few if any have examined the impact of water potential or effect on mixtures of pesticides. Novotny et al. (1999) used the filamentous white rot fungus *T. versicolor* and *Pleurotus ostreatus*, and found the latter species to be better than both *Ph. chrysosporium* and *T. versicolor*. However, they used sterile soil only, devoid of any of the natural microbial communities that would be present. This is unrealistic and gives higher degradation rates than would be achieved in practice in naturally contaminated soils.

Boyle (1995) reported an increase in growth and carbon dioxide production in natural soil supplemented with carbon amendments and observed that mineralization of [^{14}C] pentachlorophenol (PCP) (degradation to $^{14}CO_2$) was much faster in soil that had been amended with alfalfa and benomyl and inoculated with the white rot species *T. versicolor*. Another study showed that the addition of straw increased the hyphal length of white rot fungi in soil (Morgan et al., 1991). Besides strong growth capabilities, it is important that the inoculation conditions promote enzyme production. Moredo et al. (2003) investigated ligninolytic enzyme production by the white rot fungi *Ph. chrysosporium* and *T. versicolor* precultivated on different insoluble lignocellulosic materials: grape seeds, barley bran, and wood shavings. Cultures of *Ph. chrysosporium* pregrown on grape seeds and barley bran showed maximum lignin peroxidase and manganese-dependent peroxidase activities (1000 and 1232 units/L, respectively). *T. versicolor* precultivated with the same lignocellulosic residues showed maximum laccase activity (approx. 250 units/L). The utilization of these solid substrates in soil may also be advantageous as a means to distribute fungal inoculum evenly in large volumes of soil. According to Singleton (2001), growth amendments could also exert beneficial effects by sorbing pollutants and hence decreasing the amount of toxic pollutant that is bioavailable.

13.8 Use of SMCs

Composting matrices and composts are rich sources of xenobiotic-degrading microorganisms including bacteria, actinomycetes, and ligninolytic fungi, which can degrade pollutants to innocuous compounds such as carbon dioxide and water. These microorganisms can also biotransform pollutants into less toxic substances and/or immobilize pollutants within the organic matrix, thereby reducing pollutant bioavailability.

SMC is a by-product of the mushroom production, which is produced in large amounts. For every 200 g of *Pleurotus* spp. produced in Malaysia, about 600 g of spent compost is produced (Singleton, 2001) or 5 kg of SMC generated for every 1 kg of edible mushrooms according to Law et al. (2003). This resulted in 40 million tonnes of SMC in 1999. The disposal of SMC is a major problem for mushroom farmers. They either discretely burn or discard it (Singleton, 2001), and thus its exploitation as a potential bioremediation adjuvant has received significant attention (Chiu et al., 2001).

Mushroom cultivation involves the pure culture of spawn, composting and pasteurization of the substrate, and careful regulation of growing conditions. The substrates are lignocellulosic residues, such as straw, horse manure, chicken manure, and activators. The purpose of composting the substrate is to exclude microorganisms that may interfere with mushroom growth. Following mushroom harvest, SMC is likely to contain not only a large and diverse group of microorganisms but also a wide range of extracellular enzymes active against wheat straw. Singh et al. (2005) reported extraction of cellulase, hemicellulose, β-glucosidase, lignin peroxidases, and laccase from SMC. It also contains a very high organic content (20%) including cellulose, hemicellulose, and lignin (Kuo and Regan, 1999) from the unutilized lignocellulosic substrate (Singh et al., 2003).

Previous research showed some interesting findings using this type of compost as a bioremediation adjuvant: Law et al. (2003) reported that SMC of *Pleurotus pulmonarius* could remove $89.0 \pm 0.4\%$ of 100 mg PCP/L within 2 days at room temperature predominantly by biodegradation. Kuo and Regan (1999) used sterilized SMC as an adsorption medium for removal of a mixture of pesticides (carbaryl, carbofuran, and aldicarb) with a concentration range of 0–30 mg/L and found that SMC was able to adsorb carbamate pesticides from aqueous solutions successfully, which was possibly related to the increased organic matter content.

With mushroom production being the largest solid-state fermentation industry in the world (Lau et al., 2003), and with so much waste being produced, it is extremely important to find a good use for SMC. Thus, the use of this residue as a soil amendment to improve pesticide bioremediation is an interesting area for research. Furthermore, there is no information on the effect of the addition of SMC on soil enzymes, soil respiration, and soil populations, and how these metabolic parameters are affected by the presence of pesticides and water availability.

We have examined the use of SMCs by mixing with unsterile soil at a rate of 5 g SMC to 95 g unsterile sandy loam soil treated with 10 mg/kg soil on a mixture of pesticides (simazine, trifluralin, dieldrin) using a moisture adsorption curve to achieve −0.7 and 2.8 MPa water potentials (Fragoeiro and Magan, 2008; Bastos and Magan, 2009). These treatments were stored for 42 and 84 days at 15°C. The amount of CO_2 produced was measured using Gas Chromatography (GC) analysis, as well as the total ligninolytic activity, while HPLC with UV detection was used to analyze the amounts of each pesticide remaining in each of the treatments. The soil amended with the SMC had the highest microbial activity based on respiration measurements, regardless of water potential and pesticide treatments. Figure 13.3 shows that in all cases the SMC amended treatments (− and + pesticides) had the highest ligninolytic activity after 42 and 84 days. Table 13.6 shows the effect of SMC on the differential breakdown of mixtures of three pesticides under different water potential treatments and time. This demonstrates that there was a significant enhanced breakdown of all three pesticides by the SMC inoculation after both 42 and 84 days for both water potential regimes. This demonstrates that

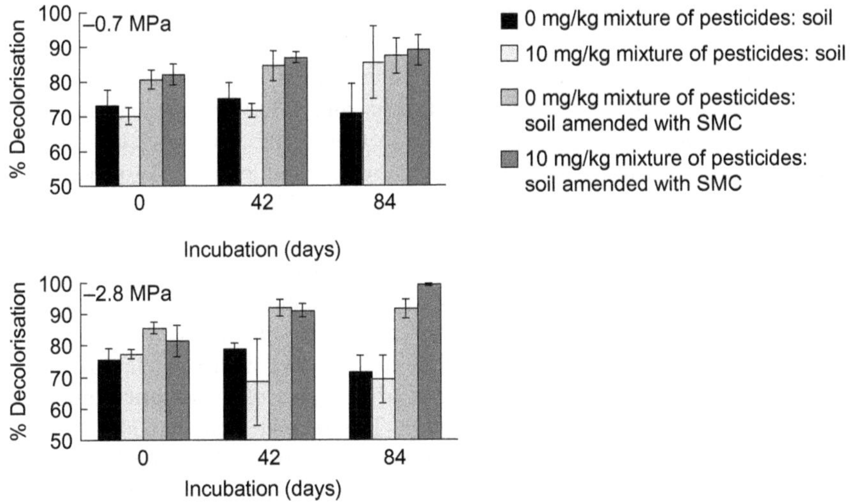

Figure 13.3 The effect of SMC on relative lignolytic enzyme production when a mixture of three pesticides (simazine, trifluralin, and dieldrin) was present in soil microcosms for 42 and 84 days at two different water potentials.

Table 13.6 Comparison of Percentage Pesticide Concentration Remaining in Soil Amended with SMC and Supplemented with a Pesticide Mixture at 10 mg/kg Soil, After 42 and 84 days' Incubation at 15°C, Under Two Different Water Regimes. LSD, Least Significant Difference (P = 0.05)

Time (days)	Water Potential (MPa)	(% Remaining)					
		Simazine		Trifluralin		Dieldrin	
		Soil + SMC	Control	Soil + SMC	Control	Soil + SMC	Control
42	− 0.7	13.2*	97.5	40.0	24.6	39.5	20.9
	− 2.8	37.5*	78.8	27.4	42.9	25.4*	60.0
		L.S.D. = 15.20		L.S.D. = 42.28		L.S.D. = 22.10	
84	− 0.7	18.2*	72.5	21.8	37.6	19.9*	46.5
	− 2.8	58.1*	70.1	18.4*	35.8	14.5*	59.8
		L.S.D. = 14.17		L.S.D. = 16.28		L.S.D. = 21.76	

*Means there was a significant difference between the control and the treatment amended with SMC. Pesticides were analyzed using HPLC with UV detection.

SMC contains a mixture of ligninolytic enzymes including laccases, actinomycetes, and, in some cases, basidiomycete hyphae. This may be an effective approach for practical applications in bioremediation systems. The advantages include the maintenance of an aerobic soil structure, moisture retention at effective rates, and preventing anaerobic conditions from occurring.

13.9 Conclusions and Future Strategies

Application of fungal technology for the cleanup of contaminants has shown promise since the 1980s when some bacteria and the white rot basidiomycete *Ph. chrysosporium* were found to be able to metabolize a number of important environmental pollutants. This led to strains being commercialized for treatments of contaminated soil-based xenobiotic contaminants. There have been many studies that have shown that both bacteria and filamentous fungi *in vitro* and *in situ* in soil microcosms have demonstrated the efficacy for degradation. However, much work has been concentrated on single xenobiotic compounds, despite the fact that in contaminated land, xenobiotic compounds are almost always present in mixtures. It is thus surprising that so few studies have addressed this important aspect, that is, the differential degradation of mixtures of xenobiotic compounds in natural soil ecosystems (Gouma, 2009). The second area that has often not been addressed is the impact of environmental conditions. Most research is carried out under largely optimal conditions for growth/colonization by the microbial inoculants. It should be noted that a wide range of bacteria and fungi are able to grow over a much wider range of water potentials than that allowing plant growth (limited to $\geq$1.4 MPa). Thus, beyond the wilting point of plants, bioremediation can continue to function effectively (Fragoeiro and Magan, 2008; Bastos and Magan, 2009). This may become very important when we examine the different interacting factors that will affect bioremediation by fungi, and indeed all microorganisms. Climate change aspects will have to be borne in mind. Thus, elevated CO_2 concentration, and slightly increased temperature conditions may well have subtle but significant impact on functioning of terrestrial ecosystems and any introduced microorganisms for bioremedial reasons. Practically no work has been done to address these aspects.

The other area that will hasten development of remediation approaches is the molecular unraveling of the genomes of specific microbial species. For example, the *Ph. chrysosporium* genome work has resulted in the elucidation of specific cytochrome P450 monooxygenases that may be differentially expressed in the presence of xenobiotic compounds. Knowledge of these gene clusters and their relative expression can now be quantified and integrated with data on ecological and physiological factors, which can give a more complete approach to understanding the mechanisms of action and functioning of fungal bioremediation systems. This could result in a more integrated "systems" approach to facilitate a much better exploitation of bioremedial fungi for treating xenobiotics in terrestrial ecosystems.

Finally, from a practical point of view, there is significant interest in organic-based bioremediation amendments for degrading organic-based contaminants including hydrocarbons, PAHs, petroleum, diesel, BTEX, and semivolatile organic compounds. Thus biopiles, where control of environmental parameters including aeration, temperature, and moisture content can be controlled, have been utilized effectively. This can stimulate the rapid microbial population development of indigenous and introduced inoculants. This can provide both more rapid degradation rates and improve the economics of treatments. Another alternative has been the use of *ex situ* bioremediation approaches, which has involved the excavation of contaminated soils which are then placed in specific treatment area that is lined in order to prevent cross-contamination with the underlying ground. Windrows have also been used to stockpile excavated soils, which can be turned for aeration and also be aerated if necessary. This does require the inoculant formulation to be viable and to have the capacity for rapid colonization of the soil windrows. In some cases, nutrient addition has been made to contaminated soil to enhance natural microbial populations to grow and enhance the breakdown of xenobiotic compounds.

References

Alcalde, M., Bulter, T., Arnold, F., 2002. Colorimetric assays for biodegradation of polycyclic aromatic hydrocarbons by fungal laccases. J. Biomol. Screening. 7, 547–553.

Alleman, B., Logan, B., Gilbertson, R., 1992. Toxicity of pentachlorophenol to six species of white rot fungi as function of chemical dose. Appl. Environ. Microbiol. 58, 4048–4050.

Arisoy, M., 1998. Biodegradation of chlorinated organic compounds by white rot fungi. Bull. Environ. Contam. Toxicol. 60, 1711–1718.

Armstrong, D.E., Chesters, G., Harris, R.F., 1967. Atrazine hydrolysis in soil. Soil Sci. Soc. Am. 31, 61–66.

Baran, S., Bielinska, J., Oleszuk, P., 2004. Enzymatic activity in an airfield soil polluted with polycyclic aromatic hydrocarbons. Geoderma. 110, 221–232.

Barr, D., Aust, S., 1994. Mechanisms white rot fungi use to degrade pollutants. Environ. Sci. Technol. 28, 78–87.

Bastos, A.C., Magan, N., 2009. *Trametes versicolor*: potential for atrazine bioremediation in calcareous clay soil, under low water availability conditions. Int. Biodeterior. Biodegrad. 63, 389–394.

Behki, R.M., Khan, S.U., 1994. Degradation of atrazine, propazine and simazine by *Rhodococcus* strain B-30. J. Agric. Food Chem. 42, 1237–1241.

Behki, R.M., Topp, E., Dick, W., Germon, P., 1993. Metabolism of the herbicide atrazine by *Rhodococcus* strains. Appl. Environ. Microbiol. 59, 1955–1959.

Bending, G.D., Friloux, M., Walker, A., 2002. Degradation of contrasting pesticides by white rot fungi and its relationship with ligninolytic potential. FEMS Microbiol. Lett. 212, 59–63.

Bezalel, L., Hadar, Y., Cerniglia, C., 1996. Mineralization of polycyclic aromatic hydrocarbons by the white rot fungus *Pleurotus ostreatus*. Appl. Environ. Microbiol. 62, 292–295.

Bhat, M.A., Tsuda, M., Horiike, K., Nozaki, M., Vaidyanathan, C.S., Nakazawa, T., 1994. Identification and characterisation of a new plasmid carrying genes for degradation of 2, 4-ichlorophenoxyacetate from *Pseudomonas cepacia* CSV90. Appl. Environ. Microbiol. 60, 307–312.

Boyle, D., 1995. Development of a practical method for inducing white rot fungi to grow into and degrade organopollutants in soil. Can. J. Microbiol. 41, 345–353.

Cameron, M.D., Timofeevski, S., Aust, S.D., 2000. Enzymology of *Phanerochaete chrysosporium* with the respect to the degradation of recalcitrant compounds and xenobiotics. Appl. Microbiol. Biotechnol. 54, 751–758.

Canet, R., Birnstingl, J., Malcolm, D., Lopez-Real, J., Beck, A., 2001. Biodegradation of polycyclic aromatic hydrocarbons (PAHs) by native microflora and combinations of white-rot fungi in a coal-tar contaminated soil. Bioresour. Technol. 76, 113–117.

Chandhry, G.R., Ali, A.N., 1988. Bacterial metabolism of carbofuran. Appl. Environ. Microbiol. 54, 1414–1419.

Chiu, S., Law, S., Ching, M., Cheung, K., Chen, M., 2001. Themes for mushroom exploitation in the 21st century: sustainability, waste management, and conservation. J. Gen. Appl. Microbiol. 46, 269–282.

Chivukula, M., Renganathan, V., 1995. Phenolic azo dye oxidation by laccase from *Pyricularia oryzae*. Appl. Environ. Microbiol. 61, 4374–4377.

Cook, A.M., 1987. Biodegradation of s-triazine xenobiotics. FEMS Microbiol. Rev. 46, 93–116.

Duran, N., Esposito, E., 2000. Potential applications of oxidative enzymes and phenoloxidase-like compounds in wastewater and soil treatment: a review. Appl. Catal. B Environ. 28, 83–99.

Elyassi, A. 1998. Bioremediation of the pesticides dieldrin, simazine, and trifluralin using tropical and temperate white rot fungi. PhD Thesis, Cranfield University, Bedford, U.K. MK43 0AL.

Field, J., Jong, E., Feijo-Costa, G., Bont, J., 1993. Screening for ligninolytic fungi applicable to the biodegradation of xenobiotics. Trends Biotechnol. 11, 44–49.

Fragoeiro, S., 2005. Use of Fungi in Bioremediation of Pesticides (PhD Thesis). Cranfield University, Bedford, UK. Mk43 0AL.

Fragoeiro, S., Magan, N., 2005. Enzymatic activity, osmotic stress and degradation of pesticide mixtures in soil extract liquid broth inoculated with *Phanerochaete chrysosporium* and *Trametes versicolor*. Environ. Microbiol. 7, 348–355.

Fragoeiro, S., Magan, N., 2008. Impact of *Trametes versicolor* and *Phanerochaete crysosporium* on differential breakdown of pesticide mixtures in soil microcosms at two water potentials and associated respiration and enzyme activity. Int. Biodeterior. Biodegrad. 62, 376–383.

Gadd, G. (Ed.), 2001. Fungi in Bioremediation. Cambridge University Press, Cambridge, UK.

Gianfreda, L., Rao, M.A., 2004. Potential of extra cellular enzymes in remediation of polluted soils: a review. Enzyme Microb. Technol. 35, 339–354.

Glenn, J.K., Gold, M.H., 1983. Decolourization of special polymeric dyes by the lignin degrading basidiomycete *Phanerochaete chryosporium*. Appl. Environ. Microbiol. 45, 1741–1747.

Gold, M., Glenn, J., Alic, M., 1988. Use of polymeric dyes in lignin biodegradation assays. Methods Enzymol. 161, 74–78.

Gouma, S., 2009. Biodegradation of Mixtures of Pesticides by Bacteria and White Rot Fungi (PhD thesis). Cranfield University. Bedford, U.K. MK43 0AL.

Grigg, B.C., Bischoff, M., Turco, R.F., 1997. Co-contaminant effects on degradation of triazine herbicides by a mixed microbial culture. J. Agric. Food Chem. 45, 995–1000.

Haggblom, M.M., 1992. Microbial breakdown of halogenated aromatic pesticides and related compounds. FEMS Microbiol. Rev. 103, 29–72.

Hatakka, A., 1994. Lignin-modifying enzymes from selected white rot fungi: production and role in lignin degradation. FEMS Microbiol. Rev. 13, 125–135.

Head, I.M., Cain, R.B., Suett, D.L., 1992. Characterisation of a carbofuran-degrading bacterium and investigation of the role of plasmids in catabolism of the insecticide carbofuran. Arch. Microbiol. 158, 302–308.

Hernando, M.D., Ejerhoon, M., Fernandez-Alba, A.R., Chisti, Y., 2003. Combined toxicity effects of MTBE and pesticides measured with *Vibrio fischeri* and *Daphnia magna* bioassays. Water Res. 37, 4091–4098.

Hestbjerg, H., Willumsen, P., Christensen, M., Andersen, O., Jacobsen, C., 2003. Bioaugmentation of tar-contaminated soils under field conditions using *Pleurotus ostreatus* refuse from commercial mushroom production. Environ. Toxicol. Chem. 22, 692–698.

Hickey, W., Fuster, D., Lamar, R., 1994. Transformation of atrazine in soil by *Phanerochete chrysosporium*. Soil Biol. Biochem. 26, 1665–1671.

Hohener, P., Hunkeler, D., Hess, A., Bregnard, T., Zeyer, J., 1998. Methodology for the evaluation of engineered *in situ* bioremediation: lessons from a case study. J. Microbiol. Methods. 32, 179–192.

Karns, J.S., Mulbry, W.W., Nelson, J.O., Kearny, P.C., 1986. Metabolism of carbofuran by a pure bacterial culture. Pestic. Biochem. Physiol. 25, 211–217.

Khadrani, A., Siegle-Murandi, F., Steinman, R., Vrousami, T., 1999. Degradation of three phenylurea herbicides (chlortorulon, isoproturon and diuron) by *Micromycetes* isolated from soil. Chemosphere. 38, 3041–3050.

Kuo, W., Regan, R., 1999. Removal of pesticides from rinsates by adsorption using agricultural residues as medium. J. Sci. Health B. 34, 431–447.

Lang, E., Gonser, A., Zadrazil, F., 2000a. Influence of incubation temperature on activity of ligninolytic enzymes in sterile soil by *Pleurotus* sp. and *Dichomitus squalens*. J. Basic Microbiol. 40, 33–39.

Lang, E., Kleeberg, I., Zadrazil, F., 2000b. Extractable organic carbon and counts of bacteria near the lignocellulose-soil surface interface during the interaction of soil microbiota and white rot fungi. Bioresour. Technol. 75, 57–65.

Lau, K., Tsang, Y., Chiu, S., 2003. Use of spent mushroom compost to bioremediate PAH-contaminated samples. Chemosphere. 52, 1539–1546.

Law, W., Lau, W., Lo, K., Wai, L., Chiu, S., 2003. Removal of biocide pentachlorophenol in water system by the spent mushroom compost of *Pleurotus pulmonarius*. Chemosphere. 52, 1531–1537.

Lin, J., Chang, D., Sheng, G., Wang, H., 1991. Correlations among several screening methods used for identifying wood decay fungi that can degrade toxic chemicals. Biotechnol. Tech. 5, 275–280.

Mackay, D., Frasar, A., 2000. Bioaccumulation of persistent organic chemicals: mechanisms and models. Environ. Pollut. 110, 375–391.

Magan, N., 2007. Ecophysiology: impact of environment on growth, synthesis of compatible solutes and enzyme production. In: Boddy, L., Frankland., J.C. (Eds.), Ecology of Saprotrophic *Basidiomycetes*. Elsevier Ltd, Amsterdam, Holland.

Mandellbaum, R.T., Allan, D.L., Wackett, L.P., 1995. Isolation and characterisation of a *Pseudomonas* sp. that mineralises the s-triazine herbicide atrazine. Appl. Environ. Microbiol. 61, 1451–1457.

Margesin, R., Walder, G., Schinner, F., 2000a. The impact of hydrocarbon remediation (diesel oil and polycyclic aromatic hydrocarbons) on enzyme activities and microbial properties of soil. Acta Biotechnol. 20, 313–333.

McFarland, M., Salladay, D., Ash, D., Baiden, E., 1996. Composting treatment of alachlor impacted soil amended with the white rot fungus *Phanerochaete chrysosporium*. Hazard. Waste Hazard. Mater. 13, 363–373.

Memic, M., Vrtacnik, M., Vatrenjak-Velagic, V., Wissiak Grm, K.S., 2005. Comparative biodegradation studies of pre-emergence broadleaf and grass herbicides in aqueous medium. Int. Biodeterior. Biodegrad. 55, 109–113.

Meysami, P., Baheri, H., 2003. Pre-screening of fungi and bulking agents for contaminated soil bioremediation. Adv. Environ. Res. 7, 881–887.

Moredo, N., Lorenzo, M., Domiguez, A., Moldes, D., Cameselle, C., Sanroman, A., 2003. Enhanced ligninolytic enzyme production and degrading capability of *Phanerochaete chrysosporium* and *Trametes versicolor*. World J. Microbiol. Biotechnol. 19, 665–669.

Morgan, P., Lewis, S., Watkinson, R., 1991. Comparison of abilities of white-rot fungi to mineralize selected xenobiotic compounds. Appl. Microbiol. Biotechnol. 14, 691–696.

Mougin, C., Pericaud, C., Malosse, C., Laugero, C., Asther, M., 1996. Biotransformation of the insecticide lindane by the white rot basidiomycete *Phanerochaete chrysosporium*. Pestic. Sci. 47, 51–59.

Nadeau, L.J., Menn, F.M., Breen, A., Sayler, G.S., 1994. Aerobic degradation of 1,1.1-trichloro-2,2-bis(4-chlorophenyl)ethane (DDT) by *Alcaligenes eutrophus* A5. Appl. Environ. Microbiol. 60, 51–55.

Novotny, C., Erbanova, P., Sasek, V., Kubatova, A., Cajthaml, T., Lang, E., et al., 1999. Extracellular oxidative enzyme production and PAH removal in soil by exploratory mycelium of white rot fungi. Biodegradation. 10, 159–168.

Novotny, C., Rawal, B., Bhatt, M., Patel, M., Sazek, V., Molitoris, H., 2003. Screening of fungal strains for remediation of water and soil contaminated with synthetic dyes. In: Sasek, V., et al., (Eds.), The Utilization of Bioremediation to Reduce Soil Contamination: Problems and Solutions. Kluwer Academic Publishers, Dordrecht, The Netherlands, pp. 143–149.

Nyanhongo, G., Gubitz, G., Sukyai, P., Leitner, C., Haltrich, D., Ludwig, P., 2007. Oxidoreductases from *Trametes* spp. in biotechnology: a wealth of catalytic activity. Food Technol. Biotechnol. 45, 250–268.

Paszczynski, A., Crawford, R., 2000. Recent advances in the use of fungi in environmental remediation and biotechnology. In: Bollag, J.-M., Stotzky, G. (Eds.), Soil Biochemistry, 10. Marcel Dekker, New York, NY, pp. 379–422.

Patel, M.N., Gopinathan, K.P., 1986. Lysozyme-sensitive bio emulsifier for immiscible organophosphorus pesticides. Appl. Environ. Microbiol. 52, 1224–1226.

Pemberton, J.M., Fisher, P.R., 1977. 2, 4-D plasmids and persistence. Nature. 268, 732–733.

Pointing, S., 2001. Feasibility of bioremediation by white rot fungi. Appl. Microbiol. Biotechnol. 57, 20–33.

Radkte, C., Cook, W., Anderson, A., 1994. Factors affecting antagonism of the growth of *Phanerochaete chrysosporium* by bacteria isolated from soil. Appl. Microbiol. Biotechnol. 41, 274–280.

Reddy, C., Mathew, Z., 2001. Bioremediation potential of white rot fungi. In: Gadd, G. (Ed.), Fungi in Bioremediation. Cambridge University Press, Cambridge, UK.

Rodriguez Couto, S., Toca Herrera, J.L., 2006. Industrial applications of laccases: a review. Biotechnol. Adv. 24, 500–513.

Ryan, T., Bumpus, J., 1989. Biodegradation of 2,4,5-trichlorophenoxyacteic acid in liquid culture and in soil by the white rot fungus *Phanerochaete chrysosporium*. Appl. Microbiol. Biotechnol. 31, 302–307.

Sandmann, E.R.I.C., Loos, M.A., 1988. Aromatic metabolism by a 2,4-D degrading *Arthrobacter* sp. Can. J. Microbiol. 34, 125–130.

Sasek, V., 2003. Why mycoremediations have not yet come to practice. In: Sasek, V., et al., (Eds.), The Utilization of Bioremediation to Reduce Soil Contamination: Problems and Solutions. Kluwer Academic Publishers, Dordrecht, The Netherlands, pp. 247–276.

The utilisation of bioremediation to reduce soil contamination: problems and solutions. In: Sasek, V., Glaser, J.A., Bouveye, P. (Eds.), Earth and Environmental Sciences 19. Nato Science Series. Kluwer Academic Publishers, Dordrecht, The Netherlands.

Serdar, C.M., Gibson, D.T., Munnecke, D.M., Lancaster, J.H., 1982. Plasmid involvement in parathion hydrolysis by *Pseudomonas diminyta.* Appl. Environ. Microbiol. 44, 246–249.

Sethunathan, N., Yoshida, Y., 1973. A *Flavobacterium* that degrades diazinon and parathion. Can. J. Microbiol. 19, 873–875.

Singh, A., Abdullah, N., Vikineswary, S., 2003. Optimization of extraction of bulk enzymes from spent mushroom compost. J. Chem. Technol. Biotechnol. 78, 743–752.

Singh, B.K., Walker, A., Denis, J., Wright, D.J., 2006. Bioremedial potential of fenamiphos and chlorpyrifos degrading isolates: influence of different environmental conditions. Soil Biol. Biochem. 38, 2682–2693.

Singleton, I., 2001. Fungal remediation of soils contaminated with persistent organic pollutants. In: Gadd, G. (Ed.), Fungi in Bioremediation. Cambridge University Press, Cambridge, UK.

Skipper, H., 1998. Bioremediation of contaminated soils. In: Sylvia, D., et al., (Eds.), Principles and Applications of Soil Microbiology. Prentice Hall, Inc., New Jersey, U.S.A.

Staszczak, M., Zdunek, E., Leonowicz, A., 2000. Studies on the role of proteases in the white-rot fungus *Trametes versicolor*: effect of PMSF and chloroquine on ligninolytic enzymes activity. J. Basic Microbiol. 1, 51–53.

Tam, A.C., Bekhi, R.M., Khan, S.U., 1987. Isolation and characterization of an *S*-ethyl-*N,N* dipropylthiocarbamate-degrading *Arthrobacter* strain and evidence for plasmid associated *S*-ethyl-*N,N*-dipropylthiocarbamate degradation. Appl. Environ. Microbiol. 53, 1088–1093.

Thurston, C.F., 1994. The structure and function of fungal laccases. Microbiology. 140, 19–26.

Trejo-Hernandez, M., Lopez-Munguia, A., Ramirez, R., 2001. Residual compost of *Agaricus bisporus* as a source of crude laccase for enzymic oxidation of phenolic compounds. Process Biochem. 36, 635–639.

Van Eerd, E.L., Hoagland, R.E., Zablotowicz, R.E., Hall, C.J., 2003. Pesticide metabolism in plants and microorganisms. Weed Sci. 51, 472–495.

Vidali, M., 2001. Bioremediation an overview. Pure Appl. Chem. 73, 1163–1172.

Xu, G.M., Zheng, W., Li, Y.Y., Wang, S.H., Zhang, J.S., Yan, Y.C., 2008. Biodegradation of chlorpyrifos and 3,5,6-trichloro-2-pyridinol by a newly isolated *Paracoccus* sp. strain TRP. Int. Biodeterior. Biodegrad. 62, 51–58.

Yadav, J., Reddy, C., 1993. Degradation of benzene, toluene, ethylbenzene and xylene (BTEX) by the lignin degrading basidiomycete *Phanerochaete chrysosporium.* Appl. Environ. Microbiol. 59, 756–762.

Yadav, J.S., Loper, J.C., 2000. Cytochrome P450 oxidoreductase gene and its differentially terminated cDNAs from the white rot fungus *Phanerochaete chrysosporium.* Curr. Genet. 37, 65–73.

Zhang, J., Chiao, C., 2002. Novel approaches for remediation of pesticide pollutants. Int. J. Environ. Pollut. 18, 423–433.

Zouari-Mechini, H., Mechini, T., Dhouib, A., Sayadi, S., Martinez, A., Martinez, M.J., 2006. Laccase purification and characterization from *Trametes trogii* isolated in Tunisia: decolorization of textile dyes by the purified enzyme. Enzyme Microb. Technol. 39, 141–148.

14 Microbial Bioremediation of Industrial Effluents

Deviprasad Samantaray, Swati Mohapatra, and Bibhuti Bhusan Mishra

Department of Microbiology, Orissa University of Agriculture and Technology, Bhubaneswar, Odisha, India

14.1 Introduction

The contamination of air, water, and soil with toxic chemicals causes high risks for the ecosystem both directly and indirectly. Rapid industrialization and explosive development of chemical and mining industries vis-a-vis population explosion have resulted in global deterioration of environmental quality. The environment is sensitive to heavy metals due to their longevity and toxicity (Aravindhan et al., 2007). Extensive utilization of minerals for human need finds application in various industries. Therefore, mining activities along with rapid industrialization are generally considered indices of progress in any country. India is endowed with various types of minerals. Because of commercial importance of minerals, policy planners have emphasized on rapid mining process to overcome the need of the time by industrialization. However, as a fallout of extensive mining and industrial activities, heavy metal contaminated land and water has become a serious environmental health issue in India. In this regard, industrial wastes are the major source of contamination of toxic metals like Hg, Zn, Cr and Al.

Chromium is one of the toxic chemicals considered to be a more hazardous pollutant even at low concentration. Chromium compounds are widely used in leather tanning, steel production, and alloy formation, as metal corrosion inhibitors, and in paints as pigment and various other applications. Chromium generally occurs in two oxidation states, Cr^{3+} and predominantly Cr^{6+}, in air, water, and soil (Cheung and Gu, 2006; Daulton et al., 2007). Hexavalent chromium is 100 times more poisonous and 1000 times mutagenic than Cr^{3+}; hence, it has been listed as a priority pollutant and a human carcinogen by the United States Environmental Protection Agency (USEPA) (Cheung and Gu, 2006). Hexavalent chromium is highly soluble in water and mobile through the ecosystem, while Cr^{3+} is insoluble and forms a precipitate with organics in nature (Bajgai et al., 2012).

Microbial Biodegradation and Bioremediation. DOI: http://dx.doi.org/10.1016/B978-0-12-800021-2.00014-5

Many of today's environmental problems, as well as their potential solutions, are intimately interwoven with the microbial component of the global ecosystem. Increasingly, scientists have recognized that microorganisms occupy a key position in the orderly flow of materials and energy through the global ecosystem by virtue of their metabolic activities to transform organic and inorganic matter. The bioactive potentiality of microorganisms in nature is a ready answer for degradation and recycling of the hazardous compounds being added to our environment because of the industrial and chemical boom. They can reduce toxicity of various pollutants and wastes that are detrimental to the valuable gift of nature. Many microorganisms such as bacteria, fungi, and algae have been recognized for their ability to resist either the toxic effect of hexavalent chromium or the biotransformation of Cr^{6+}; thus, it becomes less toxic or nontoxic to them via sequestration mechanisms such as reduction, complexation, alkylation, and precipitation (Iihan et al., 2004; Ertugrul et al., 2009). Through these mechanisms, microorganisms are also able to bioabsorb and bioaccumulate hexavalent chromium in their cells with the help of numerous binding sites present on their cell wall.

Ironically, some of these mechanisms make an environment susceptible to heavy metal toxicity. For example, reduction of toxic heavy metal ions to relatively less toxic (Cr^{6+} to Cr^{3+}) makes the heavy metal ions mobile through the water in the soil; therefore, chances of its presence in nonpolluted sites and the probability of its getting into runoff water increase. Toxic metals classified as environmental pollutants cannot be degraded, but their oxidation state can be changed to another less toxic state by microorganisms. Most of the microorganisms are antipolluters that metabolize toxic chemical substances and convert recacitrant compounds to its simpler form present in the pollutant. The virtually omnipresent microorganisms are powerful tools for bioremediation. It is an important and unique biological process, which is globally recognized as a cost-effective and eco-friendly technology. Thus, bioremediation of hexavalent chromium aims at extracting the metals to make them unavailable to flow into the ecosystem, or extract to mobilizing them for reuse or safe disposal (Crawford and Crawford, 1995). The various properties of microorganisms like reduction, adsorption, and bioaccumulation of heavy metals give the potential for a cheap alternative method of heavy metal removal from soil and industrial wastewaters. Both living and dead biomaterials are capable of removing heavy metal ions from the heavy metal–contaminated sites through diverse mechanisms (Vindhan, 2004).

Conventional treatment of Cr^{6+} waste involves a two-stage process such as chemical reduction of Cr^{6+} to Cr^{3+} followed by precipitation of Cr^{3+} by using lime, caustic soda, or sodium bicarbonate (Cushnie, 1985). Even though the process is quite effective, the large volume of sludge generated and the release of dangerous gases and cost of the chemical-reducing agents make it imperative to look into safe and cheaper alternatives. The biological system seems to be a more suitable approach. Therefore, bioremediation has become as a cost-effective, efficient, and environmentally friendly alternative for removing heavy metals from industrial effluents. The advantage of bioremediation is that this process does not require using aggressive and concentrated chemicals, and metal ions bound biomass could be reused after elution (Chojnacka, 2007). Bioremediation is a suitable alternative to conventional methods, but the presence of co-contaminants of chromium such as Cu, Fe, Hg, Ni and Co may limit its application (Singh and Tripathi, 2007). Microbial populations in heavy metal–polluted environments harbour microorganisms that have adapted to the toxic heavy metals and become "metal resistant" (Kasan and Baecker, 1989). Bioreduction of Cr^{6+} occurs directly due to microbial metabolism or indirectly by bacterial metabolites (Losi et al., 1994).

Many scientists have investigated and demonstrated the feasibility of using biological processes for the treatment of Cr^{6+}-contaminated sites and industrial effluents by either pure culture or a consortium of Cr^{6+}-reducing bacteria (Romanenko et al., 1976; Bopp and Ehrlich, 1988; Lupton et al., 1992; Turick and Apel, 1997; Camargo et al., 2003).

14.2 Chromium Production

Chromium, the 24th element in the periodic table, was first discovered in Siberian red lead ore by French Chemist Nicholas Louis Vanquelin in 1978. It was named "Chrom" from the Greek word "$\chi\rho\omega\mu\alpha$" because of its brilliant hues. This first-row transition metal finds a variety of uses in industries exploiting its color, strength, hardness, corrosion resistance, and oxidizing capabilities (Darrim, 1956). Chromium is extracted from chromite ore, which has large deposits in South Africa, the Philippines, Southern Zimbabwe, and Turkey (Mathews and Morning, 1980). South Africa is the world's largest producer of ferrochrome. The country holds about 70% of the world's total chrome reserves, most of it derived from the Bushveld Igneous Complex (BIC) ores. South Africa produces an estimated 7,417,329 tons of chromium. India is the third leading chromite ore producer globally with an output of about 3.5–4 megatons (MT) per year. Chromite ore is mainly produced in the state of Odisha, with a large portion of the chromite produced consumed by local ferrochrome-producing companies (Ferrochrome Facts, 2007). Odisha is rich with various types of minerals, chromite being chief among them. The Sukinda mining area of Jajpur in Odisha has 97% of India's chromite deposits and has been declared one of the most polluted places in the world (Blacksmith Institute Report, 2007) due to chromium pollution. The total production of Odisha in 2004–05 was 3,123,386 MT, of which Jajpur's deposits alone contribute 3,035,201 MT (ENVIS Newsletter, 2006).

14.3 Chromium Toxicity

Chromium plays a key role in the biological system, but beyond a certain level, it is toxic (Balamurugan et al., 2004), mutagenic (Gili et al., 2002), carcinogenic (Codd et al., 2003) and teratogenic (Asmatullah et al., 1998). Moreover, the metal contamination imparts many adverse effects on human beings such as brain damage, reproductive failure, nervous system failure, and tumor formation. Chromium and its compounds are widely used in different industries and then enter into the ecosystem (Figure 14.1) through effluent.

Generally chromium is present in the environment in two different oxidation states like Cr^{3+} and Cr^{6+} (Devi et al., 2012). The naturally occurring trivalent chromium is less toxic and nonbioleachable. It is an essential micronutrient in animal physiology, playing a role in glucose and lipid metabolism (Anderson, 1989; Mertz, 1993). It is involved in peripheral action of insulin, normal glucose utilization, and stimulation of enzyme systems (Mertz, 1969), and possibly in the stabilization of nucleic acids (Huff et al., 1964). The Cr^{6+} is water soluble, toxic, and bioleachable, as well as mutagenic, carcinogenic, and teratogenic; it is a powerful epithelial irritant as many researchers have reported (Petrilli and Flora, 1977; Gale, 1978; Gruber and Jennette, 1978; Langand, 1983; IARC, 1990; Daulton et al., 2007). Hexavalent chromium is present in the effluents from electroplating, paint, pigment, cement, mining, dyeing, fertilizer, and photography industries. At high concentrations, all compounds of chromium are toxic. Ingestion of chromium may cause epigastric pain, nausea, vomiting, and severe diarrhea. According to USEPA, the tolerance limit of Cr^{6+} in drinking water is 0.05 mg/L. Hexavalent chromium is carcinogenic in nature (Devi et al., 2012). Hence, it is highly imperative to treat the industrial effluent containing Cr^{6+} before its discharge.

Hexavalent chromium is a strong oxidant, and the ion can pass through cell membranes (Figure 14.2) many times faster than the trivalent form, which is a

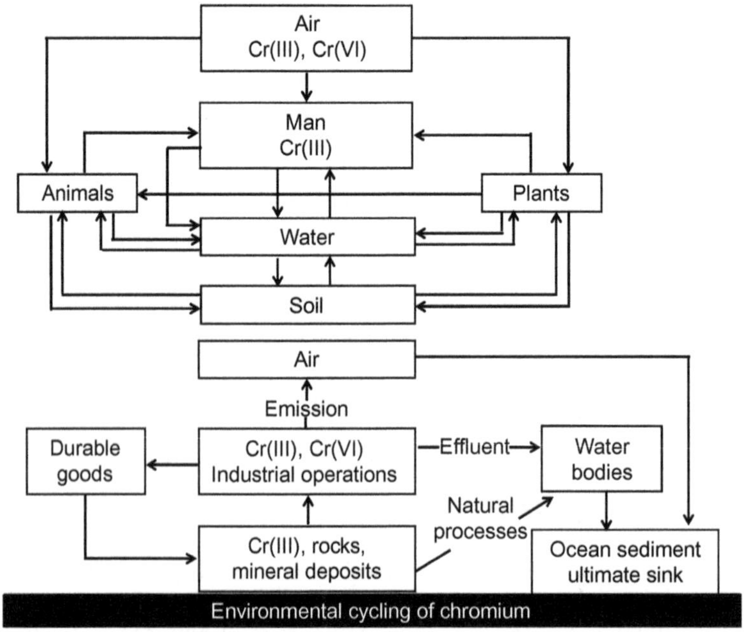

Figure 14.1 Chromium cycle in the environment.

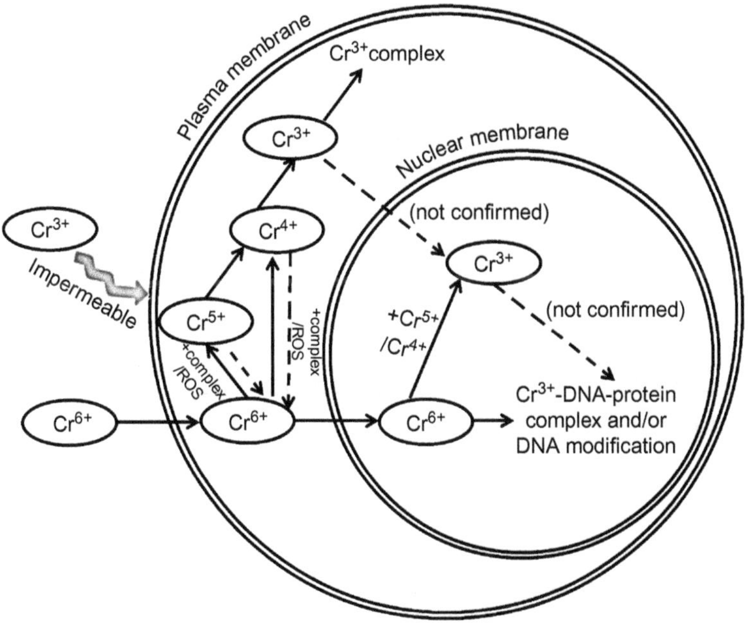

Figure 14.2 Vincent schematic diagram of hexavalent chromium toxicity.

major reason for its carcinogenicity. Intracellularly, it is then reduced to the trivalent form by various reducing agents like ascorbic acid, sodium sulfite, glutathione, Nicotinamide adenine dinucleotide phosphate hydrogen (NADPH), and Nicotinamide Adenine Dinucleotide Hydrogen (NADH) (Petrilli and Flora, 1978). The trivalent form binds and reacts with nucleic acid and other cell components by producing free radicals (Medeiros et al., 2003). A low concentration of chromium has been reported to stimulate plant growth; however, chromium concentration of 5–60 mg/kg of soil can damage plant roots (Pratt, 1966). Hexavalent chromium has been shown to affect growth, photosynthesis, morphology, and enzyme activities in algae and is toxic in concentrations ranging from 20 to 10,000 ppb as suggested by Schroll (1978), Silverberg et al. (1977), and Towill et al. (1978).

Microorganisms require a very low concentration of chromium for their growth and development, but a high concentration is toxic for them. Many microorganisms can accumulate chromium (Dursun et al., 2003; Pas et al., 2004), but the negative effects in bacterial cells such as cell elongation, cell enlargement, and reduction in cell division lead to cell growth inhibition as reported by Paran (1983) and Theodotou et al. (1976). Hexavalent chromium in the range of 0.05–5 mg/L of medium is generally toxic to microorganisms, though an internal concentration of chromium is species dependent (Babich et al., 1982). When Cr^{6+} concentration increases from 0.1 to 0.4 mg/L in aqueous systems, diatoms have been found to be replaced by algae and cyanobacteria. In *Escherichia coli* strain (NR 9064), high concentrations of chromium led to formation of DNA–DNA cross-links and decreased polymerase activity (Snow, 1994). However, in fungi, chromium toxicity leads to reduced growth of mycelia (Babich et al., 1982). It is also toxic even in low concentration (1 mg/kg of soil), which reduced soil microbial transformations like nitrification (Ross et al., 1981). Cr^{6+} has been shown to be mutagenic to *E. coli*, *Bacillus subtilis*, and *Salmonella typhimurium* as reported by Nishoka (1975), Petrilli and Flora (1977) and Venitt and Levy (1974), respectively, generally causing breaks in DNA strands. The genotoxic effects of hexavalent chromium on bacterial cells include frame shift mutation and base pair substitution (Petrilli and Flora, 1978).

14.4 Bioremediation of Chromium Toxicity: The Green Chemistry

Conventional techniques of Cr^{6+} effluent treatment is quiet effective, but not economical and eco-friendly. Thus application of microorganisms is highly advantageous in this regard. Aerobic microorganisms such as *Pseudomonas auroginosa*, *Alcaligenes eutrophus*, *Waustersia eutropha*, *Pseudomonas fluorescens*, *Pseudomonas synxantha*, *E. coli*, *Bacillus megaterium*, *Bacillus* sp., and *Pseudomonas maltophila*, and anerobes like *Pseudomonas dechromaticans*, *Pseudomonas chromatophila*, *Aeromonas dechromatica*, *B. subtilis*, *Bacillus cereus*, *Pseudomonas auroginosa*, *Pseudomonas ambigua*, *Micrococcus roseus*, *Enterobacter cloacae*, and *Desulfovibrio desulfuricans* are involved (Cheung and Gu, 2006) in detoxification of hexavalent chromium. However, biomonitoring of hexavalent chromium is also possible using phenotypic responses of a unique blue color pigment producing bacterium *Vogesella indigofera*, following exposure to Cr^{+6} (Cheung and Gu, 2002).

Microorganisms develop different resistance mechanisms to chromium for survival in Cr-contaminated sites. The microbial response depends on the nature of the toxic elements. The resistance mechanism can be exclusion by permeability barriers, exclusion by active transport, intracellular sequestration by binding proteins of the cell, extracellular sequestration, and detoxification by chemical modification of the metal from toxic to nontoxic forms. The reduction in metal sensitivity to cellular targets can be by mutation to decrease metal sensitivity, increased production of damaged cell components, increased efficiency of repair of damaged cell components, plasmid encoded mechanism, etc.

Chromium resistance in bacteria is either chromosomal or plasmid mediated (Peitzsch et al., 1998; Juhnke et al., 2002). Plasmid-associated resistance has been observed in *Streptococcus lactis* (Efstathiou and Mckay, 1977), *Pseudomonas* sp. (Summers and Jacoby, 1978), *Alcaligenes eutrophus* (Nies and Silver, 1989; Cervantes and Silver, 1992; Peitzsch et al., 1998) etc. Some chromium-reducing bacteria can grow by reducing hexavalent chromium and simultaneously detoxifying the environment. Most hexavalent chromium-reducing bacteria reported so far are gram negative (Baldi et al., 1990; Francis et al., 2000). Bacterial chromium reduction can be direct, enzymatic, or indirect by bacterial metabolites. Enzymatic reduction by some bacteria is mainly by soluble or membrane-bound enzyme systems (Figure 14.3). Membrane-associated chromate reductase activity was first reported by Wang et al. (1989) in *E. cloacae* (H01), which reduces hexavalent chromium to trivalent form by precipitation. *Shewanella putrefaciens* (MR-1) also shows cytoplasmic membrane-associated chromate reductase activity in anaerobic conditions using NADH and formate as electron donors for the enzyme (Myers et al., 2000). Rahman et al. (2007) reported that *Pseudomonas* sp. (C-171) showed resistance to 2000 ppm of Cr^{6+} in the form of $K_2Cr_2O_7$. In this bacterium, the growth rate and reduction of chromium was found to be inversely proportional to the Cr^{6+} supplementation whereas, slight elongation of bacterial cell due to accumulation of chromium hydroxide has been observed.

Chromium resistance in bacteria is mediated through two different mechanisms such as efflux mediated and reduction of Cr^{6+} to Cr^{3+}. Chromate efflux by *chrA* transporter has been established in *Pseudomonas aeruginosa* and *Cupriavidus metallidurans*, consisting of an energy-dependent process driven by the membrane potential. Most characterized enzymes for chromate reduction belong to the NADPH-dependent flavoprotein family of reductase. Expression of components of the machinery for repair of DNA damage and systems related to the homeostasis of iron and sulfur are also mechanisms of bacterial resistance to chromate (Ramirez et al., 2008). A study on reduction of Cr^{6+} by using *Pseudomonas putida* (PRS-2000) reveals that chromate reductase activity is associated with soluble protein and not with the membrane fraction. Crude enzyme activity is heat labile, and sulfate or nitrate does not affect reduction (Ishibashi et al., 1990). Evidence reveals that Cr^{6+} reduction is dependent upon pH, temperature, inoculum concentration, and Cr^{6+} concentration (Camargo et al., 2003). An alkaline pH and 30°C is the optimum

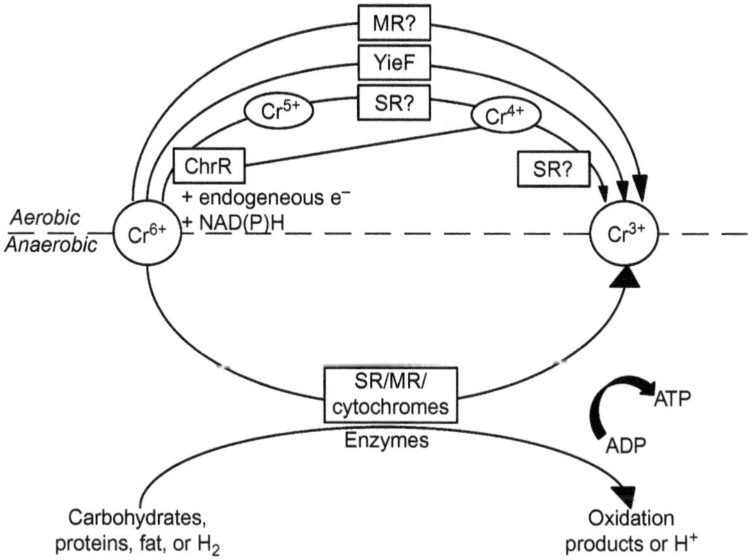

Figure 14.3 Mechanism of enzymatic hexavalent chromium reduction.

bioparameter for growth of chromium-resistant or -reducing bacteria such as *B. cereus*, *Bacillus thuringensis*, and *Arthrobacter crystallopoites*. As Cr^{6+} reduction is enzyme mediated, changes in pH affect the degree of ionization of enzymes, changing the protein conformation and affecting enzyme activity (Farrell and Ranallo, 2000). Generally, most chromium-resistant bacteria can carry out Cr^{6+} reduction at an optimum temperature range of 30–37°C (Losi et al., 1994).

However, soluble chromium reductase activity has been observed in *E. coli* (Shen and Wang, 1993), *Pseudomonas* sp. (CRB5) (McLean and Beveridge, 2001), and *Bacillus coagulans* (Philip et al., 1998). Indirect chromium reduction is due to changes in pH, redox potential during growth, and production of metabolites.

Indirect reduction of Cr^{6+} by bacterial isolates in the medium resulted in the production of off-white residues, which were the sign of chromate reduction. Bacterial conversion of Cr^{6+} to Cr^{3+} is due to production of metabolite (Smillie et al., 1981; Fude et al., 1994; Rahman et al., 2007; Mishra et al., 2010) like H_2S in the medium. The H_2S produced by the bacteria reduces Cr^{6+} to Cr^{3+}, and the trivalent chromium reacts with H_2S to form chromium sulfide, which is not stable at aqueous solution; it is deposited in the form of chromium hydroxide precipitate (off-white) in the medium. Hexavalent chromium reduction by bacteria is also due to production of acidic metabolic by-products from aerobic and anaerobic respiration. These metabolites decrease pH and redox potential (Beveridge, 1989; McLean and Beveridge, 2001), which favors conversion of Cr^{6+} to Cr^{3+} and production of chromium oxides and hydroxide. Chemical thermodynamics predict that low redox potential results in precipitation of these chromium oxides and hydroxide.

14.5 Case Study

A similar study was also undertaken to understand and elucidate the complex microbial activity in the Sukinda Valley, because of the alarmingly high level of chromium pollution in that particular area. In our experiment, an attempt was made to detoxify hexavalent chromium by chromium-resistant bacterial isolates from this area. Random soil, sediment, and water samples were collected aseptically from four different sites: Kalarangi, South Kaliapani, Kamardha, and the Dumsala canal of the Sukinda mining area of Jajpur district of Odisha. Physiochemical parameters like temperature, pH, moisture content, total chromium, and hexavalent chromium content of the samples were estimated. The average pH of soil, sediment, and water was 8.59, 6.99, and 7.84, respectively. The total chromium content was 4.24 g/kg, 5.7 g/kg, and 1.12 mg/L in soil, sediment, and water samples, respectively. The hexavalent chromium content of soil, sediment, and water was (Mishra et al., 2010) 39 mg/kg, 46.74 mg/kg, and 0.689 mg/L, respectively. Four bacterial isolates, namely, *Micrococcus luteus*, *P. putida*, *Serratia marcescens*, and *Acinetobacter calcoaceticus*, tolerated hexavalent chromium beyond 500 ppm were selected for reduction at different pH, temperatures, times of incubation, and concentrations of hexavalent chromium.

For screening of hexavalent chromium-reducing bacteria, the 24-h Cr^{6+} reduction test was conducted in a nutritive media (LB) at pH 7.0 (optimum for bacteria) and 30°C (optimum for bacteria isolated from environment). The result reveals that out of all four isolates, *A. calcoaceticus* reduced 67.14% Cr^{6+}, which was the highest of all. The percentages of hexavalent chromium reduction under similar conditions by other isolates like *S. marcescens*, *M. luteus*, *P. putida* were 65.02%, 53.14%, and 50.72%, respectively. However, no reduction of hexavalent chromium was observed in a control set without bacteria, which indicates the reduction of Cr^{6+} due to the presence of bacteria. The reduction of Cr^{6+} by these bacterial isolates in the medium resulted in the production of off-white residues, which were the sign of chromate reduction.

However, in a nonnutritive (Samantaray and Mishra, 2012) medium, *A. calcoaceticus* reduced Cr^{6+} by 38.1% at 30°C/24 h/pH 7.0 and all other bacterial isolates like *S. marcescens*, *M. luteus*, *P. putida* reduced 37.05%, 31.02%, and 26.05%, respectively, but no reduction was observed in the control. This difference in trend of reduction in a nonnutritive medium in comparison to a nutritive medium may be due to a decrease in physiological and metabolic activities of the isolates (Losi et al., 1994; Camargo et al., 2003) and viability after some time and possible inhibition of biomass activity by prolonged chromate toxicity in a nonnutritive medium. Comparative analysis of Cr^{6+} reduction in nutritive and non-nutritive medium indicated that *A. calcoaceticus* possesses the higher potential among the isolates and selected for further studies.

Effect of hexavalent chromium concentration on the growth of viable cell numbers of *A. calcoaceticus* indicates that the viable cell count was higher, i.e., 9.6×10^7 CFU/mL at 100 ppm of Cr^{6+}, and then the trend was decreased (Samantaray and Mishra, 2011) up to 800 ppm in comparison to the control. However, the viable cell count was 8.4×10^7 CFU/mL at 50 ppm of Cr^{6+}, which is higher than the control, i.e., 8.1×10^7 CFU/mL, which indicates requirement of hexavalent chromium as a substrate for their optimal growth and development. Pei et al. (2009) and Zakaria et al. (2007) found that growth of *Acinetobacter haemolyticus* was higher at 90 ppm and reduced to 48% at 110 ppm in LB medium, which is due to apparent hexavalent chromium toxicity. Thus, 100 ppm of hexavalent chromium was selected and supplemented in the LB medium during the period of experimentation.

Effect of inoculum size on Cr^{6+} reduction was also studied and it was found that, increase in the rate of Cr^{6+} reduction with increase in inoculum size up to a limit. The optimum reduction of Cr^{6+} (74.62%) was observed with 5% inoculum volume at an increasing trend from 1–5% and decreased (Samantaray and Mishra, 2011) further above 5%. Thus, 5% inoculum was supplemented during the period of study. A similar result was observed by Rahman et al. (2007), who found that maximum reduction was recorded at 30% (v/v) inoculums among 10% and 20% (Wang et al., 1989; Rahman et al., 2000; Pei et al., 2009). They reported that the higher the cell density, the greater the percentage of reduction. This may be due to the fact that, increased bacterial cells in terms of inoculum size increases the rate of H_2S production thus fasten the rate of Cr^{6+} reduction.

The highest hexavalent chromium-reducing bacterial isolate, *A. calcoaceticus*, was selected for parametric studies. Hourly Cr^{6+} reduction results reveal that

A. calcoaceticus could reduce 85% of hexavalent chromium optimally at pH 8.0, in LB within 24 h. For most of the isolates, the optimum pH for growth correlates with the highest rate of hexavalent chromium reduction. The trend increases with increase in time, i.e., up to 24 h. Thus, pH 8.0 was kept constant for Cr^{6+} reduction. For most of the isolates, the optimum pH for growth correlates with the highest rate of hexavalent chromium reduction (Camargo et al., 2003). The relationship between pH and Cr^{6+} reduction was not surprising because chromate is the dominant chromium species in aqueous environments at pH 6.5–9.0 (McLean and Beveridge, 2001). The optimum pH for growth of Cr^{6+}-resistant bacteria was reported at 7–7.8 (Losi et al., 1994), but hexavalent chromium forms are soluble over a wide pH range and generally mobile in soil–water systems (Losi et al., 1994). Similar results were also obtained by Wang et al. (1990), reporting that Cr^{6+} reduction in *E. cloacae* occurred at pH 6.5–8.5 and was inhibited at pH 5–9. As Cr^{6+} reduction is enzyme mediated, changes in pH will affect the degree of ionization of the enzyme, changing the protein conformation and affecting the enzyme activity.

The hexavalent chromium reduction profile monitored at different temperatures ranging from 20°C to 37°C in LB for 24 h at 8.0 hexavalent chromium reduction by *A. calcoaceticus* in LB indicates that the percentage of reduction (85%) is higher at an optimum temperature of 30°C. The decreasing trend of Cr^{6+} reduction was also observed with an increase in temperature for which it was kept constant at 30°C in order to study the effect of metal on Cr^{6+} reduction. As observed, the highest reduction was reported at an optimum temperature of 30°C and the percentage of reduction decreased with an increase in temperature. This is possibly because of decreased enzyme activity with increase in temperature. This could be due to loss of viability or metabolic activity of the cells on prolonged incubation at higher temperature (Aravindhan et al., 2007). Similar results were also obtained by Camargo et al. (2003). Losi et al. (1994) reported an optimum temperature of 30–37°C for chromate reduction. However, Wang et al. (1990) reported that no chromate reduction was observed at 4°C and 60°C. Temperature is an important selection factor for bacterial growth and affects enzymatic reactions necessary for chromate reduction.

Consortium study was undertaken to know effectiveness of the synchronized use of the two isolates for Cr^{6+} reduction. However, *S. marcescens* was found to inhibit the growth of *A. calcoaceticus* in LA medium. Thus, consortia hexavalent reduction was not possible by these two desired bacterial isolates. This might be due to the production of red coloured water soluble pigment prodigiosin by *S. marcescens.* Khanafari et al. (2006) reported that the red prodigiosin pigment produced by *S. marcescens* has antimicrobial, immunosuppressive, and anti-proliferate activity. In our study, *S. marcescens* was capable of producing red water-soluble pigment in Nutrient Agar (NA), Nutrient Broth (NB), Luria-Bertani Agar (LA), and Luria-Bertani Broth (LB), respectively. However, pigment production was also observed in a wide range of pH 4–14, Cr^{6+} concentration up to 1000 ppm, and in a temperature range of 20–37°C.

Natural habitats are generally characterized by the coexistence of a large number of toxic and nontoxic substances for which it is imperative to study multiple metal effects on the physiology and biochemistry of microorganisms (Verma and Singh,

1995). *A. calcoaceticus* showing the highest hexavalent chromium reduction was tolerant to a broad range of heavy metals like Fe^{2+}, Cu^{2+}, Ni^{2+}, Hg^{2+}, and Co^{2+} up to concentrations of 1000, 900, 1000, 100, and 300 ppm, respectively. Among all these metals tested, the highest tolerance was observed toward Fe^{2+} and Ni^{2+}. These observations assume great significance because effluents from any metal related to industry have several metal ions or contaminants. Tolerance to other metals has an added advantage of withstanding the presence of other metal ions while performing the desired activity. There are reports of the use of Cr^{6+} reducing microorganisms for treatment of other waste materials (Lovely, 1995). Thus, these locally isolated strains possess huge credentials for detoxification of hexavalent chromium from industrial effluent and chromium contaminated sites.

Effects of metal on hexavalent chromium reduction by *A. calcoaceticus* indicate that, in 1 ppm of copper, 89.39% Cr^{6+} reduction was observed (Samantaray and Mishra, 2012) at 30°C/24 h/pH 8. In the presence of iron, 68.44% hexavalent chromium was reduced, and the rate of reduction decreased in the presence of nickel as compared to control. Although the organism showed more tolerance to iron, no change in chromium reduction was observed. The increase in reduction in the presence of copper may be due to enhanced enzyme activity of chromate reductase (Pal and Paul, 2004; Faisal and Hasnain, 2004; Elangovan et al., 2006) as it also acts as a micronutrient for optimal growth of the bacteria.

14.6 Conclusion

This case study revealed that 89.39% Cr^{6+} reduction was observed by *A. calcoaceticus* at 30°C/24 h/pH 8 and in the presence of 1 ppm copper in a nutritive (LB) medium. Thus, it is concluded that *A. calcoaceticus* may be used in the bioremediation of hexavalent chromium toxicity. Understanding the potential of microorganisms in recycling of metals may lead to improved processes for bioremediation of metal-contaminated areas. Hexavalent chromium toxicity is a major concern, thus there is a high level of interest in developing methods aimed at detoxifying chromium-contaminated areas at minimal costs with fewer side effects. The process, which is in its nascent laboratory stage, is now moving on to the developmental stage. Slowly, microorganisms are proving to be the right tools for environmental pollution control. Our current state of knowledge about the state of affairs in chromium-contaminated areas leaves us with many queries, answers to which can help us to find a solution to control chromium pollution.

References

Anderson, R.A., 1989. Essentiality of chromium in humans. Sci. Total Environ. 86, 75–81.

Aravindhan, R., Sreeram, K.J., Rao, J.R., Nair, B.U., 2007. Biological removal of carcinogenic chromium(VI) using mixed *Pseudomonas* strains. J. Gen. Appl. Microbiol. 53 (2), 71–79.

Asmatullah, Q.N.N., Shakoori, A.R., 1998. Hexavalent chromium reduction induced congenital abnormalities in chick embryo. J. Appl. Toxicol. 18 (3), 167–171.

Babich, H., Schiffienbauer, M., Stotzky, G., 1982. Comparative toxicity of chromium(III) and chromium(VI) to fungi. Bull. Environ. Contam. Toxicol. 28, 452–459.

Bajgai, R.C., Georgieva, N., Lazarova, N., 2012. Bioremediation of chromium ions with filamentous yeast *Trichosporon cutaneum* R57. J. Biol. Earth Sci. 2, 70–75.

Balamurugan, K., Rajaram, R., Ramasami, T., 2004. Caspase-3: its potential involvement in Cr(III) induced apoptosis of lymphocytes. Mol. Cell Biochem. 259 (1–2), 43–51.

Baldi, F., Vaughan, A.M., Olson, G.J., 1990. Chromium(VI)-resistant yeast isolated from a sewage treatment plant receiving tannery wastes. Appl. Environ. Microbiol. 56, 913–918.

Beveridge, T.J., 1989. The role of cellular design in bacterial metal accumulation and mineralization. Annu. Rev. Microbiol. 43, 147–171.

Blacksmith Institute Report, 2007. The world's worst polluted places, A project of Blacksmith Institute, pp. 16–17.

Bopp, L.H., Ehrlich, H.L., 1988. Chromate resistance and reduction in *Pseudomonas fluorescens* strain LB300. Arch. Microbiol. 150 (5), 426–431.

Camargo, F.A., Bento, F.M., Okeke, B.C., Frankenberger, W.T., 2003. Chromate reduction by chromium-resistant bacteria isolated from soils contaminated with dichromate. J. Environ. Qual. 32, 1228–1233.

Center for Environmental Studies, 2006. Envis Newsl. 5 (1), 1–6, <Available from: http://www.cesorissa.org/pdf/newsletter5.pdf>.

Cervantes, C., Silver, S., 1992. Plasmid chromate resistance and chromate reduction. Plasmid. 27, 65–71.

Cheung, K.H., Gu, J., 2006. Mechanism of hexavalent chromium detoxification by microorganisms and bioremediation application potential: a review. Int. Biodeterior. Biodegradation. 59, 8–15.

Cheung, K.H., Gu, J.D., 2002. Bacterial colour response to hexavalent chromium, Cr^{+6}. J. Microbiol. 40, 234–236.

Chojnacka, K., 2007. Bioaccumulation of Cr^{+3} ions by blue-green alga *Spirulina* sp. Part I. A comparison with biosorption. Am. J. Agric. Biol. Sci. 2 (4), 218–223.

Codd, R., Irwin, J.A., Lay, P.A., 2003. Sialoglycoprotein and carbohydrate complexes in chromium toxicity. Curr. Opin. Chem. Biol. 17 (2), 213–219.

Coleman, R.N., Paran, J.H., 1983. Accumulation of hexavalent chromium by selected bacteria. Environ. Technol. Lett. 4, 149–156.

Crawford, R.L., Crawford, D.L. (Eds.), 1995. Bioremediation: Principles and Applications. Cambridge University Press, UK.

Cushnie Jr., G.C., 1985. Electroplating Wastewater Pollution Control Technology. Noyes Publications, Park Ridge, NJ.

Darrim, M., 1956. Chromium compounds—their industrial use. In: Udy, M.J. (Ed.), Chromium. Reinhold, New York, NY, pp. 251–262.

Daulton, T.L., Little, B.J., Jones-Meehan, J., Blom, D.A., Allard, L.F., 2007. Microbial reduction of chromium from the hexavalent to divalent state. Geochim. Cosmochim. Acta. 71, 556–565.

Devi, B.D., Thatheyus, A.J., Ramya, D., 2012. Bioremeoval of hexavalent chromium, using *Pseudomonas fluorescens*. J. Microbiol. Biotechnol. Res. 2 (5), 727–735.

Dursun, A.Y., Ulsu, G., Cuci, Y., Aksu, Z., 2003. Bioaccumulation of copper (II), lead (II), and chromium (VI) by growing *Aspergillus niger*. Process Biochem. 38 (12), 1647–1651.

Efstathiou, J.D., Mckay, L.L., 1977. Inorganic salts resistance associated with a lactose-fermenting plasmid in *Streptococcus lactis*. J. Bacteriol. 130, 257–265.

Elangovan, R., Abhipsa, S., Rohit, B., Ligy, P., Chandraraj, K., 2006. Reduction of Cr (VI) by a *Bacillus* species. Biotechnol. Lett. 28 (4), 247–252.

Elsevier Engineering Information, 2007. Ferrochrome Facts, Source: <http://www.estain-lesssteel.com>.

Ertugrul, S., San, N.O., Donmez, G., 2009. Treatment of dye (Remazol Blue) and heavy metals using yeast cells with the purpose of managing polluted textile wastewaters. Ecol. Eng. 35, 128–134.

Faisal, M., Hasnain, S., 2004. Microbial conversion of Cr(VI) in to Cr(III) in industrial effluent. Afr. J. Biotechnol. 3 (11), 610–617.

Farrell, S.O., Ranallo, R.T., 2000. Experiments in Biochemistry. A Hands-On Approach. Saunders College Publication, Orlando, FL.

Francis, C.A., Obraztsova, A.Y., Tebo, B.M., 2000. Dissimilatory metal reduction by the facultative anaerobe *Pantoea agglomerans* SP1. Appl. Environ. Microbiol. 66, 543–548.

Fude, L., Harris, B., Urrutia, M.M., Beveridge, T.J., 1994. Reduction of Cr(VI) by a consortium of sulfate-reducing bacteria (SRB III). Appl. Environ. Microbiol. 60, 1525–1531.

Gale, T.F., 1978. Embryonic effect of chromium trioxide in hamsters. Environ. Res. 16, 101–109.

Gili, P., Medeiros, A., Lorenzo-Lorenzo-Louis, P.A., de la Rosa, E.M., Munoz, A., 2002. On the interaction of compounds of chromium(VI) with hydrogen peroxide. A study of chromium(VI) and (V) peroxides in the acid–basic pH range. Inorg. Chim. Acta. 331 (1), 16–24.

Gruber, J.E., Jennette, K.W., 1978. Metabolism of the carcinogen chromatc by rat liver microsomes. Biochem. Biophys. Res. Commun. 82, 700–706.

Huff, J.W., Sastry, K.S., Gordon, M.P., Wacker, W.E.C., 1964. The action of metal ions on tobacco mosaic virus ribonucleic acid. Biochemistry. 3, 501–506.

IARC, 1990. Chromium, Nickel, and Welding, Monograph on the Evaluation of the Carcinogenic Risk to Humans, 49, 677.

Iihan, S., Nourbakhsh, N.M., Kilicarslan, S., Ozdag, H., 2004. Removal of chromium, lead and copper ions from industrial wastewaters by *Staphylococcus saprophyticus*. Turk. Electron. J. Biotechnol. 2, 50–57.

Ishibashi, Y., Cervantes, C., Silver, S., 1990. Chromium reduction in *Pseudomonas putida*. Appl. Environ. Microbiol. 56 (7), 2268–2270.

Juhnke, S., Peitzsch, N., Hubner, N., Grosse, C., Nies, D.H., 2002. New genes involved in chromate resistance in *Ralstonia metallidurans* strain (CH34). Arch. Microbiolol. 179 (1), 15–25.

Kasan, H.C., Baecker, A.A.W., 1989. Activated sludge treatment of coal gasification effluent in a petro-chemical plant—II. Metal accumulation by heterotrophic bacteria. Water Sci. Technol. 21 (4/5), 297–303.

Khanafari, A., Assadi, M.M., Fakhr, F.A., 2006. Review of prodigiosin, pigmentation in *Serratia marcescens*. Online J. Biol. Sci. 6 (1), 1–13.

Langand, S., 1983. The carcinogenicity of chromium compounds in man and animals. In: Burrows, D. (Ed.), Metabolism and Toxicity. CRC Press, Inc., Boca Raton, FL, pp. 13–30.

Losi, M.E., Amrhein, C., Frankenberger, W.T., 1994. Environmental biochemistry of chromium. Rev. Environ. Contam. Toxicol. 36, 91–121.

Lovely, D.R., 1995. Bioremediation of organic and metal contaminants with dissimilatory metal reduction. J. Ind. Microbiol. 14, 85–93.

Lupton, F.S., DeFilippi, L.J., Goodman, J.R., 1992. Bioremediation of chromium(VI) contaminated solid residues. US Patent 5155042.

Mathews, N.A., Morning, J.L., 1980. Metals and Minerals. Washington D.C. U.S. Bureau of mines, mineral yearbook 1978–79. Ref Type: Serial (Book, Monograph) [1], pp. 193–205.

McLean, J., Beveridge, T.J., 2001. Chromate reduction by a *Pseudomonas* isolated from a site contaminated with chromated copper arsenate. Appl. Environ. Microbiol. 67 (3), 1076–1084.

Medeiros, M.G., Rodrigues, A.S., Batoreu, M.C., Laires, A., Rueff, J., Zhitkovich, A., 2003. Elevated levels of DNA–protein crosslink and micronuclei in peripheral lymphocytes of tannery workers exposed to trivalent chromium. Mutagenesis. 18 (1), 19–24.

Mertz, W., 1969. Chromium occurrence and function in biological systems. Physiol. Rev. 49, 163–239.

Mertz, W., 1993. Chromium in human nutrition: a review. J. Nutr. 123, 626–633.

Mishra, V., Samantary, D.P., Dash, S.K., Mishra, B.B., Swain, R.K., 2010. Study on hexavalent chromium reduction by chromium-resistant bacterial isolates of Sukinda mining area. Our Nat. 8 (1), 63–71.

Myers, C.R., Carstens, B.P., Antholine, W.E., Myers, J.M., 2000. Chromium(VI) reductase activity is associated with the cytoplasmic membrane of anaerobically grown *Shewanella putrefaciens* MR-1. J. Appl. Microbiol. 88, 98–106.

Nies, D.H., Silver, S., 1989. Plasmid-determined inducible efflux is responsible for resistance to cadmium, zinc, and cobalt in *Alcaligenes eutrophus*. J. Bacteriol. 171, 896–900.

Nishioka, H., 1975. Mutagenic activities of metal compounds in bacteria. Mutat. Res. 31, 185–189.

Pal, A., Paul, A.K., 2004. Aerobic chromate reduction by chromium-resistant bacteria isolated from serpentine soil. Microbiol. Res. 159 (4), 347–354.

Pas, M., Milacic, R., Drasar, K., Pollak, N., Raspoor, P., 2004. Uptake of chromium(III) and chromium(IV) in compounds in the yeast cells structure. Biometals 17 (1), 25–33.

Pei, Q.H., Shahir, S.A.S., Raj, A.S.S., Zakaria, Z.A., Ahmad, W.A., 2009. Chromium(VI) resistance and removal by *Acinetobacter haemolyticus*. World J. Microbiol. Biotechnol. 25, 1085–1093.

Peitzsch, N., Gunther, E., Nies, D.H., 1998. *Alcaligenes eutrophus* as a bacterial chromate sensor. Appl. Environ. Microbiol. 64 (2), 453–458.

Petrilli, F.L., Flora, S.D., 1977. Toxicity and mutagenicity of hexavalent chromium on *Salmonella typhimurium*. Appl. Environ. Microbiol. 33, 805–809.

Petrilli, F.L., Flora, S.D., 1978. Metabolic deactivation of hexavalent chromium mutagenicity. Mutat. Res. 54, 139–147.

Philip, L., Iyengar, L., Venkobachar, C., 1998. Cr(VI) reduction by *Bacillus coagulans* isolated from contaminated soils. J. Environ. Eng. ASCE. 124, 1165–1170.

Pratt, P.F., 1966. Chromium. In: Chapman, H.D. (Ed.), Diagnostic Criteria for Plants and Soils. Quality Printing Co. Inc., Abilene, TX, pp. 136–141.

Rahman, M.U., Gul, S., Haq, Z.U., 2007. Reduction of chromium by locally isolated *Pseudomonas* sp. C-171. Turk. J. Biol. 31, 161–166.

Ramirez-Diaz, M.I., Diaz-Perez, C., Vargas, E., Riveros-Rosas, H., Campos-Garcia, J., Cervantes, C., 2008. Mechanisms of bacterial resistance to chromium compounds. Biometals. 21 (3), 321–332.

Romanenko, V.I., Kuznetsov, S.I., Korenkov, V.I., 1976. Method of biological purification of industrial effluents from chromates and bichromates. US Patent 3941691.

Ross, D.S., Sjogren, R.E., Bartlett, R.J., 1981. Behavior of chromium in soils 4. Toxicity to microorganisms. J. Environ. Qual. 10, 145–148.

Samantaray, D.P., Mishra, B.B., 2011. Parametric studies on hexavalent chromium reduction by *Acinetobacter calcoaceticus*. J. Res. 29 (1&2), 05–07.

Samantaray, D.P., Mishra, B.B., 2012. Effect of metal on hexavalent chromium reduction by *Acinetobacter calcoaciticus*. Bioscan. 7 (4), 627–629.

Schroll, H., 1978. Determination of the absorption of Cr^{+6} and Cr^{+3} in an algal culture of *Chlorella pyrenoidosa* using CR-51. Bull. Environ. Contam. Toxicol. 20, 721–724.

Shen, H., Wang, Y., 1993. Characterization of enzymatic reduction efflux by means of the *chrA* chromate resistance protein from hexavalent chromium by *Escherichia coli* ATCC 33456, *Pseudomonas aeruginosa*. Appl. Environ. Microbiol. 59, 3771–3777.

Silverberg, B.A., Wong, P.T.S., Chau, Y.K., 1977. Effect of tetramethyl lead on freshwater green-algae. Arch. Environ. Contam. Toxicol. 5, 305–313.

Singh, S.N., Tripathi, R.D. (Eds.), 2007. Environmental Bioremediation Technologies. *Springer-Verlag*, Berlin, Heidelberg.

Smillie, R.H., Hunter, K., Loutit, M., 1981. Reduction of chromium (VI) by bacterially produced hydrogen sulphide in a marine environment. Water Res. 15, 1351–1354.

Snow, E.T., 1994. Effect of chromium on DNA replication *in vitro*. Environ. Health Perspect. 102 (3), 41–44.

Summers, A.O., Jacoby, G.A., 1978. Plasmid-determined resistance to boron and chromium compounds in *Pseudomonas aeruginosa*. Antimicrob. Agents Chemother. 13, 637–640.

Theodotou, A., Stretton, R.J., Norbury, A.H., Massey, A.G., 1976. Morphological effects of chromium and cobalt complexes on bacteria. Bioinorg. Chem. 5, 235–239.

Towill, L.E., Shriner, C.R., Drury, J.S., 1978. Reviews of the Environmental Effects of Pollutants. III Chromium. The National Academy Press, Washington, DC, p. 100.

Turick, C.E., Apel, W.A., 1997. A bioprocessing strategy that allows for the selection of Cr (VI)-reducing bacteria from soils. J. Ind. Microbiol. Biotechnol. 18, 247–250.

Venitt, S., Levy, L.S., 1974. Mutagenicity of chromates in bacteria and its relevance to chromate carcinogenesis. Nature 250, 493–494.

Verma, S.K., Singh, S.P., 1995. Multiple chemical resistance in the cyanobacteria, *Nostoc muscorum*. Bull. Environ. Contam. Toxicol. 54, 614–619.

Vindhan, R.M., 2004. Bioaccumulation of chromium from tannery wastewater: an approach for chrome recovery and reuse. Environ. Sci. Technol. 38, 300–306.

Wang, P.C., Mori, T., Komori, K., Sasatsu, M., Toda, K., 1989. Isolation and characterization of an *Enterobacter cloacae* strain that reduces Cr(VI) under anaerobic conditions. Appl. Environ. Microbiol. 55, 1965–1969.

Wang, P.C., Mori, T., Toda, K., Ohtake, H., 1990. Membrane-associated chromate reductase activity from *Enterobacter cloacae*. J. Bacteriol. 172, 1670–1672.

Zakaria, Z.A., Zakaria, Z., Surif, S., Ahmad, W.A., 2007. Hexavalent chromium reduction by *Acinetobacter haemolyticus* isolated from heavy-metal contaminated wastewater. J. Hazard. Mater. 146 (1–2), 30–38.

15 Phycoremediation Coupled with Generation of Value-Added Products

Lowell Collins, Devin Alvarez and Ashvini Chauhan

Environmental Biotechnology Laboratory, School of the Environment, Florida A&M University, Tallahassee, FL, USA

15.1 Introduction

Intense demands for energy have driven the need to identify and optimize alternate energy sources that are renewable and sustainably manageable. Phycoremediation is the process of using algal species to sequester, remove, break down, or metabolize pollutants from soil and water. The algal biomass that results from phycoremediation can be fed into systems that process green waste, also known as biomass and biodegradable organic municipal waste and both waste products have a strong potential for the sustainable production of biofuels (Demirbas, 2009; Gibbons and Hughes, 2009). In addition, anaerobic fermentation of such organic-rich biodegradable waste can result in the production of biogas, which is an extremely versatile and environmentally benign fuel which releases significantly lower amounts of greenhouse gases and other particulate matter relative to conventional and nonrenewable fuels. More recently, the "biorefinery" concept came into being, which is focused on the process of "refining" the algal biomass into different products. Algal biomass cultivation alongside industrial processes provides the additional advantages of certain algae species that can remove waste-borne pollutants; at the same time, the photosynthetic microorganisms can sequester CO_2.

Thus phycoremediation can (a) remediate excess or residual nutrients from municipal treated wastewater and effluents that are specifically high in organic matter; (b) remove or biocatalyze xenobiotic and recalcitrant compounds; (c) perform secondary or tertiary treatment of acidic and metal wastewaters; (d) can sequester CO_2 from the atmosphere; and (e) detect toxic or recalcitrant compounds by using algae-based biosensors. Despite many advantages associated with green waste technologies, the high cost of algal biofuels (US$300–2600 per barrel) relative to petroleum (US$40–80 per barrel) strongly suggests that further improvements will be required to make it both commercially scalable and cost effective (Pienkos and Darzins, 2009; Hannon et al., 2010; Singh et al., 2011). Therefore, to

Microbial Biodegradation and Bioremediation. DOI: http://dx.doi.org/10.1016/B978-0-12-800021-2.00015-7

take full advantage of bio-products being formed by photosynthetic cyanobacteria and/or microalgae, biorefinery techniques will need to be further improved. Also, it is well known that almost 99% of the environmental biodiversity remains unknown primarily because of our inability to culture most microorganisms. Thus, better techniques to isolate and culture the under-explored algal and cyanobacterial biodiversity can also potentially yield hyper-lipid-producing strains.

Another successful solution can potentially rest in manipulating lipid accumulation in microalgae by appropriately modifying their genomes through the available genetic engineering (GE) techniques (Misra et al., 2013). Toward this end, three potential approaches to accomplish overproduction of lipids in microalgae have been proposed. The first approach, called biochemical engineering (BE), is based on causing a physiological stress, most often with nitrogen, that results in nutrient starvation that funnels metabolic fluxes within the microalgal cells to accumulate lipids. The second approach, called the GE approaches, is dependent on our understanding the lipid metabolic pathways, and then specifically focusing attention on the rate-limiting enzymes, so as to create channeling of metabolites to lipid biosynthesis by overexpressing one or more key enzymes in recombinant microalgal strains. The third approach, called the transcription factor engineering (TFE) approach, relies on enhancing the production of a particular metabolite by means of overexpressing Transcription factors (TFs) that are responsible for regulating a specific metabolic pathway involved in the accumulation of a target such as lipids. TFE is an emergent strategy because it can potentially bypass the inhibitive effects of BE and the limitation of "secondary bottlenecks" as are typically commonly observed in GE approaches. For example, an engineered *Arabidopsis thaliana* was shown to overproduce the transcription factor WRINKLED1 (WRI1), which is involved in the regulation of seed oil biosynthesis (Sanjaya et al., 2011). In addition, this group reduced the expression of APS1 that is shown to encode the ADP-glucose pyrophosphorylase that participates in starch biosynthesis. Thus, the resulting AGPRNAi-WRI1 lines were shown to accumulate less starch and more hexoses, with the production of 5.8 times more oil in vegetative tissues than in plants with WRI1 or AGPRNAi alone (Sanjaya et al., 2011).

In fact, recently there has been significant progress made on overexpressing or knocking-out genes from microalgae found to have potential for biofuel production (Hu et al., 2008; Courchesne et al., 2009; Radakovits et al., 2010). However, prior to the application of metabolomics, genomics, transcriptomics, and proteomic techniques, a deeper understanding of lipid-generating biosynthetic pathways in microalgae needs to be obtained so as to better design the most appropriate approach (Hu et al., 2008). Toward this end, an increasing number of microalgal genome sequences are being available that can result in the application of "omic" approaches to unravel algal lipid metabolism and identify more precisely the gene targets for development of engineered strains with higher lipid contents (Rodriguez-Moya and Gonzalez, 2010; Yu et al., 2011; Georgianna and Mayfield, 2012). In fact, at the Joint Genome Institute, which is funded by the US Department of Energy, more than 60 algae genome–sequencing projects are ongoing. Despite this effort, the majority of genome projects have focused largely on

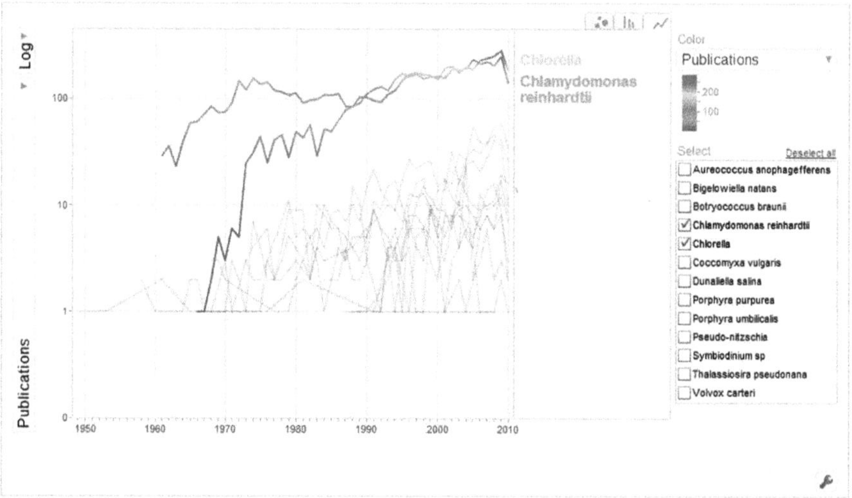

Figure 15.1 Increased publications related to *Chlorella* and *Chlamydomonas* algal species.

Chlorella and *Chlamydomonas* (Figure 15.1), indicating the vast research potential that needs to be explored in this area.

Thus, the goal of this chapter is to (1) discuss the trends in the use of microalgae and cyanobacteria coupled with wastewater remediation and the generation of value-added products, (2) provide an update on the developments being made in the field of bioprospecting for high lipid–producing microorganisms from different environmental niches, and (3) address the recent developments made in the successful application of genomics of lipid-producing microalgae and cyanobacteria that will result in sustainable production of algal biofuels.

15.1.1 Phycoremediation and Generating Value-Added Bio-Products

Compounds, organic and inorganic, have been accumulating in the environment as a result of agricultural, industrial, and domestic activities. Uncontrolled proliferation of untreated human wastes and agricultural runoff are the most common sources of water and soil pollution. The environmental and ecological consequences resulting from human pollution and untreated wastes have accumulated and must be addressed if further negative consequences are to be avoided. In order to mitigate the effects of industrial and municipal pollution, the practice of treating wastewater must increase significantly and on a global scale. Due to shortcomings in many of our current water treatment practices, the quality of freshwater is becoming as much of a concern as scarcity. Waste cleaning practices help to reduce the contamination of waters and soils from nutrient overload and other toxic contaminants. Untreated wastewater has nutrients in solution that create ecosystem eutrophication or deliver toxic compounds to the environment. Eutrophic conditions stimulate ecosystem imbalances that, if not addressed, can escalate to create conditions that threaten human health. In addition to the potential negative effects of nutrient overload, the effects of heavy metal and chemical contamination of waters and soils are difficult to repair.

The need for cost-effective means of removing or neutralizing hazardous compounds is growing daily. A database of known compounds at the Chemical Abstract Service (CAS) is currently growing from the known 66 million registered compounds, with the addition of about 12,000 novel compounds every day (CAS, 2012). Discharging these compounds into water bodies or spraying them out onto soils has drastically affected the environment and will continue to do so. Soil quality and water quality have already been affected globally by industry and human development. Increased urbanization and industrialization has given rise in particular to municipal waste-related environmental problems. Municipal wastewater is a significant source of environmental degradation from domestic products and pharmaceuticals as well as eutrophication from nutrient overloading.

Microalgae and cyanobacteria respond positively to a wide range of organic and inorganic pollutants. These organisms are able to remove compounds from solution through bioaccumulation of compounds into the cell body as well as onto the cell surface, in addition to mineralization activity related to cell metabolism. Algae and cyanobacteria are able to grow in water with low or high nutrient value as well as in harsher ecosystems like high to hyper-salinity or extreme pH. In addition to the highly versatile survival mechanisms of each organism, some microalgae species have the ability to utilize both autotrophy and heterotrophy, making them mixotrophic in the right conditions. In the presence of high levels of pollutants, algae and cyanobacteria experience physiological changes (Fogg, 2001) as well as genetic responses (Gonzalez et al., 2012) which allow algae to rapidly multiply, or bloom. This adaptation makes microalgae and cyanobacteria ideal for the remediation of environmental pollutants as well as the sequestration of CO_2. Phycoremediation is the utilization of specifically micro or macro algae species for the removal, or biodegradation, of extra environmental compounds. Phycoremediation is an important tool for the treatment of contaminated soil and water, and it encompasses numerous applications which include but are not limited to the stabilization of acidic contamination, removal of metals in contaminated waters, the removal of excessive nutrients from water, and the degradation or sequestration of toxic compounds from water and soil.

In order for phycoremediation to control pollutants efficiently, the use of specific species or a combination of species must be selected that are able to treat or interact favorably with the contamination. Qualities such as tolerance of extreme temperatures, potential for value-added products (such as lipids, carbohydrates, and/or proteins), mixotrophy, and sedimentation behavior are key elements to investigate. Researchers have studied the remediation of wastewaters polluted by a range of sources to investigate the task(s) performed by different algae species at each stage of the remediation process (Abdel-Raouf et al., 2012). Table 15.1 gives these studies with the kind of wastewater each study investigated. These studies demonstrate that for each task involved in remediating a given waste, there can be multiple species capable of performing that task.

Table 15.1 Range of Phycoremediation

Sector	Organism	Reference
Municipal Sewage		
N/P removal	*Gloeocapsa gelatinosa*, *Euglena viridis*, and *Synedra affinis*	Sengar et al. (2011)
Endocrine disruptors	*Anabaena cylindrica*, *Chlorococcus*, *S. platensis*, *Chlorella*, *S. quadricauda*, *Anabaena vax*	Shi and Wang (2010)
Urine	*Spirulina Plantensis*	Ying et al., (2011)
Agriculture		
Poultry waste	*C. Vulgaris*	Murugesan et al. (2010a,b)
Organic pesticides	*Extensive catalog*	Subashchandrabose et al. (2013)
Pig manure	*Scendesmus intermedius*, *Nannochloris* sp.	Jimenez-Perez et al. (2004)
Industry		
Heavy metals	*Anabaena variabilis*	Parameswari et al. (2010)
	Spirogya	Gupta et al. (2001)
Leather plant	*C. Vulgaris*	Rao et al. (2011)
Alginate	*Chroococcus turgidus*	Sivasubramanian et al. (2009)
Food Processing		
Olive Mill	*Scenedesmus obliqus*	Hodaifa et al. (2013)

It should be noted that the successful cultivation of algae and cyanobacteria depends on more than just the nutrients present in a given ecosystem. A combination of complex environmental factors and their interactions influence cell development in algae and cyanobacteria. Factors such as the pH of the water or soil (Azov and Shelef, 1987), light availability and intensity, temperature, and the myriad biotic factors at play in water and soil all influence algal cell development.

The use of phycoremediation to treat wastewaters offers solutions to some of the problems associated with conventional treatment techniques. Some of the disadvantages of conventional treatment techniques relate to variability in chemical efficacy, depending on the compound or nutrient to be removed. Another disadvantage of conventional treatment techniques involving chemical remediation is that the chemicals used to treat the water are released into the environment. Some of these chemicals remain active, leading to differing degrees of environmental damage from exposure to "treated" wastewaters. Furthermore, De la Noue et al. (1992) pointed out that when nutrient-rich water (treated or untreated) is discarded, there is an economic loss that coincides with the loss of valuable nutrients that could be recovered, recycled, and repurposed. Algae cell bodies retain nutrients from the process of phycoremediation, and the resulting biomass makes the nutrients available for collection and recycling (Pizarro et al., 2006). Collected nutrients in the form of algal biomass can be used for a variety of economically valuable purposes, including, but not limited to, soil restoration, organic fertilizers, fermentative processes, and other agricultural/industrial processes.

15.1.2 Municipal Wastewater

The environmental changes from municipal waste discharge are caused by the addition of nutrients that disrupt the ecology (Prased, 1982; Geddes, 1984). Rapid algal cell development from high amounts of nitrogen and phosphorus disrupts higher trophic levels (Sawayama et al., 1998). Municipal wastewater can contain concentrations of total nitrogen and phosphorus between 10 and 100 mg/L (De la Noue et al., 1992). In order to prevent the kind of environmental damages caused by municipal wastewater, the waste needs to be treated thoroughly (Sawayama et al., 2000). All of the environmentally harmful components of wastewater currently require costly chemical treatments during water treatment (Gasperi et al., 2008). The chemical treatment of water sources leaves the treatment agents active in solution, which can lead to environmental deposition.

Municipal sewage treatment involves at least two phases, but often includes three. The three standard phases of treatment include primary treatment, which is the removal of solids from the water; secondary treatment, which removes the majority of suspended and dissolved organic solids from the water; and tertiary treatment, which more thoroughly removes organic and inorganic compounds. A major disadvantage of the implementation of these treatment processes is that the relative cost of each phase of treatment compounds exponentially for each phase following primary treatment (Oswald, 1988b). For example, a complete tertiary treatment phase that removes ammonia, nitrate, and phosphate would cost approximately 4 times more than the primary treatment phase. Another disadvantage of this treatment process is that, even after chemical treatment, sludge is left over that is difficult to handle, store, or utilize. Getting rid of waste sludge incurs additional costs beyond the costs of treatment itself. A final disadvantage of the current model for waste treatment is that it discards valuable nutrients that have the potential to be recycled (Pizarro et al., 2006).

Microalgae species' metabolic processes remove nitrogen and phosphorus and adsorb toxic metals to the cell wall (Mallick, 2002; Ahluwalia and Goyal, 2007). In addition to removing harmful compounds during water treatment, algae produce clean atmospheric oxygen and can buffer pH. At the height of algal activity, carbonate and bicarbonate ions will react to supply carbon dioxide for the cells. This process leaves an excess of hydroxyl ions, and the addition of these ions raises the pH of water up to or above a pH of 9. The cultivation of algal biomass generates oxygen, and this source of O_2 can lessen the need to mechanically aerate wastewater holding ponds, partially offsetting the cost of aeration. This algae-generated aeration is valuable because the oxygenation of treatment ponds stimulates the bioremediation of organic and inorganic compounds through the metabolic activities of aerobic bacteria (Munoz and Guieysse, 2006). Certain microalgae species of *Chlorella* and *Scenedesmus* have been found to remove up to 80% of (and in some cases completely remove) ammonia, nitrite, and total phosphorus from secondary treated wastewater (Martinez et al., 2000; Zhang et al., 2008; Ruiz-Marin et al., 2010). A large body of research has been published on *Chlorella* and *Scenedesmus* because of the species' hardiness in sewage water (Bhatnagar et al., 2010; Wang et al., 2010). Furthermore, the use of microalgae for wastewater treatment has proven to be a more effective and efficient method over other aquatic plants, such as water hyacinth (*Eichornia crassipes*), cattails (*Typha*), and phragmites (*Phragmites australis*) (Werblan et al., 1978; Wolverton, 1982; Finlayson and Chick, 1983; Finlayson et al., 1987).

15.1.3 Industrial Wastes and Effluents

With respect to industrial wastes, the application of phycoremediation is based on the ability of certain algal species to remove heavy metal pollutants (e.g., cadmium, zinc, and chromium), as well as many organic chemicals (e.g., hydrocarbons, herbicides, and pesticides) (Mallick, 2002; Ahluwalia and Goyal, 2007; De-Bashan and Bashan, 2010). The nitrogen and phosphorus contents of industrial wastes are typically much lower than those of municipal wastes, signifying slower algal growth in the former. However, one example of potentially significantly increased algal growth in industrial effluent comes from the carpet mill industry (Chinnasamy et al., 2010).

Conventional treatment of industrial effluents includes the heavy use of chemicals to achieve pH balance and control bacteria. The process leads to the creation of particulate mass, which requires removal in the form of sludge. The waste sludge that forms is hazardous in most cases and must therefore be removed and disposed of with additional treatment, or stored. The process of phycoremediation reduces the formation of this environmentally hazardous sludge that requires storage in landfills, or specialized waste dumps. Even when the compounds that require treatment are particularly unavailable to mineralization, the conversion of the pollutant into a less hazardous compound is possible with phycoremediation.

The breaking down of complex toxic molecules is one application of phycoremediation. Algal biomass has been used to convert estrogenic bisphenol into a non-estrogenic monohydroxy bisphenol (Hirooka et al., 2005). Another example of phycoremediation for conversion is the application of *Scendesmus obliquus* in utilizing 1-napthalenesulfonic acid as a metabolite, converting it to 1-hydroxy-2-napthalenesulfonic acid, 1-naphthol and 1-naphthyl Β-D-glucopyranoside (Kneifel et al., 1997). Often industrial wastes include a specific group of organic chemicals called phenolic compounds, which are toxic to most organisms at low concentrations (Pimentel et al., 2008). Phenols are extremely difficult to remove from environments once they are discharged into the area (Araña et al., 2001). The algal species *Ankistrodesmus braunii* and *Scenedesmus quadricauda* are both able to degrade different phenolic compounds, most of which are reduced by 70% within

Table 15.2 Cyanobacterial/Microalgal Degradation of Organic Pollutants of Industrial Origin

Organic Pollutant	Cyanobacterium/Microalga	Reference
Phenol	*Ochromonas danica*	Semple and Cain (1996)
Tributyltin (TBT)	*C. vulgaris Chlorella* sp., *Chlorella miniata*	Tsang et al. (1999)
Benzoapyrene (BaP)	*Selanastrum capricornutum*	Tam et al. (2002)
Phenanthrene (PHE)	*S. capricornutum*	Warshawsky et al. (1988)
Naphthalene	*Agmenellum quadruplicatum*, *C. vulgaris*	Schoeny et al. (1988)
1-Naphthalenesulfonic acid	*Scenedesmus obiquus*	Chan et al. (2006)
1-Methylnaphthalene	*A. quadruplicatum Oscillatoria* sp., *Anabaena* sp.	Cerniglia et al. (1979)
2-Methylnaphthalene	*A. quadruplicatum*, *Oscillatoria* sp., *Anabaena* sp.	
2,4,6-Trinitrotoluene	*Anabaena* sp.	Pavlostathis and Jackson (1999)
Dibenzofuran	*Ankistrodesmus* sp.	Todd et al. (2002)
Dibenzo-*p*-dioxin	*Scenedesmus* sp.	
Bisphenol	*Chlorella fusca*	Hirooka et al. (2005)
Bisphenol A	*Pseudokirchneriella subcapitata*, *Scenedesmus acutus*, *Coelastrum reticulatum*	Nakajima et al. (2007)
Dimethyl phthalate	*Closterium lunula*	Yan and Pan (2004)
Sinapic acid	*Stichococcus bacillaris*	DellaGreca et al. (2003)
Azo compounds	*C. vulgaris*	Jinqi and Houtian (1992)

10 days (Pinto et al., 2002). Some species of cyanobacteria are able to break down phenols as well (Hirooka et al., 2006) (Table 15.2).

The bioaccumulation of compounds in algal biomass has also been applied to take up Polychlorinated biphenyl (PCBs); the compound was taken up into the cell organelles, as well as attached to the surface of the cell (Jabusch and Swackhamer, 2004). Algae are good accumulators of industrial compounds like organochlorides and tributyl (Payer and Runkel, 1978; Wright and Weber, 1991). Additionally, the acidic nature of some industrial wastes can be buffered by algal growth. The growth of algae in wastewater can balance the acidity which can help reduce harmful degradation of waste infrastructure. In a study of dairy effluent, which is generally acidic, the pH was buffered from around five to around eight by *Nosotc* sp. (Murugesan et al., 2010a,b).

15.1.4 Agricultural Waste

Herbicides and insecticides can be collectively referred to as pesticides, which have an indiscriminant toxic effect on ecosystems. Even pesticides designed to interact with specific components of agriculture have unintended negative ecological impacts as a waste product. The use of pesticides and fertilizers in modern agriculture has allowed food production to keep up with growing global population. However, the environmental consequences of the widespread application of pesticides have been observed and described in terms of trophic degradation. The chemicals in pesticides remain active long after their application and are eventually absorbed into the surrounding ecosystem.

Environmental damage from agricultural waste can be curved with the application of phycoremediation treatments. The cultivation of algal biomass can clean up agricultural waste and can be accomplished using relatively inexpensive techniques (Singh and Walker, 2006; Caceres et al., 2010). Species of algae such as *Isochrysis galbana*, *Dunaliella teriolecta*, *Phaeodactylum tricornutum*, *Phaeodactylum subcapitata*, and *Synechococcus* are able to remove pesticides through cellular adsorption (Weiner et al., 2004). The main factors in the removal of pesticides are the cellular volume, the surface area of the cell, and the concentration of the pesticide(s) in solution. Unfortunately, the algae that tend to interact when the pesticide concentrations are low enough to be nontoxic; in such cases, the environmental damage has usually already been done.

Triazine is one group of herbicides that has been shown to be degraded by the algal species *Chlorella vulgaris* at a remarkable rate; biomass metabolized the compound within 12 h (Gonzalez-Barreiro et al., 2006). Pesticide treatments could be applied with *Chlorella saccharophil*, which is able to remove an organophosphorus insecticide, pyridaphenthion, from solution in less than 5 days (Jonsson et al., 2001). Phycoremediation can also help mitigate environmental exposure to fluometuron, which is used in cotton fields (El-Rahman Mansy and El-Bestawy, 2002). Soil contamination is difficult to remediate with current techniques; however, with phycoremediation the contamination can be treated during exposure, or directly after, depending on how the runoff is contained. Algal strains have additionally been proven to catabolize DDT in the soil and have been documented as highly efficient for this purpose (Megharaj et al., 2000) (Tables 15.3 and 15.4).

15.1.5 Petrochemical Waste

Petrochemical waste is another widespread waste and is particularly difficult to deal with once a site is contaminated. Certain algae are able to degrade oily wastes that contain hydrocarbon compounds (Cerniglia et al., 1980; Carpenter et al., 1989). The ability of some algal species to degrade crude oil, as well as conventional motor oil, has been well established (Walker et al., 1975a,b). Environmentally dangerous petrol waste compounds, referred to as PAHs, are released from the incomplete combustion of coal and petroleum oil. Many of the compounds that are part of the PAH group of pollutants are dangerous because they are carcinogenic to both humans and animals (Dejmek et al., 2000). Freshwater algal species, as well as cyanobacteria, will take additional time during the growth phase in the presence of crude oil, which allows for greater biomass production and removal of additional PAHs (Gamila and Ibrahim, 2004). The

Table 15.3 Substances of Agricultural Importance Influenced by Cyanobacterial/Microalgal Degradation (Subashchandrabose et al., 2012)

Organic Chemical	Cyanobacteria/Microalgae	Reference
DDT	*Aulosira fertilissima*, *Chlorococcum* sp., *Anabaena* sp., *Nostoc* sp.	Lal et al. (1987), Megharaj et al. (2000)
γ-Hexachlorocyclohexane	*Anabaena* sp., *Anabaena* sp. (*pRL634*)	Kuritz and Wolk (1995)
Methyl parathion	*C. vulgaris*, *Scenedesmus bijugatus*, *Nostoc linckia*, *Nostoc muscorum*, *Oscillatoria animalis*, *Phormidium foveolarum*	Megharaj et al. (1994)
Metflurazon	*C. fusca*	Thies et al. (1996)
Norflurazon	*C. fusca*	Thies et al. (1996)
Fluometuron	*Ankistrodesmus cf.*, *Nannoselene*, *Selenastrum*, *Capricornutum*	Zablotowicz et al. (1998)
Atrazine	*Ankistrodesmus* sp., *Selenastrum* sp.	Zablotowicz et al. (1998)
α-Endosulfan	*Scenedesmus* sp., *Chlorococcum* sp., *Scenedesmus* sp.	Sethunathan et al. (2004)
Diclofop-methyl	*C. vulgaris Chlorella pyrenoidosa*, *S. obliquus*	Cai et al. (2007)
Dichlorprop-methyl	*C. pyrenoidosa*, *C. vulgaris*, *S. obliquus*	Li et al. (2008)
Fenamiphos	*Pseudokirchneriella*, *Subcapitata*, *Chlorococcum* sp.	Cáceres et al. (2008)

Table 15.4 Biosorptive Potential of Cyanobacteria/Microalgae for Organic Chemical Substances (Subashchandrabose et al., 2012)

Organic Chemical	Cyanobacterium/Microalga	Reference
Fenitrothion Chlorpyrifos DDT	*Anabaena* sp., *Aulosira fertilissima*	Lai et al. (1987)
Pyrene	*C. vulgaris*, *S. quadricauda*	Lei et al. (2002)
Anthracene	*Chlorella protothecoides*	Yan et al. (2002)
PCBs congeners	*C. reinhardtii*	Jabusch and Swackhamer (2004)
Naphthalene Phenanthrene Pyrene	*Prototheca zopfii*	Ueno et al. (2008)

metabolic processes of certain marine cyanobacteria species can break down phenanthrene, which is carcinogenic in mammals (Narro et al., 1992). In petrochemical wastes, dissolved solids are particularly harmful if released into the environment. A cyanobacterium, *Oscillatoria quadrippunctulata*, is able to absorb the dissolved

Table 15.5 World Microalgae Production

Species	Producer	Products	Production (tons/year)
Spirulina (*Arthrosphira*)	Hainan Simai Pharma Co. (China)	Powders, extracts, tablets, powders, extracts	3000
	Earthrise Nutritionals (California, USA)	Tablets, powders, bevrages	
	Cyanotech Corp. (Hawaii, USA)	Extracts, tablets, chips	
	Myanmar Spirulina Factory (Myanmar)	Pasta, liquid extract	
Chlorella	Taiwan Chlorella Man. Co. (Taiwan)	Tablets, powders, nectar	2000
	Klotze (Germany)	Powders	
D. salina	Cognis Nutrition and Health (Australia)	Powders, beta-carotene	1200
Aphanizomenon flosaquae	Blue Green Foods (USA)	Capsules, crystals	500
	Vision (USA)	Powder, capsules, crystals	

Source: Adapted from Pulz and Gross (2004); Spolaore et al. (2006); Hallmann (2007).

solids in petrochemical wastes, which include phenols, sulfides, and other aromatic compounds (Joseph and Joseph, 2001). The algal species *Spirulina platensis* has the ability to remediate dissolved solids from petroleum effluent as well (Veeralakshmi et al., 2007).

15.1.6 Phycoremediation and Value-Added Products from Algal Biomass

The potential of value-added products makes the process of phycoremediation even more attractive. Algae store lipids within the cell that are useful in pharmaceuticals, nutraceuticals, and biodiesel production. Even more fermentative products are possible; ethanol and methane have both been generated from algal biomass. The algal cell body contains carbohydrates, as well as proteins, which are present in high enough ratios to have value for use in nutrients for soil restoration and animal feed (Spolaore et al., 2006) (Table 15.5).

15.1.7 Production Techniques

There are two main techniques for the mass production of algal biomass: open pond cultivation and closed photobioreactors. Each of these techniques has both advantages and disadvantages. Open pond cultivation is done outdoors. This means exposure to weather and outside biological influences. Closed photobioreactors, on the other hand, are sealed and more closely control cultivation conditions (such as the use of synthetic light sources) to generate algal biomass. Open pond systems are less energy intensive and generally do not require additional light sources (beyond solar input), making them more suitable for mass culture. Photobioreactors require close monitoring in order to maintain ideal conditions. Although photobioreactors require more energy input, they can generate very specific biomass.

Open pond systems must be located in zones with ideal climate and conditions because they are subject to weather and seasonal variability. The ponds are also sensitive to outside contamination by environmental populations or free-floating organisms. Competing organisms (such as protozoa) or other grazing organisms (such as midge fly larva) can cause significant damage to open pond cultures. A major benefit to using open ponds, apart from the availability of solar energy, is that it makes mass cultivation possible on otherwise unused or agriculturally unviable land. Controlling pond conditions can be difficult; controlling outside biological contamination is even more difficult (Table 15.6). In open pond systems, the water conditions must be maintained while balancing environmental contamination by other species, dilution from rain, and evaporative loss. One example of an algal species that does well in extreme conditions is *Dunaliella salina*, which is grown as a monoculture for its beta-carotene content. This species survives in extreme halophilic conditions that do not permit competitive species to survive.

Open ponds and photobioreactors are both subject to culture collapse from biological interferences, or otherwise unknown causes. This problem has yet to be resolved and, unfortunately, culture collapses are common. Open pond systems have the benefit of large-scale productions but suffer to a greater degree from environmental difficulties than do photobioreactors. Either system, if deployed

Table 15.6 Advantages and Limitations of Open Ponds and Photobioreactors

Production System	Advantages	Limitations
Raceway pond	Low cost, easy to clean, uses nonagricultural land, low energy inputs, easy maintenance	Culture collapse, large land area, poor biomass productivity, poor mixing, inefficient light, and CO_2 utilization
Tubular PBR	Large illuminated surface area, suitable for outdoor culture, good biomass productivities, prevents culture collapse	Fouling, large space requirement, pH, DO, and CO_2 gradient control are obstacles
Flat plate PBR	High biomass productivities, easy to sterilize, low oxygen buildup, readily tempered, good light path, prevents culture collapse	Difficult scale-up, temperature control, hydrodynamic stress, fouling, cost
Column PBR	Compact, high mass transfer, low energy consumption, good mixing with low shear stress, easy to sterilize	Cost, sophisticated construction, command and control of key growth factors (pH, CO_2, light, shear forces)

Source: Adapted from Brennan et al. (2009).

outdoors, can be subject to evaporative loss, sharp temperature variation, and/or inconsistent exposure to sunlight from weather interferences.

Closed photobioreactors are primarily employed by industry seeking specialized products from the algae cell, such as pharmaceutical or nutraceutical products. The closed reactor allows for much tighter control of light exposure, temperature, salinity, and pH, as well as tighter control of the movement of the culture through the system. The control of these conditions allows for the cultivation of a single species for long periods of time with little to no risk of contamination. Currently, closed systems are either cycled using an aeration technique called an airlift or cycled mechanically using a pump. The airlift system allows maximum exposure for CO_2 and O_2 exchange within the liquid because it uses the gas to manipulate the growth media. The perturbation of the culture is important, both to circulate gasses and to expose as much of the culture as possible to the light source.

Photobioreactors can be a range of shapes and styles. Flat panel reactors contain the culture between two plates of clear material and allow for maximum exposure, but have volume limitations. Column bioreactors are the most common and have the most efficient methods for mixing and control over conditions. Tubular systems range from a network of transparent piping to biocoil reactors, which are cylinders wrapped in smaller transparent piping.

15.1.8 Value-Added Potential

Culturing methods dictate the products that are made available from the biomass. Biodiesel is by far the most sought-after value-added product of algae cultivators. However, the production of sufficient amounts of biomass (or higher-yielding biomass) at a cost that is economically viable has not yet been achieved. A more immediately feasible value-added product is biohydrogen, which can be produced from the biogas generated by algal biomass (which is also a source of CO_2 fixation) (Melis and Happe, 2001). Other immediately viable products that can be generated by algal biomass are useful for agricultural purposes. Algae are rich in proteins and carbohydrates that can be used in a wide range of animal feeds. Environmental algal biomass is relatively inexpensive to produce and can be incorporated into pellet feeds for cattle, poultry, fish, and swine. Benefits to livestock from consuming algae-based feed include healthier skin and coat, improved weight control, and higher fertility rates (Pulz and Gross, 2004). In aquaculture, algae pellets are currently used as feedstock for oysters, salmon, and shrimp. An additional application of algal biomass in agriculture is for soil restoration and organic fertilizer. The creation of biochar from algal biomass has promise for use as a fertilizer, as well as for CO_2 sequestration as a long-term carbon sink. Biochar has the potential to sequester up to 84% of carbon emissions (Gaunt and Lehmann, 2007) (Figure 15.2).

With respect to human consumption, algal biomass is a rich source of PUFAs (polyunsaturated fatty acids), which are critical to human physiology and physical development (Hu et al., 2008). The PUFA industry is currently supplied primarily by fish oil extracts. Fish oils, however, are particularly sensitive to oxidation and are not viable for people with dietary restrictions that exclude fish and fish products.

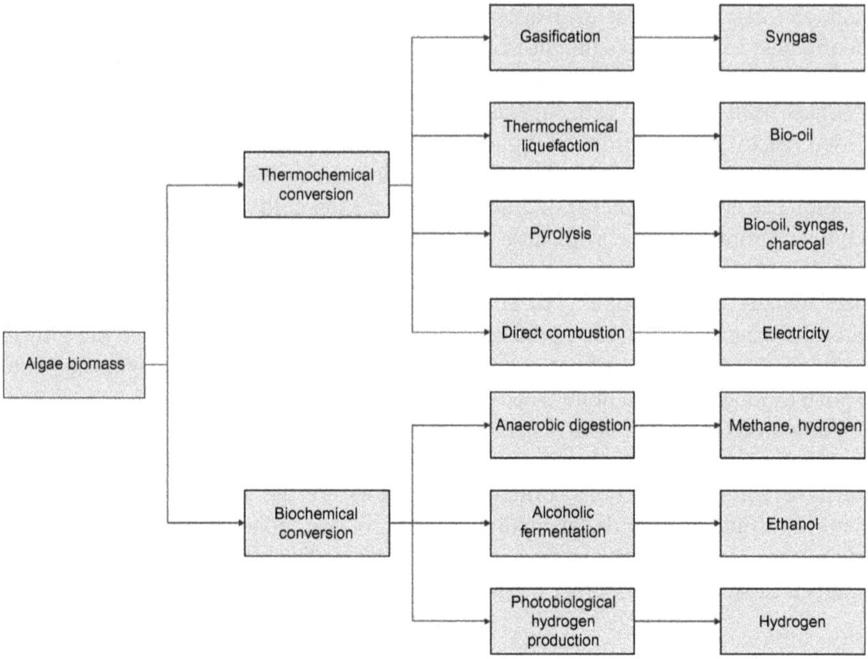

Figure 15.2 Process options for algae biomass.

Cultivation within fish bodies also raises the risk of accumulation of toxins within the PUFAs. Cultivation within fish bodies also creates the potential for the PUFAs to combine with other fatty acids, creating a need for a separation process, which means additional time and money spent (Pulz and Gross, 2004). As an alternative to fish sources, microalgae are a viable primary source of PUFAs and have been supplied to chickens to raise omega-3 levels in their eggs (Pulz and Gross, 2004).

Phycoremediation has been proven to be an effective method of treatment for waste streams in a multitude of scenarios. Algal biomass can be cultivated in a variety of environmental and industrial contexts to meet production needs. The techniques involved may vary depending on the requirements of the species, the demands of the situation, and the products desired. As of yet, many of the issues preventing efficient mass cultivation of specific species are unresolved. Species-specific cultivation in large-scale production will require innovative practices that are economically viable. Phycoremediation, as an alternative to conventional treatment methods, offers a reduction in both cost and pollution. Phycoremediation additionally provides incentives due to the potential for a wealth of value-added products from algal biomass. With innovative techniques, algal-based products may increase in value and, eventually, lead to a more sustainable existence for a growing human population. Techniques for isolating, observing, and modifying algal cells will be explored in the following sections.

15.2 What Is Bioprospecting?

The discovery and commercial applications of new products from the biological world is called bioprospecting. Thus, bioprospectors look to the natural world to discover commercial value in products or to explore functions of organisms: animals, plants, fungi, bacteria, archaea—and any others yet to be discovered. Throughout history, researchers have often turned to the natural world for answers to difficult problems. Even more recently, the prospecting efforts in exotic ecosystems have led to the discovery of enzymes, like TAQman, which has driven the biotechnological revolution further (Brock, 1997). The kinds of value that can be found in the biological world are as diverse as the varieties of organisms that exist. Viable categories of value include but are not limited to food, pharmaceuticals, nutraceuticals, cosmeceuticals, fuels, enzymes, and many other biological products related to biotechnology. In such a vast field, it is important to know where to begin. A list of some of the bioprospecting efforts and their respective values, products, and functions is provided in Table 15.7.

15.2.1 How to Bioprospect and Needs Assessments for Informed Prospecting

Bioprospecting efforts can encompass an almost limitless number of aspects pertaining to the Earth's biodiversity. As such, it is important to start with a general framework. Refer to Figure 15.3 for a microbial bioprospecting workflow. Establishing an assessment of needs is the first step in any prospecting effort. Two key questions to consider are (a) what value(s), function(s), or product(s) are we looking for? and (b) What type(s) of organism(s) might possess these value(s), function(s), or product(s)? In some cases, answers to these questions may be easily discovered; in others, some guesswork may be necessary. Performing a needs assessment enables researchers to design an informed prospecting effort. At the minimum, the type(s) of life form(s) being sought should be determined. The organism(s) desired have a preferred habitat, and the bioprospector must identify and locate the correct habitat to explore.

The following questions should be addressed by bioprospectors when conducting a needs assessment: Where does the desired organism live? What is the life cycle of the organism(s)? Is any particular habitat possibly going to produce organism(s) more suitable to the product(s), value(s), and function(s)? For example, if seeking an organism to act in a remediation capacity for any number of wastewaters (industrial, agricultural, municipal), it is prudent to first prospect in the wastewaters themselves. It is very likely that a suitable organism is already present in the appropriate environment. For example, in an ongoing project in the Chauhan laboratory at the School of the Environment, Florida A&M University, wastewater holding tanks were explored to isolate lipid-producing microalgae and cyanobacteria. This high-throughput bioprospecting, performed by using flow-assisted cell sorting (FACS) and BODIPY-stained samples, resulted in hundreds of isolates that are being screened for lipid production (Figure 15.4).

Table 15.7 Worldwide Bioprospecting Efforts

Value, Product, Function	Location	Organism	References
Biotechnology			
Taqman enzyme[a]	Yellowstone	*Thermus Aquaticus*	Brock (1997)
Antifreeze proteins	North Pole	Marinomonas, Psuedomonas sp.	Wang (2012a)
Thermostable isomerase	Arctic Circle	*Thermoanaerobacter mathranii*	Wang (2012a)
Pharmaceuticals, Nutraceuticals, Cosmacueticals			
L-DOPA	Maharashtra, India	*Mucuna monosperma* Wight	Inamdar et al. (2012)
Leukemia inhibition	Western Canada	Vascular plants	Deeg (2012)
Isoflavonoids	China	*Erythrina arborescens*	Wang (2012a)
Docsahexaenoic acid	USA	*Crypthecodinium cohnii*	Spolaore et al. (2006)
Antimicrobial			
Antifungalmycin	China	Streptomyces padanus *702*	Wang (2012a,b)
Eugenol	West Indies	*Pimenta dioica*	Cowan (1999)
Dienone	Amazonia	*Verongia fistularis*	Cowan (1999)
Fuels			
Methane	Ubiquitous	*Termites/Methanogens*	He (2013)
Hydrogen	Icelandic Hotspring	*Thermoanaerobacter thermohydrosulfuricus*	Koskinen et al. (2008)
Biodiesel	Africa/India	*Jatropha curcas*	Jain and Sharma (2010)
Remediation			
Toluene degradation	Scott Base, Antarctica	*Psuedomonas* spp.	Farell (2003)
Lead biosorption	China	*Cladophora fascicularis*	Deng (2007)
Uranium biosorption	France	*Mucor miehei*	Guibal et al. (1992)
Acid mine alkalization	Spain	*Sulfate-reducing bacteria*	Garcia et al. (2001)

[a]Not specific bioprospecting, effort, but a result of basic research conducted by Thomas Brock.

Another question to ask when conducting a needs assessment is which habitats most closely resemble the downstream cultivation conditions. Some production strategies will not include cultivation, but instead chemical synthesis or expression vectors, as is the case for an easily cultivated organism like *Escherichia coli*. Ideally, one can find an organism that is already optimized for the intended future applications.

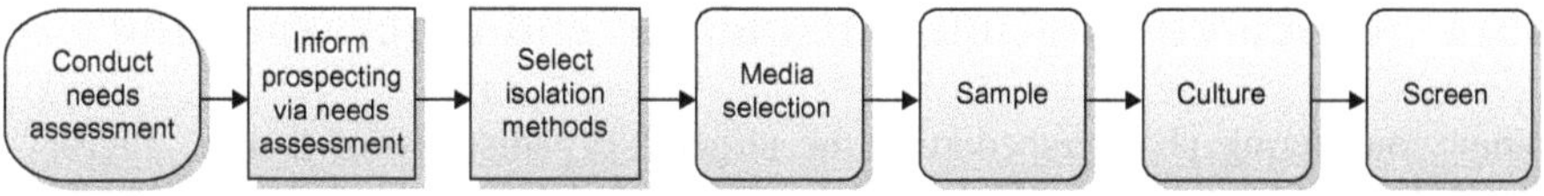

Figure 15.3 Microbial bioprospecting workflow.

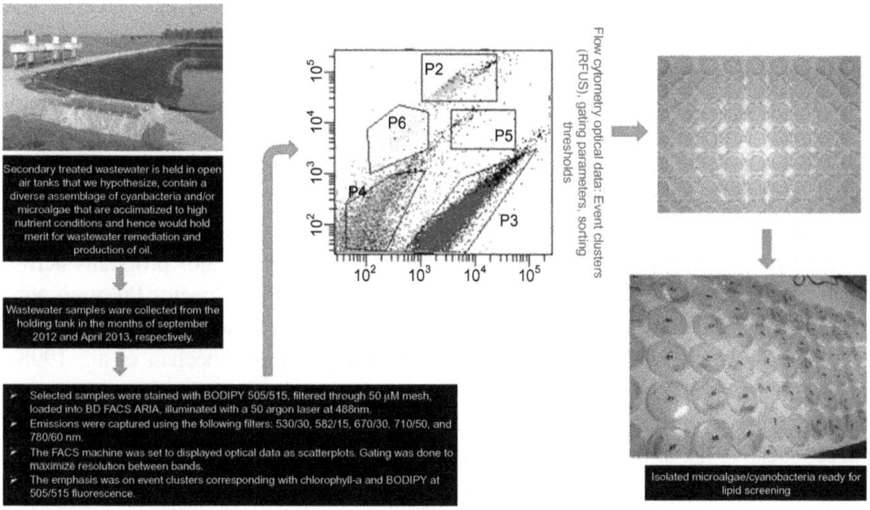

Figure 15.4 Schematic overview of microalgae and/or cyanobacteria strain isolation from wastewater holding ponds using FACS.

Once investigators know which organisms are viable to work with and where they might be found, they must determine the optimal approach for studying them. Organism-dependent isolation methods and media selections should be determined beforehand. It is critical to remember that all isolation methods and media selections will introduce bias (which can be a useful tool, as in the case of enrichments, or phototaxis) and will have certain limitations that must be openly acknowledged.

A range of microorganisms that were known to exist in ecology but could not be cultured now can be observed with genetic informatics, because of recent biotechnological advances in genetics. A host of viable organisms are not culturable for a variety of reasons; including complex alleopathic environmental relationships (mutualistic, communalistic, parasitic) and unknown micro-eco-niches. However, this does not need to be a deterrent to bioprospecting efforts. Domestication has always been, and continues to be, an evolving process. Presently, several workarounds are being advanced, including host expression, chemical synthesis, media culture, and polyculturing.

15.3 Phycoremediation, Microalgae, and Bioprospecting

When considering phycoremediation the targeted organism is, of course, microalgae. This includes single-celled, filamentous, and colonial eukaryotes, as well as the prokaryotic cyanobacteria. When attempting to isolate microalgae, the challenges to consider are cell fragility (cell wall shearing and single-cell vulnerability), media limitations, allopathic relationships, and eco-niche specificity. Some microalgae, like many of the *Chlorella* species, are hardy and can survive and thrive in a range of conditions, while others are quite frail.

Microalgae are extremely diverse; the number of known species is only 1% of what is believed to exist. Microalgae are classified by their color and include both prokaryotic and eukaryotic cellular systems. Microalgae include photoautotrophs, heterotrophs, mixotrophs, and organochemotrophs. They are found in the air, soil, and water. Accounting for about 50% of Earth's primary productivity, they are a foundational element to all biological processes on the planet (Anderson, 2005).

Algal prospecting efforts have been carried out in several major programs across the globe. The first, and most notable, effort was the Aquatic Species Program conducted by the United States Department of Energy, which isolated and evaluated over 3000 species from every major region of the United States (Sheehan et al., 1998). Bioprospectors do not always search for new species. In fact, fruitful prospecting efforts can come from mining the existing culture collections for new value, functions, and products. A list of the major culture collections is provided in Table 15.8.

Microbial evolution is a continuous process, and new specific needs are constantly arising. As such, new prospecting efforts are also needed. When structuring your effort and choosing isolation methods, it is helpful to know the sizes and other defining morphological characteristics as given in Table 15.9. For example, pigments can be helpful when using fluorescence properties to discriminate, as is the case when using FACS.

15.4 Isolation Methods

Isolation methods are well described in existing literature. For a good review and explanation of the methods, the reader is referred to *Algal Culturing Techniques* (edited by Anderson, 2005). Before choosing appropriate isolation methods, it must be decided if unialgal (one algae species, any number of bacterial and fungal species) or axenic (one type of algae species, nothing else) cultures are desired. Isolation methods have traditionally been manual; however, automatic isolation methods have been and are still being developed. See Table 15.10 for a list of isolation methods.

15.4.1 Filtration

Filtration methods for species isolation have employed the use of nets and gels. The filters introduce size bias that enables the prospector to separate the desired

Table 15.8 Major Algal Culture Collections

Collection	Web site	Phone	E-mail
UTEX	http://web.biosci.utexas.edu/utex/	1-512-471-4019	utalgae@uts.cc.utexas.edu
CCAP	http://www.ccap.ac.uk/	n/a	CCAP@sams.ac.uk
NCMA	https://ncma.bigelow.org/	(001) 207 315 2567	ncma@bigelow.org
Roscoff	http://www.sb-roscoff.fr/Phyto/RCC/index.php?option = com_frontpage&Itemid = 1	33.2.98.29.25.62	rcc@sb-roscoff.fr
CSIRO	http://www.csiro.au/Organisation-Structure/National-Facilities/Australian-National-Algae-Culture-Collection.aspx	61.3.6232.5117	susan.blackburn@csiro.au
SCCAP	http://www.sccap.dk/	45.3532.2303	sccap@bio.ku.dk
NIVA	http://www.niva.no/en/fagomraader/laboratorietjenester/algekultursamling	47.22.18.51.00	niva@niva.no
CAUP	http://botany.natur.cuni.cz/algo/caup.html	420.221.951.648	skaloud@natur.cuni.cz
Plymouth	http://www.mba.ac.uk/culture-collection/	44(0)1752.633207	sec@mba.ac.uk
CICCM	http://cultures.cawthron.org.nz/	64.3.548.2319	info@cawthron.org.nz
CCAC	http://www.ccac.uni-koeln.de/geninform.htm	49.221.470.8097	barbara.melkonian@uni-koeln.de
ACOI	http://acoi.ci.uc.pt/contacts.php	003.512.398.552.30	liliamas@ci.uc.pt
Murdoch	http://collections.ala.org.au/public/show/co103	02.6246.4431	info@ala.org.au
WFCC	http://www.wfcc.info/	n/a	philippe.desmeth@belspo.be
Extensive Culture Collection Lists			
WFCC	http://www.wfcc.info/ccinfo/collection/by_region/Europe/		
Phyco	http://www.phycology.net/Content/PNetContent.cfm?MID = 135		
Utex	http://web.biosci.utexas.edu/utex/otherResources.aspx		

Table 15.9 Microalgae Classes and Pigments

Division/Class	Principal Pigments
Prokaryota	
Cyanophycota	
Cyanophyceae	Chlorophyll a, β-carotene, c-phycoerythrin, allophycocyanin, c-phycocyanin
Prochlorophycota	Chlorophyll a, b, β-carotene, zeaxanthin, cryptoxanthin
Eukaryota	
Rhodophycota	
Rhodophyceae	Chlorophyll a, d, r-phycocyanin, allphycocyanin, c-phycoerythrin, α and β-carotene
Chromophycota	
Chrysophyceae	Chlorophyll a, c1, c2, β-carotene
Prymnesiophyceae (Haptophyceae)	Chlorophyll a, c1, c2, β-carotene
Xanthophyceae	Chlorophyll a, c, β-carotene, diatoxanthin, diadinoxanthin, heteroxanthin, vaucheriaxanthin ester
Eustigmatophyceae	Chlorophyll a, c, β-carotene, violxanthin, vaucheriaxanthin
Bacillariophyceae	Chlorophyll a, c, β-carotene, fucoxanthin, diatoxanthin, diadinoxanthin
Dinophyceae	Chlorophyll a, c2, peridinin, neoperidinin, β-carotene
Phaeophyceae	Chlorophyll a, c1, c2, β-carotene, fucoxanthin
Raphidophyceae	Chlorophyll a, c, diadinoxanthin, dinoxanthin, heteroxanthin, fucoxanthin
Cryptophyceae	Chlorophyll a, c2, α-carotene, diatoxanthin, phycoerythrin, phycocyanin
Euglenophycota	
Euglenophyceae	Chlorophyll a, b, β-carotene, astaxanthin, antheraxanthin, diadinoxanthin, neoxanthin
Chlorophycota	
Chlorophyceae	Chlorophyll a, b, α-, β-, γ-carotene, lutein, siphonoxanthin, siphonein
Charphyceae	Chlorophyll a, b, α-, β-, γ-carotene, various xanthophylls
Prasinophyceae	Chlorophyll a, b, β-carotene, siphonein, siphonoxanthin

The author searched extensively for general class-specific size ranges, but none were found. The only broad size classifications were found in Bellinger and Sigee (2010) were as follows: pico (0.2–2 μm), nano (2–20 μm), micro (20–200 μm), and macro (>200 μm).
Source: Adapted from South and Whittick (1987).

Table 15.10 Isolation Methods

Method	Application	Advantage	Limitation	Reference
Filtration	Size discrimination	Low cost, effective	Filter imperfections	Perez et al. (1985)
Enrichment	Media selection pressure to control isolation outcome	Uncover Viable but not cultivable (VBNCs), find organisms for tailored downstream conditions	Limiting by nature (VBNCs)	Droop (1954, 1959, 1966, 1971), Lee and Soldo (1992), Nichols (1973), Pringsheim (1912, 1946), Guillard (1973, 1975, 1995)
Micropipette	Single cell isolation; select one to several cell bodies manually	Hyperspecific, locate exotics, can be used to produce axenic cultures	Labor intensive, cell wall shearing, microscope requirements, more involved training	Guillard (1973), Richmond (2004), Rogerson et al. (1986), Droop (1969), Guillard and Morton (2003)
Streaking	Unialgal/axenic culture development	Low cost, effective, minimal training	Repeated iterations required, limited to organisms that grow on agar	Gerloff et al. (1950)
Dilution	Produce unialgal culture by placing one cell in a liquid culture. Reduce number of organisms present in sample	Simple in field, isolate exotics and new species	Labor intensive, requires multiple iterations	Rippka et al. (2000), Waterbury et al. (1979)
Cell Spray	Produce unialgal/axenic cultures	production of unialgal/axenic cultures without need for excessive repetition	Limited to organisms that grow on agar, spray may shear wall of some species, requires hood or clean space for effective application	Sosnowski (2013)
Centrifugation	To produce density gradient discrimination	Contaminant removal	Cannot be used on its own to produce unialgal/axenic strains	

(*Continued*)

Table 15.10 (Continued)

Method	Application	Advantage	Limitation	Reference
Sonication and vortexing	Break up colonies, mats, remove epiphytic algae and bacteria	Effective at producing managable units for isolation, or freeing epiphytes	Cannot be used on its own to produce unialgal/axenic strains, sonication and vortexing can damage some species relative	Sutherland (1976), Azma (2010)
Phototaxis	Isolate species with affinity or lack thereof for light	Isolate unialgal/axenic motile strains	Strains must exhibit phototaxis, multiple swimmers of different species	Guillard (1973), Bold (1942), Meeuse (1963), Paasche (1971)
Antibiotics	Produce axenic cultures, remove cyanobacteria	Effective removal of persistent bacterial contaminants	Limited understanding of application with respect to algae. Not a cure all, may cause mutation and other variance with algae species, known and unknown	See Table 15.11
Flow cyotometry	Single cell isolation	Reduces labor cost, highthroughput	Fluorescence required. High equipment and training cost. System lasers and PMTs	Hyka (2011), Reckermann (2000)
Optical Tweezers	Single Cell isolation	Automated axenic cell isolation	No published application for algae isolation	Pilat et al. (2013)
Microfluidics	Single Cell Isolation	Automated axenic cell isolation	No published application for algae isolation	Kim et al. (2010)

All techniques mentioned here can be found in Anderson (2005) and Bux (2013), additional research references provided for interested parties.

species from other contaminants, e.g., viruses, bacteria, fungal spores, hyphae, and other algae. However, filtration alone is rarely a way to isolate species and is best used in combination with other methods. To use filtration most effectively, carefully consider the size of the species sought in comparison with other contaminants. Keep in mind that filters are known to have some irregularity and that contaminants may not completely be eliminated.

15.4.2 Enrichment and Media Selection

The media selection process is critical to the prospecting effort. The media choice (and culture conditions) will dictate the boundaries of the possible culturable organisms. Enrichment culturing is part of the media selection process, as well a useful isolation technique. Enrichment cultures include culture medias, peat moss extract, macronutrients (e.g., phosphorous and nitrates), trace elements (e.g., iron and silica), salinity, vitamins (e.g., B1, B12, and other polyketides), and for mixotrophs that require bacteria, organic carbon sources like grain, potato, etc. Culture enrichments can make VBNC organisms accessible. Experimentation can yield new methods of enrichments.

Enrichments can also be used to select for organisms that have the qualities an investigator is interested in. If a nitrogen-fixing cyanobacteria is desired, a media deplete with nitrogen is useful. If looking for an organism that will grow in an environment with heavy metals, high nutrient concentrations, high or low pH, or other unique conditions, the prospector may find it useful to enrich the culture using methods to encourage these conditions. Antibiotics are also useful in enrichment strategies; however, they will be discussed further in their own section. Many media formulations can be found by Anderson (2005) and in the UTEX culture collection (http://web.biosci.utexas.edu/utex/). Enrichment is additionally a useful method for eliminating fungal contaminants (Anderson, 2005).

15.4.3 Micropipette Isolation

Micropipette isolation is a single-cell selection method. A micropipette is made by heating and pulling a Pasteur pipette to create an opening small enough to collect single cells. Anderson (2005) recommends the following microscopes for this work: dissecting, inverted, and compound. Dissecting scopes work best with dark field illumination; transmitted light will also work. An inverted microscope with epifluorescent functionality is useful for isolation work with cell pigmentation. While compound microscopes are considered the most difficult of the three to use, they are widely available and break down easily, making them optimal for field sampling trips. When using a micropipette, two methods for cell collection are used: natural capillary action or suction created by the prospector and a tube connected to the pipette. Following collection, cells are moved to a sterile droplet. To reduce the possibility of contamination, several cycles of capture and suspension may be needed before the cell is placed in its final culture medium. Microalgae cells can be fragile, and investigators must be careful to avoid cell wall rupture from shearing forces.

15.4.4 Streaking

Streaking on solid media is an effective method for obtaining isolated colonies. However, not all algae species will grow on agar plates. An inoculating loop is loaded with a sample and streaked across an agar plate. The streaking promotes a cell density gradient such that, near the end of the streak, isolated colonies form. These colonies may be collected and streaked again, or placed in liquid media.

15.4.5 Serial Dilution

Dilution approaches for isolation may be used as a precursor to other isolation methods, or on their own to obtain single-cell isolates. Dilution techniques work well for rich populations. The goal of serial dilution is to deposit a single cell into a liquid medium, thereby establishing a single-cell isolate. Early pioneers Kufferath (1928, 1929), Droop (1954), and Throndsen (1978) discuss this method. Anderson (2005) recommends five to six repetitions of 1:10 dilutions in most cases. Knowing the starting cell concentration makes the process far more effective because the number of needed dilutions can be calculated with a higher degree of accuracy. This method is often employed in attempts to discover random species from field samples (particularly when coupled with multiple enrichment approaches). Bacteria and other contaminants, often more prevalent than algae, prevent this technique from producing axenic cultures. However, coupled with automated sorting, streaking, various enrichment strategies, and/or antibiotics, this technique can be extremely useful in producing axenic single-cell isolates.

15.4.6 Density Gradient Separation

Density gradient separation using centrifugation is not capable of producing strict species isolation. It is useful for creating discriminating bands within a sample, which can then be used to separate groups of cells for further isolation work.

15.4.7 Sonication and Vortexing

The use of sonication and vortexing are useful for breaking up colonial algal groups and for removing epiphytes from cell walls. Several researchers advocate the use of sonication and vortexing combined with mild surfactants to remove bacteria and other epiphytes from samples. When appropriately coupled with other methods, sonication and vortexing can improve axenic culture isolation. Furthermore, to isolate axenic cultures of epiphytes, sonication and/or vortexing may be necessary. However, care must be taken to select the appropriate sonication frequency. Too high a frequency may rupture target cells; too low a frequency may be ineffective (Sutherland, 1976; Azma, 2010).

15.4.8 Phototaxis

Phototaxis is useful when attempting to isolate flagellate species that exhibit a motile affinity for light (Bold, 1942; Paasche, 1971). The basic approach is to establish a light gradient, dark on one side, and light on the other. The sample is loaded, and the organism(s) swim toward or away from the light. It is difficult to establish unialgal cultures when more than one motile phototaxic species is present. However, if one is a stronger swimmer, temporal gradient may be useful for obtaining unialgal cultures when more than one motile phototaxic species is present (Anderson, 2005). After migration has induced separation, cells are then transferred to culture media, or combined with other techniques to obtain axenic or unialgal isolates (Anderson, 2005).

15.4.9 Use of Antibiotics

The use of antibiotics is very specific when attempting to produce axenic cultures of algae. The extreme variability of antibiotics means that what works for one species may not work for another. However, there are some general guidelines based on what researchers have done; see Table 15.11.

Anderson (2005) outlines three general strategies surrounding the use of antibiotics for isolation methods. The first approach is best used for purifying unialgal cultures where the variety of bacteria present is smaller than that of the multialgal culture. Following antibiotic exposure, the algal cells are isolated using a micropipette and washed several times in sterile water before placement in media. It is important to analyze cell viability under a microscope prior to selection, as antibiotic exposure can slow and/or halt cell growth altogether, depending on species.

The second approach by Anderson (2005) uses a strong antibiotic mix and a series of successive dilutions with a dense and growing culture of algae in addition to a small amount of organic matter to promote bacterial growth (as many antibiotics are effective only when bacterial growth is occurring). The dilution occurs in twofold processes, such that antibiotic concentration is halved each time, while the algae concentration remains constant. Dilutions are made every 24–48 h and added to fresh growth medium without antibiotics and a bacterial test medium. Approach II is preferable over approach I because it requires no extra manipulation and leaves a dense cell culture.

The third approach exposes algal culture to individual antibiotic treatments in a serial fashion. This method is typically less toxic to the algae, provides workarounds for resistant strains, and ensures a broad application of antibiotics that aim to kill multiple bacterial types (ideally all).

A general understanding of antibiotics is useful when designing purification strategies. Anderson (2005) gives the following rules of thumb: (a) many antibacterials act as cell wall inhibitors and will only function on actively growing bacteria, requiring that some organic matter be added to the medium; (b) combining cell wall inhibitors with protein synthesis inhibitors is counterproductive because one works best when growth occurs, and the other stops growth (Anderson, 2005). Other antimicrobial agents have also been used to purify cultures, e.g., Tellurite

Table 15.11 Antibiotic Approaches

Strategy	Cocktails	Advantage	Limitation	Reference
Approach I, II, III				
Time exposure to antibiotics, followed by single cell isolation, and repeated exposures if needed	100, 25, 25 mg/L penicillin G, dihydrostreptomyacin sulfate, gentamicin	Treatment can be gentle	Time cost	Guillard (1973), Richmond (2004), Hoshaw and Rosowski (1979), Oppenheimer (1955)
Dilute strong antibiotic mix with actively growing dense algal culture every 24–48 h	2500, 200, 200, 400 mg/L benzyl penicillin G sulfate, chloramphenicol, neomycin, actidione	Time cost	Risk of cell damage due to high concentrations	Richmond (2004), Droop (1967), Cottrell and Suttle (1993), Divan et al. (1993), Berland and Maestrini (1969), Leon et al. (1986), Lehman (1976)
Sequential transfer through different antibiotics	Variable	Less toxic, workaround mutations	Investigation of tolerance and bacterial susceptibility	Cottrell and Suttle (1993), Hoshaw and Rosowski (1979)

For a more complete discussion on the topic see Anderson (2005).
Source: Adapted from Anderson (2005).

and other strong oxidizers, molecular iodine in alcohol, sodium hypochlorite, Argyrol, ionic detergents, and enzymatic lysozymes.

15.4.10 Flow Cytometry

Flow cytometry can be used to establish unialgal and axenic cultures of algae. The technique makes use of the inherent optical properties of algae. A sample is prepared for the flow cytometer which, using hydrodynamic focusing, focuses the sample into a concentrated stream. This stream passes through a laser which provides optical data, including emissions spectra, size, and inner cell complexity. The system can sort using the user-defined preferences based on the optical data. Using an electrical charge, the sample stream is sorted into tubes, Petri dishes, or well plates. There are a number of both limitations and advantages to using flow cytometry and fluorescent automatic cell sorting. Limitations are that the machines are costly, and the training needed to properly use them is extensive. The main advantage is that they reduce considerably the time cost associated with isolation work. Several authors have reported successful unialgal and axenic isolation using these tools (Reckermann, 2000).

15.4.11 Optical Tweezers and Microfluidics

Optical tweezers and microfluidic devices are two emerging isolation strategies. Optical tweezers use an infrared beam which is focused onto a dielectric medium, causing polarization (Anderson, 2005; Pilat et al., 2013). This polarization can be strong enough to hold cells and other particles in place, and the polarization may be used to separate algal cells. Considering that algae cells are negatively charged, an "optical tweezer" approach for isolation is worth researching.

Microfluidics has been used to sort viruses and bacteria and is more efficient when combined with optical trapping (optical tweezers) (Anderson, 2005; Kim et al., 2010). A particle sorter has been used to separate small cells, macromolecules, and colloids when combined in the manner with optical tweezing (Anderson, 2005). With established effectiveness in sorting small particles, microfluidics and optical trapping warrant further consideration as isolation strategies (Anderson, 2005).

15.5 Culturing the Target Strain(s)

Culture conditions include the media choice and impactful environmental conditions, such as temperature, pH, and photon flux density. Culturing on a production-level scale is beyond the scope of this chapter, but is of little concern during the prospecting effort, unless the downstream applications of the remediation effort are already understood. If that is the case, the culture conditions could be used as an enrichment strategy to isolate species well suited to predesigned remediation strategies. Additionally, the climactic conditions of the sampling location should be considered when designing culture conditions. If sampling in extreme cold or heat, care must be taken to provide optimal conditions for psychrophilic and thermophilic organisms. However, it may be more important to isolate an organism capable of providing a specific value, product, or function. If this is the case, the remediation/production strategy may be designed around optimizing the organisms' growth and function.

15.5.1 Screening

Once isolated and cultured, any organism(s) will be screened for potential and/or targeted value(s), product(s), and function(s). See Table 15.12 for a summary of value-added products from microalgae.

Advancements in genomics, metabolomics, proteomics, and transcriptomics are enhancing studies for both culturable and nonculturable organisms. New methods allow for a better understanding of algal ecology and new strategies may be discovered while prospecting for novel species. Observing species with genetic sensing,

Table 15.12 Microalgae Value, Products, and Functions

Value/Product/ Function	Phylum	Genus	References
Feed			
Livestock	Cyanobacteria, Rhodophyte, Chlorophyte	*Spirulina platensis*, *Porphyridium* sp., *Chlorella* sp.	Spolaore et al. (2006), Herber-McNeill and Van Elswyk (1998), Shields and Lupatsch (2012), Stamey et al. (2012), Moate et al., (2013), Bhaskar (1986)
Aquaculture	Cyanobacteria, Chlorophyta, Eustigmatophyceae, Heterokonta, Bacillariophyta, Haptophyta, Dinophyta	Multiple	Becker (2007), Spolaore et al. (2006), Herber-McNeill and Van Elswyk (1998), Shields and Lupatsch (2012), Brown et al. (1997), Duerr et al. (1998), Guedes and Malcata (2012), Muller-Fuega et al. (2003), Anon, 2010)
Human	Cyanobacteria, Chlorophyta	*Spriulina*, *Chlorella*	Becker (2007), Spolaore et al. (2006), Kay and Barton (1991), Belay et al. (1993), Benneman (1992), Gantar and Svircev (2008), Mokady et al. (1989)
Energy			
Hydrogen	Chlorophyta	*Chlamydomonas reinhardtii*	Gaffron and Rubin (1942), Melis and Happe (2001)
Methane	Algal biomass digestion via methanogenesis		Gunaseelan (1997), Yen (2007) Golueke et al. (1957)
Biodiesel	Many investigated, see references		Chisti (2007), Li et al. (2008); Thomas and Kim (2013)
Ethanol	Cyanobacteria	Engineered strain	Algenol, Deng et al. (1999), Woods et al. (2004)

(*Continued*)

Table 15.12 (Continued)

Value/Product/ Function	Phylum	Genus	References
Nutracuetical			
Astaxanthin	Chlorophyta	*Haematococcus pluvialis*	Spolaore et al. (2006), Olaziola (2000), Duffose (2005), Borowitzka (2013)
β-carotene	Chlorophyta	*Dunaliella*	Spolaore et al. (2006), Duffose (2005), Borowitzka (2013)
Pharmaceuticals			
Antiviral	Diverse, very open to bioprospecting, see references		Thomas and Kim (2013), Borowitzka (1995), Schwartz et al., (1990)
Phycoremediation			
Municipal wastewater	See Phycoremediation section		
Agricultural wastcwater			
Industrial wastewater			
Oil and derivatives			
Pesticides			
Endocrine disruptors			

without the need for culturing, gives investigators insight into previously unobserved ecologies. New techniques may allow investigators to move to chemical synthesis or host expression for improving novel species. Once a species is selected, new exploratory methods and tools, such as the "omics" mentioned in the following section, are opening up new solutions to engineering algal cells, in addition to GE and directed evolution.

15.6 Information Garnered from the Whole Genome Sequencing of Lipid-Producing Microalgae

Recent advancements in the next-generation sequencing technologies have produced whole genome sequences of several microalgal strains sequences, along with expressed sequence tags and transcriptomes (Grossman, 2005; Radakovits et al., 2010; Tirichine and Bowler, 2011). In fact, more than 10 microalgal strains have been completely sequenced (Misra et al., 2013) (Table 15.13), such as *C. vulgaris* NC64A and *Nannochloropsis gaditana*. Additionally, several mitochondria or/and plastid genomes from microalgae have also been sequenced, such as those for *D. salina* CCAP19/18 (Smith et al., 2010), *Botryococcus braunii* (Weiss et al., 2010; 2011), and others (Misra et al., 2013).

Despite the whole genome sequencing, the information that has been accruing on microalgae genomes, most research has not address algal lipid production potential. Information and genomic techniques that apply to the industry of bio-fuel development are still in development. In fact, the more recent approach of algal transcriptome profiling appears to be more successful at revealing the suite of genes being expressed under a given environmental condition such as nutrient starvation (Guarnieri et al., 2011; Rismani-Yazdi et al., 2011; Misra et al., 2013). Transcriptomics, which is the technique of quantifying the relative abundance of mRNA levels in a particular strain of interest, has the strong potential to provide a comprehensive snapshot of the suite of genes being actively expressed; this information can then be used to follow a specific approach of BE, GE, TFE, or a combination thereof (Liu and Benning, 2012).

Toward this end, the first de novo transcriptomics study was reported on *Dunaliella tertiolecta*, which revealed a suite of enzymes involved in the biosynthesis and catabolism of starch, fatty acids, and Triacylglycerols (TAGs) (Rismani-Yazdi et al., 2011). Several other studies have also used the transcriptomics approach to probe the TAG-generating pathway in the oleaginous microalga, *Chlorella variabilis* UTEX395, under both nitrogen-replete and -depleted conditions (Guarnieri et al., 2011). Furthermore, an excess of transcriptome information is also available for *Chlamydomonas reinhardtii* under a variety of environmental stresses. Notably, Boyle et al. (2012) performed a genome-wide expression analysis study to investigate the underlying mechanisms for the induction of TAG accumulation in *C. reinhardtii* which resulted in the identification of three critical acyltransferase genes—DGAT1, DGATT1, and PDAT1—that are most likely responsible for TAG accumulation in this microalga (Boyle et al., 2012). It was also shown by this study that nitrogen deprivation causes *Chlamydomonas* cells to switch their metabolism from converting acetate to glucose into fatty acid incorporation by down-regulating both glyoxylate cycle activity and gluconeogenesis. In another study, transcriptome sequencing was used to assess the transcriptome response of *Synechocystis* sp. strain PCC 6803 to two concentrations of exogenous n-butanol (Anfelt et al., 2013). Using the physiology and transcriptomics data, these authors collated several genes for overexpression in an attempt to improve butanol tolerance and found that overexpression of the small heat shock protein, HspA, improved tolerance to butanol significantly. Thus, a transcriptomics-directed engineering approach resulted in the creation of a more solvent-tolerant cyanobacteria strain that can potentially be a more productive biofuel-generating host. In yet another study, the authors investigated potentially novel gene targets in an attempt to genetically engineer the oleaginous microalga *C. vulgaris* (Guarnieri et al., 2013). These authors applied label-free, comparative shotgun proteomic analyses, via a transcriptome-to-proteome pipeline, in order to examine the nitrogen

Table 15.13 A Status Update on Ongoing and Completed Algal Genome Projects Worldwide

Organism	Organelle	Status	Genome Size	No. of Chromosomes	GC (%)	Protein Coding Genes	Reference	Web Link
Chlamydomonas reinhardtii strain CC-503	Nucleus	Complete	121 Mb	17	64	15,143	Merchant et al. (2007)	http://genome.jgi-psf.org/Chlre3/Chlre3.home.html
Volvox carteri strain UTEX2908	Nucleus	Complete	138 Mb	14	56	14,520	Prochnik et al. (2010)	http://genome.jgi-psf.org/Volca1/Volca1.home.html
Chlorella vulgaris strain NC64A	Nucleus	Complete	46.2 Mb	12	67	9791	Blanc et al. (2010)	http://genome.jgi-psf.org/ChlNC64A_1/ChlNC64A_1.home.html
Coccomyxa subellipsoidea strain C-169	Nucleus	Complete	48.8 Mb	20	53	9851	Blanc et al. (2012)	http://genome.jgi-psf.org/Chlvu1/Chlvu1.home.html
Dunaliella salina strain CCAP19/18	Plastid	Complete	269 Kb		32.1	34	Smith et al., 2010	
	Mitochondria	Complete	28.3 Kb		34.4	42	Joint Genome Institute	
	Nucleus	Ongoing	300 Mb					
Botryococcus braunii Berkeley	Nucleus	Ongoing	166.2 (−2.2)Mb		54.4		Weiss et al. (2010, 2011)	
Yamanaka	Nucleus	Ongoing	166.0 (−0.4)Mb					
Songkla	Nucleus	Ongoing	211.3 (−1.7)Mb					
Nephroselmis olivaceae	Plastid	Complete	200.7 Kb		42.1	127	Turmel et al. (1999)	
Mesostigma viride	Plastid	Complete	118.3 Kb		30.1	135	Lemieux et al. (2000)	
Prasinophytes *Ostreococcus lucimarinus* strain CCE9901	Nucleus	Complete	13.2 Mb	21	43	7651	Palenik et al. (2007)	http://genome.jgi-psf.org/Ost9901_3/Ost9901_3.home.html

(*Continued*)

Table 15.13 (Continued)

Organism	Organelle	Status	Genome Size	No. of Chromosomes	GC (%)	Protein Coding Genes	Reference	Web Link
Ostreococcus tauri strain OTH95	Nucleus	Complete	12.6 Mb	20	58	7892	Derelle et al. (2006)	http://genome.jgi-psf.org/Ostta4/Ostta4.home.html
Ostreococcus sp. strain RCC809	Nucleus	Ongoing	13.6 Mb			7773	Joint Genome Institute	http://genomeportal.jgi-psf.org/OstRCC809_2/OstRCC809_2.home.html
Micromonas pusilla RCC299 2.2 163	Nucleus	Complete	20.9 Mb	17	64	10,056	Worden et al. (2009)	http://genome.jgi.doe.gov/MicpuN3/MicpuN3.home.html
CCMP1545	Nucleus	Complete	21.9 Mb	19	65	10,575		http://genome.jgi-psf.org/MicpuC2/MicpuC2.home.html
Bathycoccus prasinos strain RCC1105	Nucleus	Complete	15 Mb	19	48	7847	Moreau et al. (2012)	
Rhodophytes *Cyanidioschyzon merolae* strain	Nucleus	Complete	16.5 Mb	20	55	5331	Matsuzaki et al. (2004)	http://merolae.biol.s.u-tokyo.ac.jp
Galdieria sulphuraria 15 Mb	Nucleus	Ongoing	15 Mb				Michigan State University	http://genomics.msu.edu/galdieria/
Chondrus crispus	Nucleus	Ongoing					Genoscope	http://www.genoscope.cns.fr/spip/
Cyanidium caldarium	Plastid	Complete	164.9 Kb		32.7	232	Glockner et al. (2000)	
Porphyra purpurea	Plastid	Complete	191.0 Kb		33	251	Reith and Munholland (1995)	
Stramenopiles *Phaeodactylum*	Nucleus	Complete	27.4 Mb		49	10,402	Bowler et al. (2008)	http://genome.jgi-psf.org/Phatr2/Phatr2.home.html

tricornutum strainCCP1055/1								
Thalassiosira pseudonana CCMP1335	Nucleus	Complete	34.5	24	47	11,242	Armbrust et al. (2004)	http://genome.jgi-psf.org/thaps1/thaps1.home.html
Fragilariopsis cylindrus CCMP1102	Nucleus	Ongoing	81 Mb				Joint Genome Institute	http://genome.jgi-psf.org/Fracy1/Fracy1.home.html
Pseudo-nitzchia multiseries CLN-47	Nucleus	Ongoing	218 Mb				Joint Genome Institute	http://genome.jgi.doe.gov/Psemu1/ Psemu1.home.html
Ectocarpus siliculosus strain EC32	Nucleus	Complete	195.8 Mb		53.6	16,256	Cock et al. (2010), Gobler et al. (2011)	http://genome.jgi.doe.gov/Auran1/Auran1.home.html
Aureococcus anophagefferens strain CCMP1984	Nucleus	Complete	57 Mb			11,501		
Nannochloropsis gaditana	Nucleus	Complete	29 Mb		54.2	9052	Radakovits et al. (2012),	http://nannochloropsis.genomeprojectsolutions-databases.com
Odontella sinensis	Plastid	Complete	119.7 Kb		31.8	178	Kowallik et al. (1995)	
Cryptophytes *Guillardia theta* strain CCMP2712	Nucleus	Complete	87.2 Mb		53	24,840	Curtis et al. (2012)	
Chlorarachniophytes *Bigelowiella natans* strain CCMP2755	Nucleus	Complete	94.7 Mb		45	21,708	Curtis et al. (2012)	
Haptophytes *Emiliana huxleyi* strain CCMP1516	Nucleus	Ongoing	167.7 Mb				Joint Genome Institute	http://genome.jgi-psf.org/Emihu1/Emihu1.home.html
Glaucophytes *Cyanophora paradoxa*	Plastid	Complete	135.5 Kb		30.4	192	Stirewalt et al. (1995)	

deprivation response of *C. vulgaris*, which revealed several potential targets for strain-engineering strategies such as the proteins involved in transcriptional regulation, lipid biosynthesis, cell signaling, and cell cycle progression.

Collectively, these findings validate the importance of transcriptome sequencing as an emerging and powerful technique for the identification of metabolically active genes for algal lipid production.

15.7 Bioinformatics Resources to Study Lipid Metabolic Pathways in Microalgae

Adequate mining of information from the whole genome sequences of oleaginous microalgae and cyanobacteria requires appropriate bioinformatics resources which contain information on metabolic pathways and facilitate comparative genomic analysis for functional annotation and interpretation of identified genes. For higher plants, such databases specific for lipid pathway analyses already exist (Beisson et al., 2003; Caspi et al., 2008; Sucaet and Deva, 2011; Child et al., 2012), but hardly any database addresses this need for the study of microalgae. It is hoped that with the explosion in the availability of algal whole genome sequences, there will be a concomitant progress in the bioinformatics tools and databases to study the algal communities of interest to the biofuel community. Here we recapitulate the various databases and bioinformatic tools with special reference to lipid metabolic pathways in microalgae. Also presented are specific selections of plant-specific databases that are also relevant for microalgal biofuel research.

15.7.1 Kyoto Encyclopedia of Genes and Genomes (http://www.genome.jp/kegg/)

Kyoto Encyclopedia of Genes and Genomes (KEGG) is perhaps the most extensively used resource of metabolic pathways that exist in organisms (Kanehisa et al., 2010); the database consists of a total of 220 genome sequences from eukaryotes, 2544 from bacteria, and 162 from archaeal, respectively. KEGG consists of 16 databases which can be broadly grouped under either the (a) systems information, which includes the pathway maps and functional units; (b) genomic information, which includes the organisms with complete genomes and metabolites, and (c) chemical information, which includes the enzyme nomenclature and biochemical reactions.

An exciting addition to KEGG is the ortholog prediction tool, which is very helpful to identify orthologous and paralogous groups of genes from evolutionarily related organisms (Rismani-Yazdi et al., 2011; Misra et al., 2013).

15.7.2 MetaCyc (http://metacyc.org)

This database contains information on pathways that are found to exist in more than 600 organisms ranging from bacteria, plants, and even humans

(Caspi et al., 2008). Specific to the oleaginous algae, ChlamyCyc (http://chlamycyc.mpimp-golm.mpg.de) is an option which contains a comprehensive listing of metabolic pathways and biochemical processes found to exist in *C. reinhardtii* (May et al., 2009). ChlamyCyc provides information on a total of 253 pathways, 2851 enzymes, 1419 enzymatic reactions, 1416 compounds, and 928 literature citations. The web-based portal of ChlamyCyc also houses both the known and predicted biochemical pathways from other well-known pathway databases. Users can also utilize the integrated Inparanoid (Remm et al., 2001) and OrthoMCL-DB (Chen et al., 2006) databases to obtain phylogenomic analyses. Moreover, for the functional annotation of uncharacterized or novel genes and products, a web-based version of the Basic Local Alignment Search Tool (Altschul et al., 1990) is integrated with ChlamyCyc.

15.7.3 Algal Functional Annotation Tool (http://pathways.mcdb.ucla.edu)

This is a stand-alone, web-based suite for functional annotation of multiple algal genome sequences (Lopez et al., 2011). An assembled genome sequence needs to be annotated. Annotation in genetic sequencing is the process of assigning pathways, ontology, and protein family nomenclature to the predicted proteins by utilizing different bioinformatics resources such as KEGG (Kanehisa et al., 2010), Eukaryotic Orthologous Groups of proteins (KOG) (Tatusov et al., 2000), MetaCyc (Caspi et al., 2008), Pfam (Finn et al., 2008), Panther (Thomas et al., 2003), Reactome (Matthews et al., 2009), Gene Ontology (Ashburner et al., 2000), InterPro (Hunter et al., 2009), and MapMan Ontology (Thimm et al., 2004). In addition to the high-throughput data that gives details about algae gene expression under variable environmental conditions, more information can be interpreted from data about the functionally related genes of the organisms if different search tools are used to identify the data.

15.7.4 AUGUSTUS

This is a web-based software that can be very useful for eukaryotic gene prediction using genomic sequences (Stanke et al., 2004). It can also be downloaded onto a computer and run as a stand-alone program. More recently, WebAUGUSTUS, a web interface for training AUGUSTUS and predicting genes with AUGUSTUS, was also made available (http://bioinf.uni-greifswald.de/webaugustus). AUGUSTUS comes with an optimization script that can be used to find values for the meta-parameters, like splice window sizes, which further optimizes the prediction accuracy. Currently the limitation of AUGUSTUS is the fact that the process has been built around species specific gene sets that have been used as the basis for the reference sets. Those reference sets dictate how AGUSUTUS predicts the gene sets and the current standards are based on the alge *C. reinhardtii* and *Galdieria sulphuraria*, and they are thermo-acidophilic unicellular red alga. AUGUSTUS can also predict alternative splicing and alternative transcripts in eukaryotes.

15.7.5 Arabidopsis Lipid Gene Database (http://www.plantbiology.msu.edu/lipids/genesurvey/index.html)

The core metabolic pathways present in microalgae and the model plant *A. thaliana* are conserved (Hu et al., 2008). Therefore, the homologous gene sequences, ESTs, and lipid biosynthetic pathways present in microalgae can be mapped based on the *Arabidopsis* metabolic pathways as reference. The arabidopsis lipid gene database thus includes a repository of 600 encoded plant proteins classified according to biological function, subcellular localization, and alternate splicing which have been successfully used for comparative genomic studies (Li et al., 2010;Sharma and Chauhan, 2012).

15.7.6 Miscellaneous Noteworthy Databases

Notably, genome-wide assessment of plant biofuel feedstock species can be performed by using the Biofuel Feedstock Genomic Resource (BFGR; http://bfgr.plantbiology.msu.edu). The BFGR database incorporates functional annotation of genes as well as ESTs from more than 54 lignocellulosic biofuel feedstock species (Childs et al., 2012). In addition, pDAWG (http://csbl1.bmb.uga.edu/pDAWG/) offers full information on plant phylogenome and genome sequences, and structure–function information (Mao et al., 2009) on the complete genome sequences of 7 higher plants and 12 microalgal species, including several algae and diatoms.

References

Abdel-Raouf, N., Al-Homaidan, A.A., Ibraheem, I.B.M., 2012. Microalgae and wastewater treatment. Saudi J. Biol. Sc. 19, 257–275.

Ahluwalia, S.S., Goyal, D., 2007. Microbial and plant derived biomass for removal of heavy metals from wastewater. Bioresour. Technol. 98, 2243–2257.

Altschul, S., Gish, W., Miller, W., Myers, E., Lipman, D., 1990. Basic local alignment search tool. J. Mol. Biol. 215, 403–410.

Anderson, R.A., 2005. Robert Andersen, Provasoli-Guillard National Center for Culture of Marine Phytoplankton. ME USA *Algal Culturing Techniques* Elsevier Academic Press, West Boothbay Harbor, ISBN: 978-0-12-088426-1.

Anfelt, J., Hallstrom, B., Nielsen, J., Uhlen, M., Hudson, E.P., 2013. Using transcriptomics to improve butanol tolerance of *Synechocystis* sp. strain PCC 6803. Appl. Environ. Microbiol. 79 (23), 7419–7427.

Araña, J., Tello Rendon, E., Dona Rodriguez, J.M., Herrera Melian, J.A., Gonzalez Diaz, O., Perez Pena, J., 2001. Highly concentrated phenolic wastewater treatment by the photo-Fenton reaction, mechanism study by FTIR-ATR. Chemosphere. 44, 1017–1023.

Ashburner, M., Ball, C.A., Blake, J.A., 2000. Gene ontology: tool for the unification of biology. The gene ontology consortium. Nat. Genet. 25, 25–29.
Azma, M., 2010. Improved protocol for the preparation of *Tetraselmis suecica* axenic culture and adaptation to heterotrophic cultivation. Open Biotechnol. J. 4, 36–46.
Azov, Y., Shelef, G., 1987. The effect of pH on the performance of the high-rate oxidation ponds. Water Sci. Technol. 19 (12), 381–383.
Becker, E.W., 2007. Micro-algae as a source of protein. Biotechnol. Adv. 25, 207–210.
Beisson, F., Koo, A.J.K., Ruuska, S., 2003. Arabidopsis genes involved in acyl lipid metabolism. A 2003 census of the candidates, a study of the distribution of expressed sequence tags in organs, and a web-based database. Plant Physiol. 132, 682–697.
Belay, A., Ota, Y., Miyakawa, K., Shimamatsu, H., 1993. Current knowledge on potential health benefits on *Spirulina*. J. Appl. Phycol. 5, 235–241.
Benneman, J.R., 1992. Microalgae aquaculture feeds. J. Appl. Phycol. 4, 233–245.
Berland, B.R., Maestrini, S.Y., 1969. Action de quelques antibiotiques sure le developpement de cinq diatomees en culture. J. Exp. Mar. Biol. Ecol. 3, 62–75.
Bhaskar, B.V., 1986. Meat meal and algae (*Spirulina*) as ingredients of calf starter rations. Agric. Wastes. 15, 51–58.
Bhatnagar, A., Bhatnagar, M., Chinnasamy, S., Das, K., 2010. *Chlorella minutissima*—a promising fuel alga for cultivation in municipal wastewaters. Appl. Biochem. Biotechnol. 161, 523–536.
Bold, H.C., 1942. The cultivation of algae. Bot. Rev. 8, 69–138.
Borowitzka, M.A., 1995. Microalgae as sources of pharmaceuticals and other biologically active compounds. J. Appl. Phycol. 7, 3–15.
Borowitzka, M.A., 2013. High-value products from microalgae—their development and commercialization. J. Appl. Phycol. 25, 743–756.
Boyle, N.R., Page, M.D., Liu, B., 2012. Three acyltransferases and a nitrogen responsive regulator are implicated in nitrogen starvation-induced triacylglycerol accumulation in *Chlamydomonas*. J. Biol. Chem. 287, 15811–15825.
Brennan, L., Owende, P., 2009. Biofuels from microalgae—a review of technologies for production, processing, and extracts of biofuels and co-products. Renew. Sustain. Energy Rev. 14, 557–577.
Brock, T.D., 1997. The value of basic research: discovery of *Thermusaquaticus* and other extreme thermophiles. Genetics. 146 (4), 1207–1210.
Brown, M.R., Jeffery, S.W., Volkman, J.K., 1997. Nutritional properties of microalgae for mariculture. Aquaculture. 151 (1–4), 315–331.
Bux, F. (Ed.), 2013. Biotechnological Applications of Microalgae: Biodiesel and Value Added Products. CRC Press, LLC Boca Raton, Florida. USA.
Cáceres, T., Megharaj, M., Naidu, R., 2008. Toxicity and transformation of fenamiphos and its metabolites by two micro algae Pseudokirchneriella subcapitata and Chlorococcum sp. Sci. Total Environ. 398, 53–59.
Caceres, T., Megharaj, M., Venkateswarlu, K., Sethunathan, N., Naidu, R., 2010. Fenamiphos and related organophosphorus pesticides: environmental fate and toxicology. Rev. Environ. Contam. Toxicol. 205, 117–162.
Cai, X., Liu, W., Jin, M., Lin, K., 2007. Relation of diclofop-methyl toxicity and degradation in algae cultures. Environ. Toxicol. Chem. 26, 970–975.
Carpenter, M., Robertson, J., Skierkowski, P., 1989. Biodegradation of an oily bilge waste using algae. Biotreatment. The Use of Microorganisms in the Treatment of Hazardous Materials and Hazardous Wastes. The Hazardous Materials Control Research Institute, pp. 141–150.

CAS—Chemical Abstract Services, 2012. www.cas.org.
Caspi, R., Foerster, H., Fulcher, C.A., 2008. The MetaCyc database of metabolic pathways and enzymes and the BioCyc collection of pathway/genome databases. Nucleic Acids Res. 36, D623–D631.
Cerniglia, C.E., Gibson, D.T., Van Baalen, C., 1979. Algal oxidation of aromatic hydrocarbons: formation of 1-naphthol from naphthalene by Agmenellum quadruplicatum, strain PR-6. Biochem. Biophys. Res. Commun. 88, 50–58.
Cerniglia, C.E., Gibson, D.T., Van Baalen, C., 1980. Oxidation of napthalene by cyanobacteria and microalgae. J. Gen. Microbiol. 116, 495–500.
Chan, S.M.N., Luan, T., Wong, M.H., Tam, N.F.Y., 2006. Removal and biodegradation of polycyclic aromatic hydrocarbons by Selenastrum capricornutum. Environ. Toxicol. Chem. 25, 1772–1779.
Chen, F., Mackey, A.J., Stoeckert Jr., C.J., Roos, D.S., 2006. OrthoMCL-DB: querying a comprehensive multi-species collection of ortholog groups. Nucleic Acids Res. 34, D363–D368.
Childs, K.L., Konganti, K., Buell, C.R., 2012. The biofuel feedstock genomics resources: a web-based portal and database to enable functional genomics of plant biofuel feedstock species. Database Article ID bar.061, 1–9. doi:10.1093/database/bar061.
Chinnasamy, S., Bhatnagar, A., Hunt, R.W., Das, K.C., 2010. Microalgae cultivation in a wastewater dominated by carpet mill effluents for biofuel applications. Bioresour. Technol. 101, 3097–3105.
Chisti, Y., 2007. Biodiesel from microalgae. Biotechnol. Adv. 25, 294–306.
Cottrell, M.T., Suttle, C.A., 1993. Production of axenic cultures of *Micromonaspusilla* (Prasinophyceae) using antibiotics. J. Appl. Phycol. 29, 385–387.
Courchesne, N.M., Parisien, A., Wang, B., Lan, C.Q., 2009. Enhancement of lipid production using biochemical, genetic and transcription factor engineering approaches. J. Biotechnol. 141, 31–41.
Cowan, M., 1999. Plant products as antimicrobial agents. Clin. Microbiol. Rev. 12, 564–582.
De la Noue, J., Laliberte, G., Proulx, D., 1992. Algae and wastewater. J. Appl. Phycol. 4, 247–254.
De-Bashan, L.E., Bashan, Y., 2010. Immobilized microalgae for removing pollutants: review of practical aspects. Bioresour. Technol. 101, 1611–1627.
Deeg, K., 2012. Growth inhibition of human acute lymphoblastic CCRF-CEM leukemia cells by medicinal plants of the West-Canadian Gwich'in Native Americans. Nat. Prod. Bioprospect. 2, 35–40.
Dejmek, J., Solansky, I., Benes, I., Lenicek, J., Sram, R.J., 2000. The impact of polycyclic aromatic hydrocarbons and fine particles on pregnancy outcome. Environ. Health Perspect. 108, 1159–1164.
DellaGreca, M., Pinto, G., Pollio, A., Previtera, L., Temussi, F., 2003. Biotransformation of sinapic acid by the green algae Stichococcus bacillaris 155LTAP and Ankistrodesmus braunii C202.7a. Tetrahedron Lett. 44, 2779–2780.
Demirbas, A., 2009. Biorefineries: current activities and future developments. Energy Convers. Manage. 50 (11), 2782–2801.
Deng, L., 2007. Sorption and desorption of lead(II) from wastewater by green algae *Cladophora fascicularis*. J. Hazard. Mater. 143, 220–225.
Deng, M.D., Coleman, J.R., 1999. Ethanol synthesis by genetic engineering in cyanobacteria. Appl. Environ. Microbiol. 65, 523–528.

Divan, C.L., Schnoes, H.K., 1982. Production of axenic *Gonyaulax* cultures by treatment with antibiotics. Appl. Environ. Microbiol. 44, 250–254.

Droop, M.R., 1954. A note on the isolation of small marine algae and flagellates for pure culture. J. Mar. Biol. Assoc. U.K. 33, 511–541.

Droop, M.R., 1959. Water Soluble factors in the nutrition of Oxyrrhis marina. J. Mar. Biol. Assoc. U.K. 38, 605–620.

Droop, M.R., 1966. Ubiquinone as a protozoan growth factor. Nature. 212, 1474–1475.

Droop, M.R., 1967. A procedure for routine purification of algal cultures with antibiotics. Br. Phycol. Bull. 3, 295–297.

Droop, M.R., 1969. Algae. In: Norris, J.R., Ribbon, D.W. (Eds.), Methods in Microbiology, vol. 3B. Academic Press, New York, NY, pp. 269–313.

Droop, M.R., 1971. Terpenoid quinones and steroids in the nutritions of Oxyrrhis marina. J. Mar. Biol. Assoc. U.K. 51, 455–470.

Duerr, E.O., Molnar, A., Sato, V., 1998. Cultured microalgae as aquaculture feeds. J. Mar. Biotechnol. 6/2, 65–70.

Duffose, L., 2005. Microorganisms and microalgae as sources of pigments for food use: a scientific oddity or an industrial reality? Food Sci. Technol. 16, 389–406.

El-Rahman Mansy, A., El-Bestawy, E., 2002. Toxicity and biodegradation of fluometuron by selected cyanobacterial species. World J. Microbiol. Biotechnol. 18, 125–131.

Farell, R.L., 2003. Toluene-degrading antartic *Psuedomonas* strains from fuel-contaminated soil. Biochem. Biophys. Res. Com. 312, 235–240.

Finn, R.D., Tate, J., Mistry, J., Coggill, P.C., Sammut, S.J., Hotz, H.R., et al., 2008. The Pfam protein families database. Nucleic Acids Res. 36, D281–D288.

Finlayson, C.M., Chick, A.J., 1983. Testing the potential of aquatic plants to treat abattoir effluent. Water Res. 17, 415–422.

Finlayson, M., Chick, A., Von Oertzen, I., Mitchell, D., 1987. Treatment of piggery effluent by an aquatic plant filter. Biol. Wastes. 19, 179–196.

Fogg, G.E., 2001. Adaptation to Environmental Stresses E Rai, LalChand E Gaur, JaiPrakash R Algal Adaptation to Stress—Some General Remarks, A Fogg, G.E Springer BerlinHeidelberg.

Gaffron, H., Rubin, J., 1942. Fermentative and Photochemical Production of Hydrogen in Algae, 26. The Rockefeller University Press, New York, NY, 219–240.

Gamila, H.A., Ibrahim, M.B.M., 2004. Algal bioassay for evaluating the role of algae in bioremediation of crude oil isolated strains. Bull. Environ. Contam. Toxicol. 73, 883–889.

Gantar, M., Svircev, Z., 2008. Microalgae and cyanobacteria: food for thought. J. Phycol. 44, 260–268.

Garcia, C., Moreno, D.A., Ballester, A., Blazquez, M.L., Gonzalez, F., 2001. Bioremediation of an industrial acid mine water by metal-tolerant sulphate reducing bacteria. Miner. Eng. 14, 997–1008.

Gasperi, J., Garnaud, S., Rocher, V., Moilleron, R., 2008. Priority pollutants in wastewater and combined sewer overflow. Sci. Total Environ. 407, 263–272.

Gaunt, J., Lehmann, J., 2007. Presentation at Power-Gen Renewable Energy and Fuels from Plant to Power Plant (Las Vegas).

Geddes, M.C., 1984. Limnology of lake Alexandrina River, Muarry, South Australia and the effect of nutrients and light on the phytoplankton. Aust. J. Mar. Fresh Water Res. 35 (4), 399–416.

Georgianna, D.R., Mayfield, S.P., 2012. Exploiting diversity and synthetic biology for the production of algal biofuels. Nature. 488, 329–335.

Gerloff, G.C., Fitzgerald, G.P., Skoog, F., 1950. The isolation, purification, and culture of blue-green algae. Am. J. Bot. 37, 216–218.

Gibbons, W.R., Hughes, S.R., 2009. Integrated biorefineries with engineered microbes and high-value co-products for profitable biofuels production. In Vitro Cell. Dev. Biol. 45 (3), 218–228.

Golueke, C.G., Oswald, W.J., Gotaas, H.B., 1957. Anaerobic digestion of algae. Appl. Microbiol. 5, 47–55.

Gonzalez, R., Garcia-Balboa, C., Rouco, M., Lopez-Rodas, V., Costas, E., 2012. Adaptation of microalgae to lindane: a new approach for bioremediation. Aquat. Toxicol. 109, 25–32.

Gonzalez-Barreiro, O., Rioboo, C., Herrero, C., Cid, A., 2006. Removal of triazine herbicides from freshwater systems using photosynthetic microorganisms. Environ. Pollut. 144, 266–271.

Grossman, A.R., 2005. Paths towards agal genomics. Plant Physiol. 137, 410–427.

Guarnieri, M.T., Nag, A., Smolinski, S.L., Darzins, A., Seibert, M., Pienkos, P.T., 2011. Examination of triacylglycerol biosynthetic pathways via de novo transcriptomic and proteomic analyses in an unsequenced microalgae. PLoS One. 6, 1–13.

Guarnieri, M.T., Nag, A., Yang, S., Pienkos, P.T., 2013. Proteomic analysis of *Chlorella vulgaris*: potential targets for enhanced lipid accumulation. J. Proteomics. 20 (93), 245–253.

Guedes, A.C., Malcata, F.X., 2012. Nutritional value and uses of microalgae in aquaculture. In: Muchlisin, Z.A. (Ed.), Aquaculture. InTech Open Acess Publisher. Rijeka, Croatia, pp. 59–78. doi:10.5772/1516.

Guibal, E., Rouplh, C., Cloirec, P.L., 1992. Uranium biosorption by a filamentous fungus mucor miehei pH effect on mechanisms and performances of uptake. Water Res. 26, 1139–1145.

Guillard, R.R.L., 1973. Methods for microflagellates and nanoplankton. In: Stein, J.R. (Ed.), Handbook of Phycological Methods: Culture Methods and Growth Measurements. Cambridge University Press, Cambridge, pp. 69–85.

Guillard, R.R.L., 1975. Culture of phytoplankton for feeding marine invertebrates. In: Smith, W.L., Chanley, M.H. (Eds.), Culture of Marine Invertebrate Animals. Plenum Press, New York, NY, pp. 26–60.

Guillard, R.R.L., 1995. Culture methods. In: Hallegraef, G.M., Anderson, D.M., Cembella, A.D. (Eds.), Manual on Harmful Marine Microalgae. IOC Manuals and Guides No. 33 UNESCO, Paris, pp. 45–62.

Guillard, R.R.L., Morton, S.L., 2003. Culture methods. In: Hallegraeff, G.M., Anderson, D. M., Cembella, A.D. (Eds.), Manual on Harmful Marine Microalgae. UNESECO, Paris, pp. 45–62.

Gunaseelan, V.N., 1997. Anaerobic digestion of biomass for methan production: a review. Biomass Bioenergy. 13, 83–114.

Gupta, V.K., Shrivastava, A.K., Jain, N., 2001. Biosorption of chromium (VI) from aqueous solution by green algae *Spirogya* species. Water Res. 35, 4079–4085.

Hannon, M., Gimpel, J., Tran, M., Rasala, B., Mayfield, S., 2010. Biofuels from algae: challenges and potential. Biofuels. 1, 763–784.

He, S., 2013. Comparative metagenomic and metatranscriptonomic analysis of hindgut paunch microbiota in wood-and dung feeding higher termites. PLoS One. Available from: http://dx.doi.org/doi:10.1371/journal.pone.0061126.

Herber-McNeill, S.M., Van Elswyk, M.E., 1998. Dietary marine algae maintains egg consumer acceptability while enhancing yolk color. Poult. Sci. 77, 493–496.

Hirooka, T., Nagase, H., Uchida, K., Hiroshige, Y., Ehara, Y., Nishikawa, J., 2005. Biodegradation of bisphenol A and disappearance of its estrogenic activity by the green alga *Chlorella fusca* var. vacuolata. Environ. Toxicol. Chem. 24, 1896–1901.

Hirooka, T., Nagase, H., Hirata, K., Miyamoto, K., 2006. Degradation of 2,4-dinitrophenol by a mixed culture of photoautotrophic microorganisms. Biochem. Eng. J. 29, 157–162.

Hodaifa, G., Sanchez, S., Eugenia Martinez, M.P., Orpez, R., 2013. Biomass production of *Scendesmus obliquus* from mixtures of urban and olive mill wastewaters used as culture medium. Appl. Energy. 104, 345–352.

Hoshaw, R.W., Rosowski, J.R., 1979. Methods for microscopic algae. In: Stein, J.R. (Ed.), Handbook of Phycological Methods: Culture Methods and Growth Measurements. Cambridge university Press, Cambridge, pp. 53–68.

Hoshaw, R.W., Rosowski, J.R. (Eds.), 1980. Handbook of Phycological Methods: Culture Methods and Growth Measurements. Cambridge University Press, Cambridge, pp. 53–68.

Hu, Q., Sommerfeld, M., Jarvis, E., Ghirardi, M., Posewitz, M., Seibert, M., et al., 2008. Microalgal triacylglycerols as feedstocks for biofuel production: perspectives and advances. Plant J. 54, 621–639.

Hunter, S., Apweiler, R., Attwood, T.K., Bairoch, A., Bateman, A., Binns, D., et al., 2009. InterPro: the integrative protein signature database. Nucleic Acids Res. 37, D211–D215.

Hyka, P., 2011. Flow cytometry for the development for the development of biotechnological processes with microalgae. Biotechnol. Adv. 31, 2–16.

Inamdar, S., Joshi, J., Jadhav, J., Bapat, V., 2012. Innovative use of intact seeds of Mucuna monosperma Wight for improved yield of L-DOPA. Nat. Prod. Bioprospect. 2, 16–20.

Jabusch, T.W., Swackhamer, D.L., 2004. Subcellular accumulation of polychlorinated biphenyls in the green alga *Chlamydomonas reinhardtii*. Environ. Toxicol. Chem. 23, 2823–2830.

Jain, S., Sharma, M.P., 2010. Prospects of biodiesel from Jatropha in India: a review. Renew. Sust. Energy Rev. 14, 763–771.

Jimenez-Perez, M.V., Sanchez-Castillo, P., Romera, O., Fernandez-Moreno, D., Perez-Martinez, C., 2004. Growth and nutrient removal in free and immobilized planktonic green algae isolated from pig manure. Enzyme Microb. Technol. 34, 392–398.

Jinqi, L., Houtian, L., 1992. Degradation of azo dyes by algae. Environ. Pollut. 75, 273–278.

Jonsson, C.M., Paraiba, L.C., Mendoza, M.T., Sabater, C., Carrasco, J.M., 2001. Bioconcentration of the insecticide pyridaphenthion by the green algae *Chlorella saccharophila*. Chemosphere. 43, 321–325.

Joseph, V., Joseph, A., 2001. Microalgae in petrochemical effluent: growth and biosorption of total dissolved solids. Bull. Environ. Contam. Toxicol. 66, 522–527.

Kanehisa, M., Goto, S., Furumichi, M., Tanabe, M., Hirakawa, M., 2010. KEGG for representation and analysis of molecular networks involving diseases and drugs. Nucleic Acids Res. 38, D355–D360.

Kay, R.A., Barton, L.L., 1991. Microalgae as food and supplement. Crit. Rev. Food Sci. Nutr. 30, 555–573.

Kim, H.S., Weiss, T.I., Devarenne, T.P., Han, A., 2010. A highthroughput microfluidic light controlling platform for biofuel producing photosynthetic microalgae analysis. In: 14th Intl Conf on Mini Systems for Chem and Life Sciences.

Kneifel, H., Elmendorff, K., Hegewald, E., Soeder, C.J., 1997. Biotransformation of 1-naphthalenesulfonic acid by the green alga *Scenedesmus obliquus*. Arch. Microbiol. 167, 32−37.

Koskinen, P.E.P., Lay, C., Beck, S.R., Tolvanen, K.E.S., Kaksonen, A.H., Orlygsson, J., et al., 2008. Bioprospecting thermophilic microrganisms from Icelandic hot springs for hydrogen and ethanol production. Energy Fuels. 22, 134−140.

Kufferath, H., 1928. La culturedes algues. Rev. Algol. 4, 127−346.

Kuritz, T., Wolk, C., 1995. Use of filamentous cyanobacteria for biodegradation of organic pollutants. Appl. Environ. Microbiol. 61, 234−238.

Lal, S., Lal, R., Saxena, D.M., 1987. Bioconcentration and metabolism of DDT, fenitrothion and chlorpyrifos by the blue-green algae Anabaena sp. and Aulosira fertilissima. Environ. Pollut. 46, 187−196.

Lee, J.J., Soldo, A.T., 1992. Protocols in Protozoology. Society for Protozoology, Lawrence Kansas (unpaginated).

Lehman, J.T., 1976. Ecological and nutritional studies on Dinobryon Ehrenb.: seasonal periodicity and the phosphate toxicity problem. Limnol. Oceanogr. 21, 646−658.

Lei, A., Wong, Y., Tam, N., 2002. Removal of pyrene by different microalgal species. Water Sci. Technol. 46, 195−201.

Leon, C., Kumazawa, S., Mitsui, A., 1986. Cyclic appearance of aerobic nitrogenase activity during synchronous growth of unicellular cyobacteris. Curr. Microbiol. 13, 149−153.

Li, Y., Horsman, M., Wu, N., Lan, C.Q., Dubois-Calero, N., 2008. Biofuels from microalgae. Biotechnol. Process. 24, 815−820.

Li, L., Li, H., Li, J., Xu, S., Yang, X., Li, J., et al., 2010. A genome-wide survey of maize lipid-related genes: candidate genes mining, digital gene expression profiling and co-location with QTL for maize kernel oil. Sci China Life Sci. 53, 690−700.

Liu, B., Benning, C., 2012. Lipid metabolism in microalgae distinguishes itself. Curr. Opin. Biotechnol. 24, 1−10.

Lopez, D., Casero, D., Cokus, S.J., Merchant, S.S., Pellegrini, M., 2011. Algal functional annotation tool: a web-based analysis suite to functionally interpret large gene lists using integrated annotation and expression data. BMC Bioinformatics. 12, 282−292.

Mallick, N., 2002. Biotechnological potential of immobilized algae for wastewater N, P and metal removal: a review. BioMetals. 15, 377−390.

Mao, F., Yin, Y., Zhou, F., Chou, W., Zhou, C., Chen, H., et al., 2009. pDAWG: an integrated database for plant cell wall genes. Bioenerg. Res. 2, 209−216.

Martinez, M.E., Sanchez, S., Jimenez, J.M., El Yousfi, F., Munoz, L., 2000. Nitrogen and phosphorus removal from urban wastewater by the microalga *Scenedesmus obliquus*. Bioresour. Technol. 73, 263−272.

Matthews, L., Gopinath, G., Gillespie, M., Caudy, M., Croft, D., de Bono, B., et al., 2009. Reactome knowledge base of human biological pathways and processes. Nucleic Acids Res. 7, D619−D622.

May, P., Christian, J.O., Kempa, S., Walther, D., 2009. Chlamy-Cyc: an integrative systems biology database and web-portal for *Chlamydomonas reinhardtii*. BMC Genomics. 10, 209−220.

Meeuse, B.J.D., 1963. A simple method for concentrating photoactic flagellates and separating them from debris. Arch. Mikrobiol. 45, 423−424.

Megharaj, M., Kantachote, D., Singleton, I., Naidu, R., 2000. Effects of long-term contamination of DDT on soil microflora with special reference to soil algae and algal transformation of DDT. Environ. Pollut. 109, 35−42.

Megharaj, M., Madhavi, D.R., Sreenivasulu, C., Umamaheswari, A., Venkateswarlu, K., 1994. Biodegradation of methyl parathion by soil isolates of microalgae and cyanobacteria. Bull. Environ. Contam. Toxicol. 53, 292–297.

Melis, A., Happe, T., 2001. Hydrogen production. Green algae as a source of energy. Plant Physiol. 127, 740–748.

Misra, N., Panda, P.K., Parida, B.K., 2013. Agrigenomics for microalgal biofuel production: an overview of various bioinformatics resources and recent studies to link OMICS to bioenergy and bioeconomy. OMICS. 17 (11), 537–549.

Moate, P.J., Williams, S.R., Hannah, M.C., Eckard, R.J., Auldust, M.J., Ribaux, B.E., et al., 2013. Effects of feeding alagal meal high in docosahexaenoic acid on feed intake, milk production, and methane emissions in dairy cows. J Dairy Sci. 96, 3177–3188.

Mokady, S., Abramovici, A., Cogan, U., 1989. The safety evaluation of *Dunaliella baradawil* as a potential food source supplement. Food Chem. Toxicol. 27, 221–226.

Muller-Feuga, A., Robert, R., Cahu, C., Robin, J., Divanach, P., 2003. In: Stottrup, J.G., McEvoy, L.A. (Eds.), Uses of Microalgae in Aquaculture. Live Feeds in Marine Aquaculture. Blackwell Science Ltd, Oxford, pp. 253–299.

Munoz, R., Guieysse, B., 2006. Algal–bacterial processes for the treatment of hazardous contaminants: a review. Water Res. 40, 2799–2815.

Murugaseen, S., Venkatesh, P., Dhamotharan, R., 2010a. Phycoremediation of poultry wastewater by microalga. Biosci. Biotech. Res. Comm. 3, 2.

Murugesan, S., Venkatesh, P., Dhamotharan, R., 2010b. Phycoremediation of poultry wastewater by micro alga. Biosci. Biotech. Res. Comm. 3 (2), 142–147.

Nakajima, N., Teramoto, T., Kasai, F., Sano, T., Tamaoki, M., Aono, M., et al., 2007. Glycosylation of bisphenol A by freshwater microalgae. Chemosphere. 69, 934–941.

Narro, M.L., Cerniglia, C.E., Van Baalen, C., Gibson, D.T., 1992. Metabolism of phenanthrene by the marine cyanobacterium *Agmenellum quadruplicatum* PR-6. Appl. Environ. Microbiol. 58, 1351–1359.

Nichols, H.W., 1973. Growth media: freshwater. In: Stein, J.R. (Ed.), Handbook of Phycologocial Methods: Culture Methods and Growth Measurements. CambridgeUniversity Press, Cambridge, pp. 7–24.

Olaziola, M., 2000. Commercial Production of astaxanthin from *Haematococcus pluvialis* using 25,000 liter outdoor photobioreactors. J. Appl. Phycol. 12, 499–506.

Oppenheimer, C.H., 1955. The effect of bacteria on the development and hatching of pelagic fish eggs, and the control of such bacteria by antibiotics. Copeia. 1, 43–49.

Oswald, W.J., 1988b. Micro-algae and wastewater treatment. In: Borowitzka, M.A., Borowitzka, L.J. (Eds.), Micro-algal Biotechnology. Cambridge University Press, pp. 305–328.

Paasche, E., 1971. A simple method for establishing bacteria free cultures of phototactic flagellates. J. Cons. Int. Explor. Mer. 33, 509–511.

Parameswari, E., Lakshmanan, A., Thilagavathi, T., 2010. Phycoremediation of heavy metals in polluted water bodies. Electron. J. Environ. Agric. Food Chem. 9, 808–814.

Pavlostathis, S.G., Jackson, G.H., 1999. Biotransformation of 2,4,6-trinitrotoluene in Anabaena sp. cultures. Environ. Toxicol. Chem. 18, 412–419.

Payer, H.D., Runkel, K.H., 1978. Environmental pollutants in freshwater algae from open-air mass cultures. Arch. Hydrobiol. Beih. 11, 184–198.

Perez, M.J., Vincente, C., Legaz, M.E., 1985. An improved method to isolate lichen algae by gel filtration. Plant Cell Rep. 4, 210–211, Plant J. 54, 621–639.

Pienkos, P.T., Darzins, A., 2009. The promise and challenges of microalgal-derived biofuels. Biofuel Biprod. Bior. 3, 431–440.

Pilat, Z., Jezek, J., Sery, M., Tritilek, M., Nedbal, L., Zemanek, P., 2013. Optical trapping of of microalgae at 735-nm 1064 nm: photodamage assessment. J. Photochem. Photobiol. B. 121, 27–31.

Pimentel, M., Oturan, N., Dezotti, M., Oturan, M.A., 2008. Phenol degradation by advanced electro-chemical oxidation process electro-Fenton using a carbon felt cathode. Appl. Catal. B. Environ. 83, 140–149.

Pinto, G., Pollio, A., Previtera, L., Temussi, F., 2002. Biodegradation of phenols by microalgae. Biotechnol. Lett. 24, 2047–2051.

Pizarro, J., Santander, E., Herrera, L., 2006. Nutrients measured on water bottle samples at station Bm_1997-08-30_1, Bajo Molle, Chile. Universidad Arturo Prat, Iquique, doi:10.1594/PANGAEA.547826.

Prased, D.Y., 1982. Effect of phosphorus on decomposition of organic matter in fresh water. Indian J. Environ. Health. 24 (3), 206–214.

Pringsheim, E.G., 1912. Die Kultur von Algen in Agar. Beiter. Biol. Pfl. 11, 305–332.

Pringsheim, E.G., 1946. Pure Cultures of Algae. Cambrdige University Press, London, 119 pp.

Pulz, O., Gross, W., 2004. Valuable products from biotechnology of microalgae. Appl. Microbiol. Biotechnol. 57, 287–293.

Radakovits, R., Jinkerson, R.E., Darzins, A., Posewitz, M.C., 2010. Genetic engineering of algae for enhanced biofuel production. Eukaryot. Cell. 9, 486–501.

Rao, P.H., Kumar, R.R., Raghvan, B.G., Subramanian, V.V., Sivasubramanian, V., 2011. Application of phycoremediation technology in the treatment of wastewater from a leather-processing chemical manufacturing facility. Water SA. 37, 7–14.

Reckermann, M., 2000. Flow sorting in aquatic ecology aquatic flow cytometry: achievements and prospects. In: Reckermann, M., Colijn, F. (Eds.), Scientia Marina, vol. 64, No. 2, pp. 235–246.

Remm, M., Storm, C.E., Sonnhammer, E.L., 2001. Automatic clustering of orthologs and inparalogs from pairwise species comparisons. J. Mol. Biol. 314, 1041–1052.

Richmond, A., 2004. Strategies for bioprospecting microalgae for potential commercial applications. In: Barclay, W. (Ed.), Handbook of Microalgal Culture: Applied Phycology and Biotechnology, second ed. Wiley-Blackwell. NJ, USA.

Rippka, R., Coursin, T., Hess, W., 2000. *Prochlorococcus marinus* Chisholm et al. 1992 aubsp. Pastoris eubsp. Nov. strain PCC 9511, the first axenic chlorophyll a2/b2 containing cyanobacterium (Oxyphotobacteria). Int. J. Syst. Evol. Microbiol. 50, 1833–1847.

Rismani-Yazdi, H., Haznedaroglu, B.Z., Bibby, K., Peccia, J., 2011. Transcriptome sequencing and annotation of the microalgae *Dunaliella tertiolecta*: pathway description and gene discovery for production of next-generation biofuels. BMC Genomics. 12, 148–165.

Rodriguez-Moya, M., Gonzalez, R., 2010. Systems biology approaches for the microbial production of biofuels. Biofuels. 1, 291–310.

Rogerson, A., DeFreitas, A.S.W., McInnes, A.C., 1986. Observations on wall morphogenesis in *Coscinodiscus asteromphalus* (Bacillariophyceae). Trans. Am. Microsc. Soc. 105, 59–67.

Ruiz-Marin, A., Mendoza-Espinosa, L.G., Stephenson, T., 2010. Growth and nutrient removal in free and immobilized green algae in batch and semi-continuous cultures treating real wastewater. Bioresour. Technol. 101, 58–64.

Sanjaya, Durrett, T.P., Weise, S.E., Benning, C., 2011. Increasing the energy density of vegetative tissues by diverting carbon from starch to oil biosynthesis in transgenic Arabidopsis. Plant Biotechnol. J. 9 (8), 874–883.

Sawayama, S., Rao, K.K., Hall, D.O., 1998. Nitrate and phosphate ions removal from water by *Phormidium laminosumim* mobilized on hollow fibres in a photobioreactor. Appl. Microbiol. Biotechnol. 49, 463–468.

Sawayama, S., Hanada, S., Kamagata, Y., 2000. Isolation and characterization of phototrophic bacteria growing in lighted up flow anaerobic sludge blanket reactor. J. Biosci. Bioeng. 89 (4), 396–399.

Semple, K.T., Cain, R.B., 1996. Biodegradation of phenols by the alga *Ochromonas danica*. Appl. Environ. Microbiol. 62, 1265–1273.

Schoeny, R., Cody, T., Warshawsky, D., Radike, M., 1988. Metabolism of mutagenic polycyclic aromatic hydrocarbons by photosynthetic algal species. Mutat. Res. 197, 289–302.

Schwartz, R.E., Hirsch, C.F., Sesin, D.F., Flor, J.E., Chartrain, M., Fromtling, R.E., et al., 1990. Pharmaceuticals from cultured algae. J. Ind. Microbiol. 5, 113–124.

Sengar, R.M.S., Singh, K.K., Singh, S., 2011. Application of phycoremediation technology in the treatment of sewage water to reduce pollution load. Indian J. Sci. Res. 2, 33–39.

Sethunathan, N., Megharaj, M., Chen, Z.L., Williams, B.D., Lewis, G., Naidu, R., 2004. Algal degradation of a known endocrine disrupting insecticide, α-endosulfan, and its metabolite, endosulfan sulfate, in liquid medium and soil. J. Agric. Food Chem. 52, 3030–3035.

Sharma, A., Chauhan, R., 2012. In silico identification and comparative genomics of candidate genes involved in biosynthesis and accumulation of seed oil in plants. Comp. Funct. Genomics.1–14, Article ID 914843.

Sheehan, J., Dunahay, T., Benemann, J., Roessler, P., 1998. A Look Back at the US Department of Energy's Aquatic Species Program—Biodiesel from Algae. Report NREL/TP-580–24190. National Renewable Energy Laboratory, Golden, CO.

Shi, W., Wang, L., 2010. Removal of estrone, 17α-ethinylestradiol, and 17β-estradiol in algae and duckweed-based wastewater treatment systems. Environ. Sci. Pollut. Res. 17, 824–833.

Shields, R.J., Lupatsch, I., 2012. Algae for aquaculture and animal feeds. Schwerpunkt. 21, 23–37.

Singh, A., Nigam, P.S., Murphy, J.D., 2011. Mechanism and challenges in commercialisation of algal biofuels. Bioresour. Technol. 102, 26–34.

Singh, B.K., Walker, A., 2006. Microbial degradation of organophosphorus compounds. FEMS Microbiol. Rev. 30, 428–471.

Sivasubramanian, V., Subramanian, V.V., Raghavan, B.G., Ranjithkumar, R., 2009. Large scale phycoremediation of acidic effluent from an alginate industry. Sci. Asia. 35, 220–226.

Smith, D.R., Lee, R.W., Cushman, J.C., Magnuson, J.K., Tran, D., Polle, J.E.W., 2010. The *Dunaliella salina* organelle genomes: large sequences, inflated with intronic and intergenic DNA. BMC Plant Biol. 10, 83–97.

Sosnowski, T.R., 2013. Spraying of cell colloids in medical atomizers. Chem. Eng. Trans. 32, 2257–2262.

South, G.R., Whittick, A., 1987. Introduction to Phycology. Blackwell Scientific Publications, London, UK, viii, 341 pp.

Spolaore, P., Joannis-Cassan, C., Duran, E., Isambert, A., 2006. Commercial applications of microalgae. J. Biosci. Bioeng. 101, 87–96.

Stamey, J.A., Shepherd, D.M., de Veth, M.J., Corl, B.A., 2012. Use of algae or algal oil rich in n-3 fatty acids as a feed supplement for dairy cattle. J. Dairy Sci. 95, 5269–5275.

Stanke, M., Steinkamp, R., Waack, S., Morgenstern, B., 2004. AUGUSTUS: a web server for gene finding in eukaryotes. Nucleic Acids Res. 32, W309–W312.
Subashchandrabose, S.R., Balasubramanian, R., Mallavarapu, M., Kadiyala, V., Naidu, R., 2013. Mixotrophic cyanobacteria and microalgae as distinctive biological agents for pollutant degradation. Environ. Int. 51, 59–72.
Subashchandrabose, S.R., Venkateswarlu, K., Ramakrishnan, B., Naidu, R., Megharaj, M., 2012. Centre for Environmental Risk Assessment and Remediation, University of South Australia, SA5095, Australia, and Cooperative Research Centre for Contamination Assessment and Remediation of Environment.
Sucaet, Y., Deva, T., 2011. Evolution and application of pathway resources and databases. Brief Bioinform. 12, 530–544.
Sutherland, J.W., 1976. Ultrasonication—an enrichment technique for microcyst-forming bacteria. J. Appl. Bacteriol. 41, 185–188.
Tam, N.F.Y., Chong, A.M.Y., Wong, Y.S., 2002. Removal of tributyltin (TBT) by live and dead microalgal cells. Mar. Pollut. Bull. 45, 362–371.
Tatusov, R.L., Galperin, M.Y., Natle, D.A., Koonin, E.V., 2000. The COG database: a tool for genome-scale analysis of protein functions and evolution. Nucleic Acids Res. 28, 33–36.
Thies, F., Backhaus, T., Bossmann, B., Grimme, L.H., 1996. Xenobiotic biotransformation in unicellular green algae (involvement of cytochrome P450 in the activation and selectivity of the pyridazinone pro-herbicide metflurazon). Plant Physiol. 112, 361–370.
Thimm, O., Bläsing, O., Gibon, Y., Nagel, A., Meyer, S., Krüger, P., et al., 2004. MAPMAN: a user driven tool to display genomics data sets onto diagrams of metabolic pathways and other biological processes. Plant J. 37, 914–939.
Thomas, N.V., Kim, S.K., 2013. Beneficial effects of marine algal compounds in cosmeceuticals. Mar. Drugs. 11, 146–164.
Thomas, P.D., Campbell, M.J., Kejariwal, A., Mi, H., Karlak, B., Daverman, R., et al., 2003. PANTHER: a library of protein families and sub families indexed by function. Genome Res. 13, 2129–2141.
Throndsen, J., 1978. The dilution-culture method. In: Sournia, A. (Ed.), Phytoplankton Manual. UNESCO, Paris, pp. 218–224.
Tirichine, L., Bowler, C., 2011. Decoding algal genomes: tracing back the history of photosynthetic life on earth. Plant J. 66, 45–57.
Todd, S.J., Cain, R.B., Schmidt, S., 2002. Biotransformation of naphthalene and diaryl ethers by green microalgae. Biodegradation. 13, 229–238.
Tsang, C.K., Lau, P.S., Tam, N.F.Y., Wong, Y.S., 1999. Biodegradation capacity of tributyltin by two Chlorella species. Environ. Pollut. 105, 289–297.
Ueno, R., Wada, S., Urano, N., 2008. Repeated batch cultivation of the hydrocarbon-degrading, micro-algal strain Prototheca zopfii RND16 immobilized in polyurethane foam. Can. J. Microbiol. 54, 66–70.
Veeralakshmi, M., Kamaleswari, J., Murugesan, S., Kotteswari, M., 2007. Phycoremediation of petrochemical effluent by Cyanobacterium. Indian Hydrobiol. 10 (1), 101–108.
Walker, J.D., Colwell, R.R., Petrakis, L., 1975a. Degradation of petroleum by an alga, *Prototheca zopfii*. Appl. Environ. Microbiol. 30, 79–81.
Walker, J.D., Colwell, R.R., Vaituzis, Z., Meyer, S.A., 1975b. Petroleum-degrading a chlorophyllous alga *Prototheca zopfii*. Nature. 254, 423–424.
Wang, F., 2012a. New isoflavonoids from erythrina arborescens and structure revision of anagyroidisoflavone A. Nat. Prod. Bioprospect. 3, 238–242.

Wang, Y., 2012b. Antifungalmycin, an antifungal macrolide from *Streptomycespadanus* 702. Nat. Prod. Bioprospect. 2, 41–45.

Wang, L., Li, Y.C., Chen, P., Min, M., Chen, Y.F., Zhu, J., et al., 2010. Anaerobic digested dairy manure as a nutrient supplement for cultivation of oil-rich green microalgae *Chlorella* sp. Bioresour. Technol. 101, 2623–2628.

Warshawsky, D., Radike, M., Jayasimhulu, K., Cody, T., 1988. Metabolism of benzo(a)pyrene by a dioxygenase enzyme system of the freshwater green alga Selenastrum capricornutum. Biochem. Biophys. Res. Commun. 152, 540–544.

Waterbury, J.B., Watson, S.W., Guillard, R.R.L., Brand, L.E., 1979. Widespread occurrence of a unicellular, marine, planktonic, cyanobacterium *Synechococcus*. In: Platt, T., Li, W.K.W. (Eds.), Photosynthetic Picoplankton. Can. Bull. Fish. Aquat. Sci., 214, 71–120.

Weiner, J.A., DeLorenzo, M.E., Fulton, M.H., 2004. Relationship between uptake capacity and differential toxicity of the herbicide atrazine in selected microalgal species. Aquat. Toxicol. 68, 121–128.

Weiss, T.L., Johnston, J.S., Fujisawa, K., Sumimoto, K., Okada, S., Chappell, J., et al., 2010. Phylogenetic placement, genome size, and GC content of the liquid-hydrocarbon-producing green microalga *Botryococcus braunii* strain Berkeley (SHOWA) *Chlorophyta*. J. Phycol. 46, 534–540.

Werblan, D., Smith, R.J., Van der Valk, A.G., Davis, C.B., 1978. Treatment of waste from a confined hog feeding unit by using artificial marshes. In: Mickim, H.L. (Ed.), Proceedings of International Symposium on Land Treatment of Wastewater, Hannover, New Hapshire, pp. 1–13.

Wolverton, B.C., 1982. Hybrid wastewater treatment system using anaerobic microorganisms and red (*Phragmites communis*). Econ. Bot. 36, 373–380.

Woods, R.P., Coleman, R.P., Deng, M.D., 2004. U.S. Patent 6,699,696 B2, Washington, DC.

Wright, P.J., Weber, J.H., 1991. Biosorption of inorganic tin and methylin compounds by estuarine macroalgae. Environ. Sci. Technol. 25, 287–294.

Yan, H., Pan, G., 2004. Increase in biodegradation of dimethyl phthalate by Closterium lunula using inorganic carbon. Chemosphere. 55, 1281–1285.

Yan, X., Yang, Y., Li, Y., Sheng, G., Yan, G., 2002. Accumulation and biodegradation of anthracene by Chlorella protothecoides under different trophic conditions. Chin. J. Appl. Ecol. 13, 145–150.

Yen, H.W., 2007. Anerobic co-digestion of algal sludge and waste paper to produce methane. Bioresour. Technol. 98, 130–134.

Ying, J., Wang, C., Zhang, C., 2011. Effect of immobilization on growth of microalgae and removal of nitrogen from urine. In proceedings of International conference on Remote Sensing, Environment and Transportation Engineering (RSETE), 2011. 4939–4942. Available from: http://dx.doi.org/doi:10.1109/RSETE.2011.5965420.

Yu, W.L., Ansari, W., Schoepp, N.G., Hannon, M.J., Mayfield, S.P., Burkart, M.D., 2011. Modifications of the metabolic pathways of lipid and triacylglycerol production in microalgae. Microb. Cell Fact. 10, 91–102.

Zablotowicz, R.M., Schrader, K.K., Locke, M.A., 1998. Algal transformation of fluometuron and atrazine by N-dealkylation. J. Environ. Sci. Health B. 33, 511–528.

Zhang, E.D., Wang, B., Wang, Q.H., Zhang, S.B., Zhao, B.D., 2008. Ammonia-nitrogen and orthophosphate removal by immobilized *Scenedesmus* sp isolated from municipal wastewater for potential use in tertiary treatment. Bioresour. Technol. 99, 3787–3793.

16 Feasibility of Using Bioelectrochemical Systems for Bioremediation

Song Jin[a,b] *and Paul H. Fallgren*[b]

[a]Department of Civil and Architectural Engineering, University of Wyoming, Laramie, WY 82071, USA, [b]Advanced Environmental Technologies, LLC, Fort Collins, CO 80525, USA

16.1 Introduction

Bioelectrochemical systems (BESs) represent one of the emerging approaches for cleaning up environmental contaminants and simultaneously generating energy. BESs rely on microorganisms to catalyze the oxidization of organic and inorganic electron donors and deliver electrons to the anode, which subsequently transfers electrons to the cathode through a conductive circuit. In past decades, BES has been widely studied for a number of applications in various bioelectrochemical cells, with the best examples being "microbial X cells" (MXCs) (Shroder, 2011) and cell-free "biofuel cells." In "MXCs," "X" stands for the specific applications for a group of BES. For example, electrons can be captured directly for current generation or ionic migration microbial fuel cells (MFCs) or microbial desalination cells (MDCs) (Liu et al., 2004; Lovley, 2006; Logan, 2008; Cao et al., 2009; Mehanna et al., 2010), chemical production microbial chemical cells (MCCs) (Clauwaert et al., 2007; Rozendal et al., 2009; Butler et al., 2010; Rabaey et al., 2010), or supplemented by external power input for producing fuel, including hydrogen and methane gas microbial electrolysis cells (MECs) (Logan et al., 2008; Cheng et al., 2009). Some of the common types of BESs (MXCs and biofuel cells) are listed in Table 16.1. More recently, BESs have even been proposed for remediation with less consideration on power generation where such studies have been reported for wastewater treatment (He et al., 2013), and for groundwater, sediments, and soil bioremediation (Gregory and Lovley, 2005; Jin et al., 2010; Zhang et al., 2010; Morris et al., 2011; Huang et al., 2011a; Kumar et al., 2012; Morris and Jin, 2012; Yan et al., 2012; Jin et al., 2013). For bioremediation applications, the BES electrodes serve as the perpetual intermediate electron acceptors/donors for contaminant-degrading microbes to promote rapid degradation of pollutants in anaerobic subsurface environments (Aulenta et al., 2009; Lovley and Nevin, 2011;

Microbial Biodegradation and Bioremediation. DOI: http://dx.doi.org/10.1016/B978-0-12-800021-2.00016-9

Table 16.1 Examples of Different Types of BESs

BES Type	Description
Microbial fuel cell (MFC)	Closed fuel cell system consisting of at least one anode and one cathode, where microbially mediated electron transfer occurs from oxidation of an electron donor and reduction of an electron acceptor. Primary application is to treat wastes with focus on electricity production.
Microbial electrolysis cell (MEC)	BES consisting of at least one anode and one cathode, where anodic microorganisms oxidize organic compounds and the resulting H^+ is reduced to H_2, which is assisted by the application of an external voltage. The cathode reaction may also result in formation of CH_4 depending on microorganisms that are present.
Microbial desalination cell (MDC)	Closed fuel cell-like system consisting of at least one anode and one cathode, where microbial oxidation of organic compounds produces electrical potentials at the electrodes that drive ions through ion exchange membranes.
Microbial chemical cell (MCC)	Closed fuel cell-like system consisting of at least one anode and one cathode, where microbially catalyzed reactions at electrodes result formation of target chemicals (e.g., hydroxides, hydrogen peroxide) depending on conditions and feedstock.
External-voltage BES	BES consisting electrodes where voltage is applied from an external source for enhancing microbial reactions. Applications in environmental remediation (e.g., TCE reduction), wastewater treatment, carbon sequestration, and chemical production.
Open-type BES	BES consisting of electrodes and configurations that are not confined by boundaries, except for the boundary of the application area. Lack of unit boundaries allows for unconventional configurations customized for specific applications. Applications include *in situ* environmental remediation and wastewater treatment.
Sediment fuel cell (SFC)	Open- or closed-type BES with at least one anode inserted in sediments composed of organic compounds, and at least one cathode at the surface exposed to atmosphere. Primary application is electricity generation from organic-rich sediments.
Microbial electrochemical snorkel (MES)	Open-type BES which consists of a single conductive unit where one end facilitates anode-type reactions and the other end facilitates cathode-type reactions. No load or power source is connected, allowing free and direct

(*Continued*)

Table 16.1 (Continued)

BES Type	Description
	transfer of electrons from the anodic end to the cathodic end. Primary application is for enhancing biodegradation of organic compounds in wastewater and other matrices.
Biofuel cell (BFC) or enzymatic fuel cell	Fuel cell system consisting of at least one anode and one cathode, where enzyme-mediated electron transfer occurs from oxidation of an electron donor and reduction of an electron acceptor. Primary application is to treat wastes or utilize an organic fuel source for electricity production.

Morris and Jin, 2012). Biofuel cells (BFCs) are enzymatically based BESs, which usually rely on extracted, cell-free enzymes attached on electrodes (anode and/or cathode) to drive electron transfer for electricity production. Organic fuels such as methanol, ethanol, or wastewater organics have been reported as electron donors in BFCs (Davis and Higson, 2007; Falk et al., 2012; Holzinger et al., 2012); however, there are no reports of BFCs being applied in environmental remediation.

The remediation process in a BES carries several advantages over traditional bioremediation processes for contaminant removal. First of all, by monitoring the electrical current and electrode potential, a BES offers an alternative method for continuous monitoring of microbial activities and control of the supply of electron donors (Williams et al., 2010). Second, no electron-donating compounds need to be injected for reductive remediation such as dechlorination. This eliminates the need for compound shipping, transport, storage, injection, and secondary groundwater contamination phenomena (Aulenta et al., 2008; Rabaey and Keller, 2008). Third, the "passive aeration" mechanisms of the BESs can maintain and even increase the concentration of dissolved oxygen in otherwise anaerobic matrices, such as hydrocarbon-contaminated sediments (Morris and Jin, 2012), helpful to more effective biodegradations under higher oxidizing potentials. Finally, a more effective usage of the supplied electrons is expected in a BES, since the microorganisms responsible for the bioremediation process can be grown directly at the surface of the electrodes or in their proximity. This mechanism helps enrich capable microbial populations within a short period of time, shortening the lag phase of bioremediation.

This chapter summarizes recent developments in applications of BESs in environmental remediation, including (1) enhanced bio-oxidation (degradation) of organic and inorganic contaminants, (2) dechlorination, and (3) harmful metal reduction and recovery. Potential limitations and challenges for environmental remediation by BESs are also discussed.

16.2 BES Configurations, Microbial Processes, and Remediation

16.2.1 BES Designs

Many different configurations have been developed for BESs (Logan and Regan, 2006); however, for environmental remediation research, a commonly utilized design is a two-chamber reactor, consisting of two compartments usually separated by an ion exchange membrane or simple separating materials. Sometimes the chambers are also connected by a plain salt bridge (Morris et al., 2009a) or consolidated into one unit (Erable et al., 2011). Two-chamber BESs tend to be favored for remediation research where the degradation of specific waste is usually examined individually in anodic or cathodic chambers for oxidative or reductive reactions, respectively.

In anodic biodegradation studies, a typical design is a two-chamber BES, such as those MFCs that use unsustainable and/or expensive cathode materials, including noble metals, to catalyze the cathodic reduction of a terminal electron acceptor like oxygen or ferricyanide (Luo et al., 2009). A proton exchange membrane is also an integral component of these MFCs. Such designs make it impractical to scale up applications for remediation, due to the high material costs and system fragility in the environment, such as contaminated soils and groundwater. Incorporating salt bridges or having no proton link may lower the cost required for MFC construction (Morris et al., 2009a), but such systems typically produce weak currents, reflecting their high internal resistance and waste of electrons on heat. Moreover, the effect of the distance between two ends of the salt bridge on the internal resistance is unclear, affecting the field applicability of MFCs. In remediation studies, in which benthic MFCs were also frequently adopted, the anode is inserted into river or marine sediment rich in organic matter, and the cathode into the overlying oxic water (He et al., 2013). Although no separator is needed in a benthic MFC, studies with such reactors have emphasized the importance of close proximity between the electrodes to promote optimal proton transfer and maximum current output. Transforming BESs to a practical process for subsurface bioremediation is likely to require an improved design to make it feasible.

In studies using the cathode to remove contaminants, the design of the BESs has gained less attention since in most cases the potential difference between the electrodes was set by using a potential workstation or power supply (Mu et al., 2009). Solar or wind power could be an alternative to a bioanode to sustainably generate the electricity necessary to supply the electrons for the cathodic remediation.

16.2.2 Microbiology and Mediators

Many studies have demonstrated the use of an anode as the primary electron acceptor to promote biodegradation of organic contaminants in an anode chamber. Although most of these studies have successfully used mixed cultures as the inoculum for the acclimation of anode respiration bacteria, identifying and selecting microbial communities that can utilize certain contaminants is still a technical challenge. In contrast, bacteria enriched on or adjacent to the cathode that can use the electrons for dissimilatory reductions have been identified as (1) perchlorate by *Dechloromonas*, *Azospira*, and *Dechlorospirillum* sp. (Thrash et al., 2007); (2) nitrate by *Geobacter* sp. (Gregory et al., 2004; Park et al., 2005); (3) hexavalent uranium by *Geobacter* sp. (Gregory and Lovley, 2005); (4) hexavalent chromium by *Trichococcus* sp. and *Pseudomonas* sp. (Tandukar et al., 2009); and (5) trichloroethene (TCE) by *Desulfitobacterium* sp. (Aulenta et al., 2007, 2009).

Optimization of BES applications in environmental remediation will probably benefit from an understanding of how anodic/cathodic respiration bacteria exchange electrons with the electrode surface. Thus far, primarily anodic processes have been studied with respect to electron transfer mechanisms. The proposed mechanisms include (1) direct electron transfer via redox components located on the outer surface of the microorganism (e.g., cytochromes, pilus-like appendages) and (2) mediated electron transfer based on diffusible redox mediators. Most of the understanding of direct electron transfer by bacteria to electrodes came from studies of dissimilatory metal-reducing bacteria such as *Geobacter* and *Shewanella* species (more details in Lovley, 2011). Although the number of publications about biocathode studies has been increasing significantly in recent years, little is known about the biochemical mechanisms of microbial electron uptake from a cathode. Some reports have shown that mature bioanodes may be turned into biocathodes upon changing the operating conditions (Rozendal et al., 2008; Cheng et al., 2010), suggesting similar roles of electrochemical active bacteria during electron-releasing and electron-accepting reactions. However, a recent investigation based on gene expression and deletion analysis of *Geobacter sulfurreducens* indicated that the mechanisms for electron transfer from electrodes differ significantly from the mechanisms for electron transfer to electrodes (Strycharz et al., 2011).

Although some *Geobacter* spp. have been reported to directly access electrons off the electrode surface, electron mediators such as antraquinone-2,6-disulfonate (AQDS) and methyl viologen (MV) were still present in most of the BES remediation studies. Except for AQDS and MV, many electro-active substrates, such as quinones (Thrash et al., 2007), phenazines (Rabaey et al., 2005), and humic substances (Lovley et al., 1996, 1999), can be used in a nondegradative manner by a bacteria's electron donors and/or acceptors (Thrash and Coates, 2008). Table 16.2 lists types of mediators, accompanied by their function as either an electron acceptor or donor and corresponding potential. Either dissolved in the bulk liquid or physically "anchored" to the electrode surface, these (charge carrying, ampholytic, etc.) compounds can be selectively oxidized or reduced by microorganisms without being consumed. They facilitate the shuttling of electrons from the carbon electrode to the functional bacteria and some mediators are naturally occurring. However, the addition of commercial electron mediators may not be optimal for environmental remediation as it adds cost to the treatment and requires an additional step to remove them from the treated matrix.

Table 16.2 Mediators (Electron Shuttles) Commonly Used in BESs

Electron Shuttle	Function	Reduction Potential, E^0 (V)	Reference
MV	e-donor	−0.450	Aulenta et al. (2009)
Cobalt sepulchrate	e-donor	−0.350	Emde and Schink (1990)
Neutral red	e-donor	−0.325	Park et al. (1999)
AQDS	e-donor	−0.184	Thrash et al. (2007) and Aulenta et al. (2010)
Iron	e-acceptor/donor	0.76	Yunker and Radovich (1986)

16.2.3 Potential and Power Supply

In a BES, the possible metabolic energy (such as ATP) gain from a given metabolic process can be calculated by

$$\text{ATP} = \sum (E^0_{\text{acceptor}} - E_{\text{donor}}) \times Q$$

Here, E(V) is the standard potential of the electron donor or acceptor and Q(C) the current multiplied by time t(s) (Huang et al., 2011b). Thus, an optimal applied potential can provide an appropriate selective pressure for adaptation of microorganisms. This selective and evolutionary pressure can lead to the enhancement of the ability of microorganisms for electrochemical interaction with electrodes as well as improved current production, which is associated with clear differences in the properties of the outer surfaces of the cells. In most of the remediation studies using BESs, as shown in Tables 16.3 and 16.4, the set potential at the electrode usually takes the place of the potential of a chemical electron donor or acceptor. The anode or cathode potential is commonly set in two different ways: (1) two-electrode systems set the working electrode versus a counter electrode and (2) three-electrode systems poise the working electrode potential versus a constant potential reference electrode.

Setting the anode potential has been shown to promote enrichment of the biofilm, reduce the startup time of the system, and increase the subsequent power density due to better acclimation of the exoelectrogenic bacteria (Finkelstein et al., 2006; Busalmen et al., 2008; Wagner et al., 2010). Optimized biocathode potentials of 242 and 345 mV have been shown to reduce the time for startup and enhance the performance of aerobic biocathodes (Liang et al., 2009; TerHeijne et al., 2010b; Huang et al., 2011b). For example, with −300 mV set potential, the startup time of an MFC for Cr(VI) reduction was reduced to 19 days, the reduction of Cr(VI) was improved to 19.7 mg/L day, and the maximum power density was increased to 6.4 W/m^3 compared to the control (26 days, 14.0 mg/L day, and 4.1 W/m^3) (Huang et al., 2011b). Aulenta et al. (2009) also reported that the rate and extent of TCE dechlorination, as well as the competition for the available electrons, were highly dependent on the set cathode potential.

Table 16.3 Examples of BES Configurations and Anodic Pollutant Degradation

BES Configuration	Pollutant	Pollutant Removal Rate	Reference
Two-chamber MFC	Phenol	63.6 mg/(L-day)	Luo et al. (2009)
	Diesel fuel	6.9 mg/(L-day)	Morris et al. (2009a,b)
	1,2-Dichloroethane	102 mg/(L-day)	Pham et al. (2009)
	Naphthalene	~10.5 μM/day	Zhang et al. (2010)
Open-cell BES	Phenol	~7 mg/(L-day)	Huang et al. (2011a,b)
	Petroleum hydrocarbons	183 mg/(kg-day)	Morris and Jin (2012)

It is noteworthy that the value of the applied potential must be controlled at a certain level to eliminate the effect of nonbiological reactions. When the poised cathode potential is lower than −414 mV, which is the theoretical cathode potential of H_2 evolution, the H_2 electrochemically generated through the electrolysis of water at the cathode surface is likely to play a role in the microbial reduction of pollutant. Thrash et al. (2007) observed a H_2 evolution in their reactor in which perchlorate reduction was obtained with a biocathode at −500 mV poised potential.

16.2.4 Applications of BES in Environmental Remediation

BESs can couple pollutant removal with energy production, providing a beneficial use of degraded contaminants. The contaminant remediation can be accomplished using both electrodes of a BES. The first is through the microbial oxidation action adjacent to the anode under anaerobic or anoxic conditions. In this case, organic or inorganic contaminants of lower oxidizing potentials (more susceptible to oxidation) are used as a carbon source as well as electron donors by the electrogenesis that generates electrons and protons. The second is through the electrochemical or bioelectrochemical reduction at or adjacent to the cathode of the BES. Compounds of higher oxidizing potentials (more susceptible to reduction) can be reduced.

16.3 Anodic Remediation

Although limited groundwater contaminants, including diesel (Morris et al., 2009a); phenol (Zhang et al., 2010); 1,2-dichloroethane (Pham et al., 2009); toluene, benzene, naphthalene (Zhang et al., 2010); and nitrobenzene (Li et al., 2010), have been tested as anode substrates in BESs, almost all of the these studies confirmed that the contaminant removal at or adjacent to the anode can be accelerated under closed circuit conditions as compared to open circuit conditions, as shown in Table 16.3. This suggests that bioelectrochemical activities of anode bacteria could be enhanced when using the anode as an electron acceptor. Similar to the traditional bioremediation technologies, acclimation of an efficient microbial community in the anode chamber, especially on the surface of the anode, was considered the most important step in this process. It has been reported that limited kinds of simple compounds such as acetate can be directly used as a carbon source by electrochemically active bacteria (Lovley and Nevin, 2008), which could become a constraint for the power generation of BESs when using complex compounds as substrates. Therefore, having a more diverse microbial community may help to treat various contaminants and promote the current production in the BESs.

16.4 Cathodic Remediation

While organic and inorganic pollutants can be biodegraded through oxidation at/or adjacent to anodes, pollutants can also be degraded through reduction in the cathode. In this example, pollutants of higher oxidizing potentials are used as the electron acceptor to take electrons from the cathodic electrode or cathodic microorganisms. Therefore, different from the anode oxidation process, which is highly microbiologically dependent, the oxidized pollutant degradation at or near the cathode of BESs could be realized through direct electrochemical or bioelectrochemical reduction. Table 16.4 lists examples of cathodic reduction of common environmental contaminants.

16.4.1 Heavy Metal Reduction

Direct electrochemical reduction at the cathode of MFCs has been applied to removal and recovery of metals, such as Cr(VI) (Li et al., 2008, 2009; Wang et al., 2008) and copper (TerHeijne et al., 2010a,b; Wang et al., 2010), from mining and metallurgical wastewaters. For example, Wang et al. (2008) studied the use of a two-chamber MFC to treat Cr(VI)-containing wastewater, using Cr(VI) as the cathodic electron acceptor. Cr(VI) was completely removed after 150 h operation, with an initial concentration of 100 mg/L. The maximum power density of this system was 150 mW/m^2. TerHeijne et al. (2010a,b) proposed a metallurgical MFC to achieve the combination of copper removal/recovery and electricity generation from waste streams. The cathodic recovery of copper compared to the produced electricity was 84% and 43%, respectively, under anaerobic and aerobic conditions. Similar to the traditional electrochemical technology, a low pH is necessary for the performance of the metal-reducing cathode. For example, the pH of catholyte in the Cu^{2+} recovery study was set at 2.0 since Cu^{2+} may precipitate as CuO or Cu_2O and may not be available for reduction at a pH $\geq$4.5. Li et al. (2008) also reported that low pH had a positive effect on power generation when Cr(VI) was used as an electron acceptor in the BES. Therefore, an effective pH separator, such as a bipolar membrane (TerHeijne et al., 2010a,b), is thus needed between the anode and cathode to prevent the pH in the catholyte from increasing and the anolyte pH from dropping.

Table 16.4 Examples of Cathodic Pollutant Reduction in BESs

Cathode Chamber Pollutant	Electrical Current Source	Cathode Chamber Catalyst/Biocatalyst	Pollutant Reduction Rate	Reference
TCE	External potential application	Mixed culture (+MV)	0.026 μM/h	Aulenta et al. (2007)
	External potential application	Mixed culture	–	Aulenta et al. (2009)
	External potential application	*Geobacter lovleyi*	–	Aulenta et al. (2009)
Tetrachloroethene	External potential application	Mixed culture	0.96 μmol/h	Lohner and Tiehm (2009)
	External potential application	*G. lovleyi*	0.37 μmol/h	Strycharz et al. (2008)
Uranium (VI)	External potential application	*Geobacter sulfurreducens*	0.095 μmol/h	Gregory and Lovley (2005)
Chromium (VI)	Microbial anodic electron transfer	–	8.12 mg/(L-h)	Li et al. (2008)
	Microbial anodic electron transfer	Rutile	0.97 mg/(L-h)	Li et al. (2009)
	Microbial anodic electron transfer	Mixed culture	0.43 mg/(L-h)	Tandukar et al. (2009)
	Microbial anodic electron transfer	Mixed culture	3.61 mg/(L-h)	Huang et al. (2011b)
Nitrate-Nitrogen	Microbial anodic electron transfer	Mixed culture	6.083 g/(L-h)	Clauwaert et al. (2007)
	Microbial anodic electron transfer	Mixed culture	0.484 mM/h	Virdis et al. (2009)
	External potential application	Mixed culture	3 mg/(L-h)	Gregory et al. (2004)
	External potential application	*Geobacter metallireducens*	0.22 mg/(L-h)	Gregory et al. (2004)
	External potential application	Mixed culture	0.74 mg/(L-h)	Park et al. (2005)
	Microbial anodic electron transfer	Mixed culture	0.26 mg/(L-h)	Lefebvre et al. (2008)
Nitrite-Nitrogen	Microbial anodic electron transfer	Mixed culture	4.37 mg/(L-h)	Puig et al. (2011)
Perchlorate	Microbial anodic electron transfer	Mixed culture	0.83 mg/(L-h)	Thrash et al. (2007)
	Microbial anodic electron transfer	*Azospira* sp. (+AQDS)	4.13 mg/(L-h)	Thrash et al. (2007)
	Microbial anodic electron transfer	Mixed culture	1 mg/(L-h)	Butler et al. (2010)

Bioelectrochemical reduction of Cr(VI) was also conducted using BESs with biocathodes. Tandukar et al. (2009) reported that when initial Cr(VI) was below 80 mg/L, Cr(VI) reduction occurred data rate of 0.46 mg Cr(VI)/g VSS · h (where VSS is volatile suspended solids). Chromium was not detected in the catholyte, which indicated the complete reduction of chromium as $Cr(OH)_3$ precipitate (Tandukar et al., 2009). Huang et al. (2011b) also studied the reduction of Cr(VI) and energy production with a biocathode MFC, using indigenous bacteria from groundwater contaminated with Cr(VI) as inocula. With an initial Cr(VI) concentration of 39.2 mg/L, a Cr(VI) reduction rate of 2.4 ± 0.2 mg/g VSS h, and a maximum power production of 2.4 ± 0.1 W/m^3, at a current density of 6.9 A/m^3 was achieved. Phylotypes closely related to the Cr(VI) reducers *Trichococcus pasteurii* and *Pseudomonas aeruginosa* were found in the cathode. Separating the Cr(III) from the biomass remains a major issue in the feasibility of this technology.

16.4.2 Dechlorination

Chlorinated solvents, such as TCE and perchloroethene (PCE), are almost entirely anthropogenic, and many are highly toxic or carcinogenic. Some of them can be used as a terminal electron acceptor by microorganisms under highly reducing conditions. TCE is reduced stepwise to *cis*-dichloroethene (*cis*-DCE) and vinyl chloride (VC), with some organisms capable of transforming VC to completely harmless ethane. Studies show that a negatively polarized solid-state electrode (i.e., −450 mV vs standard hydrogen electrode (SHE)), in the presence of exogenous (i.e., MV) or self-produced (yet unidentified) redox mediators, could serve as an electron donor for the reductive dechlorination of TCE to ethene by a mixed culture (Aulenta et al., 2007, 2009, 2010). In these studies, however, *cis*-DCE and VC were typically the main TCE dechlorination products, being dechlorinated only at low rates. Recently, Strycharz et al. (2008) demonstrated that an electrode could be used to stimulate the reductive dechlorination of chloroethene by *Geobacter lovleyi*, even in the absence of redox mediators. Aulenta et al. (2009) investigated the reaction, however, and found that *G. lovleyi* failed to dechlorinate chloroethenes beyond *cis*-DCE. The possibility to stimulate the complete transformation of chloroethenes into nonchlorinated (and nontoxic) end products is a major challenge of the microbial reductive dechlorination as a bioremediation process.

Recently, electrochemical methods have been attempted for reduction of TCE at presence or absence of microbes. When electrons were provided by inserting a cathodic electrode into both water and saturated soils, TCE was rapidly reduced in a few days (Jin et al., 2008; Jin and Fallgren, 2009). In such systems, the role of microbial populations is not clear, though presumably certain species might have benefited from the presence of electrons released by the cathode. This direct reduction of TCE was attributed to electron shuttles naturally existing in soils and groundwater.

16.4.3 Perchlorate Reduction

Perchlorate is an emerging contaminant of concern due to its mobility in the environment and its inhibitory effect on iodine uptake in the thyroid gland. To date, two groups have looked at the use of BESs to stimulate microbial perchlorate reduction. Butler et al. (2010) constructed a MFC with a denitrifying biocathode for perchlorate reduction and utilized the system to identify putative biocathode-utilizing perchlorate-reducing bacteria. They confirmed that perchlorate reduction can be realized without the need for additional electron mediators or applied electrode potential. It has also been investigated the microbial perchlorate reduction in a BES. Using *Dechloromonas* and *Azospira* species as biocatalysts in the cathode chamber, 90 mg/L perchlorate was reduced readily. However, AQDS is necessary as the electron shuttle for the reduction process in this study. After long-term operation of the BESs, they isolated a novel strain and claimed that it can utilize the cathode as electron donor directly without any electron shuttles.

When the cathode is used as the electron donor, the bioelectrochemical perchlorate reduction system overcomes many issues faced by chemical stimulation of microbial reduction. For example, *ex situ* treatment by chemical stimulation can be plagued by overgrowth of the organisms, leading to increased cost, biofouling, and treatment failure. In addition, chemical addition to treatment streams can lead to the abiotic production of carcinogenic downstream disinfection by-products (Thrash and Coates, 2008).

16.4.4 Nitrate/Nitrite Reduction

Elevated nitrate contamination in groundwater is an increasing problem in the USA and around the world. Nitrate contamination comes from nitrogenous fertilizers and detergents, causing health risks such as blue baby syndrome (Greer and Shannon, 2005). Many groups have reported that nitrate can be biologically reduced in the cathode chamber of BESs (Park et al., 2005; Clauwaert et al., 2007; Virdis et al., 2010; Puig et al., 2011), according to the following equations:

$$NO_3^- + 2e^- + 2H^+ \rightarrow NO_2^- + H_2O \qquad (E^0 = +0.433\text{V vs SHE})$$

$$NO_2^- + e^- + 2H^+ \rightarrow NO + H_2O \qquad (E^0 = +0.350\text{V vs SHE})$$

$$NO + e^- + H^+ \rightarrow 1/2N_2O + 1/2H_2O \qquad (E^0 = +1.175\text{V vs SHE})$$

$$1/2N_2O + e^- + H^+ \rightarrow 1/2N_2 + 1/2H_2O \qquad (E^0 = +1.355\text{V vs SHE})$$

It was also shown that nitrate and nitrite could both be used interchangeably by cathode respiration bacteria for autotrophic nitrogen reduction (Puig et al., 2011). The accumulation of one of the intermediates listed above would reduce the Coulombic efficiency at the cathode as it decreases the total oxidation capacity of the electron acceptor that is present. Morris et al. (2009b) compared a two-chambered BES to a similar abiotic fuel cell (steel fuel cell, SFC) for nitrate reduction as an alternative to the current method. The study used petroleum compounds from a refinery wastewater as the organic substrate in the two-chambered BES coupled with nitrate reduction in the cathode. The SFC contained no organic substrate or microbial inoculation in the anode chamber. Instead, a steel wool anode was used as the electron donor and the cathode chamber was filled with sterile groundwater, where higher nitrate reduction occurred with bacteria present in the cathode chamber.

More recently, Chen et al. (2013) identified that denitrifying bacteria appear to be able to consume electrons released from a zero-valent iron (ZVI) mediated electrode. In this study, partially passivated ZVI served as a conductor or extended electrode in a direct current (DC) electrical circuit. Denitrifying bacteria appeared to harvest electrons from the circuit and reduce nitrate into N_2 and NH_4^+. This observation raised an interesting question about the electrons from a DC current and the microbial oxidation of organic and inorganic compounds.

16.5 Current State and Challenges

This review summarized results and observations from studies conducted in the laboratory for applying BES technologies for environmental remediation. This indicates that the current status of applying BES technologies for environmental remediation is primarily being investigated in laboratory-scale studies. Very few researchers have integrated and tested a BES technology at field-scale levels. Since 2010, field trials with an open system–type BES with a floating cathode configuration (Figure 16.1) have been installed and operated at a leaking underground storage tank site, where groundwater is contaminated with diesel fuel, gasoline, and

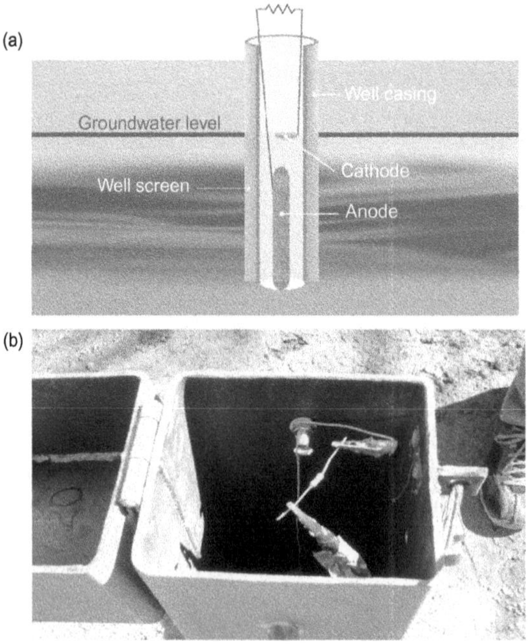

Figure 16.1 Field-scale design of an open-type BES with a floating cathode (a) and its implementation at a leaking underground storage tank site (b).

benzene (Jin et al., 2013). This type of BES was installed in several wells spanning the edges of the contaminant plume to the source zone. After 1 year, results indicate that diesel fuel levels decreased to 40% of the original level at the edge of the plume and to approximately 65–60% of the original level at the source zone. Further operation was disrupted due to draught and receding groundwater levels. This is one example of a major challenge in implementing BES technologies in the field, which is changing environmental conditions that can affect the performance of the BES. The BES design would need to account for any possible change in the environment, such as possibly installing deeper wells to account for lowering water table, for example.

Electrode material cost is another challenge due to surface area requirements. The materials used in laboratory-scale BES studies may not be economically feasible to use when scaling-up to the field. Another challenge is determining the radius of influence (ROI) of the BES, where open-type systems are most applicable for *in situ* situations. This determination would require larger scale laboratory studies or small-scale field tests. The ROI is an important design parameter that would need to be maximized using affordable electrode materials to implement effective and economical BESs for full-scale remediation of contaminated sites. Based on the range of contaminants that can be removed from the environment in reported BES studies, as well as results from field-scale tests, the application of BES as a bioremediation technology is feasible at full scale and is gaining interest as a zero- to low energy–utilizing remediation technology.

References

Aulenta, F., Catervi, A., Majone, M., Panero, S., Reale, P., Rossetti, S., 2007. Electron transfer from a solid-state electrode assisted by methyl viologen sustains efficient microbial reductive dechlorination of TCE. Environ. Sci. Technol. 41, 2554–2559.

Aulenta, F., Canosa, A., Majone, M., Panero, S., Reale, P., Rossetti, S., 2008. Trichloroethene dechlorination and H_2 evolution are alternative biological pathways of electric charge utilization by dechlorinating culture in a bioelectrochemical system. Environ. Sci. Technol. 42, 6185–6190.

Aulenta, F., Canosa, A., Reale, P., Rossetti, S., Panero, S., Majone, M., 2009. Microbial reductive dechlorination of trichloroethene to ethene with electrodes serving as electron donors without the external addition of redox mediators. Biotechnol. Bioeng. 103, 85–91.

Aulenta, F., Di Maio, V., Ferri, T., Majone, M., 2010. The humic acid analogue antraquinone-2,6-disulfonate (AQDS) serves as an electron shuttle in the electricity-driven microbial dechlorination of trichloroethene to *cis*-dichloroethen. Bioresour. Technol. 101, 9728–9733.

Busalmen, J.P., Esteve-Nunez, A., Feliu, J.M., 2008. Whole cell electrochemistry of electricity-producing microorganisms evidence an adaptation for optimal exocellular electron transport. Environ. Sci. Technol. 42, 2445–2450.

Butler, C.S., Clauwaert, P., Green, S.J., Verstraete, W., Nerenberg, R., 2010. Bioelectrochemical perchlorate reduction in a microbial fuel cell. Environ. Sci. Technol. 44, 4685–4691.

Cao, X.X., Huang, X., Liang, P., Xiao, K., Zhou, Y.J., Zhang, X.Y., et al., 2009. A new method for water desalination using microbial desalination cells. Environ. Sci. Technol. 43, 7148–7152.

Chen, L., Jin, S., Fallgren, P.H., Liu, F., Colberg, P.J.S., 2013. Passivation of zero-valent iron by denitrifying bacteria and the impact on trichloroethene reduction in groundwater. Water Sci. Technol. 67, 1254–1259.

Cheng, K.Y., Ho, G., Cord-Ruwisch, R., 2010. Anodophilic biofilm catalyzes cathodic oxygen reduction. Environ. Sci. Technol. 44, 518–525.

Cheng, S., Xing, D.F., Call, D.F., Logan, B.E., 2009. Direct biological conversion of electrical current into methane by electromethanogenesis. Environ. Sci. Technol. 43, 3953–3958.

Clauwaert, P., Rabaey, K., Aelterman, P., De Schamphelaire, L., Ham, T.H., Boeckx, P., et al., 2007. Biological denitrification in microbial fuel cells. Environ. Sci. Technol. 41, 3354–3360.

Davis, F., Higson, S.P.J., 2007. Biofuel cells—recent advances and applications. Biosens. Bioelectron. 22, 1224–1235.

Emde, R., Schink, B., 1990. Enhanced propionate formation by *Propionibacterium freudenreichii* subsp. *freudenreichii* in a 3-electrode amperometric culture system. Appl. Environ. Microbiol. 56, 2771–2776.

Erable, B., Etcheverry, L., Bergel, A., 2011. From microbial fuel cell (MFC) to microbial electrochemical snorkel (MES): maximizing chemical oxygen demand (COD) removal from wastewater. Biofouling. 27, 319–326.

Falk, M., Blum, Z., Shleev, S., 2012. Direct electron transfer based enzymatic fuel cells. Electrochim. Acta. 82, 191–202.

Finkelstein, D.A., Tender, L.M., Zeikus, J.G., 2006. Effect of electrode potential on electrode-reducing microbiota. Environ. Sci. Technol. 40, 6990–6995.

Greer, F.R., Shannon, M., 2005. Infant methemoglobinemia: the role of dietary nitrate in food and water. Pediatrics. 116, 784–786.

Gregory, K.B., Lovley, D.R., 2005. Remediation and recovery of uranium from contaminated subsurface environments with electrodes. Environ. Sci. Technol. 39, 8943–8947.

Gregory, K.B., Bond, D.R., Lovley, D.R., 2004. Graphite electrodes as electron donors for anaerobic respiration. Environ. Microbiol. 6, 596–604.

He, Y.R., Xiao, X., Li, W.W., Cai, P.J., Yuan, S.J., Yan, F.F., et al., 2013. Electricity generation from dissolved organic matter in polluted lake water using a microbial fuel cell (MFC). Biochem. Eng. J. 71, 57–61.

Holzinger, M., Le Goff, A., Cosnier, S., 2012. Carbon nanotube/enzyme biofuel cells. Electrochim. Acta. 82, 179–190.

Huang, D.Y., Zhou, S.G., Chen, Q., Zhao, B., Yuan, Y., Zhuang, L., 2011a. Enhanced anaerobic degradation of organic pollutants in a soil microbial fuel cell. Chem. Eng. J. 172, 647–653.

Huang, L., Chai, X., Cheng, S., Chen, G., 2011b. Evaluation of carbon-based materials in tubular biocathode microbial fuel cells in terms of hexavalent chromium reduction and electricity generation. Chem. Eng. J. 166, 652–661.

Jin, S., Fallgren, P., 2009. Electrically induced reduction of trichloroethene in clay. J. Hazard. Mater. 153, 127–130.

Jin, S., Fallgren, P., Morris, J., Edgar, E., 2008. TCE degradation in groundwater by supplemented electrons. Chem. Eng. J. 140, 642–645.

Jin, S., Morris, J.M., Fallgren, P.H., 2010. Influential fuel cell systems including effective cathodes and use with remediation efforts. US Patent No. 7858243 B2.

Jin, S., Fallgren, P., Nelson, C., 2013. *In-situ* remediation of petroleum-contaminated groundwater using a bioelectrochemical system with no energy input. Presented at the Second International Symposium on Bioremediation and Sustainable Environmental Technologies, Jacksonville, FL.

Kumar, A.K., Reddy, M.V., Chandrasekhar, K., Srikanth, S., Mohan, S.V., 2012. Endocrine disruptive estrogens role in electron transfer: bio-electrochemical remediation with microbial mediated electrogenesis. Bioresour. Technol. 104, 547–556.

Lefebvre, O., Al-Mamun, A., Ng, H.Y., 2008. A microbial fuel cell equipped with a biocathode for organic removal and denitrification. Water Sci. Technol. 58, 881–885.

Li, J., Liu, G.L., Zhang, R.D., Luo, Y., Zhang, C.P., Li, M.C., 2010. Electricity generation by two types of microbial fuel cells using nitrobenzene as the anodic or cathodic reactants. Bioresour. Technol. 101, 4013–4020.

Li, Y., Lu, A.H., Ding, H.R., Jin, S., Yan, Y.H., Wang, C.Q., et al., 2009. Cr(VI) reduction at rutile-catalyzed cathode in microbial fuel cells. Electrochem. Commun. 11, 1496–1499.

Li, Z.J., Zhang, X.W., Lei, L.C., 2008. Electricity production during the treatment of real electroplating wastewater containing Cr^{6+} using microbial fuel cell. Proc. Biochem. 43, 1352–1358.

Liang, P., Fan, M.Z., Cao, X.X., Huang, X., 2009. Evaluation of applied cathode potential to enhance biocathode in microbial fuel cells. J. Chem. Technol. Biotechnol. 84, 794–799.

Liu, H., Ramnarayanan, R., Logan, B.E., 2004. Production of electricity during wastewater treatment using a single chamber microbial fuel cell. Environ. Sci. Technol. 38, 2281–2285.

Logan, B.E., 2008. Microbial Fuel Cells. John Wiley & Sons, New York, NY.

Logan, B.E., Regan, J.M., 2006. Microbial fuel cells—challenges and applications. Environ. Sci. Technol. 40, 5172–5180.

Logan, B.E., Call, D., Cheng, S., Hamelers, H.V.M., Sleutels, Y.H.J.A., Jeremiasse, A.W., et al., 2008. Microbial electrolysis cells for high yield hydrogen gas production from organic matter. Environ. Sci. Technol.(42), 8630–8640.

Lohner, S.T., Tiehm, A., 2009. Application of electrolysis to stimulate reductive PCE dechlorination and oxidative VC biodegradation. Environ. Sci. Technol. 43, 7098–7104.

Lovley, D.R., 2006. Microbial fuel cells: novel microbial physiologies and engineering approaches. Curr. Opin. Biotechnol. 17, 327–332.

Lovley, D.R., 2011. Reach out and touch someone: potential impact of DIET (direct interspecies energy transfer) on anaerobic biogeochemistry, bioremediation, and bioenergy. Rev. Environ. Sci. Biotechnol. 10, 101–105.

Lovley, D.R., Nevin, K.P., 2008. Chapter 23: Electricity production with electricigens. In: Wall, J.D., Harwood, C.S., Demain, A.L. (Eds.), Bioenergy. ASM Press, Washington, DC, pp. 295–306.

Lovley, D.R., Nevin, K.P., 2011. A shift in the current: new applications and concepts for microbe-electrode electron exchange. Curr. Opin. Biotechnol. 22, 441–448.

Lovley, D.R., Coates, J.D., Blunt-Harris, E.L., Phillips, E.J.P., 1996. Humic substances as electron acceptors for microbial respiration. Nature. 382, 445–448.

Lovley, D.R., Fraga, J.L., Coates, J.D., Blunt-Harris, E.L., 1999. Humics as an electron donor for anaerobic respiration. Environ. Microbiol. 1, 89–98.

Luo, H., Liu, G., Zhang, R., Jin, S., 2009. Phenol degradation in microbial fuel cells. Chem. Eng. J. 147, 259–264.

Mehanna, M., Saito, T., Yan, J.L., Hickner, M., Cao, X.X., Huang, X., et al., 2010. Using microbial desalination cells to reduce water salinity prior to reverse osmosis. Energy Environ. Sci. 3, 1114–1120.

Morris, J.M., Jin, S., 2012. Enhanced biodegradation of hydrocarbon-contaminated sediments using microbial fuel cells. J. Hazard. Mater. 213–214, 474–477.

Morris, J.M., Jin, S., Crimi, B., Pruden, A., 2009a. Microbial fuel cell in enhancing anaerobic biodegradation of diesel. Chem. Eng. J. 146, 161–167.

Morris, J.M., Fallgren, P.H., Jin, S., 2009b. Enhanced denitrification through microbial and steel fuel-cell generated electron transport. Chem. Eng. J. 153, 37–42.

Morris, J.M., Jin, S., Fallgren, P.H., Cui, K., Ren, Z., 2011. Enhanced biodegradation of hydrocarbon-contaminated sediments using a modified microbial fuel cell. Presented at the International Symposium on Bioremediation and Sustainable Environmental Technologies, Reno, NV.

Mu, Y., Rozendal, R.A., Rabaey, K., Keller, J., 2009. Nitrobenzene removal in bioelectrochemical systems. Environ. Sci. Technol. 43, 8690–8695.

Park, D.H., Laivenieks, M., Guettler, M.V., Jain, M.K., Zeikus, J.G., 1999. Microbial utilization of electrically reduced neutral red as the sole electron donor for growth and metabolite production. Appl. Environ. Microbiol. 65, 2912–2917.

Park, H.I., Kim, D.K., Choi, Y., Pak, D., 2005. Nitrate reduction using an electrode as direct electron donor in a biofilm-electrode reactor. Proc. Biochem. 40, 3383–3388.

Pham, H., Boon, N., Marzorati, M., Verstraete, W., 2009. Enhanced removal of 1,2-dichloroethane by anodophilic microbial consortia. Water Res. 43, 2936–2946.

Puig, S., Serra, M., Vilar-Sanz, A., Cabre, M., Baneras, L., Colprim, J., et al., 2011. Autotrophic nitrite removal in the cathode of microbial fuel cells. Bioresour. Technol. 102, 4462–4467.

Rabaey, K., Keller, J., 2008. Microbial fuel cell cathodes: from bottleneck to prime opportunity? Water Sci. Technol. 57, 655–659.

Rabaey, K., Boon, N., Hofte, M., Verstraete, W., 2005. Microbial phenazine production enhances electron transfer in biofuel cells. Environ. Sci. Technol. 39, 3401–3408.

Rabaey, K., Butzer, S., Brown, S., Keller, J., Rozendal, R.A., 2010. High current generation coupled to caustic production using a lamellar bioelectrochemical system. Environ. Sci. Technol. 44, 4315–4321.

Rozendal, R.A., Jeremiasse, A.W., Hamelers, H.V.M., Buisman, C.J.N., 2008. Hydrogen production with a microbial biocathode. Environ. Sci. Technol.(42), 629–634.

Rozendal, R.A., Leone, E., Keller, J., Rabaey, K., 2009. Efficient hydrogen peroxide generation from organic matter in a bioelectrochemical system. Electrochem. Commun. 11, 1752–1755.

Shroder, U., 2011. Discover the possibilities: microbial bioelectrochemical systems and the revival of a 100-year-old discovery. J. Solid State Electrochem. 15, 1481–1486.

Strycharz, S.M., Woodard, T.L., Johnson, J.P., Nevin, K.P., Sanford, R.A., Loffler, F.E., et al., 2008. Graphite electrode as a sole electron donor for reductive dechlorination of tetrachloroethene by *Geobacter lovleyi*. Appl. Environ. Microbiol. 74, 5943–5947.

Strycharz, S.M., Glaven, R.H., Coppi, M.V., Gannon, S.M., Perpetua, L.A., Liu, A., et al., 2011. Gene expression and deletion analysis of mechanisms for electron transfer from electrodes to *Geobacter sulfurreducens*. Bioelectrochemistry. 80, 142–150.

Tandukar, M., Huber, S.J., Onodera, T., Pavlostathis, S.G., 2009. Biological chromium(VI) reduction in the cathode of a microbial fuel cell. Environ. Sci. Technol. 43, 8159–8165.

TerHeijne, A., Liu, F., van der Weijden, R., Weijma, J., Buisman, C.J.N., Hamelers, H.V.M., 2010a. Copper recovery combined with electricity production in a microbial fuel cell. Environ. Sci. Technol.(44), 4376–4381.

TerHeijne, A., Strik, D.P.B.T.B., Hamelers, H.V.M., Buisman, C.J.N., 2010b. Cathode potential and mass transfer determine performance of oxygen reducing biocathodes in microbial fuel cells. Environ. Sci. Technol. 44, 7151–7156.

Thrash, J.C., Coates, J.D., 2008. Review: direct and indirect electrical stimulation of microbial metabolism. Environ. Sci. Technol. 42, 3921–3931.

Thrash, J.C., van Trump, J.I., Weber, K.A., Miller, E., Achenbach, L.A., Coates, J.D., 2007. Electrochemical stimulation of microbial perchlorate reduction. Environ. Sci. Technol. 41, 1740–1746.

Virdis, B., Rabaey, K., Yuan, Z.G., Rozendal, R.A., Keller, J., 2009. Electron fluxes in a microbial fuel cell performing carbon and nitrogen removal. Environ. Sci. Technol. 43, 5144–5149.

Virdis, B., Rabaey, K., Rozendal, R.A., Yuan, Z.G., Keller, J., 2010. Simultaneous nitrification, denitrification and carbon removal in microbial fuel cells. Water Res. 44, 2970–2980.

Wagner, R.C., Call, D.I., Logan, B.E., 2010. Optimal set anode potentials vary in bioelectrochemical systems. Environ. Sci. Technol. 44, 6036–6041.

Wang, G., Huang, L.P., Zhang, L.P., 2008. Cathodic reduction of hexavalent chromium [Cr(VI)] coupled with electricity generation in microbial fuel cells. Biotechnol. Lett. 30, 1959–1966.

Wang, Z., Lim, B., Lu, H., Fan, J., Choi, C., 2010. Cathodic reduction of Cu^{2+} and electric power generation using a microbial fuel cell. Bull. Korean Chem. Soc. 31, 2025–2030.

Williams, K.H., Nevin, K.P., Franks, A., Englert, A., Long, P.E., Lovley, D.R., 2010. Electrode-based approach for monitoring *in situ* microbial activity during subsurface bioremediation. Environ. Sci. Technol. 44, 47–54.

Yan, Z., Song, N., Cai, H., Tay, J.H., Jiang, H., 2012. Enhanced degradation of phenanthrene and pyrene in freshwater sediments by combined employment of sediment microbial fuel cell and amorphous ferric hydroxide. J. Hazard. Mater. 199–200, 217–225.

Yunker, S.B., Radovich, J.M., 1986. Enhancement of growth and ferrous iron oxidation rates of *T. ferrooxidans* by electrochemical reduction of ferric iron. Biotechnol. Bioeng. 28, 1867–1875.

Zhang, T., Gannon, S.M., Nevin, K.P., Franks, A.E., Lovley, D.R., 2010. Stimulating the anaerobic degradation of aromatic hydrocarbons in contaminated sediments by providing an electrode as the electron acceptor. Environ. Microbiol. 12, 1011–1020.

17 Microbial Bioremediation: A Metagenomic Approach

Muthuirulan Pushpanathan[a], *Sathyanarayanan Jayashree*[a], *Paramasamy Gunasekaran*[b] *and Jeyaprakash Rajendhran*[a]

[a]Department of Genetics, Centre for Excellence in Genomic Sciences, School of Biological Sciences, Madurai Kamaraj University, Madurai, India, [b]Thiruvalluvar University, Vellore, India

17.1 Introduction

Environmental pollution is the buildup and accumulation of toxic heavy metals in the air, water, and land that reduce the ability of the contaminated sites to support life. The rise in human population density and anthropogenic activity has led to degradation of the Earth's surface through misuse of environmental resources and improper disposal of wastes. In addition, the advancements in science and technology as well as the increase in industry have led to an increase in the dumping of wastes, ranging from raw sewage to nuclear waste, into the environment, which poses a serious problem for the survival of humanity. The conventional methods of waste disposables such as digging hole and dumping wastes, heat incineration, and chemical decomposition of contaminants were found to be more complex, uneconomical, and also lack public acceptance (Karigar and Rao, 2011). Microbial bioremediation is an alternative, cost-effective and eco-friendly technology that provides sustainable ways to clean up contaminated environments. Recently, a wide variety of organisms such as bacteria, fungi, algae, and plants with efficient bioremediating properties were successfully employed for efficient removal of toxicants from the polluted environments (Vidali, 2001; Leung, 2004).

The microbial bioremediation process mainly depends on the biodegradation of pollutants by the enzymatic activity of microbial enzymes, which leads to the bioconversion of toxic pollutants to nontoxic or less toxic substances. The use of indigenous microbes for bioremediation is highly advantageous, because their growth is highly

Microbial Biodegradation and Bioremediation. DOI: http://dx.doi.org/10.1016/B978-0-12-800021-2.00017-0

influenced by pH, temperature, oxygen, soil structure, moisture, and appropriate level of nutrients, poor bioavailability of contaminants and presence of other toxic compounds, which limit the use of other exogenous organisms for the treatment of polluted sites (Dana and Bauder, 2011). Traditional culture-based approaches have resulted in identification, biochemical and genetic characterization of superior metal-resistant and/or -accumulating or -transforming microorganisms. Molecular approaches have also provided insight into the microbial interaction with toxic heavy metals, which revealed that interaction of microbes with toxic metals is preferential for potential treatment of polluted sites (Gadd, 2010). The molecular approaches involving 16S rRNA gene-based identification of microbes and analysis of genes involved in bioremediation provide in-depth knowledge of microbes involved in bioremediation as well as their functional capabilities in remediating the polluted environment. The culture-based techniques have provided only limited information about the indigenous microbial diversity in contaminated sites. The detection and enumeration of microbial diversity by using conventional 16S rDNA gene cloning and sequencing represent only the predominant bacteria phyla prevailing in the contaminated sites. The success of bioremediation failed for reasons such as lack of information on the bacterial species prevailing in the contaminated sites; lack of understanding of the metabolic capabilities, especially the factors controlling the growth and activity of microorganisms in the contaminated environment; and lack of understanding of how indigenous microbial communities respond to changes in environmental conditions (Lovley, 2003).

Microorganisms exist on every part of our planet and contribute to its geochemistry, the cycling of elements, and the breakdown of wastes. This ability of microorganisms relies on their huge genetic and metabolic diversity, which affords great potential for their application toward biotechnological purposes. Due to limitations associated with existing culturable techniques, a new approach known as "metagenomics" has emerged to explore the genetic resource of both culturable and unculturable microorganisms from any environment (Handelsman et al., 1998). A metagenome represents a mixture of microbial genomes extracted directly from an environmental sample. The metagenomic approach circumvents the isolation and cultivation of individual microorganisms, and the metagenomic DNA extracted directly from the environment is used to identify the genes involved in bioremediation. Prior to screening desired gene of interest, the environmental sample is subjected to enrichment procedure inorder to increase the abundance ratio of microbes with desired function. The enriched culture is then used for further metagenome screening procedure to identify genes based on their function (Knietsch et al., 2003). With the recent development of efficient cloning vectors such as bacterial artificial chromosomes (BACs) and cosmids, together with improved DNA isolation techniques and robust high-throughput screening methodologies, it is now possible to identify and express large fragments of DNA and subsequently screen large clone libraries using a functional-based approach (Lorenz and Eck, 2005). The present chapter reviews the metagenomic-based strategies for bioremediation of contaminated sites that will provide new perspectives on environmental pollution.

17.2 Microbial Bioremediation: Culture-Independent Approach

Microorganisms are the most abundant and diverse group of life on Earth. They play a major role in the biogeochemical cycling of compounds highly essential for the functioning of the ecosystem. The characterization of microbial communities in contaminated sites remains unexplored due to huge microbial diversity and the uncultivable status of the vast majority of organisms prevailing in contaminated sites. Metagenomic approaches, together with high-throughput sequencing technologies, have revealed the diversity, adaptation, and evolution of microorganisms surviving in contaminated sites. Molecular analysis of environmental communities have revealed that only $<1\%$ of the total number of prokaryotic species present in given environmental samples are readily cultivable under standard laboratory conditions, and therefore, a majority of microbes in the environment are not readily accessible for basic research of biotechnological applications (Handelsmann, 2004). The recent development of culture-independent molecular tools has provided insight into the microbial ecology of contaminated sites, which could be exploited for remediation of polluted environments.

17.3 Genome and Target Gene Enrichment

The active subset of microbial populations in the contaminated environments could be assessed by genome enrichment followed by metagenomic analysis. Stable isotope probing (SIP) technology can be used for selective enrichment of macromolecules such as DNA, RNA, and phospholipid-derived fatty acids (Neufeld et al., 2007). SIP is based on the incorporation of isotopes (^{13}C, ^{15}N and ^{18}O) labelled substrates into the environmental sample, followed by metagenomic DNA extraction and separation of labelled (heavier) and unlabelled (lighter) nucleic acid by density gradient centrifugation. The "heavier" labeled fraction was used for construction of a metagenomic DNA library. The labeled DNA from the extracted DNA pool can be separated by immune capture. Furthermore, microbial communities exposed to high concentrations of xenobiotics for longer durations are expected to be naturally enriched in target genes of interest. Enrichment of microbes in selective media that contains pollutant as the sole energy source would selectively increase the active subset of pollutant degrading microbial populations, which could be used for construction of metagenomic DNA library (Entcheva et al., 2001). Therefore, enrichment of target genes prior to library construction would increase the number of positive hits during subsequent screening procedures.

17.4 Metagenome Extraction and Library Construction from Contaminated Sites

In culture-independent methods, the metagenomic DNA was extracted directly from the contaminated environment. There are several methods for extracting DNA from environmental samples. Commercial kits are also available for extraction and purification of environmental DNA from the soil environment, which are easy to use with considerable reproducibility. Most researchers have reported the advantageous features of their respective methods over the previously reported methods. However, none of the methods reported hitherto are universally applicable, and every type of environmental sample requires optimization of DNA extraction methods (Rajendhran and Gunasekaran, 2008). Extraction of total metagenomic DNA from the contaminated sites is necessary to represent all microbial genomes. Two different strategies have been used for isolation of metagenomic DNA from the environmental samples: cell recovery and direct lysis. In general, the metagenome extraction processes involve physical or chemical methods. The mechanical disruption of cells by bead beating recovered more diversity compared to that accomplished by chemical methods, whereas the chemical methods resulted in isolation of high-molecular-weight DNA suitable for construction of a large insert metagenomic DNA library (Cowan et al., 2005). Construction of small inserts metagenomic DNA libraries make use of plasmid cloning vectors that requires smaller DNA fragments to be cloned. However, if the gene encoding chemical compounds are larger, the metagenomic clone libraries with large inserts are desirable. Large insert metagenomic DNA libraries makes use of cloning vectors such as cosmid, fosmid or Bacterial artificial chromosomes (BAC) that requires high molecular weight to be cloned (Hildebrand et al., 2004). Therefore, the choice of vectors and the type of construction of metagenomic DNA libraries are highly dependent on the size of the genes that encode molecules or chemical compounds that have to be screened.

17.5 Metagenomic Strategies for Accessing Biodegradative Genes from Contaminated Sites

The strategies and outcome of bioremediation in open systems or confined environments are highly dependent on a variety of physicochemical and biological factors that need to be assessed and monitored continuously. In particular, microorganisms are key players in bioremediation applications, yet their catabolic potential and their dynamics *in situ* remain unexplored. Several screening strategies have been employed for accessing biodegradative genes from metagenome of polluted environment, which include a function-based and sequence-based screening methods (Figure 17.1).

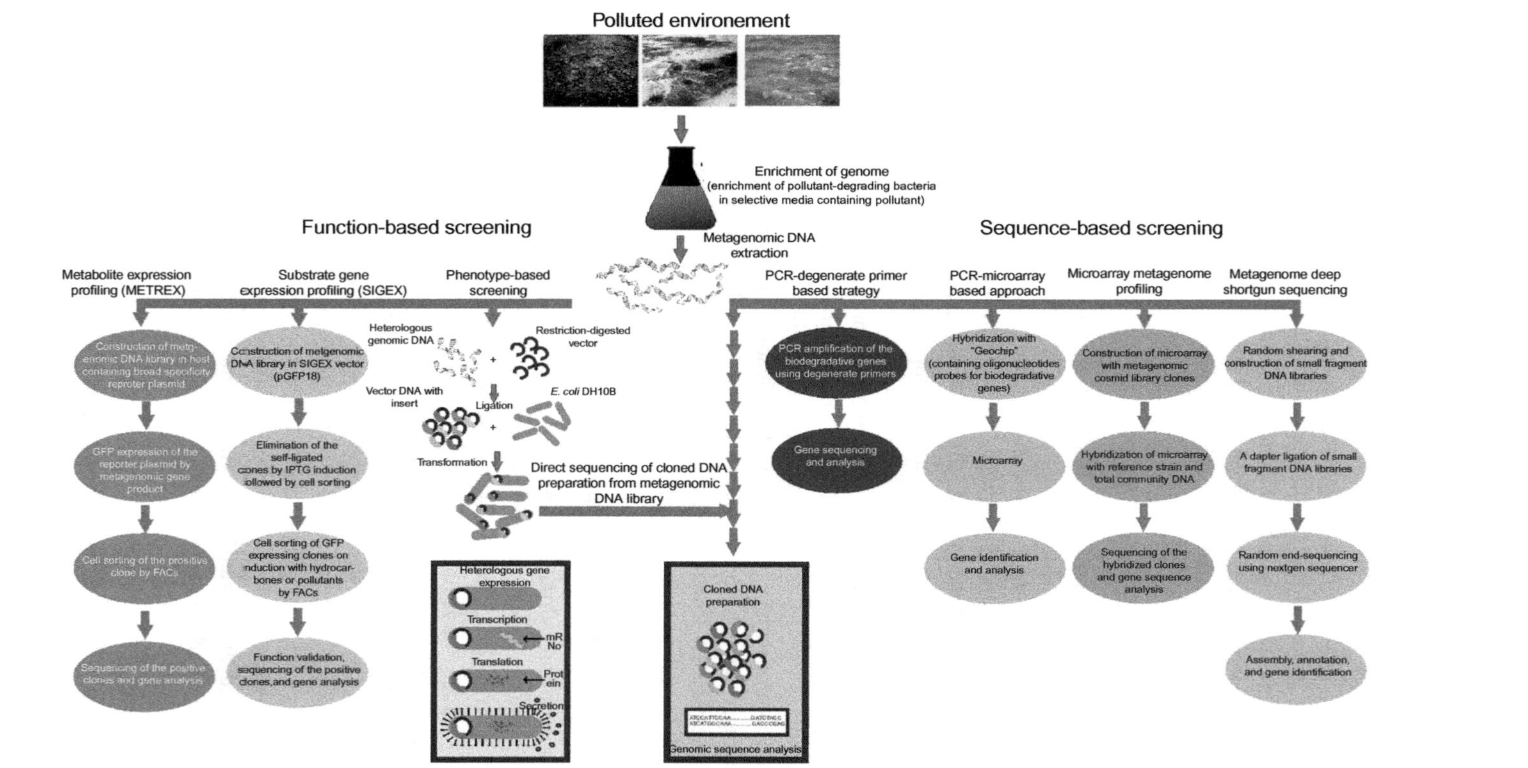

Figure 17.1 Metagenomic strategies for accessing biodegradative genes from polluted sites: Metagenomic screening strategies for accessing biodegradative genes from polluted sites include (a) function-based screening: phenotype-based screening, substrate-induced gene expression profiling, metabolite expression profiling, and (b) Sequence-based screening: PCR-degenerate primer-based screening, microarray metagenome profiling, metagenome shotgun sequencing technologies.

17.5.1 Function-Based Screening of Biodegradative Genes from Contaminated Sites

The function-based approach is not dependent on sequence information or sequence similarity to known genes. This is the only approach that has the potential to discover new classes of genes with known or new functions. Function-based screening methods of biodegradative genes include *phenotype-based screening*, *substrate-induced gene expression profiling* (SIGEX), and *metabolite expression profiling* (METREX).

17.5.1.1 Phenotype-Based Screening

In this approach, a metagenomic DNA library was constructed using metagenomes isolated directly from the polluted sites, and the library was screened for clones exhibiting the desired phenotype, particularly with respect to the genes encoding catabolic enzymes or chemical compounds involved in remediation of the polluted sites (Henne et al., 2000; Rondon et al., 2000). A major advantage of this approach is that the positive clone will certainly harbor complete genes necessary for expression of the desired phenotype. Three different types of function-based approaches have been employed for screening of metagenomic libraries: (1) direct detection of specific phenotypes of individual clones, (2) heterologous complementation of host strains or mutants, and (3) induced gene expression (Simon and Daniel, 2009). The function-based approach has resulted in identification of biodegradative enzymes from metagenomic DNA libraries such as nitrilases (Robertson et al., 2004), extra dioldioxygenases (EDO) (Suenaga et al., 2007), naphthalene dioxygenases (NDO) (Ono et al., 2007), and styrene monooxygenase (Van Hellemond et al., 2007).

17.5.1.2 Metabolite-Regulated Expression Profiling

METREX is a high-throughput intracellular screening strategy that has greater sensitivity in screening poorly expressed gene products by using fluorescence-activated cell sorters (FACs) that can rapidly screen millions of clones. Earlier, it was used to screen molecules that mimic quorum-sensing signal molecules (Williamson et al., 2005). The same strategy can be used to screen biodegradative genes from contaminated sites. This strategy will make use of the broad-specificity reporter plasmid with genes that stimulate transcription of the desired genes and the inducible promoter linked with the reporter green fluorescent protein. Since, the host cells contains both metagenomic DNA and the reporter plasmid, the poorly expressed gene products could be rapidly indentified based on GFP expression of the reporter plasmid and sorted using fluorescence activated cell sorter (FACs). The positive clones harboring insert DNA are sequenced and annotated to assign its function.

17.5.1.3 Substrate-Induced Gene Expression Profiling

SIGEX is a high-throughput strategy, which uses an operon trap vector for screening of substrate-inducible genes. Initially, the metagenomic DNA library was constructed with the SIGEX vector (pGFP18), and the self-ligated clones showing GFP fluorescence on Isopropyl β-D-1-thiogalactopyranoside (IPTG) induction were eliminated. Further, the nonfluorescing cells harboring the insert DNA were sorted on IPTG induction and subsequently used for screening of genes inducible by substrate. This strategy could also be used to screen the biodegradative genes of interest by circumventing colorimetric assays and conventional plating methods. It does not require the modified substrates that are used in colorimetric screenings, which are toxic, cause side effects, and are more expensive than unmodified substrates. Recently, the application of this approach has resulted in identification of a phenol-degradative operon (the *pox* operon) (Hino et al., 1998), genes involved in benzoate and catechol pathways (Rothmel et al., 1990; Cowles et al., 2000), and genes encoding enzymes for aromatic-hydrocarbon transformation such as BZO032 (decarboxylase), BZO062 (dehydrogenase), BZO071 (cytochrome P450), BZO135 (tyrosine lyase), and NAP3 (hydroxylase) (Uchiyama et al., 2005). The application of SIGEX has limitations such as that this strategy is sensitive to the structure and orientation of genes with desired traits. The catabolic genes that are expressed constitutively are missed during screening procedure. In addition, SIGEX cannot detect any active clones with desired catabolic genes cloned in the direction opposite to gfp.

17.5.2 Sequence-Based Screening of Biodegradative Genes from Contaminated Sites

The sequence-based approach is dependent on sequence information, in which the biodegradative genes are screened based on their conserved nucleotide sequences. Sequence-based screening methods of biodegradative genes include polymerase chain reaction (PCR)-degenerate primer-based screening, microarray metagenome profiling, and metagenome shotgun sequencing technologies.

17.5.2.1 PCR-Degenerate Primer-Based Screening

In this approach, the desired gene fragments were identified based on the conserved nucleotide sequences of known catabolic enzymes. Degenerate primers or hybridization probes were designed based on the consensus sequence of the known catabolic enzymes, which could be used for direct amplification of the gene encoding the enzymes of similar function from the environmental metagenome. Subsequently, the identified genes will be validated by their expression in culturable heterologous expression systems (Hildebrand et al., 2004). The major drawback of this approach is that the degenerate primers or probes designed based on existing gene sequences of desired function resulted in a bias in genes similar to already-existing genes and also eliminated the homologous genes encoding novel catabolic enzymes. The other major drawback of this approach is the partial recovery of target gene fragments by PCR reaction, whose expression requires the additional steps to recover whole genes necessary for their functional activity (Uchiyama and Watanabe, 2006).

17.5.2.2 PCR-Based and Microarray Approaches

In recent years, massive parallel screening of metagenomic libraries for biodegradative genes has been possible using microarray technology. This approach makes use of a "Geochip" that is aimed at providing direct linkages between biogeochemical processes and functional activities of microbial communities present in different environments. The chip contains 24,243 oligonucleotide probes covering >10,000 genes involved in metal reduction/resistance and degradation of organic contaminants (He et al., 2007). This approach has limited the identification of novel genes or proteins from the environment as because the primers and target-specific probes used in the screening procedure were designed only based on the known gene families or proteins available in the database.

17.5.2.3 Microarray Metagenome Profiling

In this approach, instead of hybridizing metagenomic DNA to microarrays containing a known set of probes, the metagenomic DNA itself can be used to construct the microarray (Sebat et al., 2003). For this, 642 cosmid libraries derived from groundwater samples were used to build microarray containing ~1 kb PCR products amplified from the cosmid insert plus a set of 16S rDNA controls. The microarray was hybridized with DNA of 14 reference strains isolated from the groundwater sample and from groundwater total community. The clones hybridized specifically to community genomic DNA and not with reference strains are suggested to be derived from microbes that were failed to isolate in pure culture. The insert DNA of the hybridized clone was sequenced, and it could be assigned to functions of ecological importance. This approach does not require any prior knowledge about the microbial communities prevailing in the environments and affords great potential to uncover novel biodegradative genes from metagenomic sources.

17.5.2.4 Metagenome Deep Shotgun Sequencing

Metagenome deep shotgun sequencing of the contaminated sites provided a global picture of the microbial ecosystems and a wealth of genomic information on unculturable microbes inhabiting these sites and their metabolic pathways. This approach involves isolation of metagenomic DNA from the polluted sites and construction of small insert metagenomic DNA library by random shearing of metagenomic DNA followed by random end-sequencing employing a Next-Generation sequencer such as 454 Life Sciences Genome Sequencer FLX. The data that contains sequence of each fragments are assembled into several hundred contigs and scaffolds. The assembled sequence reads are then annotated to assign into their functions. Deep shotgun sequencing of environmental metagenomes has resulted in identification of a 79-kb catabolic gene cluster for anaerobic benzoate and hydroxybenzoate degradation in the metagenome of the anoxic seafloor of the Black Sea (Kube et al.,

2005). The approach circumvents cloning and library construction and achieves large biodegradative gene clusters from environmental metagenomes.

A large set of databases have been developed for sequence-based approaches and most of the metagenomic datasets in this database are being generated from polluted environments. There are tools available for metagenome analysis including the Integrated Microbial Genome/Microbiome (IMG/M) (Markowitz et al., 2008), which serves as a platform for analyzing the functional capability of microbial communities. This is based on the metagenome sequences of the microbial communities in the environment that allow the comparative analysis of genome and metagenome sequences available on the Joint Genome Institute (JGI)-Integrated Microbial Genomes system, MetaBioME (Sharma et al., 2009). MetaBioME is a resource specifically aimed at novel enzyme discovery from metagenomic data, and Orphelia is a tool for discovering Open Reading Frames (ORFs) in short metagenomic sequences of unknown phylogeny (Hoff et al., 2009). The major drawback of metagenomic data analysis is the lack of reference sequences and genomes. The increased data and tools available should gradually minimize this problem.

17.6 Microbial Community Profiling of Contaminated Sites by Direct Sequencing

For the past few decades, researchers have primarily relied on 16S rRNA sequences for identification and classification of bacteria. Recent development of new sequencing technologies allows us to explore the total microbial community prevailing in the contaminated sites and has revolutionized field of metagenomics by circumventing the need for cloning and library construction (Figure 17.2). Pyrosequencing technology is the first alternative to the conventional Sanger method for DNA sequencing. It has potential advantages over Sanger sequencing technology such high accuracy, flexibility, high throughput parallel processing, time consuming and cost effective (Droege and Hill, 2008). To date, the pyrosequencing approach has been used widely to explore the microbial communities of different environmental samples. The application of this approach in contaminated sites will provide in-depth knowledge of the indigenous microbial community prevailing in that polluted environments that could be exploited for different bioremediation strategies such as *in situ* and *ex situ* bioremediation. Investigation of microbial community composition and diversity at a diesel-contaminated site using a pyrosequencing approach revealed the occurrence of high relative abundances of bacterial phyla such as *Proteobacteria*, *Firmicutes*, *Actinobacteria*, *Acidobacteria*, and *Chloroflexi* (Sutton et al., 2013). Polybrominated diphenyl ethers (PBDEs) are hydrophobic, semivolatile, and toxic compounds that are highly resistant to microbial degradation and possess strong adsorption on sediments. These residues are bioaccumulated and biomagnified in the environment and have been grouped into

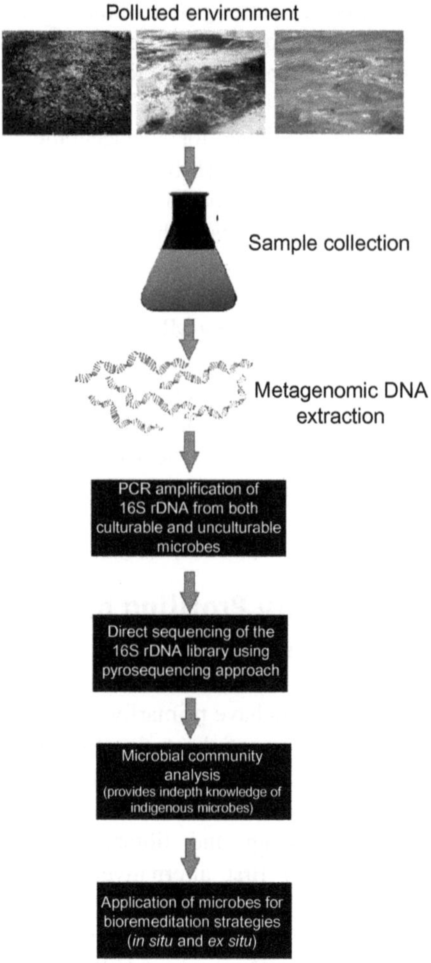

Figure 17.2 Microbial community analysis of the polluted sited by direct sequencing approach: Pyrosequencing of 16S rDNA amplified from both the culturable and unculturable microbes of the polluted sites provides in-depth knowledge of indigenous microbes prevailing in that environment that could be explored for *in situ* or *ex situ* bioremediation strategies.

new persistent organic pollutant (POPs). Recently, the high-throughput multiplex barcoded pyrosequencing approach has revealed that the microbial community structures and compositions involved in PBDEs degradation are manipulated under different electron donor–amending conditions (Xu et al., 2012). Investigation of indigenous bacterial communities in permafrost soils contaminated by oil spills revealed the presence of bacterial phyla such as *Proteobacteria*, *Acidobacteria* (esp. Acidobacteriaceae), *Actinobacteria* (esp. Intrasporangiaceae), *Bacteroidetes* (esp. Sphingobacteria and Flavobacteria), and *Chloroflexi* (esp. Anaerolineaceae), which were found to be common in all oil spill sites (Yang et al., 2009).

17.7 Conclusion

Microorganisms plays an essential role in the functioning of the ecosystem. Bioremediation technologies utilize the metabolic potential of microorganisms to clean up contaminated sites. The traditional method of pollutant degradation involves the isolation of one or more organisms capable of degrading target pollutants in the environment. These conventional isolation methods have explored only a small fraction of the diverse pollutant-degrading microorganisms in the environment. Only $<1\%$ of microbes in a soil environment can be cultured by conventional culturable methods, and thus, a large fraction of a diverse group of microorganisms in the environment defy cultivation in pure culture. A comprehensive understanding of both the community structure and its function is required to achieve an effective and reliable strategy to clean up contaminants from polluted sites. Metagenomic approaches, together with emerging high-throughput sequencing technologies have provided insight into the complex interactions between different microbial communities and their metabolic potentials, which could be exploited in various bioremediation strategies for efficient removal pollutants from the environment.

References

Cowan, D., Meyer, Q., Stafford, W., Muyanga, S., Cameron, R., Wittwer, P., 2005. Metagenomic gene discovery: past, present and future. TRENDS in Biotechnology. 23, 321–329.

Cowles, C.E., Nichols, N.N., Harwood, C.S., 2000. BenR, a XylShomologe, regulates three different pathways of aromatic acid degradation in *Pseudomonas putida*. J. Bacteriol. 182, 6339–6346.

Dana, L.D., Bauder, J.W., 2011. A General Essay on Bioremediation of Contaminated Soil. Montana State University, Bozeman, MT.

Droege, M., Hill, B., 2008. The Genome Sequencer FLX System—longer reads, more applications, straight forward bioinformatics and more complete data sets. J. Biotechnol. 136, 3–10.

Entcheva, P., Liebl, W., Johann, A., Hartsch, T., Streit, W.R., 2001. Direct cloning from enrichment cultures, a reliable strategy for isolation of complete operons and genes from microbial consortia. Appl. Environ. Microbiol. 67, 89–99.

Gadd, G.M., 2010. Metals, minerals and microbes: geomicrobiology and bioremediation. Microbiology. 156, 609–643.

Handelsman, J., 2004. Metagenomics: application of genomics to uncultured microorganisms. Microbiol. Mol. Biol. Rev. 68, 669–685.

Handelsman, J., Rondon, M.R., Brady, S.F., Clardy, J., Goodman, R.M., 1998. Molecular biological access to the chemistry of unknown soil microbes: a new frontier for natural products. Chem. Biol. 5, 245–249.

He, Z., Gentry, T.J., Schadt, C.W., Wu, L., Liebich, J., Chong, S.C., et al., 2007. GeoChip: a comprehensive microarray for investigating biogeochemical, ecological and environmental processes. ISME J. 1 (1), 67–77.

Henne, A., Schmitz, R.A., Bomeke, M., Gottschalk, G., Daniel, R., 2000. Screening of environmental DNA libraries for the presence of genes conferring lipolytic activity on *Escherichia coli*. Appl. Environ. Microbiol. 66 (7), 3113–3116.

Hildebrand, M., Waggoner, L.E., Lim, G.E., Sharp, K.H., Ridley, C.P., Haygood, M.G., 2004. Approaches to identify, clone, and express symbiont bioactive metabolite genes. Nat. Prod. Rep. 21, 122–142.

Hino, S., Watanabe, K., Takahashi, N., 1998. Phenol hydroxylase cloned from *Ralstonia eutropha* strain E2 exhibits novel kinetic properties. Microbiology. 144, 1765–1772.

Hoff, K.J., Lingner, T., Meinicke, P., Tech, M., 2009. Orphelia: predicting genes in metagenomic sequencing reads. Nucleic Acids Res. 37, 101–105.

Karigar, C.S., Rao, S.S., 2011. Role of microbial enzymes in the bioremediation of pollutants: a review. Enzyme Res. 2011, Article ID 805187, 1–11.

Knietsch, A., Waschkowitz, T., Bowien, S., Henne, A., Daniel, R., 2003. Construction and screening of metagenomic libraries derived from enrichment cultures: generation of a gene bank for genes conferring alcohol oxidoreductase activity on *Escherichia coli*. Appl. Environ. Microbiol. 69, 1408–1416.

Kube, M., Beck, A., Meyerdierks, A., Amann, R., Reinhardt, R., Rabus, R., 2005. A catabolic gene cluster for anaerobic benzoate degradation in methanotrophic microbial Black Sea mats. Syst. Appl. Microbiol. 28, 287–294.

Leung, M., 2004. Bioremediation: techniques for cleaning up a mess. J. Biotechnol. 2, 18–22.

Lorenz, P., Eck, J., 2005. Metagenomics and industrial applications. Nat. Rev. Microbiol. 3, 510–516.

Lovley, D.R., 2003. Cleaning up with genomics: applying molecular biology to bioremediation. Nat. Rev. Microbiol. 1, 35–44.

Markowitz, V.M., Ivanova, N.N., Szeto, E., Palaniappan, K., Chu, K., Dalevi, D., et al., 2008. A data management and analysis system for metagenomes. Nucleic Acids Res. 36, 534–538.

Neufeld, J.D., Dumont, M.G., Vohra, J., Murrell, J.C., 2007. Methodological considerations for the use of stable isotope probing in microbial ecology. Microb. Ecol. 53 (3), 435–442.

Ono, A., Miyazaki, R., Sota, M., Ohtsubo, Y., Nagata, Y., Tsuda, M., 2007. Isolation and characterization of naphthalene-catabolic genes and plasmids from oil-contaminated soil by using two cultivation-independent approaches. Appl. Microbiol. Biotechnol. 74 (2), 501–510.

Rajendhran, J., Gunasekaran, P., 2008. Strategies for accessing soil metagenome for desired applications. Biotechnol. Adv. 26, 576–590.

Robertson, D.E., Chaplin, J.A., DeSantis, G., Podar, M., Madden, M., Chi, E., et al., 2004. Exploring nitrilase sequence space for enantio selective catalysis. Appl. Environ. Microbiol. 70, 2429–2436.

Rondon, M.R., August, P.A., Bettermann, A.D., Brady, S.F., Grossman, T.H., Liles, M.R., et al., 2000. Cloning the soil metagenome: a strategy for accessing the genetic and functional diversity of uncultured microorganisms. Appl. Environ. Microbiol. 66, 2541–2547.

Rothmel, R.K., Aldrich, T.L., Houghton, J.E., Coco, W.M., Ornston, L.N., Chakrabarty, A. M., 1990. Nucleotide sequencing and characterization of *Pseudomonas putida* catR: a positive regulator of the *catBC* operon is a member of the LysR family. J. Bacteriol. 172, 922–931.

Sebat, J.L., Colwell, F.S., Crawford, R.L., 2003. Metagenomic profiling: microarray analysis of an environmental genomic library. Appl. Environ. Microbiol. 69 (8), 4927–4934.

Sharma, V.K., Kumar, N., Prakash, T., Taylor, T.D., 2009. MetaBioME: a database to explore commercially useful enzymes in metagenomic datasets. Nucleic Acids Res. 38, 468–472.

Simon, C., Daniel, R., 2009. Achievements and new knowledge unraveled by metagenomic approaches. Appl. Environ. Microbiol. 85, 265–276.

Suenaga, H., Ohnuki, T., Miyazaki, K., 2007. Functional screening of a metagenomic library for genes involved in microbial degradation of aromatic compounds. Environ. Microbiol. 9 (9), 2289–2297.

Sutton, N.B., Maphosa, F., Morillo, J.A., Abu Al-Soud, W., Langenhoff, A.A., Grotenhuis, T., et al., 2013. Impact of long-term diesel contamination on soil microbial community structure. Appl. Environ. Microbiol. 79 (2), 619–630.

Uchiyama, T., Watanabe, K., 2006. Improved inverse PCR scheme for metagenome walking. Biotechniques. 41, 183–188.

Uchiyama, T., Abe, T., Ikemura, T., Watanabe, K., 2005. Substrate-induced gene expression screening of environmental metagenome libraries for isolation of catabolic genes. Nat. Biotechnol. 23, 88–93.

Van Hellemond, E.W., Janssen, D.B., Fraaije, M.W., 2007. Discovery of a novel styrene monooxygenase originating from the metagenome. Appl. Environ. Microbiol. 73 (18), 5832–5839.

Vidali, M., 2001. Bioremediation. An overview. Pure Appl. Chem. 73 (7), 1163–1172.

Williamson, L.L., Borlee, B.R., Schloss, P.D., Guan, C., Allen, H.K., Handelsman, J., 2005. Intracellular screen to identify metagenomic clones that induce or inhibit a quorum-sensing biosensor. Appl. Environ. Microbiol. 71 (10), 6335–6344.

Xu, M., Chen, X., Qiu, M., Zeng, X., Xu, J., Deng, D., et al., 2012. Bar-coded pyrosequencing reveals the responses of PBDE-degrading microbial communities to electron donor amendments. PLoS One. 7 (1), e30439.

Yang, S.Z., Jin, H.J., Wei, Z., He, R.X., Ji, Y.J., Li, X.M., et al., 2009. Bioremediation of oil spills in cold environments: a review. Pedosphere. 19 (3), 371–381.

18 *In Silico* Approach in Bioremediation

Puneet Kumar Singh[a], Jahangir Imam[a,b] and Pratyoosh Shukla[a,]*

[a]Enzyme Technology and Protein Bioinformatics Laboratory, Department of Microbiology, Maharshi Dayanand University, Rohtak, Haryana, India, [b]Central Rainfed Upland Rice Research Station (CRRI), Hazaribag, Jharkhand, India

18.1 Introduction

The *in silico* approach not only simplifies the problem to be carried out *in vitro* but also provides novel ways to conduct the task. *In silico* approaches are applied in many fields ranging from drug discoveries, genetic engineering, phylogenetic, and protein engineering. In microbiology massive data are generated related to microorganisms. These data are being analyzed by *in silico* techniques i.e. pathway analysis. These mechanisms could be used to degrade different pollutants as well, which will be beneficial for the environment.

The *in silico* approach provides an insight to study the biodegradation process of contaminants by microorganisms, it also helps to understand the biochemical reaction taking place inside the microorganism during bioremediation process. The kinetic constant determined by the kinetic degradation of some reactive textile dyes with the help of a mathematical approach describes the bioremediation process carried out by chemical kinetics (Kandelbaue et al., 2004; Cristovao et al., 2008). Along with the above process, an electrostatic method is also applied to effectively determine the redox potential for toxic chemical pollutants (Brown et al., 2008).

Bioinformatics offers a tool to address questions regarding involvement of molecular mechanisms in mineralization pathways. Mineralization involves the study of transcript structures and their expression through high-throughput techniques such as microarray and mass spectroscopy. We will be discussing the

*Corresponding Author: pratyoosh.shukla@gmail.com

Microbial Biodegradation and Bioremediation. DOI: http://dx.doi.org/10.1016/B978-0-12-800021-2.00018-2

different *in silico* approaches in this chapter that could play an important role in bioremediation processes. Genome-sequencing and metabolic models bring together every aspects of an organism through integrating genome explanation and biochemical comprehension to reconstruct a mathematical model (Palsson, 2006). Information of biochemical and metabolic reaction of a cell could be obtained from mathematical model (Durot et al., 2009).

The approach consists of random mutation and selection, as well as metabolic engineering based on the limited knowledge could not help to cope up with present situation. Simulation literally means mimicking the original environment into the computers (Lee et al., 2011). Here several metabolic reactions are made to run simultaneously, so that it can provide the real outcome into the cell. GSMM reconstruction is advanced as compared to other biochemical networks, such as transcriptional and translational networks and transcriptional regulatory networks. Mineralization involves the study of transcript structure and their expression through high-throughput techniques such as microarray and mass spectroscopy (Figure 18.1). We will discuss different bioinformatics ways in this chapter that could play role in bioremediation process.

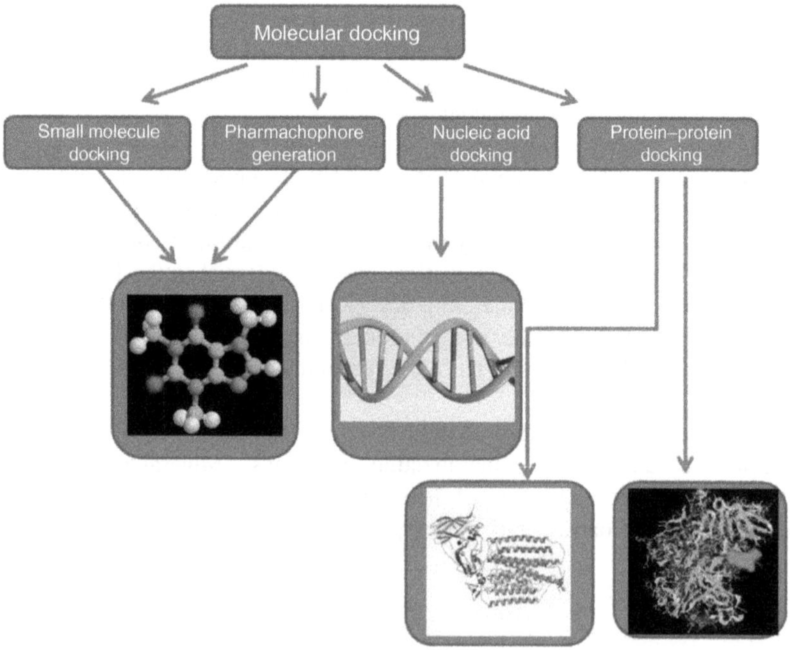

Figure 18.1 Various molecular docking approaches.

18.2 Microorganisms in Bioremediation

Microorganisms utilize their metabolic capabilities to clean up contaminants in the environment. Moreover, they do this under nonsterile condition with many other microorganisms. The microorganisms that degrade the pollutants, as well as other microorganisms, also participate in or affect the process. There are several types of pollutants in the environment which have been eradicated using microorganisms. Methane generally evolves from soil sediments and acts as a greenhouse gas. Methanotrophs reduces the emission of methane. The α-protobacterium has appeared as an important microorganism in the bioremediation of spilled oil in this pseudomonas was group with gamma, and other protobacterium. Metal toxicity is also a serious problem for the environment. α-protobacteria can detoxify metal toxicity, it contain the detoxification machineries that are considered to be useful for the metal bioremediation process. Exogenous microorganisms are introduced in the environment as one way to accelerate bioremediation (Watanabe, 2001).

18.3 Generation of a Biodegradation Pathway

Microorganisms exhibit their potential to use pollutants for their growth. The adaptation of these microorganisms is one of the key factors in reducing the concentration of pollutants (Schroll et al., 2004; van der Meer, 2008). Currently, genetically modified microorganisms are employed to overtake the charge of wild type microorganisms. They show more efficiency due to modifications in their genetic makeup. They have acquired with modified metabolic pathways, stable catabolic activity, and a diverse range of substrates (Pieper et al., 2004). Novel biodegradation pathways could be derived which would help in understanding the metabolism of the substrate, which acts as a pollutant in the environment. There are several pathway prediction tools, such as Pathway Prediction System, Meta (Ellis et al., 2008), and BNICE. The former two rely on the database of biotransformation that occurs during the reaction or cellular process, whereas the latter one generates the novel biochemical reaction.

18.3.1 Constraint-Based In Silico *Modeling of Metabolism*

Metabolic capabilities can be studied using constraint-based metabolic modeling, which also describes and investigates the metabolic flux distributions of genome-scale biochemical networks, where kinetic information is required. By transforming the stoichiometric coefficients of all reactions into one organism-specific numerical matrix (*S*), network topology and capability characteristics can be studied. Later these can be conveniently analyzed mathematically. In this way, the constraints under which the network operates could be defined. Assuming steady-state mass balance for all metabolites, this results in

$$S \times \vec{v} = 0$$

$$\vec{vlb} \leq \vec{v} \leq \vec{vub}$$

where $\vec{v}$ (mmol gDW/L/h, where DW stands for dry weight) is the flux vector describing the activity of internal and external metabolic fluxes, constrained by lower (lb) and upper (ub) bounds. Constraint-based modeling applies Flux balance analysis (FBA) as a standard technique. FBA is a widely used constraint-based modeling approach. The outcome (feasible limits of metabolic flux distribution) could be obtained by calculating the physicochemical restrictions and optimizing a metabolic purpose (e.g., maximizing growth) instead of endeavoring to calculate a strict solution using detailed kinetic models. For a constraint-based model, a stoichiometric matrix S $(m \times n)$ with m metabolites and n reactions representing a metabolic reaction network must be constructed.

18.4 Models for Bioremediation

18.4.1 Dynamic Cell Model

The dynamic cell model segregates metabolic reactions into one that occurs at equilibrium (fast reactions) or at a finite rate (slow reactions). A single-substrate isomerization reaction occurs by the fast formation of an enzyme–substrate complex ($S + E \xleftrightarrow{Q_f} ES$) and by a slow dissociation ($ES \xleftrightarrow{k,kQ_s} P + E$), where Q_f is the equilibrium constant for the fast reaction and k and kQ_s are the backward and forward rate constants for the slow reactions, respectively. With distinctive natural subsurface conditions, the combined actions of acetate kinase (AK; EC 2.7.2.1) and phosphate acetyltransferase perform the oxidation of acetate in *Geobacter* and are coupled to the reduction of Fe(III) (Lovley et al., 2004), which takes place on the extracellular membrane (Leang et al., 2003). The kinetic cell model covers the uptake of acetate and its incorporation into the biomass by means of gluconeogenesis (Gebhardt et al., 1985). Two compartments were considered intracellular and extracellular. The extracellular compartment was explained for species concentrations that represented environmental conditions, while the intracellular one was indicated for enzymatic reactions and resource allocation in cellular metabolism. Reactions are specified below: where c_i is the concentration of species i, Q_l is the equilibrium constant, k_l is the backward rate constant for reaction l, where l denotes stoichiometric coefficients, and N_p and N_r are the number of products and number of reactants, respectively. Parameters were taken from literature; since the literature does not show enzymatic forward and reverse rate constants, parameters were derived from enzyme yield, substrate affinities, and specific activities.

$$\frac{dc_i}{dt} = \sum_l \nu_{li}\left(-k_l \prod_{j=1}^{N_p} c_j^{\nu_{lj}} + Q_l k_l \prod_{j=1}^{N_r} c_j^{-\nu_{lj}}\right)$$

18.4.2 Flux Balance Model

Flux balance models were used to calculate cellular metabolic rates under a range of acetate uptake fluxes of *Geobacter sulfurreducens* metabolism by Mahadevan et al. (2006), which predicted intracellular fluxes and metabolite exchange with the environment for a given acetate uptake. The metabolic fluxes (reaction rates f) were required, where, for a network described by a stoichiometric matrix S, $S \cdot f = 0$ implied a steady state. Determination of the flux was done via optimizing a specific objective function, which is subjected to physiological constraints on the magnitude of the fluxes, as follows: lower bound $\leq f \leq$ upper bound (Mahadevan et al., 2006). A MATLAB-based FB model was implemented, and growth efficiencies were calculated from the ratio of growth rate (f_{growth}) and acetate uptake (f_{ac}) as $g_{eff} = f_{growth}/f_{ac}$. The use of constraint-based models with the FBA approach is approved with respect to microbial metabolism. There are many concerns with its application that must be considered. At present, the FBA approach and constraint-based modeling are developed in the perspective of a singular microbial community. Consequently, a modeling analysis with the FBA approach cannot be directly executed in microbial community settings. Equally, the FBA approach alone is not sufficient in extreme cases, such as in extreme nutrient-rich environments.

18.4.3 Environment Cell Model

The *Geobacter* metabolism was united to simulations of a dynamic environment by incorporating it into a reactive transport model. In the reactive transport model, the two models were connected; the reactive transport model was then used to assess the transport of substrate and biomass. In contrast, the cell model offers the cell-specific reaction rates under the environmental conditions in given instances and places. Then the reaction rates in the macroscopic reactive transport model were evaluated with the help of cell-specified rates. For dissolved constituents, the equation is as follows:

$$\phi \frac{\partial C}{\partial t} = \nabla \cdot (D * \nabla C) - \nabla \cdot (\phi v C) + \phi \sum R$$

where ϕ is porosity, t is time, C is concentration, v is pore water velocity, D is the dispersion tensor implemented with dependence on v as described by Scheidegger (1961), and R is the net reaction rate. In the implementation, the cell model was carried out by the availability of acetate as the substrate, whose spatiotemporal dynamics proceeded via the following:

$$\frac{\partial C_{ac}}{\partial t} = T_{ac} - R_{ac}^{cell} C_{BM} + R_{ferm}$$

where T_{ac} and T_{BM} denote the transport of acetate (C_{ac}) and biomass (C_{BM}), respectively, due to convection and dispersion, g_{eff} is the growth efficiency, R_{ac}^{cell} is the rate of acetate uptake, and R_{ferm} is a source of acetate from the breakdown of high-molecular-weight organics. Numerically the model is solved using sequential noniterative operator splitting. With time step "*t*," the pressure and flow field are determined, which are then used to calculate the net transport for each of the chemical species. Successively, the change in concentration due to reactions is evaluated by solving a set of coupled ordinary differential equations at each node. Reaction parameters are computed for a given environmental condition and cell state, shown by the intracellular concentrations, and are taken as constant over a time step.

18.5 Docking Approach

18.5.1 Docking

Docking is a method that predicts the preferred orientation of one molecule to a second. These two molecules can be two proteins, a protein and a drug, or a nucleic acid and a drug. Docking is concerned with the generation and evaluation of the plausible structures of intermolecular complexes. It facilitates the identification of low-energy conformation after binding of a ligand molecule with a target/receptor (Kitchen et al., 2004). The compound that binds strongly in the active site of the target molecule and which has minimum energy may act as a potential drug molecule (Meng et al., 2004).

Docking also enable us to see the interaction between molecules and ligands. These two molecules can, again, be two proteins, a protein and a drug, or a nucleic acid and a drug. A compound that interacts strongly with, or binds with, a receptor that is involved in the disease may inhibit function of disease causing agent or molecule and act as a drug. It is the identification of the low-energy binding modes of a small molecule, or ligand, in the active site of a protein, or receptor, whose structure is already known and is present in the Protein Data Bank.

Molecular Docking: Molecular docking is a key tool in structural molecular biology and computer-aided drug design. The goal of ligand–protein docking is to predict the predominant binding mode(s) of a ligand with a protein of known three-dimensional structure. Molecular Modeling is usually performed between a small molecule and a target macromolecule (Shoichet et al., 2004).

Molecular docking can be taken as a problem of "lock and key," where the main interest is in finding the optimized orientation of the "key" that will open up the "lock" (Morris et al., 2005). Here, the protein can be thought of as the "lock" and the ligand can be thought of as a "key." During the course of the process, the ligand and the protein adjust their conformation to acquire an overall orientation (Sandak et al., 1998).

There are several types of molecular docking for protein interactions (Kahraman et al., 2007).

- Protein interaction with a ligand: protein–ligand interaction
- Protein interacts with another protein: protein–protein interaction
- Protein binds to DNA: protein–DNA interactions

Of all these, protein–ligand interaction techniques are the most widely used. Docking studies predict the principal binding mode(s) of a ligand with a known three-dimensional protein structure. Docking methods make use of a scoring function that correctly positions the candidate dockings. Docking is also used to perform virtual screening of large libraries of compounds, sort the results, and elucidate the inhibition of the target by ligands, which is very useful in lead optimization. Docking of laccase with a broad range of enzymes was carried out to find enzymes for biodegradation and the result expressed a good match between the *in silico* and *in vitro* analysis (Suresh et al., 2008). There are different complexities of docking, which depend upon the molecule with which the receptor is interacting; this varies from small molecules, DNA, protein to macromolecules (Figure 18.2). Generally, to perform the docking, a library of appropriate molecules is created and then docking is performed between the molecules in the library and the target protein molecule. The results containing high binding activity are shortlisted and further investigation is carried out. There are several affirmative points that support the use of molecular docking techniques in the bioremediation process. Molecular docking offers an interesting opportunity to highlight the active strand of the enzyme that ultimately degrades the pollutants. Simulation, docking, and modeling provide detailed study of amino acid that play role in the activity of enzyme.

18.5.2 Database Approach

With an increase in the biological experimental data, the problem arises of how to store, organize, and process it. Data mining is difficult when we need to retrive data of a particular microorganism from huge amount of data. There are several databases available that provide information regarding metabolic pathways of microorganisms.

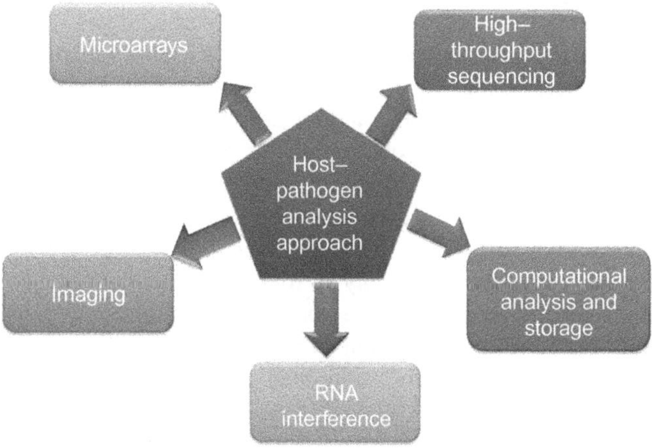

Figure 18.2 Flow diagram showing host–pathogen interaction analysis using *in vitro* and *in silico* methods.

Metarouter (Pazos et al., 2005) and Bionema (Carbajosa et al., 2009) have been introduced to bring together the data available online and present in a way that could help researchers to retrieve information. It provides information about chemical structures in a visualized format (i.e., PBD, SMILE format, molecular weight physiochemical properties).

18.6 Genomics Approach

With the advancement of technology to characterize microorganisms, the microorganisms that are involved in the bioremediation process could be easily determined by positioning them in clade by phylogenetic analysis. Along with phylogenetic, genomic is also considered a powerful computer technology used to understand the structure and function of all genes in an organism based on knowledge of the organism's entire DNA sequence. The field includes a great deal of work to determine the entire DNA sequences of organisms. This genetics pathways and functional information analysis is to elucidate its effects on, place in, and response to the entire genome networks. Such studies provide an estimate of the potential metabolic activity of the microbial community, but give little insight into the microorganisms that are responsible for bioremediation, or why particular amendments that can be evaluated for engineered bioremediation applications do or do not stimulate activity. Bioinformatics applies algorithms and statistical techniques to the interpretation, classification, and study of biological datasets. Sequence alignment is used to assemble the datasets for analysis. Comparisons of sequences, gene finding, and prediction of gene expression are the main techniques used for computational genomics; this is a field within genomics that comprises the studies of cells' and organisms' genomes. This further helps us to find genes of interest for certain diseases or other important enzymes that are helpful in medical and industrial settings. The set of tools made with the help of a bioinformatics approach allows rigorous analysis of the query sequence, which includes phylogenetic analysis, mutational identification, hydropathy regions, and CpG islands. The identification of these and other biological properties are all clues that aid the search to elucidate the specific function of the sequence. Now readily obtain millions of sequence could be reads from the genomes of their desired prokaryotic organisms. This all is possible due to the development of next-generation sequencing technologies. It is possible to reconstruct large portions of a genome from the overlapping sequence reads through sequence assembly. However, assembly is tricky because the sequence reads are generally very short and genomes often contain internally repeated segments that may block the complete reconstruction of a genome from its constituent reads. There are different approaches for addressing these challenges that involve the following:

1. More advanced assembly tools,
2. Reference genome sequences,
3. Finding the SNPs in the sequences.

Regardless of the strategy employed, there are many steps and programs involved, and the final outputs need to be annotated and interpreted with the known shortcomings of the data and methodologies in mind.

18.7 Future Prospects

Despite significant achievements in the last few decades in industrial microbial improvement, the approaches of traditional random mutation and selection and rational metabolic engineering based on local knowledge cannot meet today's needs. Recently, systems metabolic engineering based on the simulation and prediction of a mathematical model has provided a reliable method for microbial improvement. As a major tool of the systematic method, GSMM has been widely applied to guide metabolic engineering of microbial improvement. This GSMM-guided metabolic engineering is a systematic process, generally called the systematic method or the *in silico*–aided metabolic engineering approach, in industrial biotechnology. Improvements in the algorithms for bioremediation processes could simulate the environments of bioremediation sites and increase the efficiency of research in the field by reducing time and effort.

18.8 Conclusion

Genomes available in different databases could be utilized to interpret imperative conclusion which will augment the research proceeding toward bioremediation process. Likewise, docking studies of enzymes involved in bioremediation could be used to study bioremediaton at molecular level. The dynamic cell model, the flux balance model, and the environment cell model are recently introduced models that provide a pragmatic solution to the bioremediation process. *In silico* is an intact form of tools, information, and approaches with which the issues of bioremediation can be overcome. This will provide new ideas for the bioremediation process.

References

Abascal, F., Zardoya, R., Posada, D., 2009. Genetic code prediction for metazoan mitochondria with GenDecoder. Methods Mol. Biol. 537, 233–242.

Bader, M., Ohlebusch, E., 2007. Sorting by weighted reversals, transpositions, and inverted transpositions. J. Comput. Biol. 14, 615–636.

Bader, M., Abouelhoda, M., Ohlebusch, E., 2008. A fast algorithm for the multiplegenome rearrangement problem with weighted reversals and transpositions. BMC Bioinformatics. 9, 516.

Benson, D.A., Karsch-Mizrachi, I., Lipman, D.J., Ostell, J., Sayers, E.W., 2011. GenBank. Nucleic Acids Res. 39 (S1), D32–D37.

Bernt, M., Merkle, D., Ramsch, K., Fritzsch, G., Perseke, M., Bernhard, D., et al., 2007. CREx: inferring genomic rearrangements based on common intervals. Bioinformatics. 23 (21), 2957–2958.

Bergeron, A., Mixtacki, J., Stoye, J., 2006. A unifying view of genome rearrangements. Algorithms in Bioinformatics, 6th International Workshop, WABI 2006, Proceedings. Vol. 4175 of Lecture Notes in Bioinformatics. Springer, pp. 1.

Bernt, M., Chen, K.Y., Chen, M.C., Chu, A.C., Merkle, D., Wang, H.L., et al., 2011. Finding all sorting tandem duplication random loss operations. J. Discr. Algorithms. 9, 32–48.

Brown, E.N., Friemann, R., Karlsson, A., Parales, J.V., Couture, M.M.J., Eltis, L.D., 2008. Determining Rieske cluster reduction potentials. J. Biol. Inorg. Chem. 13, 1301–1313.

Bourque, G., Pevzner, P.A., 2002. Genome-scale evolution: Reconstructing gene orders in the ancestral species. Genome Res. 12, 26–36.

Carbajosa, G., Trigo, A., Valencia, A., Cases, I., 2009. Bionemo: molecular information on biodegradation metabolism. Nucleic Acids Res. 37, D598–D602.

Cristovao, R.O., Tavares, A.P.M., Ribeiro, A.S., Loureiro, J.M., Boaventura, R.A.R., Macedo, E.A., 2008. Kinetic modelling and simulation of laccase catalysed degradation of reactive textile dyes. Bioresour. Technol. 99, 4768–4774.

de Vasconcelos, A.T.R., Guimaraes, A.C.R., Castelletti, C.H.M., Caruso, C.S., Ribeiro, C., Yokaichiya, F., et al., 2005. MamMiBase: a mitochondrial genome database for mammalian phylogenetic studies. Bioinformatics. 21, 2566–2567.

Durot, M., Bourguignon, P.Y., Schachter, V., 2009. Genome-scale models of bacterial metabolism: reconstruction and applications. FEMS Microbiol. Rev. 33 (1), 164–190.

Dutilh, B.E., Jurgelenaite, R., Szklarczyk, R., van Hijum, S.A., Harhangi, H.R., Schmid, M., et al., 2011. FACIL: fast and accurate genetic code inference and logo. Bioinformatics. 27, 1929–1933.

Ellis, L.B., Gao, J., Fenner, K., Wackett, L., 2008. The University of Minnesota pathway prediction system: predicting metabolic logic. Nucleic Acids Res. 36, W427–W432.

Feijao, P.C., Neiva, L.S., de Azeredo-Espin, A.M., Lessinger, A.C., 2006. AMiGA: the arthropodan mitochondrial genomes accessible database. Bioinformatics. 22, 902–903.

Fritzsch, G., Schlegel, M., Stadler, P.F., 2006. Alignments of mitochondrial genome arrangements: Applications to metazoan phylogeny. J. Theor. Biol. 240, 511–520.

Gebhardt, N.A., Thauer, R.K., Linder, D., Kaulfers, P.M., Pfennig, N., 1985. Mechanism of acetate oxidation to CO_2 with elemental sulfur in *Desulfuromonas acetoxidans*. Arch. Microbiol. 141, 392–398.

Jameson, D., Gibson, A., Hudelot, C., Higgs, P., 2003. OGRe: a relational database for comparative analysis of mitochondrial genomes. Nucleic Acids Res. 31, 202–206.

Jones, B.R., Rajaraman, A., Tannier, E., Chauve, C., 2012. ANGES: Reconstructing ANcestral GEnomeS maps. Bioinform. Appear.

Jühling, F., Pütz, J., Bernt, M., Donath, A., Middendorf, M., Florentz, C., et al., 2012. Improved systematic tRNA gene annotation allows new insights into the evolution of mitochondrial tRNA structures and into the mechanisms of mitochondrial genome rearrangements. Nucleic Acids Res. 40 (7), 2833–2845.

Juhling, F., Morl, M., Hartmann, R.K., Sprinzl, M., Stadler, P.F., Putz, J., 2009. tRNAdb 2009: compilation of tRNA sequences and tRNA genes. Nucleic Acids Res. 37 (Database issue), D159–D162.

Kahraman, A., Morris, R.J., Laskowski, R.A., Thornton, J.M., 2007. Shape variation in protein binding pockets and their ligands. J. Mol. Biol. 368, 283–301.

Kandelbaue, A., Erlacher, A., Cavaco-Paulo, A., Guebitz, G.M., 2004. Laccase-catalyzed decolorization of the synthetic azo-dye Diamond Black PV 200 and of some structurally related derivatives. Biocatal. Biotransform. 22, 331–339.

Kitchen, D.B., Decornez, H., Furr, J.R., Bajorath, J., 2004. Docking and scoring in virtual screening for drug discovery: methods and applications. Nat. Rev. Drug Discovery. 3 (11), 935–949.

Laslett, D., Canback, B., 2008. ARWEN: a program to detect tRNA genes in metazoan mitochondrial nucleotide sequences. Bioinformatics. 24, 172–175.

Larget, B., Kadane, J.B., Simon, D.L., 2005. A Bayesian approach to the estimation of ancestral genome arrangements. Mol. Phylogenet. Evol. 36, 214–223.

Leang, C., Coppi, M.V., Lovley, D.R., 2003. OmcB, a c-type polyheme cytochrome, involved in Fe(III) reduction in *Geobacter sulfurreducens*. J. Bacteriol. 185, 2096–2103.

Lee, J.W., Kim, T.Y., Jang, Y.S., Choi, S., Lee, S.Y., 2011. Systems metabolic engineering for chemicals and materials. Trends Biotechnol. 29 (8), 370–378.

Lovley, D.R., Holmes, D.E., Nevin, K.P., 2004. Dissimilatory Fe(III) and Mn(IV) reduction. Adv. Microb. Physiol. 49, 219–286.

Lowe, T.M., Eddy, S.R., 1997. tRNAscan-SE: a program for improved detection of transfer RNA genes in genomic sequence. Nucleic Acids Res. 25, 955–964.

Lupi, R., de Meo, P.D., Picardi, E., D'Antonio, M., Paoletti, D., Castrignanò, T., et al., 2010. MitoZoa: a curated mitochondrial genome database of metazoans for comparative genomics studies. Mitochondrion. 10, 192–199.

Ma, J., Zhang, L., Bernard, B., Raney, B., Burhans, R., Kent, W., et al., 2006. Reconstructing contiguous regions of an ancestral genome. Genome Res. 16, 1557–1565.

Mahadevan, R., Bond, D.R., Butler, J.E., Esteve-Nunez, A., Coppi, M.V., Palsson, B.O., et al., 2006. Characterization of metabolism in the Fe(III)-reducing organism *Geobacter sulfurreducens* by constraint-based modeling. Appl. Environ. Microbiol. 72, 1558–1568.

Meng, E.C., Shoichet, B.K., Kuntz, I.D., 2004. Automated docking with grid-based energy evaluation. J. Comput. Chem. 13 (4), 505–524.

Matthias, B., Anke, B., Martin, M., Bernhard, M., Omar, S., Peter, F.S., 2013. Bioinformatics methods for the comparative analysis of metazoan mitochondrial genome sequences. Mol. Phylogenet Evol. 69 (2), 320–327.

Morris, R.J., Najmanovich, R.J., Kahraman, A., Thornton, J.M., 2005. Real spherical harmonic expansion coefficients as 3D shape descriptors for protein binding pocket and ligand comparisons. Bioinformatics. 21 (10), 2347–2355.

Moret, B.M.E., Wang, L.-S., Warnow, T., Wyman, S.K., 2001. New approaches for reconstructing phylogenies from gene order data. Bioinformatics. 17, 165–173.

O'Brien, E.A., Zhang, Y., Wang, E., Marie, V., Badejoko, W., Lang, B.F., et al., 2009. GOBASE: an organelle genome database. Nucleic Acids Res. 37, D946–D950.

Palsson, B., 2006. Systems Biology: Properties of Reconstructed Networks. Cambridge University Press, New York, NY.

Pazos, F., Guijas, D., Valencia, A., De Lorenz, V., 2005. MetaRouter: bioinformatics for bioremediation. Nucleic Acids Res. 33, D588–D592.

Pieper, D.H., Martins dos Santos, V.A., Golyshin, P.N., 2004. Genomic and mechanistic insights into the biodegradation of organic pollutants. Curr. Opin. Biotechnol. 15, 215–224.

Pruitt, K.D., Tatusova, T., Maglott, D.R., 2007. NCBI Reference Sequences (RefSeq): a curated non-redundant sequence database of genomes, transcripts and proteins. Nucleic Acids Res. 35, D61–D65.

Putz, J., Dupuis, B., Sissler, M., Florentz, C., 2007. Mamit-tRNA, a database of mammalian mitochondrial tRNA primary and secondary structures. RNA. 13, 1184–1190.

Sandak, B., Wolfson, H.J., Nussinov, R., 1998. Flexible docking allowing induced fit in proteins: insights from an open to closed conformational isomers. Proteins. 32, 159–174.

Scheidegger, A.E., 1961. General theory of dispersion in porous media. J. Geophys. Res. 66, 3273–3278.

Schroll, R., Brahushi, R., Dorfler, U., Kuhn, S., Fekete, J., Munch, J.C., 2004. Biomineralisation of 1, 2, 4 trichlorobenzene in soils by an adapted microbial population. Environ. Pollut. 127, 395–401.

Sheffeld, N.C., Hiatt, K.D., Valentine, M.C., Song, H., Whiting, M.F., 2010. Mitochondrial genomics in orthoptera using MOSAS. Mitochondrial DNA. 21, 87–104.

Shoichet, B.K., Kuntz, I.D., Bodian, D.L., 2004. Molecular docking using shape descriptors. J. Comput. Chem. 13 (3), 380–397.

Stoye, J., Wittler, R., 2009. A unified approach for reconstructing ancient gene clusters. IEEE/ACM Trans. Comp. Biol. Bioinf. 6, 387–400.

Suresh, P.S., Kumar, A., Kumar, R., Singh, V.P., 2008. An, *in-silco* approach to bioremediation: laccase as a case study. J. Mol. Graph. Model. 26, 845–849.

van der Meer, J.R., 2008. A genomic view on the evolution of catabolic pathways and bacterial adaptation to xenobiotic compounds. In: Diaz, E. (Ed.), Microbial Biodegradation. Genomics and Molecular Biology. Caister Academic Press, Norfolk, UK.

Watanabe, K., 2001. Microorganisms relevant to bioremediation. Environ. Biotechnol. 12, 237–241.

Wolfsberg, T.G., Schafer, S., Tatusov, R.L., Tatusova, T.A., 2001. Organelle genome resources at NCBI. Trends Biochem. Sci. 26, 199–203.

Wyman, S.K., Jansen, R.K., Boore, J.L., 2004. Automatic annotation of organellar genomes with DOGMA. Bioinformatics. 20 (17), 3252–3255.

Zhao, H., Bourque, G., 2009. Recovering genome rearrangements in the mammalian phylogeny. Genome Res. 19, 934–942.

Zheng, C., Sankoff, D., 2011. On the Pathgroups approach to rapid small phylogeny. BMC Bioinformatics. 12 (Suppl. 1), S4.

19 Microalgae in Bioremediation: Sequestration of Greenhouse Gases, Clearout of Fugitive Nutrient Minerals, and Subtraction of Toxic Elements from Waters

K. Uma Devi, G. Swapna and S. Suneetha

Department of Botany, Andhra University, Visakhapatnam, Andhra Pradesh, India

19.1 Introduction

Bioremediation with microbes usually involves cleanup of metal pollutants from various sources (Rajendran et al., 2003; Umrania, 2006; Gadd, 2010). Photosynthetic cyanobacteria and microalgae, in addition, help in sponging off the chief greenhouse gas (GHG), carbon dioxide (CO_2), which is released in large amounts during fossil fuel consumption in transport vehicles and industries—a method of biosequestration of carbon dioxide.

Microalgae are constituents of the phytoplankton that occur in all water bodies—freshwater, estuarine, or marine. They have a very efficient carbon-fixing mechanism of photosynthesis (Aizawa and Miyachi, 1986; Badger et al., 2000). They are unicellular and exist as individual cells or form into chains or colonies (Figure 19.1). They have a short generation time and multiply exponentially under suitable environmental conditions. Under conditions favoring their growth, they flourish in nature, resulting in "algal blooms" and imparting color to water. An excess of fugitive minerals (nitrates and phosphates) which are of nutritive value to plants results in eutrophication due to algal blooms in water bodies leading to the natural depletion of these minerals thus preventing their deposition to toxic levels. Microalgae also have the ability to biosorb and bioaccumulate heavy metals, some of which (e.g., iron and chromium) are of nutritive value for their growth. Therefore, they are of use in bioremediation of effluents from industries with toxic heavy metals. Being photosynthetic, algae take up carbon dioxide and, when

Microbial Biodegradation and Bioremediation. DOI: http://dx.doi.org/10.1016/B978-0-12-800021-2.00019-4

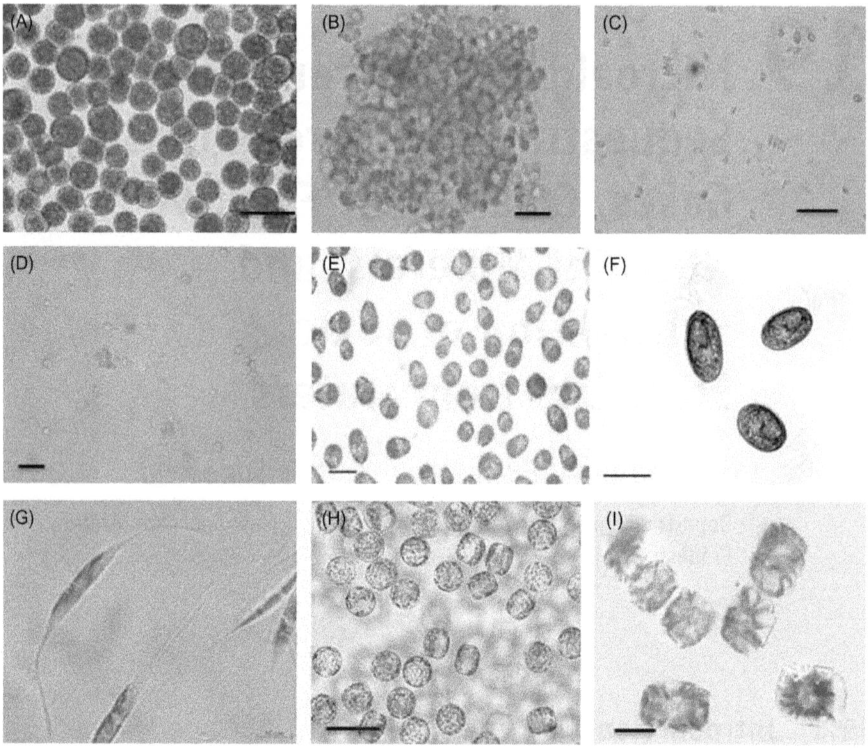

Figure 19.1 Images of microalgal species under the microscope from author's laboratory: bar represents 10 μm. (A) *Haematococcus pluvialis*, (B) *Desmodesmus* sp., (C) *Scenedesmus dimorphus*, (D) *Chlorella protothecoides*, (E) *Dunaliella salina*, (F) *Tetraselmis* sp., (G) *Cylindrotheca closterium*, (H) *Odontella aurita*, and (I) *Thalassiosira* sp.

multiplying at exponential levels, therefore sequester substantial levels of CO_2. Considering the magnitude of carbon sequestration by natural microalgal blooms emerged the concept of mass culture of microalgae for biosequestration of CO_2. The concept is bolstered by the prospect of the commercial potential of the algal biomass. The potential, promise, and accomplishments of microalgae in the bioremediation of CO_2 and other GHGs in the atmosphere, and of solid soluble pollutants in water, are presented herein.

19.2 Microalgae in Biosequestration of GHGs

At present, the concentration of CO_2 in the air is ~400 ppm/v—much more than the levels of 280 ppm/v in preindustrial times (www.globalcarbonproject.org). This increase is recognized to be the result of anthropogenic activities—primarily the burning of fossil fuels.

The effect of increased concentrations of CO_2 and other GHGs on the global climate and on human life has been recognized by researchers and called for public attention in various forums and media (e.g., Oscar award–winning documentary film "An Inconvenient Truth" by Al Gore). National and international governments, industries, and high-profile collaborative programs such as the Intergovernmental Panel on Climate Change (IPCC) and the United Nations Framework Commission on Climate Change are working toward achieving a common goal of reducing the predominant GHG, CO_2, from large point sources like power plants and other massive industrial complexes.

According to an IPCC report, the major source of CO_2 and other GHGs is fossil fuel (coal, coke, and gas)–fired power plants, which contribute up to 60% of global GHG emissions (IPCC, 2005). Physical, chemical, and biological methods are being explored for developing carbon capture and storage technologies for power plants. Biosequestration using microalgae stands out, as it can be carbon neutral and the technology can be sustainable (Table 19.1).

The photosynthetic efficiency of microalgae is reported to be even better than that of the C_4 plants. This is attributed to the presence of an efficient CO_2 concentration mechanism and the lack of photorespiration (Melis, 2009; Ort et al., 2011). Microalgae do not need fertile land for their growth. They can be grown on barren lands with shallow water ponds and minimal nutrients. The typical composition of chimney emissions (flue gas) from a coal-fired power plant chimney is ~6% oxygen, ~76% nitrogen, ~11% CO_2, ~6% water vapor, 500–800 ppm oxides of nitrogen (NO_x), 10–200 ppm oxides of sulfur (SO_X), and trace elements (Xu et al., 2003). Microalgae can utilize not only the CO_2 from these emissions but also NO_x—another GHG—(as a source of nitrate) and small amounts of SO_X. In addition, NO_x and SO_X are air pollutants that cause acid rain. Tolerance of different microalgal species to flue gas emissions ranges between 5–60% of CO_2, 50–500 ppm of NO_x, and 50–150 ppm of SO_X (Den Hende et al., 2012). There is a large scope for a high rate of algal biomass production with optimization of nutrient medium and culture conditions. There is a market for algal biomass as a source of biofuel ($US 271 million), antioxidant pigments such as carotene (with a global market of $US 247 million), astaxanthin ($US 200 million), lutein ($US 233 million), and polyunsaturated fatty acids ($US 700 million) (Spolaore et al., 2006). The proof-of-concept experiment of carbon sequestration with microalgal culture was first demonstrated through a pilot project by Dr. Isaac Berzin of MIT, Boston, MA (http://web.mit.edu). Berzin, who called microalgal cultures the "breath mints of smoke stacks," was recognized as one of the 100 most influential people of the century by *Time Magazine* (http://content.time.com). Later, the Israel Electric Corporation (IEC) in Ashkelon set up a mass culture facility for cultivation on a pilot scale of *Nannochloropsis* sp. with SO_2-free flue gas. Seambiotic Ltd. designed systems for the removal of high SO_X content in the flue gas mixture from one of IEC's coal-fired power plants using flue gas desulfurization techniques. The algae showed an increased growth rate with flue gas than pure CO_2 (US patent number US2008/0220486 A1). Similarly, the Rheinisch-Westfalisches Elektrizitatswerk (RWE) algae project in Germany, erected at the Niederaussem power plant

Table 19.1 Biosequestration of GHGs from Flue Gas Emissions in Comparison to Physical and Chemical Sequestration Methods

	Physical	Chemical	Biological	
			Phytoremediation	**Microalgae**
Method	Involves 1. CO_2 separation from main stream of exhaust gases using cryogenic or membrane separation methods. 2. Compression and transport to storage site. 3. Storage in underground geological formations or deep oceans.	Involves absorption of CO_2 into liquid solvents or solid sorbents which are later processed for the separation of CO_2. Monoethanolamine is the commonly used solvent.	Involves afforestation and more plantations. Plants utilize CO_2 for photosynthesis. More plantations result in more CO_2 recovery.	Involves 1. Feeding flue gas mixture to large-scale microalgal cultures. 2. The microalgae utilize the contents of flue gas (CO_2, oxides of nitrogen—NO_X) and small amounts of sulfur oxides (SO_X) for their growth.
Advantages	CO_2 can be used for various applications; for example, to enhance methane recovery from coal beds.	Can achieve high CO_2 recovery (up to 90%) from flue gas mixture.	Natural way of containing CO_2. Air is enriched with O_2 released during photosynthesis. Plant produce can be used for various purposes.	1. Can utilize not just CO_2 but all the GHGs in flue gas. 2. Do not compete for space or resources of agricultural crops. 3. Biomass can be utilized in the form of biofuel, animal feed, fertilizer, and nutraceutical.
Disadvantages	CO_2 separation is an expensive procedure. 1. Underground storage does not ensure safety. 2. When deposited in deep oceans, its effect on aquatic life is unknown.	No significant financial incentives to offset operational cost.	Cannot cope with the logarithmic increase in CO_2 emissions.	1. The process is climate dependent when the cultures are grown in open ponds. 2. Energy efficient biomass harvest method is not yet available. 3. Initial setup costs are very high.
References	Haszeldine (2009) and D'Alessandro et al. (2010)	Rochelle et al. (2001) and Bloom (2012)	Reyer et al. (2009)	Pedroni et al. (2001), Kumar et al. (2010), and Borowitzka and Moheimani (2013)

location, utilized desulfurized flue gases for algal growth. The project operated for 3 years, until 2011 (http://www.rwe.com/). A US patent (5659977) of Cyanotech Corporation, Hawaii, describes methods to use CO_2 from exhaust gas of a fossil fuel–fired power plant for algae production. The flue gas emissions are transferred to the bottom of a CO_2 absorption tower (6.4 m high packing material) which can provide 67 tons of CO_2 per month that is used in cultivation of microalgae and cyanobacteria. The patent also describes the utilization of heat from a fossil fuel engine to dry the algal biomass. Nature Beta Technologies Ltd., in Eilat City, Israel, reported biomass production of *Dunaliella salina* at 20 g/m^2/day when supplied with flue gas. The cost of dry biomass when cultured with flue gas mixture was estimated at USD 0.34/kg as compared to USD 17/kg when cultivated in a normal environment (Ben-Amotz et al., 2009).

While this list of pilot projects is impressive, a technological breakthrough leading to commercial success has not occurred even roughly a decade after the demonstration of the proof-of-concept project. There are many challenges to be met (Umadevi et al., 2014). Setting aside the technological challenges, the main reason for lack of commercial success was the focus on utilization of microalgal biomass as a source of biodiesel. The cost of production of biomass made it impossible to produce biofuel at prices comparative to those of fossil fuel (Stephens et al., 2010). There seems to be hope in this area, however: Exxon, a multinational company, has teamed up with the maverick scientist Craig Venter to discover the microalgal species most suitable for biodiesel production (Ratledge, 2011).

There is now a shift of emphasis from biofuel to culturing algae for nutraceutical applications. The pigments lutein, β carotene, astaxanthin, and omega-3 fatty acids; eicosapentaenoic acid (EPA); docosahexaenoic acid (DHA); and proteins with an aminogram with essential amino acids in different microalgal species are all of nutraceutical value to humans. In addition, microalgal biomass can be a very nutritious feed for poultry and cattle. Some of the microalgal species being experimented on at a commercial scale are listed in Table 19.2. Dr. Berzin, the pioneer of flue gas-fed microalgal culture as a means of CO_2 sequestration, has now reoriented his program of biodiesel from microalgae (which he targeted with his startup company, Green Fuel Technologies) and is now working in partnership with Israel-based Qualitas Health and Valicor Renewables on the culture of EPA-rich microalgal species. Since cost is not a concern in nutraceuticals, such programs are more likely to become commercially feasible parallel to the environmental benefits of mitigation of GHGs.

19.2.1 Microalgal Cultivation for Mitigation of GHGs from Power Plants and Industrial Setups

19.2.1.1 Supply of Flue Gas

It has been shown that microalgae can utilize flue gases of varied composition (Den Hende et al., 2012). The pH of the nutrient medium was found to be the major factor that affects CO_2 fixing ability of algae, and not the flue gas composition (Bark, 2012). Of the several challenges to be met to establish sustainable

Table 19.2 Commercial Algal Based Industries

Microalgal Species	Product	Production Company	Reference
Botryococcus braunii	Biofuel	NA	Metzger and Largeau (2005)
Chlorella sp.	Biofuel, lutein, and protein	Chlorella Manufacturing Co Ltd., Taiwan and Klotze, Germany	Becker (2004)
Crypthecodinium sp.	EPA and DHA (polyunsaturated fatty acids, PUFAs)	Aurora algae	www.nutraingredients-usa.com
Dunaliella salina	β carotene	Cognis-Betatene, Australia	Borowitzka (1991)
Haematococcus pluvialis	Astaxanthin	Cyanotech Corporation and Parry Nutraceuticals	Lorenz and Cysewski (2003)
Nannochloropsis sp.	Biofuel and PUFAs	Seambiotic Ltd.	US patent number US2008/0220486 A1
Scenedesmus dimorphus	Protein, bioremediation, and CO_2 fixation	NA	Terry and Stone (2002) and Ho et al. (2010)

NA, not available.

microalgal carbon sequestration systems (Umadevi et al., 2014), flue gas delivery is one. The test on the engineering side is of channeling the flue gas, cooling it down, and supplying it to the algal cultivation system with a regulated flow system. Different technologies like sumps and CO_2 absorbents were developed by commercial companies to store and supply the flue gas continuously to algal cultivation systems (Figure 19.2). Setting up microalgal culture units near the point source of flue gas emissions is not practical in many cases due to space constraints. Moreover, the atmosphere in the industrial setup or thermal power plant (TPP) is often pollution ridden and the biomass produced in such conditions can have many contaminants gained from the surrounding environment. Therefore, the option of dissolving the flue gas emissions into a water source and transporting the flue gas–concentrated water to a pristine site where algal culture facility is set up is also being examined (Pedroni et al., 2001).

19.2.2 Microalgal Mass Culture Systems

For mass culture of microalgae, two culture systems are used: photobioreactors with controlled conditions and open ponds that are exposed to the outside environment.

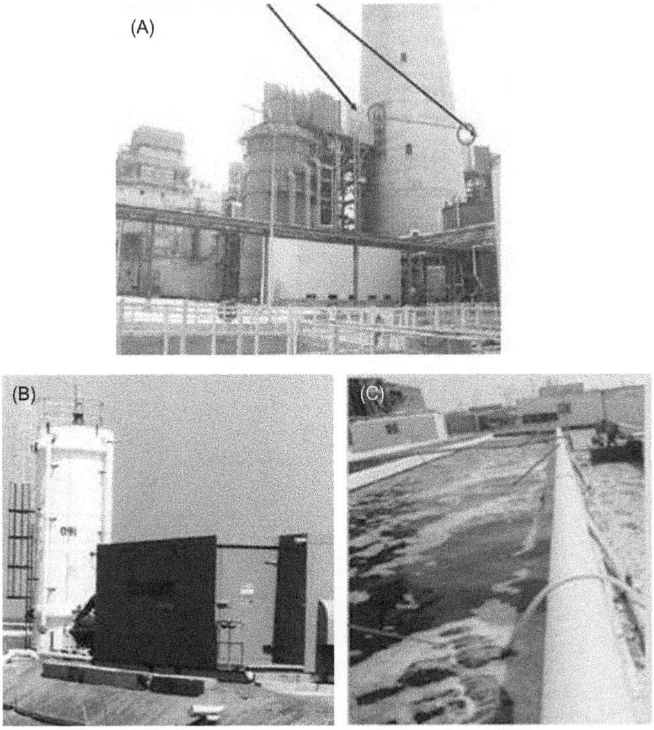

Figure 19.2 Provision for flue gas supply to algal cultivation units. (A) Outlets for tapping flue gas at the bottom of a chimney; (B) storage of CO_2 in an absorption tank filled with a patented CO_2 absorbing material in algal culture unit of Cyanotech Corporation; (C) supply of flue gas to the algal culture pond through pipes (with regulated valves) connected to a flue gas storage sump.
Source: www.cyanotech.com, www.seambiotic.com.

19.2.2.1 Photobioreactors

A photobioreactor is similar to a fermenter except that a light source is required for it since microalgae are photoautotrophic and carbon sequestration occurs only in this mode. Some species, however, can grow in mixotrophic (simultaneous organic carbon assimilation along with photoautotrophy) or heterotrophic conditions. The light source can be natural (solar) or artificial. To facilitate penetration of light, photobioreactors are made of glass, transparent plastic material, or sturdy polythene. There are different models available on the market with different materials, structures (tubular or flat plate), and shapes (Figure 19.3).

Haematococcus pluvialis was grown in a 2000 L photobioreactor using flue gases from a coal combustor and the culture was further upscaled to 25,000 L in a photobioreactor which was fed by flue gases from a propane combustor (Olaizola, 2000). The major disadvantage of photobioreactors is limitation of scalability to a large

Figure 19.3 Different structures of photobioreactors: (A) Tubular glass, (B) plastic flat plate, (C) and (D) polythene bag.
Source: (A): http://chlorelle.wordpress.com/, (B): http://silentcenter.deviantart.com/art/Photobioreactors-299621647, (C): http://www.cals.arizona.edu/spotlight/biofuels-algae-hold-potential-not-ready-prime-time, and (D): http://spectrum.ieee.org/energy/renewables/betting-on-algal-biofuels.

commercial-scale production and cost factors (Benemann, 2008; Kunjapur and Eldridge, 2010). So far photobioreactor systems have proved useful and economical for the generation of microalgal inoculum for large-scale culture in open ponds.

19.2.2.2 Open Ponds

Round or oval-shaped shallow pond divided in the middle by a grid—termed the raceway pond when oval shaped, with water circulated by a power operated paddle defines an open pond (Figure 19.4). The open ponds are easy to construct and maintain and are much cheaper than photobioreactors. The upscaling limit for open ponds is infinite. Seambiotic Ltd. (*Nannochloropsis* sp.) and Cyanotech Corporation (*Arthrospira platensis*) successfully demonstrated the production of microalgal or cyanobacterial biomass using flue gas emissions (US patent 2008/0220486

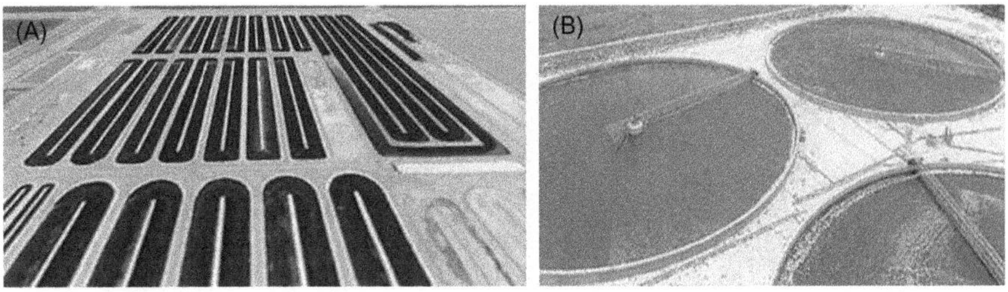

Figure 19.4 Open ponds aerated with paddle wheels: (A) Raceway pond and (B) circular pond. *Source*: (A): http://www.newswise.com/articles/study-algae-could-replace-17-of-u-s-oil-imports and (B): http://chlorelle.files.wordpress.com/2011/04/openairculture2.jpg.

A1 and US patent 5,659,977, respectively). The major drawbacks with open ponds are that biomass production in these systems is not consistent but rather climate dependent, and that there is a risk of contamination with other microalgal species and microbes.

19.2.3 Development of Inoculum for Mass Culture of Microalgae

For mass culture of microalgae either in photobioreactors or open ponds, microalgal inoculum is required to seed them. Seed culture is initially developed in small (50–100 mL) volumes in the laboratory and gradually up scaled to liter volumes, which are further scaled up in a photobioreactor or in small open ponds (Figure 19.5). This process in the laboratory requires 30–60 days depending on the size of the raceway pond to be seeded. In the laboratory, temperature and light are conditioned to favor growth. Thus, laboratory culture is both time consuming and energy intensive. Further, due to the requirement of high light intensity, the algal cultures do not attain the density attained by bacterial cultures. They are dilute and hence large culture volumes are required to obtain sufficient inoculum to seed the mass culture units. To promote dense algal growth in cultures, the nutrient mode can be shifted from an exclusively phototrophic to a mixotrophic condition. In phototrophic mode, the carbon requirement is entirely met from inorganic CO_2, while in mixotrophy (also termed photoheterotrophy) the cultures utilize both CO_2 and organic carbon supplemented in the nutrient medium. Different organic carbon sources—maltose, malt extract, glucose, acetate, glycerol, sucrose, etc.—are used. In addition, organic nitrogen sources like yeast extract and peptone are also used in mixotrophic cultures (Antia et al., 1975; Garcia et al., 2005). Mixotrophy is reported to yield high algal biomass rich in metabolites (Kobayashi et al., 1992; Liang et al., 2009; Orosa et al., 2001; Yeh and Chang, 2012). Dense growth of algae in such cultures is possible because the limiting of light due to the shadowing effect of populous cell cultures does not limit their further growth—the organic carbon substitutes for lack of phototrophy in cell dense cultures. There are very few species of algae that can grow heterotrophically, for example, *Chlorella* sp.,

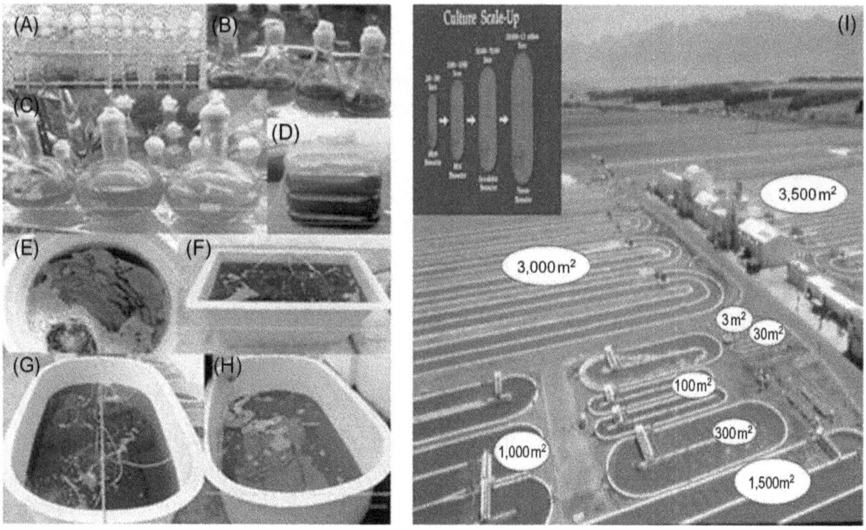

Figure 19.5 Different steps of algal culture upscaling of *Dunaliella salina* from author's laboratory. (A) 10 mL stock cultures, (B) 100 mL nutrient medium inoculated from stock cultures, (C) and (D) serial upscaling from 2 to 15 L, (E) and (F) 50 and 100 L cultures in plastic tanks supplied with continuous aeration, (G) 750 L fully grown algal culture, and (H) pigment production (red stage) due to the induction of nutrient stress. (A)–(D) are in laboratory room with controlled artificial light and temperature; (E)–(G) are in glass house with natural sunlight and no temperature control; and (I) schematic representation of large-scale algal culture upscaling in outdoor ponds of different sizes.
Source: (I): http://www.nrel.gov/biomass/pdfs/benamotz.pdf.

Chlamydomonas reinhardtii, and *Crypthecodinium cohnii* (Chen and Johns, 1996; Jiang et al., 1999; Bumbak et al., 2011), and a few more have been reported for mixotrophy, namely, *Chlorella protothecoides*, *H. pluvialis*, *Tetraselmis* sp., and *D. salina* (Ogava and Aiba, 1981; Cid et al., 1992; Orosa et al., 2001; Wan et al., 2011). Not only is the biomass production increased in mixotrophic cultures but there is a simultaneous boost in the pigment and lipid content in the cells. For example, *H. pluvialis* was reported to have more astaxanthin content when mixotrophically cultured (Cifuentes et al., 2003). *Phaeodactylum tricornutum* cultured mixotrophically was reported to have higher lipid content compared to photoautotrophic cultures (Morais et al., 2009). Mixotrophy in very large volume open-air culture units is problematic, however, as the organic carbon in the medium promotes growth of heterotrophic microbes. Therefore, the "sequential heterotrophy–dilution–photoinduction" method has been suggested (Fan et al., 2012). This technique involves initially multiplying the algal cells in heterotrophic conditions and then using this culture to seed the mass culture units with inorganic nutrient medium. A sequential mixotrophy–dilution–photoautotrophy method was successfully used in our laboratory for mass culture of *D. salina* (Suman, 2012). This method was also attempted for culture of *Scenedesmus dimorphus*, where mixotrophy (sodium acetate as carbon source) was used for high inoculum production in an indoor culture laboratory up to 20 L and photoautotrophy was used for large-scale production (1000 L) using flue gas–dissolved water. The entire upscaling process took 10 days with the two-stage method and 25 days for the exclusive photoautotrophic culture (unpublished results).

19.2.4 Harvest of Microalgal Biomass

The harvesting and drying of algal biomass are energy-intensive processes because of the very small size of microalgal cells and the very wet conditions of the harvested biomass. Biomass harvesting entails 20–30% of the total cost of algal production. The usual methods of harvesting are centrifugation, flocculation, membrane filtration, and ultrasonic separation. Centrifugation yields more biomass recovery with minimal loss but this method is financially not viable for large-scale cultivation (Grima et al., 2003). Inexpensive flocculants, both inorganic and organic, are being tested and used in the harvesting of microalgae. Inorganic flocculants include ferric chloride and alum, both of which are effective for both freshwater and marine species. However, remnants of the flocculant in the harvested biomass may not be good when it is used for human consumption or animal feed. Nontoxic organic flocculants like cationic starch have also been tested. These are suitable only for the harvesting of freshwater algae and very high doses are required to harvest large amounts of algal biomass (Vandamme et al., 2010). Chitosan, a derivative (polymer) of shrimp exoskeleton, is another effective, biodegradable organic flocculant reported to be suitable for both marine and freshwater algae (Lavoie and Noue, 1983; Morales et al., 1985). Chitosan is reported to be more effective at high microalgal cell densities (Divakaran and Pillai, 2002). Use of chemical flocculants is comparatively cheap but more research is needed to identify effective and species-specific flocculants. In selection of an algal strain for mass cultivation, traits such as generation time and cell size are critical because they ultimately affect the economics of production costs. Because the production of dense amounts of biomass reduces the harvesting cost compared to that of small-sized algae with less biomass density, Grima et al. (2003) have suggested that more research should be done to identify cheaper and easier methods of algal harvesting and drying in order to make the process cost effective. The harvested biomass is dried to remove excess water by spray drying, drum drying, freeze drying, and sun drying. Among these methods, spray drying is most convenient but again expensive.

19.3 Bioremediation of GHGs Using Microalgae: A Case Study

19.3.1 Source of Flue Gas

The Visakhapatnam steel plant (17° 42′ 0″ N, 83° 18′ 0″ E.), popularly known as "Vizag steel," is the most advanced producer of steel in India, with a production capacity of 7 Mt/year. There are 23 chimneys for release of flue gas from the various heating units. The TPP set up to meet the power requirements of the steel plant utilizes both coal and gas as fuel. Flue gas let out from the gas-fired unit of the TPP was selected to tap the flue gases for the experiments. The chimneys are equipped with sulfur scrubbing technology and its gas-fired units emit minimal/no amounts of SO_X (which is lethal to microalgae if it crosses 300 ppm; tolerance level is <150 ppm for many algae) and trace metals. The composition of flue gases from the gas-fired power plant chimney collected at four different points is CO_2 (18–20%), NO_X (75–78%), and trace elements (1–2%).

19.3.2 Preparation of Flue Gas–Enriched Water (FGW)

To test the growth of microalgae when supplied with flue gas through laboratory growth assays, the flue gas was dissolved in water as described below. This water was used for preparation of the nutrient medium for the culture of microalgae. Flue gas (25% CO_2, 73% NO_X, 2% O_2) was passed into distilled water (DW) (pH 7) from an outlet with a valve at the bottom of the chimney. The DW was contained in a 500 L plastic container. The pH of the water was periodically checked. When the pH reached 3–2.8, it did not further decrease. This water was considered as having become saturated with flue gases. The water container was transported from the Visakhapatnam steel plant to our laboratory on the university campus (~30 km). The flue gas–enriched water (FGW) was tested for the presence of sulfates that could have been formed from the SO_X in the flue gases. The standard $BaCl_2$ test (http://www.southernresearch.org/) was conducted to determine the SO_X content in FGW. The test proved negative for sulfates; no white precipitate ($BaSO_4$) was observed.

Preparation of nutrient medium using FGW: Bold's basal medium (BBM) (http://www.ccap.ac.uk/media/documents/BB_000.pdf) with half-strength nitrogen (since flue gas contains 72% NO_X) was prepared using FGW and the pH was adjusted to 6 (optimum pH for freshwater algae) using 1M NaOH solution.

19.3.2.1 Growth Assays of Microalgae in FGW

Growth of five freshwater microalgal species—*C. protothecoides* CCAP 211/7b, *S. dimorphus* UTEX 1237, *H. pluvialis* UTEX B 2505, *Neochloris oleoabundans* UTEX 1185, and *Desmodesmus* sp. (local isolate)—was tested in BBM prepared with FGW. The pH of the medium was adjusted to 6.0 which is optimal for their growth. A set of growth assays were also set up in BBM in FGW without adjusting the pH. Cultures grown in BBM prepared with DW (medium pH 6.0) were used as controls. The cultures for the experiments were set up in conical flasks. Microalgal cell inoculum was drawn from exponentially growing cultures. An inoculum volume that would constitute 20% of the final culture volume was dispensed. The culture volume in the flask was 70% of its capacity. The flasks were maintained at 90 μmol photons $m^{-2}s^{-1}$ irradiance at a 12:12 h light:dark cycle and a temperature of 22 ± 2°C. The flasks were agitated twice every day. The experiments were replicated in space (triplicate) and time (twice).

19.3.3 Measurement of Growth

Growth was measured through cell counts using a hemocytometer. Cell productivity (cells/mL/day) was calculated using the formula $C_2 - C_1/t_2 - t_1$ where C_1 refers to number of cells at the starting day (t_1) of log phase and C_2 is cell count at the end day (t_2) of log phase.

19.3.4 Results

C. protothecoides, *S. dimorphus*, and *Desmodesmus* sp. sustained the low pH (2.5) in BBM with FGW and showed significantly higher growth compared to controls (BBM in DW). The growth rate in these species was as follows: FGW (pH 6.0) > FGW (pH 2.0) > DW (pH 6.0) (Figure 19.6). *H. pluvialis* and *N. oleoabundans* were found to be intolerant to low pH FGW (pH 2.5) and showed no growth, but they showed significantly higher growth in BBM made with FGW with pH adjusted to 6.0 compared to controls (BBM in DW) (Figure 19.6).

19.3.4.1 Microalgae in Salvage of Fugitive Minerals and Toxic Heavy Metals from Contaminated Waters

The stability of an aquatic ecosystem is adversely affected by accumulation of nutrient minerals (nitrates and phosphates) and heavy metals (Cd, Pb, Se, As) through anthropogenic activity. Microalgae can be employed in removal of such minerals and heavy metals. Microalgae have high adsorption capacity and produce high biomass, and no secondary by-products are formed in remediation of wastewater (Jin-Fen et al., 2000). The special attributes of microalgae for this purpose include tolerance to extreme environments, cell wall composition, good sedimentation capacity, and capability of growing under mixotrophic conditions (Olguin, 2003).

Microalgal inoculum is introduced into the polluted medium. They grow and utilize/absorb the mineral and metal contaminants. The resulting microalgal biomass has to be harvested to clean up the water. This is difficult. Therefore, the microalgae are entrapped in low-cost polymers like chitosan, chitin, cellulose, or alginates in the form of beads and employed for uptake of these mineral salts (de la Noüe and Proulx, 1988; Garnham et al., 1992; Kaya et al., 1995; Moreno-Garrido et al., 2005). This strategy of immobilization facilitates easy harvesting of bioaccumulated mineral biomass. In addition, it has been reported that adsorption of minerals was higher in an immobilized microalgal biomass than in free living cells (Akhtar et al., 2003). Viable cells of microalgae show better adsorption than nonviable biomass (Moreno-Garrido, 2008). We are presently working on beading of microalgae and assessing the shelf life of the beads by testing the viability of the entrapped algal cells at different time intervals (Figure 19.7). Immobilization of microalgae on a flat surface has been experimented through twin layer technology (Figure 19.7). Here there are two layers: substrate and source layer. Microalgae are immobilized by self-adsorption on an ultrathin layer called the substrate layer; below this layer, medium is supplied through the source layer. Both the microalgae

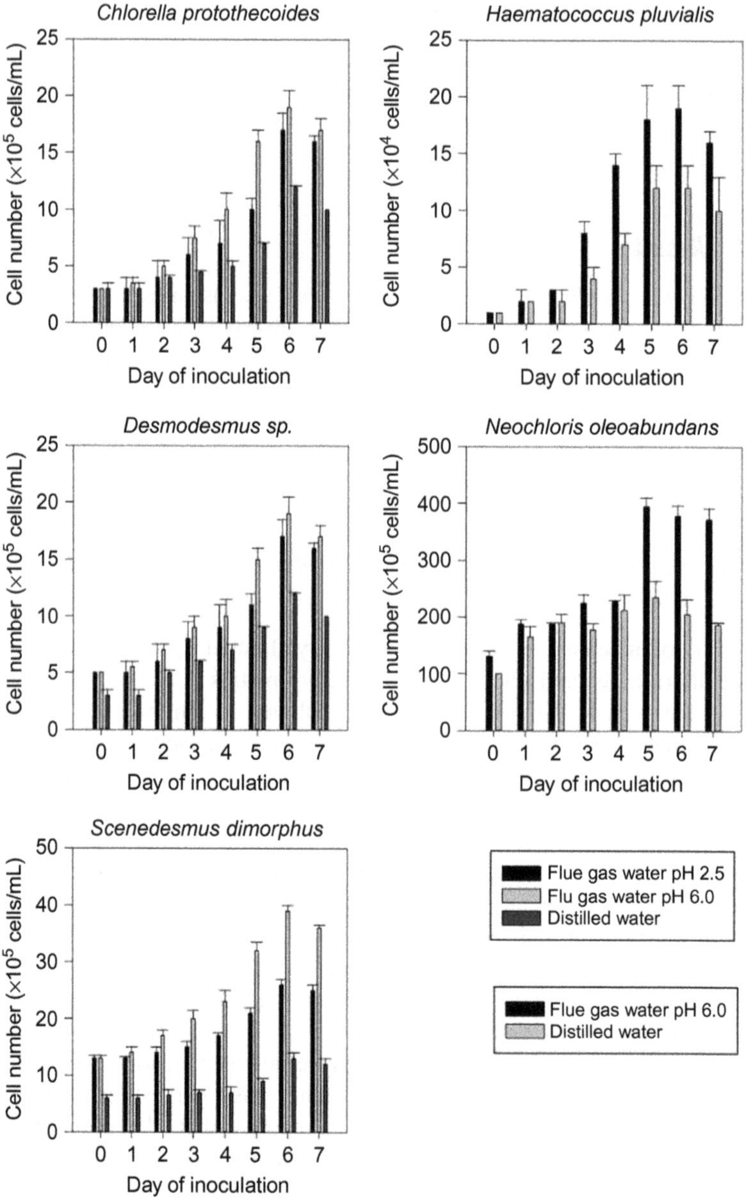

Figure 19.6 Growth of five freshwater microalgal species as measured from cell number in growth assays in nutrient medium prepared from flue gas–enriched water in comparison to nutrient medium made with DW.

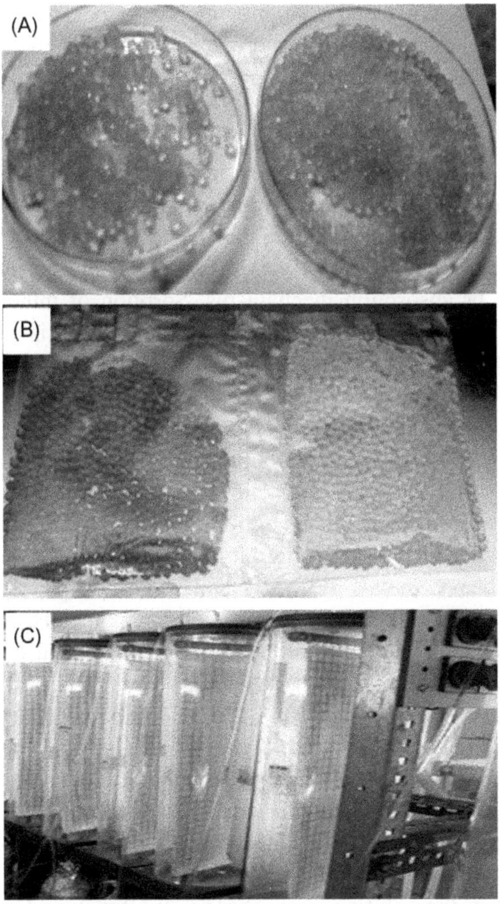

Figure 19.7 Immobilization of microalgae. (A) Entrapped green algae in sodium alginate beads; (B) packed alginate beads of a green alga and a diatom; and (C) microalgae immobilized on ultrathin twin layers.
Source: (C): http://www.melkonian.unikoeln.de/research_bioremidiation.htm.

and the growth medium are separated such that the microalgae receive nutrients through diffusion. The twin layer system was employed in subtraction of nitrogen and phosphorus from wastewater by immobilized *Chlorella vulgaris* and *Scenedesmus rubescens* (Shi et al., 2007).

Excessive input of nitrogen and phosphorus into water bodies results in eutrophication (Koelmans et al., 2001; Yang et al., 2008). Therefore, these are known as fugitive minerals. Agricultural runoff and wastes from pig farms, cattle farms, and aquaculture result in increased input of nitrates and phosphates into aquatic biosystems. Strains of *Chlorella* and *Scenedesmus* are promising candidates for treatment of such waters. *Chlorella* sp. produces biomass under autotrophic, mixotrophic, and heterotrophic conditions (Liang et al., 2009).

Among the toxic heavy metal contaminants of water, some—e.g., copper and zinc—are useful to microalgae as micronutrients. Other toxic water-polluting metals from industrial runoffs include cadmium, lead, and mercury. These are highly reactive and toxic to living cells (Cobbett and Goldsbrough, 2002). Some microalgal species are sensitive to heavy metal exposure as this results in inactivation of enzymes and disruption of the membrane integrity of the cell (Mallick and Rai, 1994). Microalgae that are tolerant to heavy metals and take up these elements from water have been identified. These algal species are isolated from habitats polluted with metals (Gekeler et al., 1988). Microalgae biosequester heavy metals through adsorption or absorption mechanisms. Sorption of minerals by these species takes place in two steps (Roy et al., 1993). The first step is a surface phenomenon: adsorption by cell wall polysaccharides and functional groups (Crist et al., 1981; Volesky, 1990). It is a rapid process. The second step is a slow process involving energy-dependent uptake of heavy metals into the cell interior (Wang et al., 2010). Principally algae remove the metals from water by an adsorption process. The negative groups interact with mineral ions and form complexes (Kaplan, 2004). Microalgae and other organisms preferentially synthesize metal-binding peptides, namely phytochelatin and metallothionein (MT), which neutralize the toxic effect caused by heavy metals (Cobbett and Goldsbrough, 2002; Perales-Vela et al., 2006). MTs are cysteine rich and can bind heavy metals. MTs act by chelating with metals such as cadmium, thereby reducing free metal ions. They are classified into three classes. Among them, two classes were reported in algae: Class II and Class III (Perales-Vela et al., 2006). Class III are secondary metabolites and products of peptides (Robinson, 1989). Torres et al. (1997) reported cadmium tolerance in *P. tricornutum* due to its ability to synthesize long-chain polypeptides of Class III MT. An increased intracellular concentration of phytochelatin was noted in *Thalassiosira weissflogii* and *Thalassiosira pseudonana* on exposure to cadmium and copper (Ahner et al., 2002). Efficiency of microalgae in metal removal depends on the pH of the solution, as it changes the functional groups like amino, amido, sulfate, and carboxyl on the cell wall (Jin-Fen et al., 2000; Donmez and Aksu, 2002). An increase in pH was found to promote removal of metal from solution (Tuzun et al., 2005; Fraile et al., 2005; Monteiro et al., 2010). Acid pretreatment of *C. vulgaris* enhanced Ni and Cu removal (Mehta and Gaur, 2001). Metal sorption potential increased due to the removal of previously bound cations from the binding site, with acid pretreatment consequently resulting in the uptake of metals (Mehta and Gaur, 2001). Donmez and Aksu (2002) reported that adsorption of Cr(IV) by *Dunaliella* decreased with an increase in salinity levels. The microalgae reported to sequester heavy metals are listed in Table 19.3.

Sivasubramanian et al. (2006) reported a successful bioremediation project with microalgae in an industrial setup. They claim to have set up the first (in 2006) phycoremediation plant in an industry: SNAP Natural & Alginate Products Pvt. Ltd, Ranipet, Tamil Nadu, India. The effluent generated by the industry is highly acidic and of very high TDS (40,000 mg/L). It is reported that the algal remediation technology helped in pH correction of the acidic industrial effluent and complete reduction of sludge. The entire effluent is evaporated using a slope tank with zero sludge

Table 19.3 Microalgal Species Employed to Biosequester Heavy Metals

Microalgal Species	Metal	Reference
Tetraselmis chuii	As	Irgolic et al. (1977)
Tetraselmis suecica	Cd	Perez-Rama et al. (2002)
Dunaliella sp.	Cr	Donmez and Aksu (2002)
Scenedesmus sp.	Hg	Inthorn et al. (2002)
Scenedesmus acutus	Pb, Cd	Inthorn et al. (2002)
Scenedesmus quadricula	Cd, Cu, Zn	Harris and Ramelow (1990)
Scenedesmus obliquus	Cr, Cu, Ni	Cetinkaya Dönmez et al. (1999)
Scenedesmus abundans	Cd, Cu	Terry and Stone (2002)
Scenedesmus incrassatulus	Cr, Cu, Cd	Pena-Castro et al. (2004)
Scenedesmus vacuolatus	Pb	Inthorn et al. (2002)
Scenedesmus obliquus	Cu	Mattuschka and Straube (1993)
Chlorella vulgaris	Hg, Cd, Pb	Inthorn et al. (2002)
	Au	Darnall et al. (1986)
	Ag	Harris and Ramelow (1990)
	Cd	Aksu (2001)
	Cu	Aksu et al. (1992)
	Cr(VI)	Nourbakhsh et al. (1994)
	Cr, Cu, Ni	Cetinkaya Dönmez et al. (1999)
	U	Sakaguchi and Nakajima (1991)
Chlorella fusca	Pb	Wehrheim and Wettern (1994)
Chlorella sorokiniana	Cd	Akhtar et al. (2003)
Chlorococcum sp.	Hg, Cd	Inthorn et al. (2002)

formation. The algal biomass produced is utilized in biofertilizer preparation and sold by the company. Sivasubramanian et al. (2006) claim that SNAP is now a zero-disposal company (http://kiachennai.com/publications.html).

References

Ahner, B.A., Wei, L., Oleson, J.R., Ogura, N., 2002. Glutathione and other low molecular weight thiols in marine phytoplankton under metal stress. Mar. Ecol. Prog. Ser. 232, 93–103.

Aizawa, K., Miyachi, S., 1986. Carbonic anhydrase and CO_2 concentrating mechanisms in microalgae and cyanobacteria. FEMS Microbiol. Lett. 39 (3), 215–233.

Akhtar, N., Saeed, A., Iqbal, M., 2003. *Chlorella sorokiniana* immobilized on the biomatrix of vegetable sponge of *Luffa cylindrica*: a new system to remove cadmium from contaminated aqueous medium. Bioresour. Technol. 88 (2), 163–165.

Aksu, Z., 2001. Equilibrium and kinetic modelling of cadmium (II) biosorption by *C. vulgaris* in a batch system: effect of temperature. Sep. Purif. Technol. 21 (3), 285–294.

Aksu, Z., Sag, Y., Kutsal, T., 1992. The biosorption of copperod by *C. vulgaris* and *Z. ramigera*. Environ. Technol. 13 (6), 579–586.

Antia, N.J., Berland, B.R., Bonin, D.J., Maestrini, S.Y., 1975. Comparative evaluation of certain organic and inorganic sources of nitrogen for phototrophic growth of marine microalgae. J. Mar. Biol. Assoc. UK 55 (3), 519–539.

Badger, M.R., von Caemmerer, S., Ruuska, S., Nakano, H., 2000. Electron flow to oxygen in higher plants and algae: rates and control of direct photoreduction (Mehler reaction) and rubisco oxygenase. Phil. Trans. R. Soc. Lond. Ser. B. Biol. Sci. 355 (1402), 1433–1446.

Bark, M., 2012. Cultivation of Eleven Different Species of Freshwater Microalgae Using Simulated Flue Gas Mimicking Effluents from Paper Mills as Carbon Source (Master's thesis). Chalmers University of Technology, Sweden.

Becker, W., 2004. Microalgae in human and animal nutrition. In: Richmond, A. (Ed.), Handbook of Microalgal Culture. Blackwell, Oxford, pp. 312–351.

Ben-Amotz, A., Polle, J.E.W., Subba Rao, D.V., 2009. The Alga *Dunaliella*: Biodiversity, Physiology, Genomics and Biotechnology. Science Publishers, Enfield, USA, 1578085454 (ISBN 13: 9781578085453).

Benemann, J.R., 2008. Open ponds and closed photobioreactors—comparative economics. Fifth Annual World Congress on Industrial Biotechnology and Bioprocessing, Chicago. Available from: <http://www.planktoleum.info/photobioreactors/BENEMANN2008-OpenPonds+ClosedPBR-ComparativeEconomics.pdf>.

Bloom, A.J., 2012. Methods for Capturing Carbon Dioxide. <http://www.eoearth.org/view/article/161514> In the Encyclopedia of Earth - A resource from CAMEL Climate adaptation and mitigation E-learning.

Borowitzka, L.J., 1991. Development of western biotechnology's algal β-carotene plant. Bioresour. Technol. 38 (2–3), 251–252.

Borowitzka, M.A., Moheimani, N.R., 2013. Sustainable biofuels from algae. Mitig. Adapt. Strat. Global Change. 18 (1), 13–25.

Bumbak, F., Cook, S., Zachleder, V., Hauser, S., Koyar, K., 2011. Best practices in heterotrophic high-cell-density microalgal processes: achievements, potential and possible limitations. Appl. Microbiol. Biotechnol. 91 (1), 31–46.

Cetinkaya Dönmez, G., Aksu, Z., Öztürk, A., Kutsal, T., 1999. A comparative study on heavy metal biosorption characteristics of some algae. Proc. Biochem. 34 (9), 885–892.

Chen, F., Johns, M.R., 1996. Heterotrophic growth of *Chlamydomonas reinhardtii* on acetate in chemostat culture. Proc. Biochem. 31 (6), 601–604.

Cid, A., Abalde, J., Herrero, C., 1992. High yield mixotrophic cultures of the marine microalga *Tetraselmis suecica* (Kylin) Butcher (Prasinophyceae). J. Appl. Phycol. 4 (1), 31–37.

Cifuentes, A.S., Gonzalez, M.A., Vargas, S., Hoeneisen, M., Gonzalez, N., 2003. Optimization of biomass, total carotenoids and astaxanthin production in *Haematococcus pluvialis* Flotow strain Steptoe (Nevada, USA) under laboratory conditions. Biol. Res. 36, 3–4.

Cobbett, C., Goldsbrough, P., 2002. Phytochelatins and metallothioneins: roles in heavy metal detoxification and homeostasis. Ann. Rev. Plant Biol. 53 (1), 159–182.

Crist, R.H., Oberholser, K., Shank, N., Nguyen, M., 1981. Nature of bonding between metallic ions and algal cell walls. Environ. Sci. Technol. 15 (10), 1212–1217.

D'Alessandro, D.M., Smit, B., Long, J.R., 2010. Carbon dioxide capture: prospects for new materials. Angew. Chem. 49 (35), 6058–6082.

Darnall, D.W., Greene, B., Henzl, M.T., Hosea, J.M., McPherson, R.A., Sneddon, J., et al., 1986. Selective recovery of gold and other metal ions from an algal biomass. Environ. Sci. Technol. 20 (2), 206–208.

de la Noüe, J., Proulx, D., 1988. Biological tertiary treatment of urban wastewaters with chitosan-immobilized *Phormidium*. Appl. Microbiol. Biotechnol. 29 (2–3), 292–297.

Den Hende, S.V., Vervaeren, H., Boon, N., 2012. Flue gas compounds and microalgae: (bio-) chemical interactions leading to biotechnological opportunities. Biotechnol. Adv. 30 (6), 1405–1424.

Divakaran, R., Pillai, V.S., 2002. Flocculation of algae using chitosan. J. Appl. Phycol. 14 (5), 419–422.

Donmez, G., Aksu, Z., 2002. Removal of chromium (VI) from saline wastewaters by *Dunaliella* species. Proc. Biochem. 38 (5), 751–762.

Fan, J., Huang, J., Li, Y., Han, F., Wang, J., Li, X., et al., 2012. Sequential heterotrophy–dilution–photoinduction cultivation for efficient microalgal biomass and lipid production. Bioresour. Technol. 112, 206–211.

Fraile, A., Penche, S., Gonzalez, F., Blazquez, M.L., Munoz, J.A., Ballester, A., 2005. Biosorption of copper, zinc, cadmium and nickel by *Chlorella vulgaris*. Chem. Ecol. 21 (1), 61–75.

Gadd, G.M., 2010. Metals, minerals and microbes: geomicrobiology and bioremediation. Microbiology. 156, 609–643.

Garcia, M.C.C., Miron, A.S., Sevilla, J.M.F., Grima, E.M., Camacho, F.G., 2005. Mixotrophic growth of the microalga *Phaeodactylum tricornutum*: influence of different nitrogen and organic carbon sources on productivity and biomass composition. Proc. Biochem. 40 (1), 297–305.

Garnham, G.W., Codd, G.A., Gadd, G.M., 1992. Accumulation of cobalt, zinc and manganese by the estuarine green microalga *Chlorella salina* immobilized in alginate microbeads. Environ. Sci. Technol. 26 (9), 1764–1770.

Gekeler, W., Grill, E., Winnacker, E.L., Zenk, M.H., 1988. Algae sequester heavy metals via synthesis of phytochelatin complexes. Arch. Microbiol. 150 (2), 197–202.

Grima, E.M., Belarbi, E.H., Fernandez, F.G.A., Medina, A.R., Chisti, Y., 2003. Recovery of microalgal biomass and metabolites: process options and economics. Biotechnol. Adv. 20 (7–8), 491–515.

Harris, P.O., Ramelow, G.J., 1990. Binding of metal ions by particulate biomass derived from *Chlorella vulgaris* and *Scenedesmus quadricauda*. Environ. Sci. Technol. 24 (2), 220–228.

Haszeldine, R.S., 2009. Carbon capture and storage: how green can black be? Science. 325 (5948), 1647–1652.

Ho, S.H., Chen, W.M., Chang, J.S., 2010. *Scenedesmus obliquus* CNW-N as a potential candidate for CO_2 mitigation and biodiesel production. Bioresour. Technol. 101 (22), 8725–8730.

Inthorn, D., Sidtitoon, N., Silapanuntakul, S., Incharoensakdi, A., 2002. Sorption of mercury, cadmium and lead by microalgae. ScienceAsia. 28, 253–261.

IPCC, 2005. In: Metz, B., Davidson, O., de Coninck, H.C., Loos, M., Meyer, L.A. (Eds.), IPCC Special Report on Carbon Dioxide Capture and Storage. Prepared by Working Group III of the Intergovernmental Panel on Climate Change. Cambridge University Press, Cambridge, UK, New York, NY.

Irgolic, K.J., Woolson, E.A., Stockton, R.A., Newman, R.D., Bottino, N.R., Zingaro, R.A., et al., 1977. Characterization of arsenic compounds formed by *Daphnia magna* and *Tetraselmis chuii* from inorganic arsenate. Environ. Health Perspect. 19, 61–66.

Jiang, Y., Chen, F., Liang, S.Z., 1999. Production potential of docosahexaenoic acid by the heterotrophic marine dinoflagellate *Crypthecodinium cohnii*. Proc. Biochem. 34, 633–637.

Jin-Fen, P., Rong-Gen, L., Li, M., 2000. A review of heavy metal adsorption by marine algae. Chin. J. Oceanol. Limnol. 18 (3), 260–264.

Kaplan, D., 2004. Absorption and adsorption of heavy metals. In: Richmond, A. (Ed.), Handbook of Microalgal Culture. Blackwell, Oxford, pp. 602–611.

Kaya, V.M., de la Noüe, J., Picard, G., 1995. A comparative study of four systems for tertiary wastewater treatment by *Scenedesmus bicellularis*: new technology for immobilization. J. Appl. Phycol. 7 (1), 85–95.

Kobayashi, M., Kakizono, T., Yamaguchi, K., Nishio, N., Nagai, S., 1992. Growth and astaxanthin formation *Haematococcus pluvialis* in heterotrophic and mixotrophic conditions. J. Ferment. Bioeng. 74 (1), 17–20.

Koelmans, A.A., Van der Heijde, A., Knijff, L.M., Aalderink, R.H., 2001. Integrated modelling of eutrophication and organic contaminant fate and effects in aquatic ecosystems: a review. Water Res. 35 (15), 3517–3536.

Kumar, A., Ergas, S., Yuan, X., Sahu, A., Zhang, Q., Dewulf, J., et al., 2010. Enhanced CO_2 fixation and biofuel production via microalgae: recent developments and future directions. Trends Biotechnol. 28, 371–380.

Kunjapur, A.M., Eldridge, R.B., 2010. Photobioreactor design for commercial biofuel production from microalgae. Ind. Eng. Chem. Res. 49 (8), 3516–3526.

Lavoie, A., Noue, J., 1983. Harvesting microalgae with chitosan. J. World Maricul. Soc. 14, 685–694.

Liang, Y.N., Sarkany, N., Cui, Y., 2009. Biomass and lipid productivities of *Chlorella vulgaris* under autotrophic, heterotrophic and mixotrophic growth conditions. Biotechnol. Lett. 31 (7), 1043–1049.

Lorenz, R.T., Cysewski, G.R., 2003. Commercial potential for *Haematococcus* microalga as a natural source of astaxanthin. Trends Biotechnol. 18, 160–167.

Mallick, N., Rai, L.C., 1994. Removal of inorganic ions from wastewaters by immobilized microalgae. World J. Microbiol. Biotechnol. 10 (4), 439–443.

Mattuschka, B., Straube, G., 1993. Biosorption of metals by a waste biomass. J. Chem. Technol. Biotechnol. 58 (1), 57–63.

Mehta, S.K., Gaur, J.P., 2001. Characterization and optimization of Ni and Cu sorption from aqueous solution by *Chlorella vulgaris*. Ecol. Eng. 18 (1), 1–13.

Melis, A., 2009. Solar energy conversion efficiencies in photosynthesis: minimizing the chlorophyll antennae to maximize efficiency. Plant Sci. 177 (4), 272–280.

Metzger, P., Largeau, C., 2005. *Botryococcus braunii*: a rich source for hydrocarbons and related ether lipids. Appl. Microbiol. Biotechnol. 66, 486–496.

Monteiro, C.M., Castro, P.M., Malcata, F.X., 2010. Cadmium removal by two strains of *Desmodesmus pleiomorphus* cells. Water Air Soil Pollut. 208 (1–4), 17–27.

Morais, K.C.C., Ribeiro, R.L.L., Santos, K.R., Taher, D.M., Mariano, A.B., Vargas, J.V.C., 2009. *Phaeodactylum tricornutum* microalgae growth rate in heterotrophic and mixotrophic conditions. Science. 8, 84–89.

Morales, J., De La Noue, J., Picard, G., 1985. Harvesting marine microalgae species by chitosan flocculation. Aquacult. Eng. 4 (4), 257–270.

Moreno-Garrido, I., 2008. Microalgae immobilization: current techniques and uses. Bioresour. Technol. 99 (10), 3949–3964.

Moreno-Garrido, I., Campana, O., Lubian, L.M., Blasco, J., 2005. Calcium alginate immobilized marine microalgae: experiments on growth and short-term heavy metal accumulation. Mar. Pollut. Bull. 51 (8), 823–829.

Nourbakhsh, M., Sag, Y., Ozer, D., Aksu, Z., Kutsal, T., Caglar, A., 1994. A comparative study of various biosorbents for removal of chromium (VI) ions from industrial waste waters. Proc. Biochem. 29 (1), 1–5.

Ogava, T., Aiba, S., 1981. Bioenergetic analysis of mixotrophic growth in *Chlorella vulgaris* and *Scenedesmus acutus*. Biotechnol. Bioeng. 23 (5), 1121–1132.

Olaizola, M., 2000. Commercial production of astaxanthin from *Haematococcus pluvialis* using 25,000 liter outdoor photobioreactors. J. Appl. Phycol. 12, 499–506.

Olguin, E.J., 2003. Phycoremediation: key issues for cost-effective nutrient removal processes. Biotechnol. Adv. 22 (1), 81–91.

Orosa, M., Franqueira, D., Cid, A., Abalde, J., 2001. Carotenoid accumulation in *Haematococcus pluvialis* in mixotrophic growth. Biotechnol. Lett. 23 (5), 373–378.

Ort, D.R., Zhu, X., Melis, A., 2011. Optimizing antenna size to maximize photosynthetic efficiency [W]. Plant Physiol. 155 (1), 79–85.

Pedroni, P., Davison, J., Beckert, H., Bergman, P., Benemann, J., 2001. A proposal to establish an international network on biofixation of CO_2 and greenhouse gas abatement with microalgae. J. Energy Environ. Res. 1 (1), 136–150.

Pena-Castro, J.M., Martınez-Jerónimo, F., Esparza-Garcıa, F., Canizares-Villanueva, R.O., 2004. Heavy metals removal by the microalga *Scenedesmus incrassatulus* in continuous cultures. Bioresour. Technol. 94 (2), 219–222.

Perales-Vela, H.V., Peña-Castro, J.M., Canizares-Villanueva, R.O., 2006. Heavy metal detoxification in eukaryotic microalgae. Chemosphere. 64 (1), 1–10.

Perez-Rama, M., Abalde Alonso, J., Herrero López, C., Torres Vaamonde, E., 2002. Cadmium removal by living cells of the marine microalga *Tetraselmis suecica*. Bioresour. Technol. 84 (3), 265–270.

Rajendran, P., Muthukrishnan, J., Gunasekaran, P., 2003. Microbes in heavy metal remediation. Ind. J. Exp. Biol. 41 (9), 935–944.

Ratledge, C., 2011. Are algal oils realistic options for biofuels? Eur. J. Lipid Sci. Technol. 113 (2), 135–136.

Reyer, C., Guericke, M., Ibisch, P., 2009. Climate change mitigation via afforestation, reforestation and deforestation avoidance: and what about adaptation to environmental change? New Forests. 38 (1), 15–34.

Robinson, N.J., 1989. Algal metallothioneins: secondary metabolites and proteins. J. Appl. Phycol. 1 (1), 5–18.

Rochelle, G.T., Bishnoi, S., Chi, S., Dang, H., Santos, J., 2001. Research needs for CO_2 capture from flue gas by aqueous absorption/stripping. Final Report for DOE Contract DE- AF26-99FT01029. Federal Energy Centre, Pittsburgh, PA, USA, pp. 154.

Roy, D., Greenlaw, P.N., Shane, B.S., 1993. Adsorption of heavy metals by green algae and ground rice hulls. J. Environ. Sci. Health A. 28 (1), 37–50.

Sakaguchi, T., Nakajima, A., 1991. Accumulation of heavy metals such as uranium and thorium by microorganisms. Miner. Bioprocess.309–322.

Shi, J., Podola, B., Melkonian, M., 2007. Removal of nitrogen and phosphorus from wastewater using microalgae immobilized on twin layers: an experimental study. J. Appl. Phycol. 19 (5), 417–423.

Sivasubramanian, V., 2006. Phycoremediation - issues and challenges. Indian Hydrobiol. 9 (1), 13–22.

Spolaore, P., Joannis-Cassan, C., Duran, E., Isambert, A., 2006. Commercial applications of microalgae. J. Biosci. Bioeng. 101, 87–96.

Stephens, E., Ross, I.L., King, Z., Mussgnug, J.H., Kruse, O., Posten, C., et al., 2010. An economic and technical evaluation of microalgal biofuels. Nat. Biotechnol. 28 (2), 126–128.

Suman, K., 2012. Bioprospecting of Biotechnologically Important Microalgae Form Hyper Saline and Marine Regions Along South Eastern Coast of India (PhD dissertation). Andhra University, Visakhapatnam, India.

Terry, P.A., Stone, W., 2002. Biosorption of cadmium and copper contaminated water by *Scenedesmus abundans*. Chemosphere. 47 (3), 249–255.

Torres, E., Cid, A., Fidalgo, P., Herrero, C., Abalde, J., 1997. Long-chain class III metallothioneins as a mechanism of cadmium tolerance in the marine diatom phaeodactylum tricornutum Bohlin. Aquatic Toxicol. 39 (3), 231–246.

Tuzun, I., Bayramoglu, G., Yalçın, E., Başaran, G., Çelik, G., Arıca, M.Y., 2005. Equilibrium and kinetic studies on biosorption of Hg(II), Cd(II) and Pb(II) ions onto microalgae *Chlamydomonas reinhardtii*. J. Environ. Manage. 77 (2), 85–92.

Umadevi, K., Swapna, G., Suman, K., 2014. Biosequestration of carbon dioxide—potential and challenges. In: Goel, M. (Ed.), Carbon Capture, Storage and Utilization. TERI Publishers, New Delhi, In press.

Umrania, V.V., 2006. Bioremediation of toxic heavy metals using acidothermophilic autotrophes. Bioresour. Technol. 97 (10), 1237–1242.

Vandamme, D., Foubert, I., Meesschaert, B., Muylaert, K., 2010. Flocculation of microalgae using cationic starch. J. Appl. Phycol. 22 (4), 525–530.

Volesky, B., 1990. Biosorption and biosorbents. In: Volesky, B. (Ed.), Biosorption of Heavy Metals. CRC Press, Boca Raton, FL, pp. 3–6.

Wan, M., Liu, P., Xia, J., Rosenberg, J.N., Oyler, G.A., Betenbaugh, M.J., et al., 2011. The effect of mixotrophy on microalgal growth, lipid content, and expression levels of three pathway genes in *Chlorella sorokiniana*. Appl. Microbiol. Biotechnol. 91 (3), 835–844.

Wang, L., Min, M., Li, Y., Chen, P., Chen, Y., Liu, Y., et al., 2010. Cultivation of green algae *Chlorella* sp. in different wastewaters from municipal wastewater treatment plant. Appl. Biochem. Biotechnol. 162 (4), 1174–1186.

Wehrheim, B., Wettern, M., 1994. Biosorption of cadmium, copper and lead by isolated mother cell walls and whole cells of *Chlorella fusca*. Appl. Microbiol. Biotechnol. 41 (6), 725–728.

Xu, X., Song, C., Wincek, R., Andresen, J.M., Miller, B.G., Scaroni, A.W., 2003. Separation of CO_2 from power plant flue gas using a novel CO_2 "molecular basket" adsorbent. Fuel Chem. Div. Prepr. 48 (1), 162–163.

Yang, X.E., Wu, X., Hao, H.L., He, Z.L., 2008. Mechanisms and assessment of water eutrophication. J. Zhejiang Univ. Sci. B. 9 (3), 197–209.

Yeh, K.L., Chang, J.S., 2012. Effects of cultivation conditions and media composition on cell growth and lipid productivity of indigenous microalga *Chlorella vulgaris* ESP-31. Bioresour. Technol. 105, 120–127.

20 Bioreactor and Enzymatic Reactions in Bioremediation

Arijit Nath[a], Sudip Chakraborty[a,b] and Chiranjib Bhattacharjee[a]

[a]Chemical Engineering Department, Jadavpur University, Kolkata, West Bengal, India, [b]Department of Chemical Engineering and Materials, University of Calabria, Cubo-42a,87036 Rende(CS), Rende (CS), Italy

20.1 Introduction

As the human population rapidly increases, it is suspected that during the next few years, the availability of fresh clean water will become severely limited all over the world (Chakraborty et al., 2013a). At the same time, the disposal of wastes from different industrial processes contaminates freshwater, resulting in threats to the ecosystem. Therefore, in the current century, because of implementation of stricter environmental legislation, the treatment of wastewater is considered to be a great challenge (Radjenovic et al., 2008). The goal of a wastewater treatment plant is to eliminate pollutant agents from the wastewater by means of physical and (bio)chemical processes (Basile et al., 2013). Wastewater treatment plants consisting of (in general) four treatment steps (1) a primarily mechanical pretreatment step, (2) a biological treatment step, (3) a chemical treatment step, and (4) a sludge treatment step (Bahja et al., 2011). Although different strategies have been developed for wastewater treatment in various process industries, but emphasis are now being given to implement the concept – "zero-effluent discharge." In this respect the membrane associated bioreactors are drawing considerable attention now a day (Mazzei et al., 2010).

20.2 Membrane-Associated Bioreactor

The recent developments in membrane technology and membrane-associated bioreactor have made the recycling of wastewater, as well as recovery and synthesis of valuable products convincing. The term *membrane-associated bioreactor* is a combination of two words: *membrane*, which is defined as a thin-walled material capable of selectively resisting the transfer of different solute molecules, and

Microbial Biodegradation and Bioremediation. DOI: http://dx.doi.org/10.1016/B978-0-12-800021-2.00020-0

bioreactor, which means a reaction vessel where a biological reaction takes place. An enzyme or microorganism is immobilized in the matrix and is used in the membrane bioreactor. In a catalytic membrane reactor, a membrane separation unit is associated with a bioreactor to separate out a reaction product (Giorno and Drioli, 2000). Shuler and Kargi (2002) proposed that the single substrate-dependent microbial growth kinetics, the Monod model, is analogous with the Michaelis–Mentent enzyme kinetics. Therefore, in the present chapter, microbial growth kinetics is considered to be analogous with the kinetics and behavior of enzyme under immobilized condition. In a membrane bioreactor, reaction and separation take place simultaneously under different operating parameters. Advantages of a membrane-associated bioreactor over a free-enzyme biochemical reaction are better product recovery along with heterogeneous reactions, reuse of enzymes, and continuous operation of the reactor. Moreover, due to possible separation of products and reactants, reversible reactions are always facilitated in this type of reactor (Nath et al., 2013). Enzyme immobilization may result in increasing enzyme stability across pH, temperatures, and storage times by providing a more suitable environment for enzymes (Chakraborty et al., 2013b). Compared with free enzymes in solution, membranes are particularly convenient immobilization supports because they provide high surface areas for improved enzyme loading and can be operated in a convective flow mode (Bhattacharyya et al., 1998). A slight disadvantage of membrane-associated bioreactors is reduction of enzyme activity at immobilized conditions due to structural and conformational changes. In contrast, it was found that the activity of the immobilized enzyme is increased after immobilization (Basile et al., 2013). The performances of membrane-associated bioreactor are considerably influenced by some operating conditions, such as substrate concentration, type of immobilized matrix material, transmembrane pressure (TMP), and stirrer speed. According to end-user requirement, different types of membrane and membrane modules are selected to increase throughput. Initially, application of membrane-based wastewater treatment was focused on tertiary treatment of secondary effluent to reuse the effluent for different purposes (Chakraborty et al., 2013b). However, the recent development (over the past 10 years) of membrane-associated reactors has emerged as an effective secondary treatment technology, which improves waste treatment as well as synthesis of valuable products from waste materials (Visvanathan et al., 2000).

20.2.1 Characteristics of Membrane-Associated Bioreactors

In membrane bioreactors, the biocatalysts are immobilized on the membrane matrix where simultaneous reaction and separation took place. In contrast, in many cases, a free-enzyme reactor assembled with a membrane separation unit, known as a catalytic membrane reactor is also used in biopharmaceutical industries (Giorno and Drioli, 2000). Immobilization of the enzyme is generally affected by three factors, namely the nature of the immobilization matrix, the types of immobilized enzymes, and the immobilization techniques (Basile et al., 2013).

20.2.1.1 Nature of the Immobilization Matrix

Two membrane characteristics are important in determining high enzyme immobilization: membrane suitability for simultaneous reaction and the separation process. The overall characteristics of a membrane are: (1) porosity, (2) morphology, (3) surface properties, (4) mechanical strength, and (5) chemical resistance. These characteristics depend on the membrane material as well as the fabrication technique. To a great extent, these properties are interrelated; e.g., a highly porous membrane structure can be maintained only if the polymer has adequate mechanical strength. Surface properties and pore morphology are related to the amount of enzyme immobilization, fouling properties, flux through the membrane, and solute separation (Scott, 1998). Furthermore, a high intramolecular gap in the membrane matrix leads to greater enzyme immobilization (adsorption process). Increasing the surface area of the membrane matrix provides a high amount of enzyme immobilization (Basile et al., 2013). Depending on the molecular weight cutoff (MWCO), membranes are classified as microfiltration (MF), ultrafiltration (UF), nanofiltration (NF), or reverse osmosis (RO) (Radjenovic et al., 2008). Figure 20.1 depicts the different MWCO membranes, which facilitates the permeation and rejection of solutes depending on their molecular weight.

Membranes are usually made up of different types of polymers, such as celluloses, polyamides, polysulfone, polyacrylonitrile, polyvinylidene difluoride, polyethylsulfone, polyethylene (PE), ethyl acetate, and polypropylene, and

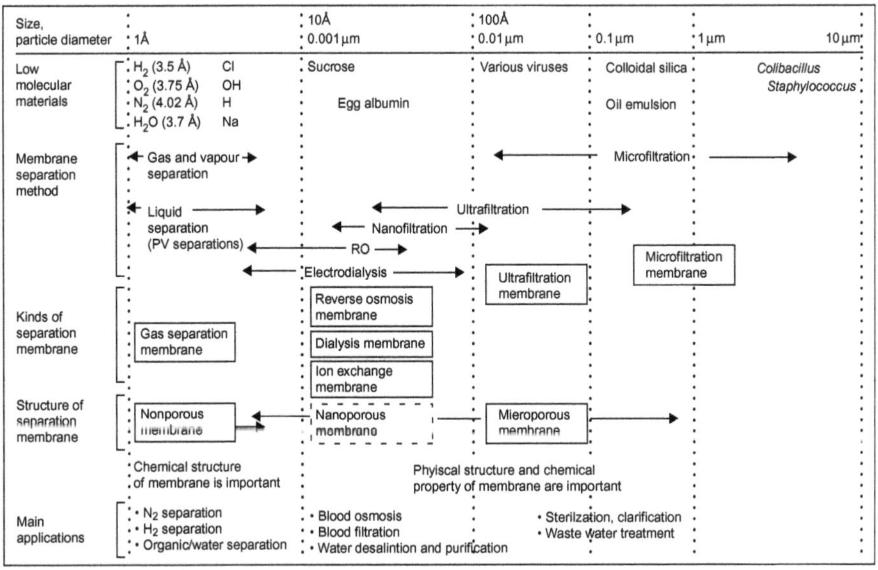

Figure 20.1 Different types of membrane depending on MWCO.
Source: Adapted from Till (2001).

occasionally ceramic materials. Depending on chemical property, a membrane may be hydrophobic or hydrophilic, which leads to high enzyme loading, low concentration polarization, and fouling. Therefore, all commercially available membranes are modified by chemical oxidation, organic chemical reaction, plasma treatment, or grafting to achieve a more hydrophilic surface (Butterfield, 1996; Butterfield et al., 2001; Wang et al., 2009). Sometimes nonspecific enzyme binding to the membrane surface occurs, which attributes the probability for enzyme denaturation on the porous matrix (Butterfield, 1996). Different commercial membranes with aldehyde or other active group functionality are available, and these are primarily cellulosic-, silica-, or poly- (ether sulfone)-based membranes. It was reported that cellulose acetate membranes can be easily hydrolyzed to cellulose by alkali treatment and subsequent oxidation to formation of functionalized membranes. By varying functionalization and casting conditions, such membranes can be selectively prepared with varying degrees of aldehyde incorporation, pore size, and thickness, which could be used for a stable and efficient immobilization media (Bhattacharyya et al., 1998).

20.2.1.2 Types of Immobilized Enzymes

Types, nature or characteristics of enzymes play a crucial role during the selection of immobilization procedure. Enzymes at immobilized condition are strongly influenced by environmental factors created by the membrane and immobilized materials. It has been found that the activity of lipase at microenvironment condition is better than the free mode. In addition, characteristics of the membrane structure, such as rigidity and hydrophobicity, affect the interaction between membrane and enzyme. Rigidity causes multipoint interaction and cross-linking between enzymes and the membrane matrix. Differences in the interaction will lead to the change in performance of enzyme to the substrate. Additionally, the concentration of enzymes also affects the immobilization. Performance of the immobilized enzyme increases up to a certain concentration of enzymes; subsequently it is saturated and decreases monotonically. This is caused by the formation of aggregates, weak interaction with the membrane surface, and oversaturation of the membrane pore (Chakraborty et al., 2013b). High concentration of enzymes leads to limited diffusion of the substrate to the enzyme and conformational changes of enzymes. Immobilized enzymes that are rendered water insoluble cannot be used in catalytic reactions or very high-molecular-weight substrates such as cellulose, starch collagen, fibroin, and DNA (Sakai-Kato et al., 2004).

20.2.1.3 Types of Immobilization Techniques

Membranes have a wide variety of surface characteristics and physicochemical properties, so they can be functionalized to provide a strong attachment for the enzymes (Bhattacharyya et al., 1998). Different types of immobilization techniques on a membrane matrix are shown in Figure 20.2. In general it was found that the amount of enzyme used and the contact time are directly proportional to the

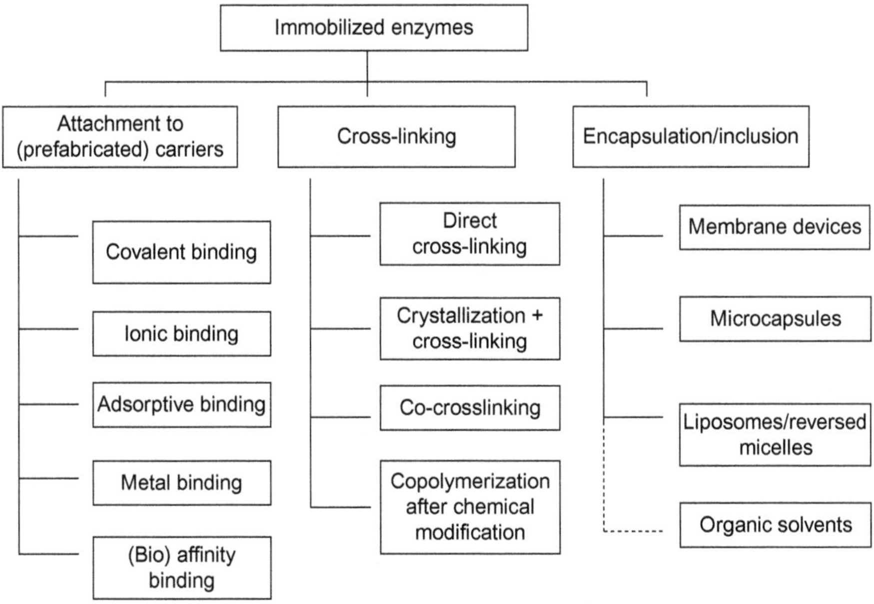

Figure 20.2 Different types of immobilization techniques on membrane matrix.
Source: Adapted from Tischer and Wedekind (2000) and Giorno and Drioli (2000).

amount of enzyme that can be immobilized until the limit (Ebrahimi et al., 2010). Once the maximum loading of enzymes is reached, no significant changes are found. This can be caused by saturation of the immobilization of enzymes on the membrane matrix (Bayramoglu et al., 2003). At immobilization conditions, sometimes an enzyme offers lower catalytic activity. Decrease of enzyme activity at immobilized conditions is caused by several factors; such as, not precisely positioning of enzymes and substrates, the attachment of the active sites of enzymes and the matrix, conformational changes of enzymes, denaturation of enzymes, and less appropriate conditions in the microenvironment. In some cases, the enzyme also needs time to adapt to the environment before it reaches maximum activity after the immobilization (Huang et al., 2011). In contrast, it was also reported that at immobilized conditions, the enzymes show better catalytic activity (Godjevargova et al., 2000).

20.2.1.3.1 Adsorption

Adsorption is the easiest technique for immobilizing the enzyme to the membrane matrix (Huckel et al., 1996; Hanefeld et al., 2009). The adsorption process usually occurs due to differences in the membrane surface charge and the charge of the enzyme that can create an interaction between membrane and enzyme (Godjevargova et al., 2000). The advantage of the adsorption technique over others

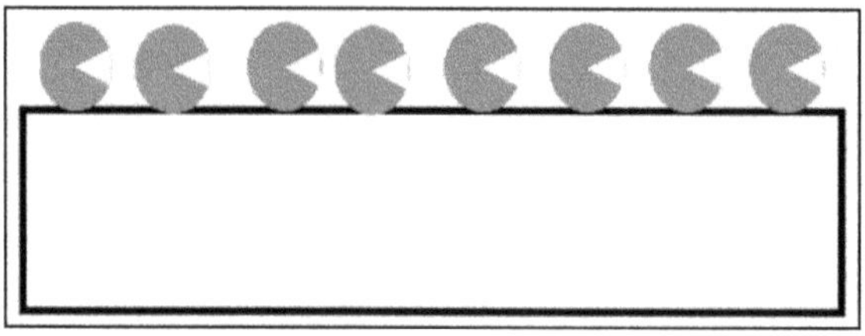

Figure 20.3 Adsorption technique for enzyme immobilization.
Source: Adapted from Basile et al. (2013).

is recovery of enzymes after completion of reaction, but desorption of enzymes offers lower activity of enzymes than original. Moreover, weak interaction between the enzyme and the membrane is also a slight disadvantage of this process (Deng et al., 2004). The adsorption mechanism is based on weak bonds such as the Van der Waal's forces and electrostatic and hydrophobic interactions between the enzyme and matrix (Hanefeld et al., 2009). Some membranes that have been used in this method are polysulfone, sulfonated polysulfone, polyethylenimine, polyacrylonitrile, PS-ZrO_2, alumina, and polypropylene. Some enzymes, namely, lipase, glucose oxydase, and β-galactosidase have been used using this immobilization technique (Pedersen et al., 1985; Deng et al., 2004; Wang et al., 2006; Cheng and Richard, 2010; Ebrahimi et al., 2010). Immobilization of enzymes by adsorption technique is depicted in Figure 20.3.

20.2.1.3.2 Entrapment

In the entrapment technique, an enzyme is immobilized at the pores of the matrix molecules. Asymmetric hollow fibers can provide an interesting support for enzyme immobilization. The enzyme can be entrapped in the sponge layer by cross-flow filtration in cases in which the pore size in the dense layer must retain the enzyme, permitting the passage of the substrate (Giorno et al., 2009). The amount of biocatalyst loaded, its distribution and activity through the support, and its lifetime are crucial issues in this process (Dave et al., 1994). Entrapment of the enzyme in the membrane matrix can be done in two ways: enzyme entrapment at an already-prepared membrane and enzyme entrapment during membrane preparation. This technique is not only used for solid membranes matrix but also applied for liquid membranes (Pal and Bhattacharya, 2002). Generally, poly(vinyl alcohol) (PVA)-tetramethoxysilane (TMOS), PVA-dimethyldimethoxysilane (DMDMOS), PVA-TMOS-DMDMOS, polystyrene, cellulose acetate (CA)-polytetrafluoroethylene (PTFE), PS, PAN, and poly(methyl methacrylate) (PMMA) have been used for these purposes. Several enzymes, such as lipase, β-glucosidase, glucose oxidase, horse radish peroxidase, and pig liver esterase (PLE), have been immobilized by this technique and are already reported (Pal and Bhattacharya, 2002; Ebrahimi et al., 2010).

Electrospun is a technique that can be used for immobilization. In this process, enzymes are immobilized on the membrane matrix during the preparation of membrane. Electrospun has huge advantages compared to other immobilization techniques: the large ratio of surface area to volume due to smaller diameter, large porosity, and easy recovery (Sakai-Kato et al., 2004). This technique has some

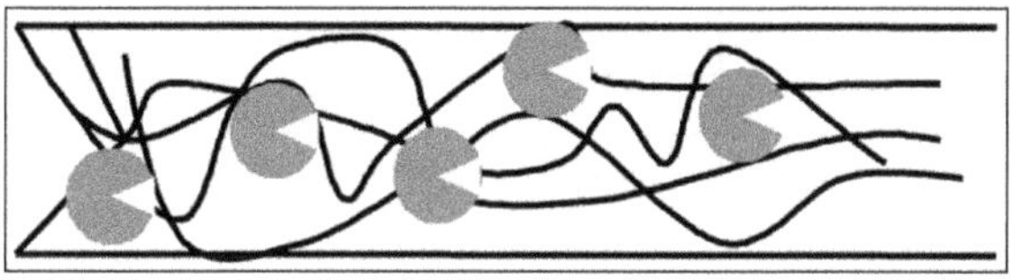

Figure 20.4 Entrapment technique for enzyme immobilization.
Source: Adapted from Basile et al. (2013).

disadvantages, i.e., when an enzyme is stuck in a membrane matrix, its activity decreases. This happens due to rigidity of the membrane structures, which cause mobility of the enzyme in the membrane. Substrate diffusion to the enzyme becomes lower, and then the activity is decreased (Sakai-Kato et al., 2004; Wang et al., 2011). Immobilization of enzymes by the entrapment technique is depicted in Figure 20.4. The simplest technique of entrapment is to circulate enzymes into the membrane reactor system, where most enzymes stick to the membrane (Ebrahimi et al., 2010). In 2006, Wang et al. successfully immobilized lipase into two kinds, and properties of membranes, i.e., PVA–PTFE and CA-PTFE to improve the stability and activity of enzymes (Godjevargova et al., 2000; Bora and Nahar, 2005, Wang et al., 2006). In several enzymes, a dual hydrophilic/hydrophobic microenvironment on the membrane could create an appropriate environment for the reaction of enzymes (Deng et al., 2004; Cheng and Richard, 2010).

20.2.1.3.3 Covalent Attachment

The covalent binding method is based on the binding of enzymes and membrane by covalent bonds (Ricca et al., 2010). In order to protect the active site, immobilization can be carried out in the presence of its substrate or a competitive inhibitor. Activity of the covalent bonded enzyme depends on the size and shape of the carrier material, the nature of the coupling method, the composition of the carrier material, and specific conditions during coupling. The main advantage of covalent attachment is that such an immobilization is very stable (Sakai-Kato et al., 2004). Covalent bond formation between the membrane and the enzyme can be done directly or through a "spacer." One major advantage of the covalent bond is strong immobilization, where the enzyme is attached to the membrane matrix by functional groups of enzymes. Covalent bonding is generally done by hydroxyl groups, sulfhydryl groups, carboxylic groups, tyrosine groups, and amino groups of enzymes (Eldin et al., 2011). In most of the chapter it was proposed that amino groups of the enzyme are bounded with the activated membrane, making the strong immobilization (Bayramoglu et al., 2003; Bora and Nahar, 2005). The use of "spacer" is interesting in this technique because it creates distance between the membrane and active sites of enzyme. This offers reduction in the denaturation process, and increases in the mobility of substrate enzyme (Huang et al., 2011). Huang et al. (2011) reported that a random bond between the enzyme and the membrane reduces the chance of bonding between the membrane and the active sites of enzymes. On the other hand, the random bond is actually responsible for blocking the active sites of enzymes (Cano and Palet, 2006). Immobilization of enzymes on a matrix by covalent bonding is depicted in Figure 20.5. These types

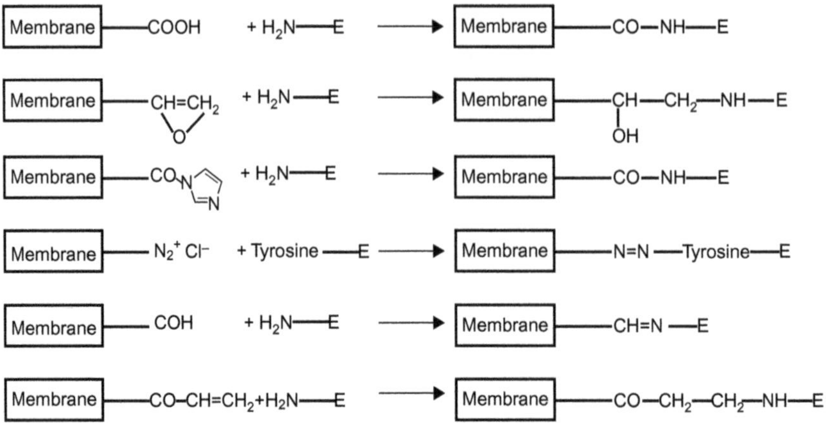

Figure 20.5 Covalent bonding for different types of membrane matrix and enzyme.

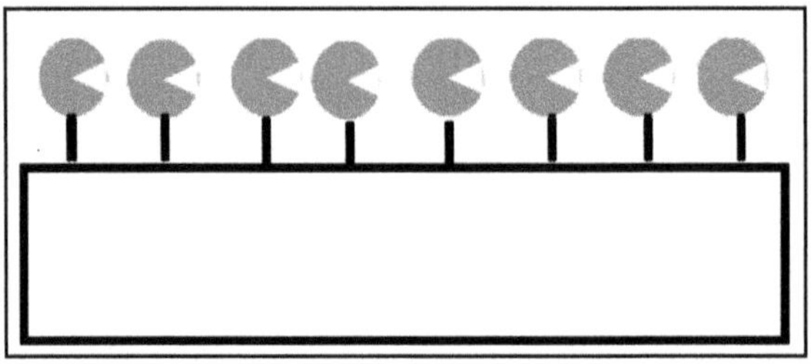

Figure 20.6 Covalent technique for enzyme immobilization.
Source: Adapted from Basile et al. (2013).

of membranes are more complex because it requires a membrane with "active" group that could react with enzyme group function. Several enzymes have been immobilized by this technique: PLE, lipase, xylanase, horse radish peroxidase, invertase, catalase, glucose oxidase, cellulase, β-galactosidase, ascorbic acid oxydase, α-amylase, and glucoamylase (Danisman et al., 2004; Cheng and Richard, 2010; Huang et al., 2011; Li et al., 2012). Different types of membranes, such as nylon, poly(2-hydroxyethyl methacrylate-co-methacrylamido-phenylalanine) (pHEMA−MAPA), poly (acrylonitrile-co-2-hydroxyethyl methacrylate) (PANCHEMA), polysulfone acrylate (PSA), regenerated cellulose (RC), poly (acrylonitrile-co-maleic acid) (PANCMA), PP, CA, polyaniline, PAN, poly(6-*O*-vinylscabacoyl D-glucose) (POVSEG), cellulose, poly (2-hydroxyethyl methacrylate-glicydil methacrylate) (pHEMA−GMA), PTFE, bromomethylated poly (2,6-dimethyl-1,4-phenylene oxide) (BPPO), PE, nylon, poly (urethane methacrylate-co-glycidyl-methacrylate) (PUA), poly(acrylonitrile-co-glycidylmethacrylate) (PAN-GM), chitosan, alumina, and silica have been used for this purpose (Bayramoglu et al., 2003; Danisman et al., 2004; Bora and Nahar, 2005; Li et al., 2012). Covalent bondings for different types of membranes and enzymes are depicted in Figure 20.6. It was reported that grafting between PTFE and polyacryl acid (PAA) and then reducing acrylate groups with HNO_2 produced active acylazide groups which can react with amino groups of enzymes (Chakraborty et al., 2013b). It was reported that existing epoxy groups on the pHEMA–GMA membrane react with the amino group of invertase and α-amylase (Bayramoglu et al., 2003). It was reported that active groups can be found on the membrane for oxidation of hydroxyl group to aldehyde group, which can react with the enzyme (Chakraborty et al., 2013b).

Covalent bond formations using a spacer are more appealing than conventional covalent processes. 1,4-Butanediol diglycidyl ether (BDE) can be used as a spacer where active epoxy groups react with enzymes. Carbodiimidazole (CDI) and 1,6-hexanediamine (HDA) for spacer with active imidazole or amino groups (Bayramoglu et al., 2003; Huang et al., 2011). Glutaraldehyde (GA) and cyanoborohydride (CBH) are meticulously used as spacers for different purposes, where active carboxylic groups of membrane easily react with enzymes (Huang et al., 2011). Other spacers are epichlorohydrin (ECH), cyanuric chloride (CC), and p-benzoquinone (pBQ). They are also used for these purposes (Huang et al., 2011). The spacer p-phenylenediamine can react with enzymes and membrane through the formation of diazonium salt. Two other kinds of spacers, biotin and avidin, can attach to each enzyme and membrane, and subsequently react together to made a spacer. It was found that direct immobilization activity is better than the use of a biotin–avidin spacer. Different techniques of covalent immobilization by a dual layer of biomimetics, derived from chitosan and gelatin, were developed. This dual layer can increase the amount of immobilized enzyme on a membrane. It was found that the presence of 1-ethyl-3-(dimethy-aminoropyl) carbodiimide hydrochloride (EDC)/*N*-hydroxy succinimide (NHS) as a coupling agent, and GA to make a strong covalent bond, allows more enzyme immobilization. The use of a biomimetic also offers high enzyme activity compared to without biomimetic. This attributes the formation of a suitable microenvironment, proper active sites position, and decreasing of the possibility of random covalent bonds (Chakraborty et al., 2013b).

In all types of covalent bond formation, it was found that increased contact time did not provide better stability. It can be caused by change in enzyme conformation or nature of bond formed with the enzyme active sites (Tominaga et al., 2004). Covalent bond formation is not limited to the organic membrane but can also occur with inorganic membranes such as alumina and silica. Improvement of thermal stability in the covalent bonding technique might be caused by limited enzyme conformation, caused in turn by multiple bonds that occur between the enzyme and the membrane (Bayramoglu et al., 2003; Huang et al., 2011). Covalent bonds lower conformational flexibility, so that energy needed to reach a state of equilibrium is higher (Danisman et al., 2004). Liu et al. (2012) found that the storage stability of covalent techniques is better than that of adsorption methods. This is caused by the presence of long chains blocking the membrane surface, so that the area that can be occupied by the enzyme is small (Deng et al., 2004). Immobilization of enzymes by the encapsulation method has been described in Figure 20.6.

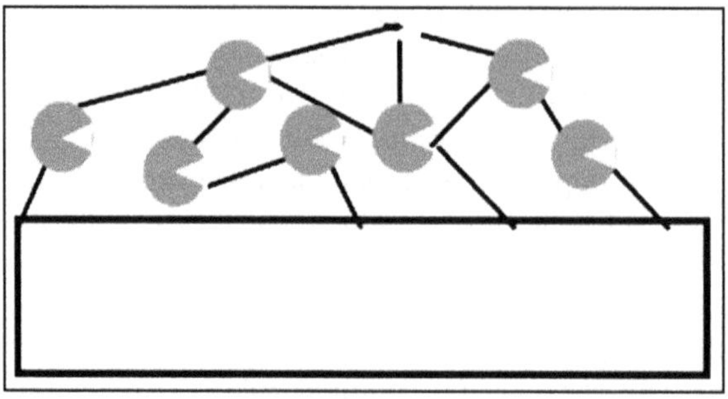

Figure 20.7 Cross-linking technique for enzyme immobilization.
Source: Adapted from Basile et al. (2013).

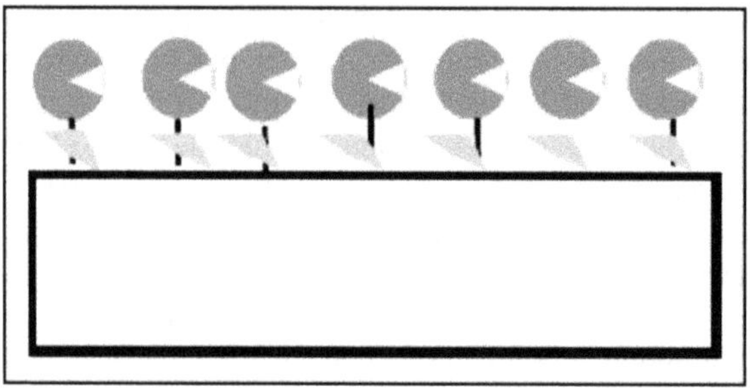

Figure 20.8 Affinity techniques for enzyme immobilization.
Source: Adapted from Basile et al. (2013).

20.2.1.3.4 Cross-Linking

Immobilization of enzymes is also possible by intermolecular cross-linking, either to other protein molecules or to functional groups on an insoluble matrix (Eldin et al., 2011). In case of cross-linking, the activity of enzymes is reduced; the process is also expensive. Since the enzyme is linked to the matrix, very little desorption is likely to occur in this method. This method is particularly attractive due to its simplicity and the strong chemical binding achieved between biomolecules (Sheldon, 2007). In Figure 20.7, cross-linking techniques of enzyme immobilization are depicted.

20.2.1.3.5 Affinity

This is a method of enzyme immobilization by a site-specific group of biomolecules on the matrix. This method allows control of the enzyme orientation in order to avoid enzyme deactivation and/or active site blocking. Several affinity methods have been described for the immobilization of enzymes. Improvement can be achieved by introducing a spacer molecule (Nouaimi et al., 2001). In this type, the aim is to create (bio-) affinity bonds between an activated support and a specific group of the protein sequence. An enzyme can contain affinity tags in its sequence (e.g., a sugar moiety) but, in some cases, the affinity tag (e.g., biotin) needs to be attached to the protein sequence by genetic engineering methods such as site-directed mutagenesis, protein fusion technology, or posttranscriptional modification (Andreescu and Marty, 2006). In Figure 20.8, cross-linking techniques of enzyme immobilization have been depicted. A comparative study for different types of immobilization procedures is described in Table 20.1.

Table 20.1 Characteristics of Different Immobilization Techniques

Binding Method	Binding Nature	Advantages	Disadvantages
Adsorption	Weak bonds	• Simple and easy • Limited loss of enzyme activity	• Desorption of enzyme • Nonspecific immobilization
Covalent binding	Chemical bindings between functional group of enzyme with matrix	No diffusion barrier Stable and short response time High enzyme activity loss	• Matrix not regenerable • Coupling with isoenzyme
Entrapment	Incorporation of the enzyme within the matrix	• No chemical reaction between the matrix, and the enzyme that could affect the activity of enzyme • Several types of enzymes can be immobilized within the same matrix	• Enzyme leakage and diffusional barrier • High concentrations of monomer and enzyme for electropolymerization
Cross-linking	Enzyme bind with matrix by cross-linker (e.g., GA)/inert molecule	• Simple	• Enzyme activity loss
Affinity	Affinity bonds between the functional group of enzyme with matrix	• Controlled and oriented immobilization	• Need of the presence of specific groups on enzyme

20.2.2 Hydrodynamics and Biochemical Reactions of Membrane-Associated Bioreactors

The membrane-associated process is a pressure-driven, fine-tuned separation technique. During the membrane filtration process, the high-molecular-weight component is rejected, while the low-molecular-weight component is passed freely through the porous channel of the membrane, known as permeate. The most important issues of the membrane separation process are low concentration polarization, fouling, which affects permeate flux, and rejection (Genkoplis, 1993). The membrane matrix, biochemical reactions, and other operating parameters also influence the permeate flux and rejection. To study the membrane bioreactor instead of the catalytic membrane bioreactor is much more challenging because reaction and separation take place at the same time. Therefore in the present chapter, hydrodynamics and biochemical reactions of membrane bioreactors have been elucidated.

Nagy and Kulcsar (2009) postulated about the mass transfer equation of membrane bioreactor (MBR), where inlet mass transfer rate, concentration profile of the boundary layer, and biocatalytic membrane layer were introduced. The principle of

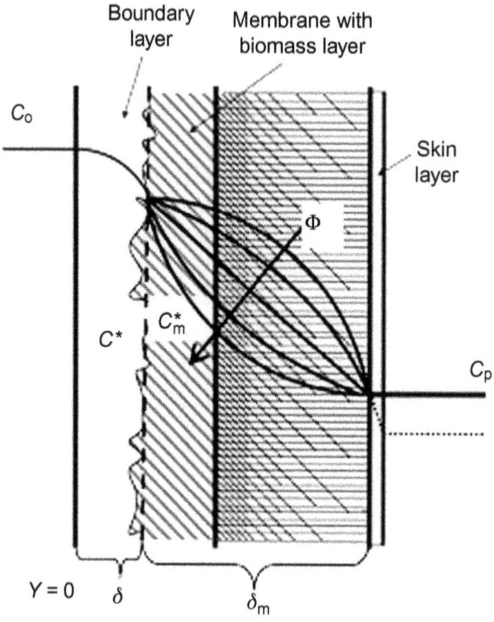

Figure 20.9 The membrane layer with immobilized enzyme thickness of δ_m, and the concentration polarization layer with thickness of δ, c^*, and c^*_m are concentrations on the liquid membrane interface (→: increasing reaction modulus).
Source: Adapted from Nagy and Kulcsar (2009).

the mass transport of substrates into the gel layer of enzyme (enzyme immobilized in membrane matrix) depends upon the mass transport parameters, namely, diffusion coefficient, convective velocity, and the bioreaction rate constant. He also proposed different cases in this respect, such as, first- and zero-order reaction kinetics, as well as the Michellis–Mentent reaction kinetics. Figure 20.9 illustrates the concentration profile of substrate and product molecules in MBR along with a boundary layer.

Several assumptions were considered in that proposed model. Those are as follows:

1. Reaction occurs at every position within the biocatalyst layer.
2. Reaction has one rate-limiting substrate or nutrient.
3. Mass transport through the biocatalyst layer occurs by diffusion and convection.
4. The partitioning of the components (substrate, product) is negligible (thus, $c^* = c^*_m$).
5. The mass transport parameters (diffusion coefficient, convective velocity, bioreaction rate coefficient) are constant or varying.
6. The effect of the concentration boundary layer should also be taken into account.

The differential mass balance equations of permeation for membrane bioreactor are as follows.

20.2.2.1 For the Michaelis–Mentent Reaction

$$\left\{\frac{d}{dy}\left(D\times\frac{dc}{dy}\right)-\frac{d(v\times c)}{dy}\right\}-\frac{v_{\max}\times c}{K_m\times c}=0 \qquad (20.1)$$

20.2.2.2 For the First-Order Biochemical Reaction

$$D_m \times \frac{d^2c}{dy^2} - \upsilon \times \frac{dc}{dy} - k_1 \times c = 0 \qquad (20.2)$$

20.2.2.3 For the Zero-Order Reaction

$$D_m \times \frac{d^2c}{dy^2} - \upsilon \times \frac{dc}{dy} - k_0 = 0 \qquad (20.3)$$

where c: concentration of solute in the solvent, D_m: diffusion coefficient, k_0: zero-order enzymatic reaction rate constant, k_1: first-order enzymatic reaction rate constant, K_m: the kinetics constant of the Michaelis–Menten equation, v_{max}: maximum enzymatic reaction velocity, and υ: convective flow of solute.

20.2.3 Different Types of Membrane-Associated Bioreactor

Membrane-associated bioreactors are classified into two categories: membrane bioreactors and catalytic membrane bioreactors.

20.2.3.1 Membrane Bioreactors

In MBR, the membranes are directly submerged in the liquid contain in bioreactor. In this process, the polymeric membrane typically involves substantially more relative membrane area per unit volume of system. Here the membranes are either horizontally or vertically oriented hollow fibers containing a rectangular or tubular support structure, or vertically oriented flat sheets contained within a support structure. The mixed liquor is placed at the shell side of the membrane module, and the effluent is extracted into the lumen of the membrane. The negative pressure on the lumen or permeate side of the membrane creates the driving force across the membrane. Continuous or discontinuous aeration, stirring, membrane rotation, and TMPs are used in certain designs to reduce the concentration polarization of the membrane module (Giorno and Drioli, 2000; Mazzei et al., 2010; Basile et al., 2013).

20.2.3.1.1 Extractive Membrane Bioreactors

Extractive membrane bioreactors enhance the performance of the process by exploiting the membrane's ability to accomplish a high degree of separation where components are transported from one phase to another. The extractive membrane bioreactor can be operated in two modes, as illustrated in Figure 20.10. These technologies have been successfully used for treatment of organic pollutants such as chloroethanes, chlorobenzenes, chloroanilines, and toluene.

Mode 1: In this situation, the membrane is deep in the bio-medium tank. The feed solution is circulated through the immerse membranes, and due to the

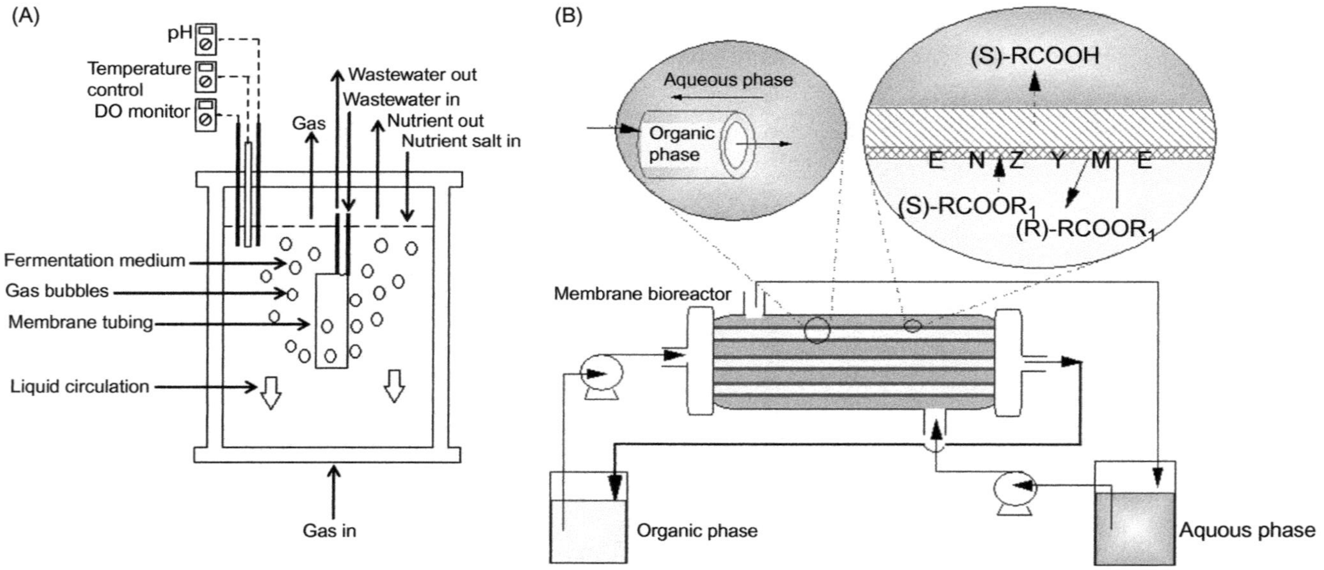

Figure 20.10 Different types of extractive membrane bioreactors: (A) Mode 1 operation and (B) Mode 2 operation.

concentration gradient, the toxic compounds are selectively passed to the surrounding bio-medium. Selective microbial cultures are cultivated in the reactor, which are responsible for degradation of waste.

Mode 2: In this type of membrane bioreactor, the wastewater containing toxic organic pollutants is circulated through the tube side of the tubular membrane and placed in the bio-medium tank. Due to the concentration gradient, the toxic pollutant is transported to the bio-medium. Within the bio-medium, the toxic pollutant is continuously degraded by selective microorganisms or stored in a particular storage tank. The treated wastewater is removed from the other end of the membrane module.

20.2.3.1.2 Bubble-Less Aeration Membrane Bioreactors

In the membrane aeration bioreactor, gas permeable membranes directly supply high-purity oxygen to a biofilm without bubble formation. The large amount of bubble-free aeration is achieved by synthetic polymeric membrane between a gas phase and a liquid phase. Here a very high air transfer rate is attained because the gas is practically diffused through the membrane. Generally a plate and frame and hollow fiber module is used for this purpose. In a conventional-activated sludge process, the efficiency of the process is controlled by the availability of air. Due to an inefficient mode of air supply, more than 80% of the oxygen is diffused as air into the atmosphere. Oxygenation by pure oxygen, as opposed to air, as an aeration medium would increase the overall mass transfer and biodegradation rate. However, conventional aeration devices have high-power requirements due to the high rate of mixing; these devices are not suitable with biofilm processes. Biofilm processes are advantageous as they enable retention of high concentrations of active bacteria. However, current research has focused on the hollow fiber arrangement with gas on the lumen side and wastewater on the shell side. The advantage of hollow fiber modules is their high surface area which gives the lower footprint. In this case, the membrane acts as a support medium for the biofilm formation; at the same time, it reduces the potential for bubble formation and the air transfer rate. In Figure 20.11, a bubble-less aeration membrane bioreactor is depicted.

20.2.3.1.3 Recycle Membrane Bioreactors

The recycle membrane bioreactor is a vessel where enzymes or microorganisms are immobilized on the membrane surface. The substrates are placed either in the reactor or pumped into the reaction vessel. The substrates are converted in the presence of enzymes or microorganisms and low-molecular-weight components (end products of the enzymatic reaction) are permeated through the membrane, while the high-molecular-weight fractions are rejected and/or recycled back into the feed tank. In these processes, the microbial species are attached to the membrane via adsorption, entrapment, micro encapsulation, etc., and the microbial biomass is separated out at the end of the run. Generally, membrane recycle reactors are operated at continuous mode. The advantages of this process are lower operating cost, reuse of enzymes, and less product inhibition of the enzymatic process. In industry, recycle membrane bioreactors are used in two different configurations: the tubular and

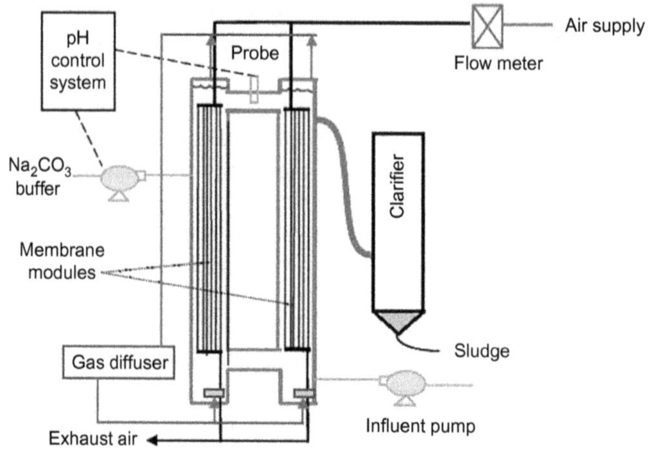

Figure 20.11 Bubble-less aeration membrane bioreactor.

beaker types. In tubular configurations, the biocatalysts are immobilized either in the shell side (annular space between the membrane fibers and the housing) or the tube side of a tubular membrane module. When the biocatalyst is immobilized inside the membrane tube, the feed substrate is pumped through the shell side. The opposite are also true when biocatalyst is immobilized on peripheral layer of the membrane tube. In the beaker type system, the reaction solution (substrate and biocatalyst) is placed in a reaction vessel. A U-shaped bundle of membrane fibers is dipped into the reaction vessel and product is continuously filtered through the membranes. Sometimes, an immobilized membrane is also attached to the end side of reaction vessel. In Figure 20.12, a recycle membrane bioreactor is depicted.

20.2.3.1.4 Membrane Separation Bioreactors

The emerging technology of biomass separation bioreactors is a combination of a suspended growth reactor for biodegradation of wastes and membrane filtration, known as membrane separation bioreactors. In this process, the solid–liquid membrane separation bioreactor occupies filtration modules as effective barriers. The membrane unit can be placed either in external loop or within the bioreactor. This configuration is popularly incorporated in activated sludge processes as a part of aerobic wastewater treatment systems. Applications of membrane separation (micro- or ultrafiltration) techniques for biosolid separation in a conventional-activated sludge process diminish the disadvantages of the sedimentation and biological treatment steps. In Figure 20.13, the developments of membrane separation bioreactors are depicted, which offers the process intensification, and lower footprint of the wastewater treatment technology. In this process, sludge retention time (SRT) is independent of hydraulic retention time (HRT). Therefore, a very long SRT is generally maintained, resulting in complete retention of slowly growing microorganisms in the system.

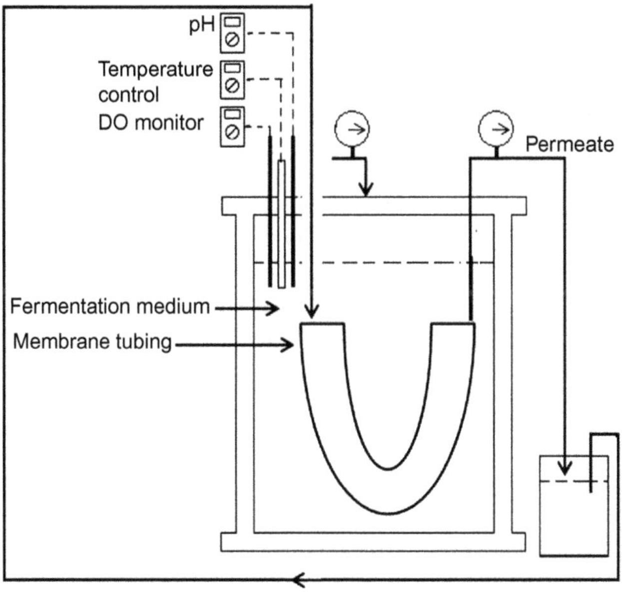

Figure 20.12 Recycle membrane bioreactor.

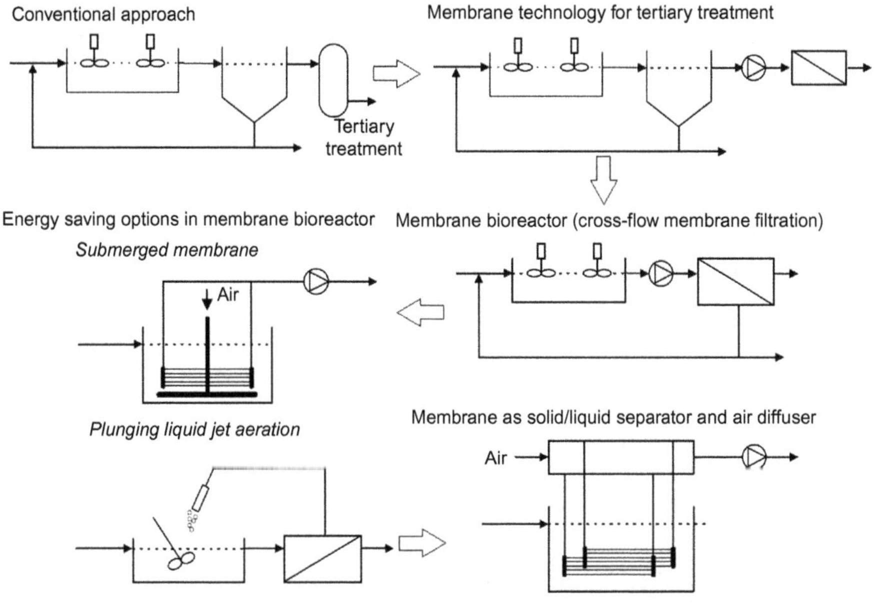

Figure 20.13 Generation of membrane separation bioreactor.
Source: Adapted from Visvanathan et al. (2000).

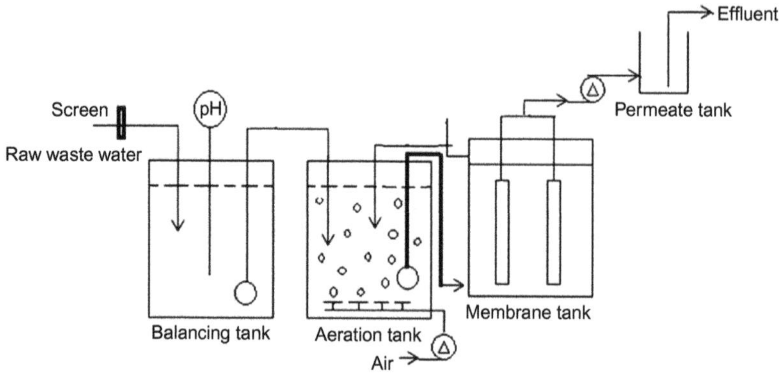

Figure 20.14 Catalytic membrane bioreactor.

20.2.3.2 Catalytic Membrane Bioreactors

When the membrane separation unit is externally attached to the reaction vessel, this is known as a catalytic membrane reactor. Here polymeric or ceramic membranes are attached in the configuration of cross-flow membrane module configuration in some cases, a dead-end membrane module is also used. The substrate and biocatalyst are placed in the reaction vessel at predetermined concentrations, and the reaction mixture is continuously pumped through the external membrane circuit. The low-molecular-weight components (end products of the enzymatic reaction) are permeated through the membrane, while the high-molecular-weight biocatalysts are rejected and recycled back into the reaction tank. Depending upon the situation, reverse configuration could also be used. To achieve desired shear across the membranes, optimum operating parameters are chosen for this system. The capital and operating costs associated with this system are comparatively higher than in an MBR system. A simplified schematic diagram of a catalytic membrane reactor is depicted in Figure 20.14. A comparative study in a comprehensive manner of external MBR systems and internal MBR systems is given in Table 20.2.

20.3 Applications of Membrane-Associated Bioreactors

The membrane associated bioreactors are now being used in different effluent treatment process with the objective of environmental remediation, as well as in some cases for the production of different value added products.

20.3.1 Treatment of Municipal Wastewater

Membrane-associated bioreactors were initially used in municipal wastewater treatment, particularly in the area of water reuse and recycling. In the mid-1990s, the development of less expensive submerged membrane bioreactors made

Table 20.2 Comparative Study of Internal MBR Systems and External MBR Systems (Table adapted from Sutton, 2006)

Comparative Factor	Internal MBR Systems	External MBR Systems
Membrane area	High membrane area per unit volume.	Low membrane area per unit volume.
Place of membrane component	Membranes are deep in the reaction tank.	Membranes are externally attached with the reaction tank.
Bioreactor and membrane component design and operation dependency	Bioreactor and membrane are attached to each other, and they are operated at same operating conditions.	Bioreactor and membrane unit are designed separately, and their operating conditions are different.
Membrane performance consistency	Alteration of membrane cleaning strategy and/or cleaning frequency is required much. It can be used for 5 years after its installation.	Less required of cleaning strategy and/or cleaning frequency of membrane. It can be used for 7 years after its installation.
Application status	Full scale application is widespread in the United States.	Many conventional technologies are replaced by this emerging technology.
Economics	Power and capital cost are low because both units are assembled in the same unit.	Power and capital cost are comparatively high because both units are attached separately.

conventional membrane bioreactors a viable alternative in municipal wastewater treatment (Kimura, 1991). The combination of membrane separation technology with anaerobic bioreactors has been mainly employed for industrial or high-strength wastewater and municipal wastewater treatment. The main advantages of this process are complete biomass retention, lower sludge production, enhanced high-quality effluent, and low operating cost (no stirring for aeration). The microbial consortium reduces the waste materials (both inorganic and organic) through adsorption and/or accumulation. In Table 20.3, applications of membrane-associated bioreactors in municipal wastewater treatment with respect to configuration of the membrane and size of operation (bench, pilot, or full scale) have been reported.

20.3.2 Treatment of Industrial Wastewater

Rapid growth of industries including pulp and paper, textiles, chemicals, pharmaceuticals, petroleum, and tanneries is resulting in a large quantity of effluent discharges. Industrial wastewater is usually characterized by high inorganic and/or organic load, extreme physicochemical nature (e.g., pH, temperature, salinity) and toxic synthetic and natural substances.

Table 20.3 Applications of Membrane-Associated Bioreactor for Domestic and Municipal Wastewater Treatment

Membrane Type	Configuration	Size of Operation	Treatment Success	Reference
Ceramic UF	Plate and frame external	Full scale, average capacity 125 m^3/day	Effluent COD <5 mg/L	Manem (1996)
Ceramic UF	Tubular external	Pilot scale, 2.4–4.8 m^3/day	COD removal >94%	Fan et al. (1996)
Ceramic UF	Tubular external	Bench scale, 0.16 m^3/day	COD removal >98%	Cicek et al. (1998)
Polymeric UF	Hollow fiber submerged	Pilot scale, capacity < 1.5 m^3/day	Effluent COD <10 mg/L	Chiemchaisri et al. (1993)
Polymeric UF	Hollow fiber submerged	Pilot scale, capacity 2.6–5.0 m^3/day	COD removal >96.5%	Cote et al. (1998)
Polymeric UF	Hollow fiber submerged	Pilot scale, capacity 6–9 m^3/day	COD removal >95%	Rosenberger et al. (2002)
Polymeric UF	Hollow fiber submerged	Pilot scale, capacity 46–74 m^3/day	COD removal >93%	Van de Roest (2002)
Polymeric UF	Plate, and frame submerged	Pilot scale, capacity 48–72 m^3/day	COD removal >91%	Van de Roest (2002)
Polymeric UF	Hollow fiber submerged	Full scale, capacity 750 m^3/day	Effluent BOD5 <1 mg/L	Garcia and Kanj (2002)
Polymeric UF	Hollow fiber submerged	Full scale, capacity 9000 m^3/day	COD removal >95%	Lorenz et al. (2002)
Polymeric MF	Hollow fiber submerged	Pilot scale, capacity 1.4–3.8 m^3/day	Effluent BOD5 <3 mg/L	Adham and Trussell (2001)
Polymeric UF	Cartridge disk external	Pilot scale, capacity 48 m^3/day	Effluent COD <5 mg/L	Ahn et al. (1999)
Polymeric UF	Tubular external	Pilot scale, capacity 360–840 m^3/day	Effluent TC <12 mg/L	Muller et al. (1995)
Stainless steel metal UF	Tubular submerged	Bench scale, reactor vol. 100 L	Most favorable efficiency	Kim and Jung (2007)
Polymeric UF	Hollow fiber External	Bench scale, reactor vol. 8 L	VS destruction >52.1%	Xu et al. (2010)
Cylindrical woven nylon UF	Submerged	Bench scale	–	Walker et al. (2009)
Kubota polymeric flat sheet	Submerged	Pilot scale, reactor vol. 3 L	COD removal >90%	Trzcinski and Stuckey (2009)
Polymeric membrane	Vibration external UF membrane	Bench scale, reactor vol. 550 L	Average total suspended solid removal 51% Average VS removal 59%	Pierkiel and Lanting (2005)
Polymeric membrane	Tubular MF Submerged	Bench scale, reactor vol. 3.8 L	–	Jeison et al. (2008)

Polymeric membrane	Flat-sheet UF submerged	Bench scale, reactor vol. 3 L	COD removal 90%	Trzcinski and Stuckey (2010)
Polymeric membrane	Flat-sheet MF submerged	Bench scale, reactor vol. 60 L	COD removal 88% Suspended solid removal >99.5%	Lin et al. (2011)
–	Flat-sheet dynamic submerged	Bench scale, reactor vol. 45 L	COD removal 57%	Zhang et al. (2010)
–	Tubular UF external	Bench scale, reactor vol. 100 L	COD removal 87% NPN fecal removal 100%	Herrera-Robledo et al. (2010)
Polymeric membrane	UF external	Bench scale, reactor vol. 10 L	COD removal 55–69%	Baek et al. (2010)
–	Hollow fiber MF external	Bench scale, reactor vol. 180 L	COD removal 88%	Lew et al. (2009)
Polymeric membrane	Nonwoven fabric UF submerged	Bench scale, reactor vol. 12.9 L	COD removal 70.13%	An et al. (2009)
–	UF external	Bench scale, reactor vol. 50 L	COD removal 88% BOD removal 90%	Nagata et al. (1989)
Polymeric membrane	Flat-sheet UF external	Bench scale, reactor vol. 15 L	COD removal >95% Total nitrogen removal 15–20% PO_4^{3-} removal 81%	Kocadagistan and Topcub (2007)
Polymeric membrane	UF external	Bench scale, reactor vol. 850 L	TOC removal >90%	Grundestam and Hellstrom (2007)
Polymeric membrane	UF external	Bench scale, reactor vol. 10 L	COD removal 58% NH_4^+ removal 67.4%	Baek and Pagilla (2006)

20.3.2.1 Treatment of Wastewater of Pulp and Paper Industry

One of the conventional technologies for treatment of kraft black liquor is evaporation, which has several disadvantages due to high temperature, high organic strength (due mainly to methanol), low suspended solids, total reduced sulfur (TRS) compounds, and turpene oils. It was reported by several researchers that membrane-associated bioreactors provide a high throughput for treatment of wastewater from the pulp and paper industry. The works in this field are reported in Table 20.4.

20.3.2.2 Treatment of Petrochemical Wastewater

In the petrochemical industry, Fischer–Tropsch Reaction Water (FTRW) is a typical wastewater characterized by high strength and consisting of short chain organic acids and other oxygenates with a low pH. Treatment of wastewater from the petrochemical industry using membrane-associated bioreactors is recognized as a safe method of waste disposal. In Table 20.4, some significant recent examples of applications of membrane-associated bioreactors for petrochemical wastewater treatment are discussed.

20.3.2.3 Treatment of Textile Wastewater

In textile industries, the primary pollutants of wastewater are spent dye and Cr^{+6}. The conventional wastewater treatment processes in the textile industry are chemical precipitation, adsorption, and membrane separation. Although the conventional filtration process was practiced for many years, it was characterized by high concentration polarization and low flux; thus membrane-associated bioreactors gained attention. In this process, microbial consortia converts the Cr^{+6} to Cr^{+3} through a bioleaching technique; in addition, their synthesized protease enzyme decomposes hides, flesh, and animal residue. The works in this field are reported in Table 20.4.

20.3.2.4 Treatment of Wastewater from the Pharmaceutical Industry

Several pharmaceutical industries discharge different organic and inorganic components which could lead to sever health hazardous due to gradual bioaccumulation of hazardous component in the environment. Different types of membrane-associated bioreactors offer new hope to environmental scientists who are concerned about the disposal of pharmaceutical industry discharge. In Table 20.4, some major examples of applications of membrane-associated bioreactors for pharmaceutical wastewater treatment are described.

20.3.3 Applications in Landfill Leachate, Human Excrement, and Sludge Digestion

In addition to municipal and industrial wastewater treatment, membrane-associated bioreactor has been widely used in the treatment of landfill leachates, which contain

Table 20.4 Applications of MBR for Treatment of Wastewater of Different Process Industry

Source Wastewater	Membrane Configuration	Size of Operation	Treatment Success	Reference
Wool scouring	UF external	Pilot scale, capacity 10 m^3/day	TOD removal, >89%	Hogetsu et al. (1992)
Various sources	UF external	Pilot scale, capacity 0.2–24.6 m^3/day	COD removal, >97%	Krauth and Staab (1993)
Pulp mill	UF external	Pilot scale, capacity 10 m^3/day	TOC removal, >85%	Minami (1994)
Automotive industry (paint line)	UF external	Full scale, capacity 113 m^3/day	COD removal, >94%	Knoblock et al. (1994)
Metal transforming	UF external	Pilot scale, capacity 0.2 m^3/day	COD removal, >90%	Zaloum et al. (1994)
Tannery wastewater	UF external	Full scale, capacity 500–600 m^3/day	COD removal, >93%	Wehrle (1994a,b)
Cosmetic industry	UF external	Full scale	COD removal, >98%	Manem (1996)
Pulp and paper	MF external	Bench scale, capacity 0.05–0.09 m^3/day	COD removal, 68–82%	Dufresne et al. (1998)
Electrical components	UF external	Full scale, capacity 10 m^3/day	COD removal, >97%	Wehrle (1999)
Fuel and lubricants	UF external	Bench scale, capacity 0.02–0.04 m^3/day	TOC removal, >95%	Scholzy and Fuchs (2000)
Kraft pulp mill	UF external	Bench scale, capacity 0.003 m^3/day	TOC removal, >93%	Berube and Hall (2001)
Kraft evaporator condensate	UF submerged	Bench scale, reactor vol. 10 L	COD removal, 97–99%	Lin et al. (2009)
TMP whitewater	UF submerged	Bench scale, reactor vol. 10 L	COD removal, 90%	Gao et al. (2010)
Petrochemical	UF submerged	Bench scale, reactor vol. 23 L	COD removal, 98%	Van Zyl et al. (2008)
Textile	MF submerged	Bench scale, reactor vol. 3.25 L	COD removal, 90%;	Baeta et al. (2012)

(*Continued*)

Table 20.4 (Continued)

Source Wastewater	Membrane Configuration	Size of Operation	Treatment Success	Reference
			Color removal, 94%	
Textile	UF submerged	Bench scale, reactor vol. 500 L	COD removal, 97%	Badani et al. (2005)
			Color removal, 70%	
Textile	UF tubular external	Bench scale, reactor vol. 20 L	COD removal, 90%	Brik et al. (2006)
			Color removal, 98%	
Textile	MF submerged	Pilot scale, reactor vol. 230 L	COD removal, 97% Color removal, 98% Total suspended solid removal, 99% Total nitrogen removal, 78%	Yigit et al. (2009)
Palm	UF submerged	Bench scale, reactor vol. 20 L	COD removal, 94%	Yuniarto et al. (2008)
Phenolic	Submerged	Bench scale, reactor vol. 4.42 L	COD removal, 67%	Viero and Sant'Anna (2008)
Pharmaceutical	Submerged	Pilot scale, reactor vol. 10,000 L	COD removal, 96%	Chang et al. (2008)
Heavy metal	Submerged	Bench scale, reactor vol. 210 L	COD removal, 508–535% Total suspended solid removal, 226–267% Total nitrogen removal, 44–53%	Katsou et al. (2011)

high concentrations of organic and inorganic molecules. Conventionally, the treatment of leachates involves different physical, biological, or membrane filtration processes, and/or a combination of them. Membrane-associated bioreactors associated with RO have been successfully utilized to degrade inorganics and heavy metals. The membrane-associated bioreactor is also used in the treatment of human excreta in domestic wastewater, also known as night soil treatment systems. This technology was necessitated by the high strength of the waste and the need for on-site treatment. Different strategies like denitrification, coagulation, filtration, and activated carbon treatment are also incorporated with membrane-associated bioreactors. In the sludge treatment process, membrane associated bioreactors have been used for treatment of different kinds of wastewater which is depicted in Table 20.5.

20.3.4 Treatment of Food and Agricultural Wastewater

The agriculture, food, farm, and dairy industries are partially or fully contributing to the economy of different countries, and the wastewater from these sources pose a serious environmental hazardous. Submerged membrane bioreactors, activated sludge processes, and catalytic membrane bioreactors are popularly used to degrade the pollution load of the aforementioned industries. Generally, mixed organisms

Table 20.5 Applications of MBR for Treatment of Landfill Leachate, Human Excrement, and Sludge Digestion

Source Wastewater	Membrane Configuration	Size of Operation	Treatment Success	Reference
Landfill leachate	UF external	Full scale, capacity 50 m^3/day	–	Manem (1996)
Landfill leachate	UF external	Full scale, capacity 264 m^3/day	COD removal, >80%	Wehrle (1997)
Landfill leachate	UF external	Full scale, capacity 250 m^3/day	COD removal, >90%	Wehrle (1998)
Human excrement	UF external	Pilot scale	BOD removal, >99%	Magara and Itoh (1991)
Human excrement	UF external	Full scale	BOD removal, >99%	Manem (1996)
Sludge digestion	MF external	Pilot scale, capacity 0.13 m^3/day	–	Pillay et al. (1994)
Landfill leachate	UF external	Bench scale, reactor volume: 50 L	COD removal, 90.7%	Zayen et al. (2010)
Sanitary landfill leachate	UF external	Bench scale, reactor volume: 44 L	COD removal, 90.4% Turbidity removal, 90.3%	Marisa and Beal (2008)
Diluted landfill leachate	UF external	Bench scale, reactor volume: 29 L	COD removal, 90%	Bohdziewicz et al. (2008)

are used for this purpose. In some cases enzyme cellulase, pectinase, and amylases are used for this degradation process. In many cases, pesticide- and herbicide-degrading microbes are added into the culture medium. The treated slurry of the reactors is popularly used in fish cultivation and gardening. The works regarding this field are reported in Table 20.6.

20.4 Case Study

The dairy industry, a major economical source in tropical and subtropical countries, generates a large volume of waste liquid effluent, namely whey, which has a high biological oxygen demand (BOD) and a high chemical oxygen demand (COD). Due to stricter environmental legislations, whey is considered to be a major challenge to the environmental scientist owing to its high pollution load. Whey is a heterogeneous mixture of different types of proteins such as immunoglobulin, α-lactalbumin, β-lactoglobulin, bovine serum albumin, lactoferrin, carbohydrates such as lactose and fat, all of which have unique physicochemical, functional, nutraceuticals, and medicinal values. Therefore, purification of different biomolecules and synthesis of diverse chemicals from casein whey will assist the dairy, food, and pharmaceutical industries. Moreover, the process is also attractive for the implementation of the concept of zero-effluent discharge with respect to dairy effluent. In the present chapter, some particular case studies on the application of catalytic membrane bioreactors, enzyme-immobilized membrane bioreactors, and microbe-immobilized membrane bioreactors have been elucidated.

20.4.1 Synthesis of Prebiotic GOS by Catalytic Membrane Bioreactors

Das et al. (2011) reported on the advantageous outcomes of catalytic membrane bioreactors compared to batch reactors with respect to GOS synthesis from de-proteinated casein whey. De-proteinated whey was prepared by cross-flow membrane module with a UF membrane. Experiments were carried out batch-wise in a cross-flow module at different operating parameters, such as feed concentration (0.1–0.4 kg/m^3) and TMP (0.15–0.2 MPa) at different volume concentration factors (VCF). To maximize the recovery of lactose, three-stage DF was carried out under each combination of operating conditions with controlled pH 7.4. Dilute lactose solution, obtained as permeate after DF operation, was again fed to the RO membrane module to enrich the concentration of lactose on the retentive side after removing the minerals as permeate. Concentrated lactose solution was diluted to different concentrations varying from 5% to 30% (w/v) with 0.025 M potassium phosphate buffer (pH 6.6). The lactose solution was treated with β-galactosidase originated from *Bacillus circulans*. A batch experiment was conducted by different enzyme concentrations, 0.05–0.5% (w/v) and temperatures (277–297 K). A fixed reaction period was maintained for 5 h. Samples from the reaction mixture were withdrawn at regular intervals of 30 min and immediately heated to stop enzymatic

Table 20.6 Applications of MBR for Treatment of Food and Agricultural Wastewater

Source Wastewater	Membrane Configuration	Size of Operation	Treatment Success	Reference
Dairy whey	UF external	Pilot scale, capacity 0.46 m^3/day	COD removal, >94%	Sutton et al. (1983)
Dairy waste	MF submerged	Lab scale, reactor vol. 20 L	Total suspended solid removal, 99%	Katayon et al. (2004)
Whey + sucrose	UF submerged	Pilot scale, reactor vol. 11 L	–	Spagni et al. (2010)
Cheese whey	MF external	Pilot scale, reactor vol. 15 L	COD removal, 98.5% BOD removal, 99.2% TSS removal, 100%	Saddoud et al. (2007)
Maize/egg processing	UF external	Full scale, capacity 500 m^3/day	COD removal >97%	Ross et al. (1992)
Brewery effluent	UF external	Pilot scale, capacity 10 m^3/day	TOC removal, >97%	Strohwald and Ross (1992)
Liquor production	UF external	Pilot scale, capacity 1.25 m^3/day	COD removal, >98%	Nagano et al. (1992)
Rendering plant	UF external	Full scale, capacity 102 m^3/day	COD removal, >95%	Wehrle (1994a,b)
Dairy products	UF external	Full scale, capacity 2000 m^3/day	COD removal, >98%	Cantor et al. (1999)
Dairy products	UF external	Full scale, capacity 2000 m^3/day	Total suspended solid removal, 99%	Wehrle (2000)
Fermentation	UF external	Bench scale, capacity 0.01 m^3/day	TOD removal, >94%	Lu et al. (2000)
Food ingredients	MF submerged	Full scale, capacity 600 m^3/day	Effluent TSS, <9 mg/L	Wehrle (2000)
Tapioca starch waste	UF external	Lab scale, reactor vol. 1 L	COD removal, >95%	Roh et al. (2006)
Slaughterhouse wastewater	UF external	Lab scale, reactor vol. 50 L	COD removal, 94.2%	Saddoud and Sayadi (2007)
Meat extract + peptone	UF submerged	Lab scale, reactor vol. 3 L	COD removal, >95%	Aquino et al. (2006)
Glucose	UF external	Pilot scale, reactor vol. 25 L	COD removal, 84.4%	Sui et al. (2007)
Glucose	UF external	Lab scale, reactor vol. 5 L	COD removal, 99.1%	Huang et al. (2008)
Volatile fatty acid	UF external	Lab scale, reactor vol. 2 L	–	Jeison et al. (2009)
Maltose + glucose + volatile fatty acid	UF external	Lab scale, reactor vol. 0.6 L	COD removal, 99.6%	Jeong et al. (2010)

(*Continued*)

Table 20.6 (Continued)

Source Wastewater	Membrane Configuration	Size of Operation	Treatment Success	Reference
Molasses	UF external	Lab scale, reactor vol. 6 L	COD removal, 81%	Wijekoon et al. (2011)
Diluted tofu processing	MF external	Lab scale, reactor vol. 5 L	–	Kim et al. (2011)
Olive mill	UF external	Lab scale, reactor vol. 15 L	COD removal, 99% Total nitrogen removal, 15–20% PO_4^{3-} removal, 81%	Stamatelatou et al. (2009)
Palm oil	UF submerged	Lab scale, reactor vol. 20 L	COD removal, 99%	Yuniarto et al. (2008)
Brewery wastewater + surplus yeast	UF external	Lab scale, reactor vol. 4.5 L	COD removal, 94%	Torres et al. (2011)
High concentration food wastewater	UF external	Pilot scale, reactor vol. 400 L	COD removal, 81.3–94.2%	He et al. (2005)
Pet food	MF submerged	Lab scale, reactor vol. 20 L	COD removal, 97%	Acharya et al. (2006)
Distillery product	Submerged	Full scale	COD removal, 75–92%	Kanai et al. (2010)
Swine manure	UF external	Lab scale, reactor vol. 6 L	COD removal, 96%	Padmasiri et al. (2007)
Sand-separated dairy manure	MF external	Lab scale, reactor vol. 100 L	COD removal, 92%; total suspended solid removal, 81.7%; NH_4^+ removal, 16.6%; total protein removal, 95.8%	Wong et al. (2009)

actions. A catalytic membrane reactor with recycle arrangement was also used for the production of GOS. The schematic diagram of the proposed process is represented in Figure 20.15. Reaction mixtures were subjected to a 50 kg/mol UF cross-flow module to separate out the enzyme, and the permeate, consisting mainly of carbohydrates, was collected in the product tank. Permeate collected from the

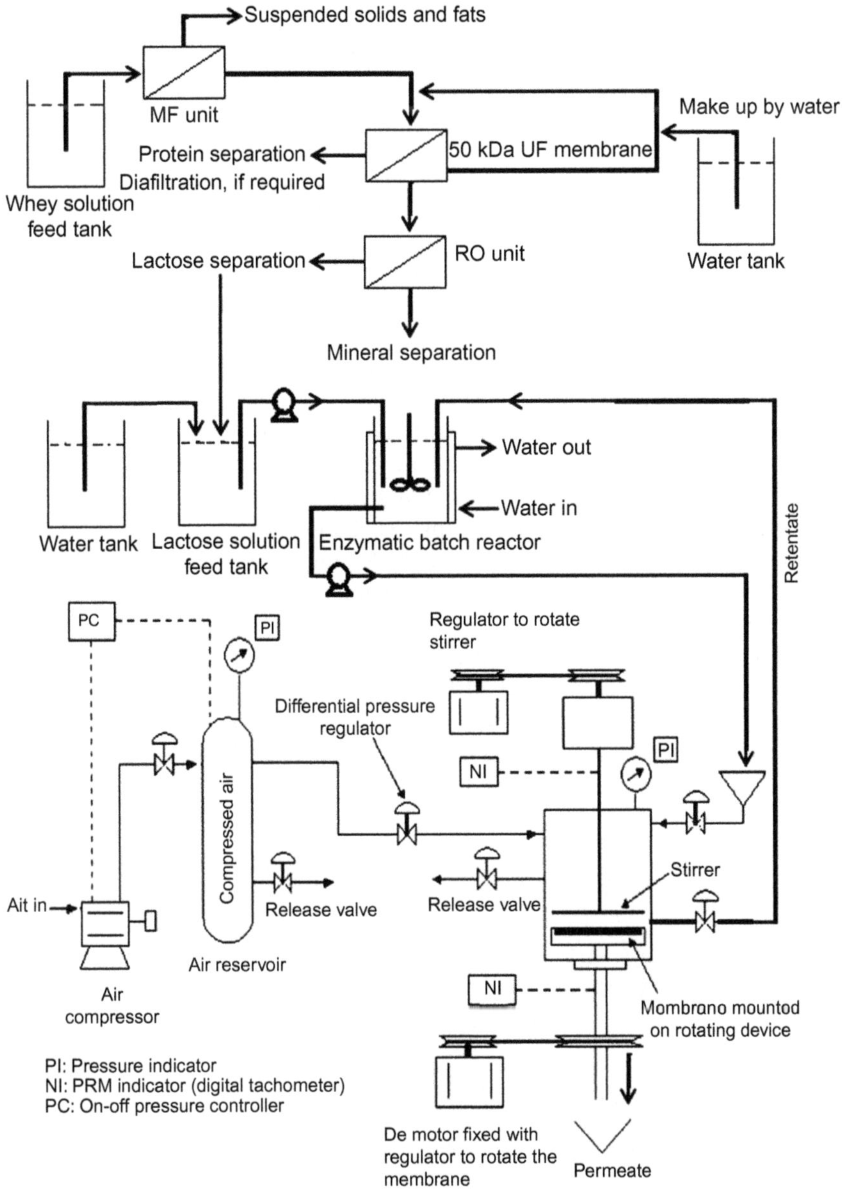

Figure 20.15 Catalytic membrane bioreactor for synthesis of prebiotic GOS from de-proteinated casein whey.

cross-flow module was mainly composed of unreacted lactose, and produced glucose, galactose, and GOS. To separate GOS from mono- and disaccharides, the permeate of 50 kg/mol UF cross-flow module was again fed to the RDMM NF membrane in conjunction with three stages of discontinuous diafiltration, at membrane speed, TMP, and temperature 0.17 rps, 1 MPa, and 298 K respectively. Around 77–78% purity of this GOS was achieved considering lactose as a substrate (200 g/L) by this proposed scheme. In a recycle membrane reactor, GOS production was found to be 33% higher than in the batch reactor, and monosaccharide concentration was 78% lower with 23% enzyme activity remaining after 5 h reaction time.

20.4.2 Synthesis of Prebiotic GOS by Enzyme Immobilized Membrane Bioreactors

Sen et al. (2012) reported that rotating disk membrane bioreactors (RDMBR) are more advantageous over batch hydrolysis of lactose followed by UF and NF in order to obtain pure GOS. The enzyme β-galactosidase was immobilized on a polymeric membrane surface by the combination of a series of mechanisms, such as adsorption of polyethyleneimine (PEI) on membrane matrix, formation of PEI and β-galactosidase aggregates, and finally cross-linking of PEI and β-galactosidase complex on the membrane surface by GA. Briefly, the membrane surface was incubated for 2 h at 25°C with 2.9 M hydrochloric acid and washed thoroughly with 0.1 M phosphate buffer (pH 8). Subsequently, the membrane was placed in 10 mL of 2.5% (w/v) gluteraldehyde solution prepared in 0.1 M phosphate buffer (pH 8) for 15 min at 25°C. Again the membrane was rinsed with deionized water and incubated in 1% (v/v) PEI prepared by 0.1 M phosphate buffer (pH 8) for 1 h at room temperature followed by a thorough water wash. Subsequently, the membrane after PEI attachment was dipped in gluteraldehyde solution. Finally, the membrane was dipped in 10 mL of 5 mg/mL of enzyme solution and incubated for 2 h, followed by a water wash. The immobilized membrane was stored in 0.1 M phosphate buffer (pH 8) before use. Figure 20.16 shows

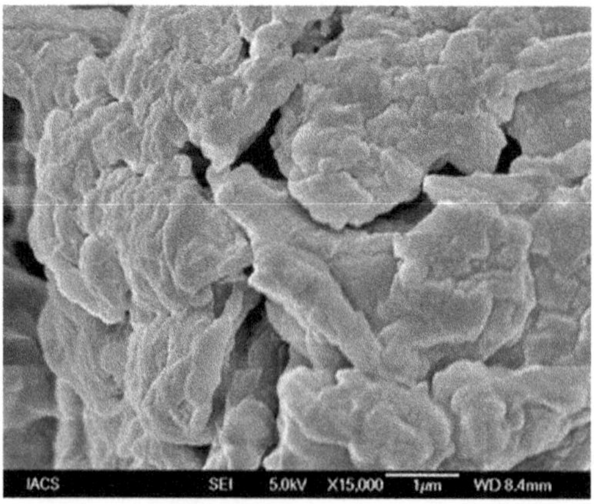

Figure 20.16 FESEM photograph of β-galactosidase immobilization on membrane surface.

that β-galactosidase was immobilized on the membrane surface in multilayer form. After immobilization, a lactose solution was fed to the immobilized membrane reactor, and the reaction was carried out for 3 h at 25°C under an initial TMP of 1 MPa. The concentration of lactose 100 kg/m^3 was prepared by 0.1 M acetate buffer (pH 6.5). The TMP was increased as the experiment proceeded to maintain the initial permeation flow rate of 3.4×10^{-8} m^3/s. To study the effect of membrane rotation on the immobilization, permeation, as well as purification of GOS, the membrane speed was varied at 0, 21, 42, 63, 84, 105, 157, 209, and 314 rps. For each of the membrane speeds, the reaction period was maintained for 3 h. The schematic diagram of the proposed process is described in Figure 20.17. The proposed membrane reactor module could serve as a single unit for the production, and purification of GOS, which ultimately lead to the lower footprint, and reduced cost of the downstream processing. A GOS purity of 80% with a yield of around 67% was achieved with an enzyme immobilized membrane. It was proposed that RDMBR is a single setup operation where problems associated with a batch mode reaction can be avoided, leading to high purity (80.2%) with high yield (67.4%) of GOS considering lactose as a substrate (100 g/L) at an optimum membrane rotation speed of 105 rps.

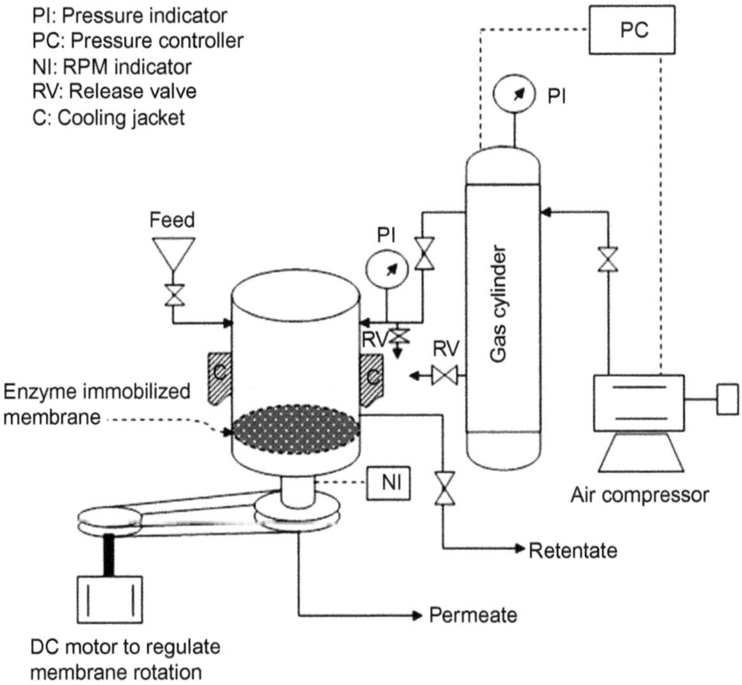

Figure 20.17 Synthesis of prebiotic GOS from lactose by membrane bioreactor. *Source*: Adapted from Sen et al. (2012).

20.4.3 Synthesis of Lactic Acid by Microbes Immobilized Membrane Bioreactors

Jung and Lovitt (2010) developed a pilot scale MBR to enhance the production of lactic acid by lactic acid bacteria (LAB). Different types of LAB, namely *Lactobacillus buchneri*, *Lactobacillus brevis*, *Oenococcus oeni*, and *Bifidobacterium longum*, were used in the experimental purposes. A 5.0 L stirred

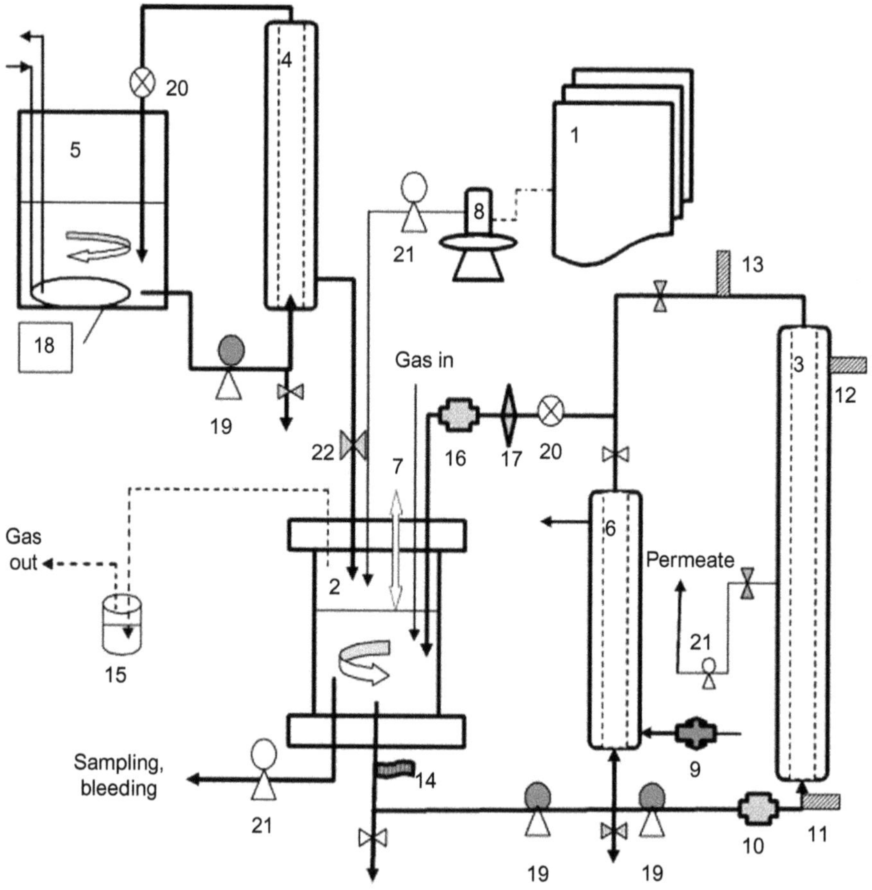

Figure 20.18 Schematic diagram of the membrane cell-recycle bioreactor (36.0 L): (1) control panel, (2) bioreactor (26.0 L), (3) production membrane (pore size, 0.2 μm; effective area, 1.0 m^2); ceramic MF, (4) feed membrane (pore size, 0.2 μm; effective area, 0.2 m^2); ceramic MF, (5) reservoir tank (100.0 L), (6) heat exchanger, (7) level controller, (8) alkaline tank and load cell (6.0 L), (9) solenoid v/v, (10) flow meter (recycling), (11) pressure gauge (P1), (12) pressure gauge (P3), (13) pressure gauge (P2), (14) thermometer and controller, (15) bubbling trap, (16) flow meter, (17) pH electrode, (18) heat exchanger, (19) two co-centrifugal magnetic pumps, (20) peristaltic pumps, (21) diaphragm v/v, (22) on/off v/v controlled by the level controller.
Source: Adapted from Jung and Lovitt (2010).

Table 20.7 Comparative Study of Different Types of Bioreactor for Biomass Formation and Lactic Acid Production

Strains	**Reactor**	μ_{max}	X	**Product Yields (g/g)**			
				$Y_{x/s}$	$Y_{l/s}$	$Y_{a/s}$	$Y_{e/s}$
L. buchneri	STR	0.20	1.83	0.07	0.44	0.1	0.22
	MBR	0.07	16.2	0.08	0.38	0.09	0.24
	MBR: STR	0.35	8.85	1.14	0.86	0.9	1.09
L. brevis	STR	0.20	2.33	0.07	0.4	0.08	0.22
	MBR	0.05	15.5	0.08	0.9	0.02	N.D
	MBR: STR	0.25	6.65	1.14	2.25	0.25	N.D
B. longum	STR	0.28	4.04	0.21	0.3	0.57	0.05
	MBR	0.13	22.2	0.18	0.25	1.25	0.19
	MBR: STR	0.46	5.5	0.86	0.83	2.19	3.8
O. oeni	STR	0.11	0.54	0.02	0.2	0.27	0.030
	MBR	0.07	12.8	0.02	0.18	0.29	0.03
	MBR:STR	0.64	23.7	1.0	0.9	1.07	

μ_{max}, maximum specific growth rate of consortium; X, biomass concentration; $Y_{x/s}$, cell product yield; $Y_{l/s}$, lactate product yield; $Y_{a/s}$, acetate product yield; $Y_{e/s}$, ethanol product yield. The table shows that the growth of LAB and volumetric productivity is much more in MBR compared with STR. Total cell concentration in perfusion culture of LAB reached 13–23 g/L, up to 14 times higher than that obtained in the STR. Overall volumetric biomass production rates in the MBR increased from 10 to 33 times that in the STR, which are again in concord with the advantage of MBR over batch process.

tank reactor (STR) (working volume 3.0 L) in a batch mode and the perfusion culture of four different LABs of the MBR (externally fitted with ceramic MF membranes) were used. The schematic diagram of the proposed MBR system is depicted in Figure 20.18. Moreover, different compositions of microbial growth medium were used to enhance the production of biomass, lactic acid, acetic acid, ethanol, and to reduce membrane fouling. In Table 20.7, maximum biomass, volumetric biomass production rate ($P_{x/t}$), and biomass production yields ($Y_{x/s}$) in the STR, and the MBR for four types of LAB is reported.

20.5 Conclusion and Scope of Future Challenge

The present work reveals the progress and development of membrane-associated bioreactors. Moreover, the chapter presents different technological viewpoints with respect to immobilization technology, fouling behavior, hydrodynamic study, and types of bioreactors. Membrane-associated bioreactors offer many advantages over conventional batch processes and enzyme or microorganism immobilized packed bed reactors in various areas such as waste treatment and the biopharmaceutical and chemical processing industries. Although different process strategies have already been developed, but due to low efficiency of enzyme, and microorganism under immobilized condition, researches are now being carried out on transgenic enzyme or microorganism. Moreover, development of new membrane materials, which could provide high-tensile stress and strength, would provide the more beneficial effect in this field. Membrane fouling is one of the major problems in enzyme catalytic membrane bioreactors and membrane bioreactors. Therefore, development of hydrophilic membranes and suitable immobilization technologies are considered to be a great challenge. Reduction of operating time, and costs of these systems is a further challenge.

References

Acharya, C., Nakhla, G., Bassi, A., 2006. Operational optimization and mass balances in a two-stage MBR treating high strength pet food wastewater. J. Envion. Eng. 132, 810–817.

Adham, S., Trussell, R.S., 2001. Membrane Bioreactors: Feasibility and Use in Water Reclamation. Water Environment Research Foundation, San Diego, CA.

Ahn, K.H., Cha, H.Y., Song, K.G., 1999. Retrofitting municipal sewage treatment plants using an innovative membrane-bioreactor system. Desalination. 124 (1–3), 279–286.

An, Y., Wang, Z., Wu, Z., Yang, D., Zhou, Q., 2009. Characterization of membrane foulants in an anaerobic non-woven fabric membrane bioreactor for municipal wastewater treatment. Chem. Eng. J. 155, 709–715.

Andreescu, S., Marty, J.L., 2006. Twenty years research in cholinesterase biosensors: from basic research to practical applications. Biomol. Eng. 23, 1–15.

Aquino, S.F., Hu, A.Y., Akram, A., Stuckey, D.C., 2006. Characterization of dissolved compounds in submerged anaerobic membrane bioreactors (SAMBRs). J. Chem. Technol. Biotechnol. 81, 1894–1904.

Badani, Z., Ait-Amar, H., Si-Salah, A., Brik, M., Fuchs, W., 2005. Treatment of textile waste water by membrane bioreactor and reuse. Desalination. 185, 411–417.

Baek, S.H., Pagilla, K.R., 2006. Aerobic and anaerobic membrane bioreactors for municipal wastewater treatment. Water Environ. Res. 78, 133–140.

Baek, S.H., Pagilla, K.R., Kim, H.J., 2010. Lab-scale study of an anaerobic membrane bioreactor (AnMBR) for dilutemunicipal wastewater treatment. Biotechnol. Bioproc. Eng. 15, 704–708.

Baeta, B.E.L., Ramos, R.L., Lima, D.R.S., Aquino, S.F., 2012. Use of submerged anaerobic membrane bioreactor (SAMBR) containing powdered activated carbon (PAC) for the treatment of textile effluents. Water Sci. Technol. 65, 1540–1547.

Basile, A., Chakraborty, S., Iulianelli, A., 2013. Membrane and Membrane Reactors for Sustainable Growth. Nova Science Publishers, Inc. Library of Congress Cataloging-in-Publication Data, New York, NY.

Bayramoglu, G.A.S., Bulut, A., Denizli, A., Arıca, M.Y., 2003. Covalent immobilisation of invertase onto a reactive film composed of 2-hydroxyethyl methacrylate and glycidyl methacrylate properties and application in a continuous flow system. Biochem. Eng. J. 14, 117–126.

Berube, P.R., Hall, E.R., 2001. Fate and removal kinetics of contaminants contained in evaporator condensate during treatment for reuse using a high-temperature membrane bioreactor. J. Pulp Paper Sci. 27 (2), 41–45.

Bhattacharyya, D., Hestekin, J.A., Brushaber, P., Cullen, L., Bachas, L.G., Sikdar, S.K., 1998. Novel poly-glutamic acid functionalized microfiltration membranes for sorption of heavy metals at high capacity. J. Membr. Sci. 141, 121–135.

Bohdziewicz, J., Neczaj, E., Kwarciak, A., 2008. Landfill leachate treatment by means of anaerobic membrane bioreactor. Desalination. 221, 559–565.

Bora, U.K.K., Nahar, P., 2005. A simple method for functionalization of cellulose membrane for covalent immobilization of biomolecules. J. Membr. Sci. 250, 215–222.

Brik, M., Schoeberl, P., Chamam, B., Braun, R., Fuchs, W., 2006. Advanced treatment of textile wastewater towards reuse using a membrane bioreactor. Process Biochem. 41, 1751–1757.

Butterfield, D.A., 1996. Biofunctional Membranes. Plenum Press, New York, NY, 117–129.

Butterfield, D.A., Bhattacharyya, D., Daunert, S., Bachas, L., 2001. Catalytic biofunctional membranes containing site specifically immobilized enzyme arrays: a review. J. Membr. Sci. 181, 29–37.

Cano, A.M.C., Palet, C., 2006. Immobilization of endo-1,4-b-xylanase on polysulfone acrylate membranes: synthesis and characterization. J. Membr. Sci. 280, 383–388.

Cantor, J., Sutton, P.M., Steinheber, R., Myronyk, M., 1999. Membrane filtration: an internal membrane bioreactor helps solve treatment plant's operational problems. Ind. Wastewater. 7 (2), 18–22.

Chakraborty, S., Sikder, J., Mukherjee, D., Mandal, M.K., Arockiasamy, D.L., 2013a. In: Piemonte, V.V., De Falco, M.M., Basile, A.A. (Eds.), Sustainable Development in Chemical Engineering, Innovative Technologies, first ed. John Wiley & Sons.

Chakraborty, S., Rusli, H., Sikder, J., Curcio, S., Drioli, E., 2013b. Immobilized biocatalytic process development and potential application in membrane separation: a review. Crit. Rev. Biotechnol. (in press). <http://dx.doi.org/10.3109/07388551.2014.923373,2013>.

Chang, C.Y., Chang, J.S., Vigneswaran, S., Kandasamy, J., 2008. Pharmaceutical wastewater treatment by membrane bioreactor process—a case study in southern Taiwan. Desalination. 234, 393–401.

Cheng, H.N., Richard, A.G. (Eds.), 2010. Green Polymer Chemistry: Biocatalysis and Biomaterials. American Chemical Society, Boston, 1–14.

Chiemchaisri, C., Yamamoto, K., Vigneswaran, S., 1993. Household membrane bioreactor in domestic wastewater treatment. Water Sci. Technol. 27 (1), 171–178.

Cicek, N., Franco, J.P., Suidan, M.T., Urbain, V., 1998. Using a membrane bioreactor to reclaim wastewater. J. Am. Water Works Assoc. 90 (11), 105–113.

Cote, P., Buisson, H., Praderie, M., 1998. Immersed membranes activated sludge process applied to the treatment of municipal wastewater. Water Sci. Technol. 38 (4–5), 437–442.

Danisman, T.T.S., Kacar, Y., Ergene, A., 2004. Covalent immobilization of invertase on microporous pHEMA–GMA membrane. Food Chem. 85, 461–466.

Das, R., Sen, D., Sarkar, A., Bhattacharyya, S., Bhattacharjee, C., 2011. A comparative study on the production of galacto oligosaccharide from whey permeate in recycle membrane reactor and in enzymatic batch reactor. Ind. Eng. Chem. Res. 50, 806–816.

Dave, B.C., Dunn, B., Valentine, J.S., Zink, J.I., 1994. Sol–gel encapsulation methods for biosensors. Analyt. Chem. 66, 1120A–1127A.

Deng, H.T.K., Liu, Z.M., Wu, J., Ye, P., 2004. Adsorption immobilization of *Candida rugosa* lipases on polypropylene hollow fiber microfiltration membranes modified by hydrophobic polypeptides. Enzyme Microb. Technol. 35, 437–443.

Dufresne, R., Lavallee, H.C., Lebrun, R.E., Lo, S.N., 1998. Comparison of performance between membrane bioreactor and activated sludge system for the treatment of pulping process wastewaters. Tappi J. 81 (4), 131–135.

Ebrahimi, M.P.L., Engel, L., Ashaghi, K.S., Czermak, P., Dimov, A., 2010. A novel ceramic membrane reactor system for the continuous enzymatic synthesis of oligosaccharides. Desalination. 250, 1105–1108.

Eldin, M., Seuror, E., Nasr, M., Tieama, H., 2011. Affinity covalent immobilization of glucoamylase onto ρ-benzoquinone-activated alginate beads: II. Enzyme immobilization and characterization. Appl. Biochem. Biotechnol. 164, 45–57.

Fan, X.J., Urbain, V., Qian, Y., Manem, J., 1996. Nitrification and mass balance with a membrane bioreactor for municipal wastewater treatment. Water Sci. Technol. 34 (1–2), 129–136.

Gao, W.J.J., Lin, H.J., Leung, K.T., Liao, B.Q., 2010. Influence of elevated pH shocks on the performance of a submerged anaerobic membrane bioreactor. Process Biochem. 45, 1279–1287.

Garcia, G.E., Kanj, J., 2002. Two years of membrane bioreactor plant operation experience at the Viejas Tribe Reservation. Paper presented at WEFTEC 2002, Chicago, IL. Water Environment Federation, Alexandria, VA.

Genkoplis, C.J., 1993. Transport Process and Unit Operations. third ed. Prentice Hall, Upper Saddle River, NJ, 632, 791 – 795.

Giorno, L., Drioli, E., 2000. Biocatalytic membrane reactors: applications and perspectives. Trends Biotechnol. 18, 339–349.

Giorno, L., Mazzei, R., Drioli, E., 2009. Biochemical Membrane Reactors in Industrial Processes. Membrane Operations: Wiley-VCH Verlag GmbH & Co. KGaA, 397–409.

Godjevargova, T.K.V., Dimov, A., Vasileva, N., 2000. Behavior of glucose oxidase immobilized on ultrafiltration membranes obtained by copolymerizing acrylonitrile and *N*-vinylimidazol. J. Membr. Sci. 172, 279–285.

Grundestam, J., Hellstrom, D., 2007. Wastewater treatment with anaerobic membrane bioreactor and reverse osmosis. Water Sci. Technol. 56, 211–217.

Hanefeld, U., Gardossi, L., Magner, E., 2009. Understanding enzyme immobilisation. Chem. Soc. Rev. 38, 453–468.

He, Y., Xu, P., Li, C., Zhang, B., 2005. High-concentration food wastewater treatment by an anaerobic membrane bioreactor. Water Res. 39, 4110–4118.

Herrera-Robledo, M., Morgan-Sagastume, J.M., Noyola, A., 2010. Biofouling and pollutant removal during long-term operation of an anaerobic membrane bioreactor treating municipal wastewater. Biofouling. 26, 23–30.

Hogetsu, A., Ishikawa, T., Yoshikawa, M., Tanabe, T., Yudate, S., Sawada, J., 1992. High rate anaerobic digestion of wool scouring wastewater in a digester combined with membrane filter. Water Sci. Technol. 25 (7), 341–350.

Huang, Z., Ong, S.L., Ng, H.Y., 2008. Feasibility of submerged anaerobic membrane bioreactor (SAMBR) for treatment of low-strength wastewater. Water Sci. Technol. 58, 1925–1931.

Huang, X.J.C., Huang, F., Ou, Y., Chen, M.R., Xu, Z.K., 2011. Immobilization of *Candida rugosa* lipase on electrospun cellulose nonfiber membrane. J. Mol. Catal B. Enzyme. 70, 95–100.

Huckel, M., Wirth, H.J., Hearn, M.T.W., 1996. Porous zirconia: a new support material for enzyme immobilization. J. Biochem. Biophys. Methods. 31, 165–179.

Jeison, D., van Betuw, W., van Lier, J.B., 2008. Feasibility of anaerobic membrane bioreactors for the treatment of wastewaters with particulate organic matter. Sep. Sci. Technol. 43, 3417–3431.

Jeison, D., Telkamp, P., van Lier, J.B., 2009. Thermophilic side stream anaerobic membrane bioreactors: the shear rate dilemma. Water Environ. Res. 81, 2372–2380.

Jeong, E., Kim, H.W., Nam, J.Y., Ahn, Y.T., Shin, H.S., 2010. Effects of the hydraulic retention time on the fouling characteristics of an anaerobic membrane bioreactor for treating acidified wastewater. Desalin. Water Treat. 18, 251–256.

Jung, I., Lovitt, R.W., 2010. A comparative study of the growth of lactic acid bacteria in a pilot scale membrane bioreactor. J. Chem. Technol. Biotechnol. 85, 1250–1259.

Kanai, M., Ferre, V., Wakahara, S., Yamamoto, T., Moro, M., 2010. A novel combination of methane fermentation and MBR-kubota submerged anaerobic membrane bioreactor process. Desalination. 250, 964–967.

Katayon, S., Noor, M.J.M.M., Ahmad, J., Ghani, L.A.A., Nagaoka, H., Aya, H., 2004. Effects of mixed liquor suspended solid concentrations on membrane bioreactor efficiency for treatment of food industry wastewater. Desalination. 167, 153–158.

Katsou, E., Malamis, S., Loizidou, M., 2011. Performance of a membrane bioreactor used for the treatment of wastewater contaminated with heavy metals. Bioresour. Technol. 102, 4325–4332.

Kim, J.O., Jung, J.T., 2007. Performance of membrane-coupled organic acid fermentor for the resources recovery form municipal sewage sludge. Water Sci. Technol. 55, 245–252.

Kim, M.S., Lee, D.Y., Kim, D.H., 2011. Continuous hydrogen production from tofu processing waste using anaerobic mixed microflora under thermophilic conditions. Int. J. Hydrogen Energy. 36, 8712–8718.

Kimura, S., 1991. Japan's aqua renaissance '90 project. Water Sci. Technol. 23 (7–9), 1573–1582.

Knoblock, M.D., Sutton, P.M., Mishra, P.N., Gupta, K., Janson, A., 1994. Membrane biological reactor system for treatment of oily wastewaters. Water Environ. Res. 66 (2), 133–139.

Kocadagistan, E., Topcub, N., 2007. Treatment investigation of the Erzurum City municipal wastewaters with anaerobic membrane bioreactors. Desalination. 216, 367–376.

Krauth, K.H., Staab, K.F., 1993. Pressurized bioreactor with membrane filtration for waste water treatment. Water Res. 27 (3), 405–411.

Lew, B., Tarre, S., Beliavski, M., Dosoretz, C., Green, M., 2009. Anaerobic membrane bioreactor (AnMBR) for domestic wastewater treatment. Desalination. 243, 251–257.

Li, Y.Q.J., Branford-White, C., Williams, G.R., Wu, J.X., Zhu, L.M., 2012. Electrospun polyacrylonitrile glycopolymer nano-fibrous membranes for enzyme immobilization. J. Mol. Catal. B. Enzyme. 76, 15–22.

Lin, H., Liao, B.Q., Chen, J., Gao, W., Wang, L., Wang, F., et al., 2011. New insights into membrane fouling in a submerged anaerobic membrane bioreactor based on characterization of cake sludge and bulk sludge. Bioresour. Technol. 102, 2373–2379.

Lin, H.J., Xie, K., Mahendran, B., Bagley, D.M., Leung, K.T., Liss, S.N., et al., 2009. Sludge properties and their effects on membrane fouling in submerged anaerobic membrane bioreactors (SAnMBRs). Water Res. 43, 3827–3837.

Liu, C., Tanaka, H., Zhang, L., Zhang, J., Huang, X., Ma, J., 2012. Fouling and structural changes of Shirasu porous glass (SPG) membrane used in aerobic wastewater treatment process for microbubble aeration. J. Membr. Sci. 421–422, 225–231.

Lorenz, W., Cunningham, T., Penny, J.P., 2002. Phosphorus removal in a membrane bioreactor system: a full-scale wastewater demonstration study. Paper presented at WEFTEC 2002, Chicago, IL. Water Environment Federation, Alexandria, VA.

Lu, S.G., Imai, T., Ukita, M., Sekine, M., Fukagawa, M., Nakanishi, H., 2000. The performance of fermentation wastewater treatment in ultrafiltration membrane bioreactor by continuous and intermittent aeration processes. Water Sci. Technol. 42 (3–4), 323–329.

Magara, Y., Itoh, M., 1991. The effect of operational factors on solid/liquid separation by ultra-membrane filtration in a biological denitrification system for collected human excreta treatment plants. Water Sci. Technol. 23 (7–9), 1583–1590.

Manem, J.A.S., 1996. Membrane bioreactors. In: Mallevialle, J., Odendaal, P.E., Wiesner, M.R. (Eds.), Water Treatment Membrane Processes. McGraw Hill, New York, NY, pp. 17.1–17.31.

Marisa, D., Beal, L.L., 2008. Anaerobic filter associated with microfiltration membrane (MAF) treating sanitary landfill leachate, 2008 IWA North American Membrane Research Conference, Amherst, MA.

Mazzei, R., Chakraborty, S., Drioli, E., Giorno, L., 2010. In: Peinemann, K.-V., Nunes, S.P., Giorno, L. (Eds.), Membrane Bioreactors in Functional Food Ingredients Production, Membrane Technology, **3**, Membranes for Food Applications. WILEY-VCH Verlag GmbH & Co. KGaA, Weinheim, Copyright _ 2010.

Minami, K., 1994. A trial of high performance anaerobic treatment on wastewater from kraft pulp mill. Desalination. 98, 273–283.

Muller, E.B., Stouthamer, A.H., Verseveld, H.W., Eikelboom, D.H., 1995. Aerobic domestic waste water treatment in a pilot plant with complete sludge retention by cross-flow filtration. Water Res. 29 (4), 1179–1189.

Nagano, A., Arikawa, E., Kobayashi, H., 1992. The treatment of liquor wastewater containing high-strength suspended solids by membrane bioreactor system. Water Sci. Technol. 26 (3–4), 887–895.

Nagata, N., Herouvis, K.J., Dziewulski, D.M., Belfort, G., 1989. Cross-flow membrane microfiltration of a bacterial fermentation broth. Biotechnol. Bioeng. 34, 447–466.

Nagy, E., Kulcsar, E., 2009. Mass transport through biocatalytic membrane reactor. Desalination. 246, 49–63.

Nath, A., Bhattacharjee, C., Chowdhury, R., 2013. Synthesis and separation of galacto-oligosaccharides using membrane bioreactor. Desalination. 316, 31–41.

Nouaimi, M., Moschel, K., Bisswanger, H., 2001. Immobilization of trypsin on polyester fleece via different spacers. Enzyme Microb. Technol. 29, 567–574.

Padmasiri, S.I., Zhang, J., Fitch, M., Norddahl, B., Morgenroth, E., Raskin, L., 2007. Methanogenic population dynamics and performance of an anaerobic membrane bioreactor (AnMBR) treating swine manure under high shear conditions. Water Res. 41, 134–144.

Pal, P.D.S., Bhattacharya, P., 2002. Multi-enzyme immobilization in eco-friendly emulsion liquid membrane reactor-a new approach to membrane formulation. Sep. Purif. Technol. 27, 145–154.

Pedersen, H., Furler, L., Venkatasubramanian, K., Prenosil, J., Stuker, E., 1985. Enzyme adsorption in porous supports: local thermodynamic equilibrium model. Biotechnol. Bioeng. 27, 961–971.

Pierkiel, A., Lanting, J., 2005. Membrane-coupled anaerobic digestion of municipal sewage sludge. Water Sci. Technol. 52, 253–258.

Pillay, V.L., Townsend, B., Buckley, C.A., 1994. Improving the performance of anaerobic digesters at wastewater treatment works: the coupled cross-flow microfiltration/digester process. Water Sci. Technol. 30 (12), 329–337.

Radjenovic, J., Matosic, M., Mijatovic, I., Petrovic, M., Barcelo, D., 2008. Membrane bioreactor (MBR) as an advanced wastewater treatment technology. Handb. Environ. Chem. 5 (S2), 37–101.

Ricca, E.C.V., Curcio, S., Basso, A., Gardossi, L., Iorio, G., 2010. Fructose production by inulinase covalently immobilized on sepabeads in batch and fluidized bed bioreactor. Int. J. Mol. Sci. 11, 1180–1189.

Roh, S.H., Chun, Y.N., Nah, J.W., Shin, H.J., Kim, S.I., 2006. Wastewater treatment by anaerobic digestion coupled with membrane processing. J. Ind. Eng. Chem. 12, 489–493.

Rosenberger, S., Kruger, U., Witzig, R., Manz, W., Szewzyk, U., Kraume, M., 2002. Performance of a bioreactor with submerged membranes for aerobic treatment of municipal waste water. Water Res. 36 (2), 413–420.

Ross, W.R., Barnard, J.P., Strohwald, N.K.H., Grobler, C.J., Sanetra, J., 1992. Practical application of the ADUF process to the full-scale treatment of maize-processing effluent. Water Sci. Technol. 25 (10), 27–39.

Saddoud, A., Sayadi, S., 2007. Application of acidogenic fixed-bed reactor prior to anaerobic membrane bioreactor for sustainable slaughterhouse wastewater treatment. J. Hazard. Mater. 149, 700–706.

Saddoud, A., Hassairi, I., Sayadi, S., 2007. Anaerobic membrane reactor with phase separation for the treatment of cheese whey. Bioresour. Technol. 98, 2102–2108.

Sakai-Kato, K., Kato, M., Ishihara, K., Toyo'oka, T., 2004. An enzyme-immobilization method for integration of biofunctions on a microchip using a water-soluble amphiphilic phospholipid polymer having a reacting group. Lab Chip. 4, 4–6.

Scholzy, W., Fuchs, W., 2000. Treatment of oil contaminated wastewater in a membrane bioreactor. Water Res. 34 (14), 3621–3629.

Scott, K., 1998. Section 2—Membrane Materials, Preparation and Characterisation, Handbook of Industrial Membranes, second ed., pp. 187–269.

Sen, D., Sarkar, A., Das, S., Chowdhury, R., Bhattacharjee, C., 2012. Batch hydrolysis and rotating disk membrane bioreactor for the production of galacto-oligosaccharides: a comparative study. Ind. Eng. Chem. Res. 51, 10671–10681.

Sheldon, R.A., 2007. Enzyme immobilization: the quest for optimum performance. Adv. Synth. Catal. 349, 1289–1307.

Shuler, M.L., Kargi, F., 2002. Bioprocess Engineering Basic Concept. second ed. Prentice-Hall Inc., Upper Saddle River, NJ, p. 2.

Spagni, A., Casu, S., Crispino, N.A., Farina, R., Mattioli, D., 2010. Filterability in a submerged anaerobic membrane bioreactor. Desalination. 250, 787–792.

Stamatelatou, K., Kopsahelis, A., Blika, P.S., Paraskeva, C.A., Lyberatos, G., 2009. Anaerobic digestion of olive mill wastewater in a periodic anaerobic baffled reactor (PABR) followed by further effluent purification via membrane separation technologies. J. Chem. Technol. Biotechnol. 84, 909–917.

Strohwald, N.K.H., Ross, W.R., 1992. Application of the ADUF process to brewery effluent on a laboratory scale. Water Sci. Technol. 25 (10), 95–105.

Sui, P., Wen, X., Huang, X., 2007. Membrane fouling control by ultrasound in an anaerobic membrane bioreactor. Front. Environ. Sci. Eng. Chin. 1, 362–367.

Sutton, P.M., 2006. Membrane Bioreactors for Industrial Wastewater Treatment: Applicability and Selection of Optimal System Configuration. Sutton & Associates, Inc., Enfield, NH.

Sutton, P.M., Li, A., Evans, R.R., Korchin, S.R., 1983. Dorr–Oliver's fixed-film, suspended growth anaerobic systems for industrial wastewater treatment and energy recovery. Proceedings of 37th Industrial Waste Conference. Purdue University, Purdue, IN, pp. 667–675.

Tischer, W., Wedekind, F., 2000. Immobilized enzymes: methods and application. In: Fessner, W.-D. (Ed.), Biocatalysis: From Discovery to Application. Springer, Berlin Heidelberg.

Till, S., Membrane Bioreactors: Wastewater Treatment Applications to Achieve High Quality Effluent, 64th Annual Water Industry Engineers and Operators' Conference All Seasons International Hotel – Bendigo, 5 and 6 September, 2001.

Tominaga, J., Kamiya, N., Doi, S., Ichinose, H., Goto, M., 2004. An enzymatic strategy for site-specific immobilization of functional proteins using microbial transglutaminase. Enzyme Microb. Technol. 35, 613–618.

Torres, A., Hemmelmann, A., Vergara, C., Jeison, D., 2011. Application of two-phase slug-flow regime to control flux reduction on anaerobic membrane bioreactors treating wastewaters with high suspended solids concentration. Sep. Purif. Technol. 79, 20–25.

Trzcinski, A.P., Stuckey, D.C., 2009. Continuous treatment of the organic fraction of municipal solid waste in an anaerobic two-stage membrane process with liquid recycle. Water Res. 43, 2449–2462.

Trzcinski, A.P., Stuckey, D.C., 2010. Treatment of municipal solid waste leachate using a submerged anaerobic membrane bioreactor at mesophilic and psychrophilic temperatures: analysis of recalcitrants in the permeate using GC–MS. Water Res. 44, 671–680.

Van de Roest, H.F., Lawrence, D.P., Van Bentem, A.G.N., 2002. Membrane Bioreactors for Municipal Wastewater Treatment. STOWA Report. IWA Publishing, London, UK.

Van Zyl, P.J., Wentzel, M.C., Ekama, G.A., Riedel, K.J., 2008. Design and start-up of a high rate anaerobic membrane bioreactor for the treatment of a low pH, high strength, dissolved organic waste water. Water Sci. Technol. 57, 291–295.

Viero, A.F., Sant'Anna Jr., G.L., 2008. Is hydraulic retention time an essential parameter for MBR performance? J. Hazard. Mater. 150, 185–186.

Visvanathan, C., Ben Aim, R., Parameshwaran, K., 2000. Membrane separation bioreactors for wastewater treatment. Environ. Sci. Technol. 30 (1), 1–48.

Walker, M., Banks, C.J., Heaven, S., 2009. Development of a coarse membrane bioreactor for two-stage anaerobic digestion of biodegradable municipal solid waste. Water Sci. Technol. 59, 729–735.

Wang, Z.G., Wj, Q., Xu, Z.K., 2006. Immobilization of lipase from *Candida rugosa* on electrospun polysulfone nanofibrous membranes by adsorption. J. Mol. Catal. B. Enzyme. 42, 45–51.

Wang, Z.G., Wan, L.S., Liu, Z.M., Huang, X.J., Xu, Z.K., 2009. Enzyme immobilization on electrospun polymer nanofibers: an overview. J. Mol. Catal. B Enzyme. 56, 189–195.

Wang, P., Wang, Z., Wu, Z., Mai, S., 2011. Fouling behaviours of two membranes in a submerged membrane bioreactor for municipal wastewater treatment. J. Membr. Sci. 382, 60–69.

Wehrle, 1994a. MBR Case Study: Bayern-Leder GmbH. <http://www.wehrle-env.co.uk/pdf/Bayern%20Leder.pdf> (21.05.03).

Wehrle, 1994b. MBR Case Study: Zweckverband TBA. <http://www.wehrle-env.co.uk/pdf/Zweckverband%20.pdf> (21.05.03).

Wehrle, 1997. MBR Case Study: EAL-Scheinberg Treatment Plant. <http://www.wehrle-env.co.uk/pdf/EAL%20.pdf> (21.05.03).

Wehrle, 1998. MBR Case Study: Trienekens GmbH. <http://www.wehrle-env.co.uk/pdf/Trienekens%20.pdf> (21.05.03).

Wehrle, 1999. MBR Case Study: Robert Bosch AG. <http://www.wehrle-env.co.uk/pdf/Robert%20Bosch%20.pdf> (21.05.03).

Wehrle, 2000. MBR Case Study: Dairy Gold Cooperative. <http://www.wehrle-env.co.uk/pdf/Dairygold.pdf> (21.05.03).

Wijekoon, K.C., Visvanathan, C., Abeynayaka, A., 2011. Effect of organic loading rate on VFA production, organic matter removal and microbial activity of a two-stage thermophilic anaerobic membrane bioreactor. Bioresour. Technol. 102, 5353–5360.

Wong, K., Xagoraraki, I., Wallace, J., Bickert, W., Srinivasan, S., Rose, J.B., 2009. Removal of viruses and indicators by anaerobic membrane bioreactor treating animal waste. J. Environ. Qual. 38, 1694–1699.

Xu, M., Wen, X., Huang, X., Li, Y., 2010. Membrane fouling control in an anaerobic membrane bioreactor coupled with online ultrasound equipment for digestion of waste activated sludge. Sep. Sci. Technol. 45, 941–947.

Yigit, N.O., Uzal, N., Koseoglu, H., Harman, I., Yukseler, H., Yetis, U., et al., 2009. Treatment of a denim producing textile industry wastewater using pilot scale membrane bioreactor. Desalination. 240, 143–150.

Yuniarto, A., Ujang, Z., Noor, Z.Z., 2008. Performance of bio-fouling reducers in aerobic submerged membrane bioreactor for palm oil mill effluent treatment. J. Teknol. UTM. 49, 555–566.

Zaloum, R., Lessard, S., Mourato, D., Carriere, J., 1994. Membrane bioreactor treatment of oily wastewater from a metal transformation mill. Water Sci. Technol. 30 (9), 21–27.

Zayen, A., Mnif, S., Aloui, F., Fki, F., Loukil, S., Bouaziz, M., et al., 2010. Anaerobic membrane bioreactor for the treatment of leachates from Jebel Chakir discharge in Tunisia. J. Hazard. Mater. 177, 918–923.

Zhang, X., Wang, Z., Wu, Z., Lu, F., Tong, J., Zang, L., 2010. Formation of dynamic membrane in an anaerobic membrane bioreactor for municipal wastewater treatment. Chem. Eng. J. 165, 175–183.

21 Microbiological Metabolism Under Chemical Stress

Rashmi Kataria[a] *and Rohit Ruhal*[b]

[a]Department of Forests Biomaterials and Technology, Swedish University of Agricultural Science (SLU), Umea, Sweden,
[b]Department of Chemistry, Umea University, Sweden

21.1 Introduction

Bacterial metabolism can be defined as the biochemical reactions that lead to generation of energy required for various processes in a cell and can include an energy-consuming biosynthetic component as shown in Figure 21.1. Microbes have developed in addition several energy-generating metabolism, which is quite different from that of eukaryotes, and this diversity makes them able to survive in almost all habitats on Earth. Some of the metabolic pathways adapted by bacteria are unique fermentation pathways (Figure 21.2), which can lead to wide range of end products, anaerobic respiration, lithotropy, photoheterotrophy, methagenosis, light-driven nonphotosynthesis energy production (archaea), and autotrophic CO_2 fixation.

Microorganisms possess great diverse enzymatic diversity and can metabolize millions of the organic compounds found on Earth. Therefore, microorganisms are used to biodegrade many biohazardous compounds. Furthermore, they are also used to treat organic and inorganic waste present as pollutants. There are substantial challenges to elucidate how microbes play ecological functions, when these chemicals or pollutants are present in the environment. These chemicals influence broad metabolic potential of these microbes e.g. gene expression of unique enzymes to degrade chemicals. It has been reported that there is a marked increase

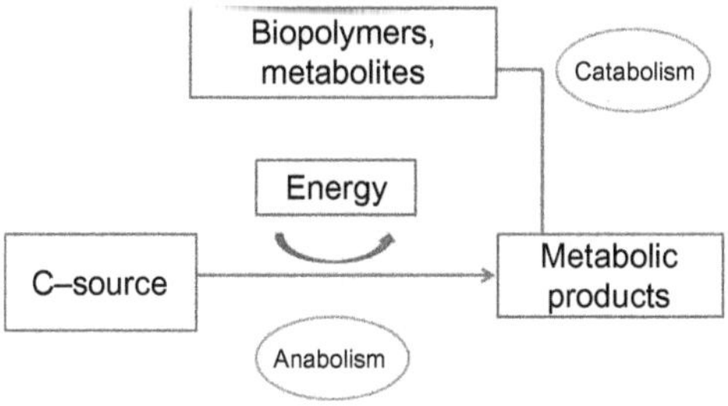

Figure 21.1 Basic metabolism of bacteria.

Microbial Biodegradation and Bioremediation. DOI: http://dx.doi.org/10.1016/B978-0-12-800021-2.00021-2

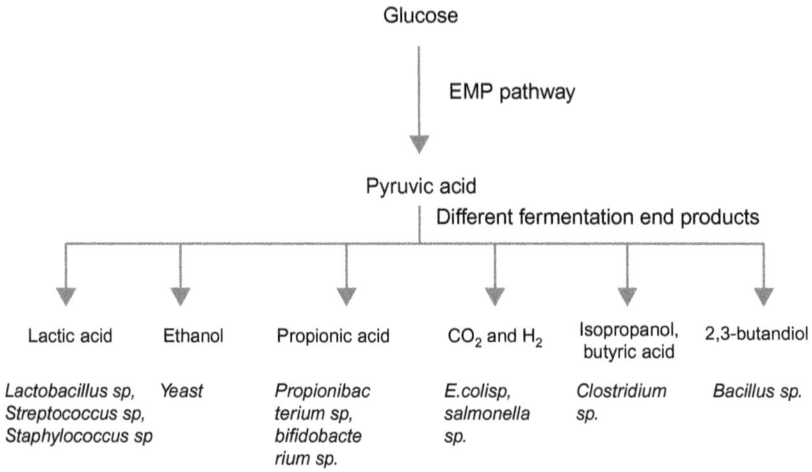

Figure 21.2 Diversity of microbial metabolism as shown by different fermentation end products of different bacteria.

in transcript involved in aromatic compound metabolism, respiration, and stress response. Thus, the objective of this chapter is to focus on the stress response of microbes toward several chemicals found in the environment. Industrial bioprocesses like fermentation and bioethanol production involve stressful environment faced by microbes, and the objective of bioprocess industries is to expand life span of microbes by reducing toxicity of their environment from several key chemicals (ethanol, butanol, and propanediol). In addition, constituents of the medium can provide chemical stress, e.g., it may include complex carbon sources provided from wastes (crude glycerol, lignocellulose waste). These carbon sources may be in the form of carboxylic acids, celluloses, or xylan (Nicolaou et al., 2010). These sugars can give osmotic stress to bacteria. Furthermore, bioprocesses that include biofuel production use hydrolysates, which include several toxic by-products. These stress-related problems are encountered in Gram-positive and -negative bacteria and several yeasts.

Bacteria encounter many chemicals in their environment, and degradation of these environmental pollutants leads to several physiochemical stresses in bacteria (Ramos et al., 2001). Further, bacteria in the environment are exposed to a number of stresses, which include lack of nutrients and water, and variations in temperature together with toxic materials (Ramos et al., 2001). Some of the compounds released in the environment are detected by bacteria as a carbon source (e.g., m-xylene), as well as stressful for their physiology (Assinder and Williams, 1990; Vlazquez et al., 2006). Thus, a bacterium has to adapt accordingly with respect to its metabolism, and it is suggested that under complex environmental conditions, there is strong selection of gene expression regulation and signal integration (Cases et al., 2003).

In the present chapter, an effort is made to understand chemical stress as it relates to bioremediation in response of bacteria.

21.2 General Bacterial Stress Responses

A bacterium adapts to changing environmental stresses, and the question arises of what exactly constitutes stress for bacteria. It can be stated that stress for bacteria may consist of the following:

- toxic compounds like pollutants in the environment: toxins, antibiotics, mutagens;
- deprivation of nutrients and water;
- Changes in optimum physiochemical parameters like very low/high temperature, oxidative stress, osmotic stress, pH fluctuations;
- interaction with host cells (especially, pathogenic bacteria).

In response to different stresses, a bacterium can have a specific response towards very high/low heat, cold, acidic, alkali, oxidative environmental conditions. SOS (named after international telegraph distress signal SOS) response is a specific stress response for any damage to DNA. Similarly, bacteria have general stress response like expression of many stress response genes, which helps bacteria in conditions of sporulation, long-term survival especially in stationary phase (Marles-Wright and Lewis, 2007). Some other responses of bacteria towards stress are – accumulation of compatible solutes, degradation of toxic compounds, and run away from stressed ambience by chemotactic movements. Bacteria can also adapt by morphological changes through differentiation and develop resistance (e.g., antibiotic resistance). The bacterial stress response toward different kind of stress is shown in Figure 21.3.

The regulation of the stress response can occur at every stage of bacterial physiology including DNA replication (supercoiling, recombination), transcription (which may include initiation or termination [sigma factors, repressors, etc.]), translation (attenuation, RNA binding factors), and protein level (metabolites changes, degradation, inhibition). The bacterial gene regulation is described by operon, regulon and modulan. Operon is defined as genes, transcribed from the same promoters

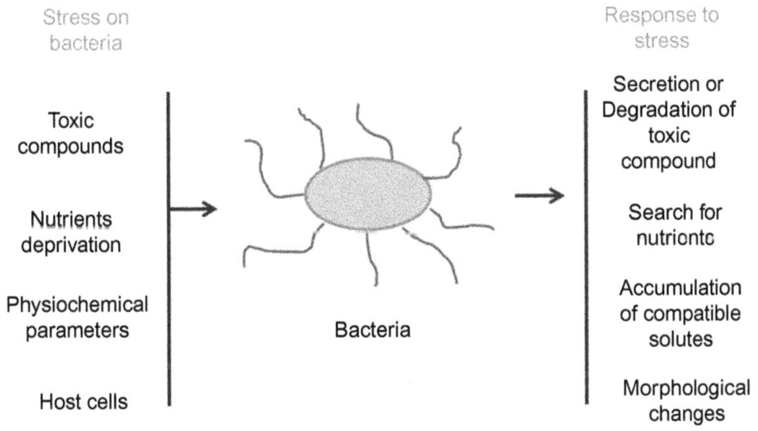

Figure 21.3 Different kinds of general stress responses in bacteria.

and is considered as transcriptional unit, while regulon is a group of operons controlled by common regulator protein, modulon represents group of operon and regulon. Genes, modulons, and regulators play a significant role in bacterial metabolism, behavior, and differentiation to particular stress responses and have the ability to coordinate and make changes in gene expression. This regulation of gene expression is mediated by regulatory transcription factors, which bond to the genes and further leads to a regulatory network. The regulation is dynamic in bacteria and depends on the extent and the kind of stress, e.g., environmental fluctuation of pH, temperature, or osmolality (Ruhal et al., 2013). It is also known that transcriptional regulatory network evolve very rapid and regulatory protein evolve much faster. Further in laboratory engineering of gene circuits helps in developing tunable circuits, which can be used for understanding bacterial survival strategies by these networks against hazardous chemicals present in environment (Acar et al., 2008). The other physiological mechanisms adopted by bacteria are two-component signal transduction, small RNA (sRNA) regulation of stress response, and stability of messenger RNA (mRNA).

21.3 Bacterial Physiological Responses to Chemical Stress

Since the objective of both this chapter and this book is to discuss bioremediation, we have focused on chemical stress in the light of bioremediation. Bioremediation can be defined simply as the conversion of toxic chemicals to less toxic or simpler molecules using microorganisms as catalysts (Pandey et al., 2009). Thus, similar to biorefinery and biofuel production (industrial processes), bioremediation also challenges bacteria to metabolically adapt against physiochemical stress developed as a response to chemicals present in their environment. To survive this situation, bacteria fine-tune their transcriptional networks in order to produce an ordered response in individuals or in their community as a whole. This phenomenon involves all stress response mechanisms of bacteria, including proteases, sigma factors, RNA modulation, and two-component signal transduction. In one example, *Pseudomonas putida* engineered to degrade chemicals like toluene, first senses the chemical stressors and then adjusts its physiology and expresses its catabolic pathway to become tolerant to these chemicals. This tolerance is a complex and multigenic trait. This can involve several simultaneous mechanisms like altered energy metabolism, molecular pumps, membrane properties, cell phenotype changes (e.g., shape and size); these can be independent from each other and are dependent on genes on chromosomes or plasmids (Ramos et al., 2002; Neumann et al., 2005; Bernal et al., 2007). In addition, it should be kept in mind that most studies were done in laboratories with single bacteria in controlled experimental conditions; hence, the situation can be different *in situ*.

21.3.1 Stress Response to Solvents

Solvents like alcohol, aldehydes, hydrocarbons, lithium chloride, urea, ethylene glycol, and phenol are lethal for bacteria. They can bind to the bacterial cell surface and ultimately lead to cell lysis and death. They have a number of effects on bacterial cells, but the most and immediate effect is reported to be on membranes (Aono and Kobayashi, 1997). Bacterial stress response to solvents has been extensively studied in numerous model bacterial organisms, including transcriptomics, proteomics, and metabolomics, and this has been utilized to generate tolerant strains (Serdessai and Bhosle, 2004). In general, solvents affect bacterial cells by making damaging biophysical changes to cell membranes, which can effect energy generation and intra- and extracellular transports. Further, they can affect DNA, causing lipid damage and leading to an enhanced stress response. Cells respond to this stress in several ways, which are discussed further.

Readjusting membrane fluidity may involve *cis–trans* isomerization of fatty acids or changes in the saturated-to-unsaturated ratio. It may also involve changes in the composition of the phospholipid head group. These responses are viewed in Figure 21.4. Following are major bacterial stress responses towards solvents:

- Changes in efflux or molecular pumps present in the membranes. These may involve a detoxification process, where bacterial cell can break toxic material into simple molecules, which can be thrown out through these pumps.
- Some of the general stress responses of bacterial cells such as expression of stress response proteins, e.g., heat shock proteins (HSPs).
- There can be accumulation of compatible solutes—sugars like trehalose, sugar alcohol glycerol, and amino acids like proline.

One of the best-studied strains regarding this stress is *P. putida* in response to phenol (Santos et al., 2004, 2007). It was reported that, during phenol stress, *Pseudomonas* sp. displayed increased expression of chaperones like DnaK, HtpG,

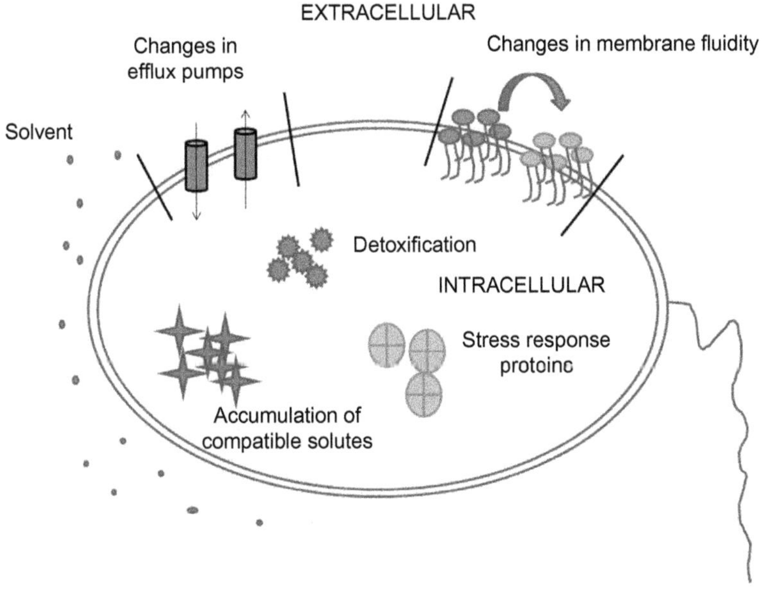

Figure 21.4 Stress responses of bacteria toward solvents.

ClpB, and the stress response protein Usp (Santos et al., 2007). Similarly, a study conducted on *P. putida* KT 2440 reveals the following upregulation and downregulation of proteins in response to phenol (Santos et al., 2004).

- Downregulation—flagellin (fliC), SurA (a peiplasmic protein required for folding of protein).
- Upregulation—oxidative stress proteins like AhpC, SodB, Tpx, and Dsb. General stress response—UspA, HtpG, GrpE, Tig. Energy metabolism—CanB, AtpH, Fpr. In addition fatty acid biosynthesis—FabB, transcriptional regulation, transport of small molecules, and proteins inhibiting cell division are upregulated.

The organic solvents with logPw (coefficient determined in octanol–water mixture) values between 1.5 and 3 are considered very toxic to bacteria (Ramos et al., 2002). Bacterial cell membranes consist of a phospholipid bilayer with proteins. Gram-negative bacteria have an outer membrane and less peptidoglycan in comparison to Gram-positive bacteria. There are phospholipid bilayers interspersed with proteins. These membranes are a kind of barrier to the external environment; they facilitate transport of proteins, ions, water etc. in and out of cell, which involves several important mechanisms like signal transduction, communication, and energy production. A number of studies have been done on the effect of solvents on cell integrity and transcriptomics networks (Sikkema et al., 1994, 1995; Bernal et al., 2007). The organic solvents can affect membranes by disturbing electrochemical potential and gradient and can inhibit membranes, which may lead to lysis and ultimately cell death (Serdessai and Bhosle, 2004). In general, solvents alter membrane fluidity and decrease the phospholipid bilayer, which can affect their biosynthesis and composition. There is *cis-trans* isomerization of phospholipids in fatty acids to counteract the increasing membrane fluidity (Weber and DeBon, 1996). It was reported that null mutation of *cti* (cis/trans isomerase gene) in *P. putida* indicates an important factor of initial damage (Junker and Ramos, 1999). Interestingly, it was observed that chain length does have an effect on chemical stress response with respect to the membrane, e.g., in *E. coli*, exposure to short-chain chemicals led to an increase in unsaturated fatty acids; alternatively, with long chains, there was an increase in saturated fatty acids (Weber and DeBon, 1996). But it is also reported that different species can respond differently to the same solvent, e.g., the increase of unsaturated fatty acids in *E. coli* and *Saccharomyces cerevisiae* (Ingram, 1976; Beaven et al., 1982), while increase of saturated fatty acids in *Bacillus subtillis*. These changes are done to maintain order and integrity of the membrane.

Another important bacterial cell response toward solvent tolerance is made with the help of efflux pumps present in the membrane. In Gram-negative bacteria, multidrug efflux pumps have an important role in solvent tolerance. These are unidirectional efflux systems, also called molecular pumps, which save cells from antibiotics primarily by expelling chemicals outside the cells. Interestingly, in Gram-negative bacteria, Resistance Nodulation Division (RND) transporters are involved with solvent tolerance, e.g., AcrAB–TolC efflux pumps in *E. coli* and TtgABC and TtgDEF pumps in *P. putida.* In *E. coli*, RND efflux pumps consist of AcrB and channel TolC

(Symmons et al., 2009). There are a number of regulators that modulate expression of AcrAB like *marR*, *marA*, *soxR*, and *soxS*. It was observed that *marR* overexpression led to increase in solvent tolerance in *E. coli*, and overexpression of other regulators led to higher activity of AcrA and TolC, which increased the tolerance of bacteria to hexane and cyclohexane (White et al., 1997; Aono et al., 1998). Furthermore, AcrAB mutants have a decreased tolerance to hexadecane (White et al., 1997). Similarly, in *P. putida*, a solvent resistance protein (SrpABC) in the form of an efflux pump was induced by solvents and alcohols but not induced in general stress conditions (pH, temperature, etc.) (Kieboom et al., 1998; Mosqueda and Ramos, 2000). Three kinds of efflux pumps in *P. putida*, known as TtgABC, TtgDEF, and TtgGHI, were reported, which were able to exclude toluene, styrene, ethylbenzene, propylbenzene, and xylenes (Duque et al., 2001).

Another cellular stress response includes HSPs, which are molecular chaperones that assist refolding of proteins by chemical stress. During chemical or any other stress, bacterial molecules like DNA, RNA, and various significant proteins are damaged and these chaperones or HSPs help in prevention of damaged protein (Han et al., 2008). In *E. coli*, HSPs have been extensively studied and are classified on the basis of size—DnaK, DnaJ, GrpE, GrpEL, GroES, and Clp family (Chung et al., 2006). These proteins help in protein refolding and prevent their aggregation. Similarly, in *B. subtillis*, different HSPs are present, which were reported to be upregulated during stress conditions. Most organisms have very specialized proteins known as *sigma factors*, which respond to stress conditions (Weber et al., 2005). *Sigma factor* encoded by *rpoS* in *E. coli* is very significant in stress conditions (Weber et al., 2005). Although HSPs do not have a direct role in solvent tolerance, a study in *C. acetobutylicum* shows overexpressions of heat shock protein GroESL, leading to reduced growth inhibition and more tolerance to butanol (Tomas et al., 2003).

21.3.2 Chemicals as Nutrient or Stressor, Their Influence on Energy Processes, and Accumulation of Compatible Solutes

The chemicals present in the environment can be both substrate and stressor. Furthermore, there will be other chemicals that may make the environment more stressful. The most widely studied strain in this area of research is *P. putida* strain. In this widely examined strain, it was found that degradation of toluene was not enough for the bacteria to become tolerant to toluene. This strain has plasmid pWW0, which has a TOL pathway that leads to the oxidation of toluene to benzoate and then catechol, but this strain still maintains its tolerance to toluene. It may be possible that intermediate compounds produced during bioremediation can be stressful for the bacteria; this observation was supported during biodegradation of polychloro-biphenyls. Thus, it is a major challenge for a bacterium to capture energy as well as to tolerate the deleterious effects of a substrate and its degradation intermediates. This was further revealed during a genome-wide global response of these strains, where it was found that the presence of 4-chlorophenyl, biphenyl led to expression not only of a catabolic pathway together with chaperones like GroEL, DnaK, and HtpG but also to accumulation of polyphosphate (Pieper and Seeger, 2008). It was also observed that exposure to biphenyl leads to oxidative stress and Reactive Oxygen Species (ROS) increase (Chavez et al., 2004). Similarly, naphthalene biodegradation also coincides with induction of antioxidant enzymes. The degradation of these chemicals can lead to oxidative stress, which may make bacterial cells adopt a mechanism to deal with them.

During *omic* studies, it was reported that during chemical stress, expression of genes related to sugar transport, glucose catabolism, and Kreb's cycle was upregulated. Thus, a bacterial cell has to adjust according to its energy requirements and fine-tune its physiology with respect to energy generation and degradation of toxic material. If it somehow becomes loosened with respect to energy, then it will not involve degradation. In a study related to toluene degradation, it was observed after proteomics analysis that energy supply was heavily disrupted due to affected proton-motive forces and that these were counteracted by downregulation of Adenosine Tri Phosphate (ATP) synthase. Furthermore, genes involved in Nicotine Adenine Di Phosphate (NADP(H)) metabolism were upregulated. Transcriptomics data suggested that genes involved in energy production and sugar consumption were upregulated, while genes involved in sugar accumulation were downregulated (Volkers et al., 2006, 2009).

Another important stress faced by bacteria is osmotic stress. The cytoplasmic membrane acts as an effective barrier for solutes present in the environments of microbes. In the bacterial cytoplasm, the concentration of intracellular solutes remains higher than in the outside in nonstress conditions, which leads water to flow down its chemical potential into the cell. This leads to turgor pressure exerted by the cytoplasmic membrane toward the cell wall. This turgor balances the variance in osmotic pressure between the intracellular and extracellular of bacterial cell (Morbach and Krämer, 2005). Turgor remains indispensable for cell envelope enlargement, growth, and division, and is maintained in general throughout the cell cycle. The measurement of turgor is considered difficult, but it is estimated that Gram-positive and -negative bacteria differ in their turgor pressure. Further, it was observed that turgor pressure of Gram-positive bacteria was higher, since it has high cytoplasmic solvents (Morbach and Krämer, 2005). Bacteria are forced to adapt to the environment with frequent fluctuations in osmolality. It can be through a decrease or increase in the external osmolality, or through osmotic hypo- or hyper-shift. A number of mechanisms are known for their tolerance to osmotic stress, but two main adaptations studied in microbes are the release of solutes out of the cytoplasm after an osmotic downshift and the accumulation of compatible solutes after an osmotic upshift (Ruhal et al., 2013). Betaine, proline, glutamine, ectoine, and trehalose were found to be effective as osmoprotectants. As the bacterium faces the challenge of osmotic stress, instant water efflux occurs, and the cell dehydrates. This may slow or stop growth. Accumulation of these kinds of solutes has an advantage for bacterial cells they assist with rehydration without disturbing other functions of the cell. Furthermore, it has been reported that at the molecular level, compatible solutes are thought to stabilize and protect enzymes by preventing

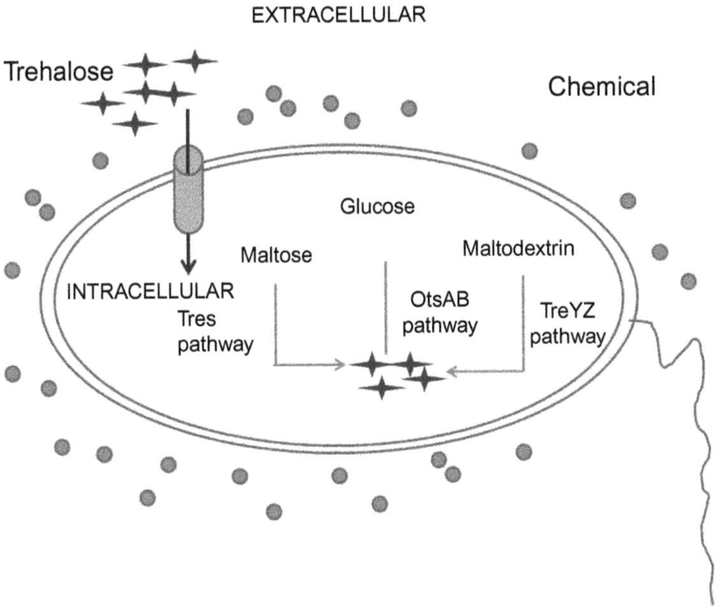

Figure 21.5 During osmotic stress due to chemicals, accumulation of trehalose is reported, which may be due to import of trehalose from extracellular of bacterial cell or biosynthesis of trehalose through different pathways utilizing different carbon sources.

dehydration of the protein. Similarly, various adaptations are motivated by microbes under temperature stress and alkaline/acidic stress (Ruhal and Choudhury, 2012a,b; Ruhal et al., 2013). One compatible solute, trehalose, is reported to accumulate in numerous bacteria under stress conditions. In particular, the effects of osmotic stress on *E. coli*, *Corneybacterium* sp., *S. cerevisiae*, and *Propionibacterium freudenreichii* have already been reported (Ruhal et al., 2013). In all of these microorganisms, the prominent role of the trehalose biosynthesis pathway has been clearly demonstrated. It can be seen in Figure 21.5. Trehalose has been implicated as a potential stress protectant that accumulates in yeasts during various stress condition. During the fermentation process, yeasts are subjected to a succession of stress conditions, such as high temperature, high sugar, and accumulation of ethanol, which affect their viability and fermentation efficiency. In yeast, a strong correlation between trehalose content and stress resistance has been demonstrated for a variety of stresses such as heat, osmotic stress, and ethanol. Furthermore, the stress response is mediated at the level of transcription, and a number of stress-induced transduction pathways concerning trehalose have already been reported (Ruhal and Choudhury, 2012a,b; Ruhal et al., 2013). Several studies have reported that trehalose has a better advantage as a protein stabilizer in comparison to other compatible solutes, because it has an unusual ability to stabilize the protein in its native conformation. A strong correlation between trehalose content and stress resistance has been demonstrated for a variety of stresses, such as heat, osmotic stress, and oxidative. Furthermore, the stress response is mediated at the level of transcription, and a number of stress-induced transduction pathways concerning trehalose were also reported (Ruhal et al., 2013).

21.4 Microbial Stress During Biofuel and Chemical Production

Another important aspect of microbial stress response and tolerance is the field of biofuel production. The development of technologies for biofuel production for alternative sources of energy from industrial waste, such as crude glycerol and lignocellulose materials, has been a subject of interest (Kataria et al., 2009; Ruhal et al., 2011). One of the challenges in these industries is the stress to microbes from toxic materials present in these carbon sources. Therefore, a procedure of detoxification by pretreatment is necessary for these wastes. Hydrolysis of lignocellulose material is done to generate different sugars used for microbial fermentation, but during this hydrolysis, some side products are formed that inhibit growth of microbes (Kataria and Ghosh, 2012; Kataria et al., 2013). These are furans like furfural and hydroxymethyl furaldehyde (HMF), which are developed by degradation of pentoses and hexoses. They have been reported to inhibit growth of microbes, reduce biocatalyst activity, and increase the lag phase of yeast as it adapts to furfural and HMF. It was reported that these toxic materials have an impact on energy metabolism, depletion of NADP/H, and cell replication. Other toxic compounds that are reported to form are acetic acid, formic acid, and levulinic acid; they are formed after degradation of hemicellulose, furfural, and HMF (Nicolaou et al., 2010) and have effect on ethanol yield; this is due to their effect on uncoupling proton-motive forces, intracellular pH drop, anion accumulation, and breakdown of the electrochemical membrane. Other toxic compounds reported are phenolic acids like 4-hydroxybenzaldehyde, which are formed due to breakdown of lignin. They have an effect on growth rate and ethanol yield and have a major effect on membranes. A number of bacteria and yeast were developed, which were tolerant to different stresses.

21.5 Conclusion

In the present chapter, an effort was made to understand how microbial cells respond to solvents, carboxylic acid, and related chemical stress. These chemicals are encountered in various bioprocesses like biofuel production, biocatalysis, and bioremediation. A number of bacteria were generated by molecular techniques after understanding of their mechanism of stress response. One of the model bacteria, *P. putida*, was studied for chemical tolerance and was found to be tolerant to toluene, butanol, decane, octane, styrene, and many other chemicals. It has also alteration in cell membranes like *cis–trans* isomerization and changes in fatty acid composition. In another model bacteria, *E. coli*, all these mechanism of tolerance was utilized to generate strains along with membrane changes, overexpression of several HSPs was observed and higher accumulations of compatible solutes to help bacteria against stress. Thus, there is need for improved genomic tools to generate synthetic programs to deal with this issue. This issue is of particular interest for the areas of biodegradation and bioremediation due to increased environmental issues caused by rapid industrialization.

References

Acar, M., Mettetal, J.T., van Oudenaarden, A., 2008. Stochastic switching as a survival strategy in fluctuating environment. Nat. Genet. 40, 471–475.

Aono, R., Kobayashi, H., 1997. Cell surface properties of organic solvents tolerant mutants of *E. coli* K12. Appl. Environ. Microbiol. 63, 3637–3647.

Aono, R., Tsukaghoshi, N., Yamamoto, M., 1998. Involvement of outer membrane protein TolC a possible member of mar-sox regulon in maintenance and improvement of organic solvents tolerance of *E. coli*. J. Bacteriol. 180, 4938–4944.

Assinder, S.J., Williams, P.A., 1990. The TOL plasmids: determinants of the catabolism of toluene and the xylenes. Adv. Microb. Physiol. 31, 1–69.

Beaven, M.J., Charpentier, C., Rose, A.H., 1982. Production and tolerance of ethanol in relation to phospholipid fatty-acyl composition in *Saccharomyces cerevisiae* NCYC 431. J. Gen. Microbiol. 128, 1447–1455.

Bernal, P., Segura, A., Ramos, J.L., 2007. Compensatory role of cis–trans isomerase and cardiolipin synthase in the membrane fluidity of *Pseudomonas putida* DOT-TIE. Environ. Microbiol. 9, 1658–1664.

Cases, I., de Lorenzo, V., Ouzounis, C.A., 2003. Transcriptional regulation and environmental adaptation in bacteria. Trends Microbiol. 11, 248–253.

Chavez, F.P., Lunsdorf, H., Jerez, C.A., 2004. Growth of polychlorinated-biphenyl-degrading bacteria in the presence of biphenyl and chlorobiphenyls generates oxidative stress and massive accumulation of inorganic polyphosphate. Appl. Environ. Microbiol. 70, 3064–3072.

Chung, H.J., Bang, W., Drake, M.A., 2006. Stress response of *E. coli*. Compr. Rev. Food Sci. Food Saf. 5, 52–64.

Duque, E.A., Segura, M.G., Ramos, J.L., 2001. Global and cognate regulators control the expression of the organic solvent efflux pumps TtgABC and TtgDEF of *Pseudomonas putida*. Mol. Microbiol. 39, 1100–1106.

Han, M.J., Yun, H., Lee, S.Y., 2008. Microbial heat shock proteins and their uses in biotechnology. Biotechnol. Adv. 26, 591–609.

Ingram, L.O., 1976. Adaptation of membrane lipids to alcohols. J. Bacteriol. 125, 670–678.

Junker, F., Ramos, J.L., 1999. Involvement of the cis–trans isomerase Cti in solvent resistance of *Pseudomonas putida* DOT-TIE. J. Bacteriol. 181, 5693–5700.

Kataria, R., Ghosh, S., 2012. Saccharification of Kans grass using enzyme mixture from *T. reesei* for bioethanol production. Bioresour. Technol. 102, 9970–9975.

Kataria, R., Choudhury, G., Ghosh, S., 2009. Potential of bioenergy production from grasses and its impact on environment. Res. J. Biotechnol. 4, 5–14.

Kataria, R., Ruhal, R., Babu, R., Ghosh, S., 2013. Saccharification of alkali treated biomass of Kans grass contributes higher sugar in contrast to acid treated biomass. Chem. Eng. J. 230, 36–47.

Kieboom, J., Dennis, J.J., Zylstra, G.J., de Bont, J.A.M., 1998. Active efflux of organic solvents by *Pseudomonas putida* S12 is induced by solvent. J. Bacteriol. 180, 6769–6772.

Marles-Wright, J., Lewis, J.R., 2007. Stress response of bacteria. Curr. Opin. Struct. Biol. 17, 755–760.

Morbach, S., Krämer, R., 2005. Osmoregulation. Handbook of Corynebacterium Glutamicum. *Taylor & Francis, Boca Raton, FL*, pp. 417–438.

Mosqueda, G., Ramos, J.L., 2000. A set of genes encoding a second toluene efflux system in *Pseudomonas putida* DOT-T1E is linked to the *tod* genes for toluene metabolism. J. Bacteriol. 181, 937–943.

Neumann, G., Veeranagouda, Y., Karegoudar, T.B., Sahin, O., Mausezahl, I., Kabelitz, N., et al., 2005. Cells of *Pseudomonas putida* and *Enterobacter* sp. adapt to toxic organic compounds by increasing their sizes. Extremophiles. 9, 163–168.

Nicolaou, S.A., Gaida, S.M., Papoutsakis, E.T., 2010. A comparative view of metabolite and substrate stress and tolerance in microbial bioprocessing from biofuels and chemicals to biocatalysis and bioremediation. Metab. Eng. 12, 307–331.

Pandey, J., Chouhan, A., Jain, R.K., 2009. Integrative approach for assessing the ecological sustainability of in-situ bioremediation. FEMS Microbiol. Rev. 33, 324–375.

Pieper, D.H., Seeger, M., 2008. Bacterial metabolism of polychlorinated biphenyls. J. Mol. Microbiol. Biotechnol. 15, 121–138.

Ramos, J.L., Gallegos, M.T., Marques, S., 2001. Response of Gram negative bacteria to certain environmental stressors. Curr. Opin. Microbiol. 4, 166–171.

Ramos, J.L., Duque, E., Gallegos, M.T., Godoy, P., Ramos-Gonzalez, M.I., Rojas, A., et al., 2002. Mechanism of solvent tolerance in Gram negative bacteria. Annu. Rev. Microbiol. 56, 743–768.

Ruhal, R., Choudhury, B., 2012a. Improved trehalose production from biodiesel waste using parent and osmotically sensitive mutant of *Propionibacterium* subsp *shermanii* under aerobic conditions. J. Ind. Microbiol. Biotechnol. 39, 1153–1160.

Ruhal, R., Choudhury, B., 2012b. Use of an osmotically sensitive mutant of *Propionibacterium freudenreichii* subspp. *shermanii* for the simultaneous productions of organic acids and trehalose from biodiesel waste based crude glycerol. Bioresour. Technol. 109, 131–139.

Ruhal, R., Aggarwal, S., Choudhury, B., 2011. Suitability of crude glycerol obtained from bio diesel waste for the productions of trehalose and propionic acid. Green Chem. 13, 3492–3498.

Ruhal, R., Kataria, R., Choudhury, B., 2013. Trends in bacterial trehalose metabolism and significant nodes of metabolic pathway in the direction of trehalose accumulation. Microb. Biotechnol. 6, 493–502.

Santos, P.M., Benndorf, D., Sa-Correia, I., 2004. Insights into *Pseudomonas putida* KT 2440 response to phenol induced stress by quantitative proteomics. Proteomics. 4, 2640–2652.

Santos, P.M., Roma, V., Benndorf, D., von Bergen, M., Harms, H., Sa-Correia, I., 2007. Mechanistic insights into the global response to phenol in the phenol-biodegrading strain *Pseudomonas* sp.M1 revealed by quantitative proteomics. OMICS. 11, 233–251.

Serdessai, Y.N., Bhosle, S., 2004. Industrial potential of organic solvent tolerant bacteria. Biotechnol. Prog. 20, 655–660.

Sikkema, J., DeBont, J.A., Poolman, B., 1994. Interactions of cyclic hydrocarbons with biological membranes. J. Biol. Sci. 269, 8022–8028.

Sikkema, J., DeBont, J.A., Poolman, B., 1995. Mechanism of membrane toxicity of hydrocarbons. Microbiol. Rev. 59, 201–202.

Symmons, M.F., Bokma, E., Koronakis, E., Hughes, C., Koronakis, V., 2009. The assembled structure of a complete tripartite bacterial multidrug efflux pump. Proc. Natl. Acad. Sci. U.S.A. 106 (17), 7173–7178.

Tomas, C.A., Welker, N.E., Papoutsakis, E.T., 2003. Overexpression of *groESL* in *Clostridium acetobutylicum* results in increased solvent production and tolerance, prolonged metabolism and changes in transcriptional program. Appl. Environ. Microbiol. 69, 4951–4965.

Velazquez, F., de Victoe, L., Valls, M., 2006. The m-xylene biodegradation capacity of *Pseudomonas putida* mt-2 is submitted to adaptation to abiotic stress: evidence from expression profiling of *xyl* genes. Environ. Microbiol. 8, 591–602.

Volkers, R.J., de Jong, A.L., Hulst, A.G., van Baar, B.L., de Bont, J.A., Wery, J., 2006. Chemostat-based proteomic analysis of toluene-affected *Pseudomonas putida* S12. Environ. Microbiol. 8, 1674–1679.

Volkers, R.J., Ballerstedt, H., Ruijssenaars, H., 2009. TrgI, toluene repressed gene I, a novel gene involved in toluene-tolerance in *Pseudomonas putida* S12. Extremophiles. 13, 283–297.

Weber, F.J., DeBon, J.A., 1996. Adaptation mechanism of microorganism to the toxic effect of organic solvent on membrane. Biochem. Biophys. Acta. 1286, 225–245.

Weber, H., Polen, T., Heuveling, J., Wendisch, V.F., Hengge, R., 2005. Genome wide analysis of the general stress response network in *E coli* sigma dependent genes, promoters and sigma factors selectivity. J. Bacteriol. 187, 1591–1603.

White, D.G., Goldman, J.D., Demple, B., Levy, S.B., 1997. Role of acrAB locus in organic solvent tolerance mediated by expression of *marA*, *soxS* or *robA* in *E. coli*. J. Bacteriol. 179, 6122–6126.

22 Bioremediation of Pesticides: A Case Study

Sujata Ray

Department of Earth Sciences, Indian Institute of Science Education and Research, Kolkata, Mohanpur, West Bengal, India

22.1 Introduction

Organochlorine pesticides, such as Dichloro-Diphenyl-Trichloroethane (DDT), are injurious to health and the environment and may cause endocrine, reproductive, and immunological dysfunctions and cancer. These chemicals are now limited in production and use in many parts of the world such as Europe and North America. While concentrations in these regions are declining slowly, these chemicals will continue to persist. Primary releases still take place from diffuse sources, such as old stockpiles. Re-emissions may occur from reservoirs such as soil, vegetation, water bodies, and contaminated sediments (Nizzetto et al., 2010). Other sources may include transportation from long distances, from countries where these chemicals continue to be used. These countries are usually developing and suffer from the lack of an economically viable alternative. For example, an economic means of controlling malaria is the application of DDT, and this pesticide is registered for restricted use in India. The use of these chemicals in developing countries makes their movement across national boundaries a matter of concern.

Transnational movement of the chemicals may occur by atmospheric long-range transport. It may even be affected by migratory birds. For example, seabird guano (excrement) has been reported to be a source of organic pollutant contamination in a remote Arctic lake (Evenset et al., 2010). This guano is found to account for 14% of the pollutant concentration in the catchment area of the lake and 80% of the contamination of the lake itself. This pathway may cause significant impact in terms of the bioconcentration of the chemicals in lakes and marine regions where the bird colonies exist. Because seabird colonies are found in coastal areas throughout the world, the continued use of pesticides in certain countries may have a significant impact on the entire human race. This impact may be particularly strong because certain species of seabirds migrate across very long distances. The arctic tern, for example, migrates from its breeding ground in the Arctic across the world to the nonbreeding area in the Antarctic.

Microbial Biodegradation and Bioremediation. DOI: http://dx.doi.org/10.1016/B978-0-12-800021-2.00022-4

Pesticides may be treated by physical, chemical, and biological processes. For example, contaminants in waste streams may be concentrated and absorbed into media such as activated charcoal. Solid contaminants may be chemically oxidized by incineration. However, physical and chemical processes of remediation are usually expensive. Moreover, they often require human exposure to hazardous chemicals and may lead to contaminant release in the environment. For example, if river sediments require physical or chemical treatment, then they must be dredged from the riverbed, which may release the pollutants absorbed in the sediments back into the water column. Incineration of hazardous solid waste may cause harmful gases to be released to the environment.

One economically feasible means of cleaning a contaminated region is the use of bioremediation. Bioremediation is the use of living organisms to promote the destruction of environmental pollutants (Perry et al., 2002). In this process, microorganisms, usually bacteria, metabolize the chemicals. The process of metabolism requires an energy source, or an electron donor, which is usually the pesticide. It also requires an electron acceptor, which is often oxygen. Other requirements are the adequate availability of nutrients such as nitrogen and phosphorus and an appropriate temperature, pH, and moisture content in soils.

When the environmental conditions are conducive, this process of biodegradation may occur without any human intervention and does not involve human exposure to the chemicals. This is known as *in situ* bioremediation. It is best accomplished by microorganisms that grow at the contaminated site itself and have adapted to its environmental conditions and to the mixture of harmful chemicals that exist there. These indigenous species together form a "consortium," and they cooperate with one another to ensure complete mineralization of the contaminants. For example, the compound 3-chlorobenzoate is degraded by one species to form benzoate, which is further degraded by another species to form acetate. Finally, acetate is degraded by a third species to form methane and carbon dioxide. Similarly, the pesticide Dalapon (2,2-dichloropropionic acid) is reported to be degraded by seven different microbiological species (Senior et al., 1976). While the addition of new species of microorganisms may enhance biodegradation, these species are often unstable because of their lack of adaptation to the contaminated site.

22.2 Challenges in Bioremediation

The growth rate of the *in situ* organisms is sometimes unacceptably slow. In this case, it is necessary to acquire knowledge of the environmental conditions that will enhance this growth rate. When the biodegradation rates are slow, then the soils may be amended by addition of nutrients such as nitrogen and phosphorus. Also, oxygen may be added by sparging air.

The desired end product of biodegradation is carbon dioxide, which is produced when complete mineralization of the chemical takes place. However, this does not

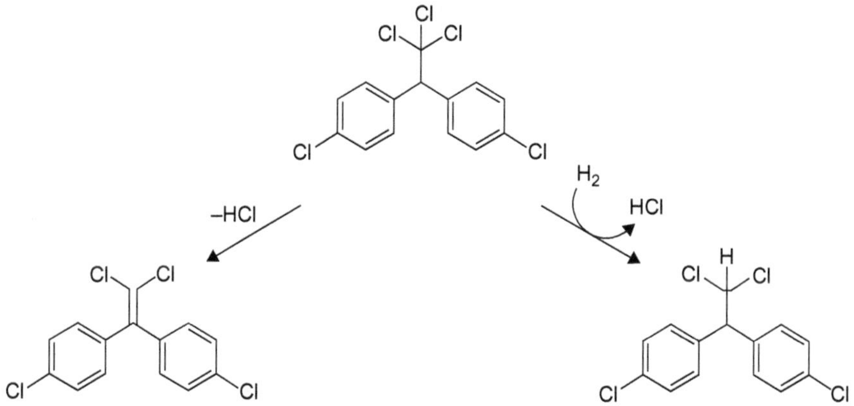

Figure 22.1 Degradation of DDT to form DDE (by elimination of HCl, left) and DDD (by reductive dechlorination, right).

always occur as the pesticide may be transformed into intermediate metabolites. Unfortunately, these intermediates may turn out to be equally harmful or more so than the parent chemical. For example, DDT is transformed into dichloro diphenyl dichloroethane (DDE) and dichloro diphenyl dichloroethelene (DDD) (Figure 22.1), both of which are considered to be as hazardous as the parent compound.

22.3 Role of Enzymes in Bioremediation

One of the tools that microorganisms use to efficiently degrade compounds such as pesticides is the use of special proteins that serve as catalysts, known as enzymes. These proteins can speed up reaction rates by as much as 10^9 times (Schwarzenbach et al., 2003). This is because enzymes can bind and hold the reacting substances in advantageous orientation with respect to one another and greatly lower the free energy of activation. Also, enzymes often contain charged structural groups that can change the electron densities of the reacting compounds. This facilitates the breaking of bonds and the creation of new ones.

Microbiological enzymes are capable of degrading pesticides through a process known as "co-metabolism." Co-metabolism is a "fortuitous" process that does not provide the microorganisms with utilizable energy. It occurs simultaneously with the biodegradation of an energy source. It occurs because microbiological enzymes are often nonspecific, in that they have the ability to bind chemicals that are structurally similar to the target compounds. Pesticides that have structural similarities to naturally occurring compounds may bind to enzymes that are capable of degrading them. The resulting biodegradation is known as co-metabolism.

22.4 Rates of Bioremediation

The rate of bioremediation of a pesticide molecule depends on the following factors: (1) the bioavailability of the pesticide, (2) the uptake rate of the pesticide molecule by the microbiological cell, (3) the rate at which the required enzymes that degrade the molecule function, and (4) the rate of growth of the cells with the pesticide molecule as a source of energy.

First, the pesticide must be made bioavailable, in that it must be in a physical state in which it can be biodegraded. In general, biodegradation occurs when the substance is dissolved in the aqueous media. This is the reason why very hydrophobic aromatic compounds are degraded much more slowly than their less hydrophobic congeners. Similarly, bioavailability is limited when the pesticide is sorbed into a nonaqueous phase, e.g., in sediments and soils. Also, certain processes may limit the delivery of necessary nutrients to the microorganism. For example, the delivery of oxygen may be limited in biofilms, where the gas must diffuse through many layers of cells to those that are attached to the holding surface.

Secondly, the pesticide must be taken up from the surrounding media by the microbiological cell. Because pesticide molecules are foreign to the cell, it is unlikely to have an apparatus that will actively transport the chemical inside. However, nonpolar pesticides may be passively transported into the cell interior by dissolving in the lipid-rich outer membrane and diffusing through it. Once the chemical reaches the enzymes that will degrade it, bioremediation proceeds rapidly. Here, the uptake of the chemical is rate limiting, and the structural features of the chemical that govern the uptake rate are important. In fact, the relative rate of biodegradation of a series of organohalides by a culture of *Pseudomonas putida* is reported to be due to the differences in permeability in the halides (Castro et al., 1985).

Thirdly, the enzyme that will degrade the chemical must become active. "Constitutively expressed" enzymes, or those that are always present in the cell, may increase in concentration when the chemical enters the cell. Some enzymes are "induced." These induced enzymes are normally absent in the cell and are synthesized only when the chemical is available for biodegradation. Some enzymes are "derepressed" or activated when the chemical enters the cell.

Finally, the rate of bioremediation depends on how quickly the microbiological population is capable of growing by utilizing the pesticide as a source of energy. The greater the affinity of the cell for the pesticide and the faster its rate of growth, the greater the rate of bioremediation.

22.5 Chemical Structure and its Impact on Bioremediation

The susceptibility of a pesticide to biodegradation is strongly influenced by its chemical structure. A number of studies have sought relationships between chemical structure and biodegradability. One such study examined the ability of microorganisms to degrade a set of 900 organic compounds with diverse structural groups

(Takatsuki et al., 1995). In this study, a chemical was considered to be easily biodegradable if the microorganisms consumed greater than 60% of the theoretical oxygen demand of the compound within 28 days. The theoretical oxygen demand of a compound is the stoichiometric amount of oxygen required for the complete conversion of a molecule of the compound to carbon dioxide. The biodegradability of the compound was correlated with its structure by establishing the following empirical relationship.

$$Y_i = a_0 + a_1 f_1 + a_2 f_2 + \cdots + a_n f_n + a_{mw}\mathrm{MW} + e_i$$

where Y_i is the probability of the chemical being susceptible to biodegradation,

a_0 is the model intercept,
a_n is the coefficient for the nth structural fragment,
f_n is the number of times the structural group n occurs in the chemical,
a_{mw} is the coefficient for the compound's molecular mass,
MW is the molecular mass of the compound, and
e_i is the error term with a mean value of zero.

The value of Y_i for each compound is determined experimentally. The values of $f_1, f_2, \ldots, f_n$ are determined from the chemical structures of the compounds. The values of the coefficients $a_1, a_2, \ldots, a_n$ and a_{mw} are determined by running the data through an optimization program that minimizes the value of the error term.

The following are a few general relationships between chemical structure and biodegradability. Groups such as carboxylic acid esters (—C(O)OR), amides (—C(O)NR_2), phosphorus acid esters, hydroxyl (—OH), formyl (—CHO), and carboxy (—COOH) are readily biodegradable anoxic conditions. These structural groups are common in nature, and it is likely that microorganisms have developed the enzymes necessary to process them.

In contrast, the presence of certain structural groups decreases the ease of biodegradation. For example, chloro and nitro groups, especially those on aromatic rings, quaternary carbons ($CR_1R_2R_3R_4$ with no R = H), and tertiary nitrogens ($NR_1R_2R_3$ with no R = H), reduce the ease of bioremediation (Schwarzenbach et al., 2003).

Organochlorine pesticides generally have low biodegradability because many contain chlorine groups on aromatic rings. Although such structural features are found in natural compounds, the enzymes required for their processing are usually present at low level of activities. When enzymatic cleavage of the aromatic carbon and halogen bond does occur, the number, position, and electronegativity of the bonds are of fundamental importance in determining the ease of bioremediation (Perry et al., 2002).

We must remember that this analysis does not take into account the fact that several functional groups present in a single compound will interact with one another and have a combined effect on its biodegradability. This will differ from the biodegradability predicted by the empirical relationship, which accounts for the presence of each group individually. For example, the presence of a carboxylic group will increase the biodegradability of an aromatic pesticide. Similarly, the presence of an additional halogen is likely to decrease the biodegradability of a chlorinated compound even further in an oxic environment. However, chlorinated compounds tend to be more easily biodegraded in anoxic conditions. This indicates that the enzymes necessary for the bioremediation of this class of compounds are active in anaerobic conditions.

This qualitative analysis of chemical structure and biodegradability may be used for the preliminary planning of bioremediation. These insights are useful in that they are the basis for the manufacture of a new class of pesticides that are both biodegradable and that accomplish the necessary functions. Pesticides that are now synthesized are usually much more biodegradable and do not persist after their functions are complete. The insights gained from the study of biodegradation of the persistent chemicals also help in the design of treatment schemes that expose the chemicals to both oxic and anoxic environments to ensure complete bioremediation.

22.6 Initial Pathways in Biodegradation of Pesticides

Depending on the chemical structure, three processes exist through which microbiological cells generally initiate the breakdown of a compound. The processes are hydrolysis, oxidation/reduction, and addition.

Hydrolysis of a pesticide occurs in an aqueous environment. When the herbicide linuron (Figure 22.2A), used by potato, rice, and wheat farmers, is hydrolyzed, it is converted to 3,4-dichloroaniline (Figure 22.2B) and *N*-methyl, *N*-methoxycarbamic acid (Schwarzenbach et al., 2003). One of the products of this reaction, *N*-methyl, *N*-methoxycarbamic acid, is converted fairly quickly to CO_2 and CH_3NHOCH_3.

In the second type of reaction, oxidation/reduction, a change occurs in the redox state of the pesticide. Whether the change is an oxidation or a reduction depends on the cosubstrates present in the environment. In a reducing environment, the pesticide DDT is principally converted to DDD. In this environment, the three chlorines in close proximity (Figure 22.2) may attract electrons from a reduced enzymatic group, e.g., a reduced metal. The reaction may be represented as follows:

$$\mathrm{Enz-Me^{red}} + \mathrm{Cl-CCl_2} - R + \mathrm{H^+} \rightarrow \mathrm{Enz-Me^{ox}} + \mathrm{Cl^-} + H - \mathrm{CCl_2} - R$$

Finally, in the process of addition, microorganisms ensure that a new group is added to the pesticide molecule. Microorganisms usually employ this strategy

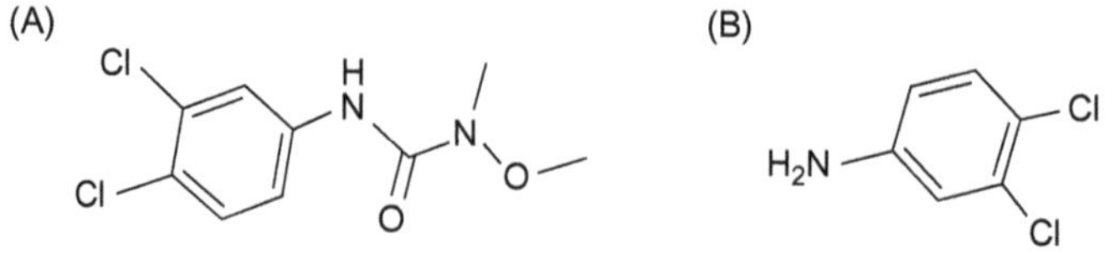

Figure 22.2 Chemical structure of (A) linuron and (B) 3,4-dichloroaniline.

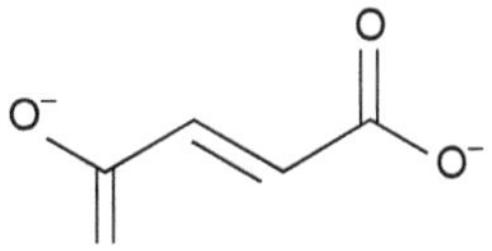

Figure 22.3 Chemical structure of fumarate.

when the environmental conditions are not favorable for any other type of reaction. This involves the investment of energy, but the addition renders the chemical far more susceptible to biodegradation. A group commonly used for microbiological addition is fumarate (Figure 22.3).

22.7 A Case Study

This case study focuses on a bioremediation experiment that was carried out on two types of soil contaminated by pentachlorophenol (PCP) using two pure cultures and a mixed one (Pu and Cutright, 2007). The two soil types were collected from the cities of Columbia and New Mexico, respectively. The pure cultures were *Arthrobacter* sp. and *Flavobacterium* sp., and the mixed culture was a 50:50 composition of the two. The study investigated the efficiency of each of the cultures in the bioremediation of PCP and illuminates many of the points that have been discussed in this chapter.

The chemical PCP was first used as a wood preservative in 1936 and was later used as a pesticide, herbicide, and disinfectant in the 1970s. However, it was found that PCP adversely affects human health, causing neurological disorders, leukemia, liver damage, and eye irritation, with long-term exposure damaging the respiratory tract, blood, kidneys, liver, immune system, eyes, and skin (Pepper et al., 1999). The use of PCP was again restricted to that of a wood preservative in the 1990s. The damaging effects of PCP and its widespread use make the study of its bioremediation important.

The biodegradability of PCP is determined to a large extent by the environmental conditions. For example, whether ring cleavage occurs before or after dechlorination is determined by whether the conditions are oxic or anoxic (Pu and Cutright, 2007). Complete mineralization of the pesticide was possible only by mixed cultures. The consortia that were most efficient in accomplishing bioremediation were those that had grown in PCP-contaminated soil (Pu and Cutright, 2007).

The study found that all the cultures degraded PCP more efficiently in the soil from New Mexico. This indicates that prevailing environmental conditions in the soil, such as pH, are critical in determining the success of bioremediation. This difference in the effectiveness of bioremediation may also be attributed to the extent to which the PCP was sorbed in the soil's organic matter. In the soil from cities of Columbia, the culture of *Flavobacterium* sp. was more capable of degrading PCP than the *Arthrobacter* sp. The mixed culture was the most effective in achieving PCP degradation. This was regardless of the PCP concentration or the soil type.

22.8 Conclusion

Microorganisms are present virtually everywhere in nature and survive in extreme conditions of pH, temperature, pressure, nutrient content, water content, and salinity. This makes them perfect tools for the process of bioremediation. However, although they are widely present, not all of the species are active in a specific environment. This may render bioremediation challenging. This chapter explores the various strategies that microorganisms use for the biodegradation of pesticides and the conditions in which this is likely to be successful. We also cite a case study to illustrate these points. This knowledge enables environmental scientists and engineers to better design bioremediation schemes.

References

Castro, C.E., Wade, R.S., Belser, N.O., 1985. Biodehalogenation: reactions of cytochrome P450 with polyhalomethanes. Biochemistry. 24, 204–210.

Evenset, A., Carroll, J., Christensen, G.N., Kallenborn, R., Gregor, D., Gabrielsen, G.W., 2010. Seabird guano is an efficient conveyer of persistent organic pollutants (POPs) to Arctic Lake ecosystems. Environ. Sci. Technol. 41, 1173–1179.

Nizzetto, L., Macleod, M., Borga, K., Cabrerizo, A., Dachs, J., Di Guardo, A., et al., 2010. Past, present and future controls on levels of persistent organic pollutants in the global environment. Environ. Sci. Technol. 44 (17), 6526–6531.

Pepper, M., Ertl, M., Gerhard, I., 1999. Long-term exposure to wood preserving chemicals containing (PCP) and lindane is related to neuro behavioural performance in women. Am. J. Ind. Med. 35, 632–641.

Perry, J.J., Staley, J.T., Lory, S., 2002. Microbial Life. Sinauer Associates, Sunderland, MA.

Pu, X., Cutright, T.J., 2007. Degradation of pentachlorophenol by pure and mixed cultures in two different soils. Environ. Sci. Pollut. Res. 14 (4), 244–250.

Schwarzenbach, R.P., Gschwend, P.M., Imboden, D.M., 2003. Environmental Organic Chemistry. John Wiley and Sons, Inc, Hoboken, NJ.

Senior, E., Bull, A.T., Slater, J.H., 1976. Enzyme evolution in a microbial community growing on the herbicide Dalapon. Nature. 263, 476–479.

Takatsuki, M., Takanayagi, Y., Kitano, M., 1995. An attempt to SAR of biodegradation. Proceedings of the Workshop, Quantitative Structure Activity Relationships for Biodegradation 67–103. National Institute of Public Health and Environmental Protection (RIVM). Bilthoven, the Netherlands.

23 Microalgae in Removal of Heavy Metal and Organic Pollutants from Soil

Madhu Priya[a,b], *Neelam Gurung*[b], *Koninika Mukherjee*[b,c] *and Sutapa Bose*[b]

[a]Department of Environmental Science, Central University of Rajasthan, Bandar Sindri, Ajmeer, India, [b]Department of Earth Sciences, Indian Institute of Science Education and Research Kolkata, Mohanpur, Nadia, West Bengal, India, [c]KIIT School of Biotechnology, Patia, Bhubaneswar, Odisha, India

23.1 Introduction

The term "microalgae" refers to aquatic microscopic plants and oxygenic photosynthetic bacteria, that is, the cyanobacteria formerly known as Cyanophyceae. They are typically found in freshwater and marine systems. They are also referred to phytoplankton or microphytes. They are unicellular and may exist individually or in chains or groups. Their size varies from species to species and ranges from a few micrometers (μm) to a few hundred micrometers. They are devoid of roots, stems, and leaves. Microalgae are photosynthetically active. Approximately 200,000–800,000 species of microalgae exist in nature, but only 50,000 of them have been described so far. Microalgae are not only taxonomically diverse but are also extremely efficient. Thus, they are of great importance in global ecology (Brown and Zeiler, 1993). A few uses of microalgae are listed below.

1. They are responsible for approximately 32% of global photosynthesis (Whittaker, 1975).
2. They produce more than 60% of the oxygen on Earth and simultaneously use up carbon dioxide during the process.
3. They are the basic food source of all kinds of organisms.
4. Microalgae are one of the most ancient organisms on Earth. Thus, many researchers have opined that they are the origin of the food chain.
5. Microalgae are capable of producing several unique products like carotenoids, antioxidants, vitamins, enzymes, polymers, peptides, toxins, sterols, and long-chain polyunsaturated fatty acids (PUFAs).
6. Due to their large photosynthetic machinery, microalgae are very rich in pigments.

Microbial Biodegradation and Bioremediation. DOI: http://dx.doi.org/10.1016/B978-0-12-800021-2.00023-6

For example, cyanobacterial phycobilisomes are of special interest because of their excellent fluorescent properties.

7. They have the potential to be used for bioenergy production if some simple nutrients such as ammonium or nitrates, phosphates, or trace amounts of certain metals are provided.
8. Certain species of algae can be used as organic fertilizer in either their raw or semidecomposed form (Riesing, 2006).
9. Microalgae including cyanobacteria are capable of carrying out specific conversions because microalgae incorporate inorganic carbon (IC) and are useful for production of isotopically labeled 13C compounds.

In recent years, another use of microalgae has attracted considerable attention. Researchers have found it to be useful in combating pollution as well. The threat of environmental pollution is now more real than ever before and one of the major types of pollution is soil pollution. Various pollutants are responsible for this but topping the list are organic pollutants and heavy metals.

23.2 Microalgae in Removal of Heavy Metals

Heavy metals are one of the many components present in soil but in recent years their presence in the environment has increased due to human activities such as mining, energy production, fuel production, electroplating, wastewater sludge treatment, and agriculture. These are conservative pollutants, that is, they are not attacked by bacterial or any other degradation process and hence become permanent additions to the environment. Subsequently, their concentrations often increase the permissible levels in soil, waterways, and sediments. The heavy metals like arsenic, mercury, chromium, nickel, lead, cadmium, zinc, and iron are toxic to life when present above a particular amount. In such cases, these contaminants are able to infiltrate deep into the layer of underground waters and pollute the groundwater as well as the surface water. Heavy metals in the soil subsequently enter the human food web through plants. On entering the food chain, these heavy metals profoundly disrupt biological processes and hence are a serious threat to human health (Krishnan et al., 2004). Moreover, they tend to bioaccumulate and hence can be transferred from one food chain to another (Pergent and Pergent-Martini, 1999). Thus, heavy metal contamination of agricultural soils has become a serious issue in crop production and human health in many developed as well as developing countries.

The heavy metals are divided into two groups:

1. Metals that are essential as nutritional requirements at trace amount for many organisms but are toxic when present in greater amounts (e.g., As, Cr, Co, Cu, Ni, Se, Va, and Zn).
2. Metals that are highly poisonous and are not known to have any nutritional value (e.g., Pb, Hg, Cd, Ur, Ag, and Be) (Inthorn, 2001).

The conventional methods for removal of heavy metals from contaminated waters include reverse osmosis, electrodialysis, ultrafiltration, ion exchange, chemical precipitation, phytoremediation, etc. However, these methods have drawbacks such as incomplete metal removal, high reagent and energy requirements, and generation of toxic sludge or other waste products that require careful disposal (Ahalya et al., 2003).

However, these xenobiotic compounds and heavy metals can be transformed, detoxified, and volatilized by microalgal metabolism, and since microalgae are non-pathogenic there is absolutely no risk of accidental release of pollutants into the atmosphere. Microalgae use the wastes as nutritional sources and degrade the pollutants enzymatically. Biosorption has also come up as a potential method of removing heavy metals from water. It has gained attention in recent years as it uses naturally abundant microalgae or by-products of fermentation industries as biosorbents. Biosorption is the ability of biological materials to accumulate heavy metals on their surfaces. It was found that various microbial sources such as moss, aquatic plants, and leaf-based adsorbents could be used as biosorbents/biomass (Niu et al., 1993; Chang et al., 1997; King et al., 2007). Microalgae exhibited high metal binding capacities (Schiewer and Volesky, 2000a,b). This property of microalgae was attributed to the presence of polysaccharides, proteins, or lipids on the surface of their cell walls which in turn contained some functional groups like amino, hydroxyl, carboxyl, and sulfate, which can act as binding sites for metals (Yu et al., 1999). Thus, heavy metals can be sequestered by the cell walls of microalgae. So microalgal biomass is considered highly effective and reliable in removing heavy metals from aqueous solutions (Volesky and Holan, 1995; Schiewer and Volesky, 2000a,b). This also makes them an ideal source of the complex multifunctional polymers which are used to sequester many different metals through adsorption or an ion-exchange process. However, the heavy metals thus accumulated affect a wide range of cellular properties of microalgae including cell viability, membrane structure and other activities. Thus, due to this tendency of absorbing toxic metals, microalgae are not usually recommended as food items. Their efficiency in scavenging metals from effluent water, from contaminants in nutrients, or from atmospheric deposition into open ponds is so high that the biomass thus produced may contain amounts at the upper limit of metal content for food use.

All organisms on this planet have evolved mechanisms to maintain equilibrium with heavy metal ions present in the surrounding medium. So the two main aims of cells are as follows:

1. to select those heavy metals essential for growth and exclude those that are not,
2. to keep essential ions at optimal intracellular concentrations (Cobbett and Goldsbrough, 2002).

Several microalgal strains have exhibited properties for removal of heavy metals. Most of the surveys that have been conducted so far are based in batch growth of the microalgal species. A marine screen was designed by Matsunaga et al. (1999) with the help of which a *Chlorella* strain capable of sustaining growth at 11.24 mg Cd^{2+}/L and 65% removal when exposed to 5.62 mg Cd^{2+}/L was characterized. While working with *Chlorella* and *Scenedesmus* strains in batch cultures at 20 mg Cr^{6+}/L, Travieso et al. (1999) found removal percentages of 48% and 31%, respectively. *Chlorella* sp. is capable of removing Al, Fe, Mg, and Mn. *Chlorella pyrenoidosa* is also extremely useful in this regard as it can remove a number of heavy metals including copper (Yan and Pan, 2002), zinc, lead, mercury, arsenic, chromium, nickel, and cadmium (Yao et al., 2011). *Chlorella minutissima* can remove Cr(VI)

(Singh et al., 2011). *Phaeodactylum tricornutum*, which shows high tolerance for Cd^{2+} ($CE_{50} = 22.3$ mg/L), also shows high removal capability (Torres et al., 1998). It has been characterized with respect to the MtIII production pattern (Torres et al., 1997). The microalgae genus *Scenedesmus* is commonly used in heavy metal removal experiments. It has been proved that it can be used for the removal of U^{6+} (Zhang et al., 1997), Cu^{2+}, Cd^{2+} (Terry and Stone, 2002), and Zn^{2+} (Aksu et al., 1998; Travieso et al., 1999; Canizares-Villanueva et al., 2001). *Spirogyra* was also found to be very useful in removing heavy metals such as Cd, Hg, Pb, As, Co, Ni (II) (Diwan, 2007), and Se. *Spirulina platensis* showed the ability to remove lead (Arunakumara et al., 2008) as well as copper and mercury (Garnikar, 2002). *Anabaena variabilis*, *Aulosira* sp., *Nostoc muscorum*, *Oscillatoria* sp., and *Westiellopsis* sp. can also remove Cr(VI) (Parameswari et al., 2010). Inthorn et al. (2001) showed that several microalgae, namely, *Scenedesmus* sp., *Chlorococcum* sp., *Chlorella vulgaris var. vulgaris*, *Fischerella* sp., *Lyngbya spiralis*, *Tolypothrix tenuis*, *Stigonema* sp., *Phormidium molle*, *Lyngbya heironymusii*, *Gloeocapsa* sp., *Oscillatoria jasorvensis*, and *Nostoc* sp., can be used for the bioremediation of heavy metals like lead, mercury, and cadmium. *Tetraselmis chuii* (Ayse et al., 2005), *Scenedesmus bijuga*, *Oscillatoria quadripunctulata* (Ajayan et al., 2011), and *Padina* sp. (Kaewsarn, 2002) can remove copper. Some of the algae used for the remediation of heavy metals are given in Table 23.1.

23.2.1 Basic Mechanism of Heavy Metal Removal

Microalgae along with related eukaryotic photosynthetic organisms and some fungi produce peptides which can bind heavy metals.

Peptides + heavy metals = organometallic complexes

These complexes are further partitioned inside vacuoles to facilitate appropriate control of the cytoplasmic concentration of heavy metal ions, thus preventing or neutralizing their potential toxic effect (Cobbett and Goldsbrough, 2002).

In contrast to this mechanism used by eukaryotes, prokaryotic cells employ ATP-consuming efflux of heavy metals or enzymatic change of speciation to achieve detoxification.

The peptides discussed can be grouped into two categories: (1) the enzymatically synthesized short-chain polypeptides named phytochelatins (class III metallothioneins), found in higher plants, algae, and certain fungi. These are named so because they were isolated from a higher plant (Phyto) and they had the capacity to bind cadmium ions (Steffens, 1990). Later, when class II metallothioneins were found to be relevant in the responses of plants to heavy metals stresses, it was proposed to change the name of PCs to class III metallothioneins. (2) The gene-encoded proteins, class II metallothioneins (identified in cyanobacteria, algae, and higher plants) and class I metallothioneins found in most vertebrates, observed in *Neurospora* and *Agaricus bisporus* (not reported in algae) (Steffens, 1990).

Table 23.1 Microalgae Used for Bioremediation of Heavy Metals

Heavy Metals	Microalgae	Reference
Cu	*Tetraselmis chuii*, *Scenedesmus bijuga*, *Oscillatoria quadripunctulata*, *Spirulina platensis*, *Padina* sp., *Scenedesmus obliquus*, *Chlorella pyrenoidosa*, and *Closterium lunula*	Ayşe et al. (2005), Ajayan et al. (2011), Garnikar (2002), Kaewsarn (2002), Yan and Pan (2002), Yao et al. (2011)
Pb	*Spirulina (Arthrospira) platensis*, *Spirogyra hyaline*, *Dunaliella*, *Scenedesmus bijuga*, *Spirulina platensis*, *Oscillatoria quadripunctulata*, *Chlorella pyrenoidosa*, *Chlorococcum* sp., *Chlorella vulgaris var. vulgaris*, *Fischerella* sp., *Lyngbya spiralis*, *Tolypothrix tenuis*, *Stigonema* sp., *Phormidium molle*, *Lyngbya heironymusii*, *Gloeocapsa* sp., *Oscillatoria jasorvensis*, *Nostoc* sp., and *Scenedesmus acutus*	Inthorn et al. (2001), Arunakumara et al. (2008), Ajayan et al. (2011), Garnikar (2002), and Yan and Pan (2002)
Ni(II)	*Oscillatoria* sp., *Spirogyra* sp., *Anabaena variabilis*, *Aulosira* sp., *Nostoc muscorum*, *Oscillatoria* sp., and *Westiellopsis* sp.	Parameswari et al. (2010) and Diwan (2007)
Se	*Spirogyra* sp., *Nostoc commune*	
Cr(VI)	*Anabaena variabilis*, *Aulosira* sp., *Nostoc muscorum*, *Oscillatoria* sp. and *Westiellopsis* sp., and *Chlorella minutissima*	Parameswari et al. (2010) and Singh et al. (2011)
Zn	*Scenedesmus bijuga*, *Oscillatoria quadripunctulata*, *Scenedesmus acutus*, *Chlorella vulgaris*, and *Chlorella pyrenoidosa*	Ajayan et al. (2011), Travieso et al. (1999), and Yao et al. (2011)
Co	*Scenedesmus bijuga*, *Oscillatoria quadripunctulata*, and *Spirogyra hyaline*	Ajayan et al. (2011)
Cd	*Scenedesmus acutus*, *Chlorella vulgaris*, *Chlorella pyrenoidosa*, *Spirogyra hyaline*, *Fischerella* sp., *Lyngbya spiralis*, *Tolypothrix tenuis*, *Stigonema* sp., *Phormidium molle*, *Lyngbya heironymusii*, *Gloeocapsa* sp., *Oscillatoria jasorvensis*, and *Nostoc* sp.	Inthorn et al. (2001) and Travieso et al. (1999)

(*Continued*)

Table 23.1 (Continued)

Heavy Metals	Microalgae	Reference
Hg	*Spirulina platensis*, *Scenedesmus* sp., *Chlorococcum* sp., *Chlorella vulgaris var. vulgaris*, *Fischerella* sp., *Lyngbya spiralis*, *Tolypothrix tenuis*, *Stigonema* sp., *Phormidium molle*, *Lyngbya heironymusii*, *Gloeocapsa* sp., *Oscillatoria jasorvensis*, *Nostoc* sp., *Scenedesmus acutus*, *Chlorella pyrenoidosa*, and *Spirogyra hyaline*	Inthorn et al. (2001), Garnikar (2002), Yao et al. (2011)
As	*Chlorella pyrenoidosa* and *Spirogyra hyaline*	Yao et al. (2011)
Al, Fe, Mg, Mn	*Chlorella* sp.	Wang et al. (2010)

23.2.2 Class III Metallothionein (MtIII) in Algae

23.2.2.1 General Structure

The general structure of MtIII was found to be (γEC) n-Gly where "n" may be anything between 2 and 11 (Steffens, 1990; Cobbett and Goldsbrough, 2002). Its molecular weight ranges from 2000 to10,000 DA (Steffens, 1990).

One important feature of this protein is that the glutamic acid residues are not-bonded with cysteine by means of an alpha-carboxyl group as in transcriptional amino acids but with a gamma-carboxyl group. Moreover, there is a gamma bond between Glu and Cys which cannot be prepared by ribosome. So, unlike most of the peptides, instead of ribosome MtIII are synthesized by an enzyme called phytochelatin synthase (PCS). It is a c-glutamyl cysteine dipeptidyl transpeptidase (E.C. 2.3.2.15) (Vatamaniuk et al., 2004) which catalyzes the transpeptidation of the c-Glu–Cys moiety of glutathione (cECG) onto a second cECG molecule to form MtIII or onto a MtIII molecule to produce an $n + 1$ oligomer. This enzyme is a tetramer of MW 95,000 with a Km for glutathione of 6.7 mM (Steffens, 1990; Cobbett and Goldsbrough, 2002). The general mechanism involved is

$$[\gamma\text{Glu} - \text{Cys}]n - \text{Gly} + [\gamma\text{Glu} - \text{Cys}] - \text{Gly} \rightarrow [\gamma\text{Glu} - \text{Cys}]\,n + 1 - \text{Gly} + \text{Gly}$$

PCS is a constitutive enzyme with no apparent gene-regulated activity.

23.2.2.2 Class III Metallothionein Biosynthesis and Regulation

MtIII complex synthesis was first discovered by Stokes et al. (1977) in the microalga *Scenedesmus acutiformis*. Ten divisions and 24 genera of algae containing MtIII were listed by Gaur and Rai (2001). It was also proved that the MtIII was the predominantly synthesized peptide in most of these algae.

It was also found that removal of heavy metal ions stopped PCS activity *in vitro*. On the other hand, *in vivo* the addition of a wide variety of heavy metal ions activates MtIII synthesis. Based on these findings, it was proposed that PCS was a heavy-metal activated enzyme (Loeffler et al., 1989). Heavy metal salts such as Cd^{2+}, Ag^{+}, Bi^{3+}, Pb^{2+}, Zn^{2+}, Cu^{2+}, Hg^{2+}, and Au^{2+} induced MtIII biosynthesis both *in vivo* and *in vitro*. Out of these Cd is the most potent activator, followed generally by Pb^{2+}, Zn, and other heavy metals. However, Vatamaniuk et al. (1999) tested the affinity of PCS from *Arabidopsis thaliana* (AtPCS) for various substrates such as glutathione and glutathiones with S group chemically blocked. Glutathione S-blocked with Cd^{2+} proved to be the best ligand. It was also found that PCS could prepare *S*-methyl-MtIII from *S*-methylglutathione even in the absence of any heavy metal ion. This indicates that though the presence of heavy metals is an important factor for the activation of PCS, glutathione with thiol blocked groups is also a very vital criterion. γ-glutamylcysteinesynthetase (γECS) and glutathione synthetase (GS) are the enzymes involved in glutathione synthesis. On genetically modifying *A. thaliana* plants with sense or antisense ECS, the mutant plants produce either high glutathione or low glutathione. The plants producing low glutathione are sensitive to Cd^{2+}. Based on these findings, researchers opined that glutathione is the primary peptide involved in binding heavy metals and the substrate for MtIII synthesis.

23.2.2.3 Sulfide Ions and Metallothionein Function

Metal–MtIII complexes contain sulfide ions (S^{2-}) (Steffens, 1990). This ion helps in stabilizing the metal–MtIII compounds (Kneer and Zenk, 1997), in turn also improving detoxification (Dameron et al., 1989). Based on the inclusion of sulfide ions in MtIII, the MtIII-complexes are divided in two categories:

- low-molecular-weight (LMW) form: metal is bound to thiol groups,
- high-molecular-weight (HMW) form: sulfide inorganic ions (S^{2-}) are incorporated in these complexes to form nanometer-sized particles (Kneer and Zenk,1997; Scarano and Morelli, 2003). The formation of the particles appears to be a matrix-mediated biomineralization process, in which the binding of the metal to c-glutamyl peptides provides the matrix (Scarano and Morelli, 2003). Results obtained in a *Schizosaccharomyces pombe* mutant sensitive to Cd^{2+} suggest that the protein sulfide oxidoreductase is responsible for maintaining an adequate equilibrium of sulfide produced during a heavy metal stress (Vande and Ow, 1999).

23.2.2.4 Sequestration and Compartmentalization to the Vacuole

Heuillet et al. (1986) first observed that the metal–MtIII complex ends up in the vacuole of the cell. They made this observation in the microalga *Dunaliella bioculata*. Later this process was characterized in detail in the yeast *S. pombe*. The hmt1 gene complements a *S. pombe* mutant that is deficient in producing HMW metal–MtIII complexes. The product of the hmt1 gene is known as HMT1 protein. It is a vacuolar transporter which is able to internalize LMW MtIII complexes in the yeast vacuole.

An ATP-dependent transporter of LMW Cd–MtIII complex exists in the higher plant *Avena sativa*. This indicates that ATP-mediated internalization of MtIII complex is a common detoxification mechanism (Haferburg and Kothe, 2007; Lloyd, 2002).

23.2.2.5 Sequestration to the Chloroplast and Mitochondria

Sometimes the MtIII–metal complex is also sequestered inside the chloroplast. This was first discovered in *Euglena gracilis*. This photosynthetic protist shows high tolerance to Cd^{2+} and high Cd^{2+}-accumulating capacity. But it lacks a specialized reservoir organelle such as a plant-like vacuole (Mendoza-Cozatl et al., 2004). Mendoza-Cozatl and Moreno-Sanchez (2005) found that in *E. gracilis* more than 60% of the accumulated Cd^{2+} is present inside the chloroplast.

Aviles et al. (2003) pretreated heterotrophic cells of *E. gracilis* with Hg^{2+}, subsequently exposing them to Cd^{2+}. They found that 79% of the total accumulated metal resided in the mitochondria. A remarkable increase in the Cys and glutathione concentration was also noted in these Cd^{2+}-treated cells. The amount of MtIII in mitochondria was around 17% of the total MtIII found in treated cells. According to Mendoza-Cozatl et al. (2004), the presence of MtIII and Cd^{2+} in *Euglena* chloroplast and mitochondria could be due to any of the following processes:

1. Cytosol is the site of synthesis of MtIII. Cd^{2+} are sequestered by MtIII in the cytosol itself, followed by transportation of the Cd–MtIII complexes into the chloroplast and mitochondria.
2. Cd^{2+} are transported as free ions inside the organelle where MtIII are synthesized. MtIII bind to Cd^{2+} inside this organelle only, forming HMW complexes.
3. Both processes occur simultaneously. MtIII are synthesized in the three cellular compartments.

The mechanisms mediated by class III metallothioneins (MtIII) in microalgae are illustrated in Figure 23.1.

In 1957, Oswald and Gootas first proposed the idea of using microalgae for the bioaccumulation of heavy metals. But this idea was not given much thought until now (Oswald, 1998). Most of the knowledge concerning algae is based on observations of higher plants. It was also observed that in aquatic ecosystems, the sedimentation of microalgae during algal blooms leads to a substantial (20–75%) decrease in the concentration of suspended heavy metals and their deposition (Luoma et al., 1998).

The algal polysaccharides (carrageenan) also help in accumulation of heavy metals. Green algae are more tolerant to metals such as zinc, lead, and copper than blue green algae and diatoms in general. Biosorption of heavy metals from aqueous solutions can be done by freshwater filamentous algae *Spirogyra hatillensis*. Cadmium, mercury, and lead are removed from aqueous solutions using marine microalgae as low-cost adsorbents. Microalgae are used to recover metals from wastewater. Wastewater can be treated using an anaerobic fixed-bed reactor in a microalgae pond. 90.2% organic nitrogen, 84.1% ammonia, and 85.5% total phosphorus can be removed by this method. *C. vulgaris* is able to remove 95.3% and 96% nitrogen and

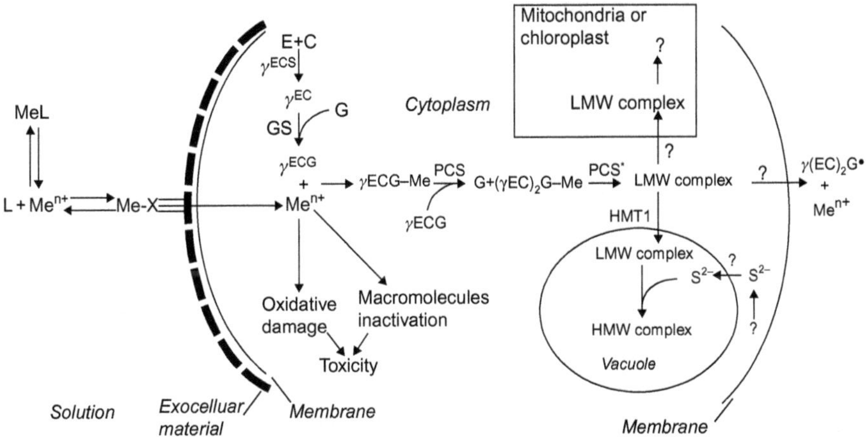

Figure 23.1 General scheme of heavy metal detoxification mechanism mediated by class III metallothioneins (MtIII) in microalgae. Abbreviations: MeL, metal complex in solution; Me^{n+}, free heavy metal ion; X, biotic exocellular ligand; E, glutamic acid; C, cysteine; G, glycine; γEC, gamma glutamylcysteine; γECG, glutathione; [γEC]2G, metallothionein $n = 2$; LMW, low molecular weight; HMW, high molecular weight; γECS, gamma glutamylcysteinesynthetase; GS, gluthathionesynthetase; PCS, phytochelatin synthase; HMT1, vacuolar ABC transporter. *When this step is repeated MtIII of longer chain can be synthesized; •MtIII is disassociated when released to medium (Perales-Vela et al., 2006).

phosphorus from secondarily treated swine wastewater. The immobilization of algal cells is one of the promising methods for wastewater treatment as it does not involve a harvesting step, which is the most difficult step in the treatment process. A gel matrix prevents free movement of the cells. Further, they have increased reaction rates because of higher cell density. They do not show any cell wash out, hence immobilized cells are preferred. Through internal immobilization of *C. vulgaris* in sodium alginate beads higher amount of nutrient are removed from raw sewage.

23.3 Organic Pollutants

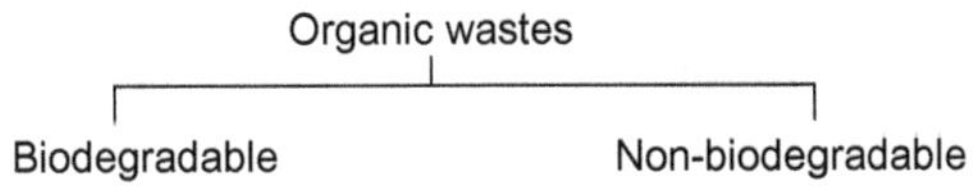

Biodegradable organic wastes include kitchen waste, garden waste, house dust after cleaning, etc. These wastes are derived from living organisms so they can be oxidized by naturally occurring microorganisms.

Nonbiodegradable organic wastes or persistent organic pollutants (POPs) are carbon-based chemicals that are resistant to environmental degradation through chemical, biological, and photolytic processes.

23.3.1 Properties of POPs

- They are semivolatile.
- They have low solubility in water.
- They have an inherent toxicity.
- These are lipophilic.

Such physical and chemical properties make them capable of long-range transport and enable them to bioaccumulate in human and animal tissue and biomagnify in food chains.

They are also referred to as persistent bioaccumulative and toxic substances, persistent toxic substance, or persistent environmental pollutants. In accordance with the Stockholm Convention on Persistent Organic Pollutants, 12 chemicals are considered to be POPs, namely, aldrin, chlordane, DDT, dieldrin, endrin, heptachlor, mirex, toxaphene, hexachlorobenzene, polychlorinated biphenyls (PCB), polychlorinated dibenzo-p-dioxins, and polychlorinated dibenzofurans (*Persistent Organic Pollutants: The Handbook of Environmental Chemistry*, Heidelore Fiedler).

23.3.2 Sources of POPs

Recently, a study was conducted by the Chinese Academy of Sciences (2013) to find the source of toxic organic pollutants present in the smog blanketing the Beijing–Tianjin–Hebei area for almost all of January 2013. The result of this study is represented below with the help of a pie chart (Figure 23.2).

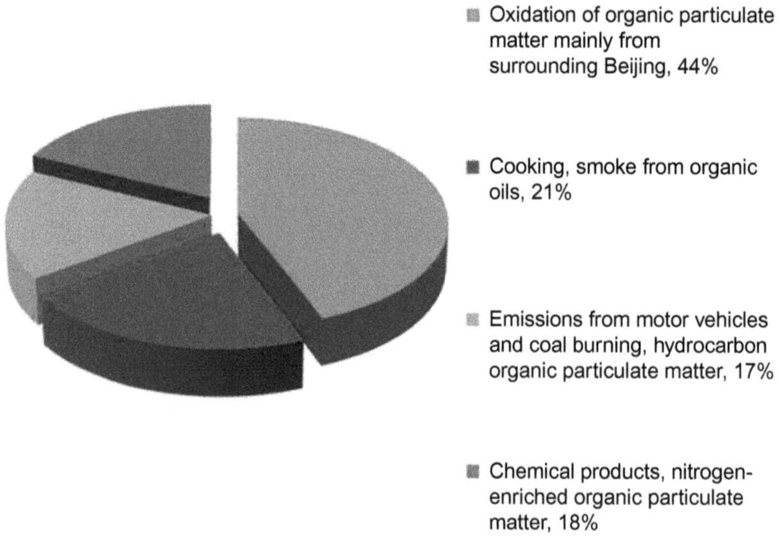

Figure 23.2 Source of toxic organic pollutants present in the smog blanketing the Beijing–Tianjin–Hebei.
Source: Chinese Academy of Sciences. Globaltimes.cn | 2013-2-17 16:30:00.

From these results, we can identify some sources of organic pollutants, namely, oxidation of organic particulate matter (especially that which is nitrogen enriched), smoke from organic oils, emissions from motor vehicles, burning of coal, etc.

Co-contamination of the environment with toxic chlorinated organic and heavy metal pollutants is another major concern. Biodegradation of chlorinated organics may be inhibited by heavy metals. In this case, heavy metals interact with enzymes directly involved in biodegradation or those involved in general metabolism. Further, heavy metals may be present in a variety of chemical and physical forms. Thus, predicting metal toxicity effects on organic pollutant biodegradation in co-contaminated soil and water environments becomes very difficult. Recent advances in bioremediation of co-contaminated environments have focused on the use of metal-resistant bacteria (cell and gene bioaugmentation), treatment amendments, clay minerals, and chelating agents to reduce bioavailable heavy metal concentrations. Phytoremediation has also shown promise as an emerging alternative cleanup technology for co-contaminated environments.

23.3.3 Biodegradation of Organic Pollutants by Microalgae

Organic pollutants are a major environmental problem worldwide, especially aromatic and heteroaromatic compounds such as phenol, pyridine, *p*-nitrophenol, trichloroethylene, and dimethyl phthalate (DMP). They are widely used in the chemical, petrochemical, and pharmaceutical industries and are therefore often found in industrial wastewater (Essam et al., 2010). These compounds are highly hazardous and have been listed by the US Environmental Protection Agency as priority pollutants. Therefore, there is a need to efficiently remove them in order to prevent their entry into the environment. Several methods have been developed to treat these toxic pollutants, such as chlorination, ozonation, adsorption, solvent extraction, membrane process, coagulation, flocculation, and biological degradation (Tchobanoglous et al., 2003). However, drawbacks of these methods include cost and their generation of secondary pollutants that may be more toxic than the parent ones. Hence, biological methods are generally preferred as they are more economical and environmentally friendly (Sogbanmu and Otitoloju, 2012). Algal–bacterial microcosms are one of the very effective methods for the biodegradation of these xenobiotic compounds. However, one of the major limitations of this promising method is the high fluctuation in influent toxicity (Munoz et al., 2009). Many algal genera have species that grow well in water containing a high concentration of organic wastes. Green algae *Chlamydomonas*, *Euglena*, diatoms, *Navicula*, *Synedra*, and the blue green algae *Oscillatoria* and *Phormidium* are emphasized to tolerate organic pollution (Palmer, 1969). At the species level, *Euglena viridis* (Euglenophyta), *Nitzschia palea* (Bacillariophyta), *Oscillatoria limosa*, *O. tenuis*, *O. princeps*, and *Phormidium uncinatum* (Cyanophyta) are reported to be present more than any other species in organically polluted waters (Palmer, 1980). Microorganisms are able to degrade organic pollutants as organic compounds can be utilized by microorganisms as the main source of carbon and other nutrients vital for their growth and survival. Microalgae are facultative chemoautotrophs (organism that obtained energy by the oxidation of organic or inorganic compounds) and it is the key reason for the degradation of organic pollutants.

Recent studies indicate that microalgae not only provide oxygen for aerobic bacterial biodegradation but can also biodegrade organic pollutants directly (Semple et al., 1999). Biodegradation of organic compounds by microalgae is the result of facultative chemoautotrophy. Organic compounds can be utilized by microorganisms as the main source of carbon and nutrients vital for their growth and survival. Liu (1992) reported that *C. pyrenoidosa*, *C. vulgaris*, and *O. tenuis* are responsible for biodegradation and decolorization of more than 30 azo compounds. The azo reducer in the algae biodegraded the azo dyes into simpler aromatic amine by breaking the azo linkage. The reduction rate depends on the molecular features and species of the algae used. Klekner (1992) observed that phenolic compounds were biodegraded by three species of blue green algae at a concentration below 1000 mg/L. *Chlorella* partly dechlorinates as well as biodegrades 2-chlorophenol at a concentration of 200 mg/L. 2,4-dinitrorophenol at a concentration of 190 mg/L was quickly degraded by *Scenedesmus*. Yan (1995) found that *C. pyrenoidosa* is able to biodegrade three kinds of phthalate esters and a new kinetic equation of organic pollutant biodegradation by microalgae was suggested. Biodegradation of DDT, BHC, and other pesticides by freshwater microalgae have also been studied. DMP can be degraded by *Dunaliella tertiolecta* and *Closterium lunula*. It has been also found out that along with providing oxygen for aerobic bacterial biodegradation, microalgae can also biodegrade organic pollutants directly (Semple et al., 1999). More than 30 azo compounds can be biodegraded and decolorized by *C. pyrenoidosa*, *C. vulgaris*, and *O. tenuis*. In this process, azo dyes are decomposed into simpler aromatic amines (Liu and Liu, 1992). Chlorella can also degrade 2,4-dinitrophenol and convert it to an isomer of dimethyl benzenediol. *Ochromonas danica* can mineralize [U-14C] phenol completely. Furthermore, microalgae can also biodegrade DDT, linear alkylbenzene sulfonate, and tributyltin.

23.3.3.1 Biodegradation of DMP by C. lunula

DMP is an artificial organic compound which is used not only in many industrial processes such as construction and automobile and chemical engineering but is also used in pesticides, medicines, and various household items. But due to its low biodegradability, it poses a great danger to the environment into which it is released. Hence, DMP is listed as a major pollutant. Fortunately, *C. lunula* has been shown to degrade DMP, producing phthalic acid (PA) as an intermediate. This intermediate product can also be further biodegraded by *C. lunula*. But if PA accumulates, it causes a sharp decrease in the pH of the microalgal culture medium, thus inhibiting both the growth of microalgae and the biodegradation of DMP. The biodegradation rate of DMP increases with the increase in initial concentration of IC. Increase in the initial concentration of IC also significantly increases the growth of the microalga.

23.3.3.2 Biodegradation of Tributyltin

The chemical formula for Tri-n-butyltin (TBT) is $C_{12}H_{27}Sn^{+}$. It has a molecular weight of 290.06. It belongs to the trialkyl organotin family. It is used in several commercial products of daily use. Some of its applications are listed below:

1. It is used as an antifouling agent in ship bottom paints.
2. It is the main active ingredient in biocides used to control a wide spectrum of organisms.
3. It helps in wood treatment and preservation.
4. It is used in textiles and industrial water systems for its antifungal properties.
5. It is also used for controlling schistosomiasis (bilharzia, bilharziosis, or snail fever) in various parts of the world.

However, such organotin compounds are very toxic. They have mutagenic as well as carcinogenic effects. They are harmful not only for humans but also for other animals and plants. They can be toxic to many marine organisms, including algae, at concentrations of 1 μg/L or less. But some types of microalgae, like the freshwater chlorophyte *Ankistrodesmus falcatus*, can tolerate a TBT concentration of 25 μg/L and degrade TBT to dibutyltin (DBT), butyltin (MBT), and inorganic tin (Maguire et al., 1984). TBT biodegradation is relatively rapid in coastal waters, where the half-life of this pollutant ranges between 4 and 14 days.TBT degrades more rapidly in the light than in the dark (Lee et al., 1987). Stimulation of algal growth may be of use in enclosed aquatic areas with organotins, since dense algal cultures can rapidly degrade these organometallic compounds. TBT is usually degraded by a sequential debutylation pathway. Hydroxylated butyltin intermediates are formed. These intermediate are unstable and lose butyne, thus giving rise to the debutylated compound (Figure 23.3).

23.3.3.3 Bioremediation of Pesticides

The effects of pesticides vary widely with concentrations used, duration of exposure, and algal species tested. Biodegradation of pesticides is usually determined by two factors. The first relates to the consortium of microbes used and the optimum conditions for their survival and activity, while the second depends on the chemical structure of the pesticides. Factors related to microbes are the number of appropriate microorganisms, the contact between microorganisms and the substrate (pesticide), pH, temperature, salinity, nutrients, light quality and intensity, available water, oxygen tension and redox potential, surface binding, presence of alternative carbon substrates, and alternative electron acceptors. The second group of factors includes chemical structure, molecular weight, functional groups of the applied pesticides, their concentration and toxicity, and their solubility in water. Some information on the interactions between pesticides and algae was compiled by Kobayashi and Rittman, showing that microalgae were capable not only of bioaccumulating pesticides, but also of biotransforming some of these environmental pollutants (Tables 23.2 and 23.3).

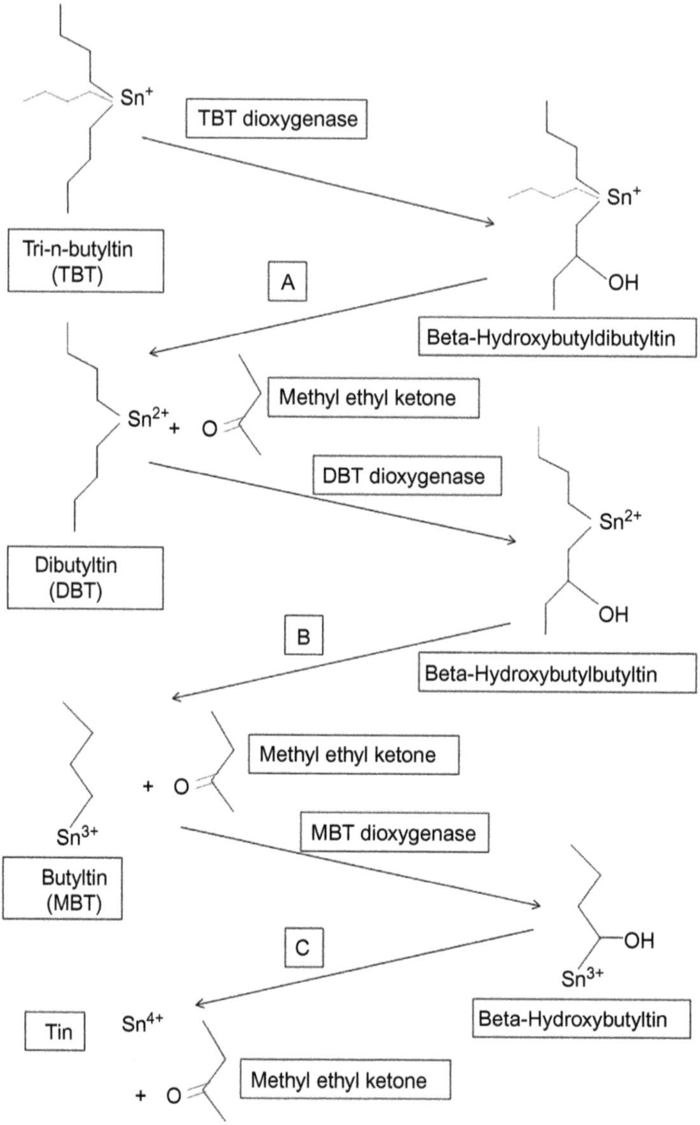

Figure 23.3 Pathway showing biodegradation of TBT by microalgae.
Source: http://umbbd.ethz.ch/tbt/tbt_image_map.html.

23.4 Conclusion

Due to the low cost of raw material, their high adsorbing capacity, the lack of secondary pollution, and other such useful characteristics, microalgae have now

Table 23.2 Microalgae in Bioaccumulation of Pesticides

Alga	Bioaccumulation	Reference
Chlamydomonas sp.	Mirex	Hollister et al., 1975
Chlorella spp.	Toxaphene, methoxychlor	Paris et al., 1975
Chlorococcum sp.	Mirex	Ware et al., 1970
Cylindrotheca sp.	DDT	
Dunaliella sp.	Mirex	
Euglena gracilis	DDT, parathion	
Scenedesmus obliquus	DDT, parathion	

Table 23.3 Microalgae in Biotransformation of Pesticides

Alga	Biotransformation	Reference
Chlamydomonas sp.	Lindane, naphthalene, phenol	Ware et al., 1970, Ellis, 1977
Chlorella sp.	Lindane and chlordimeform	
Euglena gracilis	Phenol	
Scenedesmus sp.	Phenol	

emerged as a promising tool for bioremediation. The eukaryotic algae are capable of biotransforming and biodegrading many organic pollutants and heavy metals present in the environment. Furthermore, microalgae usually enhance the degradation capacity of the local microbiota, thereby eliminating pollutants from the respective ecosystem. They are also capable of uptaking such toxic compounds from the water phase. This property of microalgae also significantly brings down the pollutant levels. However, inside their cells they must maintain nontoxic conditions. This requirement is satisfied by two principal mechanisms:

1. One prevents the indiscriminate entrance of heavy metal ions into the cell (i.e., exclusion). Extracellular polymers, mainly carbohydrates, are responsible for this mechanism.
2. The other prevents bioavailability of toxic ions once inside the cell. This is done by class III metallothioneins (MtIII). These peptides are derived from glutathione (cECG) and are a quick response of the cell to sudden and constant heavy metal stress.

Thus, it can be concluded that if the degrading capacities of microalgae can be successfully tapped, then algal technology may emerge as an effective method of bioremediation in the near future. However, it is very rare to find one single microorganism capable of completely breaking down pollutants or xenobiotic compounds. Hence a consortium of microalgae along with other microorganisms capable of degrading heavy metals and organic pollutants would be a better option.

References

Ahalya, N., Ramachandra, T.V., Kanamadi, R.D., 2003. Biosorption of heavy metals. Res. J. Chem. Environ. 7, 71–78.

Ajayan, K.V., Selvaraju, M., Thirugnanamoorthy, K., 2011. Growth and heavy metals accumulation potential of microalgae grown in sewage wastewater and petrochemical effluents. Pak. J. Biol. Sci. 14 (16), 805–811.

Aksu, Z., Egretli, G., Kutsal, T., 1998. A comparative study of copper (II) biosorption on Ca-alginate, agarose and immobilized *C. vulgaris* in a packed-bed column. Process Biochem. 33, 393–400.

Arunakumara, K.K.I.U., Zhag, X., Song, X., 2008. Bioaccumulation of Pb^{2+} and its effects on growth, morphology and pigment contents of *Spirulina (Arthrospira) platensis*. J. Ocean Univ. China. 7 (4), 397–403.

Aviles, C., Loza-Tavera, H., Terry, N., Moreno-Sanchez, R., 2003. Mercury pretreatment selects an enhanced cadmium-accumulating phenotype in *Euglena gracilis*. Arch. Microbiol. 180, 1–10.

Ayse, B.Y., Oya, I., Selin, S., 2005. Bioaccumulation and toxicity of different copper concentrations in *Tetraselmis chuii*. E.U. J. Fish. Aquat. Sci. 22 (3–4), 297–304.

Brown, L.M., Zeiler, K., 1993. Aquatic biomass and carbon dioxide trapping. Energy Convers. Manage. 34, 1005–1013.

Canizares-Villanueva, R.O., Gonzalez-Moreno, S., Domınguez-Bocanegra, A.R., 2001. Growth, nutrient assimilation and cadmium removal by suspended and immobilized *Scenedesmus acutus* cultures: influence of immobilization matrix. In: Chen, F., Jiang, Y. (Eds.), Algae and Their Biotechnological Potential. Kluwer Publishers, Dordrecht, The Netherlands, pp. 147–161. 33, 393–400.

Chang, J., Law, R., Chang, C., 1997. Biosorption of lead, copper and cadmium by biomass of *Pseudomonas aeruginosa* PU21. Water Res. 31, 1651–1658.

Cobbett, C., Goldsbrough, P., 2002. Phytochelatin and metallothioneins: roles in heavy metal detoxification and homeostasis. Annu. Rev. Plant Biol. 53, 159–182.

Dameron, C.T., Reese, R.N., Mehra, R.K., Kortan, A.R., Caroll, P.J., Steigerwald, M.L., et al., 1989. Biosynthesis of cadmium sulfide quantum semiconductor crystallites. Nature. 338, 596–597.

Diwan, S., 2007. Removal of Ni(II) from aqueous solution by biosorption using two green algal species *Oscillatoria* sp. and *Spirogyra* sp. 5th WSEAS International Conference on Environment, Ecosystems and Development, Tenerife, Spain, December 14–16.

Ellis, B.E., 1977. Degradation of phenolic compounds by freshwater algae. Plant Sci. Lett. 8, 213–216.

Essam, T., Aly, A.M., El Tayeb, O., Mattiasson, B., Guieysse, B., 2010. Characterization of highly resistant phenol degrading strain isolated from industrial wastewater treatment plant. J. Hazard. Mater. 173, 783–788.

Garnikar, D., 2002. Accumulation of copper, mercury and lead in *Spirulina platensis* studied in Zarrouk's medium. J. King Mongkut's Inst. Technol. North Bangkok. 12 (4), 333–335.

Gaur, J.P., Rai, L.C., 2001. Heavy metal tolerance in algae. In: Rai, L.C., Gaur, J.P. (Eds.), Algal Adaptation to Environmental Stresses. Physiological, Biochemical and Molecular Mechanisms. Springer-Verlag, Berlin, pp. 363–388.

Globaltimes.cn | 2013-2-17 16:30:00 <http://umbbd.ethz.ch/tbt/tbt_image_map.html>.

Haferburg, G., Kothe, E., 2007. Microbes and metals: interactions in the environment. J. Basic Microbiol. 47, 453–467.

Heuillet, E., Moreau, A., Halpren, S., Jeanne, N., Puiseux-Dao, S., 1986. Cadmium binding to a thiol molecule in vacuoles of *Dunaliella bioculata* contaminated with $CdCl_2$: electron probe microanalysis. Biol. Cell. 58, 79–86.

Hollister, T.A., Walsh, G.E., Forester, J., 1975. Mirex and marine unicellular algae: accumulation, population growth and oxygen evolution. Bull. Environ. Contain. Toxicol. 14, 753–759.

Inthorn, D., 2001. Removal of heavy metal by using microalgae. Photosynthetic Micro. Environ. Biotechnol. 310, 111–169.

Inthorn, D., Sidtitoon, N., Silapanuntakul, S., Incharoensakdi, A., 2001. Sorption of mercury, cadmium and lead by microalgae. ScienceAsia. 28, 253–261.

Kaewsarn, P., 2002. Biosorption of copper (II) from aqueous solutions by pre-treated biomass of marine algae *Padina* sp. Chemosphere. 49, 471–476.

King, P., Rakesh, N., Beenalahari, S., Prasanna Kumar, Y., Prasad, V.S.R.K., 2007. Removal of lead from aqueous solution using *Syzygium cumini* L.: equilibrium and kinetic studies. J. Hazard. Mater. 142, 340–347.

Klekner, V., Kosaric, N., 1992. Degradation of phenols by algae. Environ. Technol. 13, 493–501.

Kneer, R., Zenk, M.H., 1997. The formation of Cd–phytochelatin complexes in plant cell cultures. Phytochemistry. 44, 69–74.

Krishnan, K.K., Parmala, V., Meng, X., 2004. Detoxification of chromium (VI) in coastal waste using lignocellulosic agricultural waste. 83rd Annual Report, University of Delhi, India.

Lee, R.F., Valkirs, A.O., Seligman, P.F. Fate of tributyltin in estuarine areas. In: Proceedings of Organotin Symposuium of the Oceans 87 Conference, Marine Technology Society, Washington, DC, pp. 1411–1415.

Liu, J., Liu, H., 1992. Degradation of azo dyes by algae. Environ. Pollut. 75, 273–278.

Lloyd, J.R., 2002. Bioremediation of metals: the application of microorganisms that make and break minerals. Microbiol. Today. 29, 67–69.

Loeffler, S., Hochberger, A., Grill, E., Winnacker, E.L., Zenk, M.H., 1989. Termination of the phytochelatin synthase reaction through sequestration of heavy metals by the reaction product. FEBS Lett. 258, 42–46.

Luoma, S.N., van Geen, A., Lee, B.G., Cloern, J.E., 1998. Metal uptake by phytoplankton during a bloom in South San Francisco Bay: implications for metal cycling in estuaries. Limnol. Oceanogr. 43, 1007–1016.

Maguire, R.J., Wong, P.T.S., Rhamey, J.S., 1984. Accumulation of metabolism of Tri-n-butyltin cation by a green alga, *Ankistrodesmus falcatus*. Can. J. Fish. Aquat. Sci. 41, 537–540.

Matsunaga, T., Takeyama, H., Nakao, T., Yamazawa, A., 1999. Screening of marine microalgae for bioremediation of cadmium polluted seawater. J. Biotechnol. 70, 33–38.

Mendoza-Cozatl, D.G., Moreno-Sanchez, R., 2005. Cd^{2+} transport and storage in the chloroplast of *Euglena gracilis*. Biochim. Biophys. Acta. 1706, 88–97.

Mendoza-Cozatl, D., Loza-Tavera, H., Hernandez-Navarro, A., Moreno-Sanchez, R., 2004. Sulfur assimilation and glutathione metabolism under cadmium stress in yeast, protist and plants. FEMS Microbiol. Rev. 29, 653–671.

Munoz, R., Kollner, C., Guieysse, B., 2009. Biofilm photobioreactors for the treatment of industrial wastewaters. J. Hazard. Mater. 161, 29–34.

Niu, H., Xu, X.S., Wang, J.H., 1993. Removal of lead from aqueous solutions by Penicillium biomass. Biotechnol. Bioeng. 42, 785–787.

Oswald, W.J., 1998. Micro-algae and waste-water treatment. In: Borowitzka, M.A., Borowitzka, L.J. (Eds.), Micro-algal Biotechnology. Cambridge University Press, Cambridge, pp. 305–328.

Oswald, W., Gootas, H.B., 1957. Photosynthesis in sewage treatment. Trans. Am. Soc. Civil Eng. 122, 73–105.

Palmer, C.M., 1969. A composite rating of algae tolerating organic pollution. J. Phycol. 5, 78–82.

Palmer, C.M., 1980. Algae and Water Pollution. Castle House Publications Ltd, p. 110.

Parameswari, E., Lakshmanan, A., Thilagavathi, T., 2010. Phycoremediation of heavy metals in polluted water bodies. Electron. J. Environ. Agric. Food Chem. 9 (4), 808–814.

Perales-Vela, H.V., Pena-Casto, J.M., Canizaies-Villanieva, R.S., 2006. Heavy metal detoxification in eukaryotic algae. Chemosphere. 64, 1–10.

Paris, D.F., Lewis D.L., Barnett J.T., and Baughman G.L., 1975. Microbial degradation and accumulation of pesticides in aquatic systems. EPA Report #660/3-75-007.

Pergent, C., Pergent-Martini, C., 1999. Mercury levels and fluxes in *Posidonia oceanica* meadows. Environ. Pollut. 106, 33–37.

Riesing, T.F., 2006. Cultivating algae for liquid fuel production. Permaculture Activist. 48, pp59.

Scarano, G., Morelli, E., 2003. Properties of phytochelatin-coated CdS nanocrystallites formed in a marine phytoplanktonic alga (*Phaeodactylum tricornutum*, Bohlin) in response to Cd. Plant Sci. 165, 803–810.

Schiewer, S., Volesky, B., 2000a. Biosorption by marine algae. In: Valdes, J.J. (Ed.), Bioremediation. Kluwer Publishers, Dordrecht, The Netherlands, pp. 139–169.

Schiewer, S., Volesky, B., 2000b. In: Lovely, D.R. (Ed.), Environmental Microbe–Metal Interactions. ASM Press, Washington, DC, pp. 329–362.

Semple, K.T., Cain, R.B., Schmidt, S., 1999. Biodegradation of aromatic compounds by microalgae. FEMS Microbiol. Lett. 170 (2), 291–300.

Singh, S.K., Bansal, A., Jha, M.K., Dey, A., 2011. An integrated approach to remove Cr(VI) using immobilized *Chlorella minutissima* grown in nutrient rich sewage waste water. Bioresour. Technol. 104, 257–265.

Sogbanmu, T.O., Otitoloju, A.A., 2012. Efficacy and bioremediation enhancement potential of four dispersants approved for oil spill control in Nigeria. J. Bioremed. Biodegrad. 3, 136. Available from: http://dx.doi.org/doi:10.4172/2155-6199.1000136.

Steffens, J.C., 1990. The heavy metal-binding peptides of plants. Annu. Rev. Plant Physiol. Plant Mol. Biol. 41, 553–575.

Stokes, P.M., Maler, T., Riordan, J.R., 1977. A low molecular weight copper-binding protein in a copper tolerant strain of *Scenedesmus acutiformis*. In: Hemphil, D.D. (Ed.), Trace Substances in Environmental Health — XI. Univ. of Missouri Press, Columbia MO, pp. 146–154.

Tchobanoglous, G., Burton, F.L., Stensel, H.D., 2003. Wastewater Engineering: Treatment and Reuse. Metcalf & Eddy, Inc., New York, NY.

Terry, P.A., Stone, W., 2002. Biosorption of cadmium and copper contaminated water by *Scenedesmus abundans*. Chemosphere. 47, 249–255.

Torres, E., Cid, A., Fidalgo, P., Herrero, C., Abalde, J., 1997. Long-chain class III metallothioneins as a mechanism of cadmium tolerance in the marine diatom *Phaeodactylum tricornutum* Bohlin. Aquat. Toxicol. 39, 231–246.

Torres, E., Cid, A., Herrero, C., Abalde, J., 1998. Removal of cadmium ions by the marine diatom *Phaeodactylum tricornutum* Bohlin accumulation and long-term kinetics of uptake. Bioresour. Technol. 63, 213–220.

Travieso, L., Canizares, R.O., Borja, R., Benitez, F., Dominguez, A.R., Dupeyron, R., et al., 1999. Heavy metal removal by microalgae. Bull. Environ. Contam. Toxicol. 62, 144–151.

Vande, W.J.G., Ow, D.W., 1999. A fission yeast gene for mitochondrial sulfide oxidation. J. Biol. Chem. 274, 13250–13257.

Vatamaniuk, O.K., Mari, S., Lu, Y.-P., Rea, P.A., 1999. AtPCS1, a phytochelatin synthase from *Arabidopsis thaliana*: isolation and *in vitro* reconstitution. Proc. Natl. Acad. Sci. USA. 96, 7110–7115.

Vatamaniuk, O.K., Mari, S., Lang, A., Chalasani, S., Demkiv, L., Rea, P.A., 2004. Phytochelatin synthase, a dipeptidyltransferase that undergoes multisite acylation with γ-glutamycysteine during catalysis. Stoichiometric and site-directed mutagenic analysis of *Arabidopsis thaliana* PCS1-catalyzed phytochelatin synythesis. J. Biol. Chem. 279, 22449–22460.

Volesky, B., Holan, Z.R., 1995. Biosorption of heavy metals. Biotechnol. Prog. 11, 235–250.

Wang, L., Min, M., Li, Y., Chen, P., Liu, Y., Wang, Y., et al., 2010. Semi-continuous cultivation of *Chlorella vulgaris* for treating undigested and digested dairy manures. Appl. Biochem. Biotechnol. 162 (4), 1174–1186.

Ware, G.W., Cahill, W.P., Gerhardt, P.D., Witt, J.M., 1970. PesticidesdriftIV. On-targetdeposits from aerial application of insecticides. J. Econ. Entomol. 63, 1982–1983.

Whittaker, R.H. (Ed.), 1975. Communities and Ecosystems. Macmillan, New York, NY.

Yan, H., Pan, G., 2002. Toxicity and bioaccumulation of copper in three green microalgal species. Chemosphere. 49, 471–476.

Yan, H., Yin, C., Ye, C., 1995. Kinetic of phthalate ester biodegradation by *Chlorella pyrenoidosa*. Environ. Toxicol. Chem. 6, 931–938.

Yao, J., Li, W., Xia, F., Zheng, Y., Fang, C., Shen, D., 2011. Heavy metals and PCDD/solid waste incinerator fly ash in Zhejiang province, China: chemical and bioanalytical characterization. Environ. Monit. Assess. 184 (6), 3711–3720.

Yu, Q., Matheickal, J.T., Kaewsarn, P., 1999. Heavy metal uptake capacities of common marine macro-algal biomass. Water Res. 33, 1534–1537.

Zhang, X., Luo, S., Yang, Q., Zhang, H., Li, J., 1997. Accumulation of uranium at low concentration by the green alga *Scenedesmus obliquus*. J. Appl. Phycol. 9, 65–71.

24 Bioremediation of Aquaculture Effluents

Marcel Martinez-Porchas[a], Luis Rafael Martinez-Cordova[b], Jose Antonio Lopez-Elias[b] and Marco Antonio Porchas-Cornejo[c]

[a]Centro de Investigacion en Alimentacion y Desarrollo, Hermosillo, Sonora, Mexico, [b]Departamento de Investigaciones Científicas y Tecnológicas de la Universidad de Sonora, Hermosillo, Sonora, Mexico, [c]Centro de Investigaciones Biológicas del Noroeste, Guaymas, Sonora, Mexico

24.1 Introduction

Aquaculture has brought substantial benefits to humanity; Martínez-Porchas and Martínez-Cordova (2012) summarized some of these benefits as follows: (1) seafood produced by fisheries and aquaculture contributes 15–20% of average animal protein consumption to 2.9 billion people worldwide without considering the contribution of freshwater or brackish water species, (2) the nutritional quality of aquatic products has a high standard and represents an important source of macro- and micronutrients for people from developing countries, (3) aquaculture and fisheries are recognized as a source of employment for nearly 600 million people who have been directly employed by aquaculture or who rely on income from seafood production, and (4) aquaculture products possess high trade potential as food commodities in the international market; fish and shellfish exports from developing countries have a greater value than the combination of other important products.

In spite of these benefits, there is a stunning contrast, because aquaculture has been accused of having negative impacts on the environment with considerable repercussions. The same authors listed the main negative impacts of aquaculture as follows: (1) destruction of natural ecosystems such as mangrove forests to construct aquaculture farms, (2) salinization and/or acidification of soils, (3) pollution of water for human consumption, (4) ecological impacts on natural ecosystems because of the introduction of exotic species, (5) ecological impacts caused by inadequate medication practices, (6) changes to landscapes and hydrological patterns, (7) trapping and killing/destruction of eggs, larvae, juveniles, and adults of

Microbial Biodegradation and Bioremediation. DOI: http://dx.doi.org/10.1016/B978-0-12-800021-2.00024-8

diverse organisms, and (8) eutrophication and alterations in nitrification processes of effluent-receiving ecosystems.

Eutrophication, hypernutrification, and nitrification alterations caused by aquaculture effluents are some of the most common concerns of the activity worldwide (Herbeck and Unger, 2013). Eutrophication is a process by which an aquatic environment receives higher amounts of organic matter and nutrients that can be taken in and bioprocessed. Nitrogenous (NH_3–NH_4, NO_2, and NO_3) and phosphorous (PO_4) nutrients are observed as a rule of thumb in aquaculture effluents.

Once nitrogenous nutrients are released toward the environment, the nitrification process takes place. Nitrification is the biological process in which ammonia is oxidized into nitrite, followed by a subsequent transformation of nitrites into nitrates. This process is mainly performed by *Nitrosomonas* and *Nitrobacter* bacteria (Hargreaves, 1998); however, the environments receiving aquaculture effluents have limited capacity to process and incorporate those nutrients into food chains and biogeochemical cycles, which means that considerable amounts of nutrients tend to accumulate, causing ecological imbalances. This is commonly known as hypernutrification (Paez-Osuna and Ruiz-Fernandez, 2005).

Bioremediation has emerged as a sustainable strategy focused on alleviating the negative effects of aquaculture effluents. Different organisms such as mollusks, plants, bacteria, and others have been tested as bioremediators of aquaculture effluents. In particular, the biological capacities of some microbes have been used for the above purpose (Bender and Phillips, 2004). The aim of this chapter is to address some of the basic principles of microbial bioremediation applied to aquaculture effluents and discuss the state of the art on using microbes (bacteria and microalgae) for such purposes.

24.2 Microbes as Bioremediators

The rapid growth of microbes, their ubiquitous presence in diverse environments, and their nutritional requirements are considered useful characteristics for the aquaculture industry. For instance, Bender and Phillips (2004) argued that the "ecological success of microbial mats and their broad array of microbial activities suggest that microbial ecosystems might be useful to bioremediation of environmental pollutants."

Bacteria and other microbes have been historically used for the bioremediation of industrial effluents, particularly because of their ability to sequester heavy metals (Lefebvre and Edwards, 2010); not surprisingly, most of the scientific literature related to bioremediation using bacteria is focused on heavy metals and petroleum contaminants (Atlas, 1993; Gargouri et al., 2013). However, most of the above organisms may not be useful for bioremediation in aquaculture, because the pollutants produced by aquaculture are quite different. In shrimp farms for instance, there is evidence that from the total nitrogen inputs, only 22% was converted to harvested shrimp, 14% remained in the sediment, and 3% was assumed to be lost to the atmosphere via denitrification or volatilization of ammonia, while the rest was discharged to the environment (Jackson et al., 2003). From the total discharged nitrogen, one fraction is composed of organic N and the other by inorganic N. The inorganic fraction includes nitrite, nitrate, and ammonia, whereas the organic fraction remains in the form of proteins attached to organic suspended solids (unconsumed feed, feces, flocs) or comprising phytoplankton and/or zooplankton biomass.

In addition to the above pollutants of aquaculture effluents, other authors (Martinez-Cordova et al., 2009) have reported that apart from nitrogenous wastes, large amounts of phosphorous (PO_4), carbon, and total suspended solids are discharged by aquaculture farms. In summary, it has been calculated that 10 kg of P and 60 kg of N are discharged to the environment per ton of fish produced (Ackefors and Enell, 1994), whereas 21 kg of P and 56 kg of N are discharged per ton of shrimp produced (Martinez-Cordova et al., 2009). In this regard, Pieper and Reineke (2000) asserted that "the accumulation in the environment of highly toxic and persistent compounds emphasizes the fact that the natural metabolic diversity of the autochthonous microbes is insufficient to protect the biosphere from anthropogenic pollution." Despite the fact that ammonia and nitrite compounds can be biotransformed by the environments into less toxic compounds, the transformation rate does not match with the generation rate, which ultimately leads to accumulation of toxic compounds. Therefore, strategies such as bioremediation using microbes may be considered sustainable solutions to this problem.

The bioremediation in aquaculture using microbes is relatively recent (Chavez-Crooker and Obreque-Contreras, 2010; Martinez-Cordova et al., 2011a), in spite of some isolated efforts carried out in past decades (Bird et al., 1988). Bacteria and microalgae can play roles as effluent and sediment bioremediators through nitrification and denitrification processes; some heterotrophic bacteria can inclusively break down organic wastes (Chavez-Crooker and Obreque-Contreras, 2010), transforming them into biomass which can be used as a nutrient supply for natural or farmed organisms (Crab et al., 2007; Panigrahi and Azad, 2007).

24.2.1 Bacteria

24.2.1.1 Autotrophic Bacteria

Bioremediation of aquaculture effluents is performed under both aerobic and anaerobic conditions. Regarding aerobic conditions, the biological nitrification can be performed by some bacteria; considerable amounts of oxygen are required for this. Chemolithoautotrophic ammonia-oxidizing bacteria and ammonia-oxidizing archaea oxidize ammonia to nitrite via hydroxylamine. These ammonia-oxidizing bacteria belong to the β- and γ-Proteobacteria and use CO_2 as a carbon source (Koops and Pommerening-Roser, 2001).

In a subsequent step, the nitrite-oxidizing bacteria oxidize nitrite to nitrate, which is a significantly less toxic metabolite for aquatic organisms (Chavez-Crooker and Obreque-Contreras, 2010; Kowalchuk and Stephen, 2001); these bacteria belong to the α- and γ-Proteobacteria, phylum Nitrospirae (Fiencke et al., 2005).

A considerable number of experiments focusing on biological nitrogen removal through nitrification, denitrification, and inclusively anaerobic ammonium oxidation (anammox) have been performed, and a diversity of bacterial species have been tested (Chavez-Crooker and Obreque-Contreras, 2010). Regarding anaerobic ammonium-oxidizing (anammox) bacteria, these microbes perform an alternative denitrification pathway; in particular, 24–67% of nitrogen loss in marine sediments can be attributed to anammox (Thamdrup and Dalsgaard, 2002).

The three types of bacteria (aerobic ammonia-, nitrite-, and anaerobic ammonium oxidizing) are commonly reared and used in biofilters and bioreactors, in which high bacterial concentrations come in contact with wastewater generated by aquaculture, decreasing the ammonia loads (Tal et al., 2003, 2006). Aerobic ammonia- and nitrite-oxidizing bacteria cannot be found in high concentrations within the environment and have to be cultivated and maintained in bioreactors to reach adequate concentrations capable of performing a significant nitrification–denitrification process. In contrast, anaerobic ammonium-oxidizing bacteria can be found at high concentrations in the environment, especially in those areas where aquaculture farms discharge their effluents.

Despite the fact that anaerobic ammonium-oxidizing bacteria are a major contributor to denitrification of anoxic aquaculture-derived sediments, their proliferation and extreme concentrations in the environment are not healthy for the biota. Herein, their enhanced proliferation causes a higher oxygen demand, resulting in localized hypoxia or anoxia, killing the most susceptible aerobic life forms and having negative physiological implications in others (Camargo and Alonso, 2006). Thus, anaerobic bacteria should be used for bioremediation exclusively in bioreactors; herein, the combination of microbial nitrification with anammox has been hypothesized to be more cost-effective than other solutions (Paredes et al., 2007; Chavez-Crooker and Obreque-Contreras, 2010). Plenty of bioreactors using aerobic and anaerobic bacteria have been designed; however, the most viable strategy for large scale and open farms seems to be those systems in which effluents are in direct contact with a series of mats colonized by autotrophic/heterotrophic bacteria consortia; these consortia exhibit a bioaugmentation of nitrifying–denitrifying of nitrogenous compounds contained in wastewater effluents (Bender and Phillips, 2004; Audelo-Naranjo et al., 2011).

The use of mats colonized by microbes is a way to simulate nature but with an enhanced capacity to treat contaminants. In nature, microbial mats are laminated heterotrophic and autotrophic stratified communities; these communities are dominated by cyanobacteria, eukaryotic microalgae like diatoms, anoxygenic phototrophic bacteria, and sulfate-reducing bacteria. These microbial communities have physiological flexibility because anoxygenic and oxygenic photosynthesis can be detected within the same consortium (Paniagua-Michel and Garcia, 2003). The bioremediation performance of these organisms can be estimated; for instance, Ebeling et al. (2006) reviewed the different ammonia removal pathways (photoautotrophic-, autotrophic bacterial-, or heterotrophic bacterial based) and developed a set of stoichiometric balanced relationships using half-reaction relationships, and discussed their impact on water quality.

Different genes codings for enzymes responsible for playing a role in metabolic pathways to biotransform toxic compounds into less toxic chemicals have been studied and identified in different bacterial strains used for bioremediation of petroleum contaminants (Pieper and Reineke, 2000). Moreover, genetically modified bacteria have been used for bioremediation; recombinant strains have been tested for such purposes. However, research on genetically modified bacteria for bioremediation of aquaculture wastes has not been as extensive.

24.2.1.2 Heterotrophic Bacteria

Some heterotrophic bacteria may play a key role in the bioremediation of aquaculture wastes. In the last section, autotrophic bacteria were addressed as the main organisms converting ammonia to nitrate. However, heterotrophic bacteria (cyanobacteria: *Microcoleus chthonoplastes*, *Spirulina* sp., *Oscillatoria* sp., *Schizothrix* sp., *Calothrix* sp., *Phormidium* sp., etc.) can also transform the ammonia nitrogen into nonharmful products; in this regard, they transform ammonia into microbial biomass (Paniagua-Michel and Garcia, 2003; Ebeling et al., 2006).

Besides transforming ammonia into biomass, some heterotrophic bacteria are capable of breaking down organic wastes such as unconsumed feed, feces, and dead organisms. These wastes serve as nutrient sources for the proliferation of heterotrophic bacteria. Furthermore, that bacterial biomass can be consumed by fish

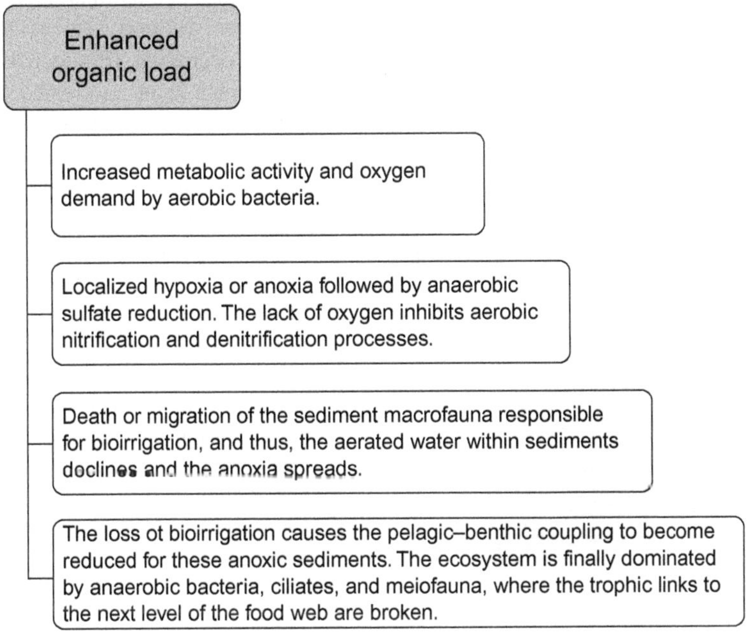

Figure 24.1 Environmental repercussions of enhanced organic load caused by aquacultural activities.
Source: Modified from Chavez-Crooker and Oberque-Contreras (2010).

and crustaceans (Figure 24.1). Thus, the proliferation of heterotrophic bacterial biomass combined with an adequate aeration/oxygenation rate may be beneficial for the productive performance of shrimp farms.

Experimental evidence has demonstrated that a high proportion of C:N ratio (20:1) should be maintained to promote the growth of heterotrophic bacteria (Ebeling et al., 2006); that heterotrophic consortium will then degrade the organic material (which is a carbon source) and assimilate ammonia directly into cellular protein. Thus, the recirculation of organic material, nutrients, and inorganic nitrogenous compounds is performed by a simple food chain within the same culture unit. In particular, the consumption of organic material by heterotrophic bacteria has paramount significance, because it eliminates a series of ecological imbalances listed in Figure 24.2.

At first, the use of heterotrophic bacteria did not have the goal of bioremediating aquaculture effluents; instead, it aimed to promote the growth of bioflocs (formed by organic matter, diatoms, algae, and heterotrophic bacteria) considering the higher rate of biomass production compared to autotrophic bacteria. This biomass comprised a food source for shrimp cultured in zero water exchange systems (Avinimelech, 1999); however, the bioremediation of wastes (N and P compounds and organic matter) represented an additional benefit. Herein, the continuous manipulation of culturing conditions (reduction of water exchange, addition of carbon sources, inoculation of nitrifying and heterotrophic bacteria, adequate aeration/oxygenation, etc.) promotes the growth of microorganisms such as microalgae, cyanobacteria, bacteria, and protozoans, which could serve as food sources for the farmed organisms, while the generation of wastes is decreased (Figure 24.3). In conclusion, an alternative flux of energy into higher trophic levels is created (Azam et al., 1983); the heterotrophic organisms thus enrich the trophic chain by converting the organic matter, dissolved organic carbon, and nitrogenous metabolites into usable energy, which finally is transferred to the upper levels (Sherr and Sherr, 2000). Under such a scenario, the culture units are converted into bioreactors in which the recirculation of nutrients and the significant reduction of waste generation are achieved.

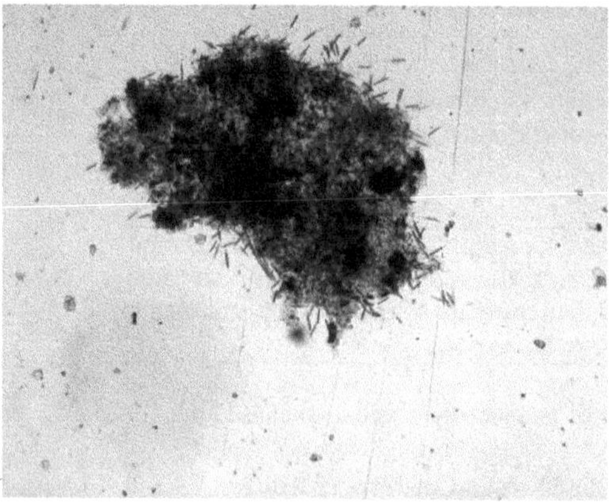

Figure 24.2 Biofloc sampled from shrimp farming tanks using the benthic diatom *Nitzchia* sp. as inducer. The biofloc contains microalgae, heterotrophic bacteria, and organic matter.
Source: Photo: Anselmo Miranda-Baeza.

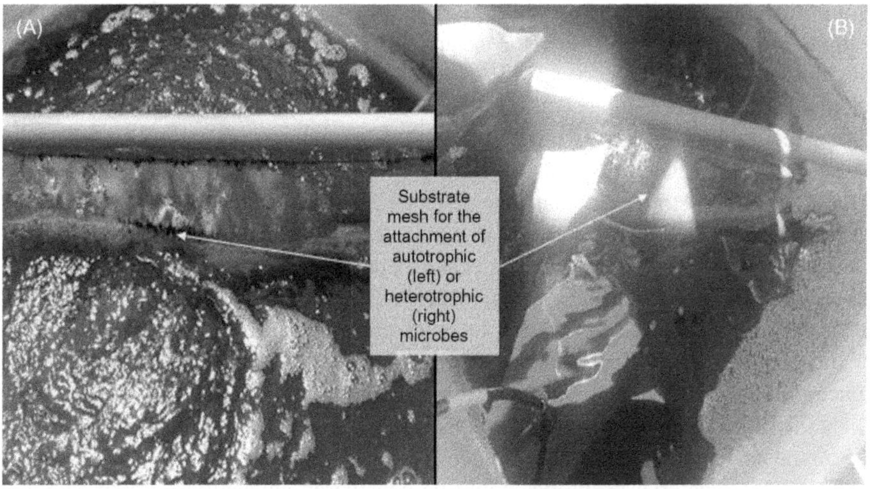

Figure 24.3 Autotrophic (A) and heterotrophic (B) microbial-based systems used for rearing shrimp (*Litopenaeus vannmei*) postlarvae. System A was provided with triple superphosphate and exposed to light conditions to enhance the proliferation of autotrophic microbes (diatoms and cyanobacteria), whereas system B was provided with molasses as the carbon source (C:N = 20:1) and exposed to continuous shade (heterotrophic bacteria and filamentous cyanobacteria).

These kinds of systems are known as Biofloc Technology (BFT) systems (Figure 24.3), and usually generate considerable amounts of total suspended solids and have low alkalinity and high CO_2 levels. Despite the fact that most of the organic matter is transformed into biofloc, the excess has to be eliminated (used for biofloc flour, fertilizer, etc.), and sodium bicarbonate may be added (100–150 mg/L as $CaCO_3$) to maintain an adequate alkalinity (Ebeling et al., 2006). Other requirements of BFT are full aeration, adequate mixing, additional carbon sources (starch, molasses, cellulose, etc.), and so forth. Additional information regarding the state of the art of BFT can be found in Avnimelech's and Emerenciano's reports (Avnimelech, 2009; Emerenciano et al., 2013).

The use of heterotrophic bacteria may be a sustainable alternative for the bioremediation of wastes generated by aquaculture; however, its particular requirements may not be adequate for all kinds of aquaculture systems. On the other hand, BFT could be performed *ex situ*, and the biofloc used as a food source for the cultured organisms or for other purposes, considering that they are a source of vitamins, essential fatty acids, carbohydrate, and protein.

Another alternative for aquaculture bioremediation could be the use of probiotic bacteria. Despite the fact that the use of probiotic bacteria in aquaculture is not focused on the elimination or breakdown of toxic compounds or organic matter, they can halt the proliferation and inclusively eradicate the presence of some pathogens. Viral and/or bacterial pathogens enter into aquaculture farms and proliferate within the culture units, increasing in number and being released in higher concentrations in the discharge effluents. Moriarty (1999) argued that "the microbial species composition in hatchery tanks or large aquaculture ponds can be changed by adding selected bacterial species to displace deleterious normal bacteria." This author reported a decrease in the abundance of luminous *Vibrio* strains, where specially selected probiotic strains of *Bacillus* species were added. An additional advantage of this strategy is that the use of antibiotics can be reduced, and the discharged effluents will not contain high concentrations of such chemicals.

Some shrimp farms, for instance, have been using aerobic bacterial consortia based exclusively on marine strains, which exhibit a faster proliferation rate and greater efficiency on the diminution or eradication of pathogens compared to most of the commercial products. Farmers use bioreactors to achieve massive bacteria concentrations (1×10^{12} CFU/mL) which are added every day to the culture units. Once the dominance of the bacterial consortia has been achieved within the ponds, the abundance and virulence of pathogens decrease.

The use of bacterial probiotics could be considered aquaculture bioremediation, but its efficacy may depend upon the thorough understanding of the nature of competition between particular species or strains of bacteria.

24.2.2 Microalgae

Microalgae are a diverse group of unicellular organisms comprising eukaryotic protists, prokaryotic cyanobacteria, and blue-green algae (Day et al., 1999); their numerous characteristics have been used in different industrial biotechnological processes. According to recent reports (Raja et al., 2008), "microalgae are the untapped resource with more than 25,000 species of which only 15 are in use."

Phytoremediation is an eco-friendly bioremediation process of removing pollutants/nutrients from an environment (soil, sediment, water) by using any green plant–based system; it is not only an energy saving but also a resource recovering system (Danaher, 2013; Kwon et al., 2013) (Figure 24.3). Most of the scientific research on phytoremediation of aquaculture wastes has been focused on the use of different algae species, whereas the use of microalgae species for the same purpose has had less attention (Hemaiswarya et al., 2011). Microalgae have shown tremendous potential to bioremediate industrial wastes such as heavy metals, dyes, toxic gases, and petroleum contaminants (Malik, 2004; Doshi et al., 2007; Lim et al., 2010; Chiu et al., 2011). However, some microalgae also have considerable potential to be used as bioremediators of aquaculture effluents (Blackburn, 2004; Hemaiswarya et al., 2011).

As stated above, bacterial consortia oxidize ammonia to the less toxic nitrate, while microalgae photosynthetically convert the dissolved inorganic nutrients (NH_3–NH_4, NO_2, NO_3, PO_4, CO_2) into "particulate nutrient packs" (Neori et al., 2004).

Martinez-Cordova et al. (2009) cultured penaeid shrimp, and the effluents generated by the system were flowed through contiguous bioremediation ponds stocked with bivalves and benthic microalgae (*Navicula* sp.); the authors associated the presence of microalgae to the decrease of total nitrogen, ammonia, nitrite, nitrate, and phosphate; finally, to prove that the effluents were adequately remediated, they used those effluents to culture another group of shrimp, with successful results in terms of production performance. Vymazaj (1988) found that microalgae species such as *Navicula* sp. were capable of removing nutrients from polluted streams with a maximum efficiency of 80% and 70% for ammonium and orthophosphates, respectively. Other diatom species may also have a similar effect on aquaculture effluents (Raja et al., 2008); however, microalgae that do not form aggregates or biofilms such as benthic species are not adequate candidates for bioremediation unless they occur at very high concentrations such as those reached in bioreactors, or those reported in systems provided with mats to promote the attachment of microalgae and bacteria. Due to the oxygen requirements during their dark phase of photosynthesis, bioremediation using microalgae should be performed *ex situ*. However, if mats and adequate aeration are provided, the bioremediation can be done *in situ* (Audelo-Naranjo et al., 2011). Additionally, sometimes silicates may be added to improve the growth of microalgae (Martinez-Cordova et al., 2011a).

Genetic improvement should also play an important role in the future development of algal industries (Raja et al., 2008). For instance, mutant strains of microalgae resistant to particular conditions can be used for bioremediation of aquacultural wastes such as NO, CO_2, and SO_2 (Chiu et al., 2011). In this context, Stevens and Purton (1997) enlisted a series of requirements that algae and microalgae species must possess to be considered potential candidates for genetic modification. The authors argued that "genetic engineering technology that allowed the manipulation of biochemical pathways leading to higher product yields and improved strains" would represent a great benefit for industrial applications. From those applications, we can highlight the bioremediation of aquaculture effluents. Li and Tsai (2009) genetically modified the microalgae *Nannochloropsis oculata*, which is an algae commonly found in shrimp ponds and other aquaculture farms. The authors inserted an algae-codon-optimized bovine lactoferin (an antimicrobial peptide) into *N. oculata* to provide an organism to defend against bacterial pathogenic infection. The survival of fish fed on the transgenic microalgae was 85% compared to 5% for fish fed on normal microalgae. If this experiment is not related to bioremediation, it provides clear and strong evidence that some characteristics of microalgae could be further modified in order to create efficient bioremediators, and that the same concept may be applied to bacteria.

24.3 Limitations of Microbial Bioremediation

In spite of the undeniable benefits, the bioremediation of aquaculture wastes using microbes has some limitations that should be considered. For instance, despite the fact that ammonia is oxidized to the less toxic nitrate by autotrophic bioremediators, pollution with nitrogenous metabolites is still being generated (Touchette and Burkholder, 2000); thus, autotrophic microbes perform a partial bioremediation, because they do not eliminate the inorganic nitrogen or the organic matter. Macroalgae (seaweed), in contrast, sequester the nutrients out of the water

(Neori et al., 2004) but do not decrease the organic load. Thus, in spite of the effectiveness of phytoremediation in removing and transforming inorganic nitrogen into less toxic compounds, the enhanced load of organic matter is still a problem (Figure 24.1).

The maintenance of nitrifying bacteria is sometimes difficult, and diverse factors such as ammonia and nitrite concentrations, carbon:nitrogen ratio, dissolved oxygen, pH, temperature, and alkalinity may have a negative effect on the nitrification–denitrification efficiency (Ebeling et al., 2006). In addition, the wastes generated by aquaculture may not meet the nutritional requirements of some microalgae species, and additional nutrients would need to be supplied during the aquaculture cycle to maintain adequate concentrations of the microalgae.

Regarding the use of genetically modified or exotic organisms (microalgae or bacteria), both strategies may require the implementation of zero water exchange systems or the use of bioreactors to prevent the escape of transgenic/exotic strains to the environment, which could represent a serious problem to the ecosystem. Snow et al. (2005) listed the main possible risks of using and introducing genetically modified organisms into the environment. However, tough bioreactors using bacteria and microalgae exhibit the most efficient bioremediation performance, they usually require high aeration/oxygenation rates and are expensive to install and operate (Ansal et al., 2010); furthermore, the operative complications would increase if used on large-scale farms.

The use of heterotrophic bacteria within the same culture units may require a constant removal of solids through some level of water exchange (Ebeling et al., 2006); though fish and crustaceans may consume considerable amounts of biofloc, the production rate of such biofloc could be higher than the consumption. In addition, heterotrophic bacteria through BFT cannot be used in earthen ponds, because the requirement of constant and vigorous mixing would create a considerable turbidity (sand + soil + biofloc) which could affect growth and survival of fish and crustaceans.

Finally, some aquaculture farms may have limited space to install bioremediation systems, which could be also expensive due to the installation of additional infrastructure. However, in the long run, the use of *in situ* bioremediation may represent an interesting solution for some farms.

There are, of course, other limitations of microbial bioremediation that will be elucidated as the research in this field becomes more extensive during the coming years; however, the benefits seem to significantly outweigh the disadvantages.

24.4 Multitrophic Bioremediation Systems: A Sustainable Alternative

Some of the limitations of microbial bioremediation rely on the fact that a single species cannot be involved in the removal and biotransformation of all the wastes generated by aquaculture. However, there are other organisms (not necessary microbes) that can play a role in the bioremediation of particular wastes that microbes cannot remove or biotransform. Herein, the combination of species from different trophic groups creates a synergistic relationship which, in turn, acts as a bioremediation system. Under the above scenario, the wastes from aquaculture are assimilated through the food web within the microcosm formed by all the organisms being co-cultured.

Integrated multitrophic aquaculture (IMTA) integrates the culture of fish or shrimp with vegetables, microalgae, shellfish, and/or seaweeds (Neori et al., 2004), while the environmental impact is sometimes eliminated or reduced to minimal levels. In other words, IMTA integrates a number of complementary and diverse organisms to optimize nutrient utilization and reduce solid wastes deposited on sediments; thus, the wastes from one organism become a source of energy for others (Chopin et al., 2008).

In a multitrophic bioremediation system using microbes, when aquaculture effluents are discharged, they usually contain high amounts of inorganic compounds and suspended solids; these solids usually trigger the turbidity level, which is a condition that affects the growth of algae and microalgae (phytoremediation). Sedimentation units may decrease the content of suspended solids, whereas organisms such as bivalves or zooplankton may consume part of the organic suspended solids (Martínez-Cordova et al., 2011b). In addition, algae, microalgae, and bacteria are oxygen consumers (particularly plants at the dark photosynthesis phase); which means that in an IMTA system, these organisms can cause dramatic oxygen drops, which creates the necessity to have adequate oxygenation and backup systems.

Finally, there are plenty of organisms from diverse taxa that may be introduced into bioremediation systems based on microbes. For instance, filter-feeding bivalves have the capacity to consume organic matter, filter-feeding *Artemia* can remove particulate N, seaweed can remove inorganic N and P, filter-feeding sponges have the capability to bioremediate the water column within 24 h, retaining up to 80% of suspended particles, etc. (Milanese et al., 2003; Marinho-Soriano et al., 2011; Martínez-Cordova et al., 2011b).

Chopin et al. (2001) asserted that "as guidelines and regulations on aquaculture effluents are forthcoming in several countries, and by adopting integrated polytrophic practices, the aquaculture industry should find increasing environmental, economic, and social acceptability and become a full and sustainable partner within the development of integrated coastal management frameworks."

24.5 Conclusion

In conclusion, the following can be stated:

1. That microorganisms play an important role in practically all the aquatic processes and their capacities may be (and in some cases are being) advantageously used for the benefit of humans in diverse fields including the bioremediation of aquaculture effluents and others.
2. The diverse types of microorganisms have different metabolic pathways which may be used for specific purposes of bioremediation.

3. Genetic modification of microbes seems to be an additional tool to maximize the effectiveness of these microorganisms as bioremediators of aquaculture effluents.
4. Considering the particular characteristics of effluents generated by aquaculture, microorganisms themselves are not capable of completely bioremediating such effluents, and the participation of other additional organisms from diverse taxa may be necessary.
5. The IMTA systems represent an important alternative for bioremediation and have proven to be efficient mostly for the culture of fish and crustaceans.
6. Bioremediation by microorganisms has some limitations and restrictions; however, the benefits outweigh those limitations.

References

Ackefors, H., Enell, M., 1994. The release of nutrients and organic matter from aquaculture systems in Nordic countries. J. Appl. Ichthyol. 10, 225–241.

Ansal, M.D., Dhawan, A., Kaur, V.I., 2010. Duckweed based bio-remediation of village ponds: an ecologically and economically viable integrated approach for rural development through aquaculture. Livestock Res. Rural Dev. 22, 129.

Atlas, R.M., 1993. Bacteria and bioremediation of marine oil spills. Oceanus (United States). 36, 2.

Audelo-Naranjo, J.M., Martinez-Cordova, L.R., Voltolina, D., Gomez-Jimenez, S., 2011. Water quality, production parameters and nutritional condition of *Litopenaeus vannamei* (Boone, 1931) grown intensively in zero water exchange mesocosms with artificial substrates. Aquac. Res. 42, 1371–1377.

Avinimelech, Y., 1999. Carbon/nitrogen ratio as a control element in aquaculture systems. Aquaculture. 176, 227–235.

Avnimelech, Y., 2009. Biofloc Technology. The World Aquaculture Society, Baton Rouge, Louisiana, USA, 176 p.

Azam, F., Fenchel, T., Field, J.G., Gray, J.S., Meyer-Reil, L.A., Thingstad, F., 1983. The ecological role of water-column microbes in the sea. Mar. Ecol. Prog. Ser. 10, 257–263.

Bender, J., Phillips, P., 2004. Microbial mats for multiple applications in aquaculture and bioremediation. Bioresour. Technol. 94, 229–238.

Bird, K.T., Bourdine, K.S., Busch, W.S., 1988. Marine algae as a tool for bioremediation of marine ecosystems. In: Dekker, M. (Ed.), Agricultural Biotechnology. Taylor & Francis, New York, NY, pp. 601–613.

Blackburn, S., 2004. Water pollution and bioremediation by microalgae: eutrophication and water poisoning. Pages 417–429. In: Richmond, A. (Ed.), Microalgal Culture: Biotechnology and Applied Phycology. Blackwell Science, Oxford, p. 566.

Camargo, J.A., Alonso, Á., 2006. Ecological and toxicological effects of inorganic nitrogen pollution in aquatic ecosystems: a global assessment. Environ. Int. 32, 831–849.

Chavez-Crooker, P., Obreque-Contreras, J., 2010. Bioremediation of aquaculture wastes. Curr. Opin. Biotechnol. 21, 313–317.

Chiu, S.Y., Kao, C.Y., Huang, T.T., Lin, C.J., Ong, S.C., Chen, C.D., et al., 2011. Microalgal biomass production and on-site bioremediation of carbon dioxide, nitrogen oxide and sulfur dioxide from flue gas using *Chlorella* sp. cultures. Bioresour. Technol. 102, 9135–9142.

Chopin, T., Buschmann, A.H., Halling, C., Troell, M., Kautsky, N., Neori, A., et al., 2001. Integrating seaweeds into marine aquaculture systems: a key toward sustainability. J. Phycol. 37, 975–986.

Chopin, T., Robinson, S., Troell, M., Neori, A., Buschmann, A., Fang, J., 2008. Ecological engineering: multi-trophic integration for sustainable marine aquaculture. In: Jorgensen, S.E., Fath, B. (Eds.), Encyclopedia of Ecology. Elsevier, Oxford, pp. 2463–2475.

Crab, R., Avnimelech, Y., Defoirdta, T., Bossier, P., Verstraet, W., 2007. Nitrogen removal techniques in aquaculture for a sustainable production. Aquaculture. 270, 1–14.

Danaher, J.J., 2013. Phytoremediation of Aquaculture Effluent Using Integrated Aquaculture Production Systems (Doctoral dissertation). Auburn University. Alabama, USA.

Day, J.G., Benson, E.E., Fleck, R.A., 1999. *In-vitro* culture and conservation of microalgae: applications for aquaculture, biotechnology and environmental research. In Vitro Cell. Dev. Biol. Plant. 35, 127–136.

Doshi, H., Ray, A., Kothari, I.L., 2007. Bioremediation potential of live and dead *Spirulina*: spectroscopic, kinetics and SEM studies. Biotechnol. Bioeng. 96, 1051–1063.

Ebeling, J.M., Timmons, M.B., Bisogni, J.J., 2006. Engineering analysis of the stoichiometry of photoautotrophic, autotrophic, and heterotrophic removal of ammonia–nitrogen in aquaculture systems. Aquaculture. 257, 346–358.

Emerenciano, M., Gaxiola, G., Cuzon, G., 2013. Biofloc Technology (BFT): a review for aquaculture application and animal food industry. Biomass Now: Cultivation and Utilization. Intech Open Science, Rijeka, Croatia, 301–328 pp.

Fiencke, C., Spieck, E., Bock, E., 2005. Nitrifying bacteria. In: Werner, D., Newton, W.E. (Eds.), Nitrogen Fixation in Agriculture, Forestry, Ecology, and the Environment. Springer, Marbirg, Germany, pp. 255–276.

Gargouri, B., Karray, F., Mhiri, N., Aloui, F., Sayadi, S., 2013. Bioremediation of petroleum hydrocarbons contaminated soil by bacterial consortium isolated from an industrial wastewater treatment plant. J. Chem. Technol. Biotechnol. Available from: http://dx.doi.org/10.1002/jctb.4188 (In press).

Hargreaves, J.A., 1998. Nitrogen biogeochemistry of aquaculture ponds. Aquaculture. 166, 181–212.

Hemaiswarya, S., Raja, R., Kumar, R.R., Ganesan, V., Anbazhagan, C., 2011. Microalgae: a sustainable feed source for aquaculture. World J. Microbiol. Biotechnol. 27, 1737–1746.

Herbeck, L.S., Unger, D., 2013. Pond aquaculture effluents traced along back-reef waters by standard water quality parameters, $\delta^{15}N$ in suspended matter and phytoplankton bioassays. Mar. Ecol. Prog. Ser. 478, 71–86.

Jackson, C., Preston, N., Thompson, P.J., Burford, M., 2003. Nitrogen budget and effluent nitrogen components at an intensive shrimp farm. Aquaculture. 218, 397–411.

Koops, H.P., Pommerening-Roser, A., 2001. Distribution and ecophysiology of nitrifying bacteria emphasizing cultured species. FEMS Microbiol. Ecol. 1255, 1–9.

Kowalchuk, G.A., Stephen, J.R., 2001. Ammonia-oxidizing bacteria: a model for molecular microbial ecology. Ann. Rev. Microbiol. 55, 485–529.

Kwon, H.K., Oh, S.J., Yang, H.S., 2013. Growth and uptake kinetics of nitrate and phosphate by benthic microalgae for phytoremediation of eutrophic coastal sediments. Bioresour. Technol. 129, 387–395.

Lefebvre, D.D., Edwards, C., 2010. Decontaminating heavy metals from water using photosynthetic microbes. In: Shah, V. (Ed.), Emerging Environmental Technologies. Springer, Dordrecht, The Netherlands, pp. 57–73.

Li, S.S., Tsai, H.J., 2009. Transgenic microalgae as a non-antibiotic bactericide producer to defend against bacterial pathogen infection in the fish digestive tract. Fish Shellfish Immunol. 26, 316–325.

Lim, S.L., Chu, W.L., Phang, S.M., 2010. Use of *Chlorella vulgaris* for bioremediation of textile wastewater. Bioresour. Technol. 101, 7314–7322.

Malik, A., 2004. Metal bioremediation through growing cells. Environ. Int. 30, 261–278.

Marinho-Soriano, E., Azevedo, C.A.A., Trigueiro, T.G., Pereira, D.C., Carneiro, M.A., Camara, M.R., 2011. Bioremediation of aquaculture wastewater using macroalgae and *Artemia*. Int. Biodeterior. Biodegrad. 65, 253–257.

Martinez-Cordova, L.R., Martinez-Porchas, M., Cortes-Jacinto, E., 2009. Camaronicultura mexicana y mundial:¿actividad sustentable o industria contaminante? Rev. Int. Contam. Ambient. 25, 181–196.

Martinez-Cordova, L.R., Lopez-Elías, J.A., Leyva-Miranda, G., Armenta-Ayon, L., Martinez-Porchas, M., 2011a. Bioremediation and reuse of shrimp aquaculture effluents to farm whiteleg shrimp, *Litopenaeus vannamei*: a first approach. Aquac. Res. 42, 1415–1423.

Martínez-Cordova, L.R., Lopez-Elías, J.A., Martinez-Porchas, M., Bernal-Jaspeado, T., Miranda-Baeza, A., 2011b. Studies on the bioremediation capacity of the adult black clam, *Chione fluctifraga*, of shrimp culture effluents. Rev. Biol. Mar. Oceanogr. 46, 105–113.

Martínez-Porchas, M., Martínez-Cordova, L.R., 2012. World aquaculture: environmental impacts and troubleshooting alternatives. Sci. World J. 2012, 389623.

Milanese, M., Chelossi, E., Mancini, R., Sara, A., Sidri, M., Pronzato, R., 2003. The marine sponge *Chondrilla nucula* Schmidt, 1862 as an elective candidate for bioremediation in integrated aquaculture. Biomol. Eng. 20, 363–368.

Moriarty, D.J., 1999. Disease control in shrimp aquaculture with probiotic bacteria. In: Proceedings of the Eighth International Symposium on Microbial Ecology, Queensland, Australia, pp. 237–243.

Neori, A., Chopin, T., Troell, M., Buschmann, A.H., Kraemer, G.P., Halling, C., et al., 2004. Integrated aquaculture: rationale, evolution and state of the art emphasizing seaweed biofiltration in modern mariculture. Aquaculture. 231, 361–391.

Paez-Osuna, F., Ruiz-Fernandez, A.C., 2005. Environmental load of nitrogen and phosphorus from extensive, semiintensive, and intensive shrimp farms in the Gulf of California ecoregion. Bull. Environ. Contam. Toxicol. 74, 681–688.

Paniagua-Michel, J., Garcia, O., 2003. *Ex-situ* bioremediation of shrimp culture effluent using constructed microbial mats. Aquacult. Eng. 28, 131–139.

Panigrahi, A., Azad, I.S., 2007. Microbial intervention for better fish health in aquaculture: the Indian scenario. Fish Physiol. Biochem. 33, 429–440.

Paredes, D., Kuschk, P., Mbwette, T.S.A., Stange, F., Muller, R.A., Koser, H., 2007. New aspects of microbial nitrogen transformations in the context of wastewater treatment: a review. Eng. Life Sci. 7, 13–25.

Pieper, D.H., Reineke, W., 2000. Engineering bacteria for bioremediation. Curr. Opin. Biotechnol. 11, 262–270.

Raja, R., Hemaiswarya, S., Kumar, N.A., Sridhar, S., Rengasamy, R., 2008. A perspective on the biotechnological potential of microalgae. Crit. Rev. Microbiol. 34, 77–88.

Sherr, B.F., Sherr, E.B., 2000. Marine microbes: an overview. In: Kirchman, D. (Ed.), Microbial Ecology of the Oceans. Wiley-Liss, New York, NY, pp. 13–46.

Snow, A.A., Andow, D.A., Gepts, P., Hallerman, E.M., Power, A., Tiedje, J.M., et al., 2005. Genetically engineered organisms and the environment: current status and recommendations. Ecol. Appl. 15, 377–404.

Stevens, D.R., Purton, S., 1997. Genetic engineering of eukaryotic algae: progress and prospects. J. Phycol. 33, 713–722.

Tal, Y., Watts, J.E., Schreier, S.B., Sowers, K.R., Schreier, H.J., 2003. Characterization of the microbial community and nitrogen transformation processes associated with moving bed bioreactors in a closed recirculated mariculture system. Aquaculture. 215, 187–202.

Tal, Y., Watts, J.E., Schreier, H.J., 2006. Anaerobic ammonium-oxidizing (anammox) bacteria and associated activity in fixed-film biofilters of a marine recirculating aquaculture system. Appl. Environ. Microbiol. 72, 2896–2904.

Thamdrup, B., Dalsgaard, T., 2002. Production of N_2 through anaerobic ammonium oxidation coupled to nitrate reduction in marine sediments. Appl. Environ. Microbiol. 68, 1312–1318.

Touchette, B.W., Burkholder, J.M., 2000. Overview of the physiological ecology of carbon metabolism in sea grasses. J. Exp. Mar. Biol. Ecol. 250, 169–205.

Vymazaj, J., 1988. The use of periphyton communities for nutrient removal from polluted streams. Hydrobiologia. 166, 225–237.

25 Aquifer Microbiology at Different Geogenic Settings for Environmental Biogeotechnology

Beyer A., Weist A., Brangsch H., Stoiber-Lipp J. and Kothe E.

Friedrich Schiller University, Institute of Microbiology, Jena, Germany

25.1 Introduction

Subsurface water holds about 30% of the world's total freshwater reserves and is the most important source of drinking water (Danielopol et al., 2003). Groundwater flows in cavities of rock bodies (karst formations), in clastic, loosely packed sediments, through pores and fissures or fractures, and within bedding planes constituting the aquifer, which is a permeable layer with high hydraulic conductivity in the saturated zone (Fitts, 2002).

Little is known about communities inhabiting aquifers with respect to different host rocks. However, it might be important to consider the geogenic setting in which the groundwater is held, since physicochemical boundary conditions clearly determine the niches for propagation and growth. With temperatures up to somewhere near 150°C, pH ranges from 4 to 9, and oxidizing to strongly reducing condition, microorganisms are versatile and may inhabit widely different habitats (Griebler and Mösslacher, 2003). Chemolithotrophy and heterotrophic growth with versatile metabolisms allow communities to develop, depending on primary production independent from light. With extensive drilling programs, aquifer microbiology becomes more and more accessible, and the advent of modern techniques allows to determine not only the microbial diversity, but of metabolic activity as well. Thus, with this chapter, we want to summarize the state of knowledge on microbial presence and activity with respect to different geogenic settings in order to allow for a better understanding of formation specific microbiology. Only such a detailed understanding will enable a tailored manipulation of subsurface systems, which seems the only way to address contamination in groundwater and prevent spreading of contaminants.

Microbial Biodegradation and Bioremediation. DOI: http://dx.doi.org/10.1016/B978-0-12-800021-2.00025-X

25.2 Groundwater: A Complex Ecosystem

Pore space is defined by porosity of a material possessing free space between the mineral grains, expressed as percentage (Rebollo et al., 1996), and depends on size and sorting of the particles as a cubic or hexagonic package. For instance, carbonate rocks are associated with the group of interparticle pore types, providing an ideal habitat for microorganisms (Lucia, 1995).

Accordingly, the subsurface is colonized by a variety of microorganisms, comprising a biomass that is almost the same as or even more than that of surface ecosystems (Gold, 1992; Dong, 2008). In terms of their origin, microbes can be autochthonous, coming from the parent rock, or allochthonous, thus being transported to the place via groundwater. Both planktonic and benthic microorganisms living freely in the water or attached to surfaces can colonize the niche (Goldscheider et al., 2006). This holds true for the different domains of life, with bacteria, archaea, and eukaryotes, such as protozoa and fungi or viruses and phages, being isolated from groundwater (Griebler and Mösslacher, 2003; Griebler and Lueders, 2009). Macroorganisms like crustaceans, nematodes, oligochaetes, and mites can be found at low prevalences depending on pore sizes (Griebler and Mösslacher, 2003), but if present, they may also contribute to biogeochemical processes (Beloin et al., 1988; Lovley and Chapelle, 1995; Goldscheider et al., 2006). Since the pores are water filled in groundwater, oxygen transport is limited. If sufficient oxygen is available in aerobic groundwater aquifers, the microbial communities are chemoheterotrophic and well adapted to oligotrophic environments (Balkwill, 1989; Madsen and Ghiorse, 1993). Here, organic matter is the most important energy source, being oxidized to carbon dioxide (Lovley and Chapelle, 1995; Gliesche, 1999). Ammonia-oxidizing bacteria, for instance, are sensitive toward water availability (and hence ionic strength) because it affects their metabolic activity through limitation of nutrients (Gleeson et al., 2010). In deep, pristine anoxic groundwater ecosystems (phreatic zone), comprising few available nutrients and a lack of light, mostly bacteria and archaea have been found (Danielopol et al., 2003). Since the input of fresh, biologically available organic carbon decreases with depth (Neff and Asner, 2001), these primary producers are often chemolithoautotrophs gaining their energy from oxidizing inorganic compounds.

Groundwater ecosystems are influenced by intensive land use, accompanied by an increase in the concentration of certain nutrients. This may lead to a change of microbial diversity, promoting organisms that are adapted to the eutrophic growth conditions and disadvantaging oligotrophs (Williamson et al., 2012). Strong impacts on community structure thus can be assumed in cases of groundwater contamination, because organic or ionic contaminants provide stress for the groundwater microflora.

25.2.1 Physicochemical Parameters

Near-surface aquifers can be influenced by climate, agriculture, and industry as well as by surface waters. This can change physicochemical parameters, e.g., temperature and pH, but also water chemistry or flow velocity (Griebler and Mösslacher, 2003). In contrast, in many aquifers with moderate depth (<100 m), the physicochemical parameters are nearly stable, providing protection for the organisms from surface influences (Goldscheider et al., 2006). For example, the groundwater temperature maintains moderate values of about 10–12°C in Central Europe (Griebler and Mösslacher, 2003; Williamson et al., 2012), or between 26.9°C and 28.2°C, corresponding to mesophilic growth conditions in India (Mahananda et al., 2010; Agarwal et al., 2012). In deep aquifers, the temperature increases continuously with depth with a gradient of about 25°C/km (Krumholz, 2000; Griebler and Mösslacher, 2003). Therefore, the temperature limit for life of more than 150°C is reached in depths of 3000–3500 m (Gold, 1992).

As oxygen is the optimal electron acceptor for the oxidation of organic compounds, partial pressure of O_2 constitutes a relevant influence on microbial community composition (Chapelle, 1993). The lack of photosynthesis in the dark limits oxygen in aquifers to, e.g., infiltration of rainwater from the surface (Chapelle, 1993). The highest O_2 values can be measured in aquifers near the surface or in aquifers which are directly connected to aerobic surface waters (Malard and Hervant, 1999).

Facultative anaerobes prefer molecular oxygen if available, but are able to use alternative electron acceptors (Chapelle, 1993). The next most energetically favorable electron acceptor is nitrate, followed by manganese(IV), iron(III), ammonia, sulfate, and methane (Lovley and Chapelle, 1995). In nature, these different redox zones are not always clearly separated, leading to parallels along the hydrological gradient (Griebler and Mösslacher, 2003). Growth of most anaerobic organisms that cannot use O_2 as an electron acceptor is inhibited in the presence of O_2 (Lovley and Chapelle, 1995).

The zonation between oxic and anoxic niches can also be seen at small scales. Within a generally aerobic groundwater body, anaerobic microenvironments can develop in minute pores or within mineralogical aggregates, in which at microscale a steep gradient of oxygen enables anaerobic metabolism (Chapelle, 1993; Goldscheider et al., 2006). This gradient is maintained mainly by aerobic use of oxygen for respiration, which can protect lower layers in biofilms or microcolonies from diffusion of harmful oxygen.

Other abiotic parameters such as hydrostatic pressure, pH, or salt concentrations usually play minor roles for the limit of microbial life, but a change of environmental conditions is associated with a change in the composition of the microbial community (Griebler and Mösslacher, 2003). In pristine groundwater systems, the relatively restricted pH range reflects the buffering capacity of the surrounding rock material, like carbonate or silicate minerals. Carbonated water from calcareous rock shows both a higher pH and a higher buffering potential. Alkalinity is usually observed in sodium-containing waters, while acidic pH occurs for example in mine water and thermal water containing free acids such as sulfuric acid resulting from sulfide oxidation, hydrochloric acid due to salt dissolution, or reduced compounds like H_2S. The enrichment of groundwater with CO_2 by microbial activity or the degradation of plant material, which leads to high concentrations of humic substances in the subsurface, may also shift pH toward acidity (Griebler and

Mösslacher, 2003). Natural pH values as low as 4.0 or 5.0 are found, while in carbonate-buffered groundwater, pH values up to 9.0 may occur. Contamination, e.g., by toxic wastes, can lead to more extreme pH conditions (Chapelle, 1993).

From pH in combination with redox potential, the solubility or speciation of prevailing ions can be predicted with E_h/pH diagrams (Hölting, 1996). Only a few ions remain in solution independent from pH, e.g., Na^+, K^+, NO^{3-}, and Cl^-. The dependency of microorganisms on physicochemical conditions allows a generalization of microbial community composition with regard to groundwater type or rock formation (Table 25.1).

25.2.2 Abundance of Microbes in Groundwater

Groundwater generally contains low amounts of bacteria, archaea, fungi, or protozoa as only limited nutrients are available due to soil filtration (Rheinheimer, 1971). Patel et al. (2007) documented that viruses and phages, the respective agents affecting bacteria, are the most abundant biological entities in aquatic environments, exceeding the abundance of bacteria by an order of magnitude (Patel et al., 2007).

The abundance of microorganisms varies with substrate concentration and composition as well as stratigraphic, hydrological, and geochemical factors (Beloin et al., 1988). Thus, high heterogeneity may be observed, while temporal and horizontal variations are usually quite low (Hunkeler et al., 2006). In pristine sediments, the population density of subsurface bacteria has been estimated to be 10- to 100-fold lower than in agricultural soil (Beloin et al., 1988). In a very permeable gravelly sandy loam layer, a higher level of biomass and activity was found in comparison to a massive clay bedrock layer (Beloin et al., 1988).

Colony forming units (CFUs) in aquifer sediments yielded 10- to 100-fold higher counts compared to the groundwater, showing the prevalence of attached growth (Kölbel-Boelke et al., 1988). In the first 5 m below the surface, the CFU decreases rapidly. In aquifers located below these 5 m, CFUs of only 100/mL and in some cases only 50/mL has been observed (Table 25.2).

25.2.3 Biofilms

In oligotrophic groundwater, prokaryotic and eukaryotic microbes often prefer biofilms as a (temporary) multicellular lifestyle, because biofilm facilitates interactions in a regulatory network and survival in adverse environmental conditions (Beveridge et al., 1997; Kostakioti et al., 2013). Only a small proportion of microorganisms are found suspended in the pore water, while most of the living microorganisms are associated with the sediment surface, forming microcolonies and biofilms (Griebler et al., 2002). The proportion of attached microbes in aquifers depends on the availability of dissolved organic carbon (DOC) and nutrients, the sediment grain-size distribution, and the sediment mineralogy (Bengtsson, 1989). Biofilms consist of a mucus matrix built from hydrated biopolymers of microbial origin, the extracellular polymeric substances (EPSs). EPS is composed of

Table 25.1 Groundwater Habitats and Respective Growth Conditions Influencing Microbial Community Composition

Subsurface Habitats	Description	Microbial Communities	Location	References
Sandstones/shales within Cretaceous rock formations	Highly to low permeable, fine-grained, organic-rich, devoid of oxygen	Heterotrophic bacteria	Western USA	Krumholz et al. (1997)
Sand/clay boundary	Subsurface sediments	Sulfate-reducing bacteria	Atlantic Coastal Plain, USA	Peter and Francis (1991) and McMahon et al. (1992)
Opalinus clay	Core	Sulfate-reducing bacteria	Mont Terri, Switzerland	Stroes-Gascoyne et al. (2007)
Sediments with lignite and clay deposits	Eocene, finely layered, interbedded with sandy material, lignite from ancient plant material	Sulfate-reducing bacteria	Texas coastal uplands aquifer system	Ulrich et al. (1998)
Six boreholes	Sulfur- and iron-containing fracture minerals, 16 groundwater samples	Methanogenic archaea, homoacetogenic bacteria, sulfate- and iron-reducing bacteria	Fennoscandian shield in Finland and Sweden	Haveman and Pedersen (2002)
Two shallow aquifers in sand and gravel deposits with clay	Oligotroph environment, saturated subsurface zones	Aerobic, nutritionally versatile bacteria: *Pseudomonas*, *Chromobacterium*, *Arthrobacter*, *Brevibacterium*; potentially facultative anaerobes and eukaryotes	Oklahoma, USA	Balkwill and Ghiorse (1985)
Deep aquifers, other subsurface sediments	Depths to 265 m	Chemoheterotrophic bacteria	South Carolina, USA	Balkwill (1989)

(*Continued*)

Table 25.1 (Continued)

Subsurface Habitats	Description	Microbial Communities	Location	References
Basalt	Groundwater wells, no evidence for migration of nutrients from the surface or depth	Anaerobic, iron- and sulfate-reducing, methanogenic, acetogenic, and fermentative bacteria/archaea	Columbia River	Stevens et al. (1993)
Granite	Groundwater in fractures, containing hydrogen and methane	Chemoautolithotrophic; homoacetogens, acetotrophic (to 112 m), and autotrophic methanogens (to 446 m)	Stripa and Äspö Hard Rock Laboratory (Sweden)	Kotelnikova and Pedersen (1998)
Rock salt	Drilling core, salt mining, Halit	Extremely halophilic archaea, *Halococcus salifodinae*	Berchtesgaden, Germany; Cheshire, England	Stan-Lotter et al. (1999)
	Salt mining	Halophilic archaea *Halobacterium* sp. BpA.1	Boulby, England	McGenity et al. (2000)
	Permian salado formation	Obligately aerobic *Bacillus sphaericus* (*permians*)	Texas, USA	Vreeland et al. (2000)
	Drilling core	Halophilic archaea *Halobacterium noricense* A1	Altaussee, Austria	Gruber et al. (2004)
Groundwater treatment/ drinking water distribution systems	Relatively high concentrations of methane, iron, manganese, dissolved organic carbon, and ammonia	Ammonia-oxidizing bacteria and archaea, nitrite-oxidizing bacteria	The Netherlands	van der Wielen et al. (2009)
Coastal groundwater system	Infiltration of bay water supplying dissolved oxygen, ammonium, organic carbon; outwash gravel, sand and silt	Ammonia-oxidizing archaea	Waquoit Bay (Waquoit, MA)	Rogers and Casciotti (2010)

Table 25.2 Cell Counts of Different Groundwater Habitats

Microorganisms	Habitat	Abundance	References
Bacteria	Karst and cave water	10^2–10^4/mL	Griebler and Mösslacher (2003)
	Sediments from cave water	10^4–10^8/mL	
	Deep groundwater from granite and gneiss systems	10^3–10^6/mL	
	Pore groundwater	10^3–10^6/mL	
	Unconsolidated sediment deposits		
	Saturated zone	10^6–10^8/mL	
	Unsaturated zone	10^4–10^8/mL	
	Groundwater of a municipal water supply		Ultee et al. (2004)
	Total count	10^3–10^4/mL	Haveman and Pedersen (2002)
	Viable count	10^2–10^3/g dry weight	Balkwill and Ghiorse (1985)
	Six boreholes (to a depth of 1390 m)	Approx. 10^5/mL	King and Parker (1988)
	Shallow aquifers (saturated subsurface zones)	10^6–10^9/g dry weight	Balkwill (1989)
	Groundwater from four wells	10^4/mL	Chapelle et al. (1987)
	Aquifer sediments	10^5–10^8/g dry weight	Goldscheider et al. (2006)
	Coastal plain sediments (182 m)	10^3–10^6/g dry weight	
	Oligotrophic aquifers	10^9/g dry weight	
Archaea		Present	Griebler and Mösslacher (2003), van der Wielen et al. (2009), Rogers and Casciotti (2010)
Protozoa			
Flagellates	Pore groundwater	1–10^2	Griebler and Mösslacher (2003)
	Unconsolidated sediment deposits (saturated zone)	10^3–10^5	
Amoebae, ciliates	Pore groundwater	Rare	
Planktonic cells	Groundwater from different types of aquifers	10^4–10^6/mL	Goldscheider et al. (2006)
Fungi		Rare	Griebler and Mösslacher (2003) and Gottlich et al. (2002)
Viruses	Pore groundwater	Present	Griebler and Mösslacher (2003)
	Marine and freshwater environments	Present	Patel et al. (2007)

carbohydrates, carbohydrate-binding proteins, adhesive fibers, and extracellular DNA, acting as stabilizer (Whitchurch et al., 2002; Qin et al., 2007). Microorganisms are embedded in EPS and by attaching to surfaces nutrients can be accumulated because of its sorption properties (Beveridge et al., 1997; Kostakioti et al., 2013). Biofilms are ubiquitous in the environment and occur at interfaces between water and solid particles or the atmosphere, e.g., in the inner walls of groundwater wells, and also in soil and sediments (Griebler and Lueders, 2009). On surface-water interfaces in pores, the respective adhesive growth would be considered microcolony formation.

25.2.4 Determining Microbial Diversity

The first step to characterize the microbial community in a given habitat is most often based on cultivation of microorganisms using defined growth media. With different cultivation methods and media, the natural conditions regarding nutrient availability, pH, and temperature are mimicked to access as large a fraction of the community as possible (Rastogi and Sani, 2011). A special medium suitable for isolation from groundwater samples should be selecting for oligotrophic heterotrophs (Reasoner and Geldreich, 1985). Isolated microorganisms from environmental samples often are not representative of the microbial community, and due to the cultivation methods used, mostly *Proteobacteria*, *Firmicutes*, *Bacteroidetes*, or *Actinobacteria* are cultivated (Reasoner and Geldreich, 1985).

The bulk of naturally occurring bacteria is not (yet) cultivable (Griebler and Mösslacher, 2003). Thus, a direct visualization is performed based on differential staining of living versus dead cells. In order to gain more, and potentially also classification-relevant, information, 16S rDNA sequencing is used, e.g., pyrosequencing, a next-generation sequencing technique used for metagenomic analyses which shows the organisms present in extracted DNA (Fakruddin et al., 2012).

Fluorescence *in situ* hybridization (FISH) is a technique for visualization of microorganisms directly in samples via epifluorescence microscopy (Andreeff and Pinkel, 1999). Although FISH is a very useful tool, it also has its limitations: nonpermeable cells or low ribosomal content affect staining; the necessity of time-consuming microscopy, specifically with samples of low numbers of microorganisms such as in groundwater, and low signals of single cells in biofilms or complex environmental samples limit its use. To overcome these limitations, other methods have been combined with FISH, e.g., filtration of water to concentrate the cells on a membrane, flow cytometry, multilabeled polyribonucleotide probes, and catalysed reporter deposition FISH (CARD-FISH) for signal enhancement (Wagner et al., 2003) may be performed.

Microarrays provide an opportunity to rapidly detect, identify, and characterize microbes in a variety of natural habitats. Here, specificity is dependent on the spotted DNA probes (Niibel et al., 2006). Depending on the question at hand, different microarrays can be applied. In different studies, a higher microbial diversity was detected using 16S rRNA-based microarrays than was shown in the corresponding clone libraries of the same samples (DeSantis et al., 2007; Rastogi et al., 2010).

Drawbacks of this method are the limited phylogenetic resolution and the fact that only known sequences and microbial groups can be detected due to the probe design (DeSantis et al., 2007). Nevertheless, microarrays are useful tools for parallel, high-throughput tracing of microorganisms and long-term monitoring of groundwater. To track changes in communities with lower resolution, polymerase chain reaction (PCR)-based restriction fragment length polymorphism (T-RFLP) or denaturing gradient gel electrophoresis (DGGE) often are used (Enwall and Hallin, 2009).

25.2.5 Assessing Microbial Activity

The first approach in assessing potentially active members of a community is the enrichment culture. Here, a microcosm with natural samples is incubated in the laboratory. If additional nutrients are given, the part of the community able to use this substrate will grow and can then be analyzed. However, the selection of substrate already influences the community, and hence, multiple substrates should be used.

The carbon source utilization of a community can be determined using a cultivation-dependent method, the BIOLOG plates (Bochner, 1989). This redox-based technique is carried out in microplates with different, separate carbon sources that are tested simultaneously. The utilization of a carbon source is indicated by the reduction of tetrazolium violet leading to color development. Subsequently, patterns can be analyzed with different statistical tools, such as principal component analysis, PCA (Garland and Mills, 1991).

BIOLOG plates have been developed for the identification of different community-level, sole-carbon-source usage, metabolic fingerprints, or community-level physiological profiles (Garland and Mills, 1991; Preston-Mafham et al., 2002). They can be used for investigating the heterotrophic community structure and for subsequent classification of communities as to their metabolic response and functional differences, where the functional diversity is defined as the number and classes of substrates used by the community (Garland and Mills, 1991; Zak et al., 1994). However, the impact of community function on the habitats largely remains obscure (Garland, 1997). BIOLOG can only show the metabolic potential of those parts of the microbial community that are able to grow under the given conditions, namely, aerobic, heterotrophic, fast-growing organisms (Heuer and Smalla, 1997). Nevertheless, BIOLOG plates have already been applied successfully to monitor changes in the indigenous microbial community of groundwater after different remediation actions (Fliermans et al., 1997).

While BIOLOG plates are mostly used to describe communities, they already provide insights into the metabolic capacities of the community present at a site. To better address the portion of microorganisms actively involved in biogeochemical cycles, RNA-dependent methods can be used (von Wintzingerode et al., 1997). However, RNA extraction often is limited, a pitfall that led to the establishment of stable isotope probing of DNA. Here, a substrate with a ^{13}C-enriched isotope signature is provided. Cells incorporating the isotope into their DNA can then be identified, if the heavy DNA is separated by density gradient centrifugation, and this portion of the environmental DNA is used for metagenome analysis. Potentially, this is also possible with autotrophic communities using labeled CO_2 (Neufeld et al., 2007). However, as with microcosms for enrichment, selection of substrate is critical.

25.3 Hydrogeobiology

Understanding groundwater flow and the properties of the surrounding material facilitates explaining physical and chemical properties of the water and growth of microorganisms. In groundwater, solute elements interact with each other and the host rock. These ions are the basis for physical parameters like viscosity, surface tension, and capillarity (Fitts, 2002). The highest water surface attraction, or capillarity, occurs in the presence of clay, due to its charge. In fine granules, the capillary forces are higher, because of the larger area of mineral surfaces (Fitts, 2002).

When atmospheric water percolates by gravity, it crosses different horizons, partly filling voids and pores in the unsaturated layer. Preferential flow through fissures of consolidated rocks or through porous clastic sediments contributes to groundwater recharge in the saturated zone. The hydrochemical parameters of the water, like acidity, temperature, and dissolved oxygen concentration, are strongly influenced by the characteristics of the passed horizons and associated microbial activity (Bengtsson, 1989; Danielopol et al., 2003).

The transition zone (ecotone) in an aquifer is the groundwater table with the capillary fringe, i.e., the contact zone between the unsaturated and saturated zones (Goldscheider et al., 2006). In this region, a highly diverse biota is located, which regulates the transfer of nutrients, energy, and particles between the different compartments. In karst aquifers, groundwater has a very short residence time, and the autochthonous community may be replaced after strong precipitation events by allochthonous organisms from surface environments (Pronk et al., 2009). In comparison, porous aquifers are characterized by smaller voids which lead to a reduced water flow and increased water residence times (Goldscheider et al., 2006).

Groundwater ecosystems can vary in size and structural complexity from small systems of a few square kilometers in alluvial aquifers to large regional aquifers, ranging from local to regional (Tóth, 1963; Lovley and Chapelle, 1995). Local groundwater bodies are well connected to the surface, causing high rates of recharge (1–30 cm/year) and high flow rates (1–100 m/year). Intermediate ones are not affected by single precipitation events with medium rates of recharge (0.01–1 cm/year) and medium flow rates (0.1–1 m/year). In regional groundwater bodies, rates of recharge, groundwater flow rates, and contact with surface environments are extremely low (Tóth, 1963; Lovley and Chapelle, 1995). This has implications for ecological dynamics, where energy and matter may remain stored for millions of years, which means that microorganisms need to be well adapted to aquifer conditions (Lovley and Chapelle, 1995). A decline in microbial diversity and numbers of cells with increasing depth thus can be observed (Rebata-Landa and Santamarina, 2006). Due to the hydrological, chemical, and geological heterogeneity of groundwater habitats (Madsen and Ghiorse, 1993), vertical layers of

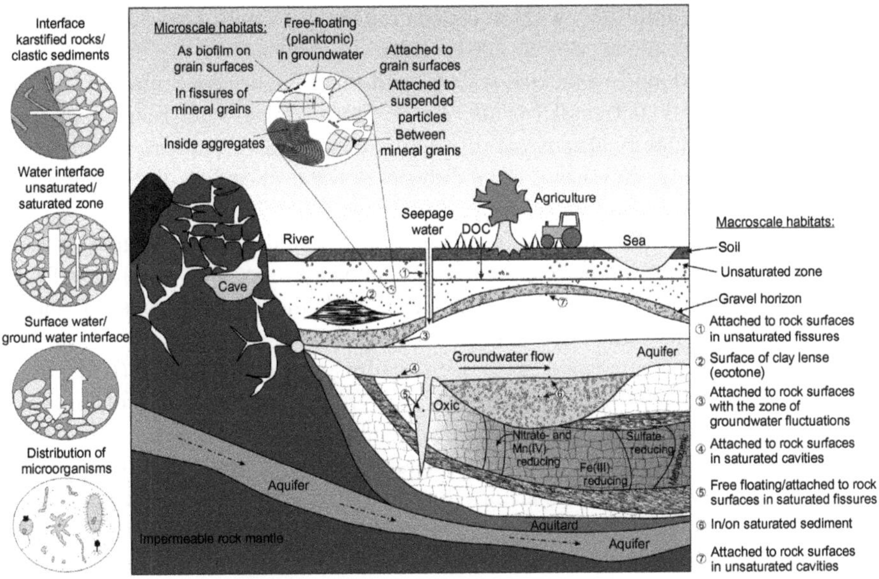

Figure 25.1 Schematic illustration of ecological macro- and microscale habitats for microorganisms and typical distribution of major terminal electron-accepting processes in groundwater aquifers and other subsurface habitats.
[modified from Lovley and Chapelle (1995), Goldscheider et al. (2006), and Griebler and Lueders (2009)].

strata are unique and complex. However, within these zones, environmental conditions can be very stable (Griebler and Lueders, 2009) (Figure 25.1).

The water flow depends on the characteristics of the host rock with porosity ranging from 0% to 60%. In granular materials such as silt and sand, the pores are connected, while in clay, the matrix porosity is lower than the fracture porosity, and in carbonate rocks, the porosity almost equals fracture porosity. The grain size, a key characteristic for unconsolidated materials, determines the amount of pore space that is available to hold water, and the water transmission through the material (Fitts, 2002). Geometrical constraints and mechanical interactions have to be considered as well. Due to the limited space available, bacterial movement and activity are confined, and the nutrient transport is limited, retarding the rate of division and reducing biodiversity (Rebata-Landa and Santamarina, 2006). Therefore, interstitial pore water provides the most dynamic habitat, while the particle surface is a more stable habitat where the carbon and nutrient supply for bacteria is determined by the net result of desorption from the particle surface and diffusion from the pore water (Bengtsson, 1989).

Several ions (Ca, Mg, Na, K, Fe, Sr or Cl, bicarbonate, and sulfate), element input through atmosphere, or weathering of rocks (Candela and Morell, 2001) delivering dissolved organic matter (DOC), contribute to the total dissolved solutes (TDS). Beside these ions, a series of trace elements (B, K, or nitrate) and gases (such as N_2, O_2, and CO_2 with a concentration <0.1 mg/L) are dissolved in groundwater (Candela and Morell, 2001). Deep groundwater mostly has a TDS of about 1000–35,000 mg/L (brackish water), with much higher concentrations found in brines (TDS > 100,000 mg/L) (Fitts, 2002). Fresh groundwater ranges between 10 and 500 mg/L, mainly because of dissolution of minerals from surrounding rock material. Specifically, clay has a large impact on water chemistry due to high ion exchange activities (Fitts, 2002; Nelson, 2002).

Humic substances, originating from microbial degradation of organic matter, form the major phase of DOC (Fitts, 2002). Typically, groundwater has a DOC content of 0.2–2 mg C/L. DOC may decrease with the depth, because of metal sorption and bacterial transformation (Fitts, 2002).

For investigating groundwater from deep aquifers, rare earth element (REE) fractionation allow for tracing flow paths, e.g., of groundwater mixing (Johannesson et al., 1997; Banks et al., 1999). REEs are preferably adsorbed onto clay minerals or Fe and Mn oxides. Johannesson and Hendry (2000) showed that light REEs (La–Nd) especially have a high affinity for adsorption onto these surfaces. In deep aquifers, surface coatings and minerals are leached, leading to a specific REE pattern. Banks et al. (1999) conclude that the REE patterns of groundwater depend on the REE contents of the host rock and weathered mineral phases, as well as the different mobilities of these elements during interaction between water and minerals, or redistribution between phases. Furthermore, the pH of the water is an important factor (Smedley, 1991). Nevertheless, in several studies, REEs have been successfully applied as natural tracers (Gosselin et al., 1992).

With these boundary conditions, microbial life can be classified as oligotrophic versus eutrophic, or anaerobic versus aerobic; with halo-, thermo-, or acidophilic niches. Furthermore, specific metal (oid) resistances (e.g., arsenic) may be observed in communities that have evolved under stable conditions in an aquifer system containing such stressors to microbial life. These abilities provide the basis for the application of microbial remediation techniques.

25.4 Application in Groundwater Remediation

Due to the growing human population, the need for drinking water has increased. However, especially in developing countries, polluted groundwater constitutes a major concern. The contamination with toxic constituents may be due to natural or anthropogenic sources (Suthar, 2010). Zouboulis and Katsoyiannis (2005) distinguish three main categories of pollution: organic compounds, inorganic pollutants, and microorganisms. The contamination of groundwater can result from the geochemistry of the geological formations, infiltration of surface water, leaks in pipelines, manure pits, etc. Biological contamination of drinking water is common in developing countries and poses a health risk, often associated with diarrheal diseases (Suthar, 2010). Members of the Enteroboacteriaceae, the so-called coliform bacteria with their prominent examples *Escherichia coli* and *Enterococcus* sp., serve as hygiene indicator bacteria for contamination with feces or manure (Aydin, 2006). Groundwater and springs generally harbor fewer bacteriological contaminants than surface water because of the filtering capacity of soil (Stambuk-Giljanovic, 2005).

Aside from microbiological contamination, indigenous bacterial communities can be taken advantage of to remediate groundwater that is contaminated with different metals such as ferrous iron, manganese, chromium, uranium, or metalloids like selenium. These contaminants can be reduced or oxidized by the microorganisms, affecting mobility. In addition, arsenic has been shown to be efficiently removed by biological oxidation, simultaneously removing dissolved iron in long-term experiments (Zouboulis and Katsoyiannis, 2005).

When waste from anthropogenic activities is disposed of in deep aquifers, metabolically active microorganisms in these formations can either disperse the contamination, causing a release of toxic substances, or decrease the toxicity of the materials by immobilization (Bachofen et al., 1998). Therefore, it is of utmost importance to monitor the fate and migration of the hazardous substances in the underground to avoid contamination of surrounding aquifers and formations. It is known that the metabolism of microbial anaerobes influences the speciation and the mobility of radionuclides, as observed during microcosm experiments which show that U and Tc can be removed from groundwater due to microbial Fe(III) reduction. The biogenic iron minerals could reduce Tc abiotically, while U was not reduced. Since reducing geological conditions is often a prerequisite for long-term safety of repositories and since the microbial activity causes geochemical changes, the latter has to be considered when planning underground waste repositories (Yoshida et al., 2006). Microbial metabolites such as surfactants, chelators, extracellular hydrolytic enzymes, and organic acids that are produced by indigenous bacteria from the deep subsurface are additional causes of changing element mobilities in groundwater bodies.

The same microbial agents can be applied in biotechnology, for example, for refinement of crude oil or mobilization of residual oil in rocks (Bachofen et al., 1998; Singh et al., 2012). Another use of microbiology tools has been introduced with tracer organisms in microbial prospecting for oil and gas, where the taxonomic composition of microbial communities from subsurface sediments is determined with a focus on hydrocarbon-degrading bacteria. These bacteria oxidize methane, ethane, propane, or butane as a carbon source and therefore are enriched in sediments above geological formations containing hydrocarbons (Rasheed et al., 2013). There is a positive correlation between microbial soil populations and the concentration of hydrocarbons in these soils. Therefore, prospective areas can be identified successfully in 90% of the cases when applied together with geoscientific and geophysical methods (Wagner et al., 2002).

The capacity to degrade organic pollutants can be used in aquifer bioremediation strategies. Organic contamination in the groundwater usually leads to the development of different redox zones. Near the pollution source, methanogenic conditions may exist, while in the plume, sulfate- and iron-reducing conditions prevail. Further downstream, at the edge of the plume, usually zones with nitrate- and manganese(IV)-reducing conditions follow (Christensen et al., 2001). This makes anaerobic or facultative anaerobic microorganisms interesting for pollutant degradation. Although degrading microbes are typically already present in contaminated aquifers, their metabolic potential sometimes is not high enough to completely mineralize the pollutants. Hence, strategies to enhance the microbial activity have been developed.

Bioremediation seems to be an efficient, economic, and environmentally sound strategy regarding removal of monoaromatic pollutants like BTEX (benzene, toluene, ethylbenzene, and xylene) from groundwater (Lynch and Moffat, 2005). In nature, aromatic compounds emanate from poorly degradable polymers like lignin, humus, and condensed tannins (Dagley, 1985). Anthropogenic sources are crude oil, gasoline, synthetic polymers, pharmaceuticals, and industrial solvents (Weelink et al., 2010). Since BTEX contaminations often are localized in anoxic conditions (Lovley, 1997), anaerobic bioremediation seems to be indicated.

The anaerobically most easily degradable BTEX component is toluene, as biodegradation is possible with CO_2, Fe(III), Mn(IV), nitrate, or even arsenate or humic substances as electron acceptors (Langenhoff et al., 1997; Chakraborty and Coates, 2004). Most of the isolated bacteria capable of toluene degradation with nitrate belong to one of the two genera *Azoarcus* (Rabus and Widdel, 1995) or *Thauera* (Anders et al., 1995), members of *Betaproteobacteria.* Coupled to the reduction of Fe(III), species of the genus *Geobacter* (*Deltaproteobacteria*) dominate (Botton et al., 2007), whereas all known sulfate reducers belong to the *Deltaproteobacteria*, such as *Desulfotignum* and *Desulfobacula* (Meckenstock, 1999; Ommedal and Torsvik, 2007). Benzene is the most persistent BTEX to anaerobic degradation, but can be degraded by *Dechloromonas* and *Azoarcus* (*Betaproteobacteria*) (Coates et al., 2001).

Another significant groundwater pollutant group is chlorinated hydrocarbons, such as tetrachlorethene (or PCE) and trichlorethene (TCE). Even polychlorinated biphenyls or phenols can be used in organohalide respiration, thereby becoming more accessible to complete mineralization (Dolfing and Beurskes, 1995). To thermodynamically favor the anaerobic degradation of most organic compounds, hydrogen, formate, and acetate need to be kept at low concentrations (Ramakrishnan, 2013). Microorganisms capable of reductive dechlorination include the genera *Dehalospirillium* (Scholz-Muramatsu et al., 1995), *Dehalobacte*r (Schumacher and Holliger, 1996), *Dehalococcoides* (Maymo-Gatell et al., 1997), and *Desulfomonile* (DeWeerd et al., 1990). *Dehalococcoides* is the only one capable of complete dechlorination of PCE and TCE to ethane, and does not seem to be present at sites where only partial dechlorination takes place (Hendrickson et al., 2002).

As *ex situ* treatment (pump-and-treat) is difficult and expensive, *in situ* degradation would be preferable. Natural attenuation by the naturally occurring microflora is often a long-term process which needs careful monitoring to prevent contaminant spreading into pristine environments. To accelerate bioremediation, optimal conditions can be engineered in enhanced bioremediation achieved by addition of electron donors or acceptors (biostimulation), or by inoculation of specific microorganisms able to degrade the pollutants (bioaugmentation; Scow and Hicks, 2005).

Bioremediation of a PCE-contaminated aquifer with biostimulation and bioaugmentation was studied in Oscoda, Michigan, with a PCE plume derived from a former dry cleaning store (Lendvay et al., 2003). The microbial population showed a heterogeneous distribution of dechlorinating species in the aquifer, and enrichment cultures with PCE and acetate resulted in a pure culture of a *Desulfuromonas*, capable of degrading PCE to *cis*-DCE. Enrichment with *cis*-dichlorethen (DCE) and lactate yielded a mixed population containing *Dehalococcoides*, which is able to completely dechlorinate DCE to ethene. For bioaugmentation of the aquifer, both were combined with reduced groundwater prior to injection, followed by addition of lactate as electron donor and phosphate and nitrate as nutrients. Within 6 weeks, dechlorination of chloroethenes to ethene was achieved. For evaluation of the inoculation, a second plot was treated with biostimulation only. The injection of lactate and nutrients resulted in dechlorination as well, although it started only after a lag period of 3 months. The degradation coincided with an increase of *Dehalococcoides* populations, strong in the bioaugmentation plot and slight in the biostimulation plot, starting right before the beginning of dechlorination. This study shows that, although the microflora present potentially is able to degrade chlorinated ethenes to ethane, bioaugmentation greatly and biostimulation slightly shortens the time that would have been needed had natural attenuation been aimed at. Therefore, bioremediation shows a high potential, but the contaminated groundwater as well as its microbiome needs to be characterized thoroughly before starting an operation.

25.5 Conclusion

During the past decade, natural attenuation and bioremediation have gained interest because of their cost-effectiveness and low impact on sites compared to geotechnological solutions. With respect to groundwater, legislative protection of the groundwater resources has to be considered. So far, there are only a few examples of successful treatments involving microbial strategies for remediation. This chapter shows why the geology, hydrogeochemistry, and microbiology of a groundwater body need to be thoroughly analyzed before a treatment is envisioned. The lack of understanding of critical components in the system will almost certainly lead to failure of the project and may even result in additional spread of contamination. Here, we have assembled data available for groundwater geobiochemistry that can be used to compare a new setting with available data, thereby aiding the acquisition of a holistic picture on the site. In addition, we provide a survey of methods addressing microbial populations and their respective activities that will aid in identification of major players at a new site. To our understanding, indigenous microorganisms, as opposed to a cure-for-all toolbox, are the key to geobiological remediation of contaminated aquifers.

References

Agarwal, B.R., Mundhe, V., Hussain, S., Pradhan, V., Yusuf, S., 2012. Study of physico-chemical parameters of groundwater around Badnapur, Dist Jalna. J. Adv. Sci. Res. 3, 94–95.

Anders, H.-J., Kaetzke, A., Kämpfer, P., Ludwig, W., Fuchs, G., 1995. Taxonomic position of aromatic-degrading denitrifying pseudomonad strains K 172 and KB 740 and their description as new members of the genera *Thauera*, as *Thauera aromatica* sp. nov., and *Azoarcus*, as *Azoarcus evansii* sp. nov., respectively, members of the beta subclass of the Proteobacteria. Int. J. Syst. Bacteriol. 45, 327–333.

Andreeff, M., Pinkel, D., 1999. Introduction to Fluorescence *In Situ* Hybridization: Principles and Clinical Applications. Wiley, New York, NY.

Aydin, A., 2006. The microbiological and physico-chemical quality of groundwater in West Thrace, Turkey. Pol. J. Environ. Stud. 16, 377–383.

Bachofen, R., Ferloni, P., Flynn, I., 1998. Microorganisms in the subsurface. Microbiol. Res. 153, 1–22.

Balkwill, D.L., 1989. Numbers, diversity, and morphological-characteristics of aerobic, chemoheterotrophic bacteria in deep subsurface sediments from a site in South-Carolina. Geomicrobiol. J. 7, 33–52.

Balkwill, D.L., Ghiorse, W.C., 1985. Characterization of subsurface bacteria associated with 2 shallow aquifers in Oklahoma. Appl. Environ. Microbiol. 50, 580–588.

Banks, D., Hall, G., Reimann, C., Siewers, U., 1999. Distribution of rare earth elements in crystalline bedrock groundwaters: Oslo and Bergen regions, Norway. Appl. Geochem. 14, 27–39.

Beloin, R.M., Sinclair, J.L., Ghiorse, W.C., 1988. Distribution and activity of microorganisms in subsurface sediments of a pristine study site in Oklahoma. Microb. Ecol. 16, 85–97.

Bengtsson, G., 1989. Growth and metabolic flexibility in groundwater bacteria. Microb. Ecol. 18, 235–248.

Beveridge, T.J., Makin, S.A., Kadurugamuwa, J.L., Li, Z., 1997. Interactions between biofilms and the environment. FEMS Microbiol. Rev. 20, 291–303.

Bochner, B.R., 1989. Sleuthing out bacterial identities. Nature 339, 157–158.

Botton, S., Van Harmelen, M., Braster, M., Parsons, J.R., Roling, W.F.M., 2007. Dominance of Geobacteraceae in BTX-degrading enrichments from an iron-reducing aquifer. FEMS Microbiol. Ecol. 62, 118–130.

Candela, L., Morell, I., 2001. Basic Chemical Principles of Groundwater. Groundwater, vol. II. EOLSS, Oxford.

Chakraborty, R., Coates, J.D., 2004. Anaerobic degradation of monoaromatic hydrocarbons. Appl. Microbiol. Biotechnol. 64, 437–446.

Chapelle, F., 1993. Ground-Water Microbiology and Geochemistry. Wiley, Oxford.

Chapelle, F.H., Zelibor Jr., J.L., Grimes, D.J., Jay, D., Knobel, L.L., 1987. Bacteria in deep coastal plain sediments of Maryland: a possible source of CO_2 to groundwater. Water Resour. Res. 23, 1625–1632.

Christensen, T.H., Kjeldsen, P., Bjerg, P.L., Jensen, D.L., Christensen, J.B., Baun, A., et al., 2001. Biogeochemistry of landfill leachate plumes. Appl. Geochem. 16, 659–718.

Coates, J.D., Chakraborty, R., Lack, J.G., O'Connor, S.M., Cole, K.A., Bender, K.S., et al., 2001. Anaerobic benzene oxidation coupled to nitrate reduction in pure culture by two strains of *Dechloromonas*. Nature 411, 1039–1043.

Dagley, S., 1985. Microbial metabolism of aromatic compounds. In: Moo-Young, M. (Ed.), The Principles of Biotechnology. Pergamon, Oxford, pp. 483–505.

Danielopol, D.L., Griebler, C., Gunatilaka, A., Notenboom, J., 2003. Present state and future prospects for groundwater ecosystems. Environ. Conserv. 30, 104–130.

DeSantis, T.Z., Brodie, E.L., Moberg, J.P., Zubieta, I.X., Piceno, Y.M., Andersen, G.L., 2007. High-density universal 16S rRNA microarray analysis reveals broader diversity

than typical clone library when sampling the environment. Microbiol. Ecol. 53, 371–383.
DeWeerd, K., Mandelco, L., Tanner, R., Woese, C., Suflita, J., 1990. *Desulfomonile tiedjei* gen. nov. and sp. nov., a novel anaerobic, dehalogenating, sulfate-reducing bacterium. Arch. Microbiol. 154, 23–30.
Dolfing, J., Beurskes, J.E.M., 1995. The microbial logic and environmental significance of reductive dehalogenation. Adv. Microb. Ecol. 14, 143–206.
Dong, H., 2008. Microbial life in extreme environments: linking geological and microbiological processes. In: Dilek, Y., Furnes, H., Muehlenbachs, K. (Eds.), Links Between Geological Processes, Microbial Activities, and Evolution of Life: Microbes and Geology. Springer, Heidelberg, pp. 237–280.
Enwall, K., Hallin, S., 2009. Comparison of T-RFLP and DGGE techniques to assess denitrifier community composition in soil. Lett. Appl. Microbiol. 48, 145–148.
Fakruddin, M.D., Chowdhury, A., Hossain, M.N., Bin Mannan, K.S., Mazumdar, R.M., 2012. Pyrosequencing—principles and applications. Int. J. Life Sci. Pharm. Res. 2, L65–L76.
Fitts, C.R., 2002. Groundwater Science. Academic Press, New York, NY.
Fliermans, C.B., Franck, M.M., Hazen, T.C., Gorden, R.W., 1997. Ecofunctional enzymes of microbial communities in ground water. FEMS Microbiol. Rev. 20, 379–389.
Garland, J.L., 1997. Analysis and interpretation of community-level physiological profiles in microbial ecology. FEMS Microbiol. Ecol. 24, 289–300.
Garland, J.L., Mills, A.L., 1991. Classification and characterization of heterotrophic microbial communities on the basis of patterns of community-level sole-carbon-source utilization. Appl. Environ. Microbiol. 57, 2351–2359.
Gleeson, D.B., Muller, C., Banerjee, S., Ma, W., Siciliano, S.D., Murphy, D.V., 2010. Response of ammonia oxidizing archaea and bacteria to changing water filled pore space. Soil Biol. Biochem. 42, 1888–1891.
Gliesche, C.G., 1999. Die Mikrobiologie des Grundwasserraumes und der Einfluß anthropogener Veränderungen auf die mikrobielle Lebensgemeinschaften. Umweltbundesamt, Berlin.
Gold, T., 1992. The deep, hot biosphere. Proc. Natl. Acad. Sci. USA 89, 6045–6049.
Goldscheider, N., Hunkeler, D., Rossi, P., 2006. Review: microbial biocenoses in pristine aquifers and an assessment of investigative methods. Hydrogeol. J. 14, 926–941.
Gosselin, D.C., Smith, M.R., Lepel, E.A., Laul, J.C., 1992. Rare earth elements in chloride-rich groundwater, Palo Duro Basin, Texas, USA. Geochim. Cosmochim. Acta 56, 1495–1505.
Gottlich, E., van der Lubbe, W., Lange, B., Fiedler, S., Melchert, I., Reifenrath, M., et al., 2002. Fungal flora in groundwater-derived public drinking water. Int. J. Hyg. Environ. Health 205, 269–279.
Griebler, C., Lueders, T., 2009. Microbial biodiversity in groundwater ecosystems. Freshwater Biol. 54, 649–677.
Griebler, C., Mösslacher, F., 2003. Grundwasser-Ökologie. Facultas, Wien.
Griebler, C., Mindl, B., Slezak, D., Geiger-Kaiser, M., 2002. Distribution patterns of attached and suspended bacteria in pristine and contaminated shallow aquifers studied with an *in situ* sediment exposure microcosm. Aquat. Microbiol. Ecol. 28, 117–129.
Gruber, C., Legat, A., Pfaffenhuemer, M., Radax, C., Weidler, G., Busse, H.J., et al., 2004. *Halobacterium noricense* sp. nov., an archaeal isolate from a bore core of an alpine Permian salt deposit, classification of *Halobacterium* sp. NRC-1 as a strain of *H. salinarum* and emended description of *H. salinarum*. Extremophiles 8, 431–439.

Haveman, S.A., Pedersen, K., 2002. Distribution of culturable microorganisms in Fennoscandian Shield groundwater. FEMS Microbiol. Ecol. 39, 129–137.

Hendrickson, E.R., Payne, J.A., Young, R.M., Starr, M.G., Perry, M.P., Fahnestock, S., et al., 2002. Molecular analysis of *Dehalococcoides* 16S ribosomal DNA from chloroethene-contaminated sites throughout North America and Europe. Appl. Environ. Microbiol. 68, 485–495.

Heuer, H., Smalla, K., 1997. Evaluation of community-level catabolic profiling using BIOLOG GN microplates to study microbial community changes in potato phyllosphere. J. Microbiol. Methods 30, 49–61.

Hölting, B., 1996. Hydrogeologie: Einführung in die Allgemeine und Angewandte Hydrogeologie. Enke, Stuttgart.

Hunkeler, D., Goldscheider, N., Rossi, P., Burn, C., 2006. Biozonosen im Grundwasser—Grundlagen und Methoden der Charakterisierung von mikrobiellen Gemeinschaften. Umwelt-Wissen Nr. 0603. Bundesamt für Umwelt, Bern.

Johannesson, K.H., Hendry, M.J., 2000. Rare earth element geochemistry of groundwaters from a thick till and clay-rich aquitard sequence, Saskatchewan, Canada. Geochim. Cosmochim. Acta 64, 1493–1509.

Johannesson, K.H., Stetzenbach, K.J., Hodge, V.F., 1997. Rare earth elements as geochemical tracers of regional groundwater mixing. Geochim. Cosmochim. Acta 61, 3605–3618.

King, L.K., Parker, B.C., 1988. A simple, rapid method for enumerating total viable and metabolically active bacteria in groundwater. Appl. Environ. Microbiol. 54, 1630–1631.

Kölbel-Boelke, J., Anders, E.M., Nehrkorn, A., 1988. Microbial communities in the saturated groundwater environment II: diversity of bacterial communities in a Pleistocene sand aquifer and their *in vitro* activities. Microb. Ecol. 16, 31–48.

Kostakioti, M., Hadjifrangiskou, M., Hultgren, S.J., 2013. Bacterial biofilms: development, dispersal, and therapeutic strategies in the dawn of the postantibiotic era. Cold Spring Harb. Perspect. Med. 3, a010306.

Kotelnikova, S., Pedersen, K., 1998. Distribution and activity of methanogens and homoacetogens in deep granitic aquifers at Aspo Hard Rock Laboratory, Sweden. FEMS Microbiol. Ecol. 26, 121–134.

Krumholz, L.R., 2000. Microbial communities in the deep subsurface. Hydrogeol. J. 8, 4–10.

Krumholz, L.R., McKinley, J.P., Ulrich, G.A., Suflita, J.M., 1997. Confined subsurface microbial communities in cretaceous rock. Nature 386, 64–66.

Langenhoff, A.A.M., Brouwers-Ceiler, D.L., Engelberting, J.H.L., Quist, J.J., Wolkenfelt, J.G.P.N., Zehnder, A.J.B., et al., 1997. Microbial reduction of manganese coupled to toluene oxidation. FEMS Microbiol. Ecol. 22, 119–127.

Lendvay, J.M., Loffler, F.E., Dollhopf, M., Aiello, M.R., Daniels, G., Fathepure, B.Z., et al., 2003. Bioreactive barriers: a comparison of bioaugmentation and biostimulation for chlorinated solvent remediation. Environ. Sci. Technol. 37, 1422–1431.

Lovley, D.R., 1997. Potential for anaerobic bioremediation of BTEX in petroleum-contaminated aquifers. J. Ind. Microbiol. Biotechnol. 18, 75–81.

Lovley, D.R., Chapelle, F.H., 1995. Deep subsurface microbial processes. Rev. Geophys. 33, 365–381.

Lucia, F.J., 1995. Rock-fabric/petrophysical classification of carbonate pore space for reservoir characterization. AAPG Bull. 79, 1275–1300.

Lynch, J.M., Moffat, A.J., 2005. Bioremediation—prospects for the future application of innovative applied biological research. Ann. Appl. Biol. 146, 217–221.
Madsen, E.L., Ghiorse, W.C., 1993. Groundwater microbiology. In: Ford, T.E. (Ed.), Aquatic Microbiology: An Ecological Approach. Blackwell, Cambridge, pp. 167–213.
Mahananda, M.R., Mohanty, B.P., Behera, N.R., 2010. Physico-chemical analysis of surface and ground water of Bargarh district, Orissa, India. Int. J. Res. Rev. Appl. Sci. 2, 284–295.
Malard, F., Hervant, F., 1999. Oxygen supply and the adaptations of animals in groundwater. Freshwater Biol. 41, 1–30.
Maymo-Gatell, X., Chien, Y.T., Gossett, J.M., Zinder, S.H., 1997. Isolation of a bacterium that reductively dechlorinates tetrachloroethene to ethene. Science 276, 1568–1571.
McGenity, T.J., Gemmell, R.T., Grant, W.D., Stan-Lotter, H., 2000. Origins of halophilic microorganisms in ancient salt deposits. Environ. Microbiol. 2, 243–250.
McMahon, P.B., Chapelle, F.H., Falls, W.F., Bradley, P.M., 1992. Role of microbial processes in linking sandstone diagenesis with organic rich clays. J. Sediment. Petrol. 62, 1–10.
Meckenstock, R.U., 1999. Fermentative toluene degradation in anaerobic defined syntrophic cocultures. FEMS Microbiol. Lett. 177, 67–73.
Neff, J.C., Asner, G.P., 2001. Dissolved organic carbon in terrestrial ecosystems: synthesis and a model. Ecosystems. 4, 29–48.
Nelson, D., 2002. Natural Variations in the Composition of Groundwater. Groundwater Foundation Annual Meeting, Springfield, OR.
Neufeld, J.D., Dumont, M.G., Vohra, J., Murrell, J.C., 2007. Methodological considerations for the use of stable isotope probing in microbial ecology. Microb. Ecol. 53, 435–442.
Niibel, U., Antwerpen, M., Strommenger, B., Witte, W., 2006. DNA microarrays for bacterial genotyping. In: Stackebrandt, E. (Ed.), Molecular Identification, Systematics and Population Structure of Prokaryotes. Springer, Berlin, pp. 387–414.
Ommedal, H., Torsvik, T., 2007. *Desulfotignum toluenicum* sp. nov., a novel toluene-degrading, sulphate-reducing bacterium isolated from an oil-reservoir model column. Int. J. Syst. Evol. Microbiol. 57, 2865–2869.
Patel, A., Noble, R.T., Steele, J.A., Schwalbach, M.S., Hewson, I., Fuhrman, J.A., 2007. Virus and prokaryote enumeration from planktonic aquatic environments by epifluorescence microscopy with SYBR Green I. Nat. Protoc. 2, 269–276.
Peter, B.M., Francis, H.C., 1991. Microbial production of organic acids in aquitard sediments and its role in aquifer geochemistry. Nature 349, 233–235.
Preston-Mafham, J., Boddy, L., Randerson, P.F., 2002. Analysis of microbial community functional diversity using sole-carbon-source utilisation profiles—a critique. FEMS Microbiol. Ecol. 42, 1–14.
Pronk, M., Goldscheider, N., Zopfi, J., 2009. Microbial communities in karst groundwater and their potential use for biomonitoring. Hydrogeol. J. 17, 37–48.
Qin, Z., Ou, Y., Yang, L., Zhu, Y., Tolker-Nielsen, T., Molin, S., et al., 2007. Role of autolysin-mediated DNA release in biofilm formation of *Staphylococcus epidermidis*. Microbiology 153, 2083–2092.
Rabus, R., Widdel, F., 1995. Anaerobic degradation of ethylbenzene and other aromatic hydrocarbons by new denitrifying bacteria. Arch Microbiol. 163, 96–103.
Ramakrishnan, B., 2013. Anaerobic/aerobic microbial degraders: game changers. J. Bioremed. Biodegrad. 4, e136.
Rasheed, M.A., Patil, D.J., Dayal, A.M., 2013. Microbial techniques for hydrocarbon exploration. In: Kutcherov, V. (Ed.), Hydrocarbon. InTech. Available from: http://dx.doi.org/doi:10.5772/50885.

Rastogi, G., Sani, R.K., 2011. Molecular techniques to assess microbial community structure, function and dynamics in the environment. In: Ahmad, I., Ahmad, F., Pichtel, J. (Eds.), Microbes and Microbial Technology: Agricultural and Environmental Applications. Springer, Heidelberg, pp. 29–57.

Rastogi, G., Osman, S., Vaishampayan, P.A., Andersen, G.L., Stetler, L.D., Sani, R.K., 2010. Microbial diversity in uranium mining-impacted soils as revealed by high-density 16S microarray and clone library. Microbiol. Ecol. 59, 94–108.

Reasoner, D.J., Geldreich, E.E., 1985. A new medium for the enumeration and subculture of bacteria from potable water. Appl. Environ. Microbiol. 49, 1–7.

Rebata-Landa, V., Santamarina, J.C., 2006. Mechanical limits to microbial activity in deep sediments. Geochem. Geophys. Geosyst. 7, Q11006.

Rebollo, M.A., Hogert, E.N., Albano, J., Raffo, C.A., Gaggioli, N.G., 1996. Correlation between roughness and porosity in rocks. Opt. Laser Technol. 28, 21–23.

Rheinheimer, G., 1971. Mikrobiologie der Gewässer. Gustav Fischer, Stuttgart.

Rogers, D.R., Casciotti, K.L., 2010. Abundance and diversity of archaeal ammonia oxidizers in a coastal groundwater system. Appl. Environ. Microbiol. 76, 7938–7948.

Scholz-Muramatsu, H., Neumann, A., Mebmer, M., Moore, E., Diekert, G., 1995. Isolation and characterization of *Dehalospirillum multivorans* gen. nov., sp. nov., a tetrachloroethene-utilizing, strictly anaerobic bacterium. Arch. Microbiol. 163, 48–56.

Schumacher, W., Holliger, C., 1996. The proton/electron ration of the menaquinone-dependent electron transport from dihydrogen to tetrachloroethene in *Dehalobacter restrictus*. J. Bacteriol. 178, 2328–2333.

Scow, K.M., Hicks, K.A., 2005. Natural attenuation and enhanced bioremediation of organic contaminants in groundwater. Curr. Opin. Biotechnol. 16, 246–253.

Singh, A., Singh, B., Ward, O., 2012. Potential applications of bioprocess technology in petroleum industry. Biodegradation. 23, 865–880.

Smedley, P.L., 1991. The geochemistry of rare earth elements in groundwater from the Carnmellis area, southwest England. Geochim. Cosmochim. Acta 55, 2767–2779.

Stambuk-Giljanovic, N., 2005. The quality of water resources in Dalmatia. Environ. Monit. Assess. 104, 235–268.

Stan-Lotter, H., McGenity, T.J., Legat, A., Denner, E.B.M., Glaser, K., Stetter, K.O., et al., 1999. Very similar strains of *Halococcus salifodinae* are found in geographically separated Permo-Triassic salt deposits. Microbiology 145, 3565–3574.

Stevens, T.O., McKinley, J.P., Fredrickson, J.K., 1993. Bacteria associated with deep, alkaline, anaerobic groundwaters in southeast Washington. Microb. Ecol. 25, 35–50.

Stroes-Gascoyne, S., Schippers, A., Schwyn, B., Poulain, S., Sergeant, C., Simonoff, M., et al., 2007. Microbial community analysis of Opalinus Clay drill core samples from the Mont Terri Underground Research Laboratory, Switzerland. Geomicrobiol. J. 24, 1–17.

Suthar, S., 2010. Contaminated drinking water and rural health perspectives in Rajasthan, India: an overview of recent case studies. Environ. Monit. Assess. 173, 837–849.

Tóth, J., 1963. A theoretical analysis of groundwater flow in small drainage basin. J. Geophys. Res. 68, 4795–4812.

Ulrich, G.A., Martino, D., Burger, K., Routh, J., Grossman, E.L., Ammerman, J.W., et al., 1998. Sulfur cycling in the terrestrial subsurface: commensal interactions, spatial scales, and microbial heterogeneity. Microb. Ecol. 36, 141–151.

Ultee, A., Souvatzi, N., Maniadi, K., Konig, H., 2004. Identification of the culturable and nonculturable bacterial population in ground water of a municipal water supply in Germany. J. Appl. Microbiol. 96, 560–568.

van der Wielen, P.W., Voost, S., van der Kooij, D., 2009. Ammonia-oxidizing bacteria and archaea in groundwater treatment and drinking water distribution systems. Appl. Environ. Microbiol. 75, 4687–4695.

von Wintzingerode, F., Gobel, U.B., Stackebrandt, E., 1997. Determination of microbial diversity in environmental samples: pitfalls of PCR-based rRNA analysis. FEMS Microbiol. Rev. 21, 213–229.

Vreeland, R.H., Rosenzweig, W.D., Powers, D.W., 2000. Isolation of a 250 million-year-old halotolerant bacterium from a primary salt crystal. Nature 407, 897–900.

Wagner, M., Wagner, M., Piske, J., Smit, R., 2002. Case histories of microbial prospection for oil and gas, onshore and offshore in northwest Europe. SEG Geophys. Ref. Ser. 11, 453–479.

Wagner, M., Horn, M., Daims, H., 2003. Fluorescence *in situ* hybridisation for the identification and characterisation of prokaryotes. Curr. Opin. Microbiol. 6, 302–309.

Weelink, S.B., Eekert, M.A., Stams, A.M., 2010. Degradation of BTEX by anaerobic bacteria: physiology and application. Rev. Environ. Sci. Biotechnol. 9, 359–385.

Whitchurch, C.B., Tolker-Nielsen, T., Ragas, P.C., Mattick, J.S., 2002. Extracellular DNA required for bacterial biofilm formation. Science 295, 1487.

Williamson, W.M., Close, M.E., Leonard, M.M., Webber, J.B., Lin, S., 2012. Groundwater biofilm dynamics grown *in situ* along a nutrient gradient. Groundwater 50, 690–703.

Yoshida, H., Yamamoto, K., Yogo, S., Murakami, Y., 2006. An analogue of matrix diffusion enhanced by biogenic redox reaction in fractured sedimentary rock. J. Geochem. Explor. 90, 134–142.

Zak, J.C., Willig, M.R., Moorhead, D.L., Wildman, H.G., 1994. Functional diversity of microbial communities: a quantitative approach. Soil Biol. Biochem. 26, 1101–1108.

Zouboulis, A.I., Katsoyiannis, I.A., 2005. Recent advances in the bioremediation of arsenic-contaminated groundwaters. Environ. Int. 31, 213–219.

26 Exploring Prospects of Monooxygenase-Based Biocatalysts in Xenobiotics

Kashyap Kumar Dubey[a], *Punit Kumar*[a], *Puneet Kumar Singh*[b] *and Pratyoosh Shukla*[b,*]

[a]Microbial Biotechnology Laboratory, Department of Biotechnology, University Institute of Engineering and Technology, Maharshi Dayanand University, Rohtak, Haryana, India, [b]Enzyme Technology and Protein Bioinformatics Laboratory, Department of Microbiology, Maharshi Dayanand University, Rohtak, Haryana, India

26.1 Introduction

Xenobiotics ("stranger to life") belong to a division of chemical compounds that are foreign in origin but are present in the environment and in living organisms. The origin of these compounds is anthropogenic (Figure 26.1) and every year thousands more xenobiotics are added to the environment. For the most part, these are not produced naturally, but their consumption makes them available to living systems. Production of xenobiotics is essential for humanity as many chemical and pharmaceutical industries use such compounds to produce medicines, plastic items, detergents, gels, biochemical kits, perfumes, laboratory research chemicals, food supplements and additives, herbicides, pesticides, and environmental pollutants (Rieger et al., 2002). Although these compounds are processed and designed to meet consumer demand, they undergo a series of chemical processes that change their property and make them consumer friendly. But these chemical processes change their physiochemical nature and eventually provide complex chemical structure which made them recalcitrance. Xenobiotics compounds are nonbiodegradable or partially biodegradable, or undergo biotransformation process and persist in the environment for long time.

Organic and inorganic xenobiotic compounds are present as contaminants in the Earth's subsurface through dumping and usage; in water through sewage and industrial discharge; and, in the air as air pollutant. Concentration of these persistent organic pollutants increases as they pass through the food chain from one tropic level to another; this process is known as biomagnification.

*Corresponding Author: pratyoosh.shukla@gmail.com

Microbial Biodegradation and Bioremediation. DOI: http://dx.doi.org/10.1016/B978-0-12-800021-2.00026-1

Figure 26.1 Addition of monooxygenases into environment.

26.1.1 Physiochemical Nature of Xenobiotics

Xenobiotics possess complex chemical structures and many nonphysiological bonds. Some of these structures have little relationship to the natural compounds from which they originate. These compounds are of high molecular mass, are insoluble or less soluble in water, have condensed aromatic rings and polycyclic structures, and are aliphatic compounds with high substitution by new groups. They are polycyclic aromatic hydrocarbons, halogen substituted aliphatic and aromatic hydrocarbons, aromatic amines, nitroaromatic compounds, azo compounds, *s*-traizines, organic sulfonic acids and synthetic polymers, and free radicals (Southorn and Powis, 1988; Spain, 1995).

26.1.2 Diseases Caused by Xenobiotics

Most xenobotics are harmful to humans. More than 20,000 compounds are classified as harmful or toxic and are associated with certain disorders (Figure 26.2), such as psychological disorders, digestive disorders, and metabolic disorders. For example, nitrosamines and aflatoxin which disrupt/modify endocrine system, estrogenic system, and embryogenic system during pregnancy. Xenobiotic compounds also caused mutagenic effects, allergies, carcinogenic effects (benzedrine, vinyl chloride), and genetic disorders. These effects are based on the nature of the compound, the time

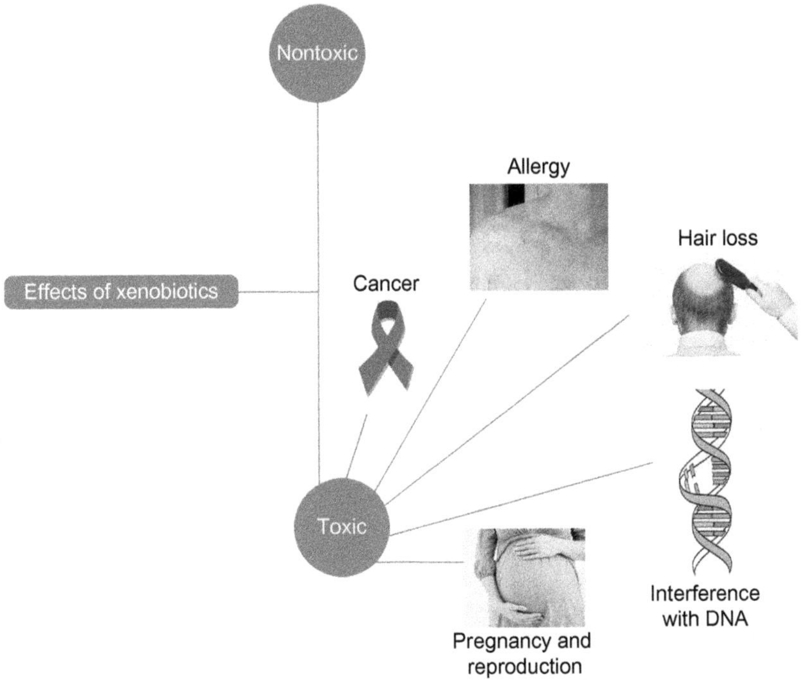

Figure 26.2 Diseases created by xenobiotics.

of exposure, and the dose. Certain xenobiotics are not toxic but after enzymatic modification they become toxic like tetrachloroethene and trichloroethene are converted into vinyl chloride that is carcinogen. Since 1971, the IARC has evaluated more than 900 xenobiotic compounds of which more than 400 have been identified as carcinogenic, probably carcinogenic, or possibly carcinogenic to humans (IARC).

26.1.3 Biodegradation of Xenobiotics

Nature has blessed us with the microorganisms which has faced the complete process of evolution and thus have capability to adapt in different physicochemical environment. A vast number of bacteria and fungi have been identified as having the capability to degrade organic waste. Biodegradation is the biologically catalyzed reduction process in complexity of chemical compounds (Alexander, 1994). Numbers of bacteria and fungi have been identified as having the capability to degrade organic waste (Lah et al., 2011). Generally, it is composed of two processes: growth and co-metabolism. In growth, microbes use organic pollutants as a source of carbon, nitrogen, and energy, which causes complete biodegradation (mineralization) of organic pollutants (Figure 26.3). This degradation of the majority of compounds takes place in aerobic conditions (Riser-Roberts, 1998). Co-metabolism is the metabolism of an organic compound in the presence of a growth substrate which is used as the primary carbon and energy source.

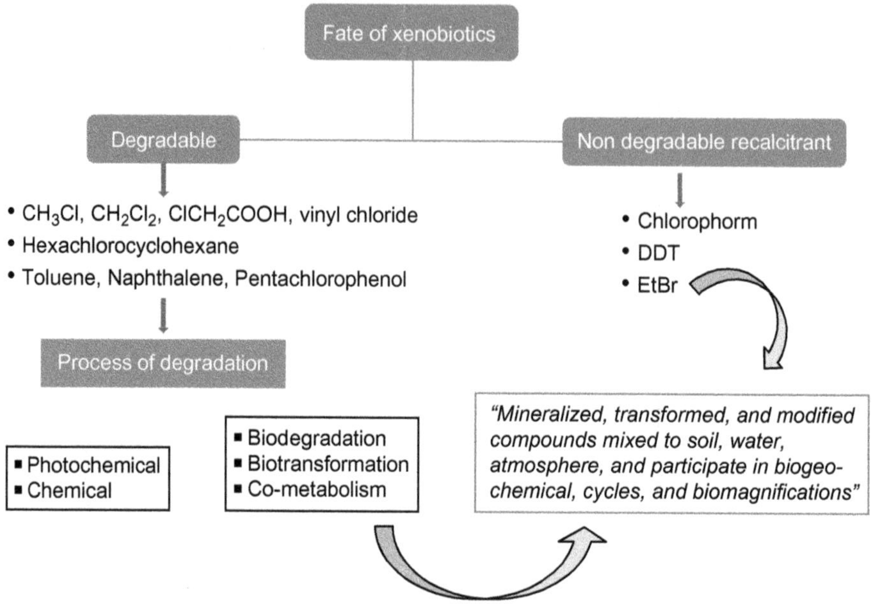

Figure 26.3 Fate of xenobiotics.

There are a vast range of enzymes present in microbes that enable them to use different chemicals as nutrients, as sources of carbon and nitrogen, or as electron acceptors, and which ultimately cause mineralization of these compounds or conversion into less complex compounds. When xenobiotics are derived from natural compounds that enzymes have faced during evolution, then these are easily targeted by enzymes. But complex molecules which are completely synthetic and possess complex chemical structures are not easily targeted by enzymes. Such complex structures undergo enzyme-assisted biotransformation processes in which their structures and physiochemical properties are altered by enzymes. During this process, a functional group may be added to a molecule by which it becomes susceptible to enzymes, as well as less toxic and more water soluble.

26.1.4 Biodegradation and Enzymes

Thus it can be understood that enzymes play a very important role in the biogeochemical cycles of elements. They are solely responsible for elimination of xenobiotics from the body or the environment through biodegradation/biotransformation processes. These enzymes have been isolated and identified in various microorganisms. These enzyme-assisted metabolic processes are known as detoxification. With the advent of new generation of xenobiotics with new physiochemical nature and structure researchers have to explore the variety of enzymes to get rid of these xenobiotics or to convert them into less harmful. These days industries are eying on these enzymes as for breakdown of xenobiotics and commercialization of these enzymes. Various research groups are engaged in study of characterization, molecular structure, domain structure, substrate specificity, optimum conditions, cofactor requirements in these enzymes. Which would be useful to convert them with broad substrate range and biotechnological applications. Thus, enzymes may have both an environmental and an economic benefit.

26.1.4.1 *Oxygenases*

The oxygenases are a superfamily of enzymes that are responsible for the metabolic incorporation of xenobiotics in a regioselective and stereoselective manner so that they undergo biodegradation or biotransformation. These enzymes are responsible for the oxidation of compounds and for creating such changes in molecules that enable them to become more water soluble or less toxic, and susceptible to other enzymes in metabolic processes. These are the category of enzymes that first interact with xenobiotics.

Based on the type of reaction they catalyze, oxidoreductases have been divided into 22 different EC-subclasses. Xu (2005) proposed classification of these oxidoreductases into four subgroups: (i) oxidases, (ii) peroxidases, (iii) oxygenases/hydroxylases, and (iv) dehydrogenases/reductases. The majority of oxidoreductases require cofactors in their catalytic activity. These cofactors are of various types such as flavins, metal ions, hemes, and pyrroloquinoline quinone (PQQ) for their catalytic activity, electron transporters (reductases), NADH, NADPH (Figure 26.4), and oxygen in molecular form, i.e., O_2. These enzymes are grouped into many protein families depending on utilization of oxygen, and broadly grouped into monooxygenases and dioxygenases (Cirino and Arnold, 2002).

$$R + NAD(P)H + O_2 + H^+ = RO + NAD(P)^+ + H_2O \text{ (Monooxygenases)}$$

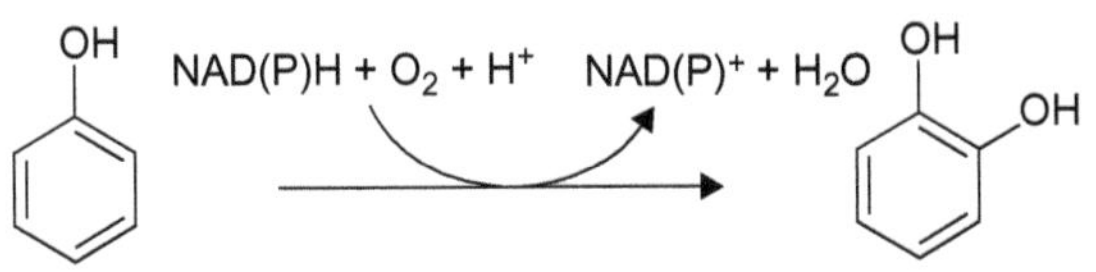

$$R + NAD(P)H + O_2 + H^+ = R(OH)_2 + NAD(P)^+ \text{ (Dioxygenases)}$$

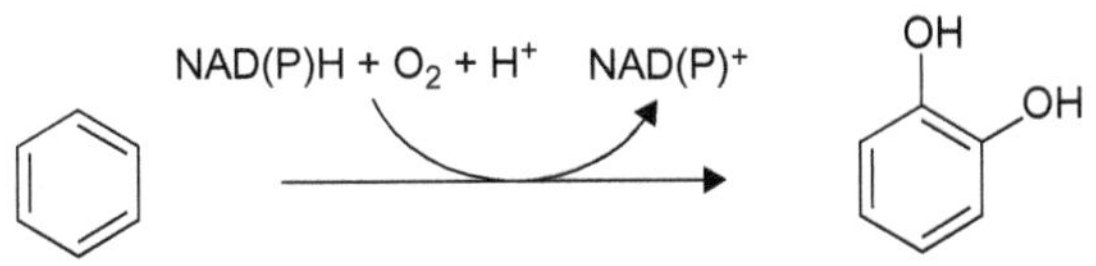

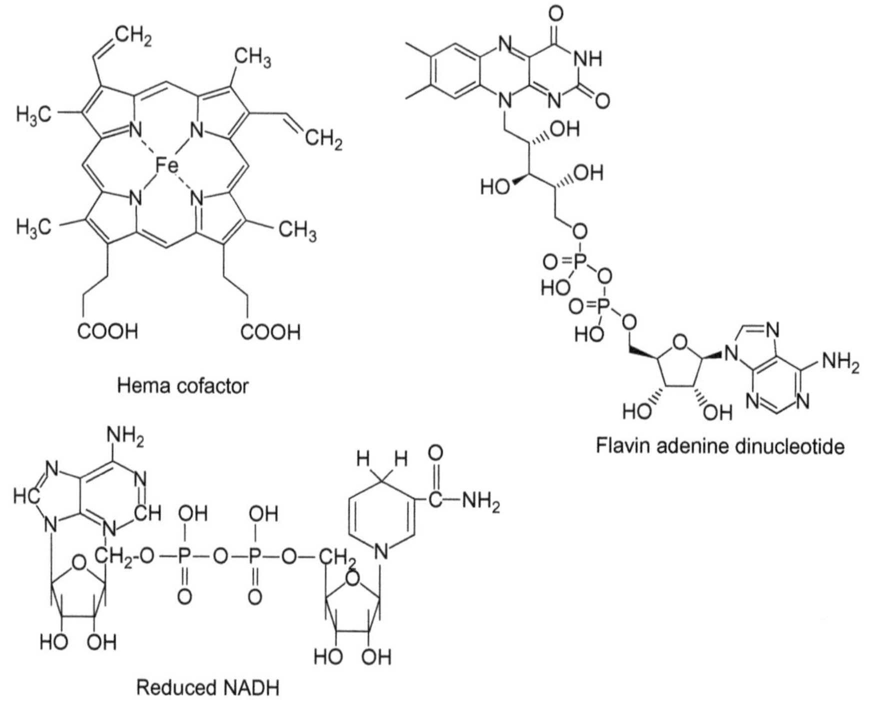

Figure 26.4 Cofactors used by monooxygenases.

In this chapter, we will discuss the capacity of monooxygenases as biocatalysts for the metabolism of xenobiotics.

From the above catalytic reactions, it is clear that O_2 is required in each process, and that in the company of oxygen, NADH or NADPH are used for the reduction/oxidation process with proteins that assist in electron transfer, such as reductases. The key chemical reactions catalyzed by these oxidative enzymes are hydroxylation, epoxidation, and sulfonation with a broad range of substrates including lipids and steroids. These reactions are also the base for the synthesis of many chemicals (Cirino and Arnold, 2002).

26.1.4.2 Monooxygenases

Monooxygenases (EC 1.13. and EC 1.14.) catalyze the transfer of single oxygen atoms into organic substrates. To carry out this catalytic activity, monooxygenases first stimulate molecular oxygen, transferring an electron to a molecule of oxygen, which is followed by oxidation of organic substrate. This activated oxygen is called reactive oxygen intermediate and it is dependent on a cofactor that is in association with monooxygenase. Monooxygenase present in internal positions acquires electrons from the substrate directly without the help of cofactor (Van Berkel et al., 2006).

On the basis of cofactor requirement for catalytic activity, monooxygenases are grouped into the following different families (Table 26.1).

1. Cofactor-dependent monooxygenases
   1.1. Heme-dependent monooxygenases
   1.2. Flavin-dependent monooxygenases
   1.3. Copper-dependent monooxygenases
   1.4. Nonheme iron-dependent monooxygenases
   1.5. Pterin-dependent monooxygenases
   1.6. Other cofactor-dependent monooxygenases
2. Cofactor-independent monooxygenases

26.1.4.2.1 Cofactor-Dependent Monooxygenases

These types of monooxygenases require a cofactor for transfer of electrons. These are classified on the basis of the cofactor required in the following sections.

26.1.4.2.1.1 Heme-Dependent Monooxygenases Heme-containing monooxygenases are cytochrome P450 monooxygenases (CYPs) which are present in almost all living organisms—mammals, plants, fungi, and bacteria (Werck-Reichhart and Feyereisen, 2000). These enzymes were discovered few decades ago. These CYPs (EC 1.14.13., EC 1.14.14., and EC 1.14.15.) are moderately abundant.

CYPs catalyze the stereo- and region-specific hydroxylation of inactivated hydroxycarbons under serene conditions and are very useful for industrial applications, and for the biotransformation of food and pharmaceutical products with high selectivity. P450s have shown potential for bioremediation of pollutants, as these enzymes can renovate chemically inert compounds to more water-soluble, hydroxylated derivatives, which could be appropriate substrates for several other enzymes (Liu et al., 2006). Bernhardt (2006) has classified CYPs into four classes on the basis of their electron transport systems: class I, class II, class III, and class IV. Class I systems possess three components: a ferredoxin (Fd), heme, and one NADH-dependent, FAD-containing ferredoxin reductase (FdR). Class I also contains soluble bacterial CYPs and membrane-bound mitochondrial CYPs. Class II CYPs possess two components: a heme domain and the reductase unit (FAD/FMN); both of these components are membrane bound. Such CYPs are frequent in eukaryotes. In class III CYPs, the components are the same as those of as class II systems, but both are situated on a single polypeptide chain. Because of this remarkable feature, there is an increased electron transport rate which further increases reaction rates. Class III CYPs are present in both eukaryotes and prokaryotes and may be membrane bound or soluble. Class IV CYPs consist of components similar to those of class I, but all the components are situated on one single polypeptide chain as in class III CYPs.

Heme b is a prosthetic group consisting of an iron ion and is coordinated by four nitrogen atoms of porphyrin. It is bounded by electrostatic and hydrogen bonds and participates in the formation of the active sites of CYPs. Structurally it is deep inside the three-dimensional structure with surrounding residues (Li and Poulos, 1997). Besides this, some members of the CYP4 subfamily of mammalian P450

Table 26.1 Classification of Monooxygenases with Their Representative Members

Monooxygenase	Examples	EC Classification	Origin	Xenobiotic Substrate
Heme-dependent monooxygenases	Cytochrome P450 BM3	1.14.13.- 1.14.14- 1.14.15.-	Prokaryotic and eukaryotic	• Alkanes • Fatty acids • Aromatic compounds
Flavin-dependent monooxygenases	• Luciferases • Type II Baeyer–Villiger monooxygenases • Salicylate 1-monooxygenase • Flavin-containing monooxygenases • *p*-hydroxybenzoate hydroxylase • Lactate monooxygenase • Salicylate 1-monooxygenase • Nitronate monooxygenases • Styrene monooxygenase • Hydroxybiphenyl 3-monooxygenase • Type I Baeyer–Villiger monooxygenases	1.13.12.- 1.14.13.-	Prokaryotic	• Long-chain aliphatic aldehydes • Aromatic ring • Styrene • Phenols
Copper-dependent monooxygenases	• Dopamine β-monooxygenase • Tyramine β monooxygenase • Peptidylglycine α hydroxylating • Monooxygenase (PHM) • Monooxygenase X • Methane monooxygenase • (pMMO) • Ammonia monooxygenase • Tyrosinase	1.14.17.- 1.14.18.1	Eukaryotic and prokaryotic	• Dopamine • Methane • Ammonia

Nonheme iron-dependent monooxygenases	• Methane monooxygenase • (sMMO) • Alkene monooxygenase • Phenol hydroxylases • Alkane hydroxylases • Toluene 4-monooxygenase • Toluene 2-monooxygenase • Xylene monooxygenase • Alkane ω hydroxylase		Prokaryotic sulfoxidation	• Hydrocarbon • Phenol and derivatives • Methane • Chlorinated solvents
Pterin-dependent monooxygenases.	• Phenylalanine • 4-Monooxygenase • Anthranilate 3-monooxygenase • Tyrosine 3-monooxygenase • Glyceryl ethyl • Monooxygenase	1.14.16.-	Prokaryotic and eukaryotic Biotransformations using whole cells	• Aliphatic substrates • Aromatic ring amino acids
Other cofactor-dependent monooxygenases	• Aclacinomycin 10-hydroxylase • *ortho*-Nitrophenol 2-monooxygenase	1.14.13.31	Prokaryotic	15-Demethoxy-rhodomycin
Cofactor-independent monooxygenases	• Tetracenomycin F1 monooxygenase • ActVA-orf6 • Quinol monooxygenase		Prokaryotic	Quinone derivatives

monooxygenases contain a covalent attachment and an ester bond to fix the heme b group. In all CYPs, a highly conserved cysteine residue present in the active site behaves as the fifth ligand of the heme iron center, thus stimulating the metal complex. CYPs catalyze a large range of oxidations reactions such as aromatic and aliphatic hydroxylation, double bond epoxidation, dealkylation, oxidative phenolic coupling, and multiple oxidation including carbon–carbon bond cleavage. With this, they are also able to carry out many other reactions such as hetero atom-dealkylations and oxidations, oxidative deaminations, dehalogenations, dehydrogenations, dehydratations, and reductions.

The Catalytic Mechanism of P450 Monooxygenases

Denisov et al. (2005) explained the reaction mechanism of action of CYPs. The reaction mechanism of CYP450A contains many reaction intermediates and a series of reduction and oxidation of iron and intermediates. In this reaction mechanism, first the binding of substrates to the enzyme takes place, leading to reduction of iron from Fe^{3+} to Fe^{2+}. After this, binding of the oxygen molecule takes place and an oxo-P450 complex is formed. In the oxo-P450 complex, Fe^{2+} is oxidized to Fe^{3+} and the terminal oxygen atom of the oxygen molecule gets negatively charged. This complex is further reduced to form a peroxo-ferric intermediate and is protonated to hydroperoxo-ferric intermediate. This is followed by second protonation at terminal oxygen and there is lysis of the O–O bond and removal of water with the formation of a new ferryl-oxo compound and with oxidation of the substrate.

In this reaction mechanism (Figure 26.5), three branch points (shunt) also exist that display a site for side reactions. These branch points are autooxidation shunt of the oxy-P450 complex, producing superoxide anions and resting state enzymes; a peroxide shunt at hydroperoxo-stage that returns intermediate to resting enzyme

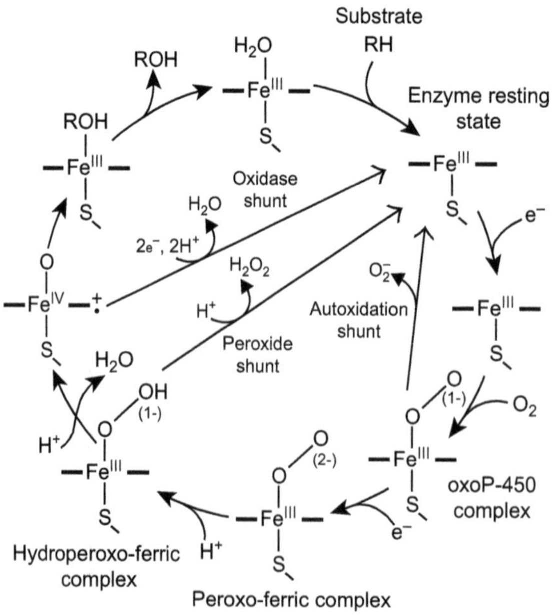

Figure 26.5 Reaction mechanism of CYP450. *Source*: Modified from Denisov et al. (2005).

forming hydrogen peroxide, and an oxidase shunt oxidizing ferryl-oxo intermediate to water without oxygenation of the substrate. These branch points are collectively referred as uncoupling and intermediate reactions to enzyme resting stage.

26.1.4.2.1.2 Flavin-Dependent Monooxygenases Flavin-dependent monooxygenases are more abundant in prokaryotes compared to CYPs, as the genetic machinery for these enzymes seems to be quite rich in prokaryotes. In contrast, many genes are found in eukaryotes; for example, *Arabidopsis thaliana* contains 29 gene homologs of flavin-containing monooxygenases. These enzymes utilize either FMN or FAD as prosthetic groups and bind strongly to the enzyme; they sometimes function as a substrate.

Two-component monooxygenases are receiving more attention because of their ability to catalyze reduction and oxidation of substrate by different polypeptide chains. They play an important role in the metabolism of aromatic and aliphatic organic compounds in microorganisms. Besides metabolism, these enzymes are also involved in the biosynthesis of antibiotics and cancer drugs like actinorhodin, rebeccamycin, violacein, enediyne, angucycline, kijanimicin, and kutzneride, and in bacterial pathogenesis.

In 1998, Kadiyala and Spain, while studying the monooxygenases associated with metabolism of *p*-nitrophenol, reported that these enzymes are composed of two components. Recently, Furuya and Kino (2013) reported two-component flavin-dependent monooxygenase from *Pseudomonas aeruginosa* metabolizing cinnamic acid derivatives. Two-component (flavin C4a and N5 locus) flavin-dependent monooxygenase, p-hydroxyphenylacetate (HPA) 3-hydroxylase was also reported by Thotsaporn et al. (2011). These two components are flavin C4a locus and flavin N5 locus. Wang and Shao (2012) reported flavin-binding monooxygenase genes (*almA*) from marine bacteria expressing flavin-binding monooxygenase (*AlmA*), which is involved in the biodegradation of long-chain n-alkanes of C_{32}. These monooxygenases are supposed to possess oil-degradation capacities. Khan et al. (2013) reported flavin-dependent monooxygenase in *Rhodococcus* sp. strain MB-P1 capable of degrading 2-chloro-4-nitroaniline.

There are two groups of flavin-dependent monooxygenases: internal and external. Very few flavin-dependent monooxygenases are internal (EC 1.13.12.): most are external (EC 1.14.13.-). These monooxygenases require NAD(P)H as an assisting coenzyme (Van Berkel et al., 2006).

Flavin-dependent monooxygenases catalyze a range of biochemical reactions, such as Baeyer–Villiger oxidations, epoxidations, and halogenations. These enzymes are also able to hydroxylate nonactivated alkanes. Biochemically these enzymes are chemo-, region-, and enantioselective.

Internal flavin-dependent monooxygenases have not been extensively explored. Very few examples of these monooxygenases are reported. Out of these, one is FMN-dependent lactate dehydrogenase monooxygenase isolated from *Mycobacterium* (Sutton, 1957). The second studied example of internal flavin-dependent monooxygenase is nitronate monooxygenases which has been shown to regulate the oxidation of nitroalkanes by generating an anionic flavin semiquinone, yielding an aldehyde and nitrite (Gadda and Francis, 2010).

On the basis of sequence structure and amino acid similarity, the external flavin-dependent monooxygenases are grouped into six subclasses (Van Berkel et al., 2006): A, B, C, D, E, and F.

Members of subclass A are encoded by a single gene and FAD is closely attached as a prosthetic group. In this situation, either NADPH or NAD(P)H is used for electron supply. These enzymes contain one dinucleotide-binding site (Rossmann fold) (Wierenga et al., 1986) and $NADP^+$ gets dissociated after the catalytic process. These members participate in the microbial degradation of aromatic compounds by *ortho-* or *para*-hydroxylation of the aromatic ring (Moonen et al., 2002). In addition to that, it shows a narrow substrate range. One example of this monooxygenase is 4-hydroxybenzoate 3-monooxygenase (EC 1.14.13.2), which was isolated from *Pseudomonas* and deliberately characterized (Entsch and van Berkel, 1995). Several other enzymes, such as 2-hydroxybiphenyl 3-monooxygenase (EC 1.14.13.44), phenol 2-monooxygenase (EC 1.14.13.7) and salicylate 1-monooxygenase, were studied by different research groups. Besides this, these monooxygenases were involved in metabolism of pharmaceutical compounds (aromatic polyketides); oxytetracycline; rifampin; mithramycin chromomycin; auricin, and griseorhodin.

Members of subclass B are also encoded by a single polypeptide chain like subclass A with FAD as the prosthetic group and NADPH for electron supply. In addition to, these contain two dinucleotide-binding domains (Malito et al., 2004); $NADP^+$ remains attached throughout the complete catalytic process. Fraaije et al. (2002) reported that these are multifunctional flavin-containing monooxygenases that are able to oxidize both carbon atoms and other (hetero) atoms. These monooxygenases are composed of three subfamilies: flavin-containing monooxygenases (FMOs), microbial *N*-hydroxylating monooxygenases (NMOs), and Baeyer–Villiger monooxygenases (Type I). FMOs are of eukaryotic origin and these are present in the livers of mammals, including humans, (FMO3) in several isoforms. FMOs are known for detoxification of drugs and several other xenobiotics present in the human body. In this way, they assist with cytochrome P450 system degradation and elimination of xenobiotics. These monooxygenases participate in catalysis of carbon-bound reactive heteroatoms: nitrogen, sulfur, phosphorus, selenium, and iodine. Suh et al. (1996) reported a different behavior of yeast FMO: that these monooxygenases were not involved in the oxidation process of nitrogen-containing compounds and catalyze only biological thiols. Stehr et al. (1998) extensively studied the microbial NMOs. These are responsible for N-hydroxylation of long-chain primary amines.

Type I BVMOs participate in Baeyer–Villiger oxidation of a ketone (or aldehyde) to an ester or lactone and detoxification of fungal toxins. These are isolated and studied in bacteria, e.g., region and enantioselective cyclohexanone monooxygenase (EC1.14.13.22) from *Acinetobacter* and steroid monooxygenase (EC 1.14.13.54) from *Rhodococcus rhodochrous*, 4-hydroxyacetophenone, cyclododecanone monooxygenase, cyclopentanone monooxygenase (EC 1.14.13.16), phenylacetone monooxygenase (EC 1.14.13.92), and monocyclic monoterpene ketone monooxygenase. Recently Khan and Cameotra (2013) reported Baeyer–Villiger monooxygenases (BVMO) from *Rhodococcus* spp. strain MB-P1, involved in the degradation of 2-hexanone.

Members of subclass "C" are FMN-containing monooxygenases, which are encoded by multiple genes expressing one or two monooxygenase components and one reductase. These are the only monooxygenase-containing FMN as a coenzyme. Reductase of this subclass may utilize NADPH and/or NADH as a coenzyme. The crystal structural study of luciferase of bacterial monooxygenase revealed that the core of these monooxygenase subunit(s) displays a TIM-barrel fold (Fisher et al., 1995). The most widely studied example of this subclass of monooxygenases is bacterial luciferases (EC 1.14.14.3), which emit light upon oxidation of long-chain aliphatic aldehydes. Other examples are 3,6-diketocamphane 1,6-monooxygenase (EC 1.14.15.2) or 2,5-diketocamphane 1,2-monooxygenase (EC 1.14.15.2), causes ring expansion of (+)- or (−)-camphor, alkanesulfonate monooxygenase (EC 1.14.14.5), all of which make use of alkanesulfonates as sulfur sources for growth. Many other monooxygenases of this subclass are reported to be involved in desulfurization of dibenzothiophene into the consequent sulfone and the degradation of nitrilotriacetate.

Members of subclass "D" are FAD dependent, encoded by two genes and NADPH and/or NADH which can be used as coenzyme by reductase. The structure of these monooxygenases has not been extensively studied, but they are supposed to be composed of α helical acyl-CoA dehydrogenase fold. These monooxygenases show their ability to degrade aromatic substrates through regioselective hydroxylation only. An extensively studied member of this subclass is bacterial 4-hydroxyphenylacetate 3-monooxygenase (EC 1.14.13.3). Other members of this subclass are 4-hydroxyphenylacetate monooxygenase and phenol monooxygenase (EC 1.14.13.7) 4-nitrophenol monooxygenase. The 4-nitrophenol monooxygenase catalyzes sequential *o*- and *p*-hydroxylations with removal of nitrite. 2,4,5-trichlorophenol monooxygenase is also involved in *p*- and *o*-hydroxylation and chloride removal. 2,4,6-trichlorophenol monooxygenase regulates dechlorinations by oxidative and hydrolytic reactions. These enzymes are reported to be involved in hydroxylation of 4-chlorophenol (Nordin et al., 2004), indole, and polyketides (Brunke et al., 2001).

Subclass E of monooxygenase is similar to the subclass D but shows an evolutionary link with subclass A, because it contains one dinucleotide-binding domain i.e., Rossmann fold. Otto et al. (2004) reported one member of this subclass i.e., styrene monooxygenase isolated from *Pseudomonas*. Styrene monooxygenases are found to be very enantioselective in oxidation of styrene into (*S*)-styrene oxide and its derivatives. Subclass F is also very similar to subclasses D and E, but its members are composed of two domains: one FAD-binding domain (Rossmann fold) and another helical domain.

These monooxygenases do not generate oxygenated products but catalyze halogenation reactions. An extensively studied member of this subclass is tryptophan 7-halogenase. Dong et al. (2005) reported that it has a structural resemblance to 4-hydroxybenzoate 3-monooxygenase.

The prokaryotic origin of these enzymes has made them suitable for biotechnological applications. Out of all the flavin-dependent monooxygenases, cyclohexanone monooxygenase (BVMOs) have been studied most extensively for their

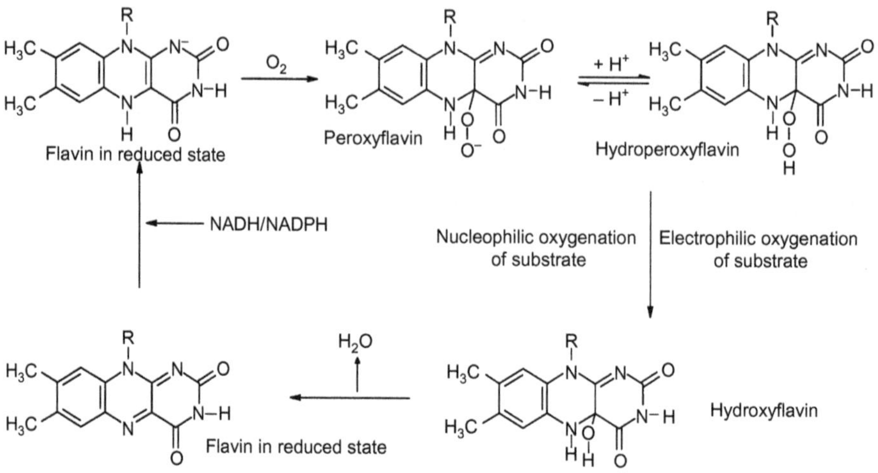

Figure 26.6 General mechanism of oxygenation reactions catalyzed by external flavoprotein monooxygenases.
Source: Modified from Van Berkel et al. (2006).

biotechnological applications because of their utility for a broad range of different substrates and their exquisite enantio and/region selectivity.

Van Berkel et al. (2006) have proposed a mechanism (Figure 26.6) of action of external and internal flavin-dependent monooxygenases. In the case of external enzymes, for the beginning of the process of oxygenation, the flavin cofactor must be in reduced form. This reduced flavin binds with oxygen and transfers its electron to oxygen and a complex of flavin radical, and superoxide is produced. This complex is called hydroperoxyflavin (peroxyflavin). This complex is unstable and breaks down into peroxide and oxidized flavin. The protonation state of the peroxyflavin determines its nucleophilic or electrophilic nature; it transfers a single atom of molecular oxygen to the substrate, and another oxygen atom forms water. The oxidized flavin is again reduced by NADH or NADPH. However, in the case of internal flavin-dependent monooxygenases (lactate monooxygenases), the flavin is reduced from substrate. The reduced flavin, reacts with molecular oxygen and oxidizes to form pyruvic acid into acetic acid and CO_2.

26.1.4.2.1.3 Copper-Dependant Monooxygenases These enzymes require copper for their catalytic activity. This is a small family of monooxygenases and is placed into EC 1.14.18.1. Most of the enzymes are reported in higher eukaryotes. Some of the reported examples of this family are dopamine β-monooxygenases (DβM), peptidylglycine α-hydroxylating monooxygenases (PHM), and tyramine β-monooxygenases. In these monooxygenases, DβM regulates the conversion of dopamine to norepinephrine and PHM catalyzes the hydroxylation of C-terminal glycine-extended peptides.

Klinman (2006) reported a detailed study of DβM and PHM and proposed a reaction mechanism of these enzymes. The structures of DβM and PHM show that two copper centers, i.e., Cu_H and Cu_M, are involved in the enzyme for catalytic activity. The crystal structure of PHM revealed that Cu_M is the site for substrate hydroxylation. During this hydroxylation process, two electrons are consumed in these two copper centers and it is believed (structural evidence is not available) that both metal centers approach each other during the catalytic cycle; however, it is difficult to understand how the electron is transferred from Cu_H to Cu_M through a bulk solvent.

During the catalysis process, Cu^{2+} is reduced to Cu^{+} by (reductant) molecules and this reduction process facilitates the binding of substrate and oxygen O_2 to the enzyme. Cu^{+} is oxidized to Cu^{2+} by the interaction of substrate and O_2 via a formal "ping-pong" mechanism in which reductant (ascorbic acid) and substrates interact with different forms of enzymes. This ping-pong mechanism also shows that both electrons required for the hydroxylation of substrate are stored on the enzyme. Depending on reduced intermediates, there are four possible O_2 activation mechanisms. In mechanism I, a single electron reduced intermediate is formed, i.e., copper superoxo, $Cu_M(+2)O_2(-1)$. In mechanism II, a double electron reduced intermediate is formed, i.e., copper peroxo, $Cu_M(+2)O_2$ (-2). In mechanism III, copper hydroperoxo $Cu_M(+2)(HO_2(-1)$ and mechanism IV leads to the formation of highly reduced copper oxo $Cu_M(+2)$ O(-1) through the reductive cleavage of copper hydroperoxo $Cu_M(+2)(HO_2(-1)$. Out of these mechanisms, mechanism I is considered as a working mechanism (Klinman, 2006). In this mechanism (Figure 26.7), ascorbic acid caused reduction of both Cu_H and $Cu_{M.}$ Both substrate and O_2 bind to the Cu_M site and form a ternary complex. An electron from Cu_M is transferred to O_2, causing activation of O_2 and the generation of a copper-superoxo intermediate. This is coupled with the transfer of hydrogen from the substrate to the charged oxygen, leading to the formation of intermediate Cu_M(II)−OOH. Simultaneously, there is an electron transfer from Cu_H to Cu_M. The Cu_M(II)−OOH undergoes reductive cleavage and produces a Cu_M(II)-oxo radical and H_2O. The Cu_M(II)-oxo radical quickly reassociates with the substrate-derived radical to produce an inner sphere alcohol product. Now ascorbic acid causes reduction of the copper center and dissociation of the product (Klinman, 2006).

Some enzymes are also reported that do not require ascorbate; one of them is methane monooxygenases (pMMO) from *Methylococcus capsulatus* and ammonia monooxygenases (AMO).

Fungal origin copper dependent monooxygenases (GH61 proteins) are cellulose-metabolizing enzymes and are involved in biomass degradation. These are reported by many research groups. Bey et al. (2013) reported copper-dependent lytic polysaccharide monooxygenases (LPMOs) from the coprophilic ascomycete *Podospora anserine*. Lo Leggio et al. (2012) and Wu et al. (2013) provided the structural information of GH61 proteins. Beeson et al. (2012) and Phillips et al. (2011) reported copper-dependent polysaccharide monooxygenase (PMO) from *Neurospora crassa* which regulates the insertion of oxygen into C−H bonds adjoining to the glycosidic linkage.

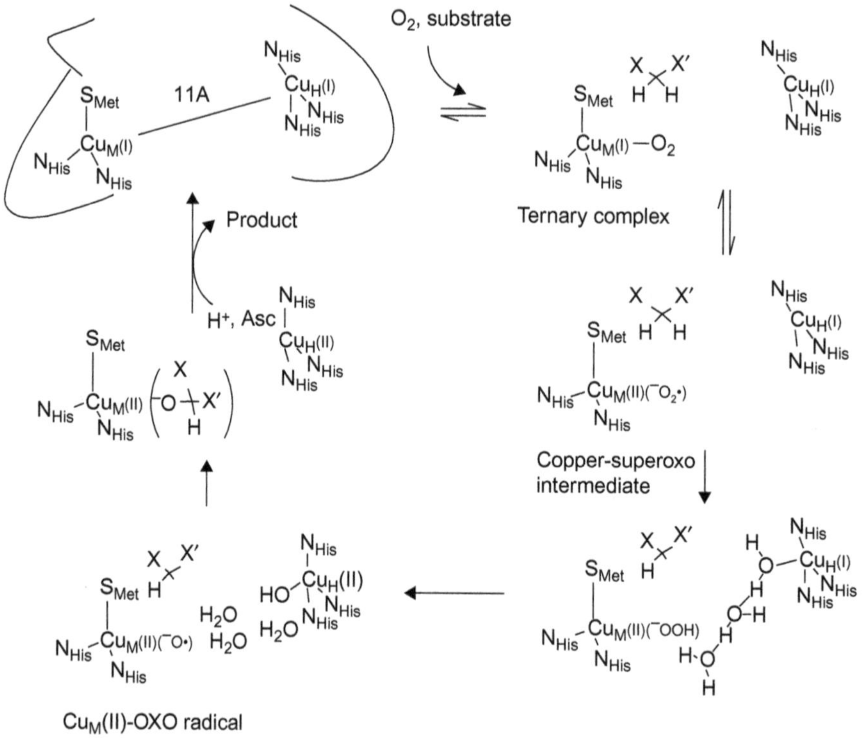

Figure 26.7 Proposed catalytic mechanism of PHM.
Source: Modified from Klinman (2006).

26.1.4.2.1.4 Nonheme Iron-Dependent Monooxygenases The enzymes of this group require two iron (Fe) atoms as cofactors for catalytic activity. These enzymes are multiple subunit enzymes that contain three subunits: one monooxygenase, one reductase, and one small regulatory polypeptide. Most of these enzymes are grouped in EC 1.14.13. Homme and Sharp (2013) reported *N*-nitrosamines degrading propane monooxygenase from *Rhodococcus* spp. Li et al. (2013) reported tetrahydrofuran and propane monooxygenases involved in the biodegradation of 1,4-dioxane.

On the basis of structural information, genetic, and biochemical data, these di-iron monooxygenases are divided into four groups: (1) soluble methane monooxygenases (sMMO) (EC 1.14.13.25); (2) amo alkene monooxygenase (EC 1.14.13.69) of bacterial strain *Rhodococcus corallinus* B-276; (3) phenol hydroxylases (EC 1.14.13.7); and (4) four-component alkane/aromatic monooxygenases (EC 1.14.15.3). Phylogenetic analysis of these monooxygenases provides the information that α- and β-oxygenase subunits are paralogous proteins and that these were evolved from gene duplication of ancient carboxylate-bridged di-iron protein, with subsequent divergence producing a catalytic α-oxygenase subunit and a structural β-oxygenase subunit. The oxidoreductase and ferredoxin components of these monooxygenases were probably obtained by

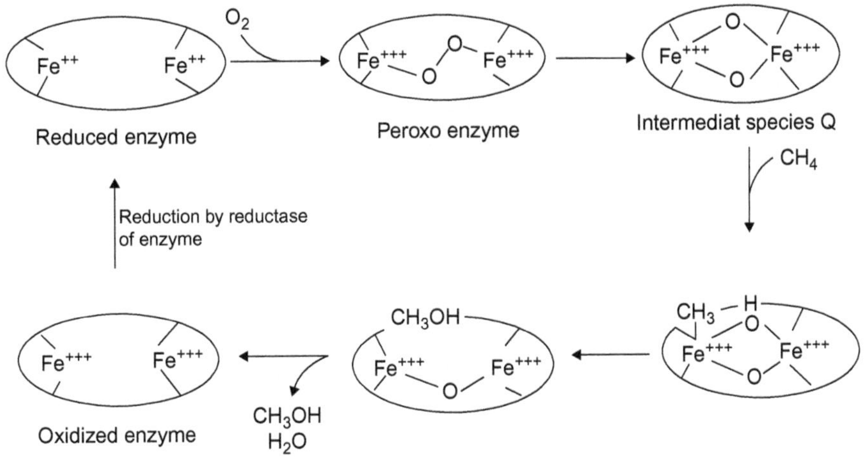

Figure 26.8 Proposed mechanism of sMMO to oxidize methane.
Source: Modified from Valentine et al. (1999).

horizontal transfer from relatives similar to distinct di-iron and Rieske center oxygenases and other catalysts. The phylogenetic analysis proposed that in all of these enzymes, the alkene/aromatic monooxygenases diverged first from the last common ancestor, followed by the phenol hydroxylases, amo alkene monooxygenase, and methane monooxygenases (Leahy et al., 2003).

Sluis et al. (2002) and Dubbels et al. (2007) isolated three-component di-iron monooxygenase "butane monooxygenase" (sBMO) from the Gram-negative β-proteobacterium *Pseudomonas butanovora*. The physical features of sBMO were surprisingly parallel to the sMMO family of soluble monooxgenases. The enzyme contained an iron-containing hydroxylase, a flavo-iron sulfur-containing NADH-oxidoreductase, and one small regulatory protein. However, the catalytic characteristics of sBMO were quantitatively dissimilar with regard to inactivation in the existence of substrate and product distribution.

Rosenzweig et al. (1995) provided a detailed study of the crystal structure of the hydroxylase component of soluble methane monooxygenase (sMMO) from *M. capsulatus* and revealed that a dinuclear iron center associated with the oxidation of methane to methanol and both iron atoms are bridged by a glutamate, a hydroxide ion, and an acetate ion, and additionally coordinated to two His residues, three Glu residues, and a water molecule. In both the oxidized and reduced structures, the di-iron core is connected through hydrogen bonds involving exogenous species to Thr residue in the active site cavity. Thr plays important role in catalytic property of enzyme (Elsen et al., 2009).

Valentine et al. (1999) discussed reaction mechanism (Figure 26.8) of sMMO on the basis of crystal structure analysis of sMMO and phenol hydroxylases. It is revealed by the study that active site iron atoms activate oxygen molecule by electron transfer and produce di-iron(IV)bis-μ-oxo-intermediate, and this intermediate hydroxylate the substrate. Biochemically these enzymes are involved in enantioslelective sulfoxidation including hydroxylation and epoxidation. The primary step of the catalytic cycle is considered the reduction of active-site iron to the diferrous (Fe^{2+}) state from diferric (Fe^{3+}) by electrons from reductase (MMOR). Now involvement of oxygen takes place: the di-iron center becomes oxidized and the first intermediate compound, known as di-iron peroxo, is produced with the help of cofactor (MMOB). The di-iron (III) peroxo intermediate consequently converts to a high-valent di-iron (IV) intermediate, known as intermediate Q and also known as bis(μ-oxo)bis(μ-carboxylato)di-iron(IV) unit. Q decays leads to its task as the active oxidizing species and hydroxylate the substrate.

$$CH_4 + O_2 + NADH + H^+ \rightarrow CH_3OH + H_2O + NAD^+$$

26.1.4.2.1.5 Pterin-Dependent Monooxygenases These catalytic monooxygenases require tetrahydrobiopterin (BH_4) as a coenzyme and one iron atom bounded by two histidine and one glutamic acid amino acids for catalytic activity. It is also hypothesized that a bacterial enzyme does not require metal for catalytic activity (Fitzpatrick, 2003). Enzymes of this family are placed in EC.1.14.16. and these are present in both prokaryotes and eukaryotes. These enzymes cause the hydroxylation of aromatic ring-possessing amino acids; members of this family are phenylalanine hydroxylase (PheH), tyrosine hydroxylase (TyrH), tryptophan hydroxylase (TrpH), and nitric oxide synthase. Another class of enzyme is alkylglycerol monooxygenase or glycerylether ether monooxygenase (EC 1.14.16.5), which is catalyzing an important step in ether lipids degradation by cleaving the ether bond of alkylglycerol derivatives (Watschinger et al., 2010). It is the only enzyme to catalyze the cleavage O-alkyl bond of ether lipids, which are essential components of brain membranes, protect the eye from cataract, interfere in or mediate signaling processes, and are required for spermatogenesis.

These enzymes contain three structural domains: a regulatory domain, a catalytic domain, and a tetramerization domain. These enzymes are iron dependent; Fe(III) in its active state is oxidized into Fe(II) and tetrahydrobiopterin works as a reductant. During the catalytic process, tetrahydrobiopterin, amino acids, and oxygen binds together for the catalytic activity; but the order of binding is random and different enzymes inherit different levels of inclination toward binding of the pterin before the amino acid.

During the catalytic process (Figure 26.9), the enzyme reductant BH_4 undergoes autoxidation and causes reduction of Fe^{3+} (ferric) into Fe^{2+} (ferrous). It binds to molecular oxygen in its active site and this ferrous form of enzyme prevails throughout the catalytic process. It is assumed that O_2 is activated by both the Fe^{2+} ion and the bound reduced BH_4, and forms Fe(IV)O ferryl-oxo reactive intermediate; this intermediate catalyzes electrophilic aromatic hydroxylation, sulfoxidation, and epoxidation of substrate. The Fe(IV)O intermediate for the pterin-dependent enzymes would be expected to be able to catalyze aliphatic hydroxylation (benzylic hydroxylation of methylated aromatic amino acids) (Fitzpatrick, 2003).

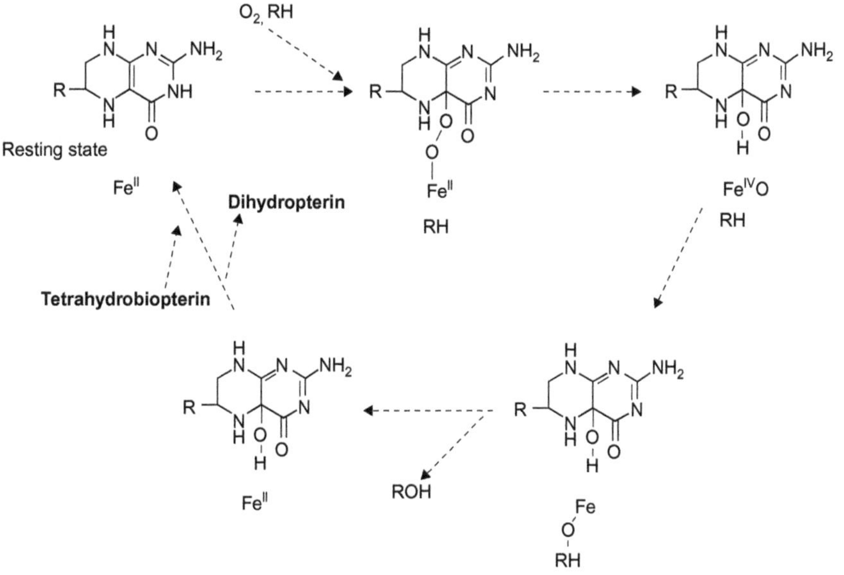

Figure 26.9 Proposed catalytic cycle of pterin-dependent monooxygenase.
Source: Modified from Fitzpatrick (2003).

26.1.4.2.1.6 Other Cofactor-Dependent Monooxygenases These types of monooxygenases have not been studied well and very few members have been isolated and identified. One member, aclacinomycin 10-hydroxylase, was isolated from *Streptomyces purpurascens* which required *S*-adenosyl-L-methionine as a cofactor. It is suggested that the cofactor contains a positive charge, which causes delocalization of electrons leading to decarboxylation of the substrate. This substrate causes activation of molecular oxygen and produces a hydroxyperoxide intermediate (Jansson et al., 2003). Xiao et al. (2012) reported a very unusual flavin-dependent monooxygenase possessing heme-binding cytochrome *b*5 domain *ortho*-nitrophenol 2-monooxygenase ONP (EC 1.14.13.31), which is involved in the oxidation of the aromatic pollutant *ortho*-nitrophenol (ONP) to form catechol via *ortho*-benzoquinone.

26.1.4.2.2 Cofactor-Independent Monooxygenases

It was a common idea that all oxygenases need metal ions and/or organic cofactors for oxygen activation, but recently several other monooxygenases were reported to catalyze the oxygenation process without the need of cofactors or metal ions. These enzymes are known as cofactors-independent monooxygenases. They present the mechanistically intriguing problem of how oxygen is activated to react with organic compounds. These monooxygenases have been grouped in the quinone-forming monooxygenases family, which is involved in the biosynthesis of numerous kinds of aromatic polyketide antibiotics in *streptomycetes* and filamentous fungi.

An initial study of such enzymes was done by Shen and Hutchinson (1993) as they studied and characterized one enzyme from *Streptomyces glauscens* named tetracenomycin F1 monooxygenase (TcmH), a member of the quinone-forming monooxygenase family involved in the oxidation of the naphthacenone tetracenomycin F1 to 5, 12-naphthacenequinone tetracenomycin. It was suggested that this enzyme is an internal cofactor-independent monooxygenase and requires only molecular oxygen for its activity; substrate was utilized as a reducing agent. Initially it was assumed that sulfhydryl groups and histidine residues played a pivotal role in catalytic activity, but the crystal structures of a second member of the quinone-forming monooxygenases enzymes, ActVA-orf6 from *Streptomyces coelicolor*, revealed that one His52 residue is important for catalytic processes and the sulfhydryl group is not important for catalytic activity (Hertweck et al., 2007); study of other crystal structures proposed the importance of tryptophan for catalytic activity. Grocholski et al. (2010) studied the crystal structure of SnoaB from *Streptomyces nogalater* and revealed that tryptophan and asparagine residues are important for catalytic activity. Other studied members of this family are ElmH from *Streptomyces olivaceus*, involved in elloramycin A biosynthesis; AknX from *Streptomyces galilaeus*, involved in biosynthesis of aclacinomycin; and FrnU from *Streptomyces roseofulvus*.

The reaction mechanism of these monooxygenases is not well elucidated and it is believed that, for catalytic processes, amino acid residues of the catalytic site participate in activation of molecular oxygen and/or substrate. For this activation, two possible mechanisms are available. In the first mechanism, there is the possibility of formation of a protein radical intermediate, and in the second mechanism there is a direct transfer of electrons to molecular oxygen from the (deprotonated) substrate to form a radical pair (Fetzner, 2002).

According to the available information on crystal structures, it is supposed that the initiation of the catalysis (Figure 26.10) process takes place by abstraction of a

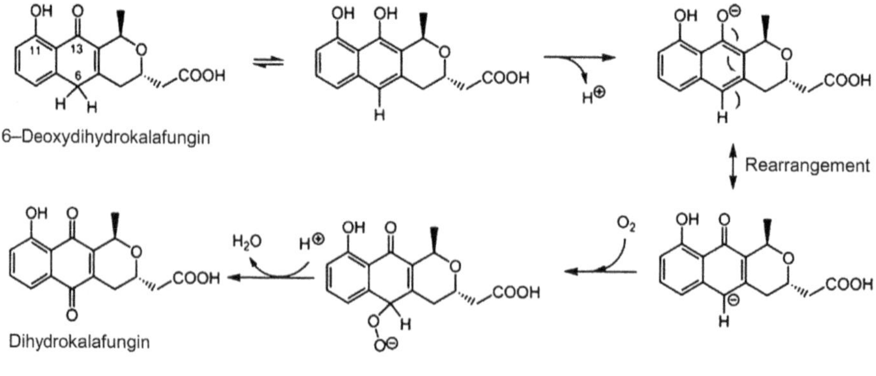

Figure 26.10 Promising mechanism for oxidation of 6-deoxydihydrokalafungin to dihydrokalafungin catalyzed by ActVA-Orf6 monooxygenase.
Source: Modified from Fetzner and Steiner (2010).

proton from the substrate by utilizing amino acids Tyr and Arg. In this process, Trp helps in substrate recognition and deprotonation. The substrate radical reacts with molecular oxygen and produces a peroxy intermediate. Protonation and subsequent dehydration of the peroxy intermediate would finally yield the *p*-quinonoid product. After this process, there is rearrangement of intermediate-producing oxidized substrate and water (Fetzner and Steiner, 2010).

Currently researchers are considering these enzymes for industrial applications because sometimes heterologous expression of cofactor-dependent enzymes creates difficulty due to troubles in biosynthesis and in insertion and folding of cofactors. These problems make industrial processes expensive, time consuming, and sometimes impossible. But cofactor-independent monooxygenases have a benefit here as they do not require any cofactor for catalytic activity and they can be synthesized by general host strains. For example, ActVA-orf6 protein and ElmH have been expressed in *Escherichia coli* successfully. Information regarding the substrate specificity and possibility of engineering in substrate specificity may be explored in all-round biocatalysts for the alteration of polyketide organization, either by biotransformation by whole-cell method or using immobilized enzymes.

26.2 Metabolism of Xenobiotics

It has been made very clear that monooxygenases are involved as biocatalysts for numerous biochemical reactions in cells, with a broad range of substrates ranging from alkanes to steroids (Pazminoa et al., 2010). Monooxygenases are involved and these biochemical reactions are resulting in biosynthesis or biodegradation of substrates. Monooxygenases are actually enzymes and it is fate of chemical nature of molecule whether it is a substrate or a product. Henceforth, the product of one biochemical reaction may be the substrate of another biochemical reaction. The biodegradation and elimination of xenobiotics is a well-understood process and takes place in three phases (Figure 26.11). In the first and second phases, compounds are modified to alter their hydrophilic nature; these become more water soluble and less toxic. Molecules are converted into cationic and anionic conjugates and are ready to be transported or eliminated by specific transporters or carriers in the third and final phase (Lee et al., 2011). Out of these three phases, monooxygenases generally are involved in the first phase and result in a change in the hydrophilic behavior of xenobiotics. Currently, researchers are focusing on the level of genetic expression of concerned genes and the biochemical characterization of enzymes after exposure to xenobiotics (Kohle and Bock, 2007; Maurice et al., 2013). Including this, researchers are trying to establish a relationship between the different phases of xenobiotic degradation. With the advent of biochemical reactions catalyzed by different monooxygenases (Table 26.2), researchers are focusing on different parameters like the age-dependent expression of monooxygenase-expressing genes and their biotechnological applications.

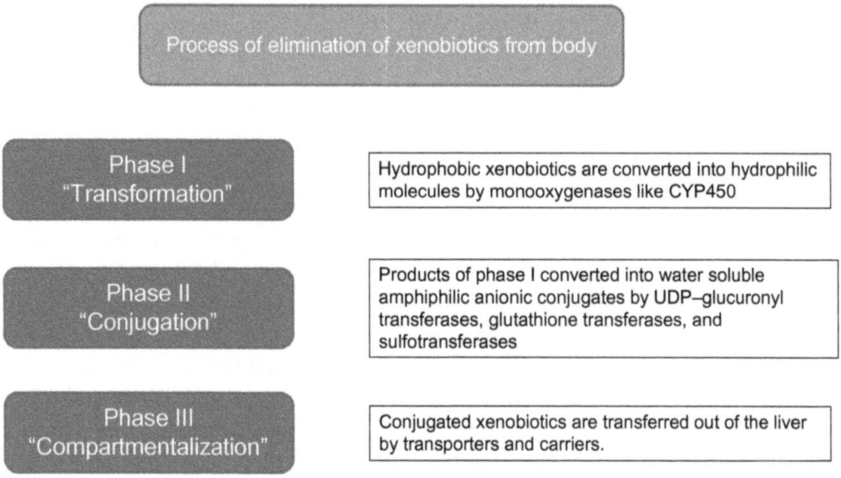

Figure 26.11 Process of elimination of xenobiotics from body.

Lee et al. (2011) reported the expression of an transporter gene expression in different life stages. This study indicated that there is a significant difference in expression of xenobiotic metabolizing enzymes for drugs and chemicals like in fetuses, newborns, children, and the aged with reference to adults.

Currently research groups are focusing on the production of transgenic plants with xenobiotic metabolizing enzymes (mainly CYP450) of bacterial, fungal, and mammalian origin. In the last two decades, a large number of experiments were performed regarding the cloning and overexpression of bacterial, mammalian, and human CYP450 isoforms in higher plants such as *Nicotiana tabaccum* (tobacco), *Solanum tuberosum* (potato), *Oryza sativa* (rice), and *A. thaliana* (Siminszky et al., 2000).

The transfer of transgenes into plants is mediated by *Agrobacterium tumefaciens* or direct DNA gene transfer. These transgenic plants show an ability to detoxify (biodegrade) organic pollutants in soil and air, and an increased tolerance, uptake, and biodegradation of explosives like nitroreductase. Additionally, transgenic plants exhibit bioaugmentation potential of microorganisms and convey the message for resourceful and environmental-friendly machinery for cleaning up polluted soil. Other advantages includes viz., carbon uptake, soil stabilization, and the possibility of the production of biofuel or fiber (Doty et al., 2007; Alvarez et al., 2008; Alvarez et al., 2009; Van Aken, 2009).

For the first time, French et al. (1999) reported the creation of transgenic tobacco plants with a bacterial-origin gene responsible for expressing pentaerythritol tetranitrate reductase, isolated from nitrate ester and a nitroaromatic explosives-degrading bacterial strain. Seeds of transgenic plants were germinated and grown with 1 mM glycerol trinitrate (GTN) or 0.05 mM trinitrotoluene, and at these concentrations germination and growth of wild-type seeds did not take place. Doty et al. (2007) developed transgenic poplar (*Populus tremula* × *Populus alba*) plants

Table 26.2 Biochemical Reactions Catalyzed by All Reported Monooxygenases

Name of Monooxygenase	EC Number	Reaction Catalyzed
β-Carotene 9′,10′-monooxygenase	1.13.11.71	All-*trans*-β-carotene + O_2 ⇔ all-*trans*-10′-apo-β-carotenal + beta-ionone
Arginine 2-monooxygenase	1.13.12.1	L-Arginine + ⇔ 4-guanidinobutanamide + CO2 + H_2O
Lysine 2-monooxygenase	1.13.12.2	L-Lysine + O_2 ⇔ 5-aminopentanamide + CO_2 + H_2O
Tryptophan 2-monooxygenase.	1.13.12.3	L-Tryptophan + O_2 ⇔ (indol-3-yl)acetamide + CO_2 + H_2O
Lactate 2-monooxygenase.	1.13.12.4	(*S*)-Lactate + O_2 ⇔ acetate + CO_2 + H_2O
Renilla-luciferin 2-monooxygenase	1.13.12.5	Renilla luciferin + O_2 ⇔ oxidized Renilla luciferin + CO_2 + light
Cypridina-luciferin 2-monooxygenase	1.13.12.6	Cypridina luciferin + O_2 ⇔ oxidized Cypridina luciferin + CO_2 + light
Photinus-luciferin 4-monooxygenase	1.13.12.7	Photinus luciferin + O_2 + ATP ⇔ oxidized Photinus luciferin + CO_2 + AMP + diphosphate + light
Watasenia-luciferin 2-monooxygenase	1.13.12.8	Watasenia luciferin + O_2 ⇔ oxidized Watasenia luciferin + CO_2 + light
Phenylalanine 2-monooxygenase	1.13.12.9	L-Phenylalanine + O_2 ⇔ 2-phenylacetamide + CO_2 + H(2)O
Oplophorus-luciferin 2-monooxygenase	1.13.12.13	Oplophorus luciferin + O_2 ⇔ oxidized Oplophorus luciferin + CO_2 + light
Nitronate monooxygenases	1.13.12.16	Ethylnitronate + O_2 ⇔ acetaldehyde + nitrite + other products
Noranthrone monooxygenases	1.13.12.20	Norsolorinic acid anthrone + O_2 ⇔ norsolorinic acid + H(2)O
2-Oxoglutarate/L-arginine monooxygenase/ decarboxylase	1.14.11.34	2-Oxoglutarate + L-arginine + O_2 ⇔ succinate + CO_2 + guanidine + (*S*)-1-pyrroline-5-carboxylate + H(2)O
Pheide a monooxygenases	1.14.12.20	Pheophorbide a + NADPH + O_2 ⇔ red chlorophyll catabolite + NADP(+)
Salicylate 1-monooxygenase	1.14.13.1	Salicylate + NADH + O_2 ⇔ catechol + NAD(+) + H(2)O + CO(2)
4-Hydroxybenzoate 3-monooxygenase	1.14.13.2	4-Hydroxybenzoate + NADPH + O_2 ⇔ protocatechuate + NADP(+) + H(2)O
Melilotate 3-monooxygenase	1.14.13.4	3-(2-Hydroxyphenyl)propanoate + NADH + O2 ⇔ 3-(2,3-dihydroxyphenyl) propanoate + NAD(+) + H(2)O
Imidazoleacetate 4-monooxygenase	1.14.13.5	4-Imidazoleacetate + NADH + O_2 ⇔ 5-hydroxy-4-imidazoleacetate + NAD (+) + H(2)O
Orcinol 2-monooxygenase	1.14.13.6	Orcinol + NADH + O_2 ⇔ 2,3,5-trihydroxytoluene + NAD(+) + H(2)O
Phenol 2-monooxygenase	1.14.13.7	Phenol + NADPH + O_2 ⇔ catechol + NADP(+) + H(2)O

(*Continued*)

Table 26.2 (Continued)

Name of Monooxygenase	EC Number	Reaction Catalyzed
Flavin-containing monooxygenase	1.14.13.8	*N*,*N*-dimethylaniline + NADPH + O_2 ⇔ *N*,*N*-dimethylaniline *N*-oxide + NADP(+) + H(2)O
Kynurenine 3-monooxygenase	1.14.13.9	L-Kynurenine + NADPH + O_2 ⇔ 3-hydroxy-L-kynurenine + NADP(+) + H(2)O
2,6-Dihydroxypyridine 3-monooxygenase	1.14.13.10	2,6-Dihydroxypyridine + NADH + O_2 ⇔ 2,3,6-trihydroxypyridine + NAD (+) + H(2)O
Trans-cinnamate 4-monooxygenase	1.14.13.11	Trans-cinnamate + NADPH + O_2 ⇔ 4-hydroxycinnamate + NADP(+) + H(2)O
Benzoate 4-monooxygenase	1.14.13.12	Benzoate + NADPH + O_2 ⇔ 4-hydroxybenzoate + NADP(+) + H(2)O
Calcidiol 1-monooxygenase	1.14.13.13	Calcidiol + NADPH + O_2 ⇔ calcitriol + NADP(+) + H(2)O
Trans-cinnamate 2-monooxygenase	1.14.13.14	Trans-cinnamate + NADPH + O_2 ⇔ 2-hydroxycinnamate + NADP(+) + H(2)O
Cholestanetriol 26-monooxygenase (cytochrome P450 27A1)	1.14.13.15	5-Beta-cholestane-3-alpha,7-alpha,12-alpha-triol + 3 NADPH + 3 O_2 ⇔ (25R)-3-alpha,7-alpha,12-alpha-trihydroxy-5-beta-cholestan-26-oate + 3 NADP(+) + 4 H(2)O
Cyclopentanone monooxygenases	1.14.13.16	Cyclopentanone + NADPH + O_2 ⇔ 5-valerolactone + NADP(+) + H(2)O
Cholesterol 7-alpha-monooxygenase	1.14.13.17	Cholesterol + NADPH + O_2 ⇔ 7-alpha-hydroxycholesterol + NADP(+) + H(2)O
4-Hydroxyphenylacetate 1-monooxygenase	1.14.13.18	4-Hydroxyphenylacetate + NAD(P)H + O_2 ⇔ homogentisate + NAD(P) (+) + H(2)O
Taxifolin 8-monooxygenase	1.14.13.19	Taxifolin + NAD(P)H + O_2 ⇔ 2,3-dihydrogossypetin + NAD(P)(+) + H(2)O
2,4-Dichlorophenol 6-monooxygenase	1.14.13.20	2,4-Dichlorophenol + NADPH + O_2 ⇔ 3,5-dichlorocatechol + NADP(+) + H(2)O
Flavonoid 3′-monooxygenase	1.14.13.21	A flavonoid + NADPH + O_2 ⇔ a 3′-hydroxyflavonoid + NADP(+) + H(2)O
Cyclohexanone monooxygenases	1.14.13.22	Cyclohexanone + NADPH + O_2 ⇔ hexano-6-lactone + NADP(+) + H(2)O
3-Hydroxybenzoate 4-monooxygenase	1.14.13.23	3-Hydroxybenzoate + NADPH + O_2 ⇔ 3,4-dihydroxybenzoate + NADP(+) + H(2)O

3-Hydroxybenzoate 6-monooxygenase	1.14.13.24	3-Hydroxybenzoate + NADH + O_2 ⇔ 2,5-dihydroxybenzoate + NAD(+) + H(2)O
Methane monooxygenase (soluble)	1.14.13.25	Methane + NAD(P)H + O_2 ⇔ methanol + NAD(P)(+) + H(2)O
Phosphatidylcholine 12-monooxygenase	1.14.13.26	1-Acyl-2-oleoyl-*sn*-glycero-3-phosphocholine + NADH + O_2 ⇔ 1-acyl-2-((*S*)-12-hydroxyoleoyl)-*sn*-glycero-3-phosphocholine + NAD(+) + H(2)O
4-Aminobenzoate 1-monooxygenase	1.14.13.27	4-Aminobenzoate + NAD(P)H + O_2 ⇔ 4-hydroxyaniline + NAD(P)(+) + H(2)O + CO(2)
3,9-Dihydroxypterocarpan 6A-monooxygenase	1.14.13.28	(6aR,11aR)-3,9-dihydroxypterocarpan + NADPH + O_2 ⇔ (6aS,11aS)-3,6a,9-trihydroxypterocarpan + NADP(+) + H(2)O
4-Nitrophenol 2-monooxygenase	1.14.13.29	4-Nitrophenol + NADH + O_2 ⇔ 4-nitrocatechol + NAD(+) + H(2)O
Leukotriene-B(4) 20-monooxygenase	1.14.13.30	(6Z,8E,10E,14Z)-(5S,12R)-5,12-dihydroxyicosa-6,8,10,14-tetraenoate + NADPH + O_2 ⇔ (6Z,8E,10E,14Z)-(5S,12R)-5,12,20-trihydroxyicosa-6,8,10,14-tetraenoate + NADP(+) + H(2)O
2-Nitrophenol 2-monooxygenase	1.14.13.31	2-Nitrophenol + NADPH + O_2 ⇔ catechol + nitrite + NADP(+) + H(2)O
Albendazole monooxygenases	1.14.13.32	Albendazole + NADPH + O_2 ⇔ albendazole *S*-oxide + NADP(+) + H(2)O
4-Hydroxybenzoate 3-monooxygenase	1.14.13.33	4-Hydroxybenzoate + NAD(P)H + O_2 ⇔ 3,4-dihydroxybenzoate + NAD(P)(+) + H(2)O
Leukotriene-*E*(4) 20-monooxygenase	1.14.13.34	(7E,9E,11Z,14Z)-(5S,6R)-6-(cystein-*S*-yl)-5-hydroxyicosa-7,9,11,14-tetraenoate + NADPH + O_2 ⇔ 20-hydroxyleukotriene *E*(4) + NADP(+) + H(2)O
Anthranilate 3-monooxygenase	1.14.13.35	Anthranilate + NADPH + O_2 ⇔ 2,3-dihydroxybenzoate + NADP(+) + NH(3)
5-*O*-(4-coumaroyl)-D-quinate 3′-monooxygenase	1.14.13.36	Trans-5-*O*-(4-coumaroyl)-D-quinate + NADPH + O_2 ⇔ *trans*-5-*O*-caffeoyl-D-quinate + NADP(+) + H(2)O
Methyltetrahydroprotoberberine 14-monooxygenase	1.14.13.37	(*S*)-*N*-methylcanadine + NADPH + O_2 ⇔ allocryptopine + NADP(+) + H(2)O
Anhydrotetracycline monooxygenase	1.14.13.38	Anhydrotetracycline + NADPH + O_2 ⇔ 12-dehydrotetracycline + NADP(+) + H(2)O
Anthraniloyl-CoA monooxygenases	1.14.13.40	2-Aminobenzoyl-CoA + 2 NAD(P)H + O2 ⇔ 2-amino-5-oxocyclohex-1-enecarboxyl-CoA + H(2)O + 2 NAD(P)(+)

(*Continued*)

Table 26.2 (Continued)

Name of Monooxygenase	EC Number	Reaction Catalyzed
Tyrosine *N*-monooxygenase (CYP 79A_1)	1.14.13.41	L-Tyrosine + O2 + NADPH ⇔ *N*-hydroxy-L-tyrosine + NADP(+) + H(2)O
Questin monooxygenases	1.14.13.43	Questin + NADPH + O_2 ⇔ demethylsulochrin + NADP(+)
2-Hydroxybiphenyl 3-monooxygenase	1.14.13.44	2-Hydroxybiphenyl + NADH + O_2 ⇔ 2,3-dihydroxybiphenyl + NAD(+) + H(2)O
(−)-Menthol monooxygenases	1.14.13.46	(−)-Menthol + NADPH + O_2 ⇔ *p*-menthane-3,8-diol + NADP(+) + H(2)O
(*S*)-Limonene 3-monooxygenase	1.14.13.47	(−)-(*S*)-Limonene + NADPH + O_2 ⇔ (−)-*trans*-isopiperitenol + NADP (+) + H(2)O
(*S*)-Limonene 6-monooxygenase	1.14.13.48	(−)-(*S*)-Limonene + NADPH + O2 ⇔ (−)-*trans*-carveol + NADP(+) + H(2)O
(*S*)-Limonene 7-monooxygenase	1.14.13.49	(−)-(*S*)-Limonene + NADPH + O2 ⇔ (−)-perillyl alcohol + NADP(+) + H(2)O
Pentachlorophenol monooxygenases	1.14.13.50	Pentachlorophenol + 2 NADPH + O_2 ⇔ 2,3,5,6-tetrachlorohydroquinone + 2 NADP(+) + chloride + H(2)O
4′-Methoxyisoflavone 2′-hydroxylase	1.14.13.53	Formononetin + NADPH + O2 ⇔ 2′-hydroxyformononetin + NADP(+) + H(2)O
Ketosteroid monooxygenase	1.14.13.54	• Ketosteroid + NADPH + O_2 ⇔ steroid ester/lactone + NADP(+) + H(2)O • Progesterone + NADPH + O_2 ⇔ testosterone acetate + NADP(+) + H(2)O
Protopine 6-monooxygenase	1.14.13.55	Protopine + NADPH + O_2 ⇔ 6-hydroxyprotopine + NADP(+) + H(2)O
Dihydrosanguinarine 10-monooxygenase	1.14.13.56	Dihydrosanguinarine + NADPH + O_2 ⇔ 10-hydroxydihydrosanguinarine + NADP(+) + H(2)O
Dihydrochelirubine 12-monooxygenase	1.14.13.57	Dihydrochelirubine + NADPH + O_2 ⇔ 12-hydroxydihydrochelirubine + NADP(+) + H(2)O
Benzoyl-CoA 3-monooxygenase	1.14.13.58	Benzoyl-CoA + NADPH + O_2 ⇔ 3-hydroxybenzoyl-CoA + NADP(+) + H(2)O
L-Lysine *N*(6)-monooxygenase	1.14.13.59	L-Lysine + NADPH + O_2 ⇔ *N*(6)-hydroxy-L-lysine + NADP(+) + H(2)O

27-Hydroxycholesterol 7-alpha-monooxygenase	1.14.13.60	27-Hydroxycholesterol + NADPH + O_2 ⇔ 7-alpha, 27-dihydroxycholesterol + NADP(+) + H(2)O
2-Hydroxyquinoline 8-monooxygenase	1.14.13.61	Quinolin-2-ol + NADH + O_2 ⇔ quinolin-2,8-diol + NAD(+) + H(2)O
4-Hydroxyquinoline 3-monooxygenase	1.14.13.62	Quinolin-4-ol + NADH + O_2 ⇔ quinolin-3,4-diol + NAD(+) + H(2)O
3-Hydroxyphenylacetate 6-monooxygenase	1.14.13.63	3-Hydroxyphenylacetate + NAD(P)H + O_2 ⇔ 2,5-dihydroxyphenylacetate + NAD(P)(+) + H(2)O
4-Hydroxybenzoate 1-monooxygenase	1.14.13.64	4-Hydroxybenzoate + NAD(P)H + O_2 ⇔ hydroquinone + NAD(P)(+) + H(2)O + CO(2)
2-Hydroxycyclohexanone 2-monooxygenase	1.14.13.66	2-Hydroxycyclohexan-1-one + NADPH + O_2 ⇔ 6-hydroxyhexan-6-olide + NADP(+) + H(2)
Quinine 3-monooxygenase	1.14.13.67	OQuinine + NADPH + O_2 ⇔ 3-hydroxyquinine + NADP(+) + H(2)O
4-Hydroxyphenylacetaldehyde oxime monooxygenases (cytochrome P-450-II-dependent monooxygenases)	1.14.13.68	(*Z*)-4-hydroxyphenylacetaldehyde oxime + NADPH + O_2 ⇔ (*S*)-4-hydroxymandelonitrile + NADP(+) + 2 H(2)O
Alkene monooxygenases	1.14.13.69	Propene + NADH + O_2 ⇔ 1,2-epoxypropane + NAD(+) + H(2)O
N-Methylcoclaurine 3′-monooxygenase (cytochrome P450 80B_1)	1.14.13.71	(*S*)-*N*-Methylcoclaurine + NADPH + O_2 ⇔ (*S*)-3′-hydroxy-*N*-methylcoclaurine + NADP(+) + H(2)O
Methylsterol monooxygenases	1.14.13.72	4,4-Dimethyl-5-alpha-cholest-7-en-3-beta-ol + NAD(P)H + O_2 ⇔ 4-beta-hydroxymethyl-4-alpha-methyl-5-alpha-cholest-7-en-3-beta-ol + NAD(P)(+) + H(2)O
(*R*)-Limonene 6-monooxygenase	1.14.13.80	(+)-(*R*)-Limonene + NADPH + O_2 ⇔ (+)-*trans*-carveol + NADP(+) + H(2)O
Vanillate monooxygenase	1.14.13.82	Vanillate + O_2 + NADH ⇔ 3,4-dihydroxybenzoate + NAD(+) + H(2)O + formaldehyde
4-Hydroxyacetophenone monooxygenase	1.14.13.84	(4-Hydroxyphenyl)ethan-1-one + NADPH + O_2 ⇔ 4-hydroxyphenyl acetate + NADP(+) + H(2)O
Isoflavone 2′-hydroxylase (CYP81E_1)	1.14.13.89	An isoflavone + NADPH + O_2 ⇔ a 2′-hydroxyisoflavone + NADP(+) + H(2)O
Phenylacetone monooxygenase	1.14.13.92	Phenylacetone + NADPH + O_2 ⇔ benzyl acetate + NADP(+) + H(2)O

(*Continued*)

Table 26.2 (Continued)

Name of Monooxygenase	EC Number	Reaction Catalyzed
Lithocholate 6-beta-hydroxylase (cytochrome P450 3A10)	1.14.13.94	Lithocholate + NADPH + O_2 ⇔ 6-beta-hydroxylithocholate + NADP(+) + H(2)O
7-α-Hydroxy-4-cholesten-3-one 12-alpha-monooxygenase	1.14.13.95	7-α-Hydroxycholest-4-en-3-one + NADPH + O_2 ⇔ 7-alpha,12-alpha-dihydroxycholest-4-en-3-one + NADP(+) + H(2)O
5-β-Cholestane-3-α,7-α-diol 12-α-monooxygenase (cytochrome P450 $8B_1$)	1.14.13.96	5-Beta-cholestane-3-alpha,7-alpha-diol + NADPH + O_2 ⇔ 5-beta-cholestane-3-alpha,7-alpha,12-alpha-triol + NADP(+) + H(2)O
Taurochenodeoxycholate 6-α-monooxygenase	1.14.13.97	Taurochenodeoxycholate + NADPH + O_2 ⇔ taurohyocholate + NADP(+) + H(2)O
Cholesterol 24-monooxygenase (cytochrome P450 $46A_1$)	1.14.13.98	Cholesterol + NADPH + O_2 ⇔ (24*S*)-24-hydroxycholesterol + NADP(+) + H(2)O
24-Hydroxycholesterol 7-α-monooxygenase	1.14.13.99	(24*R*)-Cholest-5-ene-3-beta,24-diol + NADPH + O_2 ⇔ (24*R*)-cholest-5-ene-3-beta,7-alpha,24-triol + NADP(+) + H(2)O
25-Hydroxycholesterol 7-α-monooxygenase	1.14.13.100	• Cholest-5-ene-3-beta,25-diol + NADPH + O2 ⇔ cholest-5-ene-3-beta,7-alpha,25-triol + NADP(+) + H(2)O • Cholest-5-ene-3-beta,27-diol + NADPH + O_2 ⇔ cholest-5-ene-3-beta,7-alpha,27-triol + NADP(+) + H(2)O
Senecionine monooxygenases	1.14.13.101	Senecionine + NADPH + O_2 ⇔ senecionine *N*-oxide + NADP(+) + H(2)O
Monocyclic monoterpene ketone monooxygenases	1.14.13.105	Dihydrocarvone + NADPH + O_2 ⇔ 4-isopropenyl-7-methyloxepan-2-one + NADP(+) + H(2)O
Epi-isozizaene 5-monooxygenase	1.14.13.106	(+)-Epi-isozizaene + 2 NADPH + 2 O_2 ⇔ albaflavenone + 2 NADP(+) + 3 H(2)O
Limonene 1,2-monooxygenase	1.14.13.107	(*S*)-Limonene + NAD(P)H + O_2 ⇔ 1,2-epoxymenth-8-ene + NADP(+) + H(2)O
Methanesulfonate monooxygenase	1.14.13.111	Methanesulfonate + NADH + O_2 ⇔ formaldehyde + NAD(+) + sulfite + H(2)O
3-Epi-6-deoxocathasterone 23-monooxygenase (CYP90C1, D1)	1.14.13.112	3-Epi-6-deoxocathasterone + NADPH + O_2 ⇔ 6-deoxotyphasterol + NADP(+) + H(2)O

6-Hydroxynicotinate 3-monooxygenase	1.14.13.114	6-Hydroxynicotinate + NADH + O_2 ⇔ 2,5-dihydroxypyridine + NAD(+) + H(2)O + CO(2)
Isoleucine *N*-monooxygenase	1.14.13.117	L-Isoleucine + 2 H(+) + 2 NADPH + 2 O_2 ⇔ (*E*)-2-methylbutanal oxime + CO_2 + 3 H(2)O + 2 NADP(+)
Valine *N*-monooxygenase	1.14.13.118	L-Valine + 2 H(+) + 2 NADPH + 2 O_2 ⇔ (*Z*)-2-methylpropanal oxime + CO_2 + 3 H(2)O + 2 NADP(+)
Phenylalanine *N*-monooxygenase (CYP79A2)	1.14.13.124	L-Phenylalanine + 2 NADPH + 2 O_2 ⇔ (*E*)-phenylacetaldoxime + 2 NADP(+) + 3 H(2)O + CO(2)
Tryptophan *N*-monooxygenase (CYP79B1, B2, B3)	1.14.13.125	L-Tryptophan + 2 NADPH + 2 O_2 ⇔ indole-3-acetaldoxime + 2 NADP(+) + 3 H(2)O + CO(2)
β-Carotene 3-hydroxylase	1.14.13.129	Beta-carotene + 2 NADH + 2 O_2 ⇔ zeaxanthin + 2 NAD(+) + 2H(2)O
Pyrrole-2-carboxylate monooxygenases	1.14.13.130	Pyrrole-2-carboxylate + NADH + O_2 ⇔ 5-hydroxypyrrole-2-carboxylate + NAD(+) + H(2)O
Dimethyl-sulfide monooxygenase	1.14.13.131	Dimethyl sulfide + O_2 + NADH ⇔ methanethiol + formaldehyde + NAD(+) + H(2)O
Squalene monooxygenases	1.14.13.132	Squalene + NADPH + O_2 ⇔ (3*S*)-2,3-epoxy-2,3-dihydrosqualene + NADP(+) + H(2)O
Indole-2-monooxygenase (CYP71C4)	1.14.13.137	Indole + NAD(P)H + O_2 ⇔ indolin-2-one + NAD(P)(+) + H(2)O
Indolin-2-one monooxygenase	1.14.13.138	Indolin-2-one + NAD(P)H + O_2 ⇔ 3-hydroxyindolin-2-one + NAD(P)(+) + H(2)O
3-Hydroxyindolin-2-one monooxygenase	1.14.13.139	3-Hydroxyindolin-2-one + NAD(P)H + O_2 ⇔ 2-hydroxy-2H-1,4-benzoxazin-3(4H)-one + NAD(P)(+) + H(2)O
2-Hydroxy-1,4-benzoxazin-3-one monooxygenase	1.14.13.140	2-Hydroxy-2H-1,4-benzoxazin-3(4H)-one + NAD(P)H + O_2 ⇔ 2,4-dihydroxy-2H-1,4-benzoxazin-3(4H)-one + NAD(P)(+) + H(2)O
Cholest-4-en-3-one 26-monooxygenase	1.14.13.141	Cholest-4-en-3-one + NADH + O_2 ⇔ 26-hydroxycholest-4-en-3-one + NAD (+) + H(2)O
3-Ketosteroid 9-α-monooxygenase	1.14.13.142	Androsta-1,4-diene-3,17-dione + NADH + O_2 ⇔ 9-alpha-hydroxyandrosta-1,4-diene-3,17-dione + NAD(+) + H(2)O
Trimethylamine monooxygenases	1.14.13.148	*N,N,N*-Trimethylamine + NADPH + O_2 ⇔ *N,N,N*-trimethylamine *N*-oxide + NADP(+) + H(2)O

(*Continued*)

Table 26.2 (Continued)

Name of Monooxygenase	EC Number	Reaction Catalyzed
Phenylacetyl-CoA 1,2-epoxidase	1.14.13.149	Phenylacetyl-CoA + NADPH + O_2 ⇔ 2-(1,2-epoxy-1,2-dihydrophenyl)acetyl-CoA + NADP(+) + H(2)O
Linalool 8-monooxygenase (CYP111)	1.14.13.151	Linalool + 2 NADH + 2 O_2 ⇔ (6E)-8-oxolinalool + 2 NAD(+) + 3 H(2)O
Alpha-pinene monooxygenases	1.14.13.155	(−)-Alpha-pinene + NADH + O_2 ⇔ alpha-pinene oxide + NAD(+) + H(2)O
1,8-Cineole 2-endo-monooxygenase (CYP176A)	1.14.13.156	1,8-Cineole + NADPH + O_2 ⇔ 2-endo-hydroxy-1,8-cineole + NADP(+) + H(2)O
1,8-Cineole 2-exo-monooxygenase (CYP3A4)	1.14.13.157	1,8-Cineole + NADPH + O_2 ⇔ 2-exo-hydroxy-1,8-cineole + NADP(+) + H(2)O
Amorpha-4,11-diene 12 monooxygenase (CYP71AV1)	1.14.13.158	Amorpha-4,11-diene + 3 O_2 + 3 NADPH ⇔ artemisinate + 3 NADP(+) + 4 H(2)O
(2,2,3-Trimethyl-5-oxocyclopent-3-enyl)acetyl-CoA 1,5-monooxygenase	1.14.13.160	((1R)-2,2,3-Trimethyl-5-oxocyclopent-3-enyl)acetyl-CoA + O_2 + NADPH ⇔ ((2*R*)-3,3,4-trimethyl-6-oxo-3,6-dihydro-1H-pyran-2-yl)acetyl-CoA + NADP(+) + H(2)O
2,5-Diketocamphane 1,2-monooxygenase	1.14.13.162	(+)-Bornane-2,5-dione + O2 + NADH ⇔ (+)-5-oxo-1,2-campholide + NAD(+) + H(2)O
6-Hydroxy-3-succinoylpyridine 3-monooxygenase	1.14.13.163	4-(6-Hydroxypyridin-3-yl)-4-oxobutanoate + 2 NADH + O_2 ⇔ 2,5-dihydroxypyridine + succinate semialdehyde + 2 NAD(+) + H(2)O
4-Nitrocatechol 4-monooxygenase	1.14.13.166	4-Nitrocatechol + NAD(P)H + O_2 ⇔ 2-hydroxy-1,4-benzoquinone + nitrite + NAD(P)(+) + H(2)O
4-Nitrophenol 4-monooxygenase	1.14.13.167	4-Nitrophenol + NADPH + O_2 ⇔ 1,4-benzoquinone + nitrite + NADP(+) + H(2)O
Indole-3-pyruvate monooxygenases	1.14.13.168	(Indol-3-yl)pyruvate + NADPH + O_2 ⇔ (indol-3-yl)acetate + NADP(+) + H(2)O + CO(2)
Sphinganine C(4)-monooxygenase	1.14.13.169	Sphinganine + NADPH + O_2 ⇔ phytosphingosine + NADP(+) + H(2)O
2-Methylbutanal oxime monooxygenases	1.14.13.n8	(*Z*)-2-Methylbutanal oxime + NADPH + O_2 ⇔ 2-hydroxy-2-methylbutyronitrile + NADP(+) + 2 H(2)O
Unspecific monooxygenases (cytochrome P450)	1.14.14.1	RH + reduced flavoprotein + O_2 ⇔ ROH + oxidized flavoprotein + H(2)O
Alkanal monooxygenases	1.14.14.3	RCHO + reduced FMN + O_2 ⇔ RCOOH + FMN + H(2)O + light

Alkanesulfonate monooxygenases	1.14.14.5	An alkanesufonate (*R*-CH(2)-SO(3)H) + FMNH(2) + O_2 ⇔ an aldehyde (*R*-CHO) + FMN + sulfite + H(2)O
Anthranilate 3-monooxygenase	1.14.14.8	Anthranilate + FADH(2) + O_2 ⇔ 3-hydroxyanthranilate + FAD + H(2)O
4-Hydroxyphenylacetate 3-monooxygenase	1.14.14.9	4-Hydroxyphenylacetate + FADH(2) + O_2 ⇔ 3, 4-dihydroxyphenylacetate + FAD + H(2)O
Nitrilotriacetate monooxygenases	1.14.14.10	Nitrilotriacetate + FMNH(2) + H(+) + O_2 ⇔ iminodiacetate + glyoxylate + FMN + H(2)O
Styrene monooxygenases	1.14.14.11	Styrene + FADH(2) + O_2 ⇔ (*S*)-2-phenyloxirane + FAD + H(2)O
3-Hydroxy-9,10-secoandrosta-1,3,5(10)-triene-9,17-dione monooxygenases	1.14.14.12	3-Hydroxy-9,10-secoandrosta-1,3,5(10)-triene-9,17-dione + FMNH(2) + O2 ⇔ 3,4-dihydroxy-9,10-secoandrosta-1,3,5(10)-triene-9,17-dione + FMN + H(2)O
4-(γ-L-lutamylamino)butanoyl-[BtrI acyl-carrier protein] monooxygenase	1.14.14.13	4-(Gamma-L-glutamylamino)butanoyl-[BtrI acyl-carrier protein] + FMNH(2) + O_2 ⇔ 4-(gamma-L-glutamylamino)-(2*S*)-2-hydroxybutanoyl-[BtrI acyl-carrier protein] + FMN + H(2)O
Camphor 5-monooxygenase (cytochrome P450-cam)	1.14.15.1	(+)-Camphor + reduced putidaredoxin + O_2 ⇔ (+)-exo-5-hydroxycamphor + oxidized putidaredoxin + H(2)O
Alkane 1-monooxygenase	1.14.15.3	Octane + reduced rubredoxin + O_2 ⇔ 1-octanol + oxidized rubredoxin + H(2)O
Steroid 11-β-monooxygenase (cytochrome P450 X_IB_1)	1.14.15.4	A steroid + reduced adrenodoxin + O_2 ⇔ an 11-beta-hydroxysteroid + oxidized adrenodoxin + H(2)O
Corticosterone 18-monooxygenase	1.14.15.5	Corticosterone + reduced adrenodoxin + O_2 ⇔ 18-hydroxycorticosterone + oxidized adrenodoxin + H(2)O
Cholesterol monooxygenases (cytochrome P-450)	1.14.15.6	Cholesterol + 6 reduced adrenodoxin + $3O_2$ ⇔ pregnenolone + 4-methylpentanal + 6 oxidized adrenodoxin + 4 H(2)O
Choline monooxygenase	1.14.15.7	Choline + O_2 + 2 reduced ferredoxin + 2H(+) ⇔ betaine aldehyde hydrate + H(2)O + 2 oxidized ferredoxin
Steroid 15-β-monooxygenase (CYP106A2)	1.14.15.8	Progesterone + reduced ferredoxin + O_2 ⇔ 15-beta-hydroxyprogesterone + oxidized ferredoxin + H(2)O
Spheroidene monooxygenase	1.14.15.9	Spheroidene + reduced ferredoxin + O_2 ⇔ spheroiden-2-one + oxidized ferredoxin + H(2)O
Phenylalanine 4-monooxygenase	1.14.16.1	L-Phenylalanine + tetrahydrobiopterin + O_2 ⇔ L-tyrosine + 4a-hydroxytetrahydrobiopterin

(*Continued*)

Table 26.2 (Continued)

Name of Monooxygenase	EC Number	Reaction Catalyzed
Tyrosine 3-monooxygenase	1.14.16.2	L-Tyrosine + tetrahydrobiopterin + O_2 ⇔ L-dopa + 4a-hydroxytetrahydrobiopterin
Anthranilate 3-monooxygenase	1.14.16.3	Anthranilate + tetrahydrobiopterin + O_2 ⇔ 3-hydroxyanthranilate + dihydrobiopterin + H(2)O
Tryptophan 5-monooxygenase	1.14.16.4	L-Tryptophan + tetrahydrobiopterin + O_2 ⇔ 5-hydroxy-L-tryptophan + 4*a*-hydroxytetrahydrobiopterin
Alkylglycerol monooxygenase	1.14.16.5	1-Alkyl-*sn*-glycerol + tetrahydrobiopterin + O_2 ⇔ 1-*O*-alkyl-*sn*-glycerol + dihydrobiopterin + H(2)O
Mandelate 4-monooxygenase	1.14.16.6	(*S*)-2-Hydroxy-2-phenylacetate + tetrahydrobiopterin + O_2 ⇔ (*S*)-4-hydroxymandelate + dihydrobiopterin + H(2)O
Dopamine β-monooxygenase	1.14.17.1	3,4-Dihydroxyphenethylamine + ascorbate + O_2 ⇔ noradrenaline + dehydroascorbate + H(2)O
Peptidylglycine monooxygenase	1.14.17.3	Peptidylglycine + ascorbate + O_2 ⇔ peptidyl(2-hydroxyglycine) + dehydroascorbate + H(2)O
Tyrosinase	1.14.18.1	• 2 L-dopa + O_2 ⇔ 2 dopaquinone + 2H(2)O • L-tyrosine + O_2 ⇔ dopaquinone + H(2)O
CMP-*N*-acetylneuraminate monooxygenases	1.14.18.2	CMP-*N*-acetylneuraminate + 2 ferrocytochrome b5 + O_2 + 2 H(+) ⇔ CMP-*N*-glycoloylneuraminate + 2 ferricytochrome b5 + H(2)O
Methane monooxygenases (particulate)	1.14.18.3	Methane + quinol + O_2 ⇔ methanol + quinone + H(2)O
Progesterone monooxygenase	1.14.99.4	Progesterone + AH(2) + O_2 ⇔ testosterone acetate + A + H(2)O
Steroid 17-α-monooxygenase	1.14.99.9	A C(21)-steroid + (reduced NADPH-hemoprotein reductase) + O_2 ⇔ a 17-alpha-hydroxy-C(21)-steroid + (oxidized NADPH-hemoprotein reductase) + H(2)O
Steroid 21-monooxygenase (cytochrome P-450 XXIA1)	1.14.99.10	A C(21) steroid + (reduced NADPH-hemoprotein reductase) + O_2 ⇔ a 21-hydroxy-C(21)-steroid + (oxidized NADPH-hemoprotein reductase) + H(2)O
Estradiol 6-β-monooxygenase	1.14.99.11	Estradiol-17-beta + AH(2) + O_2 ⇔ 6-beta-hydroxyestradiol-17-beta + A + H(2)O

Androst-4-ene-3,17-dione monooxygenases	1.14.99.12	Androstenedione + AH(2) + O_2 ⇔ testololactone + A + H(2)O
Progesterone 11-α-monooxygenase	1.14.99.14	Progesterone + AH(2) + O_2 ⇔ 11-alpha-hydroxyprogesterone + A + H(2)O
4-Methoxybenzoate monooxygenase	1.14.99.15	4-Methoxybenzoate + AH(2) + O_2 ⇔ 4-hydroxybenzoate + formaldehyde + A + H(2)O
Phylloquinone monooxygenases	1.14.99.20	Phylloquinone + AH(2) + O_2 ⇔ 2,3-epoxyphylloquinone + A + H(2)O
Latia-luciferin monooxygenase (demethylating	1.14.99.21	Latia luciferin + AH(2) + $2O_2$ ⇔ oxidized Latia luciferin + CO_2 + formate + A + H(2)O + light
Ecdysone 20-monooxygenase	1.14.99.22	Ecdysone + AH(2) + O_2 ⇔ 20-hydroxyecdysone + A + H(2)O
3-Hydroxybenzoate 2-monooxygenase	1.14.99.23	3-Hydroxybenzoate + AH(2) + O_2 ⇔ 2,3-dihydroxybenzoate + A + H(2)O
Steroid 9-α-monooxygenase	1.14.99.24	Pregna-4,9(11)-diene-3,20-dione + AH(2) + O_2 ⇔ 9,11-alpha-epoxypregn-4-ene-3,20-dione + A + H(2)O
2-Hydroxypyridine 5-monooxygenase	1.14.99.26	2-Hydroxypyridine + AH(2) + O_2 ⇔ 2,5-dihydroxypyridine + A + H(2)O
Juglone 3-monooxygenase	1.14.99.27	5-Hydroxy-1,4-naphthoquinone + AH(2) + O_2 ⇔ 3,5-dihydroxy-1,4-naphthoquinone + A + H(2)O
Deoxyhypusine monooxygenase	1.14.99.29	Protein *N*(6)-(4-aminobutyl)-L-lysine + AH(2) + O_2 ⇔ protein *N*(6)-((*R*)-4-amino-2-hydroxybutyl)-L-lysine + A + H(2)O
Monoprenyl isoflavone epoxidase	1.14.99.34	7-*O*-methylluteone + NADPH + O_2 ⇔ dihydrofurano derivatives + NADP(+) + H(2)O
Thiophene-2-carbonyl-CoA monooxygenases	1.14.99.35	Thiophene-2-carbonyl-CoA + AH(2) + O_2 ⇔ 5-hydroxythiophene-2-carbonyl-CoA + A + H(2)O
β-Carotene 15,15′-monooxygenase	1.14.99.36	Beta-carotene + O_2 ⇔ 2 all-*trans*-retinal
Cholesterol 25-hydroxylase	1.14.99.38	Cholesterol + AH(2) + O_2 ⇔ 25-hydroxycholesterol + A + H(2)O
Ammonia monooxygenase	1.14.99.39	Ammonia + AH(2) + O_2 ⇔ NH(2)OH + A + H(2)OAlpha-carotene + O_2 + AH(2) ⇔ alpha-cryptoxanthin + A + H(2)O
Carotene epsilon-monooxygenases (CYP 97C_1)	1.14.99.45	Alpha-carotene + O_2 + AH(2) ⇔ alpha-cryptoxanthin + A + H(2)O
Pyrimidine monooxygenase	1.14.99.46	• Uracil + FMNH(2) + O_2 ⇔ (*Z*)-3-ureidoacrylate peracid + FMN + H(2)O • Thymine + FMNH(2) + O_2 ⇔ (*Z*)-2-methylureidoacrylate peracid + FMN + H(2)O
[Tyrosine 3-monooxygenase] kinase	2.7.11.6	ATP + [tyrosine-3-monooxygenase] ⇔ ADP + [tyrosine-3-monooxygenase] phosphate

through the transfer and overexpression of mammalian cytochrome P450 2E1, a key monooxygenase responsible for metabolizing many halogenated compounds. These transgenic plants demonstrated enhanced removal rates of pollutants from hydroponic solutions and air.

Inui and Ohkawa (2005) reported the production of transgenic potato and rice with human CYP450 (CYP1A1, CYP2B6, CYP2C9, and CYP2C19) to metabolize a large number of organic compounds such as industrial chemicals, herbicides, and insecticides. These transgenic plants exhibited cross-resistance to many different structures such as herbicides and sulfonylureas. Dasgupta et al. (2011) studied the P450 SU1 gene in *Streptomyces griseolus* expressing CYP105A1, which metabolized a variety of substrates like sulfonylurea herbicides, vitamin D, coumarins, and brassinosteroids. The P450 SU1 gene was used as a negative-selection marker in plants because CYP105A1 is involved in the conversion of benign sulfonyl urea pro-herbicide R7402 into a highly phytotoxic product. For this study, the P450 SU1 gene was used to produce transgenic *Arabidopsis* plants.

There are numerous examples showing that monooxygenases are involved in metabolism (synthesis and degradation) of compounds, but this is not the only utility of these enzymes in living system. Many other applications of these enzymes are reported. For example, salicylate 1-monooxygenases are involved in biotic and abiotic stress. Kynurenine 3-monooxygenase (KMO) has been found to be associated with Huntington's disease, Alzheimer's disease, brain disorders, cancer, and inflammatory conditions (Levy et al., 2013).

Mishina and Zeier (2006) reported flavin-dependent monooxygenase (FMO) was associated with systemic acquired resistance (SAR) in *A. thaliana*. It was suggested that the expression of the FMO gene in systemic tissue is essential for the development of SAR, probably by production of a metabolite required for the transduction or amplification of a signal during the early phases of SAR establishment. In another example, Wang et al. (2013) studied the role of the CYP714D1 gene expressing a cytochrome P450 monooxygenase acting as a gibberellin (GA)-deactivating enzyme in transgenic *Populus* inhibits expression of its homologous genes CYP714, which promoted growth, biomass production, and longer xylem. Yang et al. (2004) identified TetX monooxygenase, which was associated with tetracycline resistance by oxygen-dependent degradation of the drug. More interestingly, the TetX gene was isolated from transposable elements of anaerobic *Bacteroides*.

References

Abhilash, P.C., Jamil, S., Singh, N., 2009. Transgenic plants for enhanced biodegradation and phytoremediation of organic xenobiotics. Biotechnol. Adv. 27, 474–488.

Alexander, M., 1994. Biodegradation and Bioremediation. Academic Press, New York, NY.

Alvarez, V.M., dos Santos, S.C.C., Casella, R.D., Vital, R.L., Sebastin, G.V., Seldin, L., 2008. Bioremediation potential of a tropical soil contaminated with a mixture of crude oil and production water. J. Microbiol. Biotechnol. 18 (12), 1966–1974.

Beeson, W.T., Phillips, C.M., Cate, J.H.D., Marletta, M.A., 2012. Oxidative cleavage of cellulose by fungal copper-dependent polysaccharide monooxygenases. J. Am. Chem. Soc. 134, 890–892.

Bernhardt, R., 2006. Cytochromes P450 as versatile biocatalysts. J. Biotechnol. 124, 128–145.

Bey, M., Zhou, S., Poidevin, L., Henrissat, B., Coutinho, P.M., Berrin, J.G., et al., 2013. Cello-oligosaccharide oxidation reveals differences between two lytic polysaccharide monooxygenases (family GH61) from *Podospora anserina*. Appl. Environ. Microbiol. 79, 488–496.

Brunke, P., Sterner, O., Bailey, J.E., Minas, W., 2001. Heterologous expression of the naphthocyclinone hydroxylase gene from *Streptomyces arenae* for production of novel hybrid polyketides. Antonie Van Leeuwenhoek. 79, 235–245.

Cirino, P.C., Arnold, F.H., 2002. Protein engineering of oxygenases for biocatalysis. Curr. Opin. Chem. Biol. 6, 130–135.

Dasgupta, K., Ganesan, S., Manivasagam, S., Ayre, B.G., 2011. A cytochrome P450 monooxygenase commonly used for negative selection in transgenic plants causes growth anomalies by disrupting brassinosteroid signaling. BMC Plant Biol. Available from: http://dx.doi.org/doi:10.1186/1471-2229-11-67.

Denisov, I.G., Makris, T.M., Sligar, S.G., Schlichting, I., 2005. Structure and chemistry of cytochrome P450. Chem. Rev. 105, 2253–2277.

Dong, C., Flecks, S., Unversucht, S., Haupt, C., van Pee, K.H., Naismith, J.H., 2005. Tryptophan 7-halogenase (PrnA) structure suggests a mechanism for regioselective chlorination. Science. 309, 2216–2219.

Doty, S.L., James, C.A., Moore, A.L., Vajzovic, A., Singleton, G.L., Ma, C., et al., 2007. Enhanced phytoremediation of volatile environmental pollutants with transgenic trees. Proc. Natl. Acad. Sci. 104 (43), 16816–16821.

Dubbels, B.L., Sayavedra-Soto, L.A., Arp, D.J., 2007. Butane monooxygenase of "*Pseudomonas butanovora*" purification and biochemical characterization of a terminal-alkane hydroxylating diiron monooxygenase. Microbiology. 153, 1808–1816.

Elsen, N.L., Bailey, L.J., Hauser, A.D., Fox, B.G., 2009. Role for threonine 201 in the catalytic cycle of the soluble diiron hydroxylase toluene 4-monooxygenase. Biochemistry. 48, 3838–3846.

Entsch, B., van Berkel, W.J.H., 1995. Structure and mechanism of *para*-hydroxybenzoate hydroxylase. FASEB J. 9, 476–483.

Fetzner, S., 2002. Oxygenases without requirement for cofactors or metal ions. Appl. Microbiol. Biotechnol. 60, 243–257.

Fetzner, S., Steiner, R.A., 2010. Cofactor-independent oxidases and oxygenases. Appl. Microbiol. Biotechnol. 86, 791–804.

Fisher, A.J., Raushel, F.M., Baldwin, T.O., Rayment, I., 1995. Three dimensional structure of bacterial luciferase from *Vibrio harveyi* at 2,4 A resolution. Biochemistry. 34, 6581–6586.

Fitzpatrick, P.F., 2003. Mechanism of aromatic amino acid hydroxylation. Biochemistry. 42, 14084–14091.

Fraaije, M.W., Kamerbeek, N.M., van Berkel, W.J.H., Janssen, D.B., 2002. Identification of a Baeyer–Villiger monooxygenases sequence motif. FEBS Lett. 518, 43–47.

French, C.J., Rosser, S.J., Davies, G.J., Nicklin, S., Bruce, N.C., 1999. Biodegradation of explosives by transgenic plants expressing pentaerythritol tetranitrate reductase. Nat. Biotechnol. 17, 491–494.

Furuya, T., Kino, K., 2013. Catalytic activity of the two-component flavin-dependent monooxygenase from *Pseudomonas aeruginosa* toward cinnamic acid derivatives. Appl. Microbiol. Biotechnol. Available from: http://dx.doi.org/doi:10.1007/s00253-013-4958-y.

Gadda, G., Francis, K., 2010. Nitronate monooxygenase, a model for anionic flavin semiquinone intermediates in oxidative catalysis. Arch. Biochem. Biophys. 493, 53–61.

Grocholski, T., Koskiniemi, H., Lindqvist, Y., Mantsala, P., Niemi, J., Schneider, G., 2010. Crystal structure of the cofactor-independent monooxygenase SnoaB from *Streptomyces nogalater*: implications for the reaction mechanism. Biochemistry. 49, 934–944.

Homme, C.L., Sharp, J.O., 2013. Differential microbial transformation of nitrosamines by an inducible propane monooxygenase. Environ. Sci. Technol. 47, 7388–7395.

Hertweck, C., Luzhetskyy, A., Rebets, Y., Bechthold, A., 2007. Type II polyketide synthases: gaining a deeper insight into enzymatic teamwork. Nat. Prod. Rep. 24, 162–190.

Inui, H., Ohkawa, H., 2005. Herbicide resistance in transgenic plants with mammalian P450 monooxygenase genes. Pest. Manag. Sci. 61, 286–291.

Jansson, A., Niemi, J., Lindqvist, Y., Mantsala, P., Schneider, G., 2003. Crystal structure of aclacinomycin-10-hydroxylase, a S-adenosyl-l-methionine-dependent methyltransferase homolog involved in anthracycline biosynthesis in streptomyces purpurascens. J. Mol. Biol. 334, 269–280.

Khan, F., Cameotra, S.S., 2013. Aerobic degradation of 2-hexanone by a *Rhodococcus* sp. strain MB-P1 via novel pathway. J. Petrol. Environ. Biotechnol. 4. Available from: http://dx.doi.org/doi:10.4172/2157-7463.1000151.

Khan, F., Pal, D., Vikram, S., Cameotra, S.S., 2013. Metabolism of 2-chloro-4-nitroaniline via novel aerobic degradation pathway by *Rhodococcus* sp. strain MB-P1. PLoS One. 8, e62178. Available from: http://dx.doi.org/doi:10.1371/journal.pone.0062178.

Klinman, J.P., 2006. The copper-enzyme family of dopamine β-monooxygenase and peptidylglycine α-hydroxylating monooxygenase: resolving the chemical pathway for substrate hydroxylation. J. Biol. Chem. 281, 3013–3016.

Kohle, C., Bock, K.W., 2007. Coordinate regulation of Phase I and II xenobiotic metabolisms by the Ah receptor and Nrf2. Biochem. Pharmacol. 73, 1853–1862.

Lah, L., Podobnik, B., Novak, M., Korosec, B., Berne, S., Vogelsang, M., et al., 2011. The versatility of the fungal cytochrome P450 monooxygenase system is instrumental in xenobiotic detoxification. Mol. Microbiol. 81 (5), 1374–1389.

Leahy, J.G., Batchelor, P.J., Morcomb, S.M., 2003. Evolution of the soluble diiron monooxygenases. FEMS Microbiol. Rev. 27 (4), 449–479.

Lee, J.S., Ward, W.O., Liu, J., Ren, H., Vallanat, B., 2011. Hepatic xenobiotic metabolizing enzyme and transporter gene expression through the life stages of the mouse. PLoS One. 6 (9), e24381.

Levy, M.A.C., Heyes, D.J., Lafite, P., Outeiro, T.F., Giorgini, F., Leys, D., et al., 2013. Structural basis of kynurenine 3-monooxygenase inhibition. Nature 496, 382–387.

Li, H., Poulos, T.L., 1997. The structure of the cytochrome P450BM-3 heam domain complexed with the fatty acid substrate, palmitoleic acid. Nat. Struct. Biol. 4, 140–146.

Li, M., Mathieu, J., Yang, Y., Fiorenza, S., Deng, Y., He, Z., et al., 2013. Widespread distribution of soluble di-iron monooxygenase (SDIMO) genes in arctic groundwater impacted by 1,4-dioxane. Environ. Sci. Technol. 47, 9950–9958.

Liu, L., Schmid, R.D., Urlacher, V.B., 2006. Cloning, expression, and characterization of a self-sufficient cytochrome P450 monooxygenase from *Rhodococcus ruber* DSM 44319. Appl. Microbiol. Biotechnol. 72, 876–882.

Lo Leggio, L., Welner, D., Maria, L.D., 2012. A structural overview of GH61 proteins—fungal cellulose degrading polysaccharide monooxygenases. Comput. Struct. Biotechnol. J. 2. Available from: http://dx.doi.org/10.5936/csbj.201209019.

Malito, E., Alfieri, A., Fraaije, M.W., Mattevi, A., 2004. Crystal structure of a Baeyer–Villiger monooxygenase. Proc. Natl. Acad. Sci. U.S.A. 101, 13157–13162.

Maurice, C.F., Haiser, H.J., Turnbaugh, P.J., 2013. Xenobiotics shape the physiology and gene expression of the active human gut microbiome. Cell. 152, 39–50.

Mishina, T.E., Zeier, J., 2006. The arabidopsis flavin-dependent monooxygenase FMO1 is an essential component of biologically induced systemic acquired resistance. Plant Physiol. 141, 1666–1675.

Moonen, M.J.H., Fraaije, M.W., Rietjens, Y.M.C.M., Laane, C., van Berkel, W.J.H., 2002. Flavoenzyme-catalyzed oxygenations and oxidations of phenolic compounds. Adv. Synth. Catal. 344, 1–13.

Nordin, K., Unell, M., Jansson, J.K., 2004. Novel 4-chlorophenol degradation gene cluster and degradation route via hydroxy quinol in *Arthrobacter chlorophenolicus* A6. Appl. Environ. Microbiol. 71, 6538–6544.

Otto, K., Hofstetter, K., Rothlisberger, M., Witholt, S.A., 2004. Biochemical characterization of StyAB from Pseudomonas sp. Strain VLB120 as a two-component Flavin-Diffusible Monooxygenase. J. Bacteriol. 186 (16), 5292–5302.

Pazminoa, D.E.T., Winklerb, M., Gliederb, A., Fraaije, M.W., 2010. Monooxygenases as biocatalysts: classification, mechanistic aspects and biotechnological applications. J. Biotechnol. 146, 9–24.

Phillips, C.M., Beeson, W.T., Cate, J.H., Marletta, M.A., 2011. Cellobiose dehydrogenase and a copper-dependent polysaccharide monooxygenase potentiate cellulose degradation by *Neurospora crassa*. ACS Chem. Biol. 6 (12), 1399–1406.

Rieger, P.G., Meier, H.M., Gerle, M., Vogt, U., Groth, T., Knackmuss, H.J., 2002. Xenobiotics in the environment: present and future strategies to obviate the problem of biological persistence. J. Biotechnol. 94, 101–123.

Riser-Roberts, E., 1998. Remediation of Petroleum Contaminated Soils: Biological, Physical, and Chemical Processes. Lewis, Boca Raton, FL.

Rosenzweig, A.C., Nordlund, P., Takahara, P.M., Frederick, C.A., Lippard, S.J., 1995. Geometry of the soluble methane monooxygenase catalytic diiron center in two oxidation states. Chem. Biol. 2, 409–418.

Shen, B., Hutchinson, R., 1993. Tetracenomycin F1 monooxygenase: oxidation of a naphthacenone to a naphthacenequinone in the biosynthesis of tetracenomycin C in Streptomyces glaucescens. Biochemistry 32 (26), 6656–6663.

Siminszky, B., Sheldon, B.S., Corbin, F.T., Dewey, R.E., 2000. A cytochrome P450 monooxygenase cDNA (CYP71A10) confers resistance to linuron in transgenic *Nicotiana tabacum*. Weed Sci. 48, 291–295.

Sluis, M.K., Sayavedra-Soto, L.A., Arp, D.J., 2002. Molecular analysis of the soluble butane monooxygenase from "*Pseudomonas butanovora*". Microbiology. 148, 3617–3629.

Southorn, P.A., Powis, G., 1988. Free radicals in medicine. I. Chemical nature and biologic reactions. Mayo Clin. Proc. 63, 381–389.

Spain, J.C., 1995. Bacterial degradation of nitroaromatic compounds under aerobic conditions. In: Spain, J.C. (Ed.), Biodegradation of Nitroaromatic Compounds, vol. 49. Plenum Press, New York, NY, pp. 19–35.

Stehr, M., Diekmann, H., Smau, L., Seth, O., Ghisla, S., Singh, M., et al., 1998. A hydrophobic sequence motif common to *N*-hydroxylating enzymes. Trends Biochem. Sci. 23, 56–57.

Suh, J.K., Poulsen, L.L., Ziegler, D.M., Robertus, J.D., 1996. Molecular cloning and kinetic characterization of a flavin containing monooxygenase from *Saccharomyces cerevisiae*. Arch. Biochem. Biophys. 336, 268–274.

Sutton, W.B., 1957. Mechanism of action and crystallization of lactic oxidative decarboxylase from *Mycobacterium phlei*. J. Biol. Chem. 226, 395–405.

Thotsaporn, K., Chenprakhon, P., Sucharitakul, J., Mattevi, A., Chaiyen, P., 2011. Stabilization of C4a-hydroperoxyflavin in a two-component flavin-dependent monooxygenase is achieved through interactions at flavin N5 and C4a atoms. J. Biol. Chem. 286, 28170–28180.

Valentine, A.M., Stahl, S.S., Lippard, S.J., 1999. Mechanistic studies of the reaction of reduced methane monooxygenase hydroxylase with dioxygen and substrates. J. Am. Chem. Soc. 121, 3876–3887.

Van Aken, B., 2009. Transgenic plants for enhanced phytoremediation of toxic explosives. Curr. Opin. Biotechnol. 20 (2), 231–236.

Van Berkel, W.J.H., Kamerbeek, N.M., Fraaije, M.W., 2006. Flavoprotein monooxygenases, a diverse class of oxidative biocatalysts. J. Biotechnol. 124, 670–689.

Wang, C., Bao, Y., Wang, Q., Zhang, H., 2013. Introduction of the rice CYP714D1 gene into *Populus* inhibits expression of its homologous genes and promotes growth, biomass production and xylem fibre length in transgenic trees. J. Exp. Bot. 64, 2847–2857.

Wang, W., Shao, Z., 2012. Diversity of flavin-binding monooxygenase genes (*almA*) in marine bacteria capable of degradation long-chain alkanes. FEMS Microbiol. Ecol. 80, 523–533.

Watschinger, K., Keller, M.A., Golderer, G., Hermann, M., Maglione, M., Sarg, B., et al., 2010. Identification of the gene encoding alkylglycerol monooxygenase defines a third class of tetrahydrobiopterin-dependent enzymes. Proc. Natl. Acad. Sci. U.S.A. 107, 13672–13677.

Werck-Reichhart, D., Feyereisen, R., 2000. Cytochromes P450: a success story. Genome. Biol. 1 (6), reviews 3003.1–3003.9.

Wierenga, R.K., Terpstra, P., Hol, W.G., 1986. Prediction of the occurrence of the ADP-binding beta alpha beta-fold in proteins, using an amino acid sequence fingerprint. J. Mol. Biol. 187, 101–107.

Wu, M., Beckham, G.T., Larsson, A.M., Ishida, T., Kim, S., Payne, C.M., et al., 2013. Crystal structure and computational characterization of the lytic polysaccharide monooxygenase GH61D from the basidiomycota fungus *Phanerochaete chrysosporiu*. J. Biol. Chem. Available from: http://dx.doi.org/doi:10.1074/jbc.M113.459396.

Xiao, Y., Liu, T.T., Dai, H., Zhang, J.J., Liu, H., Tang, H., et al., 2012. OnpA, an unusual flavin-dependent monooxygenase containing a cytochrome b5 domain. J. Bacteriol. 194, 1342–1349.

Xu, F., 2005. Applications of oxidoreductases: recent progress. Ind. Biotechnol. 1, 38–50.

Yang, W., Moore, I.F., Koteva, K.P., Bareich, D.C., Hughes, D.W., Wright, G.D., 2004. TetX is a flavin-dependent monooxygenase conferring resistance to tetracycline antibiotics. J. Biol. Chem. 279, 52346–52352.

Lightning Source UK Ltd.
Milton Keynes UK
UKHW02n1228290718
326400UK00003B/19/P

9 780128 000212